T0348774

Groundwater Hydrology of Springs

Groundwater Hydrology of Springs

Engineering, Theory, Management, and Sustainability

edited by

Neven Kresic
Zoran Stevanovic

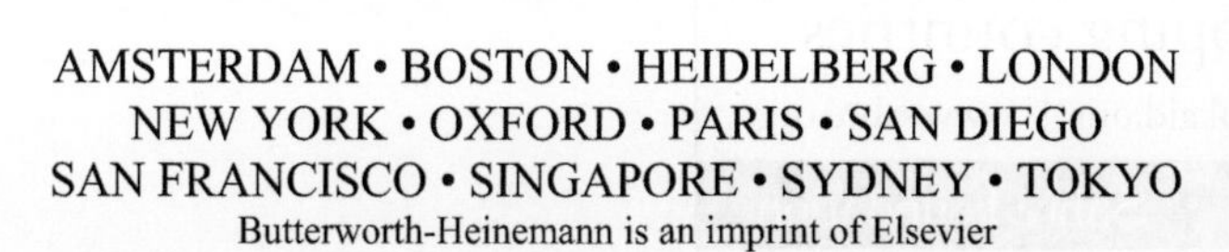
AMSTERDAM • BOSTON • HEIDELBERG • LONDON
NEW YORK • OXFORD • PARIS • SAN DIEGO
SAN FRANCISCO • SINGAPORE • SYDNEY • TOKYO
Butterworth-Heinemann is an imprint of Elsevier

Butterworth-Heinemann is an imprint of Elsevier
30 Corporate Drive, Suite 400, Burlington, MA 01803, USA
Linacre House, Jordan Hill, Oxford OX2 8DP, UK

Copyright © 2010, Elsevier Inc. All rights reserved.
Exception: Chapter 10.9 is copyright © 2010 by Gregg Eckhardt.

No part of this publication may be reproduced, stored in a retrieval system, or transmitted in any form or by any means, electronic, mechanical, photocopying, recording, or otherwise, without the prior written permission of the publisher.

Permissions may be sought directly from Elsevier's Science & Technology Rights Department in Oxford, UK: phone: (+44) 1865 843830, fax: (+44) 1865 853333, E-mail: permissions@elsevier.com. You may also complete your request online via the Elsevier homepage (http://www.elsevier.com), by selecting "Support & Contact" then "Copyright and Permission" and then "Obtaining Permissions."

Library of Congress Cataloging-in-Publication Data
Application submitted

British Library Cataloguing-in-Publication Data
A catalogue record for this book is available from the British Library.

ISBN: 978-1-85617-502-9

For information on all Butterworth-Heinemann publications
visit our Web site at: www.elsevierdirect.com

Working together to grow
libraries in developing countries

www.elsevier.com | www.bookaid.org | www.sabre.org

ELSEVIER BOOK AID International Sabre Foundation

Contents

Preface

The utilization and tapping of springwater is an ancient art. Historically, to have easy access to water, cities were often situated near large springs, while those cities without reliable water supply were destroyed or abandoned because they could not survive long sieges. The culture of springwater was emblematic of the Romans, who demonstrated their supremacy and dominance through the art of tapping and delivering springwater via spectacular aqueducts built throughout their vast empire. However, the history of capturing springs predates Roman times. In ancient Mesopotamia and Greece, springs were also preferred as drinking water sources. For example, to ensure a water supply to the ancient Mesopotamian city Nineveh, the Assyrian king Sanherib constructed the intake on the Khanis spring system and one of the first aqueducts to deliver water to the city walls. Even then, springwater was found to be a much better solution than the nearby surface waters of the Tigris.

Over the course of human history springs gradually lost their importance as primary sources of water supply in many parts of the world and for various reasons, including development of surface water resources via large reservoirs and direct groundwater extraction from aquifers using wells. Consequently, large-scale comprehensive studies of springs were relatively rare until recently, when the concept of the sustainability of water resources came into limelight worldwide. This book is a contribution to the renewed interest in both development and protection of springs for sustainable, beneficial uses.

Springs are a direct reflection of the state of groundwater in the aquifers that feed them, and they directly influence streams and other surface water bodies into which they discharge, including all dependent ecosystems. The study of springs is a borderline discipline, because they represent the transition from groundwater to surface water. Hence, they have been studied to some extent by groundwater specialists (hydrogeologists) and to some extent by surface water specialists (hydrologists). Spring management therefore includes many of the same principles that guide management of surface water and groundwater resources, the fact reflected in the selection of topics included in the book.

We are very grateful to our coauthors for the privilege of working together and learning from some of the most renowned world experts. We thank members of the Karst Commission of International Association of Hydrogeologists for their advice and support during preparation of the manuscript. Many thanks go to our friends and colleagues who provided numerous examples and photographs of springs from around the world. Finally, we thank all our students from the Department of Hydrogeology who, over the years, inspired us with challenging questions and their tireless work supporting our field investigations of countless beautiful springs, large and small.

Neven Kresic and *Zoran Stevanovic*

About the Editors

Neven Kresic Dr. Kresic is the Senior Principal with Mactec Engineering and Consulting, Inc., in Ashburn, Virginia. He is a professional hydrogeologist and professional geologist working for a variety of clients, including water utilities, industry, government agencies, and environmental law firms. Dr. Kresic is cochair of the Karst Commission of the International Association of Hydrogeologists, vice-chair of the Groundwater Management and Remediation Specialty Group of the International Water Association, past vice-president for international affairs of the American Institute of Hydrology, and cofounder of the Groundwater Modeling Interest Group of the National Ground Water Association. He is the author of five books and numerous papers on the subject of groundwater. Before coming to the United States in 1991 as a Senior Fulbright Scholar, Dr. Kresic was a professor of groundwater dynamics at Belgrade University, in the former Yugoslavia.

Zoran Stevanovic Dr. Stevanovic is Chair of Department of Hydrogeology at Belgrade University, Serbia, where he teaches a variety of courses in hydrogeology and groundwater resources engineering. He was project manager and consultant on numerous national and international research and applied engineering projects, including water resources assessment and evaluation, environmental impact studies, aquifer recharge, design of intake structures, and engineering regulation of large springs. Dr. Stevanovic serves as expert for the United Nations' Food and Agriculture Organization. He is a member of the Karst Commission of International Association of Hydrogeologists, a member of the Board on Karst and Speleology of the Serbian Academy of Sciences and Arts, an officer of the Serbian Geological Society, and past secretary of the Hydrogeology Commission of the Carpatho-Balkan Geological Association. Dr. Stevanovic is coeditor of five books and author of numerous papers and book chapters on the subject of groundwater.

List of Contributors

Ognjen Bonacci, Ph.D.
Professor of Water Resources
School of Civil Engineering and Architecture
University of Split
Split, Croatia

Gregg Eckhardt
Environmental Scientist
www.edwardsaquifer.net

Farsad Fotouhi
Vice-President, Corporate Environmental Engineering
Pall Corporation
Ann Arbor, Michigan, the United States

Nico Goldscheider, Ph.D.
Senior Lecturer and Researcher
Centre of Hydrogeology
University of Neuchatel
Neuchatel, Switzerland

Gültekin Günay, Ph.D.
Retired Professor of Karst Hydrogeology
and Geotechnics
Ankara Hacettepe University
Ankara, Turkey

Adrian Iukiewicz, Ph.D.
Enviromental Geology and Geophysics Research
Department
University of Bucharest
Bucharest, Romania

Neven Kresic, Ph.D.
Senior Principle Hydrogeologist
MACTEC Engineering and Consulting, Inc.
Ashburn, Virginia, the United States

Gerhard Kuschnig, Ph.D.
Vienna Waterworks
Vienna, Austria

Metka Petric, Ph.D.
Karst Research Institute, Scientific Research Centre
Slovenian Academy of Sciences and Arts
Postojna, Slovenia

Lukas Plan, Ph.D.
Department for Karst and Caves
Natural History Museum
Vienna, Austria

Ezat Raeisi, Ph.D.
Department of Earth Science
University of Shiraz
Shiraz, Iran

Hermann Stadler, Ph.D.
Research Hydrologist
Institute of Water Resources Management,
Hydrogeology and Geophysics
Joanneum Research
Graz, Austria

Zoran Stevanovic, Ph.D.
Professor of Hydrogeology Research
Department of Hydrogeology
School of Mining and Geology
University of Belgrade
Belgrade, Serbia

Branka Trcek, Ph.D.
Professor of Environmental Protection
Faculty of Civil Engineering
University of Maribor
Maribor, Slovenia

William B. White, Ph.D.
Professor Emeritus of Geochemistry
Materials Research Institute and Department of Geosciences
The Pennsylvania State University
University Park, Pennsylvania, the United States

Qiang Wu, Ph.D.
Institute of Water Hazard Prevention and Water Resources
China University of Mining and Technology
Beijing, China

Liting Xing, Ph.D.
Institute of Water Hazard Prevention and Water Resources
China University of Mining and Technology
Beijing, China

Wanfang Zhou, Ph.D.
Earth Resources Technology
Huntsville, Alabama, the United States

Hans Zojer, Ph.D.
Institute of Water Resources Management, Hydrogeology and Geophysics
Joanneum Research
Graz, Austria

CHAPTER

1

Sustainability and management of springs

Neven Kresic
MACTEC Engineering and Consulting, Inc., Ashburn, Virginia

1.1 INTRODUCTION

The following excerpts from the anthological work by Gunnar Brune, *Major and Historical Springs of Texas*, published in 1975 by the Texas Water Development Board, are applicable to historical significance of springs around the world and their subsequent demise, caused first by ignorance and unrestricted land use and then by the overwhelming pressure of population growth and inadequate regulations:

> *Springs were vital to the survival of Texas' earliest inhabitants, over 30,000 years ago. At an archeological site near Lewisville in Denton County, radiocarbon analysis has dated the remains of these early new-world men at 37,000+ years old, including crude sculptures, spears, and spear throwers (Newcomb, 1961). These early Americans always made their campgrounds near water, whether it was a spring, spring-fed stream, a river, or a lake. Bedrock mortars or rock mills were worn into the rock by the Indians as they ground stool, acorns, and other nuts, mesquite beans and grain. These mortars can still be seen at many Texas springs. It is also noteworthy that the Pueblo Indians of west Texas used spring water for irrigation of crops long before the arrival of the Europeans (Taylor, 1902; Hutson, 1898).*
>
> *Because the springs were so vital to the life of both the Indians and the white man, it is not surprising that many battles were fought over their possession. In 1650 when Spanish explorers first visited Big Spring in Howard County, they found the Comanche and Pawnee Indians fighting for its possession. When a network of forts was strung across Texas, they were, in nearly all cases, located near springs in order to have a reliable supply of pure water. Later the covered-wagon and stagecoach routes came to rely heavily upon springs. For example, the "Camino Real" or King's Highway, completed by the Spanish colonists about 1697 from Natchitoches, Louisiana, to San Antonio and Mexico, passed 13 major Texas springs and many more minor ones. Most of the springs in West Texas are very small in comparison to those in central and east Texas, because of the very low rainfall and recharge. Nevertheless, they often meant the difference between life and death to the early pioneers.*
>
> *Nearly all of the larger springs were used for water power by the early settlers. At least 61 were used in this way. Gritsmills, flour mills, sawmills, cotton gins, and later electric generating plants were powered by the flow of spring water.*
>
> *In the late 1800's, many medicinal or health spas sprang up around the more mineralized springs. At least 25 springs, chiefly in east Texas, were believed to be beneficial in curing various ailments. Most of these waters are high in sulfate, chloride, iron, and manganese.*

Copyright © 2010, Elsevier Inc. All rights reserved.

Many of the early settlements relied entirely on spring water. At least 200 towns were named for the springs at which they were located. About 40 still are shown on the official Texas State Highway Map, but many of the springs have dried up.

Throughout the long period during which various Indian tribes occupied Texas, spring flow remained unchanged except as affected by wet and dry climatic cycles. At the time of Columbus' epic voyages Texas abounded with springs which acted as natural spillways to release the excess storage of underground reservoirs. Early explorers described them as gushing forth in great volume and numbers. The very early accounts usually describe not springs but "fountains". This is an indication of the tremendous force with which these springs spouted forth before they were altered by modern man. As an example, less than 100 years ago Big Boiling Spring, one of the Salado Springs (Bell County) was still described as a fountain rising 5 feet high. Such natural fountains ceased to exist in Texas many years ago.

Probably the first effect upon ground-water tables and spring flow was the result of deforestation by the early white settlers. Deforested land was placed in cultivation or pasture. The deep open structure of the forest soils was altered as the organic matter was consumed and the soils became more impervious. Heavy grazing by introduced stock animals was probably especially harmful. Soon the soils were so compacted that they could take in only a small fraction of the recharge which they formerly conveyed to the underground reservoir.

This reduction of recharge affected larger areas as more and more land was placed in pasture. However, the effect upon water tables and spring flow was probably relatively small in comparison with later developments. In the middle 1800's deep wells began to be drilled. It was found that flowing wells could be brought in nearly everywhere. The "Lunatic Asylum" well in Austin, drilled to the basal Trinity Sands, threw water 40 feet high. Water from a well south of San Antonio reaching the Edwards Limestone rose 84 feet above the surface of the ground (Hill and Vaughn, 1898). Nothing could have had a more disastrous effect upon spring flows than the release of these tremendous artesian pressures through flowing wells. Most of these wells were allowed to flow continuously, wasting great quantities of water, until the piezometric heads were exhausted and the wells stopped flowing.

Although the effects of flowing wells upon spring flow were severe, there was more to come. When the wells ceased flowing, pumping began. Ground-water levels were systematically drawn down, as much as 700 feet in some areas. At first pumping for municipal and industrial use was primarily responsible. In recent years tremendous quantities of ground water have been withdrawn for irrigation, amounting to about 80 percent of the total ground water used in Texas. As a result, some streams which were formerly "gaining" streams, receiving additional water from streambed seeps and springs, are now "losing", and many streams have ceased flowing. Thousands of small springs have dried up, and the larger springs have generally suffered a decrease in flow.

Natural spring waters if taken at their source are considered to be ground water and no permit is required for their use. Once they issue forth and flow in watercourse, however, they become public surface waters. As such, a permit from the Texas Water Rights Commission is required for their use.

A spring is normally a spillway for an underground reservoir. This reservoir may be overlain by land belonging to a number of owners. If the landowners other than the spring owner choose to pump ground water heavily, lowering the water table and causing the spring to cease flowing, the spring owner has no recourse in the courts to prevent them.

An example is Comanche Springs at Fort Stockton (Pecos County). These artesian springs, issuing from a Comanchean limestone ground-water reservoir, formerly flowed as much as 66 ft^3/s, and served the Comanche and other Indians for uncounted thousands of years. From 1875 on the springs were the basis for an irrigation district which supplied water to 6,200 acres of cropland. Heavy pumping of the aquifer lowered the water table so that the spring discharge began to fall off in May 1947 (U.S. Bureau of Reclamation, 1956). The irrigation district sought an injunction in 1954 to restrain pumping which interfered with the normal flow of Comanche Springs. The injunction was denied by the courts, and the springs ceased to flow in March 1961.

Figure 1–1 is a historic hydrograph of the Comanche Springs, based on flow measurements by the United States Geological Survey (U.S. Geological Survey, 2008), showing the cyclic influence of the aquifer pumping and the final drying up of the springs. Similar examples of a catastrophic influence of aquifer pumping on springs are numerous both in the United States and around the world (Figure 1–2). Since the environmental regulations and water resource governance vary in different states and different countries, so do actions to prevent or mitigate such negative impacts of groundwater withdrawals. For example, most states in the United States have some form of required permitting for large water users, including protection of seniority rights, as illustrated by the following excerpts from a news article (Barker, 2007):

Water experts, state officials and the businesspeople and farmers at the heart of Idaho's most heated water dispute are entering their third week of testimony in a case that could dictate how the state uses one of its most precious resources in the future. The hearing is slow, plodding and arcane as attorneys for fish farmers, groundwater-pumping irrigators and the state cross-examine hydrologists about spring flows or state officials about the papers they have filed. To settle the disputes, though, officials and judges will have to resolve several points on which the water users disagree. The consequences of the decision could be massive.

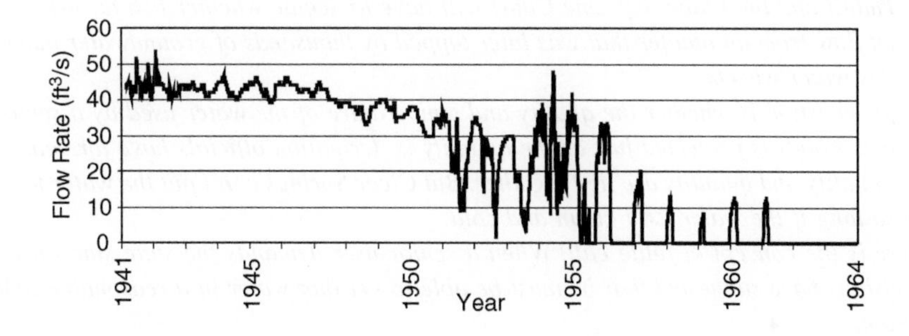

FIGURE 1–1 Mean daily discharge hydrograph of Comanche Springs. (Data from U.S. Geological Survey, 2008).

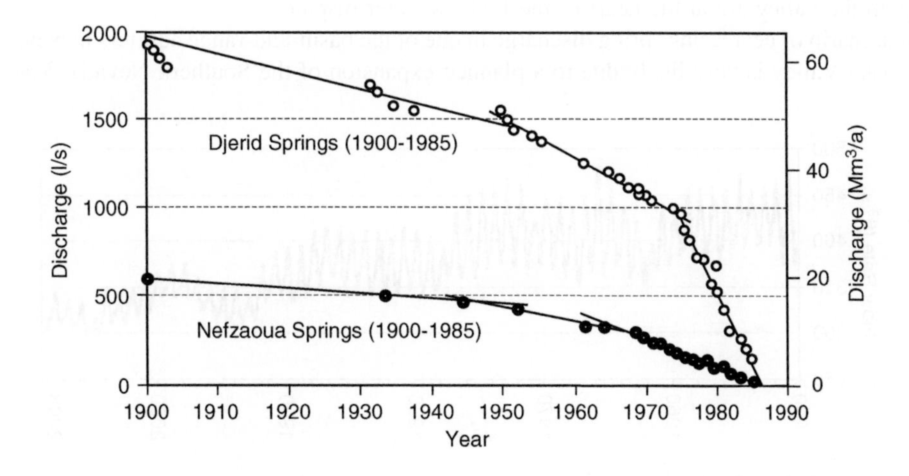

FIGURE 1–2 Progressive elimination of major spring flows in southern Tunisia during the 20th century. (From Margat, Foster, and Droubi, 2006; copyright UNESCO.)

At stake is whether the state shuts off the pumps that bring water to thousands of acres of farmland, factories and towns across central and eastern Idaho. Or will fish farmers, Clear Springs Foods and Blue Lakes Trout Farm, which have lost millions of dollars due to dropping flows from the springs on which they depend, face permanent losses?

The debate turns on legal principles that have been teased out in disputes among surface water users for nearly a century. But Schroeder and the Supreme Court may plow new ground on how the law treats disputes between people who pump their water from the ground and those who get their water from springs and rivers. The justices will have to weigh two competing sections of constitutional water law: prior appropriation—first come, first served; and the imperative to use water for its full economic development.

The primary case in surface water law upheld the constitution's prior appropriation doctrine, but the U.S. Supreme Court also ruled in favor of full economic development in 1912. In that case, Henry Schoedde was watering his crops and running a mining operation by using water wheels with buckets to take the water out of the Snake River and lift it into his canals. He sued the Twin Falls Canal Co. because its Milner Dam threatened to lower the river to a point where his water wheels wouldn't turn. The U.S. Supreme Court ruled Schoedde did not own the current in the river and could not stop others from diverting water to protect it.

Schroeder, Tuthill and the Idaho Supreme Court will have to decide whether fish farmers, who use artesian springs that flow from an aquifer that was later tapped by thousands of groundwater pumpers, are like Schoedde and his water wheels.

Another major question is whether the quality and temperature of the water used by a senior water user like Clear Springs Foods is protected just as the quantity is. Irrigation officials have for years disputed the idea that water quality and quantity are tied together. But Clear Springs can't put the water to the beneficial use it has for so long if the water isn't clean and cold.

Finally, there is the concept of futile call. When a senior user demands the state shut off a junior user, past water decisions have suggested that he must be able to get that water in a reasonable time or his call for water is moot.

Figure 1–3 is a hydrograph of one of the numerous large springs that discharge along the Snake River valley, illustrating an obvious decreasing trend caused by groundwater pumpage from the aquifer. This and other springs in the valley are at the heart of the Idaho's water dispute.

A similar scenario of decreasing spring discharge in one of the basin-and-range aquifers in Nevada and Utah underlying Snake Valley is very likely due to a planned expansion of the Southern Nevada Water Authority

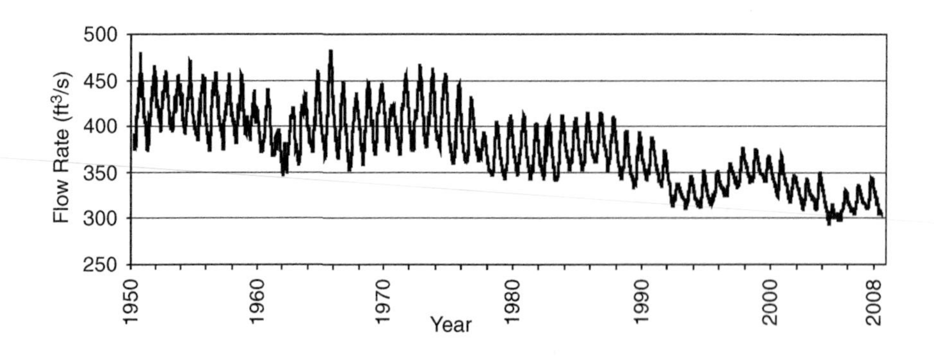

FIGURE 1–3 Mean daily discharge hydrograph of Box Canyon Springs in Twin Falls, Idaho. (Data from U.S. Geological Survey, 2008.)

(SNWA) water supply system, driven by the exploding growth of Las Vegas. This is the main reason why various environmental and citizens groups have challenged such expansion in courts and with public outreach. In addition, the state of Utah has initiated a comprehensive groundwater monitoring program in the area and completed a report summarizing likely impacts of the proposed large-scale pumping from the Snake Valley aquifer on its water resources (Kirby and Hurlow, 2005). At the time of this writing, the Utah Association of Counties is preparing to petition the federal government, requesting that four agencies of the Department of Interior, including the Bureau of Land Management and the Bureau of Indian Affairs, file protests against the plan.

A preview of what may be the ultimate impact of the proposed water diversion by SNWA is destiny of the Needle Point Spring, located in the southern part of Snake Valley. A series of wells were drilled from 1993 to 2000 to supply water for 13 center-irrigation pivots. In summer 2001, the spring, only about 1 mile from the nearest well, dried up (Figure 1–4). A number of wild horses died because they had imprinted on the spring and did not know where else to look for water. Although irrigation pumping lasts only from April to October of each year, water never flowed again from Needle Point Spring (Baker Ranches, Inc., 2008). An assessment of the historic and hydrogeologic conditions at the Needle Point Spring, including possible causes of the cessation of spring flow, is provided by Summers (2001–2005). The impact of cyclical aquifer pumping for irrigation is visible in Figure 1–5. As stated by Summers, information from drillers logs indicates that the wells are 16 in. diameter wells, designed for high-capacity irrigation pumping. Although pumping test information is not available on all the wells, the water right applications indicate pumping rates of individual wells between 500 and 1300 gpm.

The following is one of the Memo Decisions by the regulatory agency responsible for authorizing groundwater withdrawals in the vicinity of Needle Point Spring; it illustrates the circumstances that lead to the spring's demise (from Summers, 2001–2005):

> *The USA Bureau of Land Management has expressed concern about their prior rights in Needle Point Spring which may be affected by an appropriation (Water Filing 18-630) from the same well as discussed herein. Although it does not appear that this application was protested by the BLM, diversions under this application are junior in priority to the rights of the BLM and must not impair those rights. If an impairment does occur, it may be necessary for the applicant to replace or compensate for the water lost by the BLM. However, the records of the State Engineer have been reviewed. It appears that the well is about one mile distant from Needle Point Spring which reduces any likelihood that impairment will occur.*

The conclusions of the Utah Geological Survey assessment of likely impacts of the proposed water diversion from Snake Valley are as follows (Kirby and Hurlow, 2005):

- Wells proposed by the Southern Nevada Water Authority will likely adversely affect groundwater conditions in nearby Utah.
- The total drawdown of groundwater near Garrison in western Millard County could be greater than 100 ft (31 m).
- The proposed pumping may change or reverse groundwater flow patterns for much of the east-central Great Basin in Utah and Nevada. The effects may eventually propagate eastward and impact discharge at important regional springs in Wah Wah Valley and Tule Valley.
- Discharge of agriculturally and ecologically important springs will decrease.

Clay Spring is just one of the many springs in Snake Valley that could be affected by the proposed large-scale groundwater pumping (Figure 1–6). It originally attracted some of the first settlers in Snake Valley, the Clay family. Later it was used by the Civilian Conservation Corps as a water source for their camp in the 1930s. The spring pool is very deep and has not been explored yet (Baker Ranches, Inc., 2008).

FIGURE 1–4 Needle Point Spring on March 10, 1994 (top), and April 24, 2002 (bottom). (From Summers, 2001–2005.)

The photograph in Figure 1–7 illustrates the contrast of a spring ecosystem that supports a variety of wildlife including migrating birds, deer, elk, pronghorn, small mammals, and aquatic organisms, with the surrounding dry Great Basin Desert. A small drawdown in groundwater near this area could reduce or eliminate spring flow and result in the disappearance of these important oases (Baker Ranches, Inc., 2008).

The disappearance of springs due to large-scale groundwater withdrawals from aquifers is not unique to arid or semi-arid regions with low rainfall and low natural aquifer recharge. Figure 1–8 shows measured

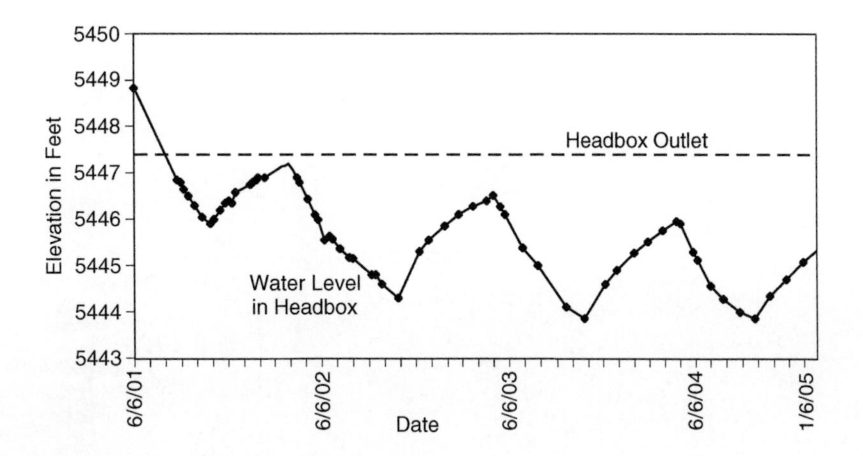

FIGURE 1–5 Needle Point Spring water levels. Spring dried up when water level in the headbox dropped below the outlet. (Modified from Summers, 2001–2005.)

FIGURE 1–6 Clay Spring emerges in South Snake Valley and helps water Burbank Meadows, a large wetland and wet meadow complex, and Pruess Lake, a warm-water fishery. (Photo courtesy of Gretchen Baker, ProtectSnakeValley.com.)

FIGURE 1–7 A spring in Spring Valley adjacent to Snake Valley. (Photo courtesy of Gretchen Baker, ProtectSnakeValley.com.)

discharge rates of Kissengen Spring in Florida. This historic large spring, located near the city of Bartow in Polk County, served as a popular recreational area for decades (Figure 1–9) and had an average flow of about 29 ft^3/s (19 million gallons per day, mgd). The cessation of spring flow, at both Kissengen Spring and many minor springs in the upper Peace River basin, was related to the regional lowering of the potentiometric surfaces of the intermediate aquifer system and the Upper Floridan aquifer from 1937 to 1950 for as much as 60 ft, mainly due to phosphate mining (Peek, 1951; Stewart, 1966; Lewelling, Tihansky, and Kindinger, 1998). Permanent cessation of the spring flow occurred in April 1960.

In addition to quantity, water quality of a once pristine spring can be diminished by unsustainable land-use practices in its drainage area, to the point of a complete loss of the source. Arguably, almost every human activity has the potential to directly or indirectly affect groundwater to a certain extent. Figure 1–10 illustrates some of the land-use activities that can result in groundwater and ultimately spring contamination (see also Chapter 8). An exponential advancement of analytical laboratory techniques in the last decade demonstrated that many synthetic organic chemicals are widely distributed in the environment, including in

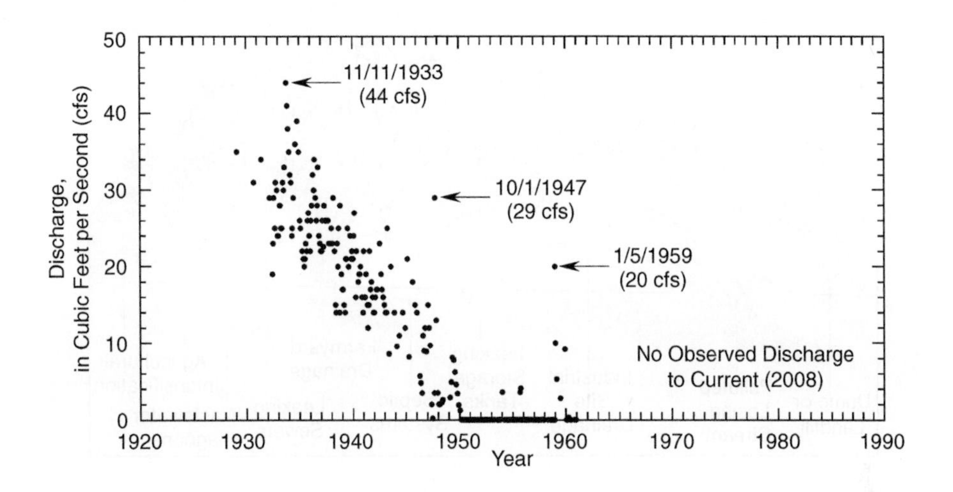

FIGURE 1–8 Periodic discharge measurements at Kissengen Spring in Polk County, central Florida. The spring remains dry through the present day. (Modified from Lewelling, Tihansky, and Kindinger, 1998.)

FIGURE 1–9 Historic photo of Kissengen Spring, which remains dry to the present day after all flow ceased in 1960. (From Peek, 1951.)

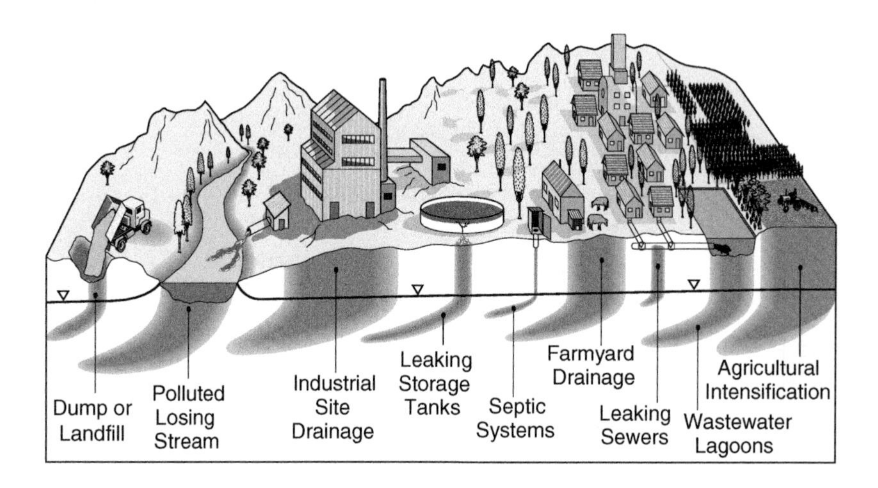

FIGURE 1–10 Land-use activities commonly generating a groundwater pollution threat. (From Foster et al., 2002–2005. Source: The World Bank, printed with permission.)

groundwater, and a considerable number of them can now be found in human tissue and organs of people living across the globe.

Not withstanding assertions that any water can be treated to become drinkable (potable), it is not surprising that multinational companies selling bottled water are frantically looking for new sources of springwater across the globe. This is understandable, since many consumers worldwide are ready to pay a premium for brands marketed as "pure springwater." In contrast, quite a few European countries, where springs of high water quality are abundant, utilize them as preferred sources of public water supply (Figure 1–11; see also Chapter 10.1) and are continuously implementing various measures for their protection. The city of Vienna, Austria, is a prime example of scientific, engineering, and regulatory efforts, at all levels, aimed at protecting its famed water supply based on springs (Chapter 10.2). This tradition in Europe can be traced back to the Romans, who preferred springwater to any other sources and perfected the art of spring capture and water distribution via spectacular aqueducts throughout their vast empire (Figures 1–12 through 1–14; see also Chapter 10.1).

As illustrated by the excerpts from Brune (1975), in Texas and the United States as a whole, interest in groundwater initially also focused on springs, particularly in the arid West. As the eastern part of the country

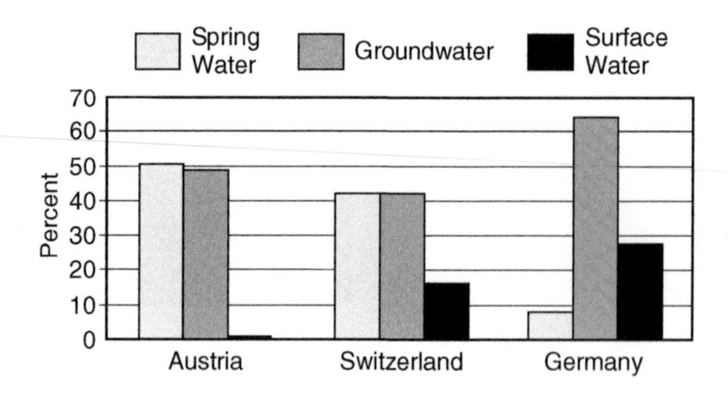

FIGURE 1–11 Utilization of springs for public water supply in Switzerland, Austria, and Germany. (Modified from Austrian Museum for Economic and Social Affairs, 2003.)

FIGURE 1–12 Part of an aqueduct built in 3rd century AD by Roman emperor Diocletian for water supply of his summer residence on the Adriatic coast (shown in Figure 1–13). The aqueduct brought water from the large karst spring shown in Figure 1–14. (Photograph courtesy of Ivo Eterovic.)

FIGURE 1–13 Bird's-eye of a restitution of Diocletian's palace in Split by the architect Ernest Hébrard. (Drawing from E. Hébrard and J. Zeiller, *Spalato, le Palais de Dioclétien*, Paris, 1912.)

developed and most of the land became privately owned, the groundwater-based public water supply shifted from springs to wells. Advances in well drilling, pump technology, and rural electrification made possible the broad-scale development of groundwater in the West beginning in the early 1900s. Large-scale irrigation with groundwater, especially after World War II, spread rapidly throughout the West, resulting in the cessation of spring flows at many locations. As a consequence, in both the eastern and western United States, the overall

FIGURE 1–14 The Jadro Spring, initially tapped by Roman emperor Diocletian in 3rd century AD, is still used for water supply of the Croatian port city of Split. (Photograph courtesy of Ivana Gabric).

utilization of springs for centralized water supply is minor compared to other parts of the world (Kresic, 2009). Where there are numerous large springs, as in the karst regions of Florida, Texas, and Missouri, for example, in many cases, such springs are located on private land or public park land and preserved for other uses, including recreation. A paragraph from Meinzer's (1927) publication on large springs in the United States illustrates this point for Florida springs:

> *Some of the springs have become well-known resorts, but otherwise not much use is made of their water. The fascinating character of these springs is indicated by the following vivid description of Silver Spring, abbreviated from a description given in a booklet published by the Marion County Chamber of Commerce. "The deep, cool water of Silver Spring, clear as air, flows in great volume out of immense basins and caverns in the midst of a subtropical forest. Seen through the glass-bottom boats, with the rocks, under-water vegetation, and fish of many varieties swimming below as if suspended in mid-air, the basins and caverns are unsurpassed in beauty. Bright objects in the water catch the sunlight, and the effects are truly magical. The springs form a natural aquarium, with 32 species of fish. The fish are protected and have become so tame that they feed from one's hand. At the call of the guides, hundreds of them, of various glistening colors, gather beneath the glass-bottom boats."*

Unfortunately, as illustrated by the example of Florida's Kissengen Spring, the preservation of large springs for beneficial uses is becoming increasingly difficult and requires a sustained, coordinated effort by the water utilities, regulatory agencies, legislature, and public (see Chapter 10.9).

1.2 CONCEPT OF SUSTAINABILITY

The term *sustainable development* was popularized by the World Commission on Environment and Development in its 1987 report, *Our Common Future* (UN Department of Economic and Social Affairs, 2009). The aim of the report, accepted by the United Nations General Assembly, was to find practical ways of addressing the environmental and developmental problems of the world. In particular, it had three general objectives:

- To reexamine the critical environmental and developmental issues and formulate realistic proposals for dealing with them.
- To propose new forms of international cooperation on these issues that will influence policies and events in the direction of needed changes.
- To raise the level of understanding and commitment to action of individuals, voluntary organizations, businesses, institutes, and governments.

In various publications, debates, interpretations, and reinterpretations over the course of years, the commission's report was, in many cases, stripped down to the following widely cited single sentence: "Sustainable development is development that meets the needs of the present without compromising the ability of future generations to meet their own needs." Since this sentence seems to focus on human generations only, it has been criticized by some as too narrow and failing to address the natural environment. However, the commission and the General Assembly did address the human and natural environments as a whole and in a holistic manner, which can be seen from key related statements of the official UN resolution 42/187 (UN Department of Economic and Social Affairs, 2009).

One commonly held and inaccurate belief, when estimating water availability and developing sustainable water supply strategies, is that groundwater use can be sustained if the amount of water removed is equal to recharge—often referred to as the *safe yield*. However, no volume of groundwater withdrawal can be truly free of any adverse consequence, especially when time is considered. The "safe yield" concept is therefore a myth, because any groundwater artificially extracted from an aquifer must come from somewhere and used to flow somewhere else. The myth falsely assumes that there are no effects on other elements of the overall water budget, including flow of any spring(s) draining the aquifer. Bredehoeft, Papadopulos, and Cooper (1982) and Bredehoeft (2002) provide illustrative discussions about the safe yield concept and the related "water budget myth."

The false premise of the "safe yield myth" and its impact on a spring and surface water stream fed by it are presented in Figure 1–15. If pumping (withdrawal) from the aquifer equals or exceeds all inflows (recharge) to the aquifer, the aquifer discharge via the spring eventually reaches zero, resulting in some adverse consequence at some point in time. Similarly, if most or all water discharging from the aquifer at the spring is captured and diverted to other users, there inevitably will be some adverse consequences on the surface-water-dependent ecosystems and users.

The direct hydrologic effects in either case will be equal to the volume of water removed from the natural system, but those effects may require some time to manifest. Because aquifer recharge, groundwater withdrawals, and amount of water captured at a spring can vary substantially over time, these changing rates can be critical information for developing both aquifer (i.e., groundwater in general) and spring management strategies.

With an increased demand for water and pressures on groundwater resources, the decades-long debate among water professionals about what constitutes "safe" withdrawal of groundwater has now changed into a debate about the "sustainable use" of groundwater. The difference is not only semantic, and confusion has occasionally resulted. For example, attempts have been made to distinguish between "safe yield" and "sustainable pumping," where the latter is defined as the pumping rate that can be sustained indefinitely without mining or dewatering the aquifer. Devlin and Sophocleous (2005) provide a detailed discussion of these and other related concepts.

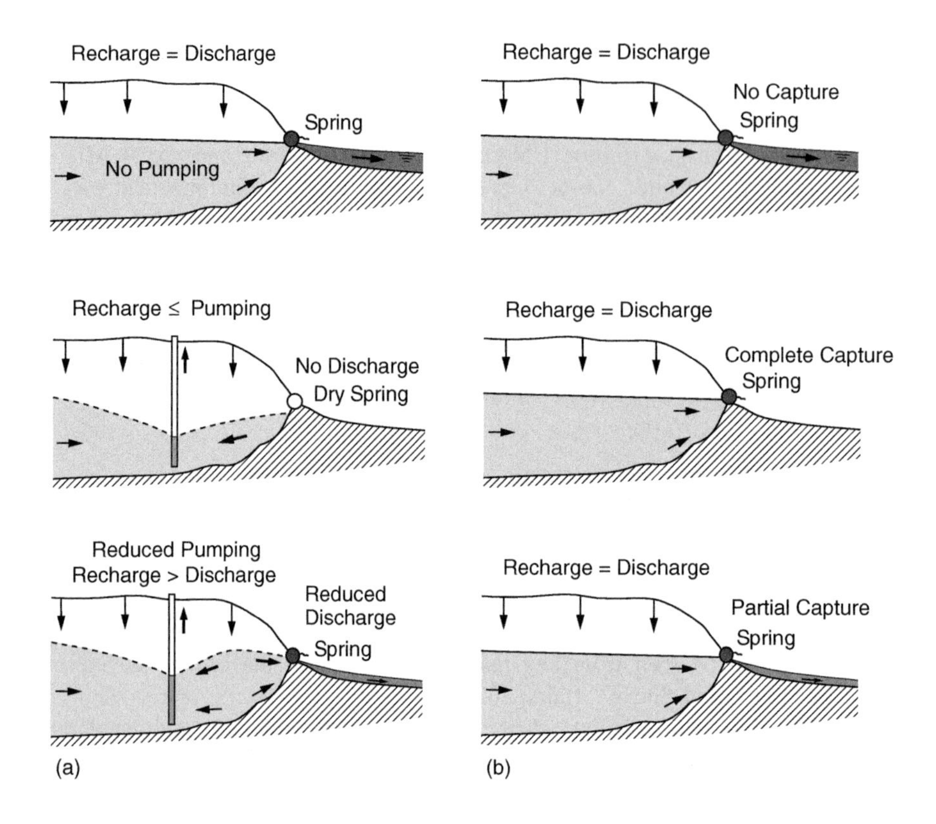

FIGURE 1–15 Spring sustainability affected by pumping from an aquifer (a) in comparison to surface stream sustainability affected by spring capture (b). Groundwater pumping or spring capture that is considered sustainable can be achieved only by accepting some consequences on surface water ecosystems and users.

What appears most difficult to understand is that the groundwater system is a dynamic one—any change in one portion of the system ultimately affects its other parts as well. Even more important, most groundwater systems are dynamically connected with surface water. As groundwater moves from the recharge area toward the discharge area (e.g., a spring or river), it constantly flows through the saturated zone that is the groundwater storage (reservoir). If another discharge area (such as a well for water supply) is created, less water flows toward the old discharge area (spring or river). This fact seems to be paradoxically ignored by those who argue that groundwater withdrawals may actually increase aquifer recharge by inducing inflow from recharge boundaries (such as surface water bodies) and therefore result in "sustainable" pumping rates. Although such groundwater management strategy may be "safe" or "sustainable" for the intended use, another question is whether it has any consequences for the sustainable use of the surface water system that now loses water rather than gains it from the groundwater system, including via springs (Kresic, 2009).

Another argument for sustainable pumping is based on managing groundwater storage. This management strategy adjusts withdrawal (pumping) rates to take advantage of natural recharge cycles. For example, during periods of high demand, some water may be withdrawn from the storage by greatly increasing pumping rates and lowering the hydraulic heads (water table) in the aquifer. During periods of low demand (low pumpage) and high natural recharge, this depleted storage would then be replenished (this is also one of the principles of spring regulation discussed in Chapter 9). However, the same question of the sustainability of this approach remains. Any portion of the natural recharge that does not contribute to the natural (nonanthropogenic) discharge has some consequences for the water users and water uses that rely on it. Depending on the volumes and rates of the denied groundwater discharge, the affected users may or may not be able to adapt to the new reality.

To sustain valued ecosystems and endangered species, segments of societies worldwide expect water to be made available, in volumes not easily quantified, to meet key habitat requirements. This relatively recent trend is accompanied by actions of environmental groups, which include legal challenges and lawsuits against various government agencies in charge of water governance. Only several decades ago, similar involvement of nongovernment groups or the public was virtually nonexistent.

The multiple aspects of groundwater sustainability are addressed in the Alicante Declaration, which since its initiation has gained wide recognition among groundwater professionals worldwide. The declaration is the action agenda that resulted from debates held in Alicante, Spain, on January 23–27, 2006, during the International Symposium on Groundwater Sustainability. This call for action for responsible use, management, and governance of groundwater is reproduced here in its entirety:

Water is essential for life. Groundwater—that part of all water resources that lie underneath land surface—constitutes more than ninety five percent of the global, unfrozen freshwater reserves. Given its vast reserves and broad geographical distribution, its general good quality, and its resilience to seasonal fluctuations and contamination, groundwater holds the promise to ensure current and future world communities an affordable and safe water supply. Groundwater is predominantly a renewable resource which, when managed properly, ensures a long-term supply that can help meet the increasing demands and mitigate the impacts of anticipated climate change. Generally, groundwater development requires a smaller capital investment than surface water development and can be implemented in a shorter timeframe.

Groundwater has provided great benefits for many societies in recent decades through its direct use as a drinking water source, for irrigated agriculture and industrial development and, indirectly, through ecosystem and stream flow maintenance. The development of groundwater often provides an affordable and rapid way to alleviate poverty and ensure food security. Further, by understanding the complementary nature of ground and surface waters, thoroughly integrated water-resources management strategies can serve to foster their efficient use and enhance the longevity of supply.

Instances of poorly managed groundwater development and the inadvertent impact of inadequate land-use practices have produced adverse effects such as water-quality degradation, impairment of aquatic ecosystems, lowered groundwater levels and, consequently, land subsidence and the drying of wetlands. As it is less costly and more effective to protect groundwater resources from degradation than to restore them, improved water management will diminish such problems and save money.

To make groundwater's promise a reality requires the responsible use, management and governance of groundwater. In particular, actions need to be taken by water users, who sustain their well-being through groundwater abstraction; decision makers, both elected and non-elected; civil society groups and associations; and scientists who must advocate for the use of sound science in support of better management. To this end, the undersigners recommend the following actions:

- ***Develop more comprehensive water-management, land-use and energy-development strategies that fully recognize groundwater's important role in the hydrologic cycle**. This requires better characterization of groundwater basins, their interconnection with surface water and ecosystems, and a better understanding of the response of the entire hydrologic system to natural and human-induced stresses. More attention should be given to non-renewable and saline groundwater resources when such waters are the only resource available for use.*

- ***Develop comprehensive understanding of groundwater rights, regulations, policy and uses****. Such information, including social forces and incentives that drive present-day water management practices, will help in the formulation of policies and incentives to stimulate socially- and environmentally-sound groundwater management practices. This is particularly relevant in those situations where aquifers cross cultural, political or national boundaries.*
- ***Make the maintenance and restoration of hydrologic balance a long-term goal of regional water-management strategies****. This requires that water managers identify options to: minimize net losses of water from the hydrologic system; encourage effective and efficient water use, and ensure the fair allocation of water for human use as well as ecological needs, taking into account long-term sustainability. Hydrological, ecological, economic and socioeconomic assessments should be an integral part of any water-management strategy.*
- ***Improve scientific, engineering and applied technological expertise in developing countries****. This requires encouraging science-based decision-making as well as "north-south" and "south-south" cooperation. Further, it is important that funds be allocated for programs that encourage the design and mass-dissemination of affordable and low-energy consuming water harnessing devices for household and irrigation.*
- ***Establish ongoing coordinated surface water and groundwater monitoring programs****. This requires that data collection become an integral part of water-management strategies so that such strategies can be adapted to address changing socio-economic, environmental, and climatic conditions. The corresponding data sets should be available to all the stakeholders in a transparent and easy way.*
- ***Develop local institutions to improve sustainable groundwater management****. This requires that higher-level authorities become receptive to the needs of local groups and encourage the development and support of strong institutional networks with water users and civic society.*
- ***Ensure that citizens recognize groundwater's essential role in their community and the importance of its responsible use****. This requires that science and applied technology serve to enhance education and outreach programs in order to broaden citizen understanding of the entire hydrologic system and its global importance to current and future generations. (Available at http://aguas.igme.es/igme/ISGWAS)*

1.3 SPRING MANAGEMENT

As pointed out by Brune (1975), the study of springs is a borderline discipline, because they represent the transition from groundwater to surface water. Hence, they have been studied to some extent by groundwater specialists (hydrogeologists) and to some extent by surface-water specialists (hydrologists). Overall, however, comprehensive studies of springs were relatively rare until recently, when the concept of sustainability of water resources came into the limelight worldwide. As illustrated by the examples from the preceding section, springs are a direct reflection of the state of groundwater in the aquifers that feed them, and they directly influence streams and other surface-water bodies into which they discharge, including all dependent ecosystems. Spring management therefore includes many of the same principles that guide management of both surface-water and groundwater resources.

In an ideal situation, when the springwater is not consumed in any way (nonconsumptive use) and there is minimal disturbance of the spring site before formation of or discharge into a surface stream, the spring management should be focused on protecting the quality and quantity of groundwater in its drainage area for the specific spring use. Some examples include recreation, fisheries (Figure 1–16), and power generation. When

(a)

(b)

FIGURE 1–16 Box Canyon Springs near Twin Falls, Idaho (a), which feed the trout farm through pipes running across the bottom of the Snake River (b). (Photographs courtesy Clear Foods, Inc.)

the springwater is used for consumptive water supply, of any type, the additional management requirements include

- Securing that clearly defined and reliable quantities of water are delivered to all users, including environmental flows, using various engineering means (see Chapters 4, 5, and 9).
- Ensuring that an adequate quality of water is delivered to all users, including drinking water treatment where applicable (Chapter 7).
- Establishing spring protection zones as required by the regulations (Chapter 8).

Two main prerequisites for establishing a realistic, workable management plan are

1. Hydrogeologic and hydrologic characterization of the spring type, drainage area, and recharge and discharge parameters, such as water quality and quantity (Chapters 2, 3, 4, and 6).
2. Reliable predictive modeling of spring discharge and water quality under natural and engineered conditions, including during possible artificial spring regulation (Chapters 5 and 9).

In addition to these management components, which are parts of supply-side management, all sources of water supply, including springs, should always include demand-side management as well. The concept of water demand management generally refers to initiatives that have the objective of satisfying existing needs for water with a smaller amount of available resources, normally through increasing the efficiency of water use. Water demand management can be considered a part of water conservation policies, which describe initiatives with the aim of protecting the aquatic environment and ensuring a more rational use of water resources.

Unfortunately, water demand management is often given a low priority or practiced reluctantly, in part due to the false premise that it involves only raising fees. The water sector is heavily subsidized in both developed and developing countries, and politicians are usually very hesitant to "wrestle" with the issue of water pricing, especially during elections. However, in addition to pricing, which is usually an effective means of demand management, many other measures, when combined, can be as effective. A very detailed discussion on demand management tools, water use, and water conservation, with examples from European countries is presented in the report *Sustainable Water Use in Europe. Part 2: Demand Management*, published by the European Environment Agency (EEA) in 1999. The following are brief excerpts from this comprehensive report, illustrating the importance of two demand-management measures, other than water pricing that can be effective in reducing pressure on water resources including springs:

> *Losses in water distribution networks can reach high percentages of the volume introduced. Thus, leakage reduction through preventive maintenance and network renewal is one of the main elements of any efficient water management policy. Leakage figures from different countries indicate the different states of the networks and the different components of leakage included in the calculations (e.g. Albania up to 75%, Croatia 30–60%, Czech Republic 20–30%, France 30%, and Spain 24–34%).*
>
> *In agriculture, the aim of the education programs is to help farmers optimize irrigation. This can be achieved through training (on irrigation techniques), and through regular information on climatic conditions, irrigation volume advice for different crops, and advice on when to start/stop adjusting irrigation volumes according to rainfall and type of soil.*

One example of a very efficient demand-side management in the Middle Ages comes from the Republic of Dubrovnik (now part of Croatia), where the capital city of Dubrovnik had a centralized water supply based on springs long before most other European capitals. After the fountain shown in Figure 1–17 was completed, a new law was enacted stating that, "The right hand of everyone caught diverting water from or plugging the

FIGURE 1–17 Great Fountain of Onofria in Dubrovnik, Croatia, designed and built by the Italian architect Onofria della Cave in 1440. Water for the fountain, which includes the water storage reservoir, was conveyed from a group of springs by a 12-km-long system of canals and aqueducts. (Photograph courtesy of Vojislav Ilic.)

conveyance canal will be cut off" and "Public servants will be sent every week to check the entire water conveyance system and determine if anything is damaged or spoiled" (Tušar, 2008). Although hardly imaginable today in the civilized world, this practice shows that those stealing or "spoiling" water designated for the use of others should bear some clearly defined consequences.

Similarly, spring (and water resource) management should have a clearly stated objective. This is true for any level of management, starting with a local water agency or water purveyor and ending at the national (federal) level. The management objective should include the establishment of threshold values for readily measured quantities such as spring flows (and associated groundwater levels in the aquifer), water quality, or changes in stream flow and surface water quality where they affect or are affected by spring discharge. When a threshold level is reached, the rules and regulations require that the amount of captured water diverted from the spring or groundwater extraction from the aquifer be adjusted or stopped to prevent exceeding that threshold.

Management objectives may range from entirely qualitative to strictly quantitative. For each particular spring, they would have a locally determined threshold value, which can vary greatly. For example, in establishing a management objective for springwater quality, one area may simply choose to establish an average value of turbidity or nitrate concentration as the indicator of whether a management objective is met, while another agency may choose to have no constituents exceeding the maximum contaminant level for public drinking water standards. Despite the great latitude in establishing management objectives, local managers of springs should remember that the objectives should support the goal of a sustainable supply for the beneficial use of the water in their particular area (California Department of Water Resources, 2003).

As discussed by Kresic (2009), spring management starts with the agreement of all stakeholders as to what constitutes its sustainable use. This agreement has to be binding and within a clearly defined regulatory framework. Whenever applicable, spring management should be seamlessly integrated with the management of groundwater (aquifer), surface water, storm water, and used water (wastewater), thus constituting an integrated water resources management (IWRM). For both to be effective or even possible, the spring management and the IWRM must rely on monitoring the water quantity, quality, and their spatial and temporal changes for all parts of the water cycle. This monitoring should include the ambient groundwater (before it is discharged at the spring), the spring discharge water, the storm water, the surface water affecting or being affected by the spring, and the wastewater generated and discharged within the spring drainage area. All monitoring data, as well as all data generated during water resource evaluation, development, and exploitation (operations and maintenance), should be stored and organized within an interactive, geographic information systems–based database.

1.3.1 Source and resource protection

Protection of aquifers and the springs that drain them (i.e., groundwater resources in general) is achieved by the prevention of possible contamination, remediation of already contaminated groundwater, and detection and prevention of unsustainable extraction. The prevention aspect includes pollution prevention programs and control measures at potential contaminant sources, land-use control, and public education. Some examples of prevention measures include (Kresic, 2009)

- Mandatory installation of devices for early detection of contaminant releases, such as leaks from underground storage tanks at gas stations and landfill leachate migration.
- Banning pesticide use in sensitive aquifer recharge areas.
- Land-use controls that prevent an obvious introduction of contaminants into the subsurface, such as from industrial, agricultural, and urban untreated wastewater lagoons.
- Land-use controls that minimize the interruption of natural aquifer recharge, such as the paving of large urban areas ("urban sprawl").
- Management of urban runoff that can contaminate both surface and groundwater resources (see, e.g., U.S. EPA, 2005).

Probably the single most important aspect of groundwater protection is public education. Unfortunately, it is also often the most underfunded or completely disregarded. Many simple means of educating the public can pay off many times more than the investment made. Some examples include public outreach with programs describing septic tank maintenance and proper use (e.g., see Riordan, 2007), disposal of toxic wastes generated in households (e.g., paints, solvents, garden pesticides), and proper disposal of unused pharmaceuticals. In terms of protecting the availability (quantity) of groundwater in the areas where it is used for the water supply, public outreach programs on water conservation are irreplaceable. Perhaps the most receptive audience to groundwater education programs are the numerous visitors of state parks that were established because of springs. Such parks should be used as "role models" for the importance of groundwater and spring protection and featured in media and school programs as much as possible.

One such park is the Barton Springs in Austin, Texas (Figure 1–18). In 2006, the U.S. Geological Survey published a scientific investigation report that summarized water quality sampling performed at the springs from 2003 to 2005. The water was found to be affected by persistent low concentrations of atrazine

FIGURE 1–18 Barton Springs in Austin, Texas, with the main spring on the far bank, which fills a "swimming hole" that many Austinites regard as sacred. On this day in the early 1990s, the pool was closed and drained because of water quality concerns. (Photograph courtesy of Gregg Eckhardt.)

(an herbicide), chloroform (a by-product of drinking water disinfection), and tetrachloroethane (a solvent). In 2008, the fight to preserve Barton Springs was the subject of *The Unforseen*, a documentary coproduced by Robert Redford, who learned to swim there as a child. The movie uses the struggle over development in the Barton Creek watershed to illustrate the many clashes between private property rights and resource protection that are occurring across the country. The film drew great reviews, but some developers said it went too far and portrays them unfairly. Environmentalists said the movie is not hard enough on those who would develop lands at the expense of common resources like Barton Springs (www.edwardsaquifer.net/barton.html).

Approaches to groundwater resources protection differ at various levels of government and may mean different things to different stakeholders, thus emphasizing the need for public education and dialogue. For an individual household that has just discovered (or was told) that the springwater it drinks has been contaminated with a dangerous carcinogenic substance for years, it would be impossible not to state that their government failed these people. Unfortunately, similar cases occur daily and worldwide. On the other hand, in many developed countries, quite a few government programs and regulations are aimed at groundwater protection.

In the United States, examples of programs that fully or in part address pollution prevention include the Source Water Assessment Program, Pollution Prevention Program, Wellhead Protection Program, aquifer vulnerability assessments, vulnerability assessments of drinking water/wellhead protection, Pesticide State Management Plan, Underground Injection Control Program, and Superfund Amendments and Reauthorization Act Title III Program (Kresic, 2009).

At the state level, this long list often translates into development and implementation of a Source Water Assessment Program, which is focused on the delineation of wellhead (or "springhead") protection areas for the existing sources of water supply. This regulatory practice is present in most developed and many developing countries (see Chapter 8). Some regulators may include a requirement for inventorying potential contaminant sources within the delineated springhead (wellhead) protection areas, including releasing this information to the public.

In the United States, the overall protection of the groundwater resource still remains a rather vague concept and is certainly not subject to any legally enforced, overall regulation by the states or the federal government. As noted by the U.S. EPA (1999): "Ground water management in this country is highly fragmented, with responsibilities distributed among a large number of federal, state, and local programs. At each level of government, unique legal authorities allow for the control of one or more of the ground water threats described in Section 3.0. These authorities need to complement one another and allow for comprehensive management of the ground water resource."

The main reason for this lack of comprehensive, overall groundwater protection regulation is that any such regulation would require a lot of political will, because it would have to include strict land-use controls and significant resources to monitor and enforce land-use practices. This, by definition, includes any agricultural, industrial, or other activity that uses land. As this is not feasible in the near future, in most cases, the real protection of both the existing sources and the resource is, in the end, left to the local communities and public water systems. They have to develop management plans, involving all local stakeholders and the public, that would minimize risks to their water supplies and their resource. Simply put, if the community (all stakeholders included) believes that a certain industry or land-use activity will not threaten its sources and the resource as a whole (both quantity and quality), everyone should be satisfied. If this is not the case, there are usually four options: (1) public education and outreach that may resolve the issue by voluntarily changing questionable land uses, (2) enactment of local regulations and ordinances that may leave some stakeholders still unhappy, (3) land acquisition and land conservation, and (4) a lawsuit (Kresic, 2009).

As pointed out by Rogers and Hall (2003), any water resource or source (e.g., a spring) management is part of the overall water governance, which includes the ability to design public policies and institutional frameworks that are socially acceptable, equitable, and environmentally sustainable. Given the complexities of water use within society, effective water governance requires the involvement of all stakeholders and must ensure that disparate voices are heard and respected in decisions on development, allocation, and management of common waters and use of financial and human resources. Governance aspects overlap the technical and economic aspects of water but include the ability to use political and administrative elements to solve a problem or exploit an opportunity (Rogers and Hall, 2003).

A recent example of groundwater governance on a grand political scale is the Groundwater Directive by the European Parliament, which refers to groundwater as "the most sensitive and the largest body of freshwater in the European Union and, in particular, also a main source of public drinking water supplies in many regions" (European Parliament and the Council of the European Union, 2006). This directive establishes specific measures to prevent and control groundwater pollution, defined as the direct or indirect introduction of pollutants into groundwater as a result of human activity. These measures include (1) criteria for the assessment of good groundwater chemical status and (2) criteria for the identification and reversal of significant and sustained upward trends of contamination and the definition of starting points for trend reversals. The directive also requires the "establishment by Member States of groundwater safeguard zones of such size as the competent national body deems necessary to protect drinking water supplies. Such safeguard zones may cover the whole territory of a Member State." Some of the more telling statements in the directive are

> *Groundwater is a valuable natural resource and as such should be protected from deterioration and chemical pollution. This is particularly important for groundwater-dependent ecosystems and for the use of groundwater in water supply for human consumption.*

> *The protection of groundwater may in some areas require a change in farming or forestry practices, which could entail a loss of income. The Common Agricultural Policy provides for funding mechanisms to implement measures to comply with Community standards.*

Protection of recharge areas, whether natural or artificial, through regulated land-use practices that cannot adversely affect either the quantity or quality of groundwater is the foundation of any successful spring (i.e., groundwater, in general) protection and management. Given many diverse human activities that can potentially lead to contamination of the subsurface, a lack of protection of recharge areas eventually decreases the availability of usable groundwater and requires expensive treatment or the substitution of a more expensive water supply, as demonstrated in many urban and rural areas worldwide.

An example of land-use change that is increasingly affecting many groundwater basins worldwide is urban development. In addition to quality impacts, urban development (pavement and buildings on former agricultural land, lining of flood control channels, and other land-use changes) has reduced the capacity of recharge areas to replenish groundwater, effectively reducing spring flows and the sustainable capacity of well fields. This, in turn, may put additional pressure on springs, as the additional demands for groundwater extraction via wells increase. Figure 1–19 is example of a decreasing trend in discharge rate of the Rainbow Springs, one of the first-magnitude Florida springs, caused by the combined effects of land-use changes and aquifer pumping.

As advised by the California Department of Water Resources (2003), to ensure that recharge areas continue to replenish high-quality groundwater, water managers and land-use planners should work together to

- Identify recharge areas so the public and local zoning agencies are aware of the areas that need protection from paving and contamination.
- Include recharge areas in zoning categories that eliminate the possibility of contaminants entering the subsurface.
- Standardize guidelines for pretreatment of recharge water, including recycled water.
- Install monitoring wells to collect data on changes in groundwater quality that may be caused by recharge.
- Consider the functions of recharge areas in land use and development decisions.

Groundwater protection and management in rural areas, including those areas where irrigation for agriculture is significant or predominant, present a special challenge. Groundwater use in such areas is a decentralized activity with many private users normally involved (Figure 1–20). The users drill their own wells, install their own equipment, follow their own pumping schedules, and use fertilizers and pesticides of their own choice, applying them in quantities and with schedules also of their own choice. In the case of major

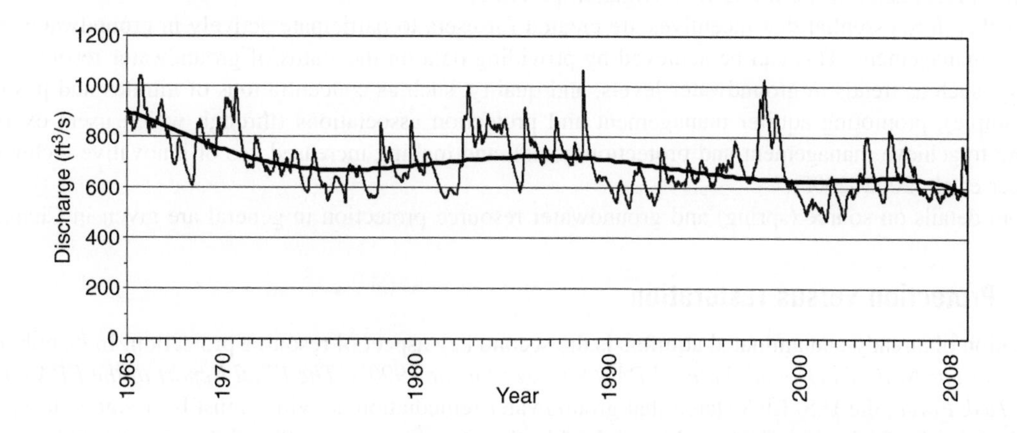

FIGURE 1–19 Discharge hydrograph of the Rainbow Springs, one of the first-magnitude springs in Florida, with the polynomial sixth order trend. (Data from U.S. Geological Survey, 2008.)

FIGURE 1–20 Intensive center-pivot irrigation using groundwater from the same aquifer supplying water to numerous springs along the right bank of the Snake River, Idaho: (1) Thousands Springs area; (2) Box Canyon Springs, also shown in Figure 1–16. (Photograph courtesy of USGS.)

aquifers, with thousands or hundreds of thousands of users, enforcement of any kind, including, for example, well discharge metering, is impossible if users have no incentive to comply. The same is true with the use of pesticides, fertilizers, or more efficient irrigation practices.

It is therefore essential that incentives are created for users to participate actively in groundwater protection and management. This can be achieved by providing data on the status of groundwater resources (both quantity, such as trends in groundwater levels, and quality, such as concentrations of nitrates and pesticides, for example), promoting aquifer management and protection associations (through which users exert peer pressure to achieve management and protection goals), and making increased use of innovative technologies (Kemper et al., 2002–2005).

More details on source (spring) and groundwater resource protection in general are given in Chapter 8.

1.3.2 Protection versus restoration

Restoration of already contaminated aquifers is the second key aspect of resource protection. In its publication *Protecting The Nation's Ground Water: EPA's Strategy for the 1990's. The Final Report of the EPA Ground-Water Task Force*, the U.S. EPA stated that groundwater remediation activities must be assigned top priority to limit the risk of adverse effects to human health, then to restore currently used and reasonably expected sources of drinking water and groundwater whenever such restorations are practicable and attainable (U.S. EPA, 1991).

The agency also stated that, "given the costs and technical limitations associated with groundwater cleanup, a framework should be established that ensures the environmental and public health benefit of each dollar spent is maximized. Thus, in making remediation decisions, USEPA must take a realistic approach to restoration based upon actual and reasonably expected uses of the resource as well as social and economic values."

Finally, given the expense and technical difficulties associated with groundwater remediation, the agency emphasized early detection and monitoring so that it can address the appropriate steps to control and remediate the risk of adverse effects of groundwater contamination to human health and the environment.

Unfortunately, the expense and technical difficulties associated with groundwater remediation have become even more apparent recently, as the societies are coping with the global economic crisis and various threats of climate change. Every aspect of sustainability is examined very carefully, including various aquifer restoration efforts, their costs and their own sustainability. The author has been involved with, and has knowledge of, various hydrogeologically complex sites, including contaminated springs, where characterization of groundwater contamination has continued for more than a decade, sometimes costing over $100 million at individual sites. At the same time and for various reasons, the actual full-scale groundwater remediation (aquifer restoration to beneficial uses) at many of these sites has not yet begun. Often, this is caused by the many remaining uncertainties as to the distribution of contaminants in the subsurface and their migration pathways (such as at many sites in karst and fractured rock environments; see Figure 1–21). These uncertainties have often resulted in the failure of remediation pilot tests; at the same time, however, in many cases, the regulators and the public alike are not ready to accept that restoration at a particular site may not be "practicable and attainable."

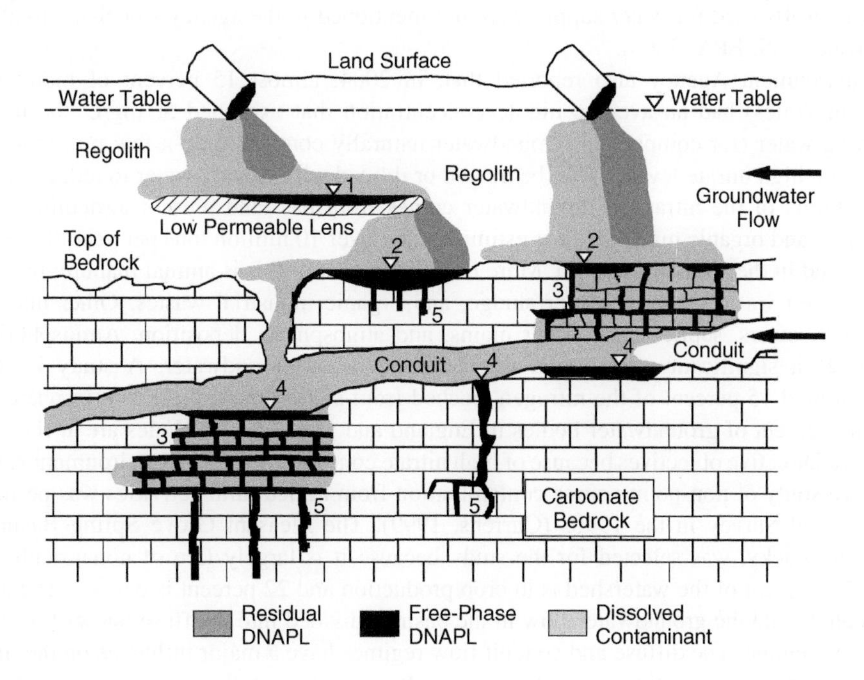

FIGURE 1–21 Distribution of potential DNAPL-accumulation sites in a hypothetical karst setting: (1) Pooling on low-permeability layer in regolith, (2) pooling on top of bedrock, (3) pooling in bedrock diffuse-flow zone, (4) pooling in conduit, (5) pooling in fractures isolated from flow. (From Wolfe and Haugh, 2001.)

Restoration of already contaminated aquifers to their natural condition is often very difficult due to the nature of contamination and complexity of hydrogeologic environments. Unfortunately, the nature of contamination can be controlled or mitigated at a local land-use (e.g., zoning) and development level only to a limited extent, whereas hydrogeologic conditions cannot be controlled at all. In some cases, to restore contaminated aquifers to their natural condition, the local socioeconomic structure as well as the laws of the society would have to be radically changed. Even then, it may take tens or hundreds of years for aquifers to return to their natural condition, assuming that the definition of natural conditions means the absence of any anthropogenic substances in groundwater (Kresic, 2009).

The use of pesticides and fertilizers is just one example of many dilemmas regulatory agencies and societies face when trying to protect both surface water and groundwater resources. The United Kingdom Environment Agency reported that pesticides were found in over a quarter of groundwater monitoring sites in England and Wales in 2004, and in some cases, they exceeded the drinking water limit. Atrazine is a weed killer used mainly to protect maize (corn) crops, and it was used in the past to maintain roads and railways. It has been a major problem, but since the nonagricultural uses were banned in 1993, concentrations in groundwater have gradually declined. As noted by the agency, banned pesticides can remain a problem for many years after they were last used (UK Environment Agency, 2007). Some other European countries have banned the use of atrazine: France, Sweden, Norway, Denmark, Finland, Germany, Austria, Slovenia, and Italy. In contrast, the U.S. Environmental Protection Agency has concluded that the risks from atrazine for approximately 10,000 community drinking water systems using surface water are low and did not ban this pesticide, which continues to be the most widely used pesticide in the United States. Incidentally, as stated by the agency, 40,000 community drinking water systems using groundwater were not included in the related study, and private wells used for water supply were not mentioned in the agency's decision to allow continuous use of atrazine (U.S. EPA, 2003).

The UK Environment Agency also reported that, in 2004, almost 15 percent of monitoring sites in England (none in Wales) had an average nitrate concentration that exceeded 50 mg/L, the upper limit for nitrate in drinking water (for comparison, groundwater naturally contains only a few milligrams per liter of nitrate). Water with high nitrate levels has to be treated or diluted with cleaner water to reduce concentrations. More than two thirds of the nitrate in groundwater comes from past and present agriculture, mostly from chemical fertilizers and organic materials. It is estimated that over 10 million tons per year of organic material is spread on the land in the United Kingdom. More than 90 percent of this is animal manure; the rest is treated sewage sludge, green waste compost, paper sludge, and organic industrial wastes. Other major sources of nitrate are leaking sewers, septic tanks, water mains, and atmospheric deposition. Atmospheric deposition of nitrogen makes a significant contribution to nitrate inputs to groundwater. A study in the Midlands concluded that around 15 percent of the nitrogen leached from soils came from the atmosphere. The agency estimates that 60 percent of groundwater bodies in England and 11 percent in Wales are at risk of failing the Water Framework Directive objectives because of high nitrate concentrations (UK Environment Agency, 2007).

An illustrative study of non-point-source contamination from agricultural activities was performed by the Kentucky Geological Survey in the 1990s (Currens, 1999). The Pleasant Grove Spring Basin in southern Logan County, Kentucky, was selected for the study because it is largely free of nonagricultural pollution sources. About 70 percent of the watershed is in crop production and 22 percent is pasture. The area is underlain by karst geology and the groundwater flow in the basin is divided into a diffuse (slow) flow regime and a conduit (fast) flow regime. The diffuse and conduit flow regimes have a major influence on the timing of contaminant maxima and minima in the spring during and after major rainfall events. Nitrate is the most widespread, persistent contaminant in the basin, but concentrations average 5.2 mg/L basinwide and generally do not exceed the drinking water maximum contaminant level (MCL) of 10 mg/L set by the U.S. EPA. Atrazine has been detected consistently and other pesticides occasionally. Concentrations of triazines (including atrazine) and alachlor have exceeded drinking water MCLs during peak spring flows. Maximum concentrations of triazines, carbofuran, metolachlor, and alachlor in samples from Pleasant Grove Spring were 44.0, 7.4, 9.6, and 6.1 μg/L, respectively. Flow-weighted average concentrations for 1992–1993 were 4.91 μg/L for atrazine-equivalent triazines and 5.0 mg/L for nitrate-nitrogen. In comparison, the maximum allowed concentration of any individual pesticide in drinking water in the European Union is 0.1 μg/L, and of all pesticides combined, it is 0.5 μg/L.

The hydrogeology of the basin is a significant controlling influence on the temporal variation of contaminant concentrations. The fast-flow conduit region is characterized by intermediate concentrations of nitrate and pesticides during low flow but substantially higher concentrations of triazines and lower concentrations of nitrate during high flow. The diffuse (slow) flow regime, which is estimated to represent slightly less than half the basin, drains into the area dominated by conduit flow. The diffuse-flow region has persistently higher concentrations of nitrate but lower, less variable concentrations of triazines. The diffuse, slow-flow area is acting as a reservoir of agricultural chemicals, maintaining a background level of triazines and nitrate during low flow in the conduit-flow regime. Triazine concentrations are significantly higher during high flow, while nitrate concentrations are diluted.

As concluded by Currens (1999), both municipal and domestic water supplies derived from groundwater can be adversely affected. Implementation of best management practices in the basin should focus on controlling animal waste, controlling crop field runoff with associated sediment and pesticide loss, and using more efficient methods of applying nutrients. A strong education program on groundwater protection is highly recommended.

These examples illustrate that, even if the local community and local or possibly even state regulators were unified in their desire to restore an aquifer to its natural condition, it is not possible to do so without changing political will and regulations at a higher (e.g., federal) level. Changing regulations, however, would be only the first necessary step. To restore a major aquifer contaminated from non–point sources to its natural condition in a meaningful period (e.g., several generations), the remediation measures would have to be extremely costly and decades long and would have to be paid and implemented by the wider society. Such efforts are therefore seldom if ever undertaken, and the restoration of aquifers to their "natural" condition is left to the natural attenuation processes, while the groundwater users are becoming accustomed to drinking treated water.

In contrast, when point-source groundwater contamination is caused by known "potentially responsible parties" (PRPs), the approaches to groundwater restoration in many cases depends on the prevailing interpretation of the existing regulations at the local and state levels. For example, many states in the United States have adopted a zero tolerance for groundwater degradation by large polluters, such as various industries and military installations; these PRPs are consequently required to restore "their" portions of contaminated aquifers to pristine natural conditions, often regardless of the underlying hydrogeologic characteristics, the risks, the associated costs, and the likely outcome.

In conclusion, prevention of groundwater and spring contamination, as well as the resource overexploitation, should not be substituted by any restoration measures.

REFERENCES

Austrian Museum for Economic and Social Affairs, 2003. Water ways. Austrian Museum for Economic and Social Affairs, Vienna.

Baker Ranches, Inc., 2008. Protect Snake Valley, Water. Available at: http://protectsnakevalley.com/water.html (Accessed December 2008).

Barker, R., 2007. Water hearing will affect users statewide. Idaho Statesman (December 10). Available at: www.idahostatesman.com/235/story/233986.html.
Bredehoeft, J.D., Papadopulos, S.S., Cooper, H.H., 1982. Groundwater—The water budget myth. In: Studies in geophysics, scientific basis of water resource management. National Academy Press, Washington, DC.
Bredehoeft, J.D., 2002. The water budget myth revisited: why hydrogeologists model. Ground Water 40 (4), 340–345.
Brune, G., 1975. Major and Historical Springs of Texas. Report 189, Texas Water Development Board, Austin.
California Department of Water Resources, 2003. California's groundwater. Bulletin 118, Update 2003. The Resources Agency, Department of Water Resources, Sacramento, State of California.
Currens, J.C., 1999. Mass flux of agricultural nonpoint-source pollutants in a conduit-flow-dominated karst aquifer, Logan County, Kentucky. Report of Investigations 1, Series XII, Kentucky Geological Survey, University of Kentucky, Lexington.
Devlin, J.F., Sophocleous, M., 2005. The persistence of the water budget myth and its relationship to sustainability. Hydrogeology Journal 13, 549–554.
European Parliament and the Council of the European Union, 2006. Directive 2006/118/EC on the protection of groundwater against pollution and deterioration. Official Journal of the European Union (December 27), L 372/19–31.
Foster, S., Garduño, H., Kemper, K., Tiunhof, A., Nanni, M., Dumars, C., 2002–2005. Groundwater quality protection; defining strategy and setting priorities. Sustainable Groundwater Management; Concepts & Tools, Briefing Note Series Note 8, The Global Water Partnership. The World Bank, Washington, DC. Available at: www.worldbank.org/gwmate.
Hill, R.T., Vaughan, T.W., 1898. The Geology of the Edwards Plateau and Rio Grande Plain Adjacent to Austin and San Antonio, Texas, with References to the Occurrence of Underground Waters. U.S. Geological Survey 18th annual report, part 2-B, pp. 103–321.
Hutson, W.F., 1898. Irrigation systems of Texas. U.S. Geological Survey Water Supply and Irrigation Paper 13, Washington, DC.
Kemper, K., Foster, S., Garduño, H., Nanni, M., Tuinhof, A., 2002–2005. Economic instruments for groundwater management: Using incentives to improve sustainability, Sustainable Groundwater Management: Concepts and Tools, Briefing Note Series Note 7, GW MATE (Groundwater Management Advisory Team). The World Bank, Washington, DC.
Kirby, S., Hurlow, H., 2005. Hydrogeologic setting of the Snake Valley hydrologic basin, Millard County, Utah, and White Pine and Lincoln Counties, Nevada—Implications for possible effects of proposed water wells. Report of Investigation 254, Utah Geological Survey, Utah Department of Natural Resources, Salt Lake City, UT.
Kresic, N., 2009. Groundwater Resources: Sustainability, Management, and Restoration. McGraw-Hill, New York.
Lewelling, B.R., Tihansky, A.B., Kindinger, J.L., 1998. Assessment of the Hydraulic Connection Between Ground Water and the Peace River, West-Central Florida. U.S. Geological Survey Water-Resources Investigations Report 97–4211, Tallahassee, FL.
Margat, J., Foster, S., Droubi, A., 2006. Concept and importance of non-renewable resources. In: Foster, S., Loucks, D.P. (Eds.), Non-renewable groundwater resources. A guidebook on socially-sustainable management for water-policy makers. IHP-VI, Series on Groundwater No. 10. UNESCO, Paris.
Meinzer, O.E., 1927. Large springs in the United States. U.S. Geological Survey Water-Supply Paper 557, Washington, DC.
Newcomb, W.W., 1961. The Indians of Texas. University of Texas Press, Austin.
Peek, H.M., 1951. Cessation of flow of Kissengen Spring in Polk County, Florida. In: Water resource studies. Florida Geological Survey Report of Investigations No. 7, Tallahassee, FL, pp. 73–82.
Rogers, P., Hall, A.W., 2003. Effective water governance. TEC Background Papers No. 7, Global Water Partnership Technical Committee (TEC), Global Water Partnership, Stockholm, Sweden.
Stewart Jr., H.G., 1966. Ground-water resources of Polk County, Florida. Florida Geological Survey Report of Investigations No. 44, Tallahassee, FL.
Summers, P., 2001–2005. Hydrogeologic Analysis of Needle Point Spring (Revised Final). Bureau of Land Management, Fillmore Field Office, Utah.
Taylor, T.U., 1902. Irrigation systems of Texas. U.S. Geological Survey Water Supply and Irrigation Paper 71, Washington, DC.
Tušar, B., 2008. Vodoopskrba u Dubrovniku [Water Supply in Dubrovnik, in Croatian]. Obrada vode (April), 54–59.
UK Environment Agency, 2007. Underground, under threat. The state of groundwater in England and Wales. Environment Agency, Almondsburry, Bristol. Available at: www.environment-agency.gov.uk.
UN Department of Economic and Social Affairs, 2009. Report of the World Commission on Environment and Development: Our Common Future (Brundtland Report). DESA, Division for Sustainable Development, The United Nations. Available at: www.un.org/esa/dsd/index.shtml.
U.S. EPA, 1991. Protecting the nation's ground water: EPA's strategy for the 1990's. The final report of the EPA Ground-Water Task Force. 21Z-1020, Office of the Administrator, Washington, DC.
U.S. EPA, 1999. Safe Drinking Water Act, Section 1429, Ground Water Report to Congress. EPA-816-R-99-016, United States Environmental Protection Agency, Office of Water, Washington, DC.

U.S. EPA, 2003. Atrazine interim reregistration eligibility decision (IRED), Q&A's—January 2003. Available at: www.epa.gov/pesticides/factsheets/atrazine.htm#q1 (Last accessed January 23, 2008).

U.S. EPA, 2005. National management measures to control nonpoint source pollution from urban areas. EPA-841-B-05-004, United States Environmental Protection Agency, Office of Water, Washington, DC.

U.S. Geological Survey, 2008. USGS Ground-Water Data for the Nation. Available at: http://waterdata.usgs.gov/nwis/gw.

Wolfe, W.J., Haugh, C.J., 2001. Preliminary conceptual models of chlorinated-solvent accumulation in karst aquifers. In: Kuniansky, E.L., (Ed.), 2001 U.S. Geological Survey Karst Interest Group Proceedings. U.S. Geological Survey Water-Resources Investigations Report 01-4011, St. Petersburg, FL, pp. 157–162.

CHAPTER 2

Types and classifications of springs

Neven Kresic
MACTEC Engineering and Consulting, Inc., Ashburn, Virginia

2.1 TYPES OF SPRINGS

Arguably the first scientific publication on springs is that of French physicist and astronomer Arago (*On Springs, Artesian Wells, and Spouting Fountains*, 1835a, 1835b), who was also first to introduce word *aquifère* (*aquifer* in English). An aquifer is a geologic formation or group of hydraulically connected geologic formations storing and transmitting significant quantities of potable groundwater. Although most dictionaries of geologic and hydrogeologic terms would have a very similar definition, it is surprising how many interpretations of the word exist in everyday practice, depending on the circumstances. The problem usually arises from the lack of common understanding of the following two terms, which are not easily quantifiable: *significant* and *potable*. For example, a spring yielding 2 gallons per minute may be very significant for an individual household with no other available source of water supply. However, if this quantity is at the limit of what the geologic formation could provide via individual springs or wells, such "aquifer" would certainly not be considered as a potential source for any significant public water supply (Kresic, 2007).

A spring is a location at the land surface where groundwater discharges from the aquifer, creating a visible flow. This discharge is caused by difference in the elevation of the hydraulic head in the aquifer and the elevation of the land surface where the discharge takes place. The opening through which groundwater discharges is called the spring's *orifice*. Springs issuing from consolidated rocks usually have a well-defined orifice, as opposed to springs in unconsolidated sediments, although the latter ones may also have a clearly visible orifice (Figure 2–1). Sometimes, the orifice can be at the bottom of a deep spring pool and not readily visible, or it may be covered by sediments and rock debris.

When the flow cannot be immediately observed but the land surface is wet compared to the surrounding area, such discharge of groundwater is called a *seep*. A *seepage spring* is a general term used to indicate diffuse discharge of water, usually from unconsolidated sediments, such as sand and gravel, or from loose soil.

A *fracture* (or *fissure*) *spring* refers to concentrated discharge of water from bedding planes, joints, cleavage, faults, and other breaks in the consolidated (hard) rock (Figure 2–2). The term *tubular springs* (or *cave springs*) is sometimes used to describe flow from relatively large openings in the rocks; such springs are characteristic of karst terrains (see Section 2.3).

Secondary springs issue from locations located away from the primary spring discharge, which is covered by colluvium (rock fall fragments) or other natural debris and therefore not visible. As emphasized in Chapter 9, when capturing such springs every attempt should be made to remove all the debris and locate the primary spring orifice(s), since the secondary discharge locations tend to migrate over time.

Copyright © 2010, Elsevier Inc. All rights reserved.

FIGURE 2–1 Small spring ("gushet") issuing from residuum soil at a creek bank. The larger of the two orifices is about 2 in. (5 cm) in diameter.

As discussed by Meinzer (1927), the water seldom issues from a single opening and may issue from a great many openings, which may be close together or scattered over a considerable area. What is considered a single spring in one locality in another locality may be regarded as a group of springs, each of which has an individual name. Some springs are designated in the singular and some in the plural, for example, Silver Spring and Thousand Springs. The idea that underlies this usage is that, if the water issues from a single opening or several openings close together, it forms a "spring," whereas if it issues from a number of openings that are farther apart, it forms "springs" (Figures 2–3 and 2–4). In fact, however, local terminology is so variable in this respect that there is no consistent distinction between "a spring" and "springs," and often there is no uniformity in usage of terms even for the same group of openings.

Although all springs (except some associated with young volcanism and hydrothermal activity, where discharge is driven by gases and temperature gradients) ultimately discharge at the land surface because of the force of gravity, they are usually divided into two main groups based on the nature of the hydraulic head in the underlying aquifer at the point of discharge:

- **Gravity springs** emerge under unconfined conditions where water table intersects land surface. They are also called *descending springs*.
- **Artesian springs** discharge under pressure due to confined conditions in the underlying aquifer. They are also called *ascending* or *rising springs*.

Geomorphology and geologic fabric (rock type and tectonic features, such as folds and faults) play the key role in the emergence of springs. When site-specific conditions are rather complicated, springs of formally different types, based on some classifications, may appear next to each other, causing confusion. For example, a lateral

FIGURE 2–2 Capture of a small spring used for watering cattle near Saltville, Virginia. The spring is issuing from an enlarged fracture in limestone.

impermeable barrier in fractured rock, caused by faulting, may force groundwater from greater depth to ascend and discharge at the surface through the overlying alluvial sediments in the stream valley (Figure 2–5). This water may have an elevated temperature due to the normal geothermal gradient in the Earth crust; such springs are called *thermal springs*. At the same time, groundwater of normal temperature may issue at a "cold" gravity spring located very close to the thermal spring, at the contact with less permeable alluvial fine-grained sediments, such as clays. Yet, a third spring may be present with its temperature varying depending on the precipitation pattern, seasonal influences, and the mixing mechanism of waters with different temperatures.

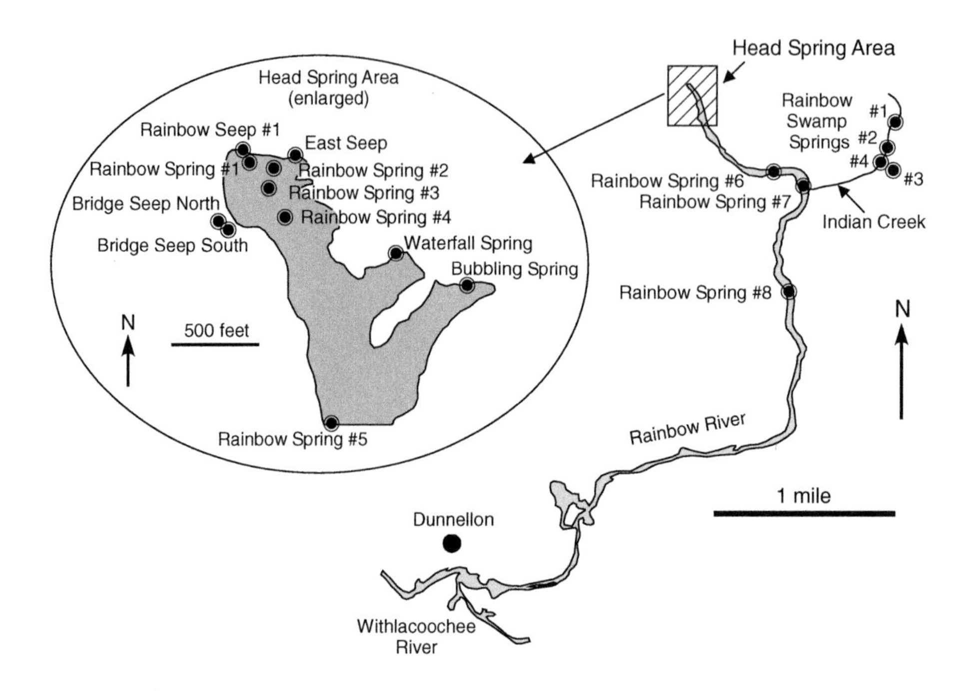

FIGURE 2–3 Locations of spring vents in the Rainbow Springs complex. (From Jones et al., 1996.)

FIGURE 2–4 Thousand Springs discharging in Hagerman Valley along the Snake River near Twin Falls, Idaho, circa 1910–1920. (Idaho Historical Society, Bisbee Collection. Printed with permission.)

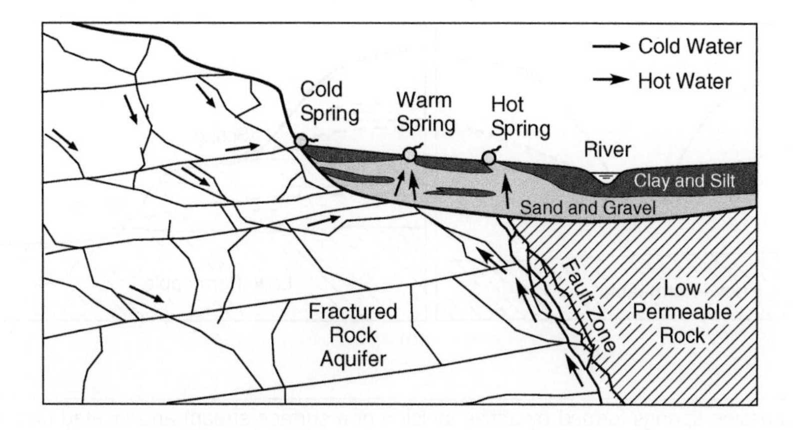

FIGURE 2–5 Three springs of different temperature in a barrier fault zone overlain by alluvium.

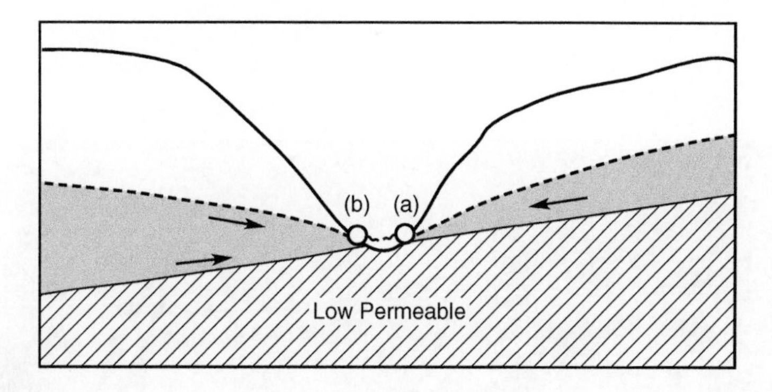

FIGURE 2–6 Contact gravity springs: (a) Descending and (b) overflowing. (Modified from Milojevic, 1966.)

In general, when the contact between the water-bearing rocks (sediments) and the impermeable underlying rock slopes toward the spring, in the direction of groundwater flow, and the aquifer is above this impermeable contact, the spring is called a *descending contact spring* (Figure 2–6a). When the impermeable contact slopes away from the spring, in a direction opposite of the groundwater flow, the spring is called an *overflowing contact spring* (Figure 2–6b).

Depression springs are formed in unconfined aquifers when the topography intersects the water table, usually due to surface stream incision (Figure 2–7a). Possible contact between the aquifer and the underlying low-permeable formation is not the main reason for spring emergence (this contact may or may not be known). When surface stream cuts through alternating layers of permeable and impermeable sediments (rocks), the original depression springs may be transformed into contact springs at various elevations above the stream channel (Figure 2–7b). Such springs are often found along depositional and erosional river terraces.

Depression springs draining local groups of interconnected fractures are common in fractured rock aquifers and are not necessarily associated with a surface stream—any erosional depression intersecting water-bearing fractures can cause emergence of a spring. This is illustrated in Figure 2–8, where discharge of water from fractures in a fault zone exposed by a road-tunnel cut is frozen due to low air temperature. Permanent,

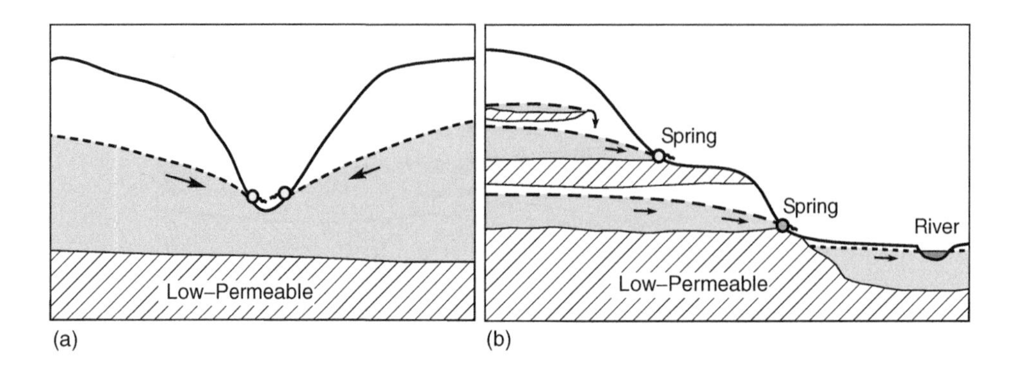

FIGURE 2–7 (a) Depression springs formed by active incision of a surface stream and located near the stream channel and (b) previous depression springs, now contact gravity springs along river terraces. (Modified from Milojevic, 1966.)

FIGURE 2–8 Tunnel and road cut in Paleozoic metamorphic rocks 5 mi south of the entrance to the Shenandoah National Park at Route 211 and the Skyline Drive near Sperryville, Virginia. The water flowing out of the fractures in the fault zone is frozen.

FIGURE 2–9 Capture of a permanent spring issuing from Paleozoic metamorphic rocks and used for the public water supply, Skyline Drive, Virginia.

small depression springs in fractured rock aquifers have been used as sources of local water supply around the world throughout human history (Figures 2–9 and 2–10).

Figure 2–11 shows some examples of *barrier springs*; the term generally refers to springs at any lateral contact between the aquifer and a low-permeable rock (sediment). Such contact can have many shapes, caused by depositional processes and tectonic movements, which form a variety of faults and folds. When groundwater is forced to ascend from greater aquifer depths along the contact due to hydrostatic pressure, the spring is called *ascending* or *artesian*. Such springs commonly have a stable water temperature that, if higher than the average air temperature at the location, makes them *thermal springs*.

Intermittent springs discharge only for a period of time, while at other times they are dry, reflecting directly the aquifer recharge pattern. They can be found in consolidated and unconsolidated rocks of all types, but the most fascinating are springs discharging from karst aquifers. As explained in more detail in Section 2.3, karstified rocks can receive a large percentage of precipitation episodes and quickly transmit this newly infiltrated water toward a previously dry spring (Figure 2–12).

Ebb-and-flow springs, or *periodic springs*, are usually found in limestone (karst) terrain. Their discharge occurs in relatively uniform time intervals (periods) and is explained by the existence of a siphon in the rock mass behind the spring (Figure 2–13). The siphon fills up and empties with regularity, regardless of the recharge (rainfall) pattern. A better known example from the United States is Periodic Spring in the Bridger-Teton National Forest near Jackson, Wyoming. Situated at the base of limestone cliffs, the spring discharges about 285 gallons per second. Springwater gushes from an opening for several minutes, stops

FIGURE 2–10 Simple capture of a small spring issuing from fractures in Paleozoic metamorphic rocks of the Piedmont physiographic province, the United States. This spring was tapped before the Civil War for the water supply of a farmhouse in Virginia. Many other similar springs are still used as sources of both potable and nonpotable water, although drilled wells are now the main form of water supply in the region.

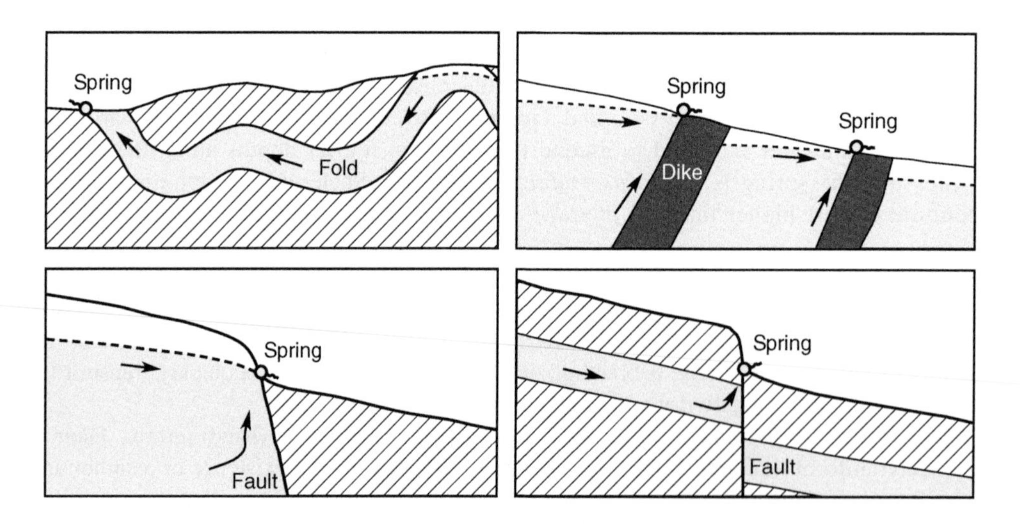

FIGURE 2–11 Several types of barrier springs issuing at lateral contacts between saturated permeable rocks (aquifers) and low-permeable rocks (barriers).

FIGURE 2–12 Intermittent karst spring Sopot in Montenegro, discharging over 200 m^3/s after a heavy summer storm (photograph courtesy of Igor Jemcov). The spring is dry most of the year (photograph courtesy of Sasa Milanovic).

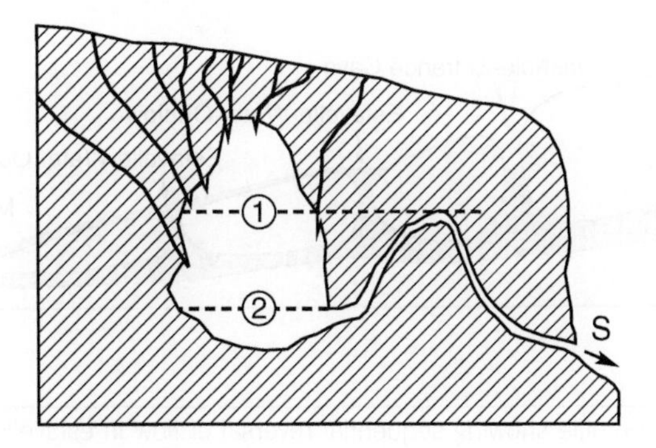

FIGURE 2–13 Schematic of an ebb-and-flow (periodic) spring. When water in the cavity reaches level (1), the siphon becomes active and the spring (S) starts flowing. When the level drops to position (2), the spring stops flowing. (From Radovanovic, 1897.)

abruptly, then begins a new cycle a short time later. Intermittent water flows range anywhere from 4 to 25 minutes and the water is clear and cold (U.S. Forest Service Intermountain Region, 2008).

Estavelle has a dual function: It acts as a spring during high hydraulic heads in the aquifer and as a surface water sink during periods when the hydraulic head in the aquifer is lower than the body of surface water (Figures 2–14 and 2–15). Estavelles are located within or adjacent to surface water features.

Paleosprings are places of former spring discharge that are now inactive. Entrances to many permanently dry caves in karst terrains are former springs (see the section on karst springs). Travertine deposits located away from permanent springs or surface streams are good indicators of former spring activity and may also indicate presence of nearby active springs (Figure 2–16).

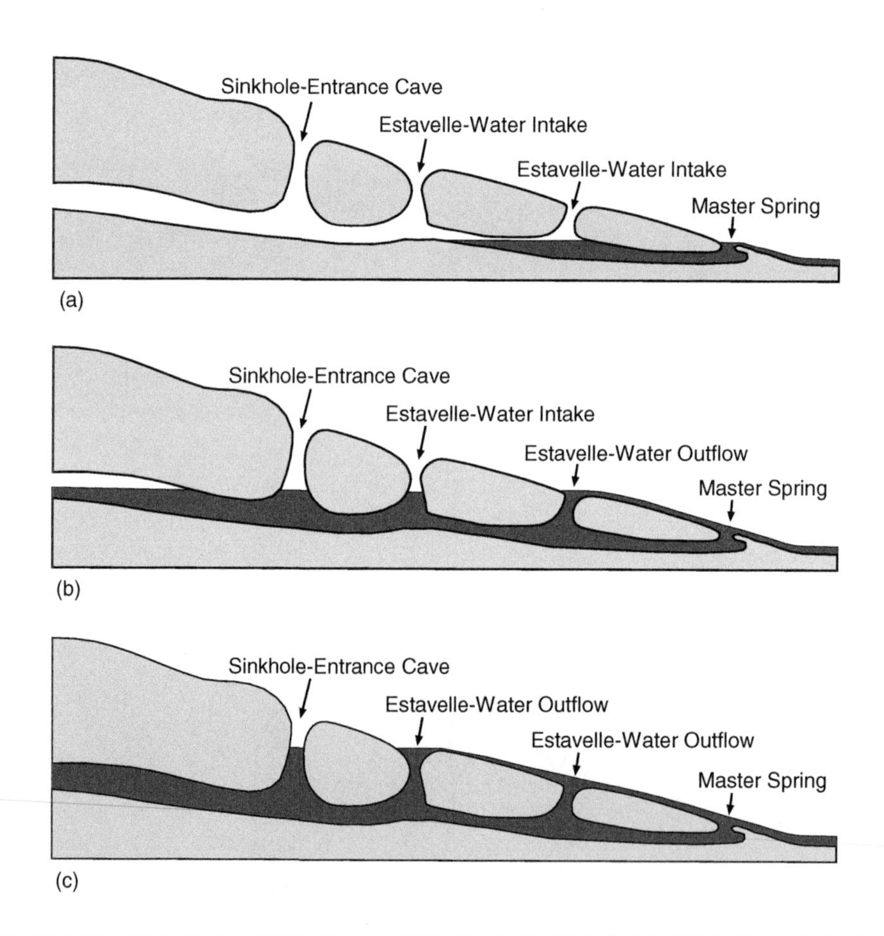

FIGURE 2–14 Hypothetical stream profile showing sequential reversal of flow in estavelles: (a) Low flow, only lower sections of karst drainage system flooded, estavelles are sinkholes (swallow holes); (b) medium flow, headward flooding of karst drainage system, lowest estavelle becomes spring (rising); and (c) high flow, storage capacity of karst drainage system exceeded, estavelles become springs, water backs up in headward caves. (From Vineyard and Feder, 1982.)

FIGURE 2–15 The Oval Sink, a reversible sinkhole (estavelle) in Springfield, Missouri. (From Vineyard and Feder, 1982; photograph by Jerry Vineyard.)

Faults play a major role in the emergence of springs, especially in fractured rock and karst aquifers. They are also not uncommon in unconsolidated and semiconsolidated sediments (Figure 2–17). In any case, faults themselves may play one of the following three roles: (1) conduit for groundwater flow, (2) storage of groundwater due to increased porosity within the fault (fault zone), or (3) barrier to groundwater flow due to a decrease in porosity within the fault.

As discussed by Meinzer (1923):

Faults differ greatly in their lateral extent, in the depth to which they reach, and in the amount of displacement. Minute faults do not have much significance with respect to ground water except, as they may, like other fractures, serve as containers of water. But the large faults that can be traced over the surface for many miles, that extend down to great depths below the surface, and that have displacements of hundreds or thousands of feet are very important in their influence on the occurrence and circulation of ground water. Not only do they affect the distribution and position of aquifers, but they may also act as subterranean dams,

FIGURE 2–16 Travertine deposits and orifices of a paleospring in Saratoga Springs, New York.

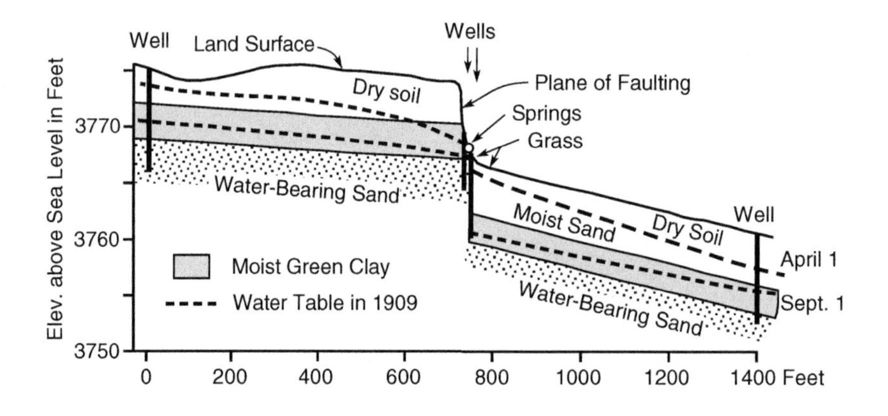

FIGURE 2–17 Section of Owens Valley, California, showing a spring produced by impounding effects of a fault. The "moist sand" on the downthrown side is equivalent to some of the "dry soil" on the upthrown side and apparently has an impounding effect. (After C. H. Lee, from Meinzer, 1923.)

impounding the ground water, or as conduits that reach into the bowels of the earth and allow the escape to the surface of deep-seated waters, often in large quantities. In some places, instead of a single sharply defined fault, there is a fault zone in which there are numerous small parallel faults or masses of broken rock called fault breccia. Such fault zones may represent a large aggregate displacement and may afford good water passages.

Excellent examples of springs produced by the rise of deep waters through fault openings are to be found along the edges of the mountain ranges of Nevada and western Utah. Many of these springs have large yields, some of them discharging several cubic feet a second. The abundance of these springs and the copious flow of some of them are the more impressive because of the aridity of the region in which they occur. The ranges of this region consist largely of tilted fault blocks, and in many places there are recent fault scarps in the alluvial slopes at the foot of the mountains. That manor of the springs along these fault lines are not merely returning to the surface water that percolates into the sediments of the adjacent alluvial slopes but yield water that ascends from deep sources along faults seems to be shown by the following facts: (1) The springs are situated along the general courses of the fault scarps, some of the groups having a more or less linear arrangement; (2) the yield of many of the springs is larger than would be expected if they were supplied from local sources, and some with the largest yields occur along narrow dry ranges that supply but little water; (3) they have relatively uniform flow throughout the year, whereas ordinary springs in the region fluctuate more with the season; (4) many of these springs yield water whose temperature is above the mean annual temperature of the region, and hot springs that are not associated with volcanic rocks are abundant; (5) many of the springs issue from deep pools that are believed to be associated with fissures.

The impounding effect of faults is caused by four main mechanisms:

- The displacement of alternating permeable and impermeable beds in such a manner that the impermeable beds are made to abut against the permeable beds, as shown in Figure 2–11.
- Due to a clayey gouge along the fault plane produced by the rubbing and mashing during displacement of the rocks, this gouge being smeared over the edges of the permeable beds. The impounding effect of faults is most common in unconsolidated formations that contain considerable clayey material (Figure 2–17).
- Cementation of the pore space by precipitation of material, such as calcium carbonate, from the groundwater circulating through the fault zone.
- Rotation of elongated flat clasts parallel to the fault plane, so that their new arrangement reduces permeability perpendicular to the fault (Kresic, 2007).

2.1.1 Submerged springs

Discharge of groundwater into surface water bodies is, in many cases, below the surface water level and not immediately visible. This discharge can be either diffuse (Figures 2–18 and 2–19) or be concentrated in the form of *submerged (subaqueous) springs* (Figure 2–20). Submerged freshwater springs discharging at the sea floor (*submarine* springs) have been known to and have intrigued people for millennia. For example, the Roman geographer Strabo, who lived from 63 BC to 21 AD, mentioned a submarine fresh groundwater spring 4 km from Latakia, Syria, near the Mediterranean island of Aradus. Water from this spring was collected from a boat, utilizing a lead funnel and leather tube, and transported to the city as a source of fresh water. Other historical accounts tell of water vendors in Bahrain collecting potable water from offshore submarine springs for shipboard and land use, Etruscan citizens using coastal springs for "hot baths" (Pausanius, ca. 2nd century AD) and submarine "springs bubbling fresh water as if from pipes" along the Black Sea (Pliny the Elder, ca. 1st century AD; from UNESCO, 2004).

Karstified rocks (see Section 2.3) make up 60 percent of the shoreline of the Mediterranean and are estimated to contribute 75 percent of its freshwater input, mostly via direct discharge to the sea (UNESCO, 2004; see Figure 2–21).

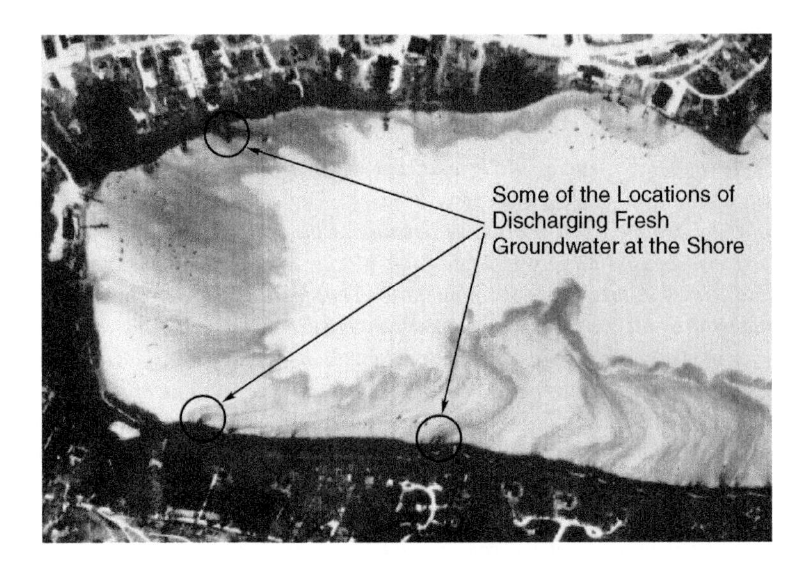

FIGURE 2–18 Aerial thermal infrared scan of Town Cove, Nauset Marsh, Cape Cod, Massachusetts. Discharging fresh groundwater is visible as dark (relatively cold) streams flowing outward from the shore over light-colored (warm) but higher-density estuarine water. Data were collected at low tide at 9:00 PM eastern daylight time on August 7, 1994. (From Barlow, 2003; photograph courtesy of John Portnoy, Cape Cod National Seashore.)

FIGURE 2–19 Bubbles of gas freed from the discharging groundwater due to pressure drop; streambed of Honey Creek in Ann Arbor, Michigan. (From Kresic, 2007; copyright Taylor & Francis Group, printed with permission.)

FIGURE 2–20 Lower outlet of Greer Spring in Oregon County, Missouri. The average flow of this first-magnitude spring is about 214 million gallons per day. (Photograph courtesy of Missouri Department of Natural Resources; available at www.dnr.mo.gov/env/wrc/springsandcaves.htm.)

FIGURE 2–21 Submarine spring Vrulja in the Plominski Bay, east coast of Istria, Croatia. The clear springwater contrasts the muddy water brought into the bay by the Boljuncica River after heavy rains. (Photograph courtesy of Andrija Rubinić.)

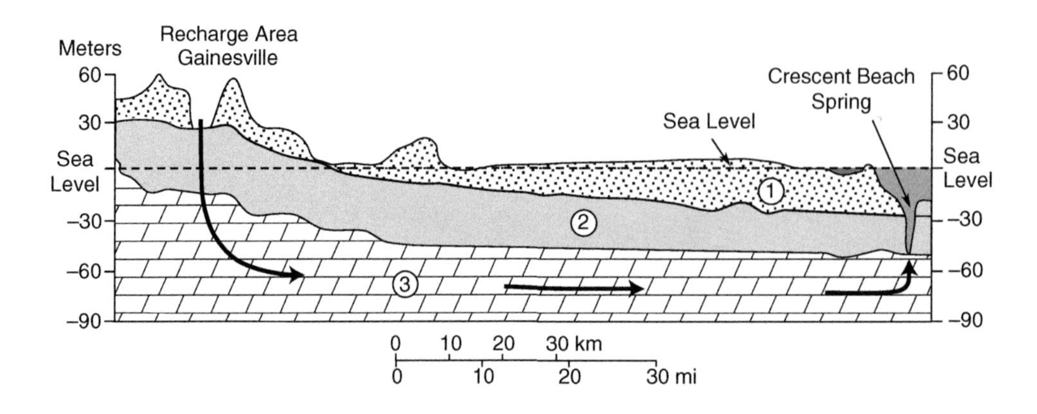

FIGURE 2–22 Idealized cross section of groundwater flow to Crescent Beach Spring, Florida: (1) Post-Miocene deposits (green clay, sand, and shell); (2) confining unit (Hawthorn Formation); (3) Upper Floridan aquifer (Eocene Ocala Limestone). (Modified from Barlow, 2003.)

Most studies of submarine springs are driven largely by potable water supply objectives. One argument for continuing efforts in that respect is that, even if the captured water is not entirely fresh, it may be less expensive to desalinate than undiluted seawater. Another argument is that the discharge of freshwater across the sea floor may be considered a waste, especially in arid regions. In such places, the detection of submarine groundwater discharge may provide new sources of drinking and agricultural water (UNESCO, 2004).

The groundwater flow toward the coast and its submarine discharge are caused by the hydraulic gradient between the inland recharge areas and the sea level (Figure 2–22). If the aquifer is confined and well protected by a thick aquitard, the groundwater flow may continue well beyond the coastline with the ultimate discharge taking place along distant submarine aquifer outcrop.

2.1.2 Thermal and mineral springs

Thermal springs can be divided into *warm springs* and *hot springs*, depending on their temperature relative to the human body temperature of 98° Fahrenheit or 37° Celsius: Hot springs have a higher and warm springs a lower temperature. A warm spring has a temperature higher than the average annual air temperature at the location of the discharge. The water temperature of both groups of thermal springs can fluctuate over time, reflecting more or less surficial influence. One such example is Granite Hot Springs near Jackson, Wyoming, with temperatures varying from 93°F in the summer to 112° in the winter. This delay is likely due to the time required for cold snowmelt water to infiltrate into the aquifer and affect the hot water coming from greater depths. In the mid-1930s, the Civilian Conservation Corps constructed a cement pool to capture the thermal heated water of the spring. The 45 × 75 ft hot pool is bordered by large granite boulders and beautiful scenic views (Figure 2–23). The spring flows from the contact of the Cambrian period Death Canyon Limestone and Flathead Sandstone. In addition to varying water temperature, the spring's fluctuating flow also indicates influence by precipitation and snowmelt.

Geysers and *fumaroles* (also called *solfataras*) are generally found in regions of young volcanic activity. Surface water percolates downward through the rocks below the Earth's surface to high-temperature regions surrounding a magma reservoir, either active or recently solidified but still hot. There the water is heated, becomes less dense, and rises back to the surface along fissures and cracks. Sometimes, these features are

FIGURE 2–23 Granite Hot Springs at Hoback Junction, 13 mi south of Jackson, Wyoming, Bridger-Teton National Forest. (Photograph courtesy of U.S. Forest Service Intermountain Region, 2008.)

called *dying volcanoes*, because they seem to represent the last stage of volcanic activity as the magma, at depth, cools and hardens.

Erupting geysers provide spectacular displays of underground energy suddenly unleashed, but their mechanisms are not completely understood. Large amounts of hot water are presumed to fill underground cavities. The water, on further heating, is violently ejected when a portion of it suddenly flashes into steam. A slight decrease in pressure or an increase in temperature causes some of the water to boil. The resulting steam forces overlying water up through the conduit and onto the ground. This loss of water further reduces pressure within the conduit system, and most of the remaining water suddenly converts to steam and erupts at the surface. This cycle can be repeated with remarkable regularity, as for example, at Old Faithful Geyser in

FIGURE 2–24 Photograph of the Old Faithful Geyser erupting in Yellowstone National Park, Wyoming, the United States. Old Faithful was named in 1870 during the Washburn-Langford-Doane Yellowstone expedition and was the first geyser in the park to be named. (From USGS, 2009b.)

Yellowstone National Park, which erupts on an average of about once every 65 minutes (Figure 2–24; U.S. Geological Survey, 2009a, 2009b).

Fumaroles, which emit mixtures of steam and other gases, are fed by conduits that pass through the water table before reaching the surface of the ground. Hydrogen sulfide (H_2S), one of the typical gases issuing from fumaroles, readily oxidizes to sulfuric acid and native sulfur. This accounts for the intense chemical activity and brightly colored rocks in many thermal areas (U.S. Geological Survey, 2009a).

Meinzer (1940) gives the following illustrative discussion regarding the occurrence and nature of thermal springs:

> *An exact statement of the number of thermal springs in the United States is, of course, arbitrary, depending upon the classification of springs that are only slightly warmer than the normal for their localities and upon the groupings of those recognized as thermal springs.*
>
> *Nearly two-thirds of the recognized thermal springs issue from igneous rocks—chiefly from the large intrusive masses, such as the great Idaho batholith, which still retain some of their original heat. Few, if any, derive their heat from the extrusive lavas, which were widely spread out in relatively thin sheets that cooled quickly. Many of the thermal springs issue along faults, and some of these may be artesian in character, but most of them probably derive their heat from hot gases or liquids that rise from underlying bodies of intrusive rock. The available data indicate that the thermal springs of the Western Mountain region derive their water chiefly from surface sources, but their heat largely from magmatic sources.*

FIGURE 2–25 Ascending thermal spring in the channel of the Rio Grande River, Big Bend National Park, Texas. The higher hydraulic head in the little pool maintains clear water as opposed to often muddy water of the river. (From Kresic, 2007; copyright Taylor & Francis Group, printed with permission.)

One such spring is shown in Figure 2–25. The spring is located in the Rio Grande River fault zone, at the contact between bedrock and alluvium, in the general area of both young and old magmatic activity.

Thermal springs are heated by the naturally occurring thermal energy within the Earth (geothermal energy). Measurements in boreholes indicate that temperature increases downward within the Earth's crust at an average rate of about 30°C/km, and from this average geothermal gradient, it has been calculated that about 4×10^{26} J of thermal energy, assuming a surface temperature of 15°C, is stored within the outer 10 km of the crust (White, 1965). Although most of the energy is stored in rocks, water and steam contained in fractures and pore spaces of the rocks are the only naturally occurring media available for transferring this energy to the Earth's surface. In the United States, average groundwater temperatures from 5 to 15 m deep are 5–7°C above the mean annual air temperature (Reed, 1983b).

Certain "hot spots" of the Earth, generally near areas of recent or Pleistocene volcanism, discharge heat at rates of 10 to more than 1000 times that of areas of "normal" heat flow of comparable size. These hot spring areas are characterized by the physical transport of most of the total heat flow in water or steam. Some of the largest and hottest spring areas have been utilized for geothermal energy. These areas are characterized by high permeability, at least locally on faults, fractures, and sedimentary layers; this high permeability permits fluid circulation, most of the total heat flow being transported upward in water or steam. The circulation has produced reservoirs of stored heat closer to the Earth's surface than is normally possible by rock conduction alone. Local near-surface thermal gradients are typically very high, but the gradient decreases greatly, and even reverses, at greater depths in any single geothermal drill hole (White, 1965).

Hot-water systems are dominated by circulating liquid, which transfers most of the heat and largely controls subsurface pressures (in contrast to vapor-dominated systems). However, some vapor may be present, generally as bubbles dispersed in the water of the shallow low-pressure parts of these systems. Most known hot-water systems are characterized by hot springs that discharge at the surface. These springs, through their chemical composition, areal distribution, and associated hydrothermal alteration, provide very useful evidence on probable subsurface temperatures, volumes, and heat contents (Renner, White, and Williams, 1975).

The temperatures of hot-water systems in North America range from slightly above ambient to about 363°C in the Salton Sea (California) system and the nearby Cerro Prieto system of Mexico. All hot-water convection systems are divided into three temperature ranges: (1) Above 150°C, these systems may be considered for generation of electricity; (2) from 90 to 150°C, these systems are attractive for space and process heating and can be utilized for electricity generation; and (3) below 90°C, these systems can supply the energy needs of increasingly popular geothermal heat pumps and many processes that now depend on fossil fuels, as shown in Table 2–1. The principal source of thermal energy in the temperature range 10–90°C within the United States is the burning of natural gas and no. 2 diesel oil; electrical resistance heating is also a common source of thermal energy in this temperature range (Reed, 1983a).

Direct temperature measurements of geothermal systems are made in either surface springs or wells. The temperatures of springs generally do not exceed the boiling temperature at existing air pressure (100°C at sea level to 93°C for pure water at an altitude of 2200 m), although some springs in Yellowstone Park and elsewhere are superheated by 1–2°C (Renner et al., 1975).

In the past, the use of hot springs and geothermal water in general in the United States was primarily for hot-water baths and pools (balneology). After 1920, however, the abundance of inexpensive natural gas for heating baths and pools caused a rapid decline in the use of natural hot water. Some use of geothermal water for space heating dates from before 1890 in such areas as Boise, Idaho, but interest in this application was rather slight until the 1970s and the first global oil crisis.

Low-temperature geothermal resources occur in two types of geothermal systems: hydrothermal convection and conduction dominated. In hydrothermal-convection systems, the upward circulation of water transports thermal energy to reservoirs at shallow depths or to the surface via thermal springs. These systems commonly occur in regions of active tectonism and above-normal heat flow, such as much of the western United States. In conduction-dominated systems, the upward circulation of fluid is less important than the existence of high vertical temperature gradients in rocks that include aquifers of significant lateral extent. These conditions occur beneath many deep sedimentary basins throughout the United States (Sorey, Natheson, and Smith, 1983a).

Most of the identified low-temperature geothermal resources associated with hydrothermal-convection systems fall into areas of isolated thermal springs and wells. In such areas, the only evidence that a

Table 2–1 Temperatures Required for Use of Low-Temperature Geothermal Water (From Lindal, 1973)

Temperature (°C)	Uses
90	Drying of stock fish; intense deicing operations
80	Space heating; greenhouse heating and milk pasteurization
70	Refrigeration (lower limit); vacuum distillation of ethanol
60	Animal husbandry; combined space and bed heating of greenhouses
50	Mushroom growing
40	Enhanced oil recovery (lower limit); soil warming
30	Water for winter mining in cold climates; balneology and deicing (lower limit)
20	Fish hatching and fish farming

geothermal reservoir exists at depth is a single thermal spring or group of closely spaced springs or a well that produces thermal water. In the western United States, thermal springs commonly occur along normal faults, whereas in the eastern United States, thermal springs occur in regions of folded and thrust-faulted rocks. Figure 2–26 shows three possible models of fluid circulation in such areas according to Sorey et al. (1983a); other models are presented by Breckenridge and Hinckley (1978) and Hobba et al. (1979).

Much of the western United States lies within the Basin and Range geologic province, which has a heat flow generally higher than normal and is characterized by extensional tectonism. The combination of range-front faults and sediment-filled basins is favorable for the occurrence of geothermal systems. Young silicic volcanic centers along the east and west margins of the province provide localized heat sources for hydrothermal-convection systems. Most thermal waters in the province result from deep circulation. Normal faults provide near-surface conduits for the circulating waters and thus control the positions of most of the identified hydrothermal-convection systems (Figure 2–26). Basin-fill sediment may act as a thermal blanket that traps heat in relatively shallow aquifers beneath large areas of some of the basins. Leakage away from fault conduits is probably the source of the thermal waters in these aquifers (Mariner et al., 1983).

In western Arkansas, identified low-temperature geothermal resources occur in areas of thermal springs in the Ouachita province, including those at Hot Springs National Park and Caddo Gap. These springs are associated with tightly folded and thrust-faulted rocks. Studies by Bedinger et al. (1979) and Steele and Wagner (1981) indicate that the chemical compositions are similar in all the springs in this province and suggest that circulation systems feeding the springs occur largely in silica-rich sandstone and chert formations. Little is known, however, about the configuration of associated low-temperature geothermal reservoirs in these areas (Sorey et al., 1983b).

Thermal springs in the eastern United States are associated with fault and fracture zones in several provinces of the Appalachian Mountains. Early descriptions of these springs dealt with their therapeutic and recreational values (Moorman, 1867; Crook, 1899; Fitch, 1927). The locations of thermal springs are controlled mostly by the structural setting and, to a lesser extent, lithology. The springs occur in areas of steeply dipping folded rocks that are transected by nearly vertical east-west-trending fracture zones. Correlation of springs with topographic lows, or gaps, apparently results from the fact that easily eroded areas correspond to zones containing many fractures, which, in turn, provide the increased vertical permeability needed to establish a hydrothermal-convection system.

The warm springs in the Appalachians issue from sandstone or carbonate rocks exposed in the steeply dipping limbs of anticlinal folds (Hobba et al., 1979). Chemical analyses of the warm-spring waters issuing from carbonate rocks exhibit consistently low concentrations of dissolved silica and high concentrations of magnesium and calcium, which indicate that the flow of warm water is restricted to the carbonate rocks. Analyses of waters from springs issuing from fractured sandstone show higher concentrations of dissolved silica and lower concentrations of magnesium and calcium, which indicates that flow is restricted to the sandstone beds (Sorey et al., 1983b).

Geochemical considerations suggest that reservoir temperatures are not substantially higher than the measured surface temperatures at most eastern thermal springs; observed temperatures range from 18 to 41°C. The occurrence of these springs in areas of average heat flow and relatively low-temperature gradients (Costain, Keller, and Crewdson, 1976; Perry, Costain, and Geiser, 1979) indicates that the depths of hydrothermal circulation are generally between 1 and 3 km.

As discussed by Duffield and Sass (2003), hot springs that are indicators of moderate- to high-temperature geothermal resources are concentrated in places where hot or even molten rock (magma) exists at relatively shallow depths in the Earth's outermost layer (the crust). Such "hot" zones generally are near the boundaries of the dozen or so slabs of rigid rock (called *plates*) that form the Earth's lithosphere, which is composed of the Earth's crust and the uppermost, solid part of the underlying denser, hotter layer (the mantle). High heat flow also is associated with the Earth's "hot spots" (also called *melting anomalies* or *thermal plumes*), whose origins are somehow related to the narrowly focused upward flow of extremely hot mantle material from very deep within the Earth. Hot spots can occur at plate boundaries (for example, beneath Iceland) or in plate

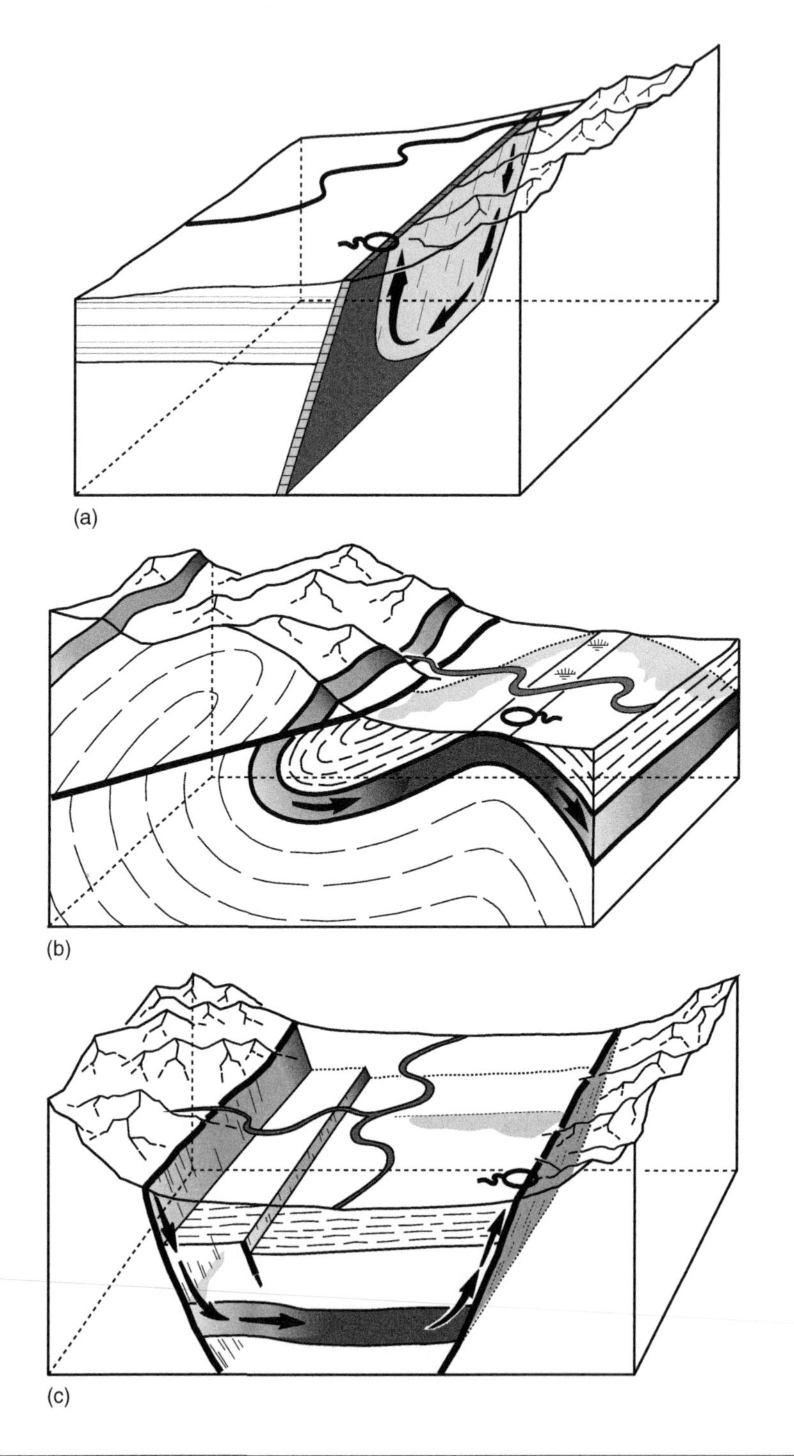

FIGURE 2–26 Conceptual models for types of hydrothermal-convection systems in which low-temperature geothermal-resource areas in the category of isolated thermal springs and wells occur: (a) Fault plane; (b) margin of anticline; and (c) deep reservoir. Arrows indicate direction of fluid circulation; dark shading shows location of reservoir containing low-temperature geothermal resources. (From Sorey et al., 1983a.)

interiors thousands of kilometers from the nearest boundary (for example, the Hawaiian hot spot in the middle of the Pacific Plate). Regions of stretched and fault-broken rocks (rift valleys) within plates, like those in East Africa and along the Rio Grande River in Colorado and New Mexico, also are favorable target areas for high concentrations of the Earth's heat at relatively shallow depths.

Zones of high heat flow near plate boundaries are also where most volcanic eruptions and earthquakes occur. The magma that feeds volcanoes originates in the mantle, and considerable heat accompanies the rising magma as it intrudes into volcanoes. Much of this intruding magma remains in the crust, beneath volcanoes, and constitutes an intense, high-temperature geothermal heat source for periods of thousands to millions of years, depending on the depth, volume, and frequency of intrusion. In addition, frequent earthquakes—produced as the tectonic plates grind against each other—fracture rocks, thus allowing water to circulate at depth and transport heat toward the Earth's surface. Together, the rise of magma from depth and the circulation of hot water (hydrothermal convection) maintain the high heat flow prevalent along plate boundaries (Duffield and Sass, 2003).

Figure 2–27 illustrates the connection between hot springs and young volcanic activity in the Cascade Range, western United States, which is part of the "Ring of Fire," a belt of volcanic arcs and oceanic trenches partly encircling the Pacific Basin and is famous for frequent earthquakes and volcanic eruptions.

Figure 2–28 shows a stretch of Hot Creek in east-central California. The creek flows through the Long Valley Caldera in a volcanically active region. This stretch of the creek, looking upstream to the southwest, has long been a popular recreation area because of the warm waters from its thermal springs. The thermal springs in Long Valley Caldera have long been known to Native Americans. Many of the hot springs have special status with Native American tribes and have been used for spiritual and medicinal purposes.

Since May 2006, springs in and near the most popular swimming areas have been "geysering" or intermittently spurting very hot, sediment-laden water as high as 6 ft (2 m) above the stream surface. At times, this geysering activity is vigorous enough to produce "popping" sounds audible from hundreds of feet away. The geysering usually lasts a few seconds and occurs at irregular intervals, with several minutes between eruptions. The unpredictability of this hazardous spring activity led the U.S. Forest Service to close parts of the Hot Creek Geologic Site in June 2006 (Farrar et al., 2007).

All the springs in Hot Creek emerge along a stream section between two faults and discharge a total of about 8.5 ft^3/s (about 240 L/s) of hot water. Figure 2–29 is a diagrammatic cross section showing the origin of hot water in the creek. A detailed explanation of the possible reasons for development of the new geysering activity in the creek (Figure 2–30) is given by Farrar et al. (2007).

Hot water circulating in the Earth's crust dissolves some of the rock through which it flows. The amounts and proportions of dissolved constituents in the water are a direct function of temperature. Therefore, if geothermal water rises quickly to the Earth's surface, its chemical composition does not change significantly and bears an imprint of the subsurface temperature. Field and laboratory studies show that this deeper, hotter temperature is commonly "remembered" by thermal waters of hot springs. Subsurface temperatures calculated from hot-springwater chemistry have been confirmed by direct measurements made at the base of holes drilled into hydrothermal systems at many locations worldwide. The technique of determining subsurface temperature from the chemistry of hot springwater is called *chemical geothermometry* (see, e.g., Duffield and Sass, 2003; Brook et al., 1979).

The term *mineral spring* (or *mineral water*, for that matter) has a very different meaning in different countries and could be very loosely defined as a spring with water having one or more chemical characteristics different from normal potable water used for public supply (see Chapter 6). For example, water can have elevated content of free gaseous carbon dioxide (naturally carbonated water), high radon content ("radioactive" water, still consumed in some parts of the world as "medicinal" water of "miraculous" effects), high hydrogen sulfide content ("good for skin diseases" and "soft skin"), high dissolved magnesium, or simply have the total dissolved solids higher than 1000 mg/L. Some water bottlers, exploiting a worldwide boom in the use of bottled springwater,

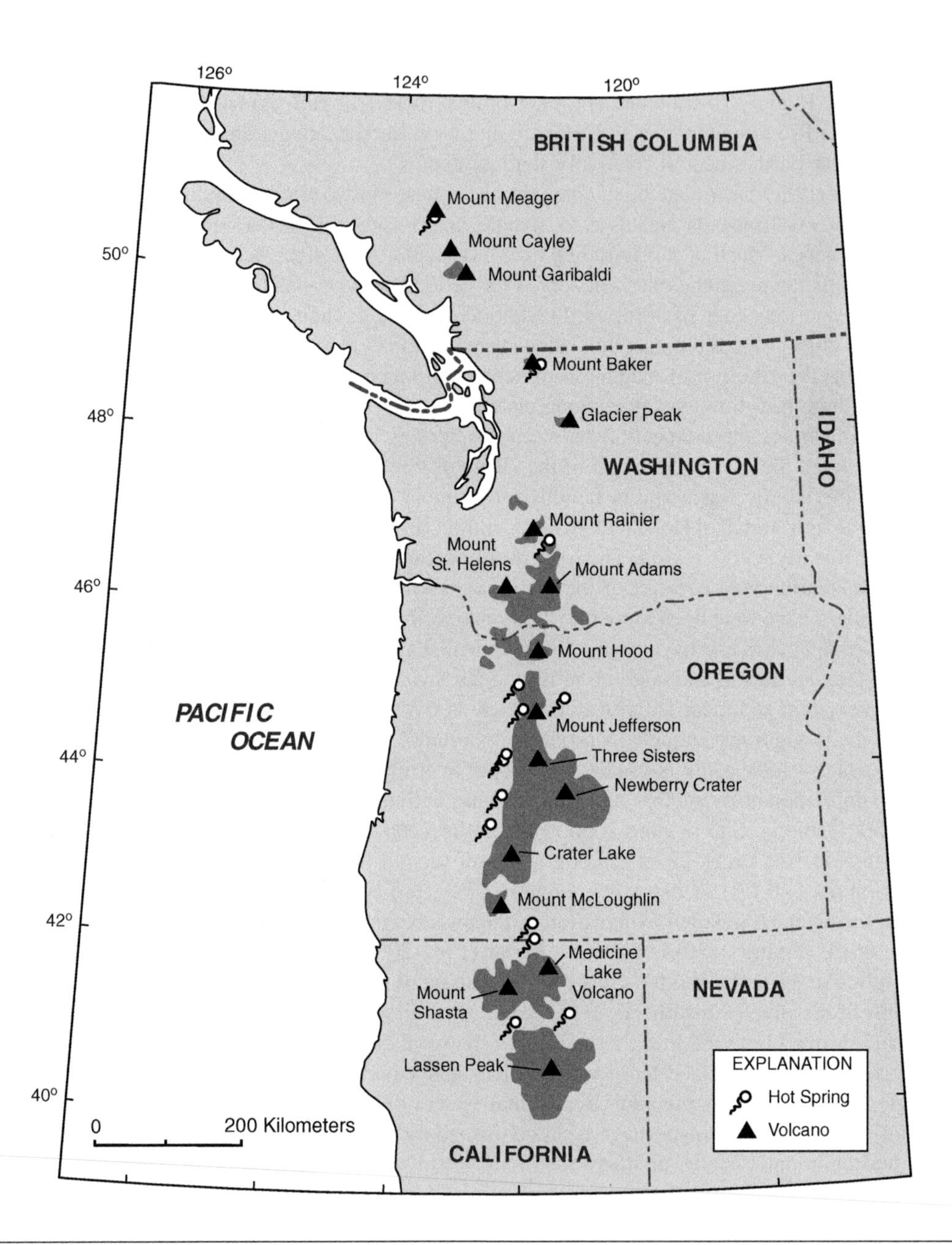

FIGURE 2–27 Principal volcanoes, geologically young volcanic rocks (dark areas) and hot springs in the Cascade Range, Pacific Northwest. (From Duffield and Sass, 2003.)

FIGURE 2–28 Hot springs in Hot Creek, which flows through the Long Valley Caldera in a volcanically active region of east-central California. (USGS photo by Chris Farrar.)

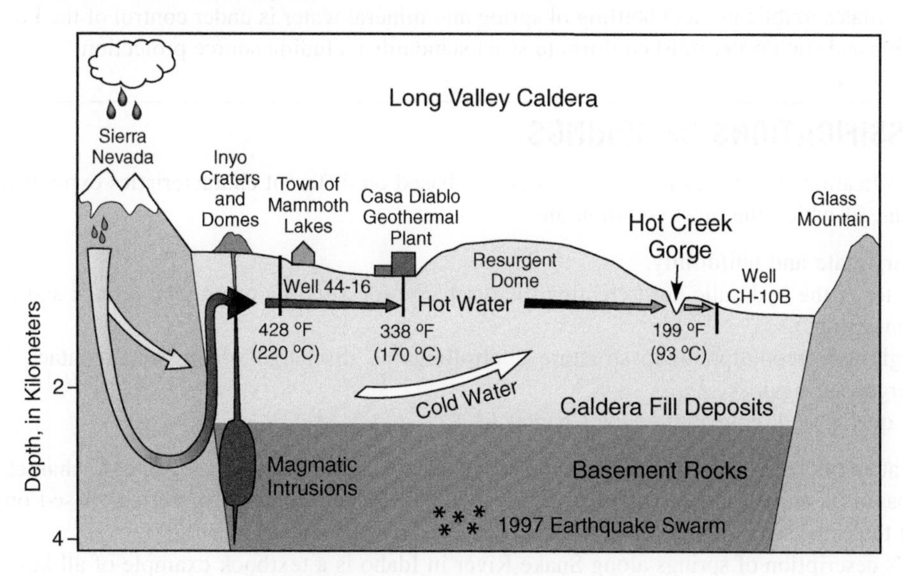

FIGURE 2–29 Diagrammatic cross section of Long Valley Caldera. The Inyo Craters and Domes and Resurgent Dome reflect volcanic activity since 760,000 years ago, when a giant eruption formed the caldera. The thermal springs in Hot Creek are fed by Sierra Nevada snowmelt that seeps underground and migrates eastward, becoming heated to temperatures as high as 428°F (220°C) in the vicinity of partially molten rock (magma) beneath the western part of the caldera. The water cools as it migrates eastward beneath the ground. The temperature at which water emerges in the springs at Hot Creek could be affected by changes in seismic activity, heat extraction, mixing with cooler water, and other factors. (From Farrar et al., 2007.)

FIGURE 2–30 The temperature in Hot Creek can change in seconds. These photos were taken 5 s before and then during a violent geysering event, in which boiling water (at 199°F or 93°C, the boiling point at this elevation) erupted above the surface. Any swimmer caught in this part of the creek would have been severely scalded. (From Farrar et al., 2007; photographs courtesy of Alix Ginter.)

label water derived from a spring as "mineral" even when it has no unusual chemical or physical characteristics. In the United States, public use and bottling of spring and mineral water is under control of the Food and Drug Administration and such water must conform to strict standards including source protection.

2.2 CLASSIFICATIONS OF SPRINGS

Various classifications of springs have been proposed, based on different characteristics explained in the previous sections, of which the most common are

- Discharge rate and uniformity.
- Character of the hydraulic head creating the discharge (descending or gravity springs and ascending or artesian springs).
- Geologic and geomorphologic structure controlling the discharge (depression, contact, barrier, and submerged springs).
- Water quality and temperature (see Chapter 6).

Various attempts have been made to group springs of different hydraulic or geologic characteristics into new types based on some focused perspective. For example, a classification of springs based on the microhabitats and the ecosystems they support is proposed by Springer and Stevens (2009).

Meinzer's description of springs along Snake River in Idaho is a textbook example of all key elements of spring characterization that can also serve as bases for spring classification (Meinzer, 1927, "Large Springs in the United States," pages 42–50). The following short excerpts from Meinzer's work illustrate these elements:

Spring Size

Many large springs issue on the north side of Snake River between Milner and King Hill, Idaho, nearly all of them in the canyon below Shoshone Falls or in short tributary canyons. According to the measurements that have been made the total discharge of these springs was 3,885 second-feet in 1902, before any irrigation developments had been made on the north side, and averaged 5,085 second-feet in 1918, after the north-side irrigation project had been developed. The great volume of water discharged by these springs can perhaps be better appreciated by recalling that in 1916 the aggregate consumption of New York, Chicago, Philadelphia, Cleveland, Boston, and St. Louis, with more than 12 million inhabitants, averaged only 1,769 million gallons a day (2,737 second-feet), or only slightly more than one-half of the yield of these Snake River springs in 1918. In fact, these springs yield enough water to supply all the cities in the United States of more than 100,000 inhabitants with 120 gallons a day for each inhabitant.... There are 11 springs or groups of springs that yield more than 100 second-feet, of which 1 yields more than 1,000 second-feet, 3 yield between 500 and 1,000 second-feet, and 7 yield between 100 and 500 second-feet. Moreover there are 5 springs that yield between 50 and 100 second-feet and numerous so-called "small" springs which would be considered huge in most localities.

Fluctuation in Discharge

The flow of these springs is relatively constant, and in this respect they differ notably from most of the large limestone springs.

Role of Geology

These springs issue chiefly from volcanic rocks or closely related deposits. The water-bearing volcanic rocks are largely basalt, but they also include jointed obsidian and rhyolite. A large part of the basin of Snake River above King Hill, Idaho, was inundated with basaltic lava during the Tertiary and Quaternary periods, and the lava rock is in many parts so broken or vesicular that it absorbs and transmits water very freely.... At the Thousand Springs the water can be seen gushing from innumerable openings in the exposed edge of a scoriaceous zone below a more compact sheet of lava rock. At Sand, Box Canyon, and Blind Canyon Springs the water ... comes from a stratum consisting largely of white sand which is overlain by a thick sheet of lava. At most of the springs there is so much talus that the true source of the water cannot be observed, but it probably issues chiefly from the large openings in scoriaceous or shattered basalt where the basalt overlies more dense rocks. The fact that most of the springs are confined to rather definite localities and issue at points far above the river indicates that the flow of the ground water to the springs is governed by definite rock structure. The great body of ground water is obviously held up in the very permeable water-bearing rocks by underlying impermeable formations. It may be that the underlying surface which holds up the water is a former land surface and that the principal subterranean streams which supply the springs follow down the valleys of this ancient surface.... At most of the springs the water issues at considerable heights above river level.

Drainage Area, Recharge, Source of Water

The lava plain lying north and northeast of these large springs extends over a few thousands of square miles and receives the drainage of a few thousand square miles of bordering mountainous country. The great capacity of the broken lava rock to take in surface water is well established and is, moreover, shown by the fact that in the entire stretch of more than 250 miles from the head of Henrys Fork of Snake River to the mouth of Malade River no surface stream of any consequence enters Henrys Fork or the main river from the north. The greater part of this vast lava plain discharges no surface water into the Snake, and a number of rather large streams that drain the mountain area to the north lose themselves on this lava plain. A part of the water that falls on the plain and adjoining mountains is lost by evaporation and transpiration, but a large part percolates into the lava rock and thence to the large springs.

Water Quality

The water of these springs does not contain much mineral matter. It is generally very clear, although the water of some of the springs, such as the Blue Lakes, has a beautiful blue color and a slight opalescence due to minute particles in suspension. So far as is known, all the springs have about normal temperatures.

Utilization and Preservation

On account of the notable height above the river at which most of these springs issue, together with the great volume of water which they discharge, they are capable of developing a great amount of water power. Large power plants have already been installed at Malade Springs and at Thousand and Sand Springs, and other large plants could be installed at other springs, especially at Clear Lakes, Box Canyon, and Bickel Springs. ... At the Crystal Springs the clear, cold water is utilized in a fish hatchery, where great quantities of trout are raised. ... The most spectacular feature of these springs is the cataracts which they form, or which they formed before they were harnessed to develop electric energy. Thousand Springs formerly gave rise to a strikingly beautiful waterfall 2,000 feet long and 195 feet high. Snowball Spring, which discharges 150 to 160 second-feet, is at the east end of the Thousand Springs. Formerly its water dashed over the rough talus slope, forming a cataract of great beauty that suggested a snowbank. The Niagara Springs, which issue from the canyon wall 125 feet above the river level, also form a spectacular cataract [Figure 2–31].

Meinzer's classification of springs based on average discharge expressed in the U.S. units is still widely used in the United States (Table 2–2). The table also includes Meinzer's classification based on the metric system. However, the classification based solely on average spring discharge, without specifying other

FIGURE 2–31 Niagara Falls Springs issuing from basalts in the valley of Snake River near Twin Falls, Idaho. (Photograph courtesy of Clear Foods, Inc.)

Table 2–2 Classification of Springs Based on Average Discharge Rate (From Meinzer, 1923)

(a) Classification Based on the Metric System		
Magnitude	**Discharge in Metric Units**	**Discharge in English Units (approximate)**
First	10 cubic meters per second or more	353 second-feet (cubic feet per second)
Second	1 to 10 cubic meters per second	35 to 353 second-feet
Third	0.1 to 1 cubic meter per second	3.5 to 35 second-feet
Fourth	10 to 100 liters per second	158 gallons per minute to 3.5 second-feet
Fifth	1 to 10 liters per second	16 to 158 gallons per minute
Sixth	0.1 to 1 liter per second	1.6 to 16 gallons per minute
Seventh	10 to 100 cubic centimeters per second	1.25 pints to 1.68 gallons per minute
Eighth	Less than 10 cubic centimeters per second	Less than 1.25 pints per minute

(b) Classification Suggested for Practical Use in the United States	
Magnitude	**Discharge**
First	100 second-feet or more
Second	10 to 100 second-feet
Third	1 to 10 second-feet
Fourth	100 gallons per minute to 1 second-foot
Fifth	10 to 100 gallons per minute
Sixth	1 to 10 gallons per minute
Seventh	1 pint to 1 gallon per minute
Eighth	Less than 1 pint per minute

discharge parameters, is not very useful when evaluating the potential for spring utilization. For example, a spring may have a very high average discharge, but it may be dry or just trickling most of the year. Practice in most countries is that the springs are evaluated based on the minimum discharge recorded over a long period, typically longer than several hydrologic years (a hydrologic year is defined as spanning all wet and dry seasons within a full annual cycle). When evaluating availability of springwater, it is important to include a measure of spring discharge variability, which should also be based on periods of record longer than one hydrologic year. The simplest measure of variability is the ratio of the maximum and minimum discharge:

$$I_v = \frac{Q_{max}}{Q_{min}} \tag{2.1}$$

Springs with an index of variability (I_v) greater than 10 are considered highly variable, and those with $I_v < 2$ are sometimes called *constant* or *steady springs*.

Meinzer proposed the following measure of variability expressed in percentage:

$$V = \frac{Q_{max} - Q_{min}}{Q_{av}} \times 100\,(\%) \tag{2.2}$$

where Q_{max}, Q_{min}, and Q_{av} are the maximum, minimum, and average discharge, respectively. Based on this equation, a constant spring would have variability less than 25 percent, and a variable spring would have variability greater than 100 percent.

An extensive study of springs in east Tennessee is illustrative of the natural variability of spring flows and the role of the underlying geology. More than 960 springs were observed and described by De Buchananne and Richardson (1956). During their reconnaissance, measurements were made of discharge from many of the springs and estimates made of flow from the remainder. The total of these measurements was about 265 million gallons per day (mgd). All but a few of the measurements were made during the relatively dry period, June through September; thus, this volume represents a near-minimum, or at least a below-average, total for the 960 springs considered. The 960 springs described by De Buchananne and Richardson are classified according to the magnitude of their flow as shown in Table 2–3.

The results of a follow-up study by Sun, Criner, and Poole (1963) are shown in Figure 2–32 for 84 large undeveloped springs in east Tennessee for which records of minimum, average, and maximum flows are available. The horizontal line shows the range of discharge, and the vertical line indicates the average discharge for the spring's period of record. Table 2–4 shows the classification of these springs based on their magnitude.

In general, the least variable springs in east Tennessee issue from shale of the Conasauga Group, and all these are of relatively small magnitude. The most variable springs issue from solution openings in formations of the Knox Group, Chickamauga Limestone, or limestones of the Conasauga Group. These cavities are of such varying size and degree of interconnection that, as the water table fluctuates from wet to dry seasons, spring discharge fluctuates in accordance with the ability of the saturated cavities to transmit the water to the spring orifices.

As discussed by Sun et al. (1963), springs are important sources of water for municipal, domestic, and farm use in east Tennessee; in 1959, 39 of the 95 municipal water-supply systems used water derived solely from springs, and 15 others used springwater as a supplemental source. Many of the springs are not developed, because of their inaccessibility or the lack of information regarding their adequacy and dependable low flows available for small-industry and community supplies. A spring discharging 450 gpm (gallons

Table 2–3 Classification of 960 Springs in Tennessee According to the Magnitude of Flow (in gallons per minute) (From Sun et al., 1963)

Discharge (gpm)	No. of Springs
<100	653
100–450	155
450–4,500	147
4,500–45,000	5
>45,000	0

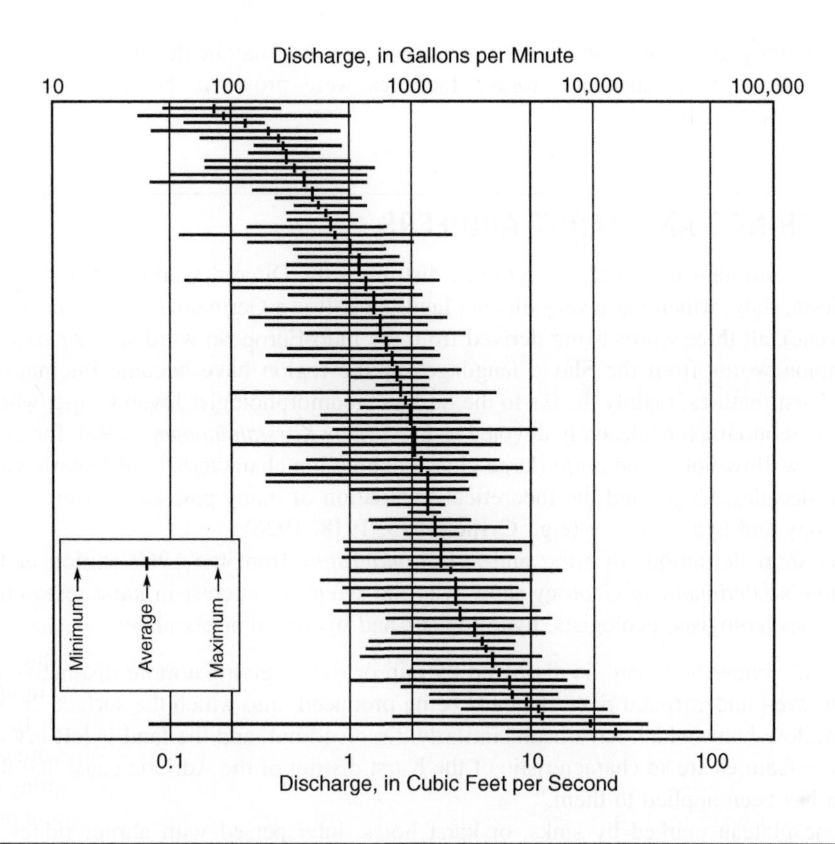

FIGURE 2–32 Minimum, average, and maximum discharges of 84 large springs in eastern Tennessee. (From Sun et al., 1963.)

Table 2–4 Classification of 84 Large Springs in East Tennessee (From Sun et al., 1963)

Magnitude	No. of Springs
First	None
Second	4
Third	62
Fourth	16
Fifth	2

per minute), or about 1 cfs (cubic foot per second), is capable of supplying a town population of 6500, if one assumes a per capita consumption of 100 gpd (gallons per day).

According to Sun et al. (1963), a number of large springs have been identified as excellent sources of water for future development in east Tennessee. However, the variability of their flow may prevent full utilization unless adequate storage facilities are provided. If there is no storage to provide water during periods of peak use and minimum yield, which usually are concurrent, a spring may be developed only to the extent of its lowest dependable flow. If adequate storage facilities were provided, however, development could approach the average annual flow.

2.3 KARST SPRINGS AND KARST AQUIFERS

Karst is a scientific term named after the geographic district of the Dinarides on the Adriatic coast, between Slovenia and Trieste, Italy, which has a very distinct landscape. It is a Germanized word for *carso* (in Italian) and *kras* (in Slovene), all three words being derived from the Indo-European word *kar* or *karra*, which means rock. Some common words from the Slavic languages of the region have become international scientific terms describing karst features, mainly thanks to the Serbian geomorphologist Jovan Cvijić, who was the first to write a scientific monograph exclusively devoted to karst (*Das Karstphänomen*, 1893), for example, *doline* (sinkhole), *ponor* (swallow hole), and *polje* (large closed depression characteristic of Dinaric karst). Over the following several decades, Cvijić laid the theoretical foundation of many past and current developments in karst geomorphology and hydrogeology (e.g., Cvijić, 1893, 1918, 1926).

The following short definitions of *karst* and *karst topography* from the 1960 edition of the American Geological Institute's *Dictionary of Geology* show why the scientific interest in karst research is shared by geomorphologists, speleologists, geologists, hydrologists, and hydrogeologists alike:

- "Limestone, no matter how hard, is dissolved by rain or rivers, giving a more distinctive type of country, caves or even underground river channels being produced, into which the surface drainage sinks by rifts and swallow-holes which have been similarly dissolved out, and the land is left dry and relatively barren. These features are so characteristic of the Karst district of the Adriatic coast that the name karst phenomena has been applied to them."
- "A limestone plateau marked by sinks, or karst holes, interspersed with abrupt ridges and irregular protuberant rocks; usually underlain by caverns and underground streams."
- "In Karst, on the eastern side of the Adriatic sea, the limestone rocks are so honeycombed by tunnels and openings dissolved out by ground waters, that much of the drainage is underground. Large sinks abound, some of them five or six hundred feet deep. Streamless valleys are common, and valleys containing streams often end abruptly where the latter plunge into underground tunnels and caverns, sometimes to reappear as great springs elsewhere. Irregular topography of this kind, developed by the solution of surface and ground waters, is known as karst topography, after the region in Austria-Hungary."

Karst features, in various but similar forms, are developed in all water soluble rocks, such as limestones, dolomites, marbles, gypsum, halite, and some conglomerates. Figures 2–33 through 2–36 illustrate some of the many fascinating features of the Slovenian karst, while Figures 2–37 through 2–40 show karst examples from the United States.

As discussed throughout this book, groundwater in karst environments has intrigued scientists, researchers, and ordinary people for millennia due to its many fascinating facets: It feeds the world's largest springs, many of which enabled the establishment of the first urban centers in human history and continue to serve as reliable sources of water supply to the present day; it creates a mysterious underground world of caves and caverns and supports living creatures often unique to specific locations; it behaves unpredictably as it can sometimes rise hundreds of feet after heavy rains in a matter of hours, giving life to numerous temporary springs and increasing the flow of permanent springs a thousandfold; and it is extremely vulnerable to both natural and anthropogenic contamination, thus seriously limiting its unrestricted use in many parts of the world (Kresic, 2009).

FIGURE 2–33 Emergence of the Notranjska River in Velika Dolina ("Big Sinkhole") of the Skocjan Caves, Slovenia. The caves and the surrounding area were placed on the UNESCO list of protected natural heritage in 1986. The river appears and disappears in several karst windows and sinks before it continues flowing underground toward the Timavo Springs near Trieste, the Adriatic coast of Italy. (Copyright Bostjan Burger, printed with permission.)

FIGURE 2–34 Disappearance of the Notranjska River into Lezeca Ponor, the Skocjan Caves. (Photograph courtesy of Borut Petric.)

Many excellent publications examine karst geomorphology, speleology, hydrology, and hydrogeology, all of which discuss karst aquifers and springs in some manner. Here is just a small sample of useful general books: Herak and Stringfield (1972), Sweeting (1972), Bögli (1980), Milanovic (1981), Bonacci (1987), White (1988), and Derek and Williams (2007). Worthington and Gunn (2009) provide a short

FIGURE 2–35 Part of the Skocjan Caves named Sumeca Jama ("Murmuring Cave"), an underground canyon with vertical sides over 120 m (360 ft) high. (Copyright Bostjan Burger, printed with permission.)

history of hydrogeology of carbonate aquifers. A good source of general information on karst hydrogeology is the Web site of the Karst Commission of the International Association of Hydrogeologists (www.iah.org/karst).

The vulnerability of karst aquifers and springs to contamination cannot be emphasized enough (see Chapter 8). This aspect of spring characterization, development, and management, for whatever the intended spring use, should be approached thoroughly and with a comprehensive plan for multidisciplinary field investigations and monitoring whenever possible. Figures 2–41 through 2–44 illustrate the karst vulnerability and the utmost importance of public education in protecting its groundwater resources.

Historically, Ricks Spring (Figure 2–41) was used as a mountain springwater source until visitors became ill after drinking from it. Ice jams on the Logan River in 1972 and dye traces led scientists to find that the water of Ricks Spring was actually coming directly from the Logan River through a fracture in the rock. Folded rocks gave way to fractures, and evidently, one of these fractures allows the river water to seep through the rock and exit at Ricks Spring (U.S. Forest Service Intermountain Region, 2008).

Karst aquifers are present on all continents and numerous oceanic islands and crop out on over 20 percent of the land surface. As the importance of water resources and their sustainability is becoming critical worldwide, karst aquifers are receiving rapidly increasing attention from the scientific, engineering, and regulatory communities, due to the many challenges related to their characterization and management. The main

FIGURE 2–36 The Boka Waterfall karst spring near Bovec, Slovenia. The spring emerges from a steep rock face, forming a waterfall that falls freely 106 m and immediately after another 30 m steeply inclined. The recharge area of the spring is the Mount Kanin alpine karst massif. (Copyright Nico Goldscheider, printed with permission.)

explanation for these challenges is straightforward—unlike all other aquifer types, karst aquifers have three types of porosity: (1) porosity of rock matrix or primary porosity, (2) common rock discontinuities such as fractures (fissures) and bedding planes or secondary porosity, and (3) solutionally enlarged voids such as channels and conduits developed from the initial discontinuities (Figures 2–45 through 2–48). Storage and transfer (flow) of groundwater in karst therefore depends on the prevailing porosity type in a specific aquifer, making any generalization without field investigations potentially erroneous.

In addition to recharge pattern and exposure of karst aquifers at the land surface (e.g., covered or outcropping), the size and discharge characteristics of karst springs are strongly related to the number, size, and degree of interconnection of interstitial pores (matrix porosity) and secondary porosity of fractures and solution openings, such as caverns, channels, and conduits. For example, the old Paleozoic limestones and dolomites of east Tennessee have little or no primary porosity, but large volumes of water are stored in the

FIGURE 2–37 Sinkholes in Monroe County, West Virginia. (Photograph courtesy of William Jones, Karst Waters Institute.)

FIGURE 2–38 Aerial photograph of the head of Muddy Creek at Piercys Mill Cave, Greenbrier County, West Virginia. (Photograph courtesy of William Jones, Karst Waters Institute.)

FIGURE 2–39 Big Spring in Carter County, the largest first-magnitude spring in Missouri, with average flow of about 289 million gallons per day. (Photograph courtesy of Missouri Department of Natural Resources; available at www.dnr.mo.gov/env/wrc/springsandcaves.htm.)

numerous solution cavities and openings along faulted, jointed, and fractured zones. The ability of such rock formations to yield a sustained flow of water to springs therefore depends on a system or network of interconnected openings through which water can infiltrate from the land surface and be transmitted to points of natural or artificial discharge. On the other hand, younger limestones, such as those constituting the Floridan aquifer in the United States, which have high matrix porosity, can store enormous volumes of groundwater outside the karst conduits, which constantly drain into the conduits, thus maintaining significant spring flows during droughts. This difference is clearly reflected in the hydraulic characteristics of spring discharge (see Chapter 4). The conduit-dominated systems without significant matrix storage feed springs of higher discharge variability, whereas the matrix-dominated systems are drained by springs that have a more uniform discharge.

In either case, however, there is a very strong tendency for developing the so-called self-organized permeability, which results in aquifer drainage by a few or just one large spring, similar to the hierarchy of surface water drainages. This mechanism is explained in detail by Worthington and Ford (2009), who present the results of related laboratory experiments and numeric modeling. The authors conclude that the invariable result of flow through limestone aquifers is that the positive feedback between increasing flow and increasing dissolution results in a high-permeability, self-organized network of channels (conduits). If the flux of water through an aquifer is minimal, such as in some confined aquifers, then the formation of channel networks is retarded. Conversely, in confined aquifers where there is a substantial flux of water as well as in unconfined aquifers, channel networks are likely to convey most of the flux of water through the aquifer after periods of 10^3–10^6 years following the onset of substantial flow through the aquifer. This range of times is short in comparison to the time that most unconfined limestone aquifers have been functioning, and so it is reasonable to infer that most such aquifers should have well-developed channel networks.

FIGURE 2–40 View of the Wakulla Spring from a cave diver's perspective. This first-magnitude spring in Wakulla County, Florida, is one of the largest and deepest freshwater springs in the world, with miles of the explored submerged cave passages. Several of the early *Tarzan* movies starring Johnny Weissmuller were filmed in the Wakulla Springs State Park. (Copyright David Rhea & Global Underwater Explorers 2006, printed with permission.)

The location, discharge rate, and other spring characteristics are very useful indicators of karst aquifer conditions and its behavior as one system. In a well-developed, integrated conduit (channel) system, as more and more water is conveyed toward a point location of a major spring, the transmissive properties of the aquifer must increase to accommodate this increasing flow. For example, a study of carbonate-rock aquifers in southern Nevada by Dettinger (1989) shows that, within 10 mi of regional springs, aquifers are an average of 25 times more transmissive than they are farther away. These are areas where flow converges, flow rates are locally high, and the conduit-type of flow likely plays a significant, if not predominant, role.

FIGURE 2–41 Ricks Spring in Wasatch-Cache National Forest, Utah. (Photograph courtesy of U.S. Forest Service Intermountain Region, 2008.)

The main consequence of the self-organized permeability of karst aquifers is that the drainage (catchment) areas of karst springs often extend beyond topographic divides, provided the underlying geology is relatively homogeneous (i.e., the carbonate sedimentary rocks extend over several surface water drainage basins). This is also the main reason why karst aquifers give rise to the world's largest springs. Except for some regional aquifers developed in extensive areas underlain by young extrusive volcanic rocks (see the next section), no other aquifer types have this characteristic of groundwater drainage being larger than the topographic drainage. Figures 2–49 through 2–51 show some examples of this unique characteristic of karst aquifers.

In addition to the lateral growth of groundwater drainage basins in karst, there is another important aspect of the karstification process: It also continues with depth, regardless of the surface water erosional basis. In other words, provided there is enough thickness of the carbonate sediments below surface water features or contact with less permeable rocks, the dissolution process results in development of karst conduits at ever-increasing depths and formation of ascending springs. In fact, some of the world's largest karst springs are of this type, as shown with examples in Figures 2–52 and 2–53 (see also Chapter 10).

FIGURE 2–42 Trash disposed in a sinkhole, Laclede County, Missouri. Dye tracing shows this sinkhole to provide recharge to Ha Ha Tonka Spring. (Photograph courtesy of Missouri Department of Natural Resources; available at www.dnr.mo.gov/env/wrc/springsandcaves.htm.)

FIGURE 2–43 Continuous stream of sewage from a college building discharging into a sink from which it finds its way to the underground water channels. Photograph used to illustrate vulnerability of groundwater to contamination in an early USGS publication for public education. (From Fuller, 1910.)

FIGURE 2–44 Spelunker examines trash around a flowstone waterfall at the bottom of the entrance to Midnight Cave near Austin, Texas, on November 20, 1993. The trash includes household garbage, used oil filters, corroded 55-gallon drums, glass pesticide bottles, partially filled turpentine cans, and automobile parts. Note the trash on the higher ledges of the cave. (From Hauwert and Vickers, 1994.)

Advances in cave diving in the last couple of decades have made some important revelations regarding major ascending springs in karst terrains. Such springs are also called *vauclusian springs*, after the famous Fontaine de Vaucluse, the source of river Sorgue in Provence, France (Figures 2–54 and 2–55):

> *The Fontaine de Vaucluse is the source of the river Sorgue and a typical karst spring. The Fontaine is a collapsed part of a cave system. In this particular case a large shaft, filled with water. At the moment the shaft is explored to a depth of 315 m. As nobody is able to dive this deep, this exploration was done using a small submarine robot called MODEXA 350. The camera of the robot showed a sandy floor at this depth, leads were not visible. The water table is most time of the year below the rim of the shaft. So the Fontaine appears as a very deep and blue lake. Small caves below lead to several springs in the dry bed of the river, just 10 m below the lake. The source is fed by the rainfall of the Plateau de Vaucluse. In spring or sometimes, after enormous rainfall, the water table rises higher than the rim. In these periods the Fontaine de Vaucluse really is a spring that produces more than 200 m^3/sec of water. (Showcaves, 2009)*

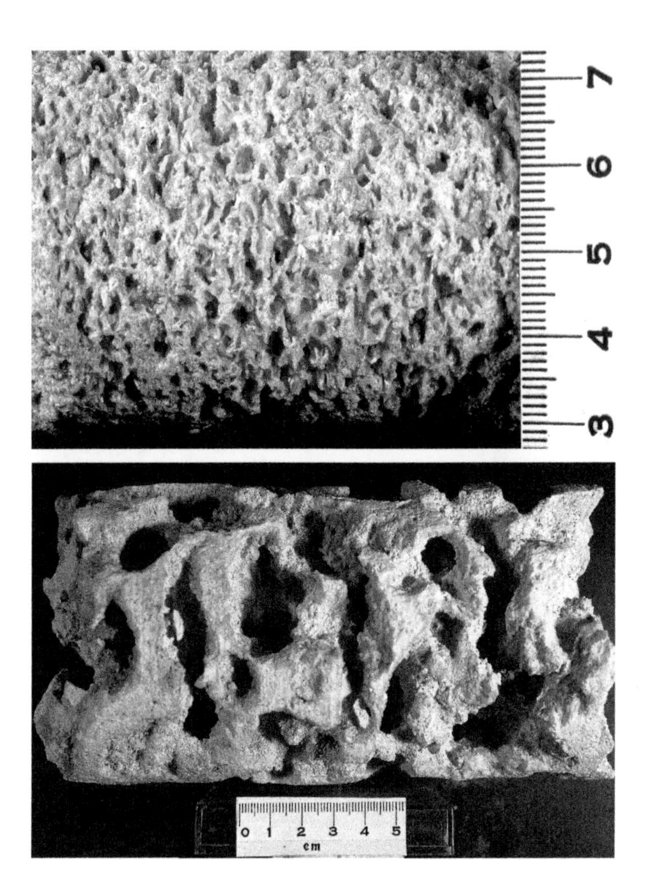

FIGURE 2–45 Primary matrix porosity of some young oolitic limestones may be over 40–50 percent, and further increased by karstification: (Top) Fort Thompson Limestone in Miami; (bottom) Miami oolite of the Biscayne aquifer, Florida, with hydraulic conductivity >1000 ft/d. (Photographs by George Sowers, printed with kind permission of Francis Sowers.)

In many cases, regardless of the size, ascending springs in limestone may issue from pools that have bottoms covered by sand, gravel, or rock colluvium (fragments) and without visible vertical conduits. However, the conduits or fractures may be masked and bridged over by clastic sediment, a possibility that cannot be excluded. For example, Meinzer (1927) gives the following account of the Bennett Spring in Dallas County, Missouri (see Figure 2–56): "It issues from a circular basin in gravel about 30 feet in diameter and gives rise to a stream that is about 1½ miles long and that empties into Niangua River at a level about 22 feet below the spring. The spring furnishes power for a mill in the village of Brice. The temperature of the water is about 58°F. The spring was visited in 1903 by Shepard, who states that at that time it boiled up with great force from a vertical cavelike opening through the limestone into a large oval basin."

As explained in more detail in Chapter 9, ascending karst springs are a primary target for spring regulation, which can increase their utilization for water supply, especially during dry seasons with higher demand. The springs can be overpumped by lowering pumps deep into the ascending channel(s), or when geologic conditions and geomorphologic conditions allow, underground dams can be constructed to control their discharge

FIGURE 2–46 Cavities of different sizes at a construction site in Paleozoic limestone, Hartsville, Tennessee. The matrix porosity is measured at 2 percent. (Photographs by George Sowers, printed with kind permission of Francis Sowers.)

and increase the hydraulic head in the aquifer. Figure 2–57 shows the general area of the Ombla Spring near Dubrovnik, Croatia, which is currently used for the water supply of the city. The spring and its drainage area have been under an extensive field investigation program for almost two decades now for the purposes of designing and constructing the largest underground dam for spring regulation ever built in karst (for more detail, see Chapters 9 and 10).

One piece of practical advice that many coauthors of this book have experienced is that, if a spring is large, "behaves" like a karst spring, and appears to have a very significant deficit in its topographic (land surface) drainage area, then it is most likely a karst spring. In other words, even when such a spring discharges

FIGURE 2–47 Underground stream passage in a Monroe County cave, West Virginia. Cave passages can vary greatly in shape and size, giving clues about the past karstification processes caused by the flowing groundwater and what can be expected at the lower, submerged horizons of the aquifer. (Photograph courtesy of William Jones, Karst Waters Institute.)

FIGURE 2–48 Turner Avenue cave passage, in the Flint Ridge section of the Mammoth Cave system. (From Palmer, 1985.)

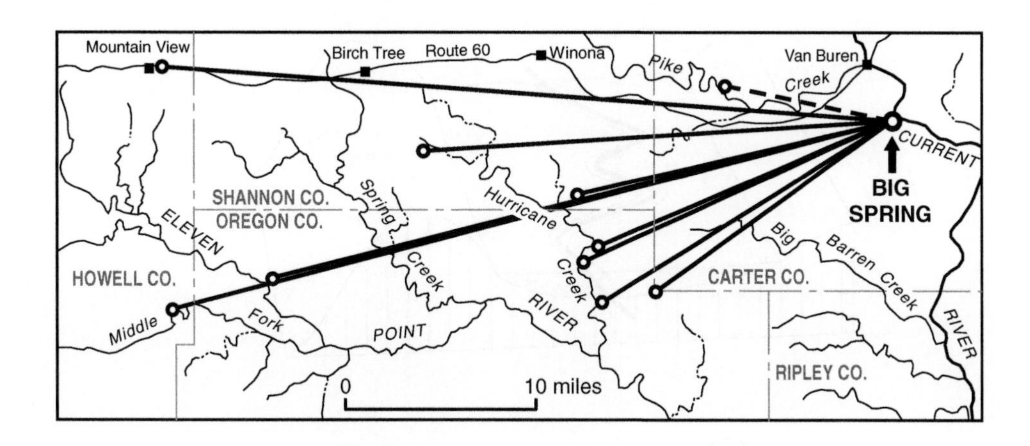

FIGURE 2–49 Water tracing experiments by Thomas J. Aley of Mark Twain National Forest, using fluorescein dyes and Lycopodium spores, show that water flows through subterranean karst channels to Big Spring in Missouri from as far as 40 mi away. (From Vineyard and Feder, 1982.)

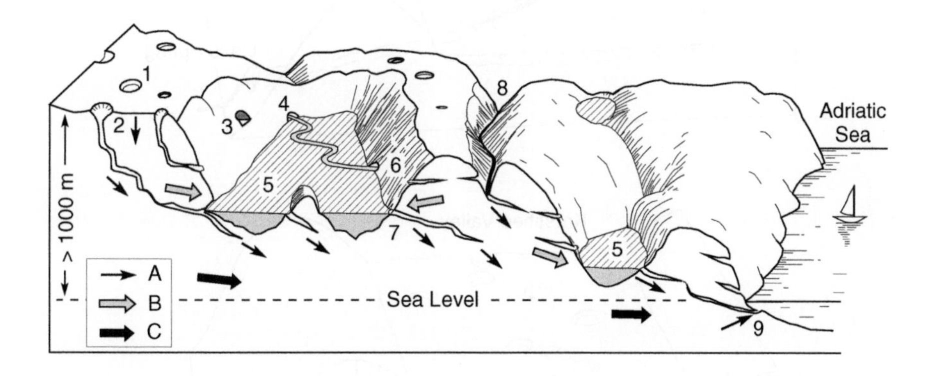

FIGURE 2–50 Karst and groundwater flow features in the thick Mesozoic carbonate platform of the Dinarides (modified from Kresic, 1987): (1) sinkhole (doline); (2) pithole (jama); (3) dry cave, former spring; (4) active spring; (5) karst polje with unconsolidated fill sediment; (6) active sink; (7) estavelle (interchangeable spring and sink); (8) deep canyon with the loosing stream; and (9) submarine spring in the Adriatic Sea. A denotes the predominant local groundwater flow direction and aquifer recharge; B denotes the local groundwater flow direction during wet season and heavy rains; C denotes the regional groundwater flow.

from a nonkarstic rock but there are sedimentary carbonates "somewhere" in its relative vicinity, one should not rule out the possibility that the primary aquifer providing water to the spring is actually a karst (limestone) aquifer.

A very illustrative example is Giant Springs in Great Falls, Montana. The spring has a very uniform discharge of about 200 mgd and forms probably the shortest river in the United States, the Roe River, which flows into the Missouri River. The springwaters originate as precipitation that falls on the Madison limestone of Little Belt Mountains over 4000 ft above the Missouri River. The limestone has all the characteristics of karst, including losing streams and dry caves. At Giant Springs, the Madison limestone is less than 400 ft below the

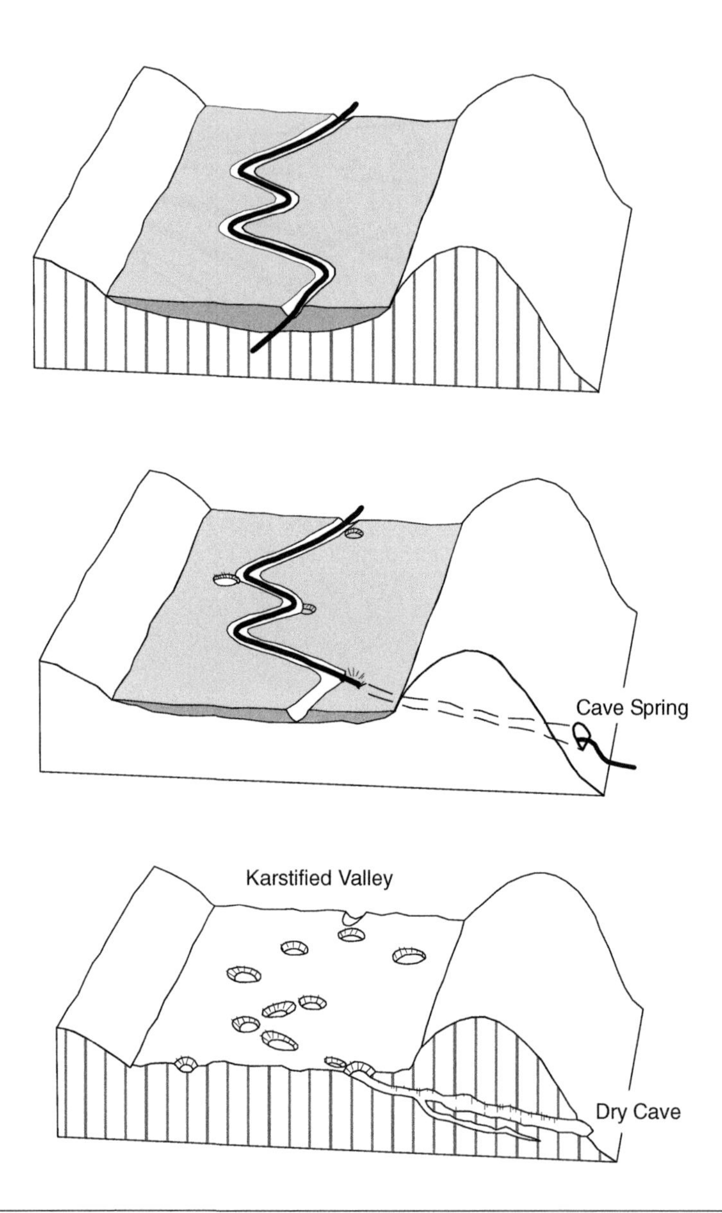

FIGURE 2–51 Schematic diagram of development of a sinking stream flowing over a thin layer of nonkarsic rock (sediment) covering a karst aquifer; the stream loses water to a spring located in a different surface watershed; as the karstification deepens, the spring may become dry and a new spring or springs develop(s) at lower elevations.

surface, and groundwater from this karstic aquifer escapes upward through fractures in the overlying Kootenai sandstone (Figure 2–58).

In conclusion, karst springs and karst aquifers represent an enormous but very vulnerable natural treasure that sustains the flow of many surface streams and ecosystems, provides habitat for unique flora and fauna, and has been used to fill the water supply needs of hundreds of millions of people around the world.

FIGURE 2–52 Ascending karst spring of the Cetina River, Croatia. (Photograph courtesy of Andrija Rubinić.)

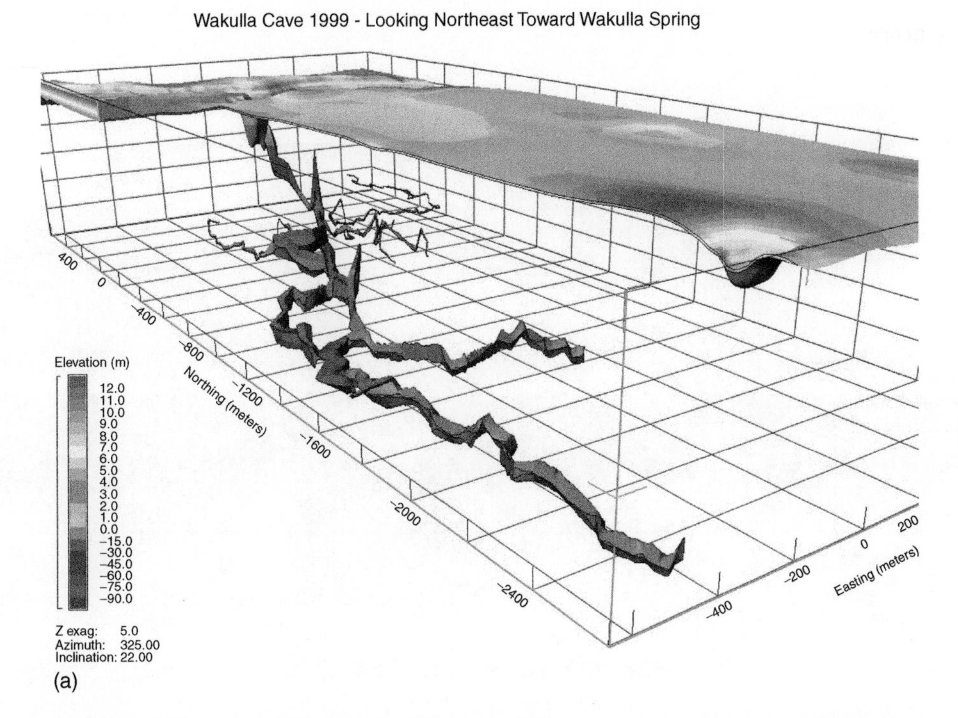

FIGURE 2–53 Two computer-generated views of the submerged Wakulla Cave passages based on the cave diving surveys. The passages all converge toward the Wakulla Spring shown in Figure 2–40. (Courtesy of Dr. Todd Kincaid, Global Underwater Explorers.)

(continued)

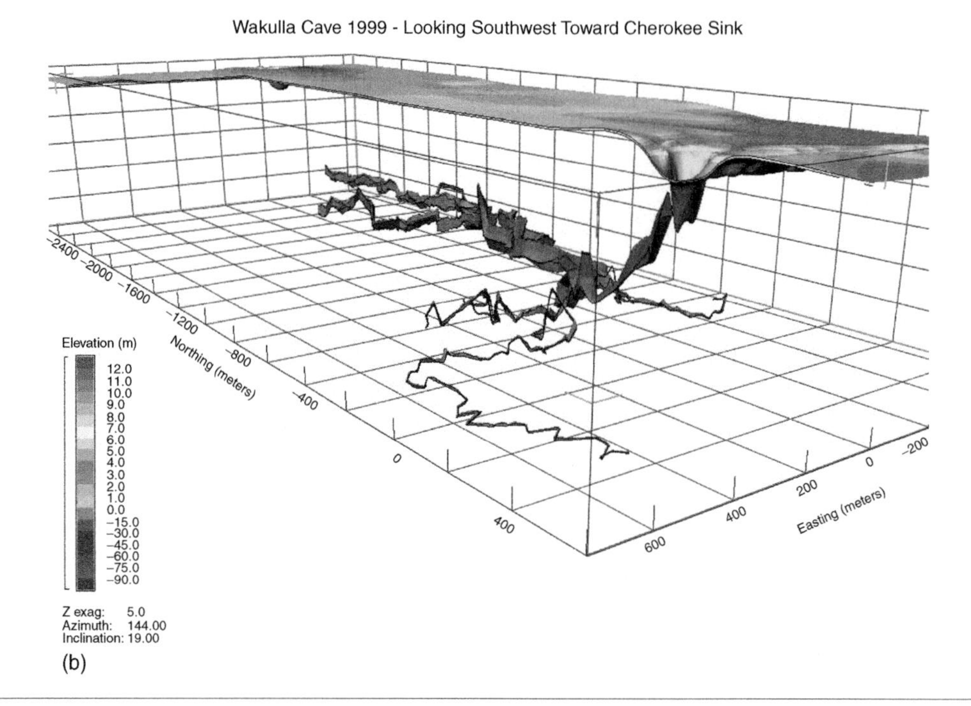

FIGURE 2–53, Cont'd

FIGURE 2–54 The main cave entrance (spring orifice) of Fontaine de Vaucluse is dry most of the year. (Copyright Bostjan Burger, printed with permission.)

FIGURE 2–55 River Sorgue in Provence, France, after receiving water from all springs in the Fontaine de Vaucluse area. (Photograph courtesy of William Jones, Karst Waters Institute.)

(a)

FIGURE 2–56 (a) Bennett Spring. (Photograph courtesy of Missouri Department of Natural Resources.) (b) Northwest-southeast longitudinal profile through Bennett Spring from data supplied by Donald N. Rimbach, Michael R. Tatalovich, and M. Grussemeyer. (From Vineyard and Feder, 1982.)

(continued)

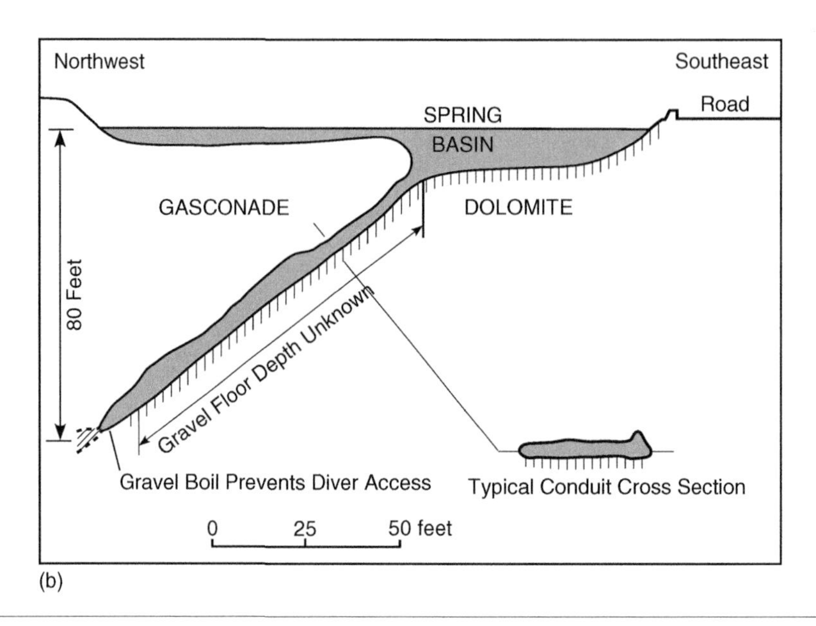

(b)

FIGURE 2–56, Cont'd

FIGURE 2–57 General view of the Ombla Spring site on the Croatian coast of the Adriatic. This overflowing ascending spring is utilized for the water supply of the city of Dubrovnik. Contact between karstified limestones and impermeable fine-grained sediments (flysch) is marked by change in vegetation. The spring issues at the bottom of the this V-shaped contact, which will be used to anchor the underground dam. (Photograph courtesy of Nicola Bilicic.)

FIGURE 2–58 Vertical fractures in Kootenai Sandstone at Giant Springs State Park near Great Falls, Montana. The fractures are extension fractures and open, providing an easy conduit for groundwater in the Madison aquifer to rise to the surface and be discharged in and next to the Missouri River. (Photograph courtesy of David Baker.)

2.4 SPRINGS IN EXTRUSIVE VOLCANIC ROCKS

As emphasized earlier, few if any aquifer types can compete with karst in terms of large, first-magnitude springs worldwide. However, in the United States, a dozen or so such springs, of equal economic and utilization importance, issue from Pliocene and younger basaltic rocks, present chiefly in the Snake River Plain in Idaho and underlying much of the Cascade Range in Oregon. Younger basic lavas also provide water to large springs in Hawaii.

The Snake River Plain regional aquifer system in southern Idaho and southeastern Oregon is a large grabenlike structure filled with basalt of Miocene and younger age. The basalt consists of a large number of flows, the youngest of which was extruded about 2000 years ago. The maximum thickness of the basalt, as estimated by using electrical resistivity surveys, is about 5500 ft (Miller, 1999). The permeability of basaltic rocks is highly variable and depends largely on the following factors: the cooling rate of the basaltic lava flow, the number and character of interflow zones, and the thickness of the flow. The cooling rate is most rapid when a basaltic lava flow enters water. The rapid cooling results in pillow basalt, in which ball-shaped masses of basalt form, with numerous interconnected open spaces at the tops and bottoms of the balls. Large springs that discharge thousands of gallons per minute issue from pillow basalt along the walls of the Snake River Canyon in the general area of Twin Falls, Idaho (see Figures 2–4 and 2–31).

Highly permeable but relatively thin rubbly or fractured lavas act as excellent conduits but have themselves only limited storage. Leakage from overlying thick, porous but poorly permeable, volcanic ash may act as the storage medium for this dual system. The prolific aquifer systems of the Valle Central of Costa Rica and Nicaragua and El Salvador are examples of such systems (Morris et al., 2003).

As discussed by Whitehead (1994), Pliocene and younger basaltic rocks are mainly flows, but in many places in the Cascade Range, the rocks contain thick interbeds of basaltic ash, as well as sand and gravel beds deposited by streams. Most of the Pliocene and younger basaltic rocks were extruded as lava flows from numerous vents and fissures concentrated along rift or major fault zones in the Snake River Plain. The lava flows spread for as much as about 50 mi from some vents and fissures. The overlapping shield volcanoes that formed around major vents extruded a complex of basaltic lava flows in some places. Thick soil, much of which is loess, covers the flows in many places. Where exposed at the land surface, the top of a flow typically is undulating and nearly barren of vegetation. The barrenness of such flows contrasts markedly with those covered by thick soil where agricultural development is intensive. The thickness of the individual flows is variable; the thickness of flows of Holocene and Pleistocene age average about 25 ft, whereas that of Pliocene age flows average about 40 ft.

In some shield-volcano eruptions, basaltic lava pours out quietly from long fissures instead of central vents and floods the surrounding countryside with lava flow on lava flow, forming broad plateaus. Lava plateaus of this type can be seen in Iceland (see Figure 6–2 in Chapter 6), southeastern Washington, eastern Oregon, and southern Idaho. Along the Snake River in Idaho and the Columbia River in Washington and Oregon, these lava flows are beautifully exposed and measure more than a mile in total thickness.

This geologic environment, in many ways, has acted hydrogeologically as karst, including the existence of sinking streams and development of integrated networks of overlapping and intersecting lava flows that drain at some of the most spectacular large springs, illustrated by photographs in Chapter 1 and this chapter (e.g., Niagara Falls Springs, Thousand Springs, Box Canyon Springs). Figure 2–59 shows one of the most striking springs in the Cascades Range in Oregon, the Roaring Springs on the upper McKenzie River, which emerge from a 50 m horizontal fracture within a single lava flow. Figure 2–60 is a schematic diagram of the large springs issuing from the Snake River basalts near Twin Falls, Idaho.

FIGURE 2–59 Roaring Springs on the South Fork McKenzie River, Oregon. (Photograph courtesy of Gordon Grant.)

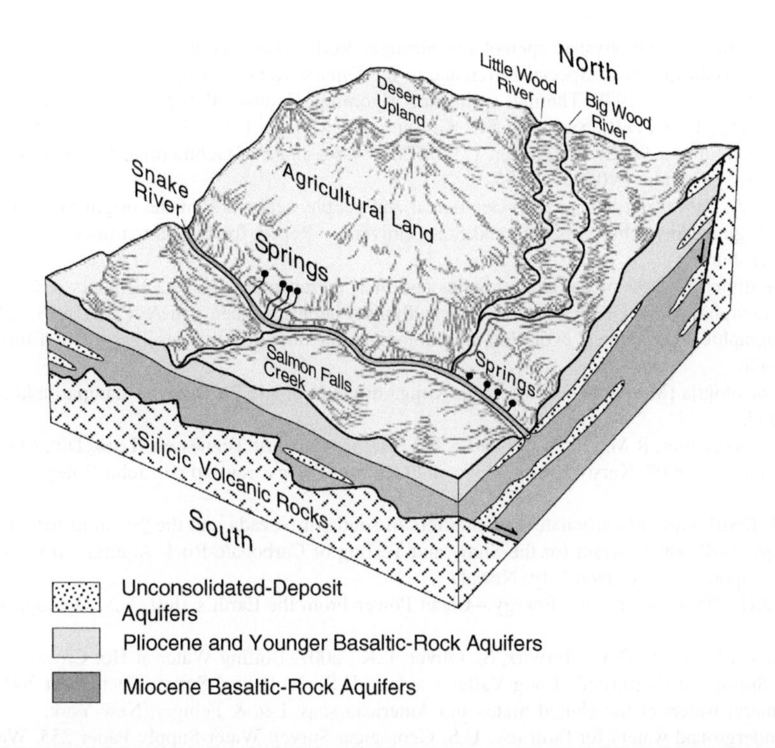

FIGURE 2–60 Basalt of Miocene and younger age fills the grabenlike trough on which the Snake River Plain has formed. Low-permeability, silica-rich volcanic rocks bound the basalt, which is locally interbedded with unconsolidated deposits. (From Miller, 1999; modified from Whitehead, 1994.)

Often present in lava flows are interconnected lava tubes at various depths below the water table, which may act like karst conduits, feeding springs of variable discharge rates that react quickly to rainfall events. For this reason, some practitioners describe such environment as *pseudokarst*.

Silicic volcanic rocks in the United States are present chiefly in southwestern Idaho and southeastern Oregon, where they consist of thick flows interspersed with unconsolidated deposits of volcanic ash and sand. Silicic volcanic rocks also are the host rock for much of the geothermal water in Idaho and Oregon. Big Springs in Fremont County, Idaho, is the source of the South Fork of the Henrys Fork River. Designated as a national natural landmark in 1980, the springs are the only first-magnitude spring in the United States that emanate from rhyolitic lava flows of the Madison Plateau.

REFERENCES

Arago, D.F.J., 1835a. Sur les puits forés, connus sous le nom de Puits Artésiens, des fontaines artésiennes, ou de fontaines jaillissantes. Bureau des Longitudes Annuaire, Paris.

Arago, D.F.J., 1835b. On springs, artesian wells, and spouting fountains. New Philosophical Journal [Edinburgh] 18, 205–246.

Barlow, P.M., 2003. Ground Water in Freshwater-Saltwater Environments of the Atlantic Coast. U.S. Geological Survey Circular 1262, Reston, VA.

Bedinger, M.S., Pearson, F.J., Reed, J.E., Sniegocki, R.T., Stone, C.G., 1979. The waters of Hot Springs National Park, Arkansas—Their nature and origin. U.S. Geological Survey Professional Paper 1044-C, pp. C1–C33.

Bögli, A., 1980. Karst hydrology and physical speleology. Springer-Verlag, New York.
Bonacci, O., 1987. Karst Hydrology with Special Reference to the Dinaric Karst. Springer-Verlag, Berlin.
Breckenridge, R.M., Hinckley, B.S., 1978. Thermal springs of Wyoming. Geological Survey of Wyoming Bulletin 60.
Brook, C.A., Mariner, R.H., Mabey, D.R., Swanson, J.R., Guffanti, M., Muffler, L.J.P., 1979. Hydrothermal Convection Systems with Reservoir Temperatures > 90°C. In: Muffler, L.J.P. (Ed.), Assessment of Geothermal Resources of the United States—1978. Geological Survey Circular 790, pp. 18–85.
Costain, J.K., Keller, G.V., Crewdson, R.A., 1976. Geological and geophysical study of the origin of the warm springs in Bath County, Virginia. Virginia Polytechnic Institute and State University Report for U.S. Department of Energy under Contract E-(40-1)-4920, Blacksburg.
Crook, J.K., 1899. The mineral waters of the United States and their therapeutic uses. Lea Brothers & Co., New York.
Cvijić, J., 1893. Das Karstphänomen. Geographischen Abhandlunged herausgegeben von A. Penck 5 (3), 218–329.
Cvijić, J., 1918. Hydrographie souterraine et evolution morphologique du karst. Receuil des Travaux de l'Institute de Géographie Alpine 6 (4), 375–426.
Cvijić, J., 1926. Geomorfologija [Morphologie Terrestre]. Knjiga druga [Volume 2]. Drzavna stamparija kraljevine Srba, Hrvata i Slovenaca, Beograd.
De Buchananne, G.D., Richardson, R.M., 1956. Ground-water resources of East Tennessee. Tenn. Div. of Geol. Bull 58, pt. 1.
Derek, F.C., Paul Williams, P., 2007. Karst Hydrogeology and Geomorphology, revised ed., John Wiley & Sons Ltd, Chichester, England.
Dettinger, M.D., 1989. Distribution of carbonate-rock aquifers in southern Nevada and the potential for their development, summary of findings, 1985–88. Program for the Study and Testing of Carbonate-Rock Aquifers in Eastern and Southern Nevada, Summary Report No. 1. Carson City, Nevada.
Duffield, W.A., Sass, J.H., 2003. Geothermal Energy—Clean Power From the Earth's Hest. U.S. Geological Survey Circular 1249, Reston, VA.
Farrar, C.D., Evans, W.C., Venezky, D.Y., Hurwitz, S., Oliver, L.K., 2007. Boiling Water at Hot Creek—The Dangerous and Dynamic Thermal Springs in California's Long Valley Caldera. U.S. Geological Survey Fact Sheet 2007–3045.
Fitch, W.E., 1927. Mineral waters of the United States and American spas. Lea & Febiger, New York.
Fuller, M.L., 1910. Underground waters for farm use. U.S. Geological Survey Water-Supply Paper 255, Washington, DC.
Hauwert, N.M., Vickers, S., 1994. Barton Springs/Edwards Aquifer: Hydrogeology and groundwater quality. Barton Springs/ Edwards Aquifer Conservation District, Austin, TX.
Heark, M., Stringfield, V.T. (Eds.), 1972. Karst; Important Karst Regions of the Northern Hemisphere. Elsevier, Amsterdam.
Hobba Jr., W.A., Fisher, D.W., Pearson Jr., F.J., Chemerys, J.C., 1979. In: Hydrology and geochemistry of thermal springs of the Appalachians. U.S. Geological Survey Professional Paper 1044-E, pp. E1–E36.
Jones, G.W., Upchurch, S.B., Champion, K.M., 1996. Origin of Nitrate in Ground Water Discharging from Rainbow Springs, Marion County, Florida. Ambient Ground-Water Quality Monitoring Program. Southwest Florida Water Management District, Brooksville, FL.
Kresic, N., 2007. Hydrogeology and Groundwater Modeling, second ed. CRC Press, Boca Raton, FL.
Kresic, N., 2009. Ground Water in Karst. In: Kresic, N. (Ed.), Theme Issue Ground Water in Karst, Ground Water, vol. 47., no. 3, pp. 319–320.
Lindal, B., 1973. Industrial and other applications of geothermal energy. In: Armstead, H.C.H. (Ed.), Geothermal energy, review of research and development. United Nations Educational, Scientific and Cultural Organization (UNESCO), Paris, pp. 135–148.
Mariner, R.H., Brook, C.A., Reed, M.J., Bliss, J.D., Rapport, A.L., Lieb, R.J., 1983. Low-Temperature Geothermal Resources in the Western United States. In: Reed, M.J. (Ed.), Assessment of Low-Temperature Geothermal Resources of the United States—1982. Geological Survey Circular 892, U.S. Department of the Interior, pp. 31–50.
Meinzer, O.E., 1923a. The occurrence of ground water in the United States with a discussion of principles. U.S. Geological Survey Water-Supply Paper 489, Washington, DC.
Meinzer, O.E., 1927. Large springs in the United States. U.S. Geological Survey Water-Supply Paper 557, Washington, DC.
Meinzer, O.E., 1940. Ground water in the United States; a summary of ground-water conditions and resources, utilization of water from wells and springs, methods of scientific investigations, and literature relating to the subject. U.S. Geological Survey Water-Supply Paper 836-D, Washington, DC, pp. 157–232.
Milanovic, P.T., 1981. Karst hydrogeology. Water Resources Publications, Littleton, CO.
Miller, J.A., 1999. Introduction and national summary. Ground-Water Atlas of the United States, United States Geological Survey, A6.
Milojevic, N., 1966. Hidrogeologija. Univerzitet u Beogradu, Zavod za izdavanje udžbenika Socijalističke Republike Srbije, Beograd.

Moorman, J.J., 1867. The mineral waters of the United States and Canada. Kelly & Piet, Baltimore.

Morris, B.L., et al., 2003. Groundwater and its susceptibility to degradation: A global assessment of the problem and options for management. Early Warning and Assessment Report Series, RS. 03-3. United Nations Environment Programme, Nairobi, Kenya.

Palmer, A.N., 1985. The Mammoth Cave region and Pennyroyal Plateau. In: Dougherty, P.H. (Ed.), Caves and karst of Kentucky. Special Publication 12, Series XI. Kentucky Geological Survey, Lexington, pp. 97–118.

Perry, W.C., Costain, J.K., Geiser, P.A., 1979. Heat flow in western Virginia and a model for the origin of thermal springs in the folded Appalachians. J. Geophys. Res. 84 (B12), 6875–6883.

Radovanovic, S., 1897. Podzemne vode; izdani, izvori, bunari, terme i mineralne vode [Ground waters; aquifers, springs, wells, thermal and mineral waters; in Serbian]. Srpska književna zadruga 42.

Reed, M.J., 1983a. Summary. In: Reed, M.J. (Ed.), Assessment of Low-Temperature Geothermal Resources of the United States—1982. Geological Survey Circular 892, U.S. Department of the Interior, pp. 67–73.

Reed, M.J., 1983b. Introduction. In: Reed, M.J. (Ed.), Assessment of Low-Temperature Geothermal Resources of the United States—1982. Geological Survey Circular 892, U.S. Department of the Interior, pp. 1–8.

Renner, J.L., White, D.E., Williams, D.L., 1975. Hydrothermal Convection Systems. In: White, D.E., Williams, D.L. (Eds.), Assessment of Geothermal Resources of the United States—1975. Geological Survey Circular 726. U.S. Department of the Interior, Washington, DC, pp. 5–57.

Showcaves, 2009. Available at: www.showcaves.com/english/fr/springs/Vaucluse.html (accessed on February 27, 2009).

Sorey, M.L., Natheson, M., Smith, C., 1983a. Methods for Assessing Low-Temperature Geothermal Resources. In: Reed, M.J. (Ed.), Assessment of Low-Temperature Geothermal Resources of the United States—1982. Geological Survey Circular 892, U.S. Department of the Interior, pp. 17–30.

Sorey, M.L., Reed, M.J., Foley, D., Renner, J.L., 1983b. Low-Temperature Geothermal Resources in the Central and Eastern United States. In: Reed, M.J. (Ed.), Assessment of Low-Temperature Geothermal Resources of the United States—1982. Geological Survey Circular 892, United States Department of the Interior, pp. 51–65.

Springer, A.E., Stevens, L.E., 2009. Spheres of discharge of springs. In: Hancock, P.J., Hunt, R.J., Boulton, A.J. (guest Eds.), Hydrogeoecology and Groundwater Ecosystems, Hydrogeology Journal, 17(1), 83–93.

Steele, K.F., Wagner, G.H., 1981. Warm springs of the Western Ouachita Mountains, Arkansas. Geothermal Resources Council Transactions 5, 137–140.

Sun, P.C.P., Criner, J.H., Poole, J.L., 1963. Large Springs of East Tennessee. Geological Survey Water-Supply Paper 1755.

Sweeting, M.M., 1972. Karst Landforms. Macmillan, London.

UNESCO (United Nations Educational, Scientific and Cultural Organization), 2004. Submarine groundwater discharge. Management implications, measurements and effects. IHP-VI, Series on Groundwater No. 5, IOC Manuals and Guides No. 44, UNESCO, Paris.

U.S. Forest Service Intermountain Region, 2008. Geologic Points of Interest by Activity; Springs/Falls. Available at: www.fs.fed.us/r4/resources/geology/geo_points_interest/activities/springs_falls.shtml (Accessed November 2008).

U.S. Geological Survey, 2009a. Geysers, Fumaroles, and Hot Springs. Available at: http://pubs.usgs.gov/gip/volc/geysers.html (Accessed January 2009).

U.S. Geological Survey, 2009b. USGS Multimedia Library. Available at: http://gallery.usgs.gov (Accessed January 2009).

Vineyard, J.D., Feder, G.L., 1982. Springs of Missouri. Water Resources Report No. 29, Missouri Department of Natural Resources, Division of Geology and Land Survey.

White, D.E., 1965. Geothermal Energy. Geological Survey Circular 519, United States Department of the Interior, Washington, DC.

White, B.W., 1988. Geomorphology and hydrology of karst terrains. Oxford University Press, New York.

Whitehead, R.L., 1994. Ground Water Atlas of the United States—Segment 7: Idaho, Oregon. U.S. Geological Survey Hydrologic Investigations Atlas HA–730–H, Washington, DC.

Worthington, S.R., Ford, D.C., 2009. Self-Organized Permeability in Carbonate Aquifers. In: Kresic, N. (Ed.), Theme Issue: Ground Water in Karst, Ground Water vol. 47, no. 3, 326–336.

Worthington, S.R., Gunn, J., 2009. Hydrogeology of Carbonate Aquifers: A Short History. In: Kresic, N. (Ed.), Theme Issue Ground Water in Karst, Ground Water vol. 47, no. 3, 462–467.

CHAPTER

Recharge of springs

3

Branka Trček[1] **and Hans Zojer**[2]

[1]University of Maribor, Faculty of Civil Engineering, Maribor, Slovenia.
[2]Joanneum Research, Institute of Water Resources Management, Graz, Austria

Knowledge of groundwater systems is a basic prerequisite for the efficient management of springs. The relationship between groundwater recharge and discharge is one of the most important aspects in the protection of groundwater resources. Groundwater and surface water are fundamentally interconnected. It is therefore often difficult to separate the two, because they recharge each other, and hence they can also contaminate each other. Recharge of groundwater may occur naturally from precipitation, surface streams and lakes, and as an anthropogenic input from irrigation and urbanization. Two types of recharge are generally distinguished: direct and indirect recharge. Direct recharge is the water added to the aquifer in excess of soil moisture deficits and evapotranspiration, by direct vertical percolation of precipitation through the unsaturated zone (Simmers, 1990). Indirect recharge results from percolation to the water table following runoff and localization in joints, as ponding in low-lying areas and lakes, or through the beds of surface watercourses (Simmers, 1990). The two distinct categories of indirect recharge are (1) that associated with the surface watercourses and (2) the localized form resulting from horizontal surface concentration of water in the absence of well-defined channels. Furthermore the recharge is divided into diffuse and discrete recharge. The former refers to seepage water through the subsurface and the latter to the groundwater recharge through a network of localized openings that are able to rapidly transport both water and contaminants.

Factors affecting the natural recharge of an aquifer are (Rushton and Ward, 1979)

- At the land surface,
  - Topography.
  - Precipitation magnitude, intensity, duration, spatial distribution.
  - Runoff, ponding of water.
  - Cropping of water, actual evapotranspiration.
- Surface waters (gaining, losing, and sinking surface streams).
- At the soil zone,
  - The nature, depth, and hydraulic properties of the soil.
  - Variability of the soil.
  - Cracking of soil on drying out or swelling due to wetting.
- At the unsaturated zone,
  - Flow mechanism through the unsaturated zone.
  - Zones with different hydraulic conductivities.

Copyright © 2010, Elsevier Inc. All rights reserved.

- At the aquifer,
  - The aquifer's ability to accept water.
  - Variation of aquifer conditions with time.

Movement of water in the atmosphere and on the land surface is relatively easy to visualize, while this not the case for the movement of groundwater. Groundwater flow paths vary greatly in length, depth, and travel time from the recharge areas to discharge areas and springs. These indicate that successful groundwater recharge estimation primarily depends on identifying the probable flow mechanisms and important features that influence the recharge for a given locality. The study of relationships between spring recharge and discharge represents the key theme of this chapter. The role of gaining, losing, and sinking streams in aquifer recharge processes is discussed at the beginning. The research methods are presented afterward, with an emphasis on artificial and environmental tracer methods. The combined use of various geochemical and hydrologic tools can considerably enhance our understanding of complex groundwater flow patterns as well as the mixing of groundwater and surface water in a particular spring.

3.1 GAINING, LOSING, AND SINKING STREAMS

After entering the aquifer in recharge areas, groundwater flows down-gradient toward the aquifer discharge areas. Left undisturbed, groundwater naturally arrives at a balance, and aquifer discharge and recharge depend on hydrometeorologic and hydrologic conditions. The direction and rate of groundwater movement are governed mainly by geologic conditions. Groundwater is discharged to the surface via springs and as base flow to streams, rivers, and lakes. Groundwater can be also recharged by surface waters, which, in general, can gain groundwater in some regions and lose water in others. It therefore follows that groundwater is responsible for maintaining the hydrologic balance of surface streams, springs, lakes, wetlands, and marshes. The volume of water flowing from surface waters to aquifers or vice versa depends on the position of the water table relative to the surface water body, that is, the hydraulic gradient between the two.

A gaining stream is a surface stream into which groundwater discharges (Figure 3–1). In this case the elevation of the water table in the vicinity of the stream is higher than the elevation of the stream water surface.

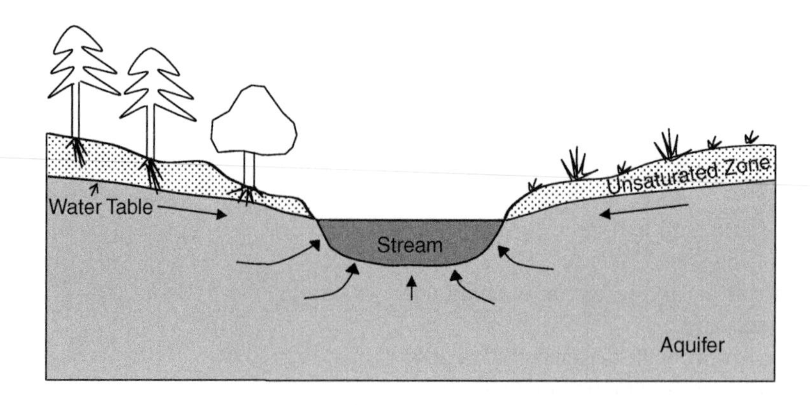

FIGURE 3–1 A gaining stream.

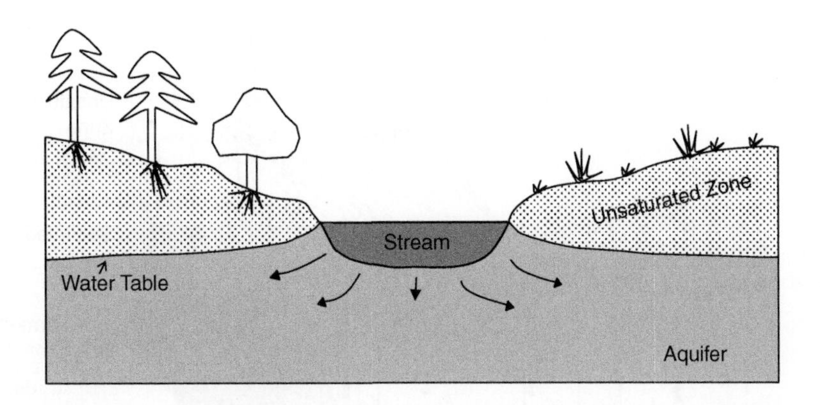

FIGURE 3–2 A losing stream.

Losing streams are surface streams that lose water through a streambed or series of openings, fractures, or swallets and hence recharge the aquifer (Figure 3–2). In this case, the altitude of the water table in the stream vicinity is lower than the altitude of the stream water surface. Losing streams are connected to aquifers by a continuous saturated zone, or they can recharge the aquifer through the unsaturated zone. In the latter case, the losing stream is called a *disconnected stream* (Figure 3–3). It can result in water table elevation below the stream if the rate of recharge through the streambed and unsaturated zone is greater than the rate of lateral groundwater flow (Figure 3–3).

A sinking stream is a spatial type of losing stream that is common in karst regions but can also occur in other geologic settings, such as lava bed plateaus. It is a stream that disappears underground at a distinct swallow hole (Figure 3–4). A karst aquifer has a complex geologic structure, where distinctive physiographic and hydrologic features develop as a result of the dissolution of soluble bedrock, such as limestone (Figure 3–4). Typical karst features include swallow holes, sinking or losing streams, caves, underground streams, and springs. Some sinking and losing streams are dry most of the year. They flow only when the karst system overflows during major storms or runoff events. In the karst, surface water and groundwater are interconnected

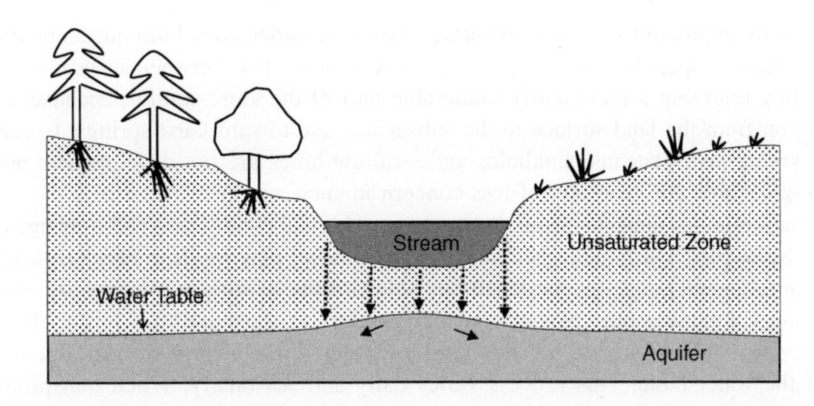

FIGURE 3–3 A disconnected stream.

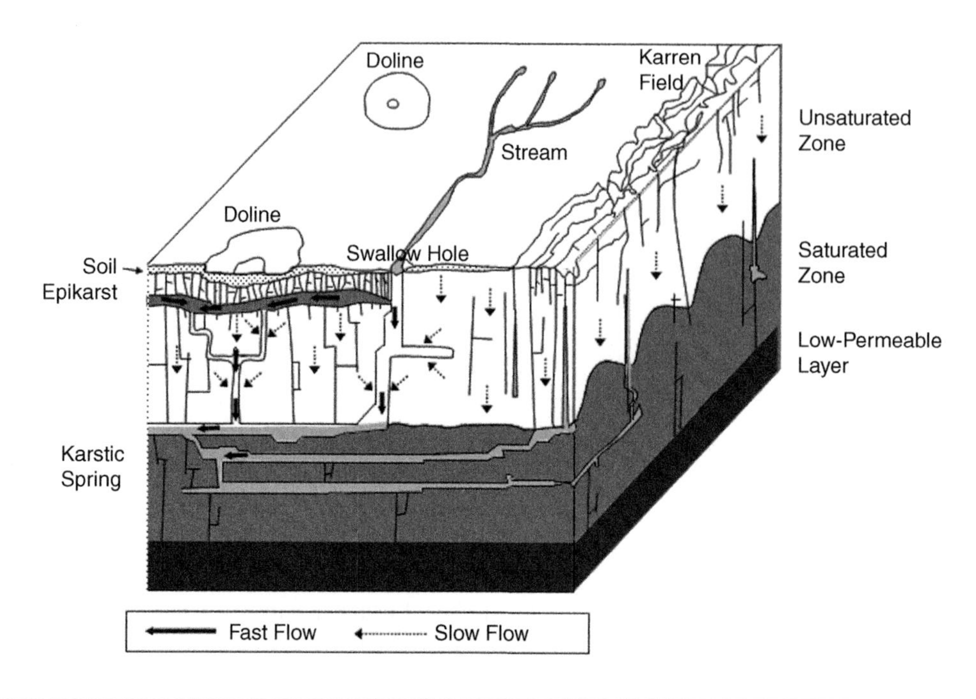

FIGURE 3–4 Karst aquifer scheme. (After COST Action 65, 1995.)

in many ways and compose a single dynamic system that transfers water between the surface and subsurface. The main characteristics of such systems include

- Distinct shapes of karst spring hydrographs indicating transport and storage conditions in the aquifer (see Chapter 4).
- Fast and large groundwater level oscillation in some boreholes, and slow and insignificant oscillation in others.
- Fast breakthrough of artificial and natural tracers.
- Variations of the groundwater chemical and isotopic composition as a function of the inflow.

Sinkholes, which are depressions in the land surface that have underground drainage, are abundant in karst areas. Regardless of their shape, all sinkholes provide a direct connection between surface runoff and groundwater. Therefore, they represent a particularly vulnerable part of the karst aquifer, because they can rapidly transport the pollution from the land surface to the subsurface and toward karst springs. Except in some rare cases, such as on young lava plateaus, sinkholes and swallow holes are not developed in nonkarstic areas; consequently, losing streams are generally of less concern in such areas.

The flow of surface streams depends primarily on meteorological conditions. Their changes affect seepage processes in streambeds, particularly near the banks. Heavy rainfall may result in rapid groundwater recharge in some geologic settings and cause increased groundwater inflow to surface streams, by which a normally losing stream becomes a gaining one. Negative recharge refers to transpiration by near-shore plants. Plant roots can penetrate into the saturated zone and transpire water directly from the aquifer, which can result in a drawdown of the water table. This process varies daily and seasonally. When transpiration effects are significant they may reduce groundwater discharge to a surface stream or even inverse the recharge, so that surface water seeps into the aquifer.

Lakes interact with groundwater as they do with surface streams. They may receive an inflow of groundwater or seep to groundwater throughout their entire bed. They may even be recharged by groundwater in part of their bed and recharge the aquifer through other parts. However, the interactions between lakes and aquifers also have some particularities. The water level fluctuations of natural lakes are much more attenuated in comparison with surface streams, while evaporation generally has greater effects on lake levels, and the lake-aquifer interactions may be much more complex. Moreover, the poorly permeable lake organic sediments can characteristically affect the distribution of seepage as well as the biogeochemical exchanges of water and solutes in comparison with surface streams. The wetlands interact with aquifers as they do to streams and lakes. They can receive groundwater inflow, recharge groundwater, or do both.

The Ljubljanica River, in the Slovenian Classic karst, is called the *river of seven names*, due to numerous sinking springs and streams in its catchment, which occupies 1100–1200 km^2 (Habič, 1976a). The elevation difference between the highest part, Mt. Snežnik (1796 m), and the Ljubljanica springs near Vrhnika, is 1505 m. Triassic, Jurassic, and Cretaceous rocks dominate in this area, while Quaternary sediments cover only karst poljes. The spring catchment is cut by several faults with the prevailing strike northwest-southeast. The altering of permeable and impermeable rocks and the tectonic structure influence both the surface runoff and groundwater flow, as illustrated in Figure 3–5. The majority of karst groundwaters discharge from higher to lower basins, via a series of sinks and karst poljes. Only a small part discharges across the karstified dolomitic barrier directly to the Ljubljanica springs.

The unique attributes of the Ljubljanica karst river basin represented great challenges for scientists and researchers in the past. The most extensive geological, hydrogeological, and geochemical investigations were performed in the framework of the Association of Tracer Hydrology between 1972 and 1975 (Gospodarič and Habič, 1976), with the intention to study the recharge-discharge relationships of the Ljubljanica springs. The results are summarized in Figure 3–5. A combined water tracing experiment, described further in Section 3.2.1, was the main investigation method for determining interconnections between surface waters and groundwaters in the Ljubljanica catchment area.

In many cases, hydrogeologic (subsurface) characteristics of a spring drainage area cannot be determined directly, and little data are available on the aquifer geometry and hydrodynamic parameters. Consequently, indirect methods are often applied to study the spring discharge-recharge relationship. Artificial and environmental tracers are a very useful indirect method described in detail in the following section. Other methods include analyses of spring discharge hydrographs as well as the numeric and stochastic modeling described in Chapters 4 and 5.

3.2 ARTIFICIAL AND ENVIRONMENTAL TRACER METHODS

The aquifer contains flow paths from the point where water enters the aquifer to the topographically lower point where it leaves the aquifer. Infiltration from rainfall and snow occurs in the recharge area, moving through the phreatic zone of the aquifer then down-gradient toward the springs in the discharge area. These flow paths can be understood by the use of both environmental and artificial tracer methods. Groundwater tracers are substances or signals contained in or added to the water and suited to observe the movement of a discrete body of water within a hydraulic system (Behrens, Hötzl, and Käss, 2001). Tracer tests are very useful tools to acquire information on the extent of recharge zones, the characteristics of groundwater flow, and the transport of contaminants (e.g., Kranjc, 1997; Zhou et al., 2002; Trček, 2003; Perrin et al., 2003). However, releasing artificial substances into the environment generally entails additional financial or logistical efforts, as well as regulatory requirements, due to the toxicity or radioactivity of some potential tracer substances. Hence, the search for naturally appearing substances suitable as tracers continues to be an important task of the applied environmental geosciences.

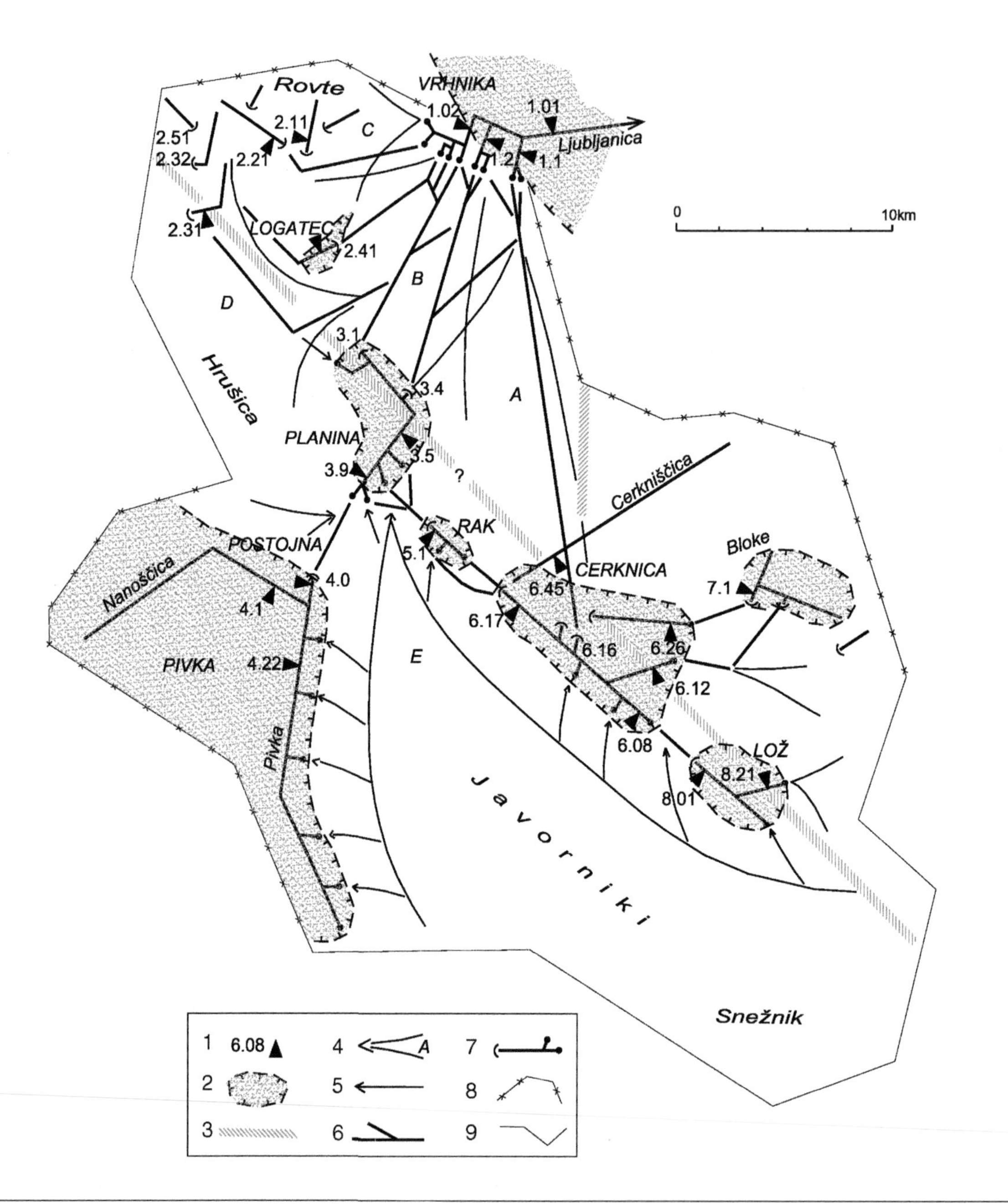

FIGURE 3–5 Schematic review of superficial and underground waters in the Ljubljanica karst river basin: (1), important gauging stations; (2), karst polje with sediments; (3), central hydrogeologic relative barrier; (4), catchment area of permanent karst springs; (5), periodic high-water discharge of karst water; (6), underground water connections; (7), superficial stream with springs and ponors; (8), superficial watershed; (9), supposed karst watershed. (After Habič, 1976a.)

3.2.1 Artificial tracers

Artificial tracers are widely used to obtain information on specific groundwater pathways, leading from the sinking points of surface streams or from human-made structures, such as wells, toward springs. The principal goals of tracer tests are to

- Prove the existence or nonexistence of flow paths between the tracer injection points and observation points, such as springs.
- Determine hydraulic parameters, such as groundwater residence time, flow rate, and dispersion.
- Assess contaminant fate and transport.
- Provide information necessary for spring management, groundwater (aquifer) protection, investigation of contaminant loads, and other water management tasks.

In some simple cases, the application of just one suitable tracer may be sufficient to provide needed answer(s). More complex studies require the application of a multitracer, a combined or a comparative tracer experiment that can provide information on hydrodynamic conditions on the entire spring catchment at equal hydraulic conditions, the transport behavior of contaminants, and hydraulic parameters of investigated systems. In the framework of the combined test, different tracers are injected into different locations at the same time. The multitracer test is distinguished from the previous one, as it is repeated at the same location but at a different time and probably different hydrologic conditions. Such a test is very useful for studies of relationships between the spring discharge and hydrologic conditions. The comparative tracer test is applied to determine if tracers are directly comparable under the conditions of groundwater flow and to investigate their interactions (Behrens et al., 1992; Käss, 1998). Such tests refer to more or less parallel injections of two or more tracers at one site under the same hydrologic and underground storage conditions.

The existing natural accesses to groundwater, like swallow holes, open features, or artificial openings (e.g., piezometers, wells), are preferably chosen for the tracer injection (Käss, 1998).

The tracer selection depends on the experiment objectives and strategy. The reference tracers are closely related to the assessment of their behavior and consequently the transport velocity and recovery.

To be useful, the tracers should have a number of properties (Käss, 1998):

- A physical and chemical structure should let the tracer travel in such a manner that allows the hydraulic parameter to be measured accurately; hence, it should be chemically stable and not be sorbed to aquifer materials or lost by filtration or any process of degradation.
- The tracer also is to be detectable in very high dilution, therefore, it should be abundant in the aquatic environment on a tolerable level.
- The tracer must be nontoxic and its use should be economic.

An ideal tracer does not exist, since most tracers have some disadvantages. Nevertheless, the tracer should always be representative for a certain parameter to be determined.

According to their physical and chemical natures and detection methods, artificial tracers are divided into three groups, which are interpreted in Table 3–1: fluorescent dyes (the most frequently used tracers), salts, and particles. The last group requires the spring sampling with a frequency adapted to the study objective and microscopic analysis. The breakthrough of the salt tracer is easily detected just by an increase of the spring-specific electrical conductivity but with low detection sensitivity. Portable fluorimeters can be used for continuous measurements of dye tracers in the spring. However, laboratory analysis of spring samples, in the case of a very sophisticated instrumentation, provides better detection accuracy, but it depends on sampling intervals.

Table 3–1 Properties of the Most Popular Groundwater Tracers (Summary after Käss 1998, Behrens et al., 2001)

Tracer Group	Tracer Name	Toxicology	Advantages	Disadvantages
Fluorescent dyes	Uranine, Rhodamine WT Sulforhodamine B Amidorhodamine G Eosin Pyranine Naphthionate Tinopal	Safe Genotoxic Ecotoxically Unsafe Safe Safe Safe Safe Safe	Extreme detection sensitivity; low or almost no background abundance in the environment; relatively easy and quick detection; well quantitatively achievable; good environmental tolerance.	Sorption in dependence with material and water pH, which varies from low for uranine, over moderate for eosin, to stronger in the sequence sulforhodamines and rhodamine B; relatively wide-shaped spectral peaks.
Particles	Spores Specific bacteria Viruses, Phages Microspheres	Safe Not evaluated Not evaluated Safe	No background abundance in the environment; suitable for identification of spreading of pathogens in hygiene issues; phages have been used for karst systems tracing on long flow conditions.	Lost by filtration or sedimentation; suitable only for fast and best close to turbulent water flow.
Salts	Sodium Potassium Lithium Strontium Chloride Bromide Iodide	Safe Safe Safe with restriction Safe with restriction Safe with restriction Safe with restriction Not evaluated	Breakthroughs are easily detected just by the increase of water electrical conductivity.	Low or moderate sorption; low to high (chlorine, sodium) background concentrations.

The heterogeneous catchments with fast- and slow-flow components require special attention due to double porosity effects. The retardation caused by the slow-flow component can amount to up to several years and more, whereby the shares of the fast- and slow-flow components can vary with the hydrologic conditions (Behrens et al., 2001). Therefore, the tracer breakthrough curves usually record only the fast flow, as the observation time of the slow flow is normally too short. The share of the fast-flow component can be calculated from the recovery rate.

The sampling methodology should be adjusted to the objectives of the tracer test. Integrative sampling is sufficient to prove underground connections, while the calculation of the recovery rate requires continuous measurements of tracer concentrations together with spring discharge measurements.

One of the critical issues in planning a tracer test is to determine what mass of tracer to inject. The optimal tracer quantity permits clear detection at the sampling point without excess expense, unacceptable water coloration and environmental loading, and additional work in the laboratory. Commonly, the tracer quantity is estimated from an empirical relationship (Käss, 1998):

$$M = Q \times L \times k \tag{3.1}$$

where

M = required tracer quantity (i.e., mass per kilogram for soluble tracers and number of particles for particulate tracer).
Q = spring discharge (m^3/s).
L = distance to the spring (km).
k = coefficient for the tracer (e.g., $k = 0.25$ for uranine and 250 for sodium chloride; other values can be found in Käss, 1998).

If the hydraulic and transport parameters are known it is possible to calculate the injection mass. The calculation is based on the advection-dispersion equation, resulting in a target concentration at the sampling point (Field, 2003).

The quantitative interpretation of tracing tests is derived from principles of solute transport. The tracer breakthrough curve observed at the spring expresses the hydrodynamic as well as the physicochemical and biological processes to which the tracers are subjected along their underground passage (Schulz, 1998). To compare the properties and the behavior of tracers in groundwater, a very detailed breakthrough curve is required. Either a continuous measurement of the tracer concentration over the breakthrough time or a very dense sampling program is necessary to reproduce the concentration changes with all the required details.

The breakthrough curve of a spring forms the base for behavior and comparison analyses of tracers. A preliminary monitoring before the tracer injection is essential. It serves for determining the tracer background concentrations used for corrections to the results of the tracer experiment. The results of measured concentrations are drawn in the form of a time-dependent tracer concentration curve. So far, the discharge rates are also available in the form of the time-dependent cumulative tracer recovery (Figure 3–6). For the comparison of tracers, it is important to normalize the concentration as a factor of the injected amount of tracers (Behrens et al., 1992).

The magnitude of the breakthrough is dominated by the tracer mass and volumetric flow (i.e., dilution). The primary features of a breakthrough curve are the premonitoring, the rising limb, the peak, and the recession (Benischke, Goldscheider, and Smart, 2007). Asymmetry is caused by dispersion, storage, and other transport processes, which is possible to explain only by a comprehensive evaluation of the hydrogeologic, hydraulic, hydrochemical, and biochemical factors of influence.

The time delay of the breakthrough curve of a conservative tracer is determined by the length of the flow path and flow velocity. A number of transport statistics can be derived from the breakthrough curve, such as

- Maximal effective flow velocity (v_{max}), determined by the time of the first tracer detection (t_{beg}).
- Dominant effective flow velocity (v_{dom}), determined by the time of the peak tracer concentration (t_{dom}).
- Mean effective flow velocity (v_{mean}), determined by the time of the gravity center of the area below the breakthrough curve (t_{mean}):

$$t_{mean} = \frac{\int_0^\infty c_t \times t \times dt}{\int_0^\infty c \times dt} \tag{3.2}$$

- Minimal effective flow velocity (v_{min}), determined by the end of the measured tracer breakthrough (t_{end}).

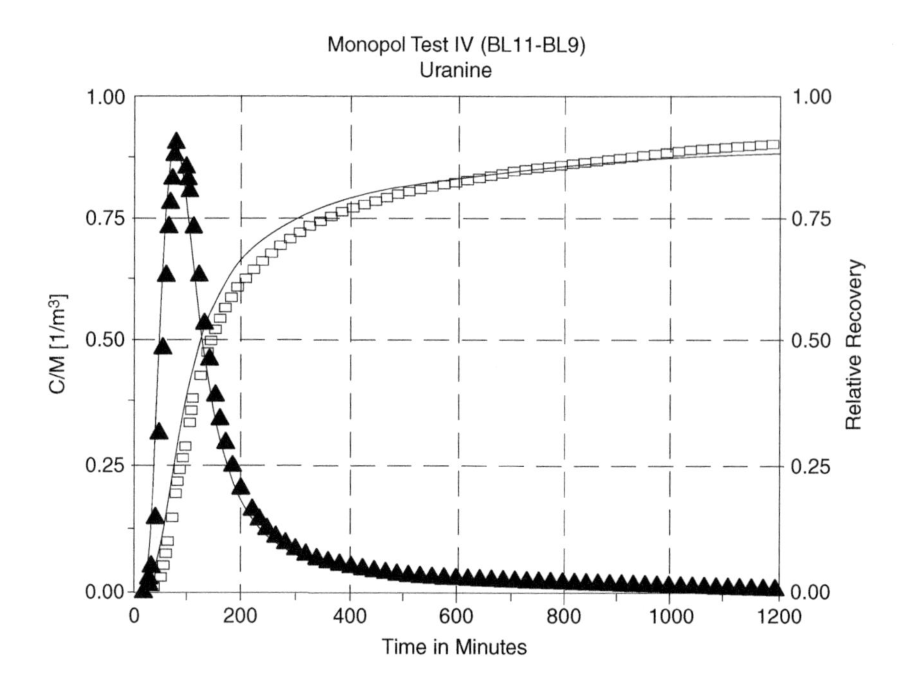

FIGURE 3–6 Breakthrough curve of uranine and recovery curve from a short-distance tracer experiment along a fractured system at the test site, Lindau: triangles and squares represent the measured values, the through-going line gives the best fit calculated with the Single Fissure Dispersion Model (SFDM). (Himmelsbach et al., 1992.)

In addition to the breakthrough curve, the spring recovery curve produces information for comparing the tracer's behavior along the underground passages, particularly by comparing the yield over different flow distances (Himmelsbach et al., 1992; Kass, 1998). The cumulative tracer recovery (mass or percent) is often plotted together with a breakthrough curve to show that the recovery curve approaches an asymptotic final value. This also allows determining the time at which half the recovered tracer has passed. It is often significantly less than 100 percent, due to tracer storage, adsorption, or decomposition.

Mass recovery requires high-resolution discharge and concentration data at each site. If the discharge is constant, the recovered tracer mass is the integral of the breakthrough curve times the discharge. For variable discharge, the recovery tracer mass, M_R, is calculated by the following equation (Käss, 1998):

$$M_R = \int_{t=0}^{\infty} Q \times c \times dt, \quad Q = A\left(\int_{t=0}^{\infty} c \times dt\right) \tag{3.3}$$

where

Q = discharge.
A = injected quantity of tracer.
c = tracer concentration.

For improved evaluation and better quantification of tracing results, appropriate models are used (Figure 3–6). The measured tracer concentrations may show more or less strong deviations from theoretical curves; hence, the parameters occurring in the analytical solutions may be determined by the best-fit methods.

The tracer test may provide excellent information on groundwater movement and contaminant transport, but it may take many months to execute and fail if poorly prepared, performed, or analyzed. In addition, each tracing test should be controlled for errors that may arise from injection, sampling, handling, analysis, and data processing (Smart, 2005).

Extensive tracer studies and field tests have been used to estimate the recharge areas of springs, the rates of groundwater movement, and the water balance of aquifers, as discussed in a previous section (Figure 3–5).

Based on data from previous investigations and available tracers, 12 swallow holes were selected for the combined water tracing experiment (Figure 3–7). On Cerkniško polje, uranine was injected into the Vodonos swallow hole (6.16), which drains water to Bistra (1.1), Lubija (1.2), and Ljubljanica springs (1.3 and 1.4), due to an unknown common runoff at the bottom of the intermittent lake. On Planinsko polje, a series of swallow holes of the Unica sinking stream were selected to ascertain if all waters from the polje flow together into the same springs or not. The Milavcovi ključi swallow hole (3.42) was marked by rhodamine, the Ribce swallow holes (3.43) by brown and green Lycopodium spores, the Dolenje Loke swallow holes (3.41) by tinopal, the Laška žaga swallow hole (3.21) by detergents, and the Strževica swallow hole by indium solution and lithium chloride. On Logaško polje, eosin was injected into the Jačka swallow hole (2.41), where the Logaščica is sinking; while on the Rovte plateau, the Hotenka sinking stream (2.31) was traced by amidorhodamine. Additionally, the Željski potok (2.32) and Pikeljščica (2.51) sinking streams were marked by red and blue Lycopodium spores, whereas the Rovtarica (2.21) and Petkovščica (2.11) sinking streams were marked by ^{51}Cr and potassium chloride, respectively. The type of tracer and its quantity were selected based on the hydraulic position of the injected point and its distance from the Ljubljanica springs.

The results of the combined tracing experiment in the Ljubljanica river catchment are summarized in Figure 3–7. They

- Certify and complete the knowledge about the underground connections between swallow holes and springs.
- Ascertain the water quality being supplied by particular swallow holes into particular springs at different hydrological situations.
- Complete the knowledge about the hydrologic and hydrogeologic karst characteristics among the Cerkniško, Planinsko, and Logaško polje and the Rovte plateau and the Ljubljanica springs (Figure 3–5).
- Compare the applicability of tested and new water tracing methods and means.
- Complete the methodology of karst water tracing.

The area of Semriach-Peggau, including the Lur cave in the Central Styrian karst, has been used for a long time as a test area for different artificial tracers. Nearly 30 tracer experiments have been performed to prove the properties and boundary conditions for applications of tracers. The karst system is characterized by the water input from precipitation and a surface stream as well as the water outflow from the karst system by two springs. It is described by Behrens et al. (1992).

The recovery rate of injected tracers into the sinking input stream, calculated as the product of discharge and tracer portions by time, increases significantly with the rising discharge of the two outflow springs, resulting in recovery rates between 10 and 100 percent. Thus, at low-water conditions, a large portion of tracers is stored for a longer time in the matrix of the aquifer, represented by microfissures and inactive karst channels. This explanation has been confirmed by the wide time variation of the peak from breakthrough curves, between 15 hours and almost three days, as well as by the runoff separation analysis of the two springs.

The portions of the tracer, found at the outflow, were used to calculate the flow rate then to quantify the whole water volume of traced water in the karst system. It shows that the water volume is nearly independent

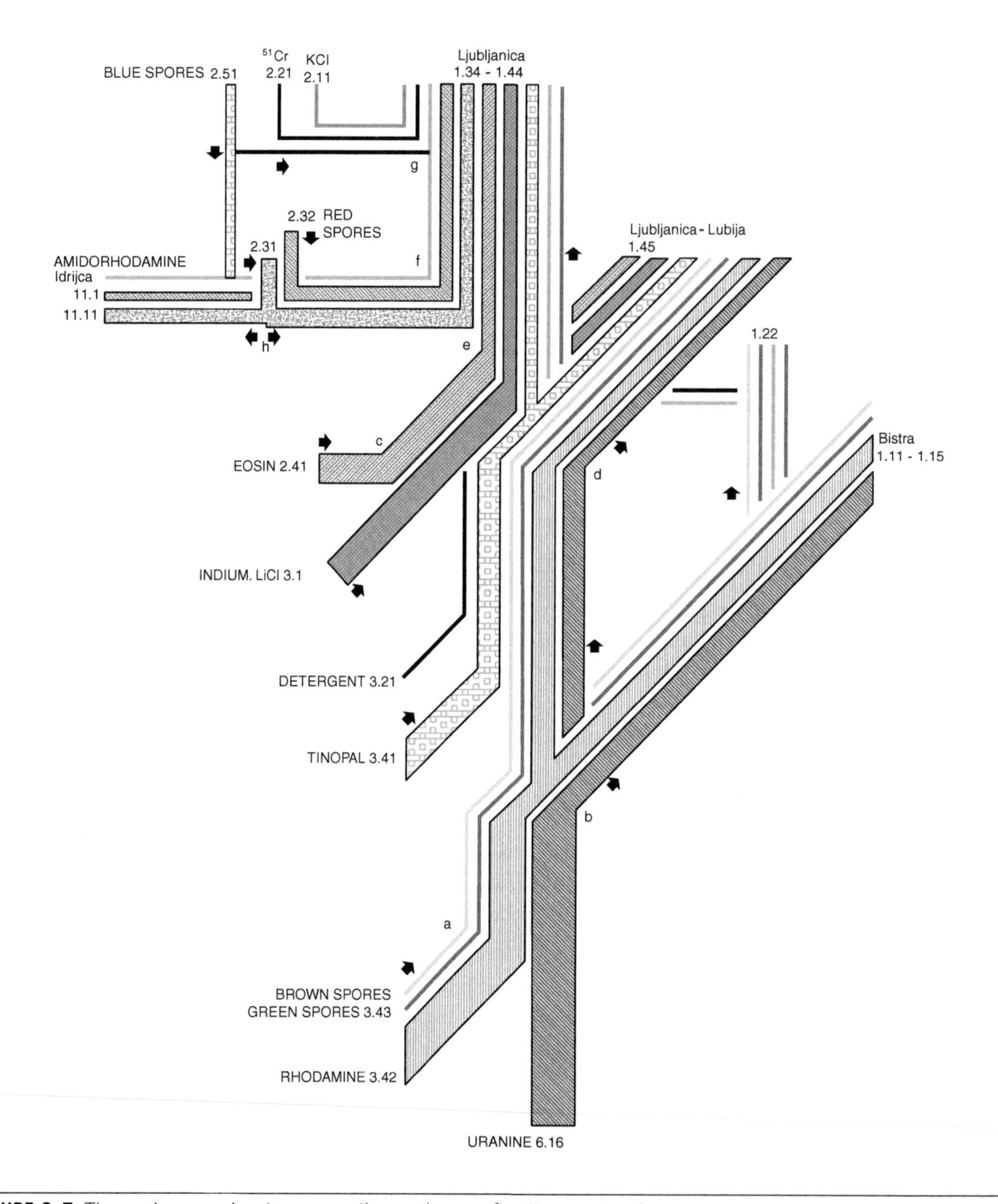

FIGURE 3–7 The underground water connections scheme after the results of a water tracing experiment in 1975: a to h are the main critical points in the underground drainage system, where the underground flows are united and dismembered. (Modified after Habič, 1976b.)

from the outflow discharge. It can therefore be concluded that the whole water volume should be considered constant, with a rising discharge. Increasing discharge can be interpreted as a function of flow velocity and not as additional space for water storage.

Investigations of storage capacity of the karst system, based on tracer experiments, result in an amount of 6.5×10^4 m^3. The residence time of water varies between 20 and 80 hours, due to hydrological conditions during the tracer experiment. On the other hand, the total volume of the karst block in both systems, vadose and phreatic, has been calculated on the base of environmental isotopes with 1.4×10^8 m^3, which is at least four magnitudes higher. It corresponds to a mean residence time of groundwater of about 50 years and a total porosity of 4.9 percent. It also indicates that, by means of tracing experiments, only a very small part of the karst water body is involved, so both numbers can be considered as realistic.

Similar investigations have been carried out in the Swiss Jura (Bauer et al., 1980). Storage calculations at the well-known, large Areuse karst spring show a volume, based on calculations of tracing experiments, of 1.9×10^6 m^3, while the total volume of the karst aquifer was calculated at some 2×10^8 m^3. The same effect has been found as described previously.

Several tracer experiments have been performed recently in Slovenia on landfills in the karst, with the aim to better understand the directions and characteristics of the groundwater flow from these sources of pollution. This was essential to prepare an efficient plan for the monitoring of groundwater in the area of influence of the landfill (Kogovšek and Petrič, 2006, 2007). The research of landfill impacts on the environment was based on the determination of tracer flow from the injection point on the landfill surface toward the main springs in the study areas.

The Mala gora landfill, near Ribnica in southeastern Slovenia, is one of the nine still active landfills on the Slovene karst. The landfill is located on well-karstified Upper Jurassic limestone, alternating with dolomite and covered with thin, often interrupted layers of brown soil (Figure 3–8). Several tracer tests were performed in the past to determine the directions and characteristics of groundwater flow in the broader studied area (Figure 3–8). To simulate dangerous or harmful landfill impacts, additional tracing at high-water conditions was performed in October 2004 (Kogovšek and Petrič, 2006). After testing the aquifer swallow capacity, a highly permeable vertical fissure at the landfill margin was selected for the tracer injection. A solution of 7 kg of uranine was injected into it and washed off with 9 m^3 of water (Figure 3–9).

The results indicated the main underground water connections to the springs, illustrated in Figure 3–8. The appearance of the tracer is forwarded by favorable hydrological conditions and, even one year after the injection, increased concentrations of uranine were measured in all springs after more intensive precipitation events (Figure 3–10). Special attention was dedicated to the Globočec spring, as the main source of water supply in the Suha krajina region. The results of tracer tests indicated that groundwater from the landfill area also flows toward Globočec during high-water conditions, although this spring is recharged mainly from other territories of the karst aquifer.

The apparent dominant flow velocity in the main direction toward the springs near Dvor was approximately 4 cm/s (Table 3–2). This indicates a very high vulnerability and a serious danger of pollution with harmful substances from the landfill. Hence, the obtained results represented valuable information for the proper planning of the groundwater resource monitoring.

The characteristics of the groundwater flow from the landfill near Sežana in the area of the Classical karst were also studied by means of a tracer test (Kogovšek and Petrič, 2007). The uranine fluorescent dye was injected into a well-permeable fissure on the karren surface near the landfill to study the directions and velocities of its flow through the karst aquifer. The monitoring of tracer concentrations over the period of one and a half years enabled making some conclusions about the dynamics of the groundwater flow in the directions

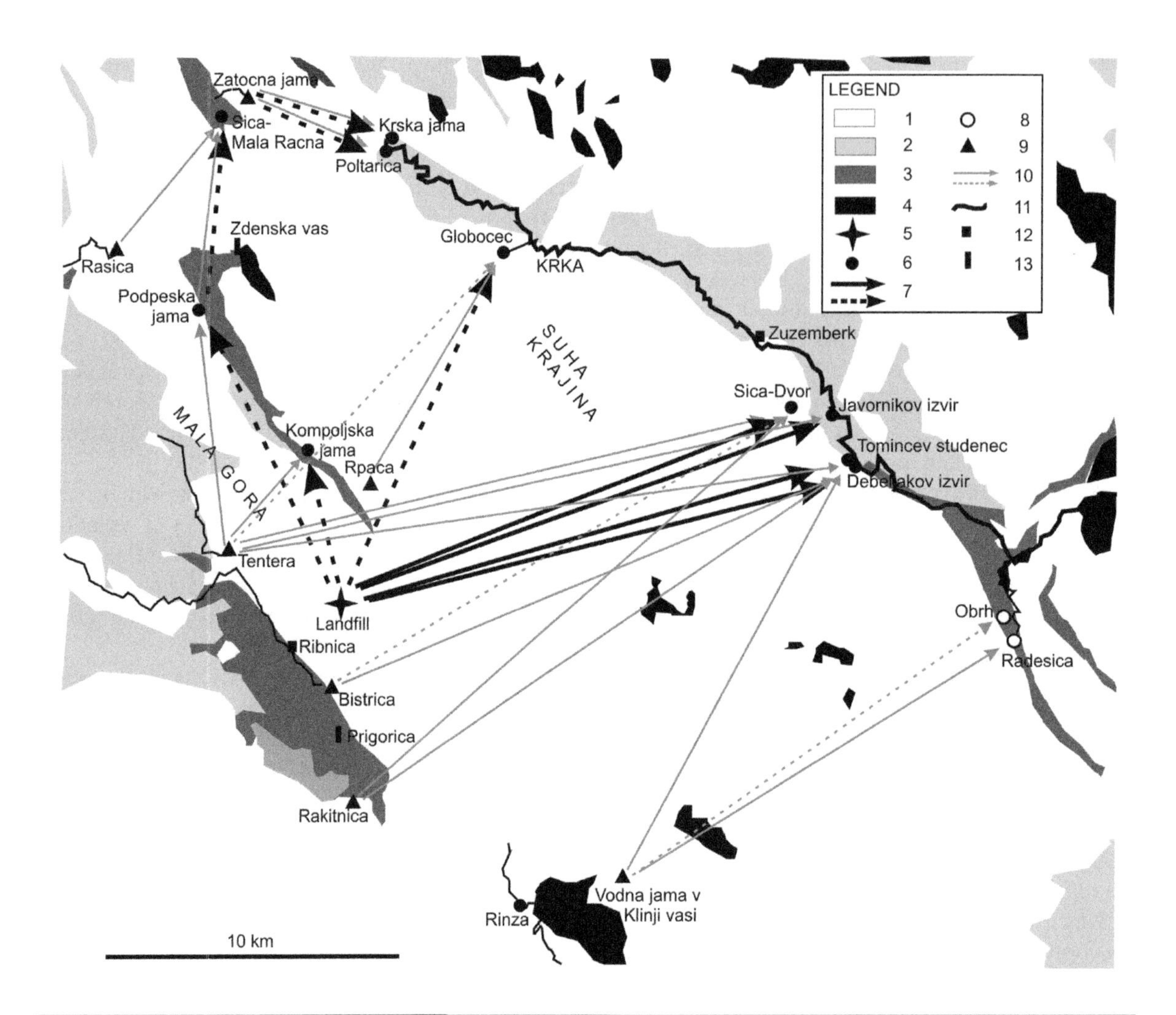

FIGURE 3–8 Hydrogeological map of the broader area of the Mala gora landfill with the results of tracer tests: (1), karst aquifer; (2), fissured aquifer; (3), porous aquifer; (4), very low-permeable rocks; (5), landfill Mala gora, injection point at the tracer test in October 2004; (6), sampling point at the tracer test in October 2004; (7), main and secondary groundwater connection proved by tracer test in October 2004; (8), spring; (9), injection point at previous tracings; (10), main and secondary groundwater connections proved by previous tracings; (11), surface flow; (12), settlement; (13), precipitation station; (14), gauging station. (Kogovšek and Petrič, 2006.)

toward the Timava, Brojnica, and Sardoč springs. Detailed observations were also studied at the Klariči pumping station, which is the main source of drinking water for the Kras region.

The main directions and high flow velocities through very permeable karst channels were proven. The calculation of the amount of recovered tracer confirmed the main flow direction toward the Timava springs (93 percent recovery rate). Based on the described results and previous experience, some guidelines for the monitoring can be emphasized.

FIGURE 3–9 Injection of an uranine solution into a fissure at the margin of the landfill. (Photo: Petrič, 2005.)

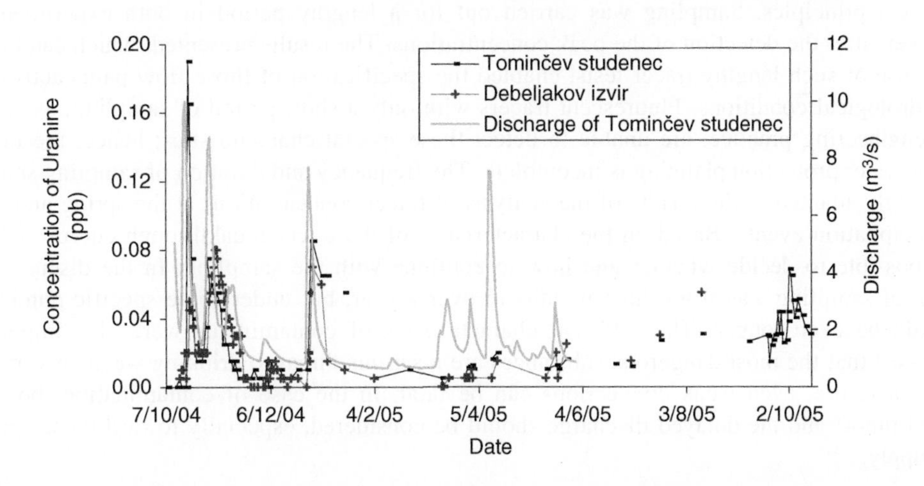

FIGURE 3–10 Concentrations of uranine in the Tominčev studenec and Debeljakov izvir springs, and discharges in Tominčev studenec. (Kogovšek and Petrič, 2006.)

The main Slovene military training area is located in the mountainous Javorniki plateau, a karstic district with no surface drainage or river valleys. Little attenuation occurs under such conditions, and as a result, there is a high risk of pollution. Tracer tests have shown that a significant proportion of the water recharging the Malni and Vipava springs comes from the vicinity of the military training area (Kogovšek et al., 1999). Consequently, any polluting activities taking place within the military training area are likely

Table 3–2 Estimation of Apparent Dominant Velocities of Groundwater Flow (Kogovšek and Petrič, 2006)

Sampling Point	Height Difference (m)	Distance (m)	t_{dom} (h)	v_{dom} (cm/s)
Tominčev studenec	395	17,800	122.5	4.0
Debeljakov studenec	397	18,045	142.0	3.5
Javornikov izvir	390	17,710	143.5	3.4
Šica-Dvor	370	16,515		3.4
Globočec	322	12,740	124.5	2.9

to affect the two springs. This is a very serious matter, as the springs have been developed to provide the water supply for the population of southwestern Slovenia. To adequately protect such water, it is necessary to obtain as much information as possible on the extent of the recharge zones and the characteristics of the groundwater flow. Hence, the presented investigation (Kogovšek et al., 1999) proves that tracer tests are a very useful tool for acquiring such information.

Other examples of tracer applications feature the danger estimation of water contamination deriving from a petrochemical storage depot and the assessment of effects resulting from the construction of a railway line on regionally important water resources (Kogovšek and Petrič, 2004). Besides the basic information about the direction and velocity of groundwater flow, the discussed studies also indicated some methodological principles. Sampling was carried out for a lengthy period in both experiments, and it continued even after the detection of the peak concentrations. The results presented, which can be obtained only by the use of such lengthy tracer tests, enabled the specification of those flow paths activated under extreme hydrological conditions. Fluorescent tracers with only a short period of sampling, as usually performed in engineering practice, are unable to detect these special characteristics; hence, the information available for water protection planning is incomplete. The frequency and duration of sampling should therefore be regularly adapted to the results of the analysis of tracer concentrations at the spring and the occurrence of precipitation events. Based on the characteristics of the tracer breakthrough curves and recovery rates, it is possible to decide whether and how to continue with the sampling. In the discussed studies, the duration of sampling was from three months to over a year, but under some specific conditions, this period should be even longer. The different characteristics of contaminants were also important. The authors stressed that the most dangerous substances are a serious threat to drinking water in very low concentrations; therefore, even weak connections can be fatal. In the case of contamination, both the first concentrated runoff and the delayed discharge should be considered, especially toward those springs used for water supply.

Tracer experiments under lysimeter conditions

The unsaturated zone of an aquifer can essentially influence the spring discharge as well as its geochemical composition. Hence, the tracer experiments were also performed in the discussed zone to get insight into its basic physicochemical characteristics and study its role in spring discharge and contamination (Figure 3–11).

In cooperation with agricultural organizations, a scientific program for the Quaternary Mur river basin south of Graz has been developed to establish a combined lysimeter configuration. The soil conditions, and especially the agricultural land use by heavy machines, are very much associated with other areas of the basin. The configuration of the technical equipment has been selected, so far, to investigate a maize monoculture as well as a crop rotation.

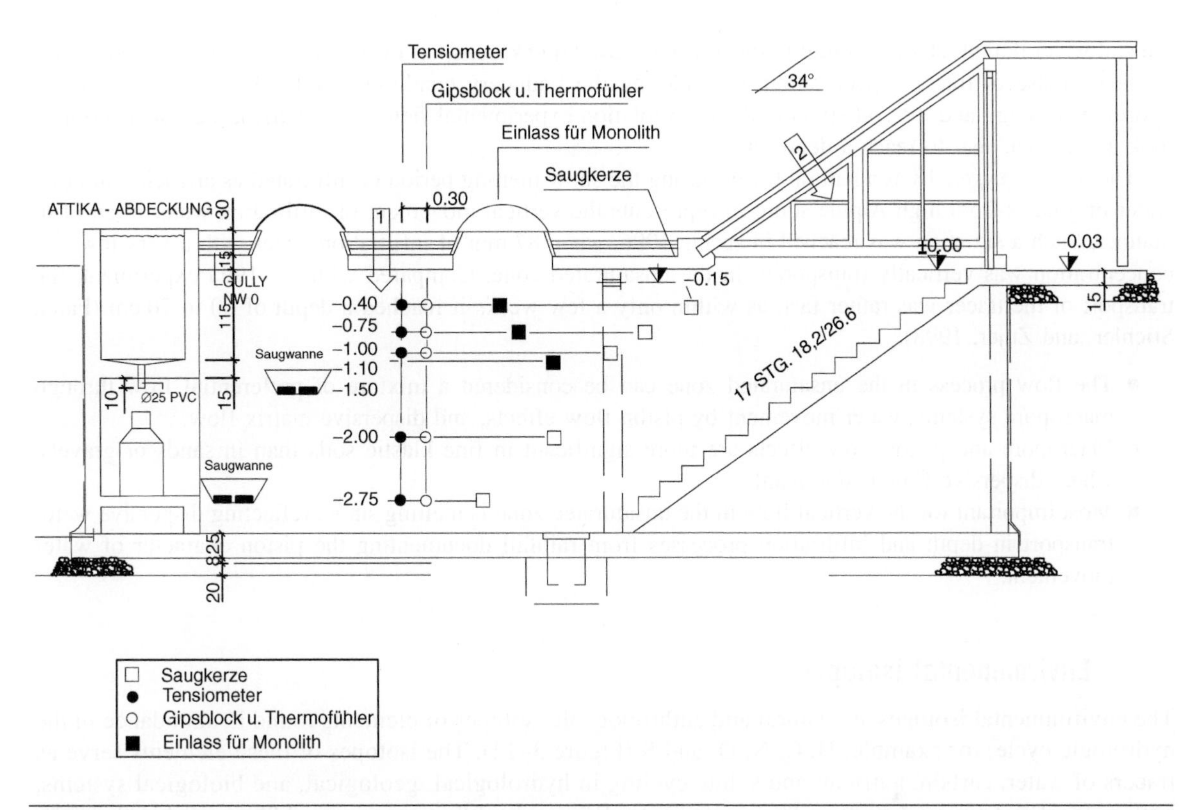

FIGURE 3–11 Original configuration of lysimeter station Wagna. Saugkerze = suction cup, Gipsblock und Thermofühler = gypsum block and thermal sensor, Einlass für Monolith = connection to monolithic block, Saugwanne = suction pan. (From Zojer, Ramspacher, and Fank, 1991.)

The movement of infiltration water depends on the permeability of the soil cover and coarse sediments:

- Installation of suction cups, horizontally penetrated through the soil at different depth.
- Installation of tensiometers.
- Installation of monolithic blocks.
- Installation of microlysimeters.

The contaminant transport in the unsaturated zone depends on different land use—exchange, retardation, and adsorption processes—in the soil zone. These investigations should help develop improved land-use practices with a reduced impact of fertilizers and pesticides into the groundwater system.

Combined tracing experiments have been carried out with the aim of comparing solute transport with water transport and the tracer movement with the behavior of contaminants. Another aspect was to verify the model conceptions concerning solute transport in the unsaturated zone (Fank, 1996). The tracer experiment was performed in springtime 1993 with 6 kg sodium bromide dissolved in 400 L of water and distributed at the surface of the whole experimental field. The experiment was accompanied by 40 mm artificial rain. Bromide is known as a very conservative tracer that is not subject to strong adsorption processes. Due to a very dry spring in 1993, no significant quick vertical movement of the tracer was recognized, and the maximum of the tracer concentration was first measured in late autumn at a 40 cm depth, after balancing the soil water deficit. Heavy rainfall in this season caused quick transport of bromide in depth proceeded by a slight decrease of tracer concentration. Considering the boundary conditions in 1993 and 1994 regarding hydrometeorology and the soil strata at the crop rotation experimental field (rape-maize), the tracer bromide took almost one year to reach a depth of 3 m.

The use of oxygen-18 as a natural tracer during the snow melting period or infiltrated as artificial rain (the water originates from high Alpine sources) represents the vertical movement of infiltration water in an ideal manner. Such a situation was reached in spring 1996, when 87 mm of infiltration water with a very low ^{18}O concentration was vertically transported in the unsaturated zone. Compared with the 1993 experiment, the transport of the tracer was rather fast, as within only a few weeks it reached a depth of 60 to 70 cm (Fank, Stichler, and Zojer, 1998):

- The flow process in the unsaturated zone can be considered a mixture of preferential flow through macropore systems, water movement by piston flow effects, and dispersive matrix flow.
- Macropore and piston flow effects are more significant in fine klastic soils than in sands or gravels, where dispersive flow is dominant.
- Most important for the vertical flow in the unsaturated zone is melting snow, reflecting dispersive water transport in depth and infiltration processes from rainfall documenting the piston character of water movement.

3.2.2 Environmental isotopes

The environmental isotopes are natural and anthropogenic isotopes of elements found in abundance in the hydrologic cycle, for example, H, C, N, O, and S (Figure 3–12). The isotopes of these elements serve as tracers of water, carbon, nutrient, and solute cycling in hydrological, geological, and biological systems, both in local and regional studies. The groundwater flow and solute transport could be traced in the aquifer from the recharge area, through the unsaturated and saturated zones to the discharge. It should be

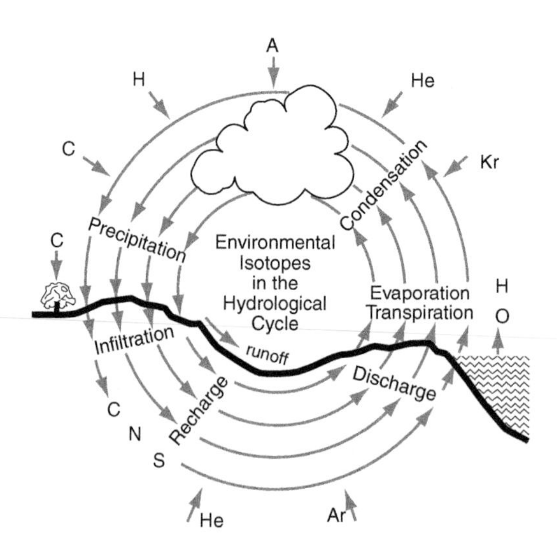

FIGURE 3–12 The hydrologic cycle and environmental isotopes.

pointed out that the groundwater isotopic composition varies at a particular aquifer sampling place as a function of the precipitation, surface water, and groundwater inflow. The solute transport could be described with two quantitative approaches: the balance method and the peak-displacement method.

The isotopic data of water from springs coupled with available information on physical hydrogeology and water chemistry help in understanding the source and mechanism of recharge, groundwater circulation and its renewability, groundwater transit time distribution, hydraulic interrelationships, the groundwater origin and its evolution due to effects of water-rock interaction, the temperature in the aquifer depth and the recharge area, as well as the source and fate of groundwater contamination. Moreover, the isotopic composition of gases can contribute additional valuable information on the groundwater origin.

Meteoric processes, for instance, modify the stable isotopic composition of water; hence, the recharge waters in a particular environment have a characteristic isotopic signature. This signature serves as a natural tracer input in hydrogeological and hydrological studies. On the other hand, the radioisotope decay provides information on circulation time and, thus, groundwater renewability.

The characteristics of the stable and radioactive environmental isotopes are described in numerous professional publications (e.g., Pearson et al., 1991; Clark and Fritz, 1997; Kendall and McDonnell, 1998; Mook, 2000), while additional valuable information can be found on the Web site of the International Atomic Energy Agency (IAEA; www.iaea.org/programmes): IAEA publications and technical reports, information on isotope reference materials, IAEA isotope databases. Therefore, only a basic summary from the cited references is given in this section.

Measurements of the stable isotopic composition of substances are conventionally reported in terms of a relative value δ:

$$\delta_x = \left(\frac{R_x}{R_{st}} - 1\right) \times 1000 \tag{3.4}$$

where R_x is the isotope ratio (e.g., $^2H/^1H$, $^{18}O/^{16}O$, and $^{13}C/^{12}C$) in the substance x, R_{st} is the isotope ratio in the corresponding international standard substance, and δ is expressed in parts per thousand (‰).

A positive δ value means that the sample contains more of the heavy isotope than the standard. A negative δ value means that the sample contains less of the heavy isotope than the standard. A $\delta^{15}N$ value of +30‰ means that there are 30 parts per thousand or 3 percent more ^{15}N in the sample than in the standard.

Various isotope standards are used for reporting isotopic compositions. Stable oxygen and hydrogen isotopic ratios are normally reported relative to the SMOW (standard mean ocean water) standard or the virtually equivalent V-SMOW (Vienna-SMOW) standard. Carbon-stable isotope ratios are reported relative to the PDB (Pee Dee Belemnite) or the equivalent V-PDB (Vienna PDB) standard. Oxygen-stable isotope ratios of carbonates are also commonly reported relative to PDB or V-PDB. Sulfur and nitrogen isotopes are reported relative to CDT (for Cañon Diablo troilite) and AIR (for atmospheric air), respectively. V-SMOW and V-PDB are virtually identical to the now-unavailable SMOW and PDB standards; therefore, the use of V-SMOW and V-PDB is preferred.

The expression $\delta_A > \delta_B$ means that substance A is enriched with an isotope (e.g., ^{18}O) or is isotopically heavier than the substance B.

The difference between the stable isotopic composition of two substances, A and B, is given by the fractionation factor:

$$\alpha_{(A-B)} = \frac{R_A}{R_B} \tag{3.5}$$

The fractionation results from two main processes:

1. **Isotopic reactions.** The isotopic exchange between different chemical substances, different phase stages, or different molecules prevails, while chemical changes are insignificant.
2. **Kinetic processes.** The differences between reaction velocities of isotope molecules mainly influence them.

The law of mass action is valid for each isotopic reaction. The isotopic reaction is described by the equilibrium constant K or the fractionation factor α that is related by the expression

$$\alpha = K^{1/n} \tag{3.6}$$

where n is the number of exchanged atoms.

This expression is usually simplified. It is considered that only one atom is exchanged in the reaction. Consequently, the fractionation factor α equals to K ($\alpha = K$).

The relationship between the fractionation factor α and δ-values of two substances, A and B, expressed relative to the same standard is described by the following equation:

$$\alpha_{(A-B)} = \frac{(1000 + \delta_A)}{(1000 + \delta_B)} \tag{3.7}$$

The term 100 ln α is often used for the expression of fractionation factors. If it is taken into account that the numerical value of α is close to 1, it follows that

$$1000 \ln \alpha_{(A-B)} \approx \delta_A - \delta_B \; 1000 \tag{3.8}$$

The approximation error is less than 0.05‰ if the difference between δ_A and δ_B is less than 10‰.

On the other hand, radioactive isotopes are age-dating isotopes. The half-life for the isotope radioactive decay forms the basis for most age-dating methods. Groundwater dating is based on the fact that the initial concentration of radioactive isotope c_0 in recharging groundwater (which should be well known) decreases with time according to the decay equation:

$$c(t) = c_0 \times e^{-\ln 2(t/T_{1/2})} \tag{3.9}$$

where $T_{½}$ is the half-life of the radioactive isotope.

The radioactive isotopes in the environment can be of either natural origin or anthropogenic. The input source functions, which describe the time-varying global fluxes of isotopes deduced from atmospheric, cosmogenic, and anthropogenic production, such as ^{3}H, ^{14}C, and ^{36}Cl, are well known. These isotopes have been measured monthly or annually in precipitation since 1950 and form the peak-shaped curves. The input functions of radioactive isotopes that resulted from nuclear power facilities or fuel reprocessing have either increased steadily (e.g., ^{85}Kr) or remain elevated (e.g., ^{129}I).

The measured spring radioactive isotope concentrations are compared with the input functions to get fairly informative age determinations over the past several decades. If more than one anthropogenic isotope is measured, the groundwater age determinations may be much more precise, because of the uncertainty regarding the input concentration c_0, which may be poorly known or vary with time.

Three temperature systems should be taken into account in the isotopic investigations of springs: low-temperature (<90°C), medium-temperature ($90 < T < 150$°C), and high-temperature (>150°C) systems. It should be stressed that the groundwater isotopic composition is mainly unaffected in low-temperature systems due to very slow reactions and processes. Therefore, the isotopic techniques have the widest range of application in hydrogeological and hydrological studies of these systems. The monitoring of the natural variation of the isotope contents in precipitation, groundwater (springs), and surface waters (if they recharge the aquifer) contributes essential information on the groundwater origin and genesis and hydrochemical and mixing processes (leakage of aquifers, mixing of groundwater with surface water) on groundwater residence times and with that on hydrodynamic properties of the study area. For two other temperature systems, the knowledge on the magnitude and temperature dependence of the isotopic fractionation factors between the minerals and fluids (gases and water) is essential for the interpretation of changes in the spring stable isotopic composition.

The environmental isotopes in low-temperature systems are common research tools of springs: isotope hydrology (tracers of the water itself) and solute isotope biogeochemistry (tracers of the water solutes). However, we should know why and when environmental isotopes are treated as convenient tracers. The most frequent examples follow:

- Waters that are recharged at different times or at different locations have different isotopic composition, which is also reflected in the spring isotopic composition.
- Groundwaters that followed different flow paths preserve distinctive fingerprints, which is also reflected in the spring isotopic composition.
- An isotopically distinctive source along a groundwater flow path has a significant influence on the spring isotopic composition, which indicates a hydrodynamic connection.
- Meteoric waters retain their distinctive fingerprints until they mix with waters of different compositions or react with minerals or other fluids.
- Groundwater solutes that originate from the atmosphere usually differ isotopically from those originating from geologic and biologic sources within the aquifer.
- Solute isotopic ratios change during the groundwater flow path due to the biological cycling of solutes and water-rock interactions; however, these changes are usually predictable and can be reconstructed from the spring isotopic compositions.

It follows that two opposite properties of isotopes are used in isotopic studies of springs. Generally, they refer to changes in groundwater isotopic composition due to physical and chemical processes, such as mixing processes and water-rock interaction, and the constancy of the isotopic composition during the groundwater flow path from the catchment area to the spring discharge. The isotopic investigations of springs most frequently include the following topics:

- Identification of the aquifer recharge-discharge relationship.
- Assessment of possible origins of waters that contribute to the spring discharge.
- Determination of flow paths from the recharge to the discharge area of the spring.
- Groundwater residence time and connected storage properties of groundwater.
- Mixing of event waters (e.g., snowmelt, storms) and preevent groundwater in the spring discharge.
- Determination of atmospheric sources that contribute to the solute isotopic composition of groundwater.
- Determination of weathering reactions along the flow paths that influence the spring isotopic composition.
- Assessment of influences of biologic cycling on the groundwater's solute isotopic composition (e.g., nutrients within an ecosystem).
- Determination of groundwater geochemical evolution.
- Identification of sources and mechanisms of groundwater contamination.
- Biodegradation processes and transport phenomena.
- Testing hydrologic models using isotopic data.

The investigation topic governs the selection of studied isotopes. It should be taken into consideration that conservative isotopes (such as ^{2}H and ^{18}O) and reactive isotopes (such as ^{87}Sr and ^{13}C) provide very different

Table 3–3 Advantages and Disadvantages of the Application of Environmental Isotopes

Advantages	Environmental isotopes indirectly provide information on groundwater residence times and mixing processes in the spring catchment and, hence, on the aquifer groundwater flow dynamic and recharge properties as well as on groundwater geochemical evolution. These data cannot be obtained by direct measurement. Long-term isotopic data reflect changes in the spring recharge-discharge relationship and complement the spring hydrographic data. Results of isotopic investigations significantly complement results of classical geological and hydrogeological methods and those of other indirect methods, such as artificial tracers, geophysical investigations, and numerical modeling.
Disadvantages	In situ analyses are not possible. Besides, the isotopic monitoring often requires long-term or frequent sampling for adequate insight into the spring hydrodynamic and geochemical characteristics. Automatic samplers offer valuable aid to this work. Isotopic investigations are quite expensive, due to sampling cost and particularly laboratory costs. The research methodology should be carefully designed to get significant results, especially in the selection of sampling places and leading parameters as well as the sampling techniques and methods. The processing and modeling of isotopic data consider a lot of assumptions, which are sometimes difficult to implement. The data interpretation should consider the uncertainties.

information, which is discussed in following subsections. The advantages and disadvantages of the application of environmental isotopes are summarized in Table 3–3, but we should be also aware that, if we have dealings with more complex hydrogeological systems, a so-called multiisotope approach is required. The selected stable and radioactive isotopes should be combined with geological, hydrogeological, and chemical parameters. Different information obtained by each studied parameter completes each other and provides insight into the groundwater flow and solute transport phenomena. For all that, the importance of the sampling method is on a par with the multiparametric approach.

Stable isotopes

Stable isotopes of oxygen and hydrogen Isotopes of elements of the water molecule oxygen (^{18}O) and hydrogen (^{2}H) have been the most frequently used in hydrological investigations of springs. Hence, they also receive the most attention in this book.

Even though the global water cycle is quite complicated, the $\delta^{18}O$ and $\delta^{2}H$ composition of its components can be predicted (Figure 3–13). Applications of stable isotope ratios of hydrogen and oxygen in groundwater are based primarily on isotopic variations in precipitation. After the infiltration of precipitation into the aquifer, only physical processes, such as diffusion, dispersion, mixing, and evaporation, alter the groundwater isotopic composition. Hence, they are ideal conservative tracers under low-temperature conditions (Clark and Fritz, 1997).

The sampled water ^{18}O and ^{2}H composition, together with hydrometric data, provides information on the movement and mixing of water masses only if the isotopic composition of each water mass significantly differs. The average groundwater ^{18}O and ^{2}H composition equals to a weighted average of the precipitation ^{18}O and ^{2}H composition as a rule, while surface water is enriched more with heavier (more positive) isotopes than precipitation, due to the evaporation process.

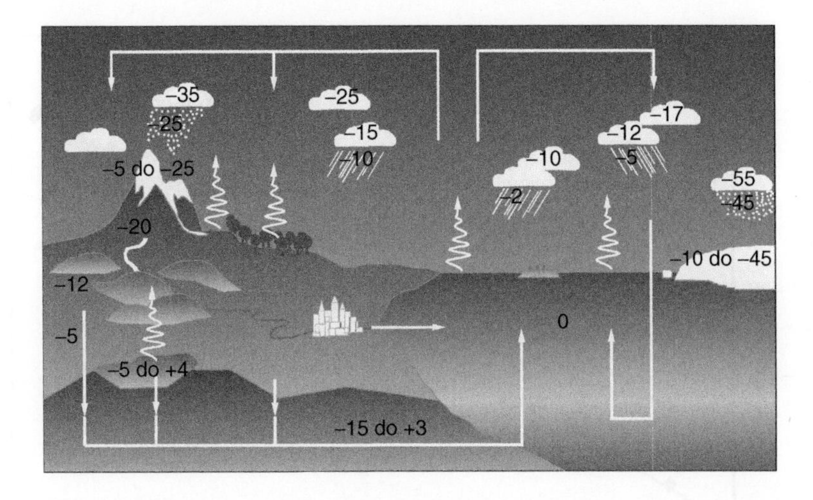

FIGURE 3–13 Typical values of ^{18}O composition (in ‰) in the global water cycle. (Modified after International Atomic Energy Agency/World Meteorological Organization, 2006.)

Craig (1961) was the first to describe the relationship of discussed isotopes in the precipitation by the global meteoric waterline (GMWL; Clark and Fritz, 1997; Rozanski, Araguds-Araguds, and Gonfiantini, 1992, 1993):

$$\delta^2H = 8 \times \delta^{18}O + 10 \tag{3.10}$$

This is illustrated in Figure 3–14. GMWL depends locally on two basic processes: the condensation conditions before the precipitation and the environmental evaporation conditions during the precipitation.

In a particular region, the local meteoric waterline (LMWL) differs from GMWL in both the slope and the intercept. The main factors that influence the LMWL deviation are described in Figure 3–14. The slope does not vary a lot, 8 $\pm$ 0.5. However, the intercept, which is also called the *deuterium excess*, may vary considerably from place to place, owing to various origins and conditions of the vapor formation, 10 $\pm$ 5‰. Therefore, the local meteoric waterline is a function of temperature during secondary evaporation as rain falls from a cloud, which results in effects of the isotopic fractionation with respect to latitude, altitude, and climate. The GMWL/LMWL equations could be calculated from large data sets of precipitation stations all over the world that are included in the GNIP Programme and available at IAEA Web pages.

At a particular location, the precipitation isotopic composition is affected by season, latitude, altitude, precipitation amount, and distance from the coast.

Precipitation that occurs in polar regions is more depleted in isotopic composition than in tropical regions. This latitude effect is due to successive rainout from the cloud during moisture transport from the tropics to the poles. A latitude effect of -0.6‰ for $\delta^{18}O$ per degree is generally observed.

The continental effect is related to the continuous depletion of the precipitation isotopic composition as the air mass moves inland from the coast. The precipitation isotopic composition changes from region to region as well as from season to season.

Heavy rainfall is depleted in isotopic composition compared to light rain. The amount of this effect should be studied locally. It could result from the evaporation of falling raindrops, the exchange with atmospheric water vapor during light rain, or the removal of different amounts of water from the atmosphere.

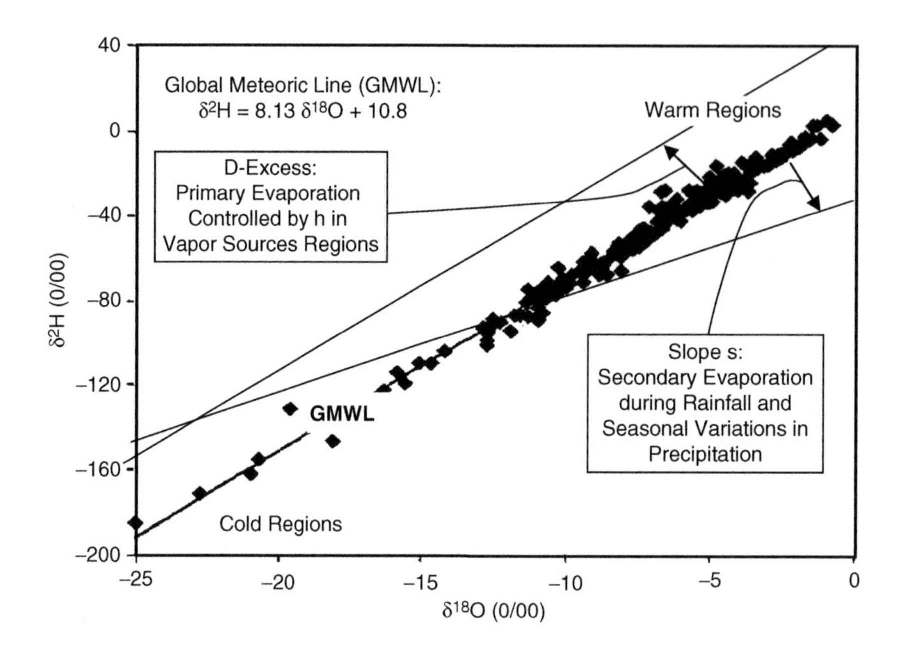

FIGURE 3–14 Global relationship between ^{18}O in ^{2}H isotopes in precipitation (data are weighted average annual values of precipitation from GNIP station) and main factors influencing the D excess and the intercept coefficient. (After Rozanski et al., 1993, and Clark and Fritz, 1997.)

The altitude effect represents the influence of altitude on the precipitation isotopic composition. In addition to air temperature, the $\delta^{18}O$ and $\delta^{2}H$ values decrease with the altitude increase. Generally the effect varies from −0.15 to −0.5‰ for $\delta^{18}O$ and from −1 to −4‰ for $\delta^{2}H$ per 100 m rise in altitude.

The seasonal effect reflects the dependence of the precipitation isotopic composition from local air temperatures, which is revealed by the lowest $\delta^{18}O$ and $\delta^{2}H$ values in colder months and the highest $\delta^{18}O$ and $\delta^{2}H$ values in warmer months, respectively.

Isotopes of the water molecule elements have the widest field of application in hydrogeological studies of springs, such as tracing the water origin, the mode of groundwater recharge, the determination of very young groundwater components, and the spring discharge response to precipitation. In principle, the regional and local investigations are based on LMWL studies and comparisons with $\delta^{18}O$ and $\delta^{2}H$ data of surface and groundwater sources.

The altitude and seasonal effects were applied in several hydrogeological investigations for the identification of the spring recharge areas and recharge-discharge characteristics, respectively. The seasonal variations were also used to determine the periods with dominant spring recharge.

In the framework of the karst hydrogeological investigations in southwestern Slovenia, extensive isotopic studies were performed in the research area that occupied about 700 km^{2} (Stichler et al., 1997). During a period of three years testing the $\delta^{18}O$ and $\delta^{2}H$ of precipitation, six main karstic springs and two sinkholes were monitored. Precipitation was sampled at five stations, which were spread over the research area at altitudes from 50 to 1070 m. Figure 3–15 illustrates the correlation between the weighted annual average of the precipitation $\delta^{18}O$ values and altitudes of corresponding sampling stations.

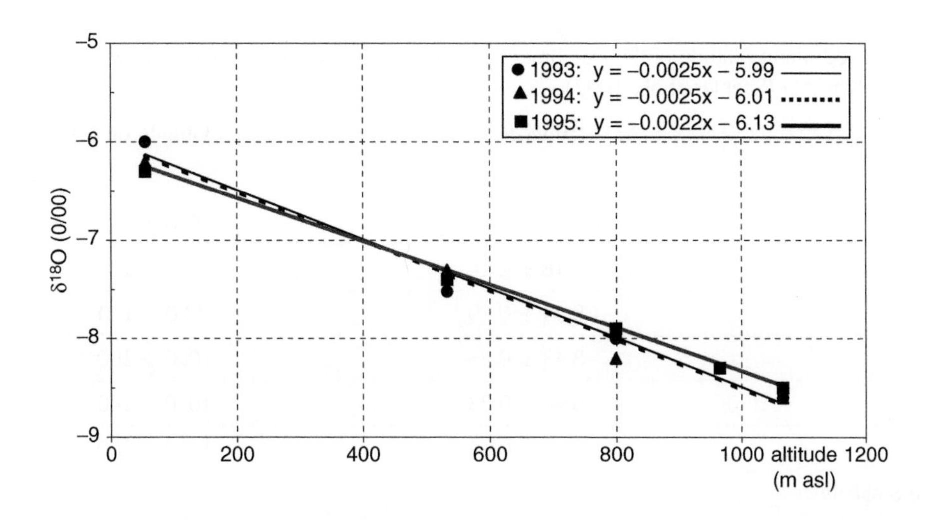

FIGURE 3–15 Correlation of δ^{18}O contents in precipitation from meteorological stations at southwestern Slovenia and altitudes. (Stichler et al., 1997.)

The weighted annual averages of the precipitation δ^{18}O were calculated by the equation

$$\delta^{18}\text{O} = \frac{\sum_{i=1}^{n} P_i \delta_i{}^{18}\text{O}}{\sum_{i=1}^{n} P_i} \tag{3.11}$$

δ^{18}O = weighted average of the precipitation's oxygen isotopic composition.
P_i = precipitation amount that fell between samples (i – 1) and (i).
$\delta_i{}^{18}$O = oxygen isotopic composition of the precipitation sample (i).

Generally, the springs and sinkholes were sampled at base flow conditions, hence their annual average δ^{18}O values were compared with Figure 3–15 and the mean altitudes of the catchment area of sampled springs and sinkholes were estimated as shown in Table 3–4.

The seasonal variation in the precipitation δ^{18}O and δ^{2}H represents an input signal that may be used for the groundwater dating. That is, precipitation infiltrates the soil and recharges the aquifer, where it is mixed with the prestored groundwater. These results in the input signal attenuation indicate a lowering of the isotopic variation amplitude. Owing to different mixing and homogenization stages, groundwater has a different ^{2}H and ^{18}O isotopic composition throughout the aquifer and, with that, a different amplitude of the isotopic seasonal variation. These differences can be applied to determine groundwater residence time, seeing that the longer the residence time, the lower is the amplitude of the groundwater isotopic seasonal variation.

Data on the isotopic investigations in southwestern Slovenia were also used for the determination of mean groundwater residence times of karst springs. This was based on differences between the heights of two amplitudes: the amplitude ^{18}O variation in the precipitation and that in the groundwater (Stichler et al., 1997).

Table 3–4 Estimated Mean Altitude of the Catchment Area of the Six Karstic Springs and Two Sinkholes in Southwestern Slovenia (Stichler et al., 1997)

	$\delta^{18}O$ (‰)	Altitude (m asl)
Name of the Spring		
Kajža	−7.52 ± 0.20	620 ± 80
Hotešk	−8.18 ± 0.24	900 ± 100
Mrzlek	−8.24 ± 0.29	920 ± 120
Hubelj	−8.43 ± 0.23	1000 ± 100
Vipava	−8.46 ± 0.33	1010 ± 140
Podroteja	−8.57 ± 0.24	1060 ± 100
Name of the Sinkhole		
Banjšice	−7.76 ± 0.17	720 ± 70
Čepovanski potok	−8.28 ± 0.14	940 ± 60

The mean groundwater residence times were calculated by the dispersion model equation (Maloszewski and Zuber, 1982, 1996):

$$T = \frac{1}{2}\pi\left(\frac{-\ln f}{P_D}\right)^{1/2} \tag{3.12}$$

T = average residence time.
f = amplitude ratio: $f = B_0/A_0$.
A_0 = amplitude of the precipitation ^{18}O variation.
B_0 = amplitude of the groundwater ^{18}O variation.
P_D = dispersion parameter.

They varied from 4.4 months in the Vipava spring to 5.8 months in the Hubelj spring. However, these values indicated that two groundwater flow components contributed to the spring discharge, having a mean transit of weeks (karst channels) and mean transit times of years (diffuse discharge from the karst aquifer).

The comparative method included differences between the precipitation and the Hubelj spring. The $\delta^{18}O$ seasonal variations were also applied to estimate the groundwater residence time (Trček, 2003, 2007). As already mentioned, the amplitude of the groundwater ^{18}O variation decreases with increasing residence time, and it becomes practically negligible after the groundwater average residence time exceeds five years (Figure 3–16).

This information was essential for age estimations of sampled groundwater in the catchment area of the Hubelj karstic spring. Polynomial and linear trends of the precipitation $\delta^{18}O$ were compared with the groundwater values and trends (Figure 3–17). Because of the typical $\delta^{18}O$ seasonal variation and its positive linear trends, the residence times of the SVR-7, SVR-4, Hubelj, and SVR-3A groundwater should be less than five years. On the other hand, the average residence time of remaining waters should be longer, owing to the

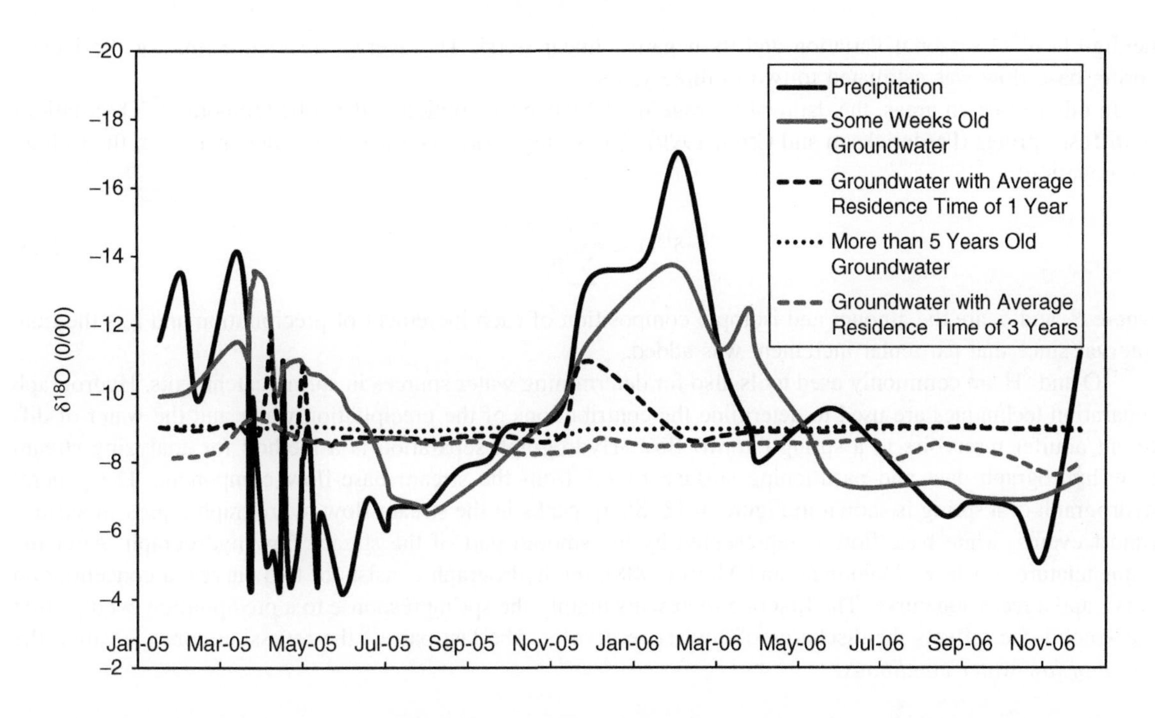

FIGURE 3–16 Comparison of the $\delta^{18}O$ variations in groundwater with different residence times and in precipitation.

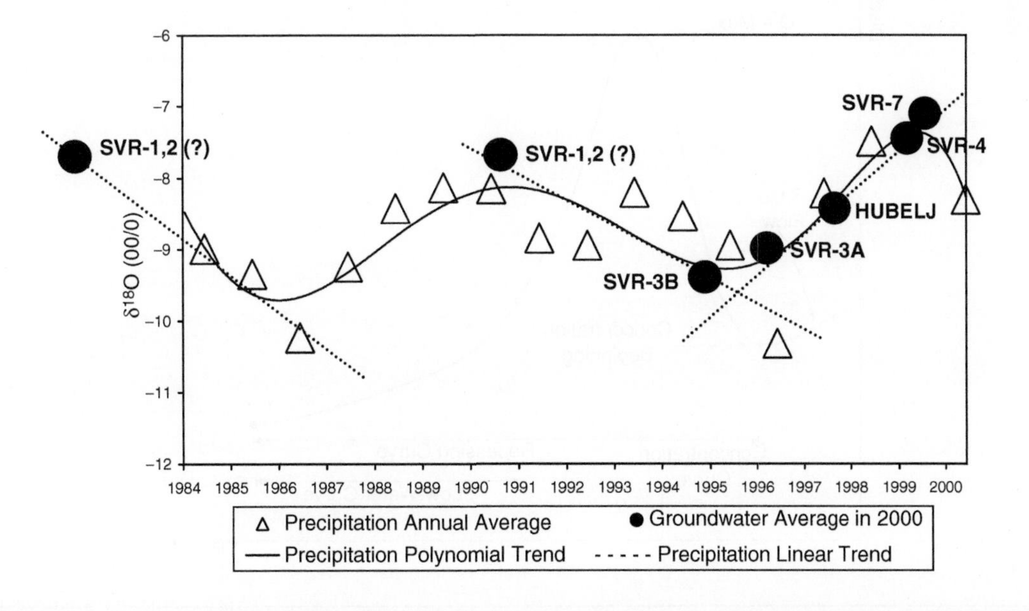

FIGURE 3–17 Trends of the average annual $\delta^{18}O$ of the Ljubljana precipitation compared with the average annual $\delta^{18}O$ of groundwater samples. (Trček, 2003.)

negligible $\delta^{18}O$ seasonal variation and its negative linear trend. The average residence time of the Hubelj spring base flow was estimated to two to three years.

In other research areas, the damped-average model has been applied to describe temporal $\delta^{18}O$ variations in diffuse springs (Frederickson and Criss, 1999). The spring residence time, τ, is calculated from the following equation:

$$\delta^{18}O = \frac{\sum P_i \delta_i e^{-t_i/\tau}}{\sum P_i e^{-t_i/\tau}} \tag{3.13}$$

where P_i and δ_i are the amount and isotopic composition of each increment of precipitation and t_i is the time interval since that particular increment was added.

^{18}O and ^{2}H are commonly used tools also for determining water sources in spring catchments. Hydrograph separation techniques are used to determine the contributions of the precipitation water and the water of different aquifer reservoirs to a spring (storm) flow. Hydrograph separation is a method for analyzing stream flow hydrograph data and partitioning surface runoff from the stream base-flow component. The general hydrograph of a spring is shown in Figure 3–18. Sharp peaks in the stream flow hydrograph represent surface runoff events, while base flow is represented by the smooth part of the stream flow hydrograph. After the nomenclature of Chow, Maidment, and Mays (1988), the hydrograph consists of two curves: a concentration curve and a recession curve. The first one represents mainly the spring response to a precipitation event, while the second one reflects the discharge of aquifer reservoirs. The last part of the recession curve is called the *curve of low-water conditions*.

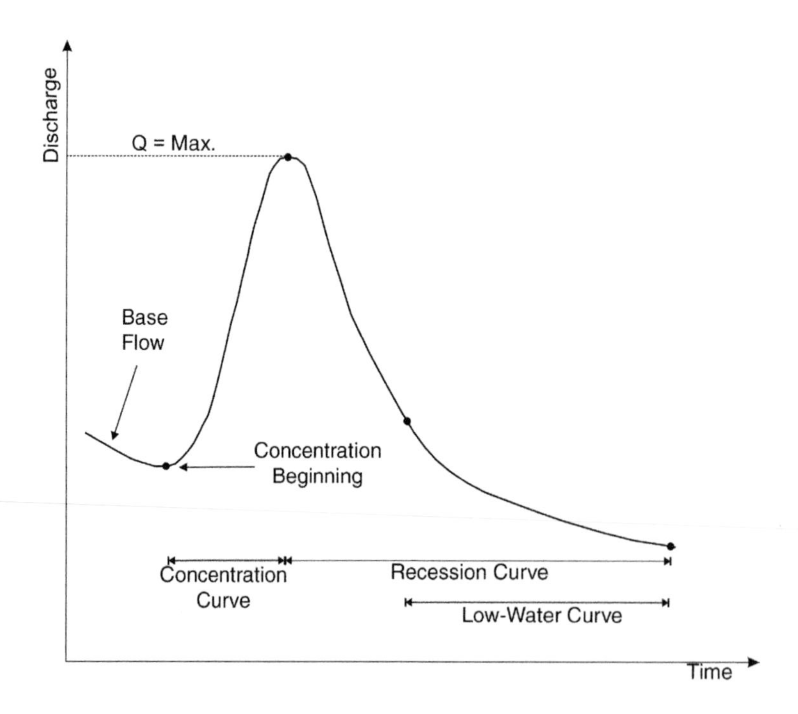

FIGURE 3–18 A general decryption of a spring hydrograph.

A single, conservative natural tracer together with hydrometric data enables the spring hydrograph separation into the relative parts of the event and preevent water components (Dincer et al., 1970; Martinec et al., 1974; Fritz et al., 1976; Sklash and Farvolden, 1979; Stichler et al., 1997; Trček, 2003; Trček, Veselič, and Pezdič, 2006). The event water component represents the precipitation water that fell within the catchment during the particular event (i.e., rain, snowmelt), while the preevent water component represents the groundwater that had been stored in the aquifer before this event (i.e., base flow).

The isotopic signatures of precipitation and groundwater also have been used several times to identify the recharge derived from snowmelt (Moser and Stichler, 1975) and Pleistocene glaciers (Siegel, 1991) and to quantify other important discharge components, such as surface water contributions (Maloszewski et al., 1990).

The instantaneous mixing ratios of the two end members could be quantified by solving a system of two linear equations of the mass balance (Sklash and Farvolden, 1979):

$$Q_t = Q_p + Q_e \quad (3.14)$$

$$Q_tC_t = Q_pC_p + Q_eC_e \quad (3.15)$$

Q = discharge.
C = isotopic or chemical composition of the conservative natural tracer.
t = symbol for the total flow.
p = symbol for the preevent water component (e.g., groundwater, base flow).
e = symbol for the event water component (e.g., rain).

The following assumptions must be considered:

- The isotopic or chemical composition of end members is significantly different.
- The water stored in the aquifer unsaturated and saturated zones does not differ significantly in its isotopic and chemical composition and thus represents the preevent water component.
- The isotopic and chemical composition of the preevent water is uniform throughout the aquifer, generally the same as that of the base flow.
- The precipitation preserves the uniform isotopic and chemical composition throughout the event.

If waters within the same flow path derive from several different sources, the simple two-component hydrograph separation technique could be rather invaluable, as it denotes only the event water contributions, which is the goal of some investigations.

Later studies demonstrated that three- and four-component hydrograph separation techniques are much more convenient for complex catchments, so that contributions of surface waters or different aquifer sources can be taken into consideration. More than one tracer enables the spring hydrograph separation into more than two components, considering that (n) components require the use of (n – 1) tracers. A combination of conservative and nonconservative natural isotopes or chemical substances is convenient for that purpose (Ogunkoya and Jenkins, 1991, 1993; Bazemore, Eshleman, and Hollenbeck, 1994; Kendall and McDonnell, 1998; Brown et al., 1999; Lee and Krothe, 2001; Trček, 2003; Trček et al., 2006). The rational and objective selection of the representative spring components is a key problem of this technique, which is also closely linked to the planning of a suitable sampling strategy. The listed studies demonstrated that the validity of selected end members should be tested to solve the uncertainty.

The relative contributions of three end members to the spring could be calculated for each spring sample by solving a system of three linear equations with three unknowns (after Ogunkoya and Jenkins, 1993):

$$Q_t = Q_e + Q_u + Q_b \tag{3.16}$$

$$Q_t C_t = Q_e C_e + Q_u C_u + Q_b C_b \tag{3.17}$$

$$Q_t D_t = Q_e D_e + Q_u D_u + Q_b D_b \tag{3.18}$$

where Q is discharge, C and D are tracers, and the subscripts are, respectively,

$t =$ the total flow.
$e =$ component 1, the precipitation water component.
$u =$ component 2, the groundwater or surface water component.
$b =$ component 3, the groundwater component (e.g., base flow).

The following assumptions must be considered:

- The isotopic and chemical composition of selected spring components are representative.
- The isotopic or chemical composition of selected end members is significantly different.
- Water that represents the base-flow water component does not differ significantly in its isotopic or chemical composition.
- The base flow has a constant isotopic and chemical composition.
- Components 1 and 2 preserve the uniform isotopic and chemical composition throughout the studied period.

However, some temporal variability of components 1 and 2 may be observed, which can be accounted for by applying the method of incremental weighted averages, as in equation (3.11). It should consider data of the precipitation that fell in the catchment by the time defined separately by each hydrograph point. The main advantage of the discussed method is that the analysis of the particular hydrograph part is not influenced by precipitation that fell later on.

The storm discharge of the already discussed Hubelj spring was studied by the three-component hydrograph separation technique to examine the role of the upper unsaturated-zone water component in summer storm flow generation. The data of two tracers, the sampled water oxygen isotopic ($\delta^{18}O$) composition and the dissolved organic carbon (DOC) concentration, were applied. The sampling design allowed the Hubelj spring storm hydrograph's separation into three end members: base-flow, upper unsaturated-zone, and event water components (Trček, 2003, 2008). However, the results indicated that, according to the epikartic hypothesis, the event and upper unsaturated-zone water components should be combined into one component, representing the fast flow that arrives from the epikarst zone. This flow is called the *epiflow* (after Kiraly, Perrochet, and Rossier, 1995). The *epiflow* is defined as the fast flow of water that was prestored in the epikarst zone and event water. Both were concentrated at the base of the epikarst zone and later drained into the karst conduit network, where they could mix with the water of the sinkholes. The separation of the Hubelj spring hydrograph into the epiflow and base-flow components is presented in Figure 3–19. At the very beginning of the hydrograph concentration, there was an epiflow breakthrough, which contained no event water (the portion of this component was negative). This breakthrough resulted in the inversion of the hydraulic gradient: The pressure of the karst conduit network became higher than that of the rock blocks. Hence, the recharge process had been tied to the karst conduit network during the hydrograph concentration, and only water that was prestored in the high permeable parts of the saturated and lower unsaturated zones was discharged into the spring. After the breakthrough, the epiflow component became negative throughout

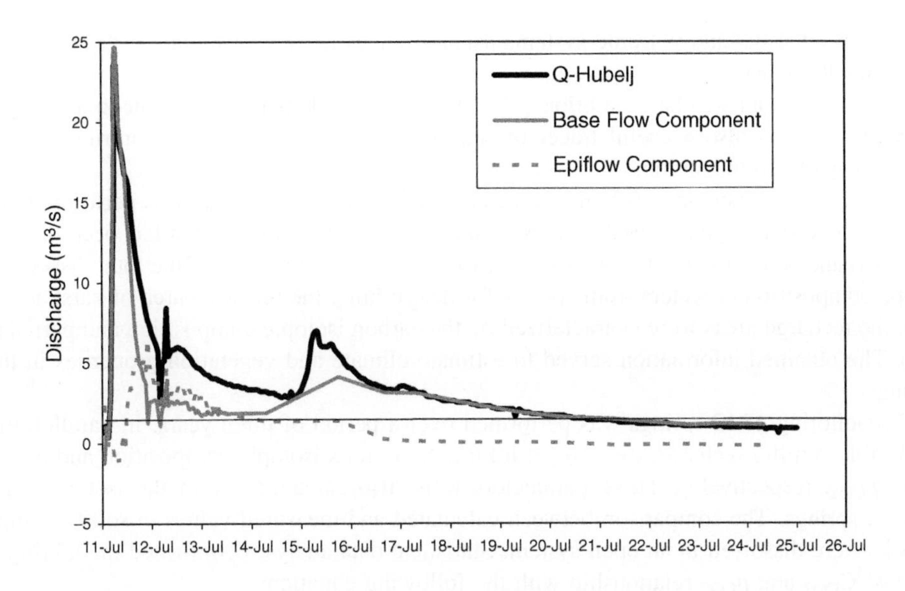

FIGURE 3–19 Hubelj spring hydrograph separation into base flow and epiflow components. (Trček, 2003.)

the hydrograph concentration, which reflects the water concentration in the epikarst zone. The epiflow component portion began to increase in the initial hydrograph recession, when the hydraulic pressure of the karst conduit network decreased. The maximum value, 84 percent, was registered at the end of the first storm cycle. The epiflow component contribution had mostly decreased later on (only the second storm cycle slightly interrupted the trend) and reached a negative value on July 21, when the inversion of the hydraulic gradient occurred again. Hence, there has been only the diffuse recharge process in the aquifer since then, and only the base flow recharged the Hubelj spring. Negative values indicate that the epikarst zone had to be internally recharged during this period. During the observed storm period, their average contributions were 41 and 59 percent, respectively.

Carbon isotopes The stable carbon isotope composition of dissolved inorganic carbon (^{13}C-DIC) is not a conservative tracer. Nevertheless, the delta ^{13}C values of DIC (δ^{13}C) can trace the carbon sources and reactions for numerous interreacting organic and inorganic species. The spring δ^{13}C composition gives information on the flow type and solute transport. It provides insight into the water geochemical evolution, rock types in the flow path surroundings, and the recharge area conditions (Hoefs, 1997; Kendall and McDonnell, 1998).

Hence, the isotope ratio $^{13}C/^{12}C$ is an important tool for quantifying water-rock interactions, identifying the proportion of different CO_2 sources in water, and determining initial geological settings of the groundwater recharge.

The spring ^{13}C isotopic composition can range quite widely. The main sources of carbon dissolved in groundwater are soil CO_2, CO_2 of geogenic origin or magmatic CO_2 (from deep crustal or mantle sources), carbonate minerals, organic matter in soils and rocks, fluid inclusions, and methane. Each of these sources has a different carbon isotopic composition and contributes to totally dissolved carbon in various proportions.

It is well known that the δ^{13}C is about 0‰ in marine carbonate rocks, −25‰ in the soil CO_2 (similar to plants), and −7‰ in the atmospheric CO_2, respectively (Kendall and McDonnell, 1998). However, the evaporate carbonates could reach the δ^{13}C values up to +10‰; the δ^{13}C of CO_2 in tropical soils may be more positive, about −11‰; the δ^{13}C of geothermal methane is about −30‰; while the δ^{13}C of the CO_2

originating from geothermal and volcanic systems usually ranges between −8 and −3‰ (Hoefs, 1997; Kendall and McDonnell, 1998).

In regions with seasonal weather conditions, the higher $\delta^{13}C$ values usually occur in a spring during the winter, hence the $\delta^{13}C$ is also a useful tracer of seasonal and discharge-related contributions of different hydrological flow paths to the spring.

The study of dissolved inorganic carbon isotope composition of karst springs was performed in the framework of the karst hydrogeological investigations in southwestern Slovenia, which has been discussed before (Trček, 1997; Urbanc et al., 1997). It was aimed at assessing the applicability of the total dissolved inorganic carbon isotope composition of waters from springs for determining the recharge area of karst aquifers. Initial conditions in the recharge areas were characterized by the carbon isotopic composition and partial pressure of the soil CO_2. The obtained information served to estimate climate and vegetation properties in the recharge areas of springs.

The $\delta^{13}C$ monitoring of 12 springs was performed over a period of three years. In parallel, the soil CO_2 was sampled in the aquifer recharge areas to get information on its isotopic composition and partial pressure ($\delta^{13}C_{CO2}$ and p_{CO2}, respectively). These parameters were also calculated from the isotopic and chemical compositions of springs. The comparison between calculated and measured values in soil CO_2 indicated that carbonate rocks were dissolved in an open system. Statistical data analyses confirmed the mixing model that describes the $\delta^{13}C_{CO2}$ and p_{CO2} relationship with the following equations:

$$\delta^{13}C_{CO_2} = -23 + 0.45\frac{1}{p_{CO_2}} \tag{3.19}$$

$$\delta^{13}C_{CO_2} = -24 + 0.48\frac{1}{p_{CO_2}} \tag{3.20}$$

It followed from the model that the $\delta^{13}C$ of the soil CO_2 varied due to mixing with the atmospheric carbon, because the $\delta^{13}C$ of the biogenic part remained constant (between −23 and −24‰). Equations (3.19) and (3.20) describe recharge areas with prevailing forest and grasslands, respectively. The differences in the isotopic composition of biogenic carbon, therefore, presented a possibility of drawing conclusions about the vegetation properties of the spring recharge areas (Trček, 1997).

On the basis of (1) correlations between measured p_{CO2} and soil temperature and between the measured $\delta^{13}C_{CO2}$ and soil temperature, and (2) influences of various vegetation types on the measured $\delta^{13}C_{CO2}$ and p_{CO2}, an attempt was made to approximately estimate the mean values of soil temperature in recharge areas of individual springs. Calculations were made by two methods:

1. Average pedotemperatures were calculated on the basis of CO_2 partial pressure, obtained from the alkalinity of individual springs and taking into account the correlation between the measured p_{CO2} and soil temperature for an approximate vegetation structure of the recharge areas.
2. Average temperatures were calculated from the DIC isotopic composition, taking into account the correlation between the measured $\delta^{13}C_{CO2}$ and soil temperature for an approximate vegetation structure of the recharge areas.

The estimated pedotemperatures are presented in Figure 3–20.

Sulfur isotopes Like carbon, sulfur is an abundant element in nature. Because of its reactive nature, many processes affect its global cycling and isotopic composition: weathering of rocks rich in sulfur, deposition of evaporates, sulfate from sea spray, and anthropogenic fluxes from burning fossil fuels. The sulfur isotopic composition ($\delta^{34}S$) varies between −5 and 25‰ in atmospheric sulfur and −10 and 35‰ in the lithospheric sulfur, respectively (Krouse and Mayer, 2000). Lithospheric sulfur originates mainly from the weathering of

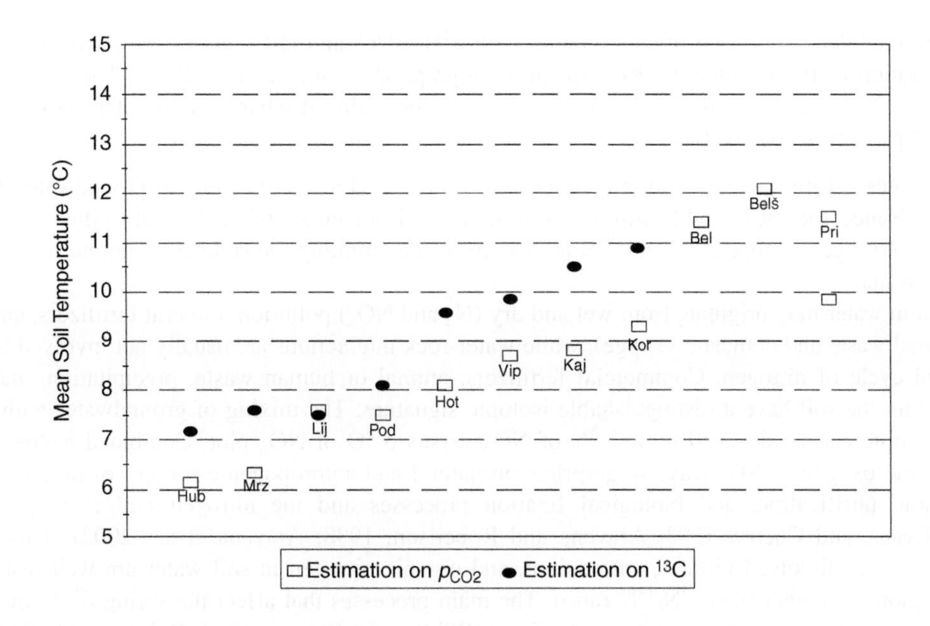

FIGURE 3–20 Estimated soil temperatures of the spring recharge areas, calculated on the basis of $\delta^{13}C$ and partial pressure of soil CO_2.

metamorphic and sedimentary rocks (marine evaporates such as gypsum and anhydrite). The $\delta^{34}S$ is about 21‰ in marine sulfur, while it ranges between −10 and 35‰ in groundwater (Krouse and Mayer, 2000).

Groundwater $\delta^{34}S$ values depend on the nature of the sulfur inputs to the water. It is found mainly in the form of sulfates and sulfides in groundwater. The main source of sulfate is the dissolution of gypsum and anhydrite, but some dissolved organic sulfur, elemental sulfur, and mineral sulfur might also be present. Hence, the stable isotopic ratio of ^{32}S and ^{34}S isotopes indicates marine, evaporitic, and volcanic sources of dissolved sulfate in a spring groundwater. The $\delta^{34}S$ values of H_2S and SO_4 of the geothermal discharge reflect their origin: The $\delta^{34}S$ of the mantle sulfur that usually originates from basalts is 0‰ and ranges between +2 and +6‰ in the crust sulfur (Clark and Fritz, 1997).

Sulfur isotopes of springs are usually used to trace natural and anthropogenic sources of sulfur, especially in agricultural catchments. A general problem with sulfur isotopes is that isotopic values of a certain compound may vary a lot.

The studies of water chemistry and the $\delta^{34}S$ values of SO_4 suggested that two flow systems exist in the karst terrain of southern Indiana: (1) a shallow flow system, dominated by surface water that infiltrates through the soil profile to the epikarst, which then flows laterally to major vertical joints and fractures to the springs, and (2) a deeper flow system driven by topographic cells forcing water downward, where it flows through and dissolves evaporate minerals from the St. Louis limestone (Lee and Krothe, 2001). The water then flows upward and discharges at topographic lows. Sulfur isotopic values $\delta^{34}S$ of SO_4 for springs in the study area ranged from +1.1 to +22.1‰, while those of H_2S ranged from −0.2 to −38.4‰. The freshwater springs had lighter values, approximately 10‰, of $\delta^{34}S$ of SO_4, and the mineral springs approximately 14‰. The latter values are similar to values of the local St. Louis gypsum (14–16‰). During low-flow conditions, the Orangeville Rise, a major freshwater spring, had the $\delta^{34}S$ of SO_4 of approximately +16‰. However, the $\delta^{34}S$ of SO_4 had decreased to +9‰ at high flow, which suggested the mixing of shallow epikarst water and deeper mineral waters.

The presented data were used in combination with δ^2H, $\delta^{13}C$, and DIC data as tracers for the four-component separation of the Orangeville Rise storm hydrograph (Lee and Krothe, 2001). The contributions of new, soil, epikarstic, and base-flow water components to the spring discharge were calculated as 10.6, 3.1, 52.3, and 34 percent, respectively.

Nitrogen isotopes Anthropogenic nitrogen inputs led to increased loads of nitrate in groundwater during the last decades. Hence, the use of stable nitrogen isotopes in environmental and ecological studies also increased considerably. Nitrogen isotopes ^{15}N and ^{14}N are useful for determining the sources of nitrate in groundwater and surface water.

Nitrogen in water may originate from wet and dry (N_2 and NO_x) pollution, mineral fertilizers, and organic matter (animal waste and domestic sewage), while water-rock interactions are usually not involved in the biogeochemical cycle of nitrogen. Commercial fertilizers, animal or human waste, precipitation, and organic nitrogen within the soil have a distinguishable isotopic signature. The mixing of groundwater with different sources of nitrate can be detected by a $\delta^{15}N$ of NO_3 versus $\delta^{18}O$ of NO_3 plot. Combined isotope analyses of nitrogen and oxygen in NO_3 leave fingerprints on natural and anthropogenic sources of nitrate, microbial denitrification, nitrification, and biological fixation processes and the nitrogen budget in groundwater (Aravena, Evans, and Cherry, 1993; Aravena and Robertson, 1998; Aravena et al., 2002). Therefore, the $\delta^{15}N$ values of the dissolved nitrates, ammonium, and organic nitrogen in soil water are well distinguished from one region to another (i.e., $^{15}N/^{14}N$ ratio). The main processes that affect the spring $\delta^{15}N$ are denitrification, temporary storage, and transport mechanisms (Böhlke and Denver, 1995; Böhlke et al., 2002).

Many studies demonstrated that effects of denitrification and assimilation can be distinguished with the use of $\delta^{15}N$ analyses combined with $\delta^{18}O$ analysis (e.g., Böttcher et al., 1990a, 1990b; Panno et al., 2001, 2008; Leis, 2002). If plant uptake alone is responsible for NO_3^- remediation, the isotopic composition of the remaining NO_3^- remains unchanged. If both denitrification and assimilation occur, the isotopic composition of the residual nitrate is enriched and the overlying plants reflect the isotopic composition of the NO_3^- source. The isotopic composition of the plants remains the same and the water becomes more enriched if denitrification is the only process occurring.

The studies of $\delta^{15}N$ and $\delta^{18}O$ of NO_3 were performed in the Leibnitzer Field in the south part of Styria, where the source of nitrate contamination in the aquifer is attributed to local, long-term agricultural land-use practices, such as spreading large amounts of liquid manure (mainly pig manure) on the soil (Leis, 2002). The main research goal was to determine which action should be taken to reduce nitrate contamination of the groundwater. Hence, the sources and fate of nitrate were investigated. The results demonstrated that the isotopic composition of nitrate is not only a powerful tool to determine its sources but can also provide hints about nitrogen transformation processes, such as nitrification and denitrification.

The sources of nitrate (NO_3^-) in groundwater of the shallow karst aquifer in southwestern Illinois's sinkhole plain were investigated using chemical and isotopic techniques (Panno et al., 2001). Water samples from 10 relatively large karst springs were collected during four different seasons and analyzed for inorganic constituents, dissolved organic carbon, atrazine, and $\delta^{15}N$ and $\delta^{18}O$ of the NO_3^- ions. The isotopic data were most definitive and suggested that the sources of NO_3^- in springwater are dominated by N fertilizer with some possible influence of atmospheric NO_3^- and, to a much lesser extent, human or animal waste. Differences in the isotopic composition of NO_3^- and some of the chemical characteristics were observed during the four consecutive seasons in which springwater samples were collected. Isotopic values for $\delta^{15}N$ and $\delta^{18}O$ of the NO_3^- ranged from 3.2 to 19.1‰ and from 7.2 to 18.7‰, respectively (Figure 3–21). The trend of $\delta^{15}N$ and $\delta^{18}O$ data for NO_3^- also indicated that a significant degree of denitrification occurs in the shallow karst hydrologic system (within the soil zone, the epikarst, and the shallow karst aquifer) prior to discharge into springs.

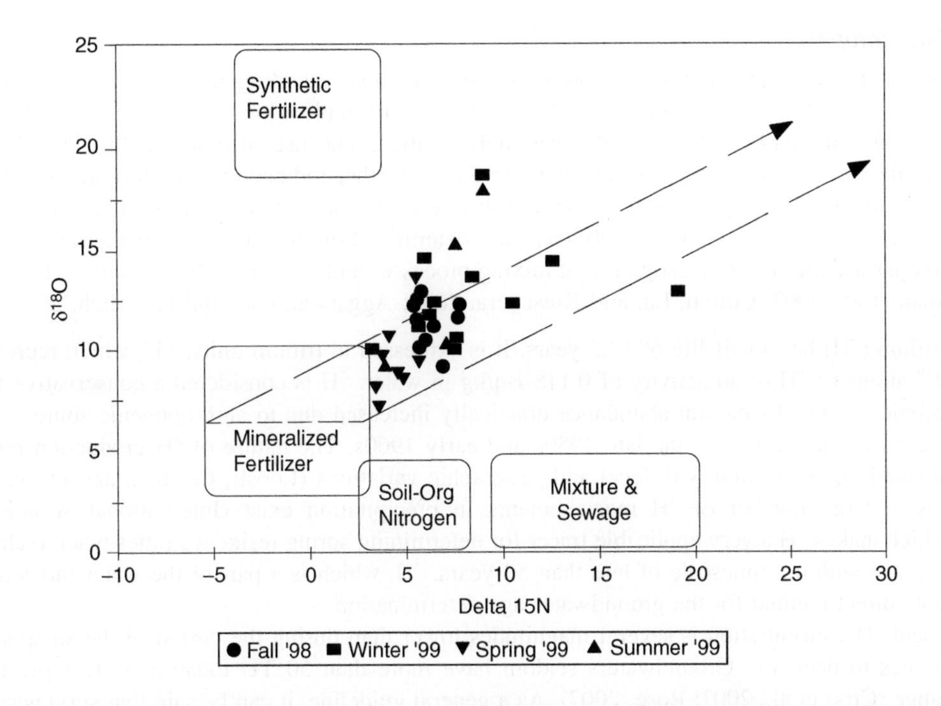

FIGURE 3–21 The isotopic compositions of NO_3^- from different potential sources and the ten springs sampled during four consecutive seasons, and lines (arrows) showing trends for denitrification processes. The horizontal dashed line near the center of the box marked "mineralized fertilizer" represents an estimate of the original isotopic composition for fertilizers used in the sinkhole plain. (Panno et al., 2001.)

Chlorine, boron, strontium, and lead isotopes The stable isotopes of chlorine, boron, strontium, and lead do not have a long tradition in studies of the spring hydrogeology; therefore, they are mentioned only briefly in this book.

The isotopic ratio $^{87}Sr/^{86}Sr$ is a valuable tracer for mixing and source studies of mineralized groundwater. Detailed analysis of the isotopic composition of Sr in stream waters, organic and mineral soil horizons, biomass, and atmospheric input at a handful of sites has provided insight into the cycling of Sr (Graustein, 1989; Aberg, Jacks, and Hamilton, 1989; Miller, Blum, and Friedland, 1993; Bailey et al., 1996; Rose and Fullagar, 2005).

Recently, the isotopic ratios $^{37}Cl/^{35}Cl$ and $^{10}B/^{11}B$ have been used in pollution studies of groundwater (Eggenkamp, 2004; Annable et al., 2007), while Pb isotopes were not generally recognized as useful tracers in the spring catchments (Erel, Morgan, and Patterson, 1991; Connelly and Thrane, 2005).

Stable isotopes in geochemical modeling Reactive solute isotopes, such as ^{13}C, ^{34}S, ^{15}N, and ^{87}Sr, can be used with chemical data in geochemical mass balance and reaction path models, such as BALANCE, PHREEQE, and NETPATH to study the geochemical processes, test hypotheses on hydrodynamic and geochemical mechanisms, and eliminate possible reaction paths. In chemical reaction modeling, usually several reaction models can be found that satisfy the data. For each model reaction path, calculations are used to predict the chemical and isotopic composition of the aqueous phase as well as the amounts of minerals dissolving or precipitated along a flow path.

Radioactive isotopes

Radioactive isotopes represent another important research method in hydrogeological investigation. As was mentioned at the beginning of Section 3.2.3, their main field of application is in groundwater dating. Only tritium (^{3}H) and carbon-14 (^{14}C) have been routinely applied. The use of chlorine-36 (^{36}Cl) has rapidly increased, while the sampling, analysis, and interpretation of other radioactive isotopes are so complicated that they are studied only in specialized research laboratories. In these studies, old groundwater is dated by ^{81}Kr and ^{129}I, groundwater ages, up to 1000 years, are determined on the basis of ^{39}Ar measurements, while uranium isotopes are useful for investigation of mixing processes and also groundwater dating (Pearson et al., 1991; Lehman et al., 2003; Collon, Lu, and Kutschera, 2004; Aggarwal, Gat, and Froehlich, 2006).

Tritium Tritium (^{3}H) has a half-life of 12.3 years. It is expressed in tritium units, TU, which represent 1 ^{3}H atom in 10^{18} atoms of ^{1}H or an activity of 0.118 Bq/kg in water. ^{3}H is considered a conservative tracer for most hydrologic studies. Its natural abundance drastically increased due to anthropogenic sources produced during nuclear weapons testing in the late 1950s and early 1960s. The nature of ^{3}H production results in a very complicated input function with local and geographic variations (Loosli, Lehman, and Däppen, 1991). Nevertheless, a large number of ^{3}H measurements in precipitation exist (International Atomic Energy Agency), which makes ^{3}H a very applicable tracer for determining spring residence times when recharge processes took place within a timescale of less than 50 years. ^{3}H, which is a part of the water molecule, is the only available direct method for the groundwater age determination.

The present ^{3}H concentration is several magnitudes lower than during the period of the weapons testing and it continues to decrease. Groundwaters seldom have more than 50 TU today and are typically in the 5–10 TU range (Criss et al., 2007; Rose, 2007). As a general guideline, it can be said that springs containing over 10 TU contain a thermonuclear test contribution while 20 TU or more would suggest a component of water recharging since 1961 (International Atomic Energy Agency, 1983).

The study of the ^{3}H composition of karst springs was performed in the framework of the karst hydrogeological investigations in southwestern Slovenia (Stichler et al., 1997). The ^{3}H values of monthly samples from the Vipava spring corresponded to the actual ^{3}H content in precipitation, which indicates a relatively short residence time of the spring groundwater (Figure 3–22). There was no significant increase or decrease of

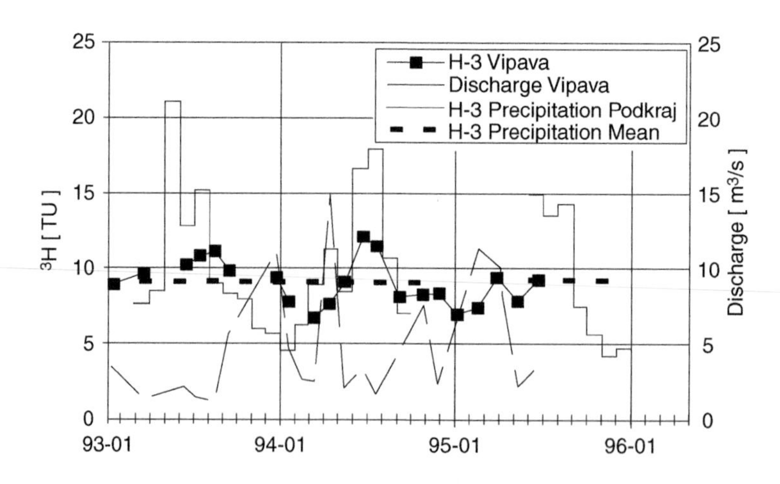

FIGURE 3–22 Discharge and ^{3}H content of monthly samples from the Vipava spring together with the monthly and weighted mean ^{3}H-values of precipitation at the meteorological station Podkraj. (Stichler et al., 1997.)

the ^{3}H content, not even during low-water periods. Only seasonal variations of the ^{3}H content in precipitation were reflected in the spring. A mean transit time of about 0.4 years was estimated from the comparison of the ^{3}H amplitudes of precipitation and spring.

As was mentioned at the beginning of Section 3.2.2, the application of ^{3}H for groundwater age determination depends on knowledge of the ^{3}H input concentration, which in fact may be poorly known or vary with time. ^{3}H and its stable daughter product ^{3}He can be measured together in spring samples to solve this problem. Studying the relation between ^{3}H and ^{3}He offers the possibility of determining groundwater ages without knowing the ^{3}H input function. The original ^{3}H input concentration can be reconstructed from the amount of ^{3}He in solution due to tritium decay when no helium has been lost from the solution (Solomon and Sudicky, 1991; Solomon et al., 1993). However, the ^{3}H/^{3}He method also presents problems, since the total ^{3}He in groundwater comes from a variety of sources. In addition to the tritiogenic origin (coming from decay of ^{3}H), ^{3}He may be also of terrigenic origin (from in situ production) as well as from the original equilibration of the water with the atmospheric gases. The last contribution can be estimated from the Ne concentration (which cannot be of terrigenic origin), while a possible terrigenic contribution is to be estimated from the concentration of ^{4}He and corrected for the atmospheric component (Schlosser et al., 1989).

Carbon isotope Carbon-14 (^{14}C) originates from neutron interactions, the decay of certain radium isotopes, alpha particle reactions, and anthropogenic sources. It has a half-life of 5730 $\pm$ 30 years. When the appropriate field measurements are collected and appropriate corrections made, ^{14}C data can provide insight into groundwater flow paths, recharge areas, and sources of recharge.

The ^{14}C atmospheric source function is well known, in contrast to the ^{14}C concentration of CO_2, which equilibrates with the water that recharges aquifers. Hence, numerous hydrochemical models have been developed to determine the initial concentrations of ^{14}C. These are based on the δ^{13}C value of groundwater DIC needed to calibrate the groundwater ^{14}C timescale, due to water-rock interactions (Mook, 1980). As was demonstrated previously, the δ^{13}C of DIC reflects groundwater chemical interaction with the aquifer rock in the open or closed dissolution systems.

The geochemical models mentioned previously include the whole groundwater chemistry (Plummer, Prestemon, and Parkhurst, 1994), so they were successfully applied to produce reliable absolute groundwater ages (e.g., Geyh, 1992, 2000). However, it should be taken into account that the results of only water samples from the same flow path can be used.

Chlorine isotope Chlorine-36 (^{36}Cl) has a half life of 3×10^5 years; therefore, it is particularly applicable to date groundwaters that are from 500,000 to a million years old.

^{36}Cl is produced primarily in the atmosphere, either by thermonuclear explosions or, in small quantities, by isotope spallation by cosmic rays converting ^{40}Ar to ^{36}Cl (Schaeffer, Thompson, and Lark, 1960; Cresswell and Bonotto, 2008). Some ^{36}Cl can be also produced in the shallow unsaturated zone from cosmic ray secondary neutrons interacting with stable ^{35}Cl (this source is mainly negligible; Bentley et al., 1982; Kaufmann et al., 1984). Other ^{36}Cl sources refer to a solution of matrix chloride in environments where in situ production of ^{36}Cl is significant. The same methodology is applied to neutron activation of stable ^{35}Cl in oceans as a result of nuclear weapons tests in the 1950s and 1960s and anthropogenic inputs from nuclear-fuel reprocessing and nuclear-power generation facilities (Cecil et al., 1992).

The chloride ion exists in most natural waters in varying concentrations, due to the dissociation of sodium chloride. Chloride ions do not adsorb onto these silicate surfaces, and therefore, they move at approximately the same rate as groundwater (Bentley, Phillips, and Davis, 1986).

Anthropogenic sources of ^{36}Cl due to nuclear weapons tests can be applied to identify a modern recharge. High ^{36}Cl concentrations in a spring indicate that recharge occurred recently.

Extensive ^{36}Cl studies were performed in the framework of a case study in northern Switzerland (Lehmann and Loosli, 1991). The results indicate that all water samples originated mainly from deep subsurface sources and thermonuclear ^{36}Cl production near the surface was neglected. The dominant path for ^{36}Cl production was neutron capture by ^{35}Cl. This process occurred on chloride dissolution in groundwater and chlorine atoms within the rock matrix. After approximately 1.5 million years, the $^{36}Cl/Cl$ ratio could reach equilibrium. It was demonstrated that its magnitude was proportional to the local neutron flux. The general result of the neutron flux calculations was that low $^{36}Cl/Cl$ ratios of about 10×10^{-15} are expected in carbonate rock, medium values in the range of 20 to 30×10^{-15} in sandstone, and high values of approximately 40 to 50×10^{-15} in crystalline rocks.

REFERENCES

Aberg, G., Jacks, G., Hamilton, P.J., 1989. Weathering rates and $^{87}Sr/^{86}Sr$ ratios: An isotopic approach. J. Hydrol. 109, 65–78.

Aggarwal, P.K., Gat, J., Froehlich, K.F.O., 2006. Isotopes in the water cycle: Past, present and future of a developing science. Springer, Dordrecht, the Netherlands.

Annable, W.K., Frape, S.K., Shouakar-Stash, O., Shanoff, T., Drimmie, R.J., Harvey, F.E., 2007. ^{37}Cl, ^{15}N, ^{13}C isotopic analysis of common agro-chemicals for identifying non-point source agricultural contaminants. Appl. Geochem. 22 (7), 1530–1536.

Aravena, R., Evans, M.L., Cherry, J.A., 1993. Stable isotopes of oxygen and nitrogen in source identification of nitrate from septic tanks. Ground Water 31, 180–186.

Aravena, R., Roberston, W., 1998. The use of multiple isotope tracers to evaluate denitrification in groundwater: A study in a large septic system plume. Ground Water 36, 975–982.

Aravena, R., Brown, C., Schiff, S.L., Elgood, R., 2002. Use of geochemical and isotope tools to evaluate nitrate attenuation in riparian wetlands in agricultural landscape in southern Ontario. Geochim. Cosmochim. Acta 66, A25.

Bailey, S.W., Hornbeck, J.W., Driscoll, C.T., Gaudette, H.E., 1996. Calcium inputs and transport in a base-poor forest ecosystem as interpreted by Sr isotopes. Water Resour. Res. 32, 707–719.

Bauer, F., Benischke, R., Bub, F.P., Burger, A., Dombrowski, H., Gospodarič, R., et al., 1980. Karsthydrologische untersuchungen mit natprlichen und künstlichen tracern im Neuenburger Jura (Schweiz). Steir. Beiträge zur Hydrogeology 32, 5–100.

Bazemore, D.E., Eshleman, K.N., Hollenbeck, K.J., 1994. The role of soil water in storm flow generation in a forested headwater catchment: Synthesis of natural tracer and hydrometric evidence. J. Hydrol. 162, 47–75.

Behrens, H., Benischke, R., Bricelj, M., Harum, T., Käss, W., Kosi, G., et al., 1992. Investigation with natural and artificial tracers in the karst aquifer of the Lurbach system (Peggau-Tanneben-Semriach, Austria). In: Association of Tracer Hydrology Transport phenomena in different aquifers (Investigations 1987–1992), Steir, Beiträge zur Hydrogeologie vol. 43. pp. 9–158.

Behrens, H., Hötzl, H., Käss, W., 2001. Application of artificial tracers in comparative tracer experiments. In: Association of Tracer Hydrology Tracers studies in the unsaturated zone and groundwater (Investigations 1996–2001) Beiträge zur Hydrogeologie vol. 52. 105–117.

Benischke, R., Goldscheider, N., Smart, C., 2007. Tracer techniques. In: Goldscheider, N., Drew, D. (Eds.), Methods in karst hydrogeology. International contribution to hydrogeology 26. Taylor and Francis, London, pp. 147–170.

Bentley, H.W., Phillips, F.M., Davis, S.N., Gifford, S., Elmore, D., Tubbs, L.E., et al., 1982. Thermonuclear ^{36}Cl pulse in natural water. Nature 300, 737–740.

Bentley, H.W., Phillips, F.M., Davis, S.N., 1986. ^{36}Cl in the terrestrial environment. In: Fritz, P., Fontes, J.C. (Eds.), Handbook of environmental geochemistry, 2b. Elsevier, New York, pp. 422–475.

Böhlke, J.K., Denver, J.M., 1995. Combined use of ground-water dating, chemical, and isotopic analyses to resolve the history and fate of nitrate contamination in two agricultural watersheds, Atlantic coastal plain, Maryland. Water Resources Researches 31, 2319–2339.

Böhlke, J.K., Wanty, R., Tuttle, M., Delin, G., Landon, M., 2002. Denitrification in the recharge area and discharge area of a transient agricultural nitrate plume in a glacial outwash sand aquifer, Minnesota. Water Resour. Res. 38 (7), 10.1–10.26.

Böttcher, J., Strebel, O., Voerkelius, S., Schmidt, H.L., 1990. Using isotope fractionation of nitrate nitrogen and nitrate oxygen for evaluation of denitrification in a sandy aquifer. J. Hydrol. 114, 413–424.

Böttcher, A., Imbeck, R., Morgante, A., Ertl, G., 1990. Nonadiabatic surface reaction: Mechanism of electron emission in the $Cs+O_2$ system. Phys. Rev. Lett. 65, 2035–2037.

Brown, V.A., McDonnell, J.J., Burns, D.A., Kendall, C., 1999. The role of event water, a rapid shallow flow component, and catchment size in summer storm flow. J. Hydrol. 217, 171–190.

Cecil, L.D., Beasley, T.M., Pittman, J.R., Michel, R.L., Kubik, P.W., Sharma, P., et al., 1992. Water infiltration rates in the unsaturated zone at the Idaho National Engineering Laboratory estimated from chlorine-36 and tritium profiles, and neutron logging. In: Kharaka, Y.K., Maest, A.S. (Eds.), Water-rock interaction. Balkema, Rotterdam, pp. 709–714.

Chow, V.T., Maidment, D.R., Mays, L.W., 1988. Applied hydrology. McGraw-Hill, New York.

Clark, I.D., Fritz, P., 1997. Environmental isotopes in hydrogeology. Lewis Publishers, New York.

Collon, P., Lu, Z.T., Kutschera, W., 2004. Tracing noble gas radionuclides in the environment. Annual Review of Nuclear and Particle Science 53, 39–67.

Connelly, J.N., Thrane, K., 2005. Rapid determination of Pb isotopes to define Precambrian allochthonous domains: An example from West Greenland. Geology 33 (12), 953–956.

COST Action 65, 1995. Final report on Hydrogeological aspects of groundwater protection in karstic areas. European Commission, Luxembourg.

Craig, H., 1961. Isotopic variations in meteoric waters. Science 133, 1702–1703.

Cresswell, R.G., Bonotto, D.M., 2008. Some possible evolutionary scenarios suggested by ^{36}Cl measurements in Guarani aquifer groundwaters. Appl. Radiat. Isot. 66 (8), 1160–1174.

Criss, R.E., Davisson, M.L., Surbeck, H., Winston, W.E., 2007. Isotopic Techniques. In: Goldscheider, N., Drew, D. (Eds.), Methods in karst hydrogeology. International contribution to hydrogeology 26. Taylor and Francis, London, pp. 123–145.

Dincer, T., Payne, B.R., Florkowski, T., Martinec, J., Tongiorgi, E., 1970. Snow-melt runoff from measurements of tritium and oxygen-18. Water Resour. Res. 6, 110–124.

Eggenkamp, H.G.M., 2004. Summary of methods for determining the stable isotope composition of chlorine and bromine in natural materials. In: De Groot, P.A. (Ed.), Handbook of stable isotope analytical techniques, vol. I. Elsevier, Amsterdam, pp. 604–622.

Erel, Y., Morgan, J.J., Patterson, C.C., 1991. Natural levels of Pb and Cd in a remote mountain stream. Geochim. Cosmochim. Acta 55 (3), 707–719.

Fank, J., 1996. Hydrogeologische Rahmenbedingungen, messtechnische Voraussetzungen und Ergebnisse der bisherigen Untersuchungen zur Frage Stoffdynamik. In: Proceedings Sixth Lysimetertagung, Lysimeter im Dienste des Grundwasserschutzes. BAL-Bundesanstalt fur alpenlandische Landwirtschaft, Gumpenstein, pp. 59–64.

Fank, J., Stichler, W., Zojer, H., 1998. Die Schneeschmelze 1996 als ^{18}O-Tracerversuch an der Lysimeteranlage in Wagna. In: Proceedings Workshop Bestimmung der Sickerwassergeschwindigkeit in Lysimetern. GSF-forschungszentrum fur Umwelt und Gesundheit, Munich, pp. 11–19.

Field, M.S., 2003. A review of some tracer-test design equations for tracer-mass estimation and sample-collection frequency. Env. Geol. 43, 867–881.

Frederickson, G.C., Criss, R.E., 1999. Isotope hydrology and time constants of the unimpounded Meramec River basin, Missouri. Chem. Geol. 157, 303–317.

Fritz, P., Cherry, J.A., Weyer, K.V., Sklash, M.G., 1976. Runoff analyses using environmental isotope and major ions. In: Interpretation of environmental isotope and hydrochemical data in groundwater hydrology. International Atomic Energy Agency, Vienna, pp. 111–130.

Geyh, M.A., 1992. ^{14}C time scale of groundwater correction and linearity. In: International Atomic Energy Agency (Ed:) Isotope techniques in water resources development. International Atomic Energy Agency, Vienna, pp. 167–177.

Geyh, M.A., 2000. An overview of ^{14}C analysis in the study of groundwater. Radiocarbon 42, 99–114.

Gospodarič, R., Habič, P., 1976. Underground water tracing—Investigations in Slovenia, 1972–1975. Institute for Karst Research SAZU, Ljubljana, Slovenia.

Graustein, W.C., 1989. ^{87}Sr/^{86}Sr ratios measure the sources and flow of strontium in terrestrial ecosystems. In: Rundel, P.W., Ehleringer, J.R., Nagy, K.A. (Eds.), Stable isotopes in ecological research. Springer-Verlag, New York, pp. 491–511.

Habič, P., 1976a. Geomorphologic and hydrographic characteristics. In: Gospodarič, R., Habič, P. (Eds.), Underground water tracing—Investigations in Slovenia, 1972–1975. Institute for Karst Research SAZU, Postojna, Slovenia, pp. 12–27.

Habič, P., 1976b. Karst hydrographic evaluations. In: Gospodarič, R., Habič, P. (Eds.), Underground Water Tracing—Investigations in Slovenia, 1972–1975. Institute for Karst Research SAZU, Ljubljana, Slovenia, pp. 197–213.

Himmelsbach, T., Hötzl, H., Käss, W., Leibundgu, C., Maloszewski, P., Meyer, T., et al., 1992. Fractured rocks—Test site Lindau/Southern Black Forest (Germany). In: Association of Tracer Hydrology (Ed:), Transport phenomena in different aquifers (Investigations 1987–1992), Steir, Beiträge zur Hydrogeologie, vol. 43. pp. 159–228.

Hoefs, J., 1997. Stable isotope geochemistry, fourth ed. Springer-Verlag, Berlin-Heidelberg.

International Atomic Energy Agency, 1983. Isotope techniques in the hydrogeological assessment of potencial sites for the disposal of high-level radioactive wastes. Technical reports series 228, International Atomic Energy Agency, Vienna, p. 151.

International Atomic Energy Agency/World Meteorological Organization, 2006. Global network of isotopes in precipitation. The GNIP database. Available at: http://www-naweb.iaea.org/napc/ih/GNIP/IHS_GNIP.html.
Käss, W., 1998. Tracing technique in geohydrology. Balkema, Rotterdam.
Kaufmann, R., Long, A., Bentley, H.W., Davis, S.N., 1984. Natural chlorine isotope variations. Nature 309, 338–340.
Kendall, C., McDonnell, J.J., 1998. Isotope tracers in catchment hydrology. Elsevier, Amsterdam.
Kiraly, L., Perrochet, P., Rossier, Y., 1995. Effect of the epikarst on the hydrograph of karst springs: A numerical approach. Bulletin d'Hydrogéologie 14, 199–220.
Kogovšek, J., Petrič, M., 2004. Advantages of longer-term tracing: Three case studies from Slovenia. Env. Geol. 47, 76–83.
Kogovšek, J., Petrič, M., 2006. Tracer test on the Mala gora landfill near Ribnica in south-eastern Slovenia. Acta Carsologica 35 (2), 91–101.
Kogovšek, J., Petrič, M., 2007. Directions and dynamics of flow and transport of contaminants from the landfill near Sežana (SW Slovenia). Acta Carsologica 36 (3), 413–424.
Kogovšek, J., Knez, M., Mihevc, A., Petrič, M., Slabe, T., Šebela, S., 1999. Military training area in Kras (Slovenia). Env. Geol. 38 (1), 69–76.
Kranjc, A., 1997. Karst hydrogeological investigations in south-western Slovenia. Acta Carsologica 26 (1), 260–353.
Krouse, H.R., Mayer, B., 2000. Sulfur and oxygen isotopes in sulphate. In: Cook, P., Herczeg, A.L. (Eds.), Environmental tracers in subsurface hydrology. Kluwer Academic Publishers, Rotterdam, pp. 195–230.
Lee, E.S., Krothe, N.C., 2001. A four-component mixing model for water in a karst terrain in south-central Indiana, USA, using solute concentration and stable isotopes as tracers. Chem. Geol. 179, 129–143.
Lehman, B.E., Loosli, H.H., 1991. Isotopes formed by underground production. In: Pearson, F.J., Balderer, W., Loosli, H.H., Lehmann, B.E., Matter, A., Peters, T., et al. (Eds.), Applied isotope hydrogeology—A case study in Northern Switzerland. Studies in environmental science 43. Elsevier, Amsterdam, pp. 239–266.
Lehman, B.E., Love, A., Purtschert, R., Collon, P., Loosli, H.H., Kutschera, W., et al., 2003. A comparison of groundwater dating with ^{81}Kr, ^{36}Cl and ^{4}He in four wells of the Great Artesian Basin, Australia. Earth and Planetary Science Letters 211 (3–4), 237–250.
Leis, A., 2002. Use of $\delta^{15}N$ and $\delta^{18}O$ isotope ratios to identify sources of nitrate in the unsaturated zone. In: Berichte des Institutes für Erdwissenschaften der Karl-Franzens-Universität Graz 6. Graz, Austria, pp. 19–21.
Loosli, H.H., Lehman, B.E., Däppen, G., 1991. Dating by radionuclides. In: Pearson, F.J., Balderer, W., Loosli, H.H., Lehmann, B.E., Matter, A., Peters, T., et al. (Eds.), Applied isotope hydrogeology—A case study in Northern Switzerland. Studies in environmental science 43. Elsevier, Amsterdam, pp. 153–175.
Maloszewski, P., Zuber, A., 1982. Determining the turnover time of groundwater systems with the aid of environmental tracers, part 1, Models and their applicability. J. Hydrol. 57, 201–231.
Maloszewski, P., Moser, H., Stichler, W., Bertleff, B., Hedin, K., 1990. Modelling of groundwater pollution by riverbank infiltration using oxygen-18 data. In: Groundwater monitoring and management, Proceedings of Dresden Symposium, March 1987. Publication 173, International Association of Hydrological Sciences, Wallingford, Oxfordshire, pp. 153–161.
Maloszewski, P., Zuber, A., 1996. Lumped parameter models for the interpretation of environmental tracer data. In: Manual on mathematical models in isotope hydrology. International Atomic Energy Agency, Vienna, pp. 9–58.
Martinec, J., Siegenthaler, H., Oescherger, H., Tongiorgi, E., 1974. New insight into the runoff mechanisms by environmental isotopes. In: Proceedings of the Symposium on Isotope Techniques in Groundwater Hydrology. International Atomic Energy Agency, Vienna, pp. 129–143.
Miller, E.K., Blum, J.D., Friedland, A.J., 1993. Determination of soil exchangeable-cation loss and weathering rates using Sr isotopes. Nature 362, 438–441.
Mook, W.G., 1980. Carbon-14 in hydrogeological studies. In: Fritz, P., Fontes, J.C. (Eds.), Handbook of environmental isotope geochemistry. Elsevier, Amsterdam, pp. 49–74.
Mook, W.G., 2000. Environmental isotopes in the hydrological cycle principles and applications. International Atomic Energy Agency, Vienna.
Moser, H., Stichler, W., 1975. Deuterium and oxygen-18 contents as an index of the properties of snow blankets. In: Snow mechanics, Proceedings of Grindelwald Symposium, April 1974. Publication 114, International Association of Hydrological Sciences, Dorking, Surrey, pp. 122–135.
Ogunkoya, O.O., Jenkins, A., 1991. Analysis of runoff pathways and flow distributions using deuterium and stream chemistry. Hydrol. Process. 5, 271–282.
Ogunkoya, O.O., Jenkins, A., 1993. Analysis of storm hydrograph and flow pathways using a three-component hydrograph separation model. J. Hydrol. 142, 71–88.
Panno, S.V., Hackley, K.C., Hwang, H.H., Kelly, W.R., 2001. Determination of the sources of nitrate contamination in karst springs using isotopic and chemical indicators. Chem. Geol. 179, 113–128.

Panno, S.V., Kelly, W.R., Hackley, K.C., Hwang, H.H., Martinsek, A.T., 2008. Sources and fate of nitrate in the Illinois River Basin, Illinois. J. Hydrol. 359, 174–188.

Pearson, F.J., Balderer, W., Loosli, H.H., Lehmann, B.E., Matter, A., Peters, T., et al., 1991. Applied isotope hydrogeology—A case study in northern Switzerland. Studies in environmental science 43. Elsevier, Amsterdam.

Perrin, J., Jeannin, P.Y., Zwahlen, F., 2003. Epikarst storage in a karst aquifer: a conceptual model based on isotopic data, Milandre test site, Switzerland. J. Hydrol. 279, 106–124.

Plummer, L.N., Prestemon, E.C., Parkhurst, D.L., 1994. An interactive code (NETPATH) for modeling NET geochemical reactions along a flow PATH, version 2.0. Water-resources investigations report 94-4169. U.S. Geological Survey, Reston, VA.

Rose, S., 2007. Utilization of decadal tritium variation for assessing the residence time of base flow. Ground Water 45 (3), 309–317.

Rose, S., Fullagar, P.D., 2005. Strontium isotope systematics of base flow in Piedmont Province watersheds, Georgia (USA). Appl. Geochem. 20 (8), 1571–1586.

Rozanski, K., Araguds-Araguds, L., Gonfiantini, R., 1992. Relation between long-term trends of ^{18}O isotope composition of precipitation and climate. Science 258, 981–985.

Rozanski, K., Araguds-Araguds, L., Gonfiantini, R., 1993. Isotopic patterns in modern global precipitation. In: Swart, P.K., Lohman, K.C., McKenzie, J., Savin, S. (Eds.), Climate change in continental isotopic records—Geophysical monograph 78. American Geophysical Union, Washington, DC, pp. 1–36.

Rushton, K.R., Ward, C.J., 1979. The estimation of groundwater recharge. J. Hydrol. 41, 345–361.

Schaeffer, O.A., Thompson, S.O., Lark, N.L., 1960. Chlorine-36 radioactivity in rain. J. Geophys. Res. 65, 4013–4016.

Schlosser, P., Stute, M., Dorr, H., Sonntag, C., Munnich, K.O., 1989. Tritogenic ^{3}He in shallow groundwater. Earth Planet. Sci. Lett. 94, 245–254.

Schulz, H.D., 1998. Evaluation and interpretation of tracing tests. In: Käss, W. (Ed.), Tracing technique in geohydrology. Balkema, Rotterdam, pp. 341–376.

Siegel, D.I., 1991. Evidence for dilution of deep, confined ground water by vertical recharge of isotopically heavy Pleistocene water. Geology 19 (5), 433–436.

Simmers, I., 1990. Aridity, groundwater recharge and water resources management. In: Lerner, D.N., Issar, A.S., Simmers, I. (Eds.), Groundwater recharge—A guide to understanding and estimating natural recharge. International contribution to hydrogeology 8. Verlag Heinz Heise, Hannover, Germany, pp. 3–20.

Sklash, M.G., Farvolden, R.N., 1979. The role of groundwater in storm runoff. J. Hydrol. 43, 45–65.

Smart, C.C., 2005. Errors and technique in fluorescent dye tracing. In: Beck, B.F. (Ed.), Sinkholes and the engineering and environmental impacts of karst. Geotechnical special publication 144, American Society of Civil Engineers, Reston, pp. 500–509.

Solomon, D.K., Sudicky, E.A., 1991. Tritium and helium-3 isotope ratios for direct estimation of spatial variations in groundwater recharge. Water Resour. Res. 27 (9), 2309–2319.

Solomon, D.K., Schiff, S.L., Poreda, R.J., Clarke, W.B., 1993. A validation of the $^{3}H/^{3}He$ method for determining groundwater recharge. Water Resour. Res. 29 (9), 2951–2962.

Stichler, W., Trimborn, P., Maloszewski, P., Rank, D., Papesch, W., Reichert, B., 1997. Environmental isotope investigations. In: Kranjc, A. (Ed.), Karst hydrogeological investigations in south-western Slovenia. Acta Carsologica vol. 26, no. 1, 213–236.

Trček, B., 1997. Carbon isotopic composition of groundwater from Trnovsko-banjška plateau. Master's thesis University of Ljubljana, Ljubljana, Slovenia.

Trček, B., 2003. Epikarst zone and the karst aquifer behaviour: A case study of the Hubelj catchment, Slovenia. Geološki zavod Slovenije, Ljubljana, Slovenia.

Trček, B., 2007. How can the epikarst zone influence the karst aquifer hydraulic behaviour? Env. Geol. 51 (5), 761–765.

Trček, B., 2008. Flow and solute transport monitoring in the karst aquifer in SW Slovenia. Env. Geol. 55 (2), 269–276.

Trček, B., Veselič, M., Pezdič, J., 2006. The vulnerability of karst springs—A case study of the Hubelj spring (SW Slovenia). Env. Geol. 49 (6), 865–874.

Urbanc, J., Trček, B., Pezdič, J., Lojen, S., 1997. Dissolved inorganic carbon isotope composition of waters. In: Kranjc, A. (Ed.), Karst hydrogeological investigations in south-western Slovenia. Acta Carsologica vol. 26, no. 1, 236–256.

Zhou, W., Beck, B.F., Pettit, A.J., Stephenson, B.J., 2002. A groundwater tracing investigation as an aid of locating groundwater monitoring stations on the Mitchell Plain of southern Indiana. Env. Geol. 41 (7), 842–851.

Zojer, H., Ramspacher, P., Fank, J., 1991. Die kombinierte Lysimeteranlage Wagna. In: Proceedings Lysimetertagung, Art der Sickerwassergewinnung und Ergebnisinterpretation. BAL-Bundesanstalt fur alpenlandische Landwirtschaft, Gumpenstein, pp. 55–62.

CHAPTER

Spring discharge hydrograph

4

Neven Kresic[1] and Ognjen Bonacci[2]
[1]MACTEC Engineering and Consulting, Inc., Ashburn, Virginia.
[2]School of Civil Engineering and Architecture, University of Split, Croatia

4.1 INTRODUCTION

The spring discharge hydrograph is the final result of various processes that govern the transformation of precipitation and other water inputs in the spring's drainage area into the single output at the spring. In many cases, the discharge hydrograph of a spring closely resembles hydrographs of surface streams, particularly if the aquifer is unconfined and reacts relatively quickly to water input(s). Springs draining karst and intensely fractured aquifers are typical examples—their discharge can increase several times or even orders of magnitude in a matter of hours after heavy rains. At the opposite spectrum are deep ascending springs (often thermal) isolated from a direct influence of surficial processes, such as infiltration of precipitation and showing only slight, delayed seasonal changes in discharge characteristics.

Figure 4–1 shows a number of possible springwater budget components that, to varying degrees, may influence generation of the spring discharge hydrograph and the spring yield (Q_S). A water budget states that the rate of change in water stored in an area, such as a spring drainage area ("springshed"), is balanced by the rate at which water flows into and out of the area:

$$\text{Water input} - \text{Water output} = \text{Change in storage} \quad (4.1)$$

Water budget equations can be written in terms of volumes (for a fixed time interval), fluxes (volume per time, such as cubic meters per day), and flux densities (volume per unit area of land surface per time, such as millimeters per day).

Common to most components of a water budget is that they cannot be measured directly but are estimated from measurements of related quantities (parameters) and estimates of other components. The exceptions are direct measurements of total precipitation, stream flows, spring discharge rates, and well-pumping rates. Other important quantities that can be measured directly and used in water budget calculations as part of various equations are the hydraulic head (water level) of both groundwater and surface water and the soil moisture content.

Water budget terms are often used interchangeably, sometimes causing confusion. The term *effective rainfall* describes the portion of precipitation that reaches surface streams via direct overland flow or near-surface flow (*interflow*). *Rainfall excess* describes that part of rainfall that generates surface runoff and does not infiltrate the subsurface. In groundwater studies, the term *effective rainfall* is sometimes

Copyright © 2010, Elsevier Inc. All rights reserved.

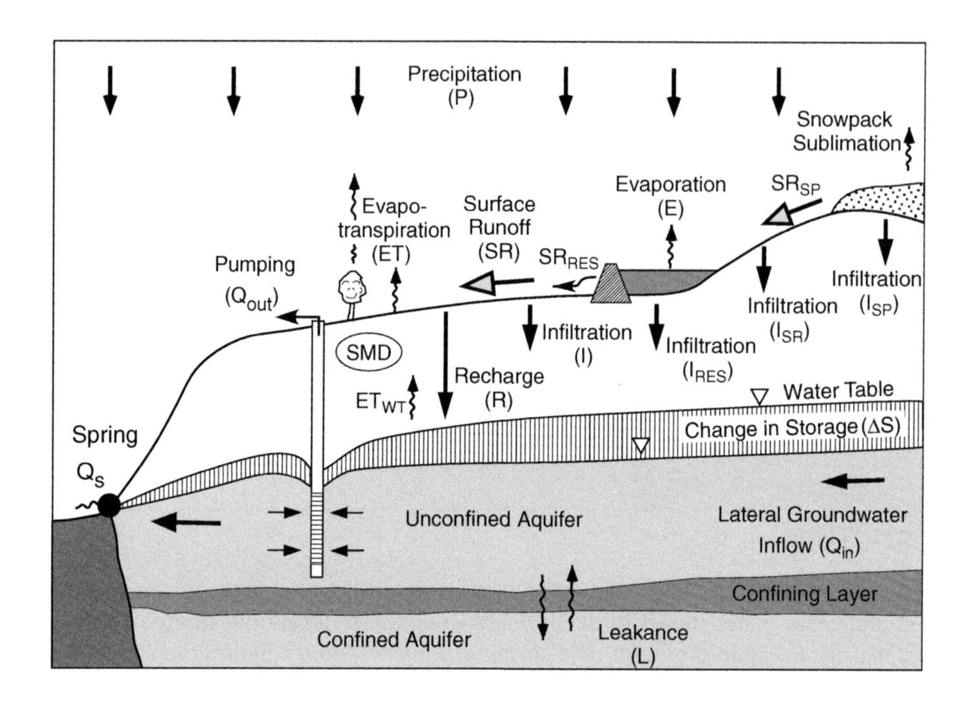

FIGURE 4–1 Elements of water budget in a spring drainage area (springshed). Explanation in text. (Modified from Kresic, 2009.)

used to describe actual infiltration, but if used in this way, it should be clearly defined. In general, *infiltration* refers to any water movement from the land surface into the subsurface. This water is sometimes called *potential recharge*, indicating that only a portion of it may eventually reach the water table (saturated zone). The term *actual recharge* increasingly is being used to avoid any possible confusion: It is the portion of infiltrated water that reaches the aquifer, and it is confirmed based on groundwater studies. The most obvious confirmation that actual groundwater recharge is taking place is a rise in the water table (hydraulic head) and an increase in the spring discharge rate. *Effective (net) infiltration*, or *deep percolation*, refers to water movement below the root zone and is often equated to actual recharge. In hydrologic studies, *interception* is the part of rainfall intercepted by vegetative cover before it reaches the ground surface and is not available for either infiltration or surface runoff. The term *net recharge* is used to distinguish between the following two water fluxes: recharge reaching the water table due to vertical downward flux from the unsaturated zone and evapotranspiration from the water table, which is an upward flux ("negative recharge"). *Areal* (or *diffuse*) *recharge* refers to recharge derived from precipitation and irrigation that occur fairly uniformly over large areas, whereas *concentrated recharge* refers to the loss of stagnant (ponded) or flowing surface water (playas, lakes, recharge basins, and sinking and losing streams) to the subsurface (Kresic, 2009).

The complexity of the water budget determination depends on many natural and anthropogenic factors present in the general area of interest, such as climate, hydrography and hydrology, geologic and geomorphologic characteristics, hydrogeologic characteristics of the surface soil and subsurface porous media, land cover and land use, presence and operations of artificial surface water reservoirs, surface water and groundwater withdrawals for consumptive use and irrigation, and wastewater management. The following are some of the relationships among the components shown in Figure 2–1 that can be utilized in quantitative water budget analyses of such a system:

$$
\begin{aligned}
I &= P - \mathrm{SR} - \mathrm{ET} \\
I &= I_{\mathrm{SR}} + I_{\mathrm{RES}} + I_{\mathrm{SP}} \\
R &= I - \mathrm{SMD} - \mathrm{ET}_{\mathrm{WT}} \qquad (4.2) \\
Q_S &= R + Q_{\mathrm{IN}} + L - \Delta S - Q_{\mathrm{OUT}}
\end{aligned}
$$

where I is infiltration in general, SR is surface water runoff, ET is evapotranspiration, I_{SR} is infiltration from surface runoff (including from sinking and losing streams), I_{RES} is infiltration from surface water reservoirs, I_{SP} is infiltration from snowpack and glaciers, R is groundwater recharge, SMD is soil moisture deficit, ET_{WT} is evapotranspiration from water table, Q_S is spring discharge Q_{IN} is lateral groundwater inflow to the aquifer feeding the spring, L is leakage back and forth between the underlying aquitard and the aquifer, Q_{OUT} is well pumpage from the aquifer, and S is change in storage of the aquifer. If the area is irrigated, yet one more component would be added to the list: infiltration of the irrigation water.

Ideally, most of these relationship would have to be established to fully quantify the processes governing the water budget of a groundwater system feeding a spring, including volumes of water stored in and flowing between three general reservoirs: the surface water, vadose zone, and saturated zone. By default, changes in one of the many water budget components cause a "chain reaction" and influence all other components. These changes take place with more or less delay, depending on both the actual physical movement of water and the hydraulic characteristics of the three general reservoirs. Figure 4–2 shows how localized recharge in one part of the system can cause a rapid response far away, followed by a more gradual change between the areas of recharge and discharge, as the newly infiltrated water starts flowing through the aquifer. The rapid response is due to propagation of the hydrostatic pressure through the system and, in this particular example, illustrates the behavior of large fractures and

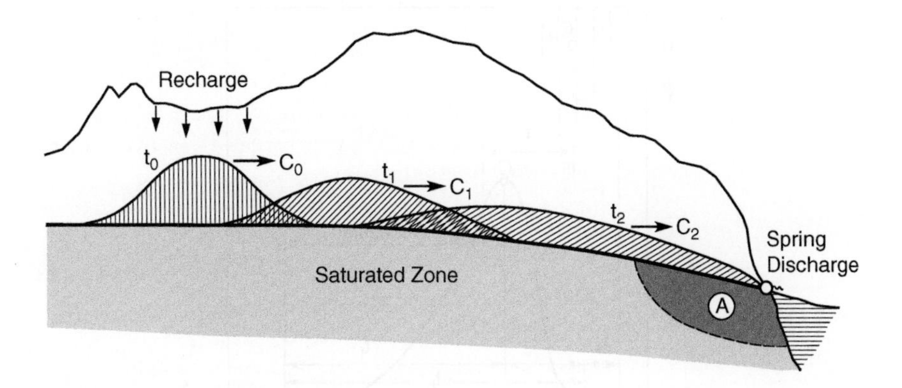

FIGURE 4–2 Formation and movement of a groundwater "wave" caused by a localized recharge event. Velocity of the wave is C_0 at time t_0, C_1 at time t_1, and C_2 at time t_2, where $C_0 > C_1 > C_2$ due to decreasing hydraulic gradients. A is the volume of "old" water discharged under pressure at the spring due to the recharge event. (Modified from Yevjevich, 1981.)

conduits in karst aquifers; the same mechanism is, to a certain extent, applicable to other aquifer types as well. In any case, it is very important to always consider groundwater systems as dynamic and constantly changing in both time and space.

As discussed by Healy et al. (2007) and applied to springs, an understanding of the water budget and underlying hydrologic processes provides a foundation for effective spring management. Observed changes in the spring's water budget over time can be used to assess the effects of climate variability and human activities on the available water resource. Comparison of spring water budgets from different areas allows the effects of factors such as geology, soil, vegetation, and land use to be quantified. Human activities affect the natural hydrologic cycle in many ways. Modifications of the land to accommodate agriculture, such as installation of drainage and irrigation systems, alter infiltration, runoff, evaporation, and plant transpiration rates. Buildings, roads, and parking lots in urban areas tend to increase runoff and decrease infiltration. In conclusion, water budget analysis provides a means for evaluating availability and sustainability of a water supply.

As mentioned earlier, the spring flow rate is one of the few water budget elements that can be measured directly. In addition, a thorough analysis of the spring discharge hydrograph provides useful information on the aquifer characteristics, such as the nature of its storage and transmissivity and the types and quantity of its groundwater reserves.

Although the processes that generate hydrographs of springs and surface streams are different, much is analogous between them, and the hydrograph terminology is the same. Figure 4–3 shows the main elements of a discharge hydrograph. The beginning of discharge after a rainfall is marked with point A, and the time between the beginning of rainfall and the beginning of discharge, called the *starting time*, with t_s. The time in which the hydrograph raises to its maximum (point C), is called the *concentration time*, t_c. The time from maximum discharge until the end of the hydrograph, when the discharge theoretically equals 0 (point E), is the *falling time*, t_f. Together, the concentration time and the falling time are called the *base time* of the hydrograph, t_b. The time between the centroid of the precipitation episode (C_P) and the centroid of the hydrograph (C_H) is called the *retardation time*, t_r. The time interval for recording the amount of precipitation and the flow rate at the spring is Δt.

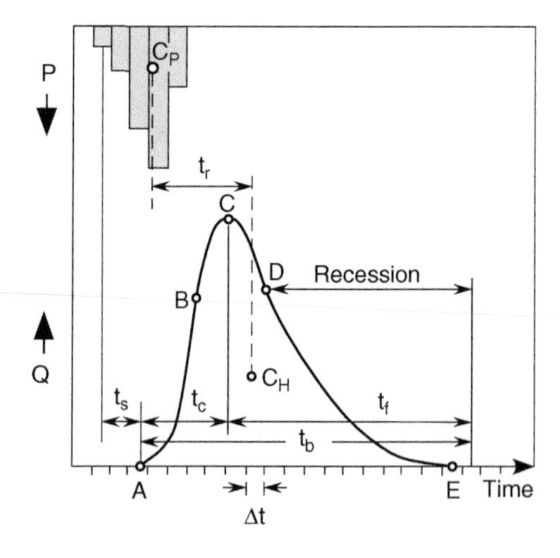

FIGURE 4–3 Components of a discharge hydrograph (explanation in text).

The shape of the hydrograph is defined by its base (*AE*), the rising limb (*AB*), the crest (*BCD*), and the falling limb (*DE*). The falling limb corresponds to the *recession period*. *B* and *C* are *inflection points*, where the hydrograph curve changes its shape from convex to concave and vice versa. For surface streams, point *D* is the end of direct runoff after the rain. In general, the part of the hydrograph from point *D* on is called the *recession curve*.

The shape of a discharge hydrograph depends on the size and shape of the drainage area, as well as the precipitation intensity. When a rainfall episode lasts longer and the intensity is lower, the hydrograph has a longer time base and vice versa: Intensive short storms cause sharp hydrographs with short time bases. The area under the hydrograph is the volume of discharged water for the recording period. The retardation time or delayed response of the spring to water input and the shape of the hydrograph are good first indicators of the recharge capacity and transmissive properties of the aquifer feeding the spring.

In reality, unless a spring or a surface stream is intermittent, the recorded hydrograph has a more complex shape, which reflects the influence of antecedent precipitation and other possible water inputs. Such hydrographs are formed by the superposition of single hydrographs corresponding to separate precipitation events (Figure 4–4) and other water inputs, such as arrival of water from a sinking stream.

The impact of newly infiltrated water on spring discharge varies with respect to the predominant type of porosity and the stage of groundwater level. In any case, the first reaction of karst, transmissive fractured rock, and some young basaltic aquifers to recharge, in most cases, is the consequence of pressure propagation through karst conduits and large fractures and not the outflow of newly infiltrated water (see Figure 4–2).

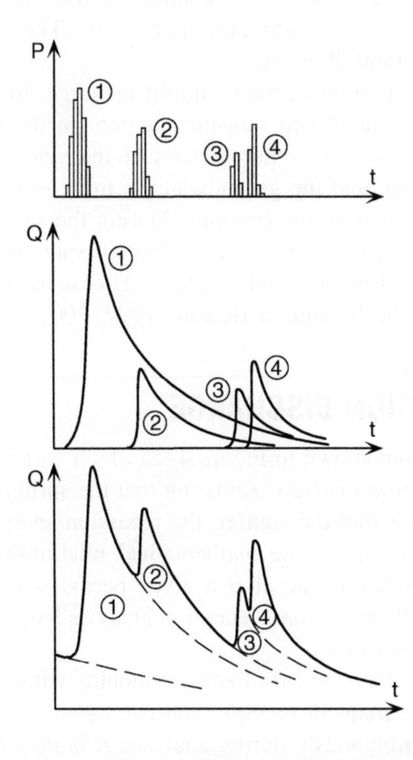

FIGURE 4–4 Complex discharge hydrograph (bottom) composed of single hydrographs (middle) as the result of several rainfall events (top). (Modified from Jevdjevic, 1956.)

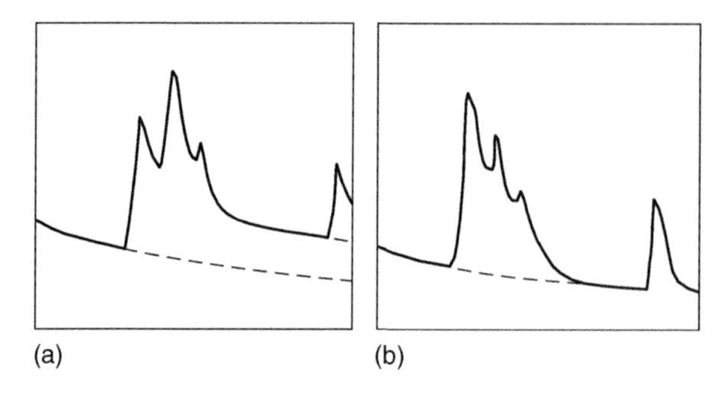

FIGURE 4–5 Possible reaction of aquifers to water inputs (e.g., infiltration of precipitation) as seen from spring discharge hydrographs: (a) Lifting of the base flow to a higher hydrograph level because of the increased storage, (b) discharge with no increase in storage. (From Kresic, 2007.)

The new water arrives at the spring with a certain delay and its contribution is a fraction of the overall flow rate. After a recharge episode and the initial response of the system are over, two cases may occur, as illustrated in Figure 4–5: (1) an increase in the volume of groundwater accumulated in storage, which is reflected in a shift of the recession curve to a higher hydrograph level, and (2) new water transmitted mostly through a well-developed network of fractures or karst channels and conduits and discharged with no significant accumulation of groundwater in the surrounding matrix porosity. The recession curve continues along the same extrapolated line as before the rainfall event.

The first case is characteristic for periods of main aquifer recharge, for example, March through June in moderate climates: Both groundwater level and moisture content in the unsaturated zone are high, while evapotranspiration loss is small compared to other periods of the year. Newly infiltrated water raises an already high hydraulic head (pressure), and the groundwater is more easily injected into the aquifer matrix porosity and narrow fissures, including from the conduits. During the summer-autumn period, the hydraulic heads and the hydraulic gradients are low, and newly infiltrated water from summer storms is transmitted quickly through well-connected large fractures and conduits. This second case may also indicate a lack of significant matrix or fissure porosity in the aquifer (Kresic, 1997, 2007).

4.2 EQUATIONS OF RECESSION DISCHARGE

Analysis of the falling hydrograph limb shown in Figure 4–6, which corresponds to a period with no significant precipitation, is called the *recession analysis*. Knowing that the spring discharge is without disturbances caused by a rapid inflow of new water into the aquifer, the recession analysis provides good insight into the aquifer structure. By establishing an appropriate mathematical relationship between spring discharge and time, it is possible to predict the discharge rate after a given period without precipitation and to calculate the volume of discharged water. For these reasons, recession analysis has been a popular quantitative method in spring discharge analysis for a long time.

The ideal recession conditions—a long period of several months without precipitation—are rare in moderate, humid climates. Consequently, frequent precipitation can cause various disturbances in the recession curve that may not be removed unambiguously during analysis. It is therefore desirable to analyze as many recession curves from different years as possible. Larger samples allow for derivation of an average recession curve as well as the envelope of minima (Figure 4–7), which enables a more accurate quantification of the expected long-term minimum discharge.

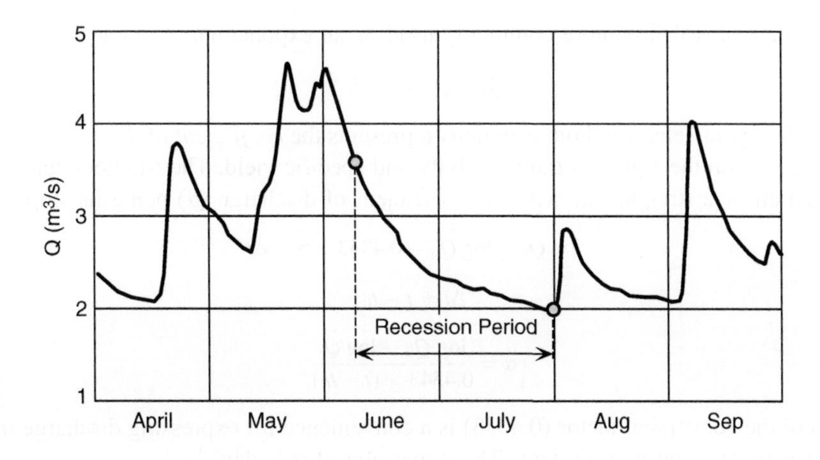

FIGURE 4–6 Spring discharge hydrograph with a recession period. (From Kresic, 2007.)

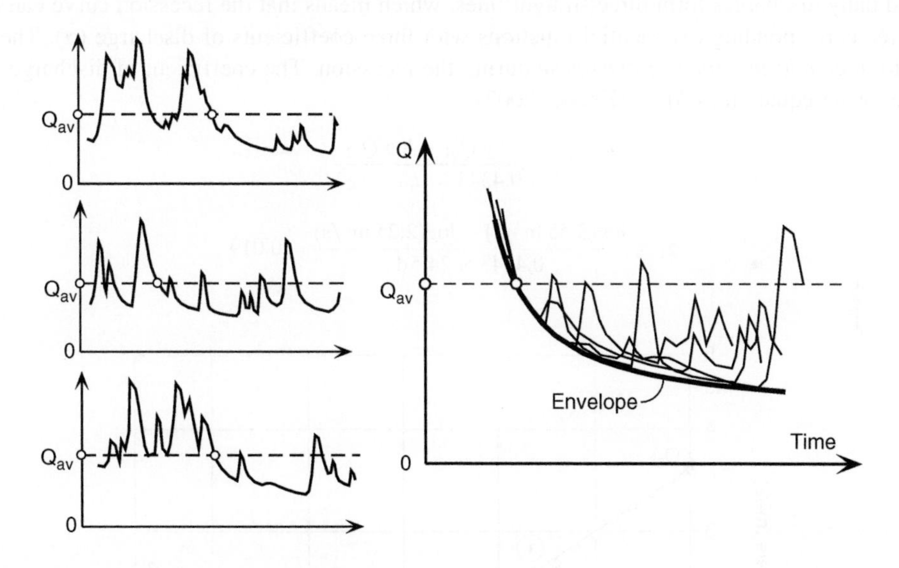

FIGURE 4–7 Three annual discharge hydrographs with main recession curves overlaid at the point of average discharge rate. The minima on the three overlaid curves are connected with the envelope, which represents a long-term recession curve. (From Kresic, 2007.)

Two well-known mathematical formulas that describe the falling limb of hydrographs and the base flow were proposed by Boussinesq (1904) and Maillet (1905). Both equations give the dependence of the flow at specified time (Q_t) on the flow at the beginning of recession (Q_0). The Boussinesq equation is of hyperbolic form:

$$Q_t = \frac{Q_0}{[1 + \alpha(t - t_0)]^2} \tag{4.3}$$

where t is the time since the beginning of recession for which the flow rate is calculated; t_0 is time at the beginning of recession usually (but not necessarily) set equal to 0.

The Maillet equation, which is more commonly used, is an exponential function:

$$Q_t = Q_0 \times e^{-\alpha(t-t_0)} \tag{4.4}$$

The dimensionless parameter α in both equations represents the *coefficient of discharge* (or *recession coefficient*), which depends on the aquifer's transmissivity and specific yield. The Maillet equation, when plotted on a semilog diagram, is a straight line with the coefficient of discharge (α) being its slope:

$$\log Q_t = \log Q_0 - 0.4343 \times \alpha \times \Delta t \tag{4.5}$$

$$\Delta t = t - t_0$$

$$\alpha = \frac{\log Q_0 - \log Q_t}{0.4343 \times (t - t_0)} \tag{4.6}$$

The introduction of the conversion factor (0.4343) is a convenience for expressing discharge in equation (4.6) in cubic meters per second and time in days. The dimension of α is day^{-1}.

Figure 4–8 is a semilog plot of time versus discharge rate for the recession period shown in Figure 4–6. The recorded daily discharges form three straight lines, which means that the recession curve can be approximated by three corresponding exponential functions with three coefficients of discharge (α). The three lines correspond to three *microregimes of discharge* during the recession. The coefficient of discharge for the first microregime, using equation (4.6), is (Kresic, 2007)

$$\alpha_1 = \frac{\log Q_{01} - \log Q_{02}}{0.4343 \times (t_{01} - t_{02})}$$

$$\alpha_1 = \frac{\log(3.55\ \text{m}^3/\text{s}) - \log(2.25\ \text{m}^3/\text{s})}{0.4343 \times 24.5\,\text{d}} = 0.019$$

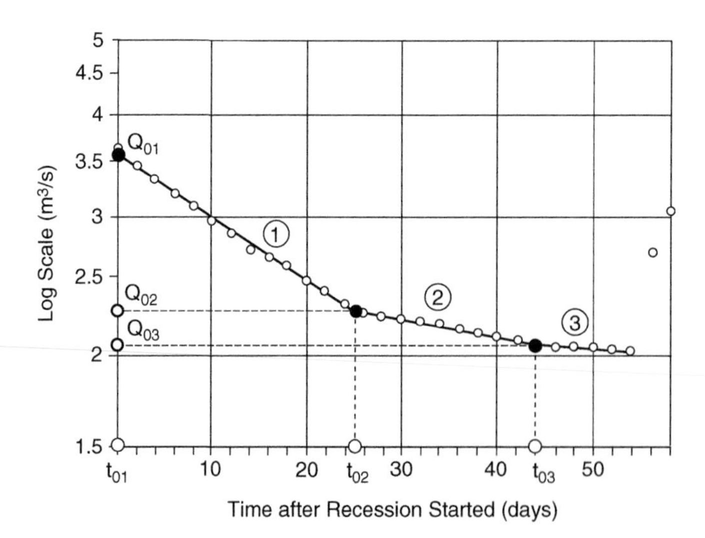

FIGURE 4–8 Semilog graph of time versus discharge for the recession period shown in Figure 4–7. The duration of the recession period is 54 days. (Modified from Kresic, 2007.)

The coefficient of discharge for the second microregime is

$$\alpha_2 = \frac{\log(2.25\,\mathrm{m^3/s}) - \log(2.06\,\mathrm{m^3/s})}{0.4343 \times (44\,\mathrm{d} - 24.5\,\mathrm{d})} = 0.0045$$

The third coefficient of discharge, or slope of the third straight line, is found by choosing a discharge rate anywhere on the line, including its extension if the actual line is short. In our case, the discharge rate after 60 days is 2.01 $\mathrm{m^3/s}$ and α_3 is then

$$\alpha_3 = \frac{\log(2.06\,\mathrm{m^3/s}) - \log(2.01\,\mathrm{m^3/s})}{0.4343 \times (60\,\mathrm{d} - 44\,\mathrm{d})} = 0.0015$$

After determining the coefficients of discharge, the flow rate at any given time after the beginning of recession can be calculated using the appropriate Maillet equation. For example, the discharge of the spring 35 days after the recession started, when the second microregime is active, is calculated as

$$Q_{35} = Q_{02} \times \mathrm{e}^{-\alpha_2(35\,\mathrm{d} - t_{02})}$$

$$Q_{35} = 2.25\,\mathrm{m^3/s} \times \mathrm{e}^{-0.0045\times(35\,\mathrm{d} - 24.5\,\mathrm{d})} = 2.146\,\mathrm{m^3/s}$$

Note that the initial discharge rate for the second microregime is 2.25 $\mathrm{m^3/s}$ and the corresponding time is 24.5 days.

Spring discharge three months (90 days) after the beginning of recession, assuming no precipitation for the entire period, may be predicted by using the characteristic values for the third microregime (see Figure 4–8), where Q_{03} is the initial discharge for that regime:

$$Q_{90} = Q_{03} \times \mathrm{e}^{-\alpha_3(90\,\mathrm{d} - t_{03})}$$

$$Q_{90} = 2.06\,\mathrm{m^3/s} \times \mathrm{e}^{-0.0015(90\,\mathrm{d} - 44\,\mathrm{d})} = 1.923\,\mathrm{m^3/s}$$

The coefficient of discharge (α) and the volume of free gravitational groundwater stored in the aquifer above spring level (i.e., groundwater that contributes to spring discharge) are inversely proportional:

$$\alpha = \frac{Q_t}{V_t} \tag{4.7}$$

where Q_t is the discharge rate at time t and V_t is the volume of water stored in the aquifer above the level of discharge (spring level).

Equation (4.7) allows calculation of the volume of water accumulated in the aquifer at the beginning of recession as well as the volume discharged during a given period of time. The calculated remaining volume of groundwater always refers to the reserves stored above the current level of discharge. The draining of an aquifer with three microregimes of discharge (as in our case) and the corresponding volumes of the discharged water are shown in Figure 4–9. The total initial volume of groundwater stored in the aquifer (above the level of discharge) at the beginning of the recession period is the sum of the three volumes that correspond to three types of storage (effective porosity):

$$V_0 = V_1 + V_2 + V_3 = \left[\frac{Q_1}{\alpha_1} + \frac{Q_2}{\alpha_2} + \frac{Q_3}{\alpha_3}\right] \times 86{,}400\,\mathrm{s}\,[\mathrm{m^3}] \tag{4.8}$$

where discharge rates are given in cubic meters per second.

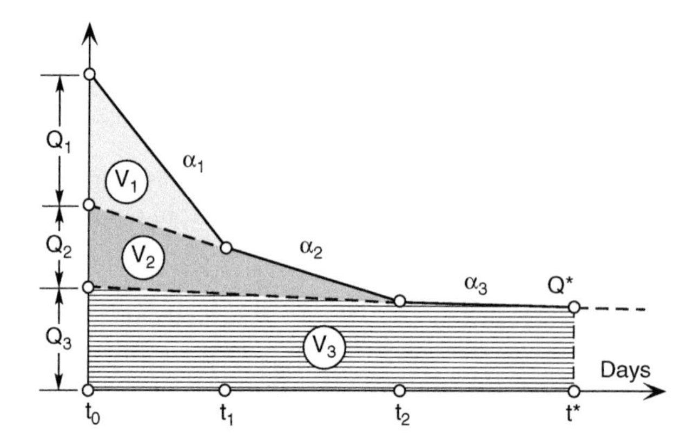

FIGURE 4–9 Schematic presentation of the recession with three microregimes of discharge, and the corresponding volumes of discharged water. (From Kresic, 2007.)

The volume of groundwater remaining in the aquifer at the end of the third microregime is the function of the discharge rate at time t^* and the coefficient of discharge α_3:

$$V^* = \frac{Q^*}{\alpha_3} \tag{4.9}$$

The difference between volumes V_0 and V^* is the volume of all groundwater discharged during the period $t^* - t_0$. In our case, the volume of groundwater stored in the aquifer at the beginning of recession is

$$V_0 = \left[\frac{(Q_{01} - Q_{02})}{\alpha_1} + \frac{Q_{02} - Q_{03}}{\alpha_2} + \frac{Q_{03}}{\alpha_3}\right] \times 86{,}400\,\text{s}[\text{m}^3]$$

$$V_0 = \left[\frac{(3.55\ \text{m}^3/\text{s} - 2.55\ \text{m}^3/\text{s})}{0.019} + \frac{(2.55\ \text{m}^3/\text{s} - 2.20\ \text{m}^3/\text{s})}{0.0045} + \frac{2.20\ \text{m}^3/\text{s}}{0.0015}\right] \times 86{,}400\ \text{s}$$

$$V_0 = 4.547 \times 10^6\ \text{m}^3 + 6.720 \times 10^6\ \text{m}^3 + 1.267 \times 10^8\ \text{m}^3 = 1.380 \times 10^8\ \text{m}^3$$

The volume of water remaining in the aquifer above the spring level at the end of recession is

$$V^* = \frac{2.03\ \text{m}^3/\text{s}}{0.0015} \times 86{,}400\,\text{s} = 1.169 \times 10^8\ \text{m}^3$$

which gives the following volume of water discharged at the spring for the duration of recession (54 days):

$$V = V_0 - V^* = 1.380 \times 10^{-8} - 1.169 \times 10^8\ \text{m}^3 = 21.1 \times 10^6\ \text{m}^3$$

Recession periods of large perennial karstic springs or springs draining highly permeable fractured rock aquifers often have two or three microregimes of discharge, as in this example. However, the recession curve can vary greatly from spring to spring and its shape can have various physical meanings.

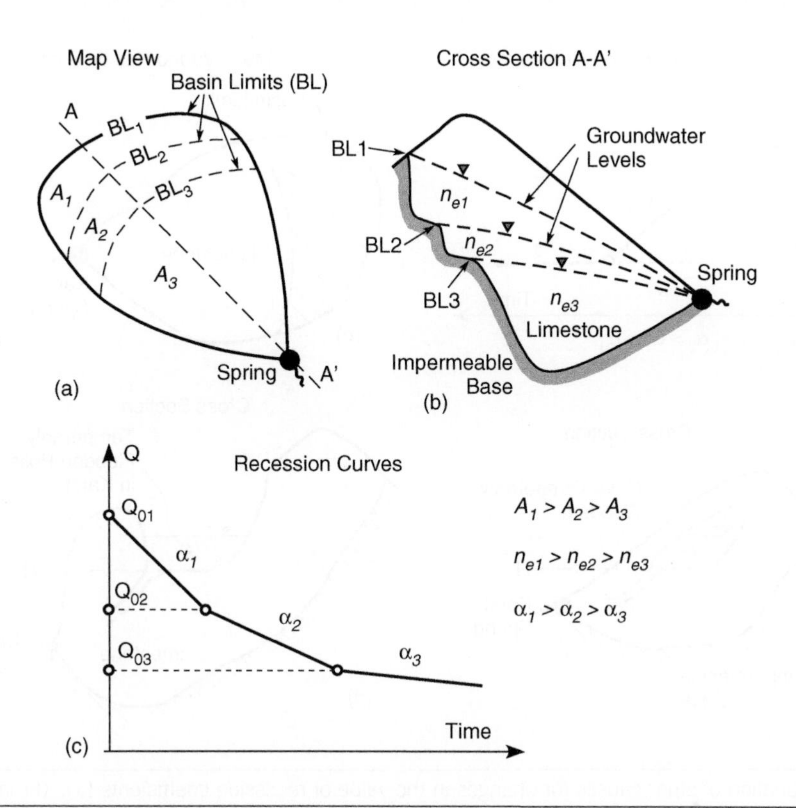

FIGURE 4–10 Explanation of common causes of changes in the value of recession coefficients (α) caused by changes in the size of catchment area (A) and effective porosity (n_e). (Modified from Bonacci, 1993.)

It has been argued that, for karst springs, the initial steep portion of the curve represents the turbulent drainage of large fractures and conduits (e.g., the first microregime in Figure 4–8), followed by a transitional portion of the curve, where the flow is less turbulent and reflects the contribution of smaller fractures and rock matrix (the second microregime), ending with the slowly decreasing curve, the so-called master recession curve, where the drainage of rock matrix and small fissures is dominant. This last portion of the curve is also the most important for predicting the future discharge in a complete absence of water inputs, that is, during prolonged droughts. However, as explained further, there may be other explanations for various breaks in the recession curve.

As discussed by Bonacci (1993), every break point in the recession line is caused by a change in a characteristic of the groundwater reservoir (aquifer). A common situation is represented in Figure 4–10. The break points in the recession curve result from the decrease in the spring drainage area and the decrease in the effective porosity of the aquifer. For example, in karsts, during high groundwater levels, the flow may be dominated by a more karstified epikarst zone, which has higher effective porosity than deeper portions of the aquifer.

Figure 4–11 presents some less common shapes of the recession curve and the changes in recession coefficients. The main difference between Figures 4–10 and 4–11 is that $\alpha_2 > \alpha_1$ in the latter; that is, the recession coefficient α increases with time. The reasons for this can be numerous. For example,

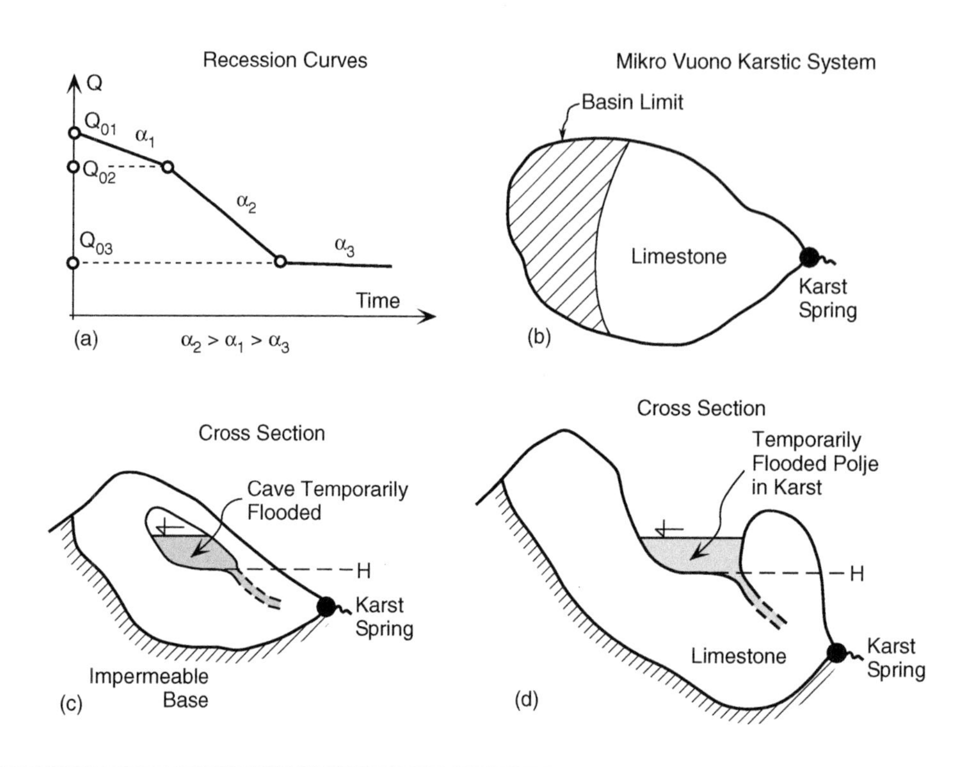

FIGURE 4–11 Explanation of other causes for changes in the value of recession coefficients (α). (From Bonacci, 1993.)

Figure 4–11b gives a schematic presentation of a catchment that consists partly of limestone and partly of schists. The outflow through limestone is much faster than that through the schists, which causes a lag in the water inflow (in the period $t_i - t_j$) from the schists, and thus a change (increase) in the slope of the recession curve in the Mikro Vuono Spring in Greece (Soulios, 1991). Figure 4–11c is a schematic representation of a temporarily flooded cave in a karst aquifer. The bottom of the main outlet of the cave is at height H. As long as the groundwater level is below that height, the discharge occurs very slowly or not at all. A sudden lowering of the groundwater level leads to water outflow from the flooded cave. Understandably, this is not particular to caves; it can occur in a wider aquifer area with locally higher effective porosity (Bonacci, 1993). An identical situation is presented in Figure 4–11d. In this case, a polje in a karst is temporarily flooded instead of an underground reservoir in a cave. The roles of the cave and the polje are identical.

The initial portion of the recession curve reflecting rapid discharge may not correspond to the simple exponential expression of the Maillet's type and may be better explained by some other functions. Deviations from exponential dependence can be easily detected if the recorded data plotted on a semilog diagram do not form a straight line(s).

Often a good approximation of the rapid drainage at the beginning of recession is the hyperbolic relation of the Boussinesq type. Its general form is

$$Q_t = \frac{Q_0}{(1 + \alpha t)^n} \tag{4.10}$$

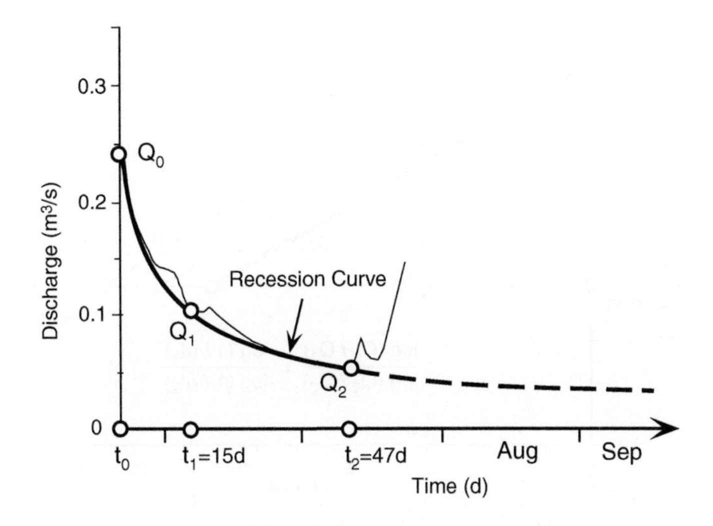

FIGURE 4–12 Recession curve of a spring with discharges used to determine the parameters of hyperbolic function. (Modified from Kresic, 2007.)

In many cases this function correctly describes the entire recession curve. On the basis of 100 analyzed recession curves of karstic springs in France, Drogue (1972) concludes that, among the six exponents studied, the best first approximations of exponent n are ½, 3⁄2, and 2.

The exact determination of exponent n and the discharge coefficient α for the function that best fits the measured data is performed graphically and by computation as follows (Kresic, 2007):

- The minimum recorded discharge at the end of recession is noted ($Q_2 = 0.057$ m^3/s for the example in Figure 4–12).
- Any discharge, Q_1, on the recession curve, that is not the result of (possible) deviation due to recent rainfall is chosen in the section between Q_2 and Q_0.
- The value of α that satisfies the following equation

$$\frac{\log(Q_0/Q_1)}{\log(Q_0/Q_2)} = \frac{\log(1+\alpha t_1)}{\log(1+\alpha t_2)} \tag{4.11}$$

is determined by trial and error, adopting an initial value for α (usually 0.5). The result can be graphically checked, as shown in Figure 4–13: The correct coefficient of discharge forms a straight line through the points defined by Q_0, Q_1, Q_2, and the corresponding times, t_0, t_1, t_2. The exact value of α for this example is 0.202:

$$\frac{\log\left(\dfrac{0.240\ \text{m}^3/\text{s}}{0.105\ \text{m}^3/\text{s}}\right)}{\log\left(\dfrac{0.240\ \text{m}^3/\text{s}}{0.060\ \text{m}^3/\text{s}}\right)} = \frac{\log(1+0.202\times 15\,\text{d})}{\log(1+0.202\times 47\,\text{d})}$$

$$0.5963 \approx 0.5929$$

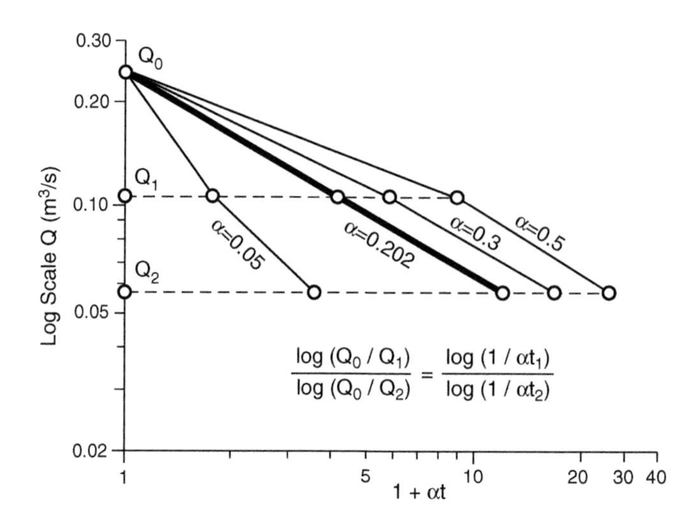

FIGURE 4–13 Graphical determination of discharge coefficient α for the example shown in Figure 4–12. (From Kresic, 2007.)

- The exponent n is calculated by substituting the determined value for α into either of the following two equations:

$$n = \frac{\log(Q_0/Q_1)}{\log(1+\alpha t_1)} \tag{4.12}$$

$$n = -\frac{\log(Q_1/Q_2)}{\log\left(\dfrac{1+\alpha t_1}{1+\alpha t_2}\right)} \tag{4.13}$$

For this example, equation (4.12) gives

$$n = \frac{\log\left(\dfrac{0.240\ \mathrm{m^3/s}}{0.105\ \mathrm{m^3/s}}\right)}{\log(1+0.202\times 15\mathrm{d})} = 0.593$$

and the recession discharge equation is then

$$Q_t = \frac{0.24\ \mathrm{m^3/s}}{(1+0.202\times t)^{0.593}}$$

The coefficient of discharge α and the exponent n have the following general relationship:

$$\alpha = \frac{\sqrt[n]{Q_0} - \sqrt[n]{Q_t}}{t \times \sqrt[n]{Q_t}} \tag{4.14}$$

As in the case of the Maillet equation, the determined hyperbolic function can be used to calculate the volume of free gravitational water stored in the aquifer above the spring level. In general, this volume at any time t since the beginning of recession is

$$V_t = \frac{Q_0}{\alpha(n-1)}\left[1 - \frac{1}{(1+\alpha t)^{n-1}}\right] \times 86{,}400\ \mathrm{s[m^3]} \tag{4.15}$$

4.2.1 Approximation with linear reservoirs

Analogous to the nonlinear nature of discharge in most natural hydrologic systems (Amorocho, 1964), the connection between the karst spring discharge (Q) during a recession period and the hydraulic head in the aquifer (H) can be approximated by the following equation (Castany, 1967; Avdagic, 1990; Bonacci, 1993):

$$Q = \lambda H^n \tag{4.16}$$

where λ is a coefficient that characterizes the conductivity and effective porosity of the aquifer and n describes the type of the connection: When $n = 1$, the discharge is linear; and when $n > 1$, it is nonlinear. The same connection can be expressed in terms of the volume (V) of water discharged:

$$V = \lambda Q^n \tag{4.17}$$

The equation of mass conservation (the continuity equation) is

$$I - Q = \frac{dV}{dt} \tag{4.18}$$

where I is water input and t is time; it follows that

$$\frac{dV}{dt} = \lambda n Q^{n-1}\frac{dQ}{dt} \tag{4.19}$$

Combining equations (4.18) and (4.19) and after integration, the discharge rate is

$$Q = \left[Q_0^{n-1} + \frac{1-n}{\lambda n}(t - t_0)\right]^{1/(n-1)} \tag{4.20}$$

Although the parameters λ and n can be identified from the discharge measurements (Q versus time), the procedure is somewhat complex and practitioners still favor approximation of aquifer (spring) discharge as outflow from a linear reservoir(s). Figure 4–14 illustrates this concept, where the volume of water (V) discharged between times t and $t + dt$ is given as

$$dV = Qdt \tag{4.21}$$

Using the reservoir (aquifer) active area (A) and the decrease in the hydraulic head (dH), this volume is

$$dV = AdH \tag{4.22}$$

Since, for the linear reservoir, $Q = \lambda H$, it follows that

$$Qdt = \lambda H dt \tag{4.23}$$

Combining equations (4.21) and (4.23) gives

$$\lambda H dt = -AdH = -dV \tag{4.24}$$

which, when integrated, results in

$$t = -\frac{A}{\lambda}\ln H + \mu \tag{4.25}$$

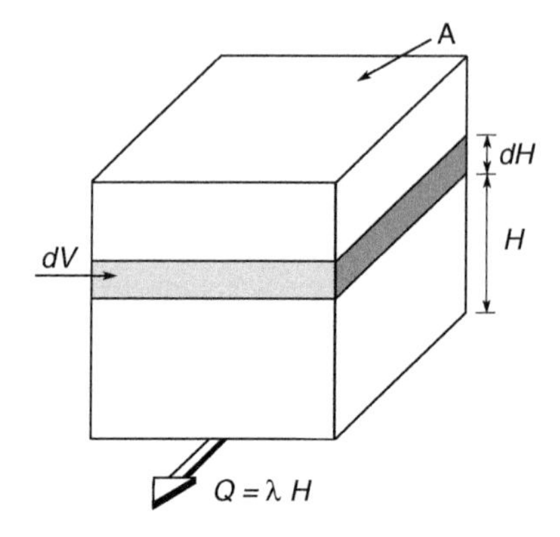

FIGURE 4–14 Schematic of discharge from a linear reservoir. (From Avdagic, 1990.)

Knowing that $H = Q/\lambda$, equation (4.25) becomes

$$Q = e^{-\lambda/A(t-\mu)} \tag{4.26}$$

The solution of equation (4.26) for the boundary conditions $t = t_0$ and $Q = Q_0$ is

$$Q = Q_0 e^{-\lambda/A(t-t_0)} \tag{4.27}$$

which is very similar to the Maillet equation (4.4). The coefficient of discharge α can be expressed with the active aquifer area (A) and the parameter dependent on the conductivity (effective porosity) of the aquifer (λ) as proposed by Castany (1967):

$$\alpha = \frac{\lambda}{A} \tag{4.28}$$

4.3 SEPARATION OF DISCHARGE COMPONENTS

Characteristics of the spring discharge hydrograph, together with the frequent monitoring of chemical and physical composition of the water as a function of time, can often provide insight into the origin and residence time of water, aquifer recharge mechanisms, and the hydraulics of the spring flow generation in both the unsaturated and saturated zones. As explained in Chapters 3 and 6, various environmental and artificial isotopes, as well as groundwater chemical constituents such as major anions, cations, and conductance, can be used to plot chemographs and evaluate processes that lead to spring discharge. As White explains, in Chapter 6, springs draining karstic carbonate aquifers are the best candidates for such an approach, as opposed to springs discharging from noncarbonate rocks, which typically have little variation in the concentration of dissolved constituents.

As discussed throughout this book, karst aquifers have a unique triple porosity, resulting in two main types of groundwater flow: the so-called slow diffuse flow or base flow and fast conduit flow (Atkinson, 1977). However, the percentages of these two components in karst spring hydrographs vary greatly as the result of many possible water inputs and different characteristics of karst aquifer. The presence of sinking streams recharging the aquifer (Figure 4–15), extensive dendritic conduit systems, and high-matrix porosity can

FIGURE 4–15 The Rak River, permanent sinking river at the entrance to Tkalca cave in classic karst of Slovenia.

produce complicated chemographs and hydrographs with multiple peaks, some of which show rapid response to rainfall and some exhibiting long delays. Separation of such hydrographs may therefore be completely ambiguous without thorough field investigations of the system. For example, Figure 4–16 shows how the same spring may react quite differently in two successive years, with similar periods of major recharge during the first half of the year, followed by long recession periods from June-July to October. Vandike (1996) provides a very detailed discussion about precipitation and recharge episodes recorded at various gauging stations in the spring drainage area and corresponding variations in spring hydrographs and chemographs.

The outlet of Maramec Spring, Missouri's fifth largest spring, is a cave opening developed in the Lower Gasconade Dolomite. Cave diving has shown that the conduit that channels water to the spring reaches depths of at least 190 ft below pool elevation. The discharge of Maramec Spring varies from a low of about 56 ft^3/s to more than 1100 ft^3/s, and averages about 155 ft^3/s (Vandike, 1996).

Dye tracing shows that Maramec Spring is recharged from a 310 mi^2 area west and south of the spring in the Dry Fork, Norman Creek, and Asher Hollow watersheds. All these watersheds are drained by losing streams that channel a significant part of their runoff into the subsurface. Hourly rainfall data collected at four locations in Dry Fork watershed, combined with hourly discharge and specific conductance data collected at Maramec Spring, show that discharge at the spring begins to increase as little as four to six hours after precipitation begins. The response time appears to be greater during relatively dry weather and less during wet weather, when antecedent soil moisture is high. The rapid increase in the Maramec Spring discharge is due to an increase in the pressure head in the karst system as the water table elevation is increased in the recharge area. The actual water supplied by the recharge does not begin to arrive at the spring for several days, and the mass center of the recharge typically reaches the spring 12 to 15 days after heavy rainfall (Vandike, 1996).

Hydrochemical separation of the spring discharge hydrograph is based on the assumption that the constitution of the water entering the aquifer is considerably different than that already in it. When recharge by rain

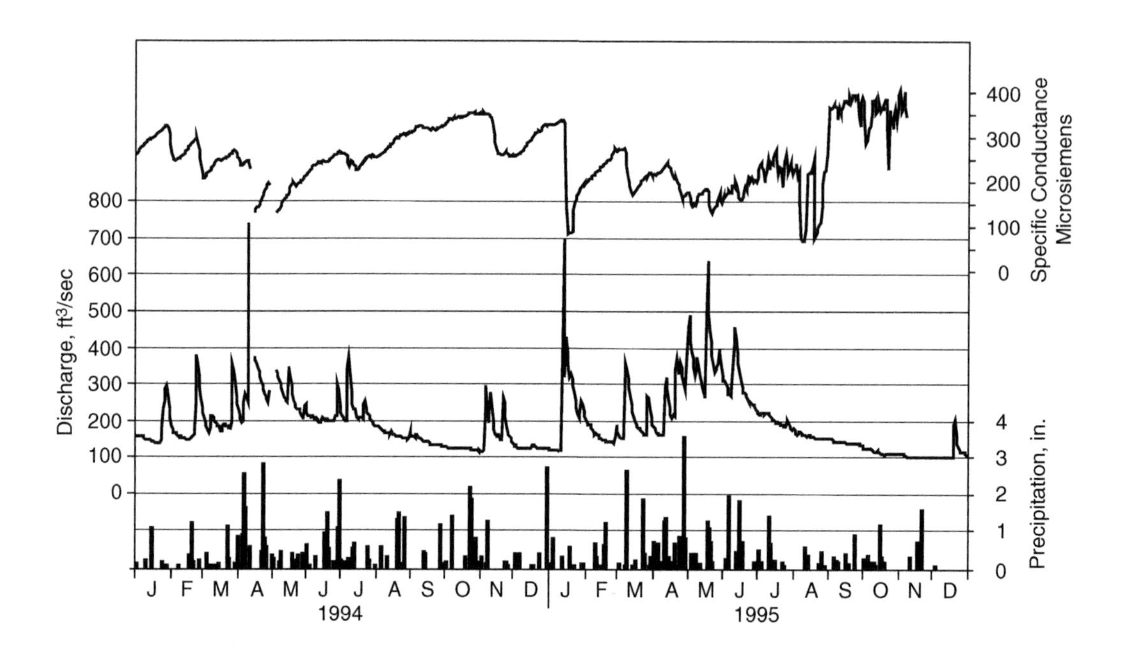

FIGURE 4–16 Average daily discharge and specific conductance at Maramec Spring, and daily precipitation at Rolla-UMR, 1994–1995. (Modified from Vandike, 1996.)

takes place, it is evident that the concentration of most cations characterizing groundwater, such as calcium and magnesium, is much lower in rainwater. Additional assumptions for the application of this method are that (after Dreiss, 1989)

- Concentration of the chemical constituents in the rainwater chosen for monitoring are uniform in both area and time.
- Corresponding concentrations in the prestorm water are also uniform over the active aquifer area and time.
- The effects of other processes in the hydrologic cycle during the rainfall event, including recharge by surface waters, are negligible.
- The concentrations and transport of constituents are not changed by chemical reactions in the aquifer.

The last condition assumes a minor dissolution of carbonate rocks during the flow of new water through the porous medium. Figure 4–17 shows that the first two assumptions regarding the calcium ion are acceptable; its concentration in the springwater drops rapidly after the heavy rains that cause an increase in discharge rate.

Assuming a simple mixing of old aquifer water (Q_{old}) and newly infiltrated rainwater (Q_{new}), the total recorded discharge of the spring is the sum of the two (after Dreiss, 1989):

$$Q_{total} = Q_{old} + Q_{new} \tag{4.29}$$

If chemical reactions in the aquifer do not cause significant and rapid changes in the concentration of calcium ions in the infiltrating rainwater (which is often true for unconfined karsts and intensely fissured aquifers, where the flow velocity is high), the calcium ion balance in the springwater is

$$Q_{total} \times C_{total} = Q_{old} \times C_{old} + Q_{new} \times C_{new} \tag{4.30}$$

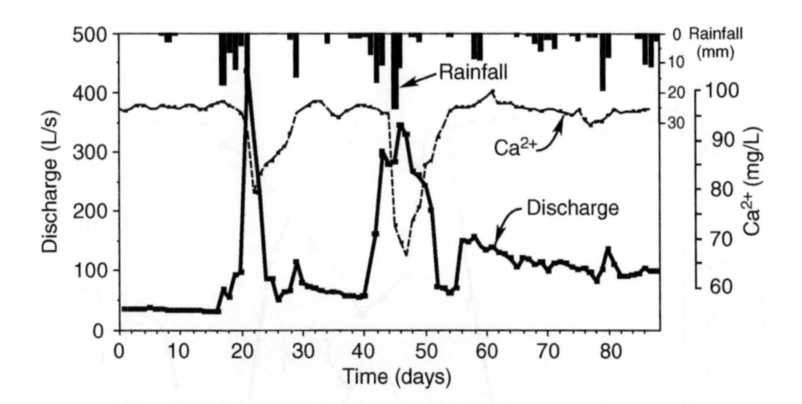

FIGURE 4–17 Spring discharge hydrograph, calcium chemograph, and rainfall in the spring drainage area. (From Kresic, 2007.)

where

- Q_{total} is the recorded spring discharge.
- C_{total} is the recorded concentration of calcium ion in the springwater.
- Q_{old} is the portion of the spring flow attributed to the "old" water (i.e., water already present in the aquifer before the rain event).
- C_{old} is the recorded concentration of calcium ion in the springwater before the rain event.
- Q_{new} is the portion of the spring flow attributed to the new water.
- C_{new} is the concentration of calcium ion in the new water.

If C_{new} is much smaller than C_{old} (which is correct in this case, since the concentration of the calcium ion in rainwater is usually less than 5 mg/L), the input mass of calcium ion is relatively small compared to its mass in the old aquifer water:

$$Q_{new} \times C_{new} << Q_{old} \times C_{old} \quad (4.31)$$

From equation (4.30), it follows, after excluding the (small) input mass, that

$$Q_{old} = \frac{Q_{total} \times C_{total}}{C_{old}} \quad (4.32)$$

Combining equations (4.29) and (4.32) gives

$$Q_{new} = Q_{total} - \frac{Q_{total} \times C_{total}}{C_{old}} \quad (4.33)$$

By applying equation (4.33), it is possible to estimate the discharge component formed by the inflow of new rainwater if the spring discharge recordings and a continuous hydrochemical monitoring are performed before, during, and after the storm event.

The result of the hydrochemical hydrograph separation for the second major rainfall event, which started approximately 40 days after the beginning of monitoring (see Figure 4–17), is shown in Figure 4–18. One day after the major increase in the flow rate, only a minor amount of new water started discharging at the spring. This lag is three days for the first major increase in the discharge of new water. The lag between the

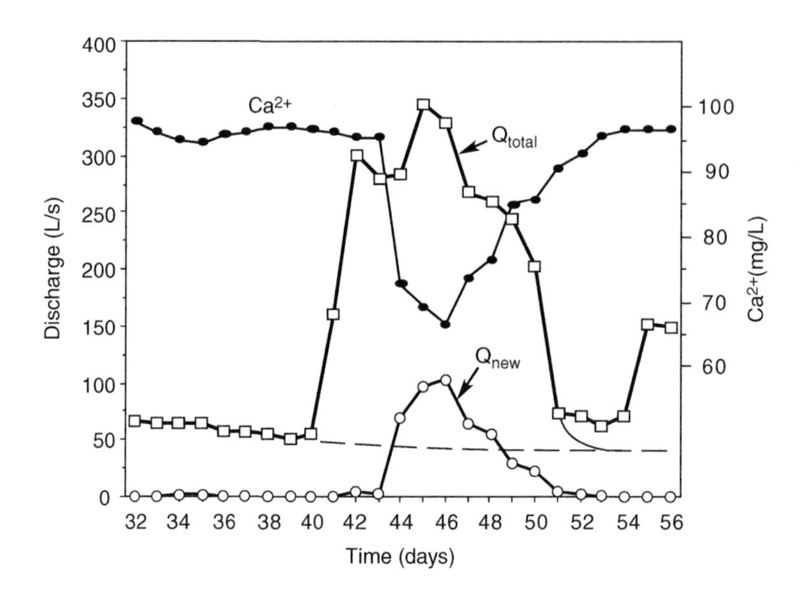

FIGURE 4–18 Spring hydrograph separation into conduit flow and diffuse flow components, that is, new rainfall and prerainfall water, respectively. (Modified from Kresic, 2007.)

maximum spring discharge and the maximum discharge of new water is again one day. New water stopped discharging at the spring 14 days after the rainfall started influencing the overall flow rate. The new water contributed only 18.3 percent to the overall flow increase at the spring. The percentage of new water in the total spring discharge for the 14-day period is 14.6 percent (Kresic, 2007).

Understanding and quantifying the various flow components in spring discharge is very important for overall spring management, utilization, and protection, as illustrated by the several examples that follow.

The findings of Desmarais and Rojstaczer (2002) are typical of the fast conduit-flow-dominated spring discharge with no delay and no significant influence of soil cover or matrix storage. Initially, the discharge from the Maynardville Limestone spring in Bear Creek Valley, Tennessee, peaks approximately one to two hours from the midpoint of summer storms. This initial peak is likely due to surface loading, which pressurizes the aquifer and results in water moving out of storage. All the storms monitored exhibited recessions that follow a master recession curve very closely, indicating that storm response is fairly consistent and repeatable, independent of the time between storms and the configuration of the rain event itself. Electrical conductivity initially increases for 0.5–2.9 days (longer for smaller storms), the result of moving older water out of storage. This is followed by a 2.1–2.5 day decrease in conductivity, resulting from an increasing portion of low-conductivity recharge water entering the spring. Stable carbon isotope data and the calcite saturation index of the springwater also support this conceptual model. Spring flow following recharge episodes is to a great extent controlled by displaced water from the aquifer rather than by direct recharge through the soil zone.

The response of a karst aquifer to threshold events in spring discharge is explained by Herman, Toran, and White (2008), who analyzed the effects of three major hurricanes, Frances, Ivan, and Jeanne, traveling up the eastern United States from the Gulf Coast and bringing large amounts of rain to central Pennsylvania. Monitoring equipment in place at Arch Spring, Blair County, Pennsylvania, captured the effects of these storms on the karstic spring flow. Together, these storms revealed a quantitative limit for the carrying capacity of the conduit system. Ivan was a much more devastating storm to the area, because rain fell on ground already saturated by Frances, but the net discharge increase at the spring was greater during the earlier Frances storm. Storm water not transported through the Arch Spring system was diverted into surface channels during these storms.

Suspended sediment collected by an automatic sampler during Frances revealed that another threshold was crossed. Concurrent with increasing discharge and high-conductance water, maximum sediment concentrations (933 mg/L) exceeded previous fluxes by an order of magnitude. The timing of the sediment pulse indicates that high sediment concentrations occur not only when the storm water reaches the spring but also when stored water is flushed out of the karst spring system. Sediment previously deposited in the conduit system is flushed only when adequate flows occur, indicating that sediment transport in karst is marked by thresholds and is a strongly nonlinear process (Herman et al., 2008).

Boyer and Kuczynska (2003) compare the temporal variability in storm flow of fecal coliform bacteria densities and *Cryptosporidium parvum* oocyst densities in agriculturally impacted karst groundwater. As emphasized by the authors, the transmission of disease in groundwater is a topic of great concern to government agencies, groundwater specialists, and the general public. *Cryptosporidium parvum* oocyst densities ranged from 0 to 1050 oocysts/L, and mean storm densities ranged from 3.5 to 156.8 oocysts/L. Fecal coliform densities ranged from less than 1 CFU/100 mL to more than 40,000 CFU/100 mL, and geometric mean storm densities ranged from 1.7 CFU/100 mL to more than 7000 CFU/100 mL. Fecal coliform densities correlated well with flow during storms, but *Cryptosporidium* oocyst densities exhibited a great deal of sample-to-sample variability and were not correlated with flow. Fecal coliform densities did not correlate positively with *Cryptosporidium* oocyst densities. Fecal coliform densities were greatest at storm peaks, when sediment loads were also greatest. As concluded by the authors, multiple transport mechanisms for fecal coliform bacteria and *C. parvum* oocysts may necessitate various agricultural land management and livestock health maintenance practices to control movement of pathogens to karst groundwater.

Goeppert and Goldscheider (2008) investigated solute and colloid transport in karst aquifers under low and high flows using tracer tests with fluorescent dyes (uranine), and microspheres of the size of pathogenic bacteria (1 μm) and *Cryptosporidium* cysts (5 μm), which were injected into a cave stream and sampled at a spring 2.5 km away. Uranine breakthrough curves (BTCs) were regular shaped and recovery approached 100 percent. Microsphere recoveries ranged between 27 and 75 percent. During low flow, the 1 μm spheres displayed an irregular BTC preceding the uranine peak. Only a very few 5 μm spheres were recovered. During high flow, the 1 μm sphere BTC was regular and more similar to the uranine curve. BTCs were modeled analytically using a conventional advection-dispersion model and a two-region nonequilibrium model. The results show that (1) colloids travel at higher velocities than solutes during low flow, (2) colloids and solutes travel at similar velocities during high flow, and (3) higher maximum concentrations occur during high flow.

4.4 PROBABILITY OF SPRING FLOWS

Analysis of spring hydrographs always includes determination of the general statistical parameters of the time series, such as average, minimum, and maximum flows for the period of record; standard deviation (variance) of the flows; coefficient of variation; flow duration curves; and frequency of characteristic flows at the minimum. Unfortunately, in the majority of real-life situations, a spring under investigation has relatively short periods of flow observation at irregular intervals. This simple fact may prevent any statistically rigorous quantitative analysis altogether. At the end, however, engineers and scientists engaged in a spring-related project still face various decisions that must include some type of quantitative analysis. Here are just some of the examples: What is the likely minimum flow in summer? Will the spring ever go dry? How big should the storage reservoir be? What is the "guaranteed" quantity of water delivered to end users? What should dimensions of the overflow structure be?

In any case, regardless of the available period of record and any specific question, most if not all answers have to be given in terms of probability. The first and most important step in the probability analysis is to evaluate the nature and occurrences (time series) of major water inputs, such as precipitation in the drainage area, and display them together with the spring hydrograph on the same timescale. This also includes long-term flow records of the nearest surface streams or other springs in similar physical settings. By visually comparing different but interconnected hydrologic time series, one gains an invaluable insight into the time-dependent processes that seem most important for the particular spring. Any short-term or long-term anthropogenic influences, such as seasonal groundwater withdrawal (pumpage) from the common aquifer or land-use changes (such as urbanization) that may affect the spring discharge should also be accounted for.

Depending on the goal(s) of the project, some characteristic flows may be more important than others. Common probability analyses include minimum and average monthly flows, minimum weekly flows for the period of highest demand (typically during August-September in moderate humid climates), and absolute minimum daily flow. There is no universally accepted limit on the data points required for probability analysis, although quite a few textbooks on statistics give a number of 30 as a "rule of thumb." As mentioned before, observation periods in practical hydrologic and groundwater studies are often short (several years or less) and this rule of thumb would invalidate many studies by default. However, in addition to the best professional judgment, probability analysis in such cases includes generation of synthetic time series using stochastic models (see Chapter 5) and reconstruction of the likely historic time series using regression with another, related time series (e.g., precipitation) that has a longer observation period (Chapter 5).

Figure 4–19 illustrates important characteristics of spring hydrographs that should be considered when performing probability analyses. Some of the questions that can be easily answered from the long record of the Crater Lake Spring discharge include

- Is there a regular pattern of spring discharge from one hydrologic year to another? (Yes.)
- Is the discharge strongly influenced by precipitation? (Yes.)
- Is there a general delay between major precipitation seasons and spring discharge? (Yes, the highest precipitation is in November-December, followed by the secondary peak in March-April; the highest flow rates are between mid-May and mid-July.)
- Is there a long-term periodicity in spring discharge? (Yes, approximately 14 years between wet and dry periods.)
- Is there a long-term trend unrelated to natural periodicity? (No, the spring behaved like a perfectly reliable clock for almost 30 years now.)

It is obvious that, if the period of record included just one or two randomly selected years from the graph in Figure 4–19, any answer to these questions would be much less certain (less probable).

A very good public-domain general text, *Hydrologic Frequency Analysis* (Engineering Manual 1110-2-1415), is available on the U.S. Army Corps of Engineers Web site at http://140.194.76.129/publications/eng-manuals. It covers all major aspects of the subject and provides guidance in fitting frequency distributions and construction of confidence limits. Techniques are presented that could reduce the errors caused by small sample sizes. Also, some types of hydrologic data are noted that usually do not fit any theoretical distributions.

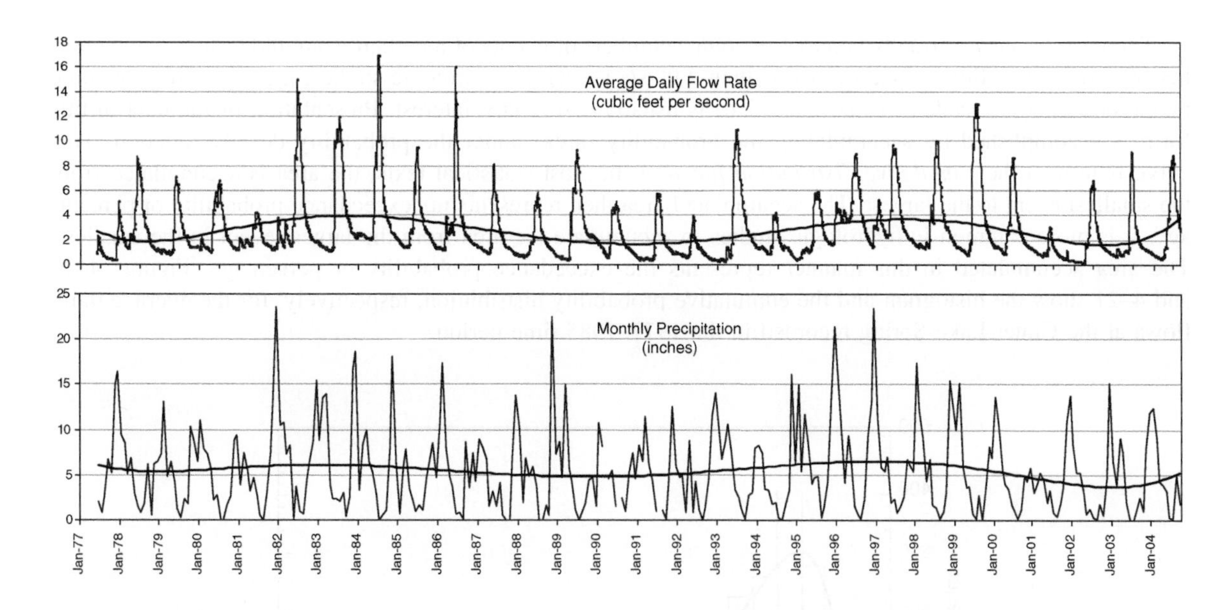

FIGURE 4–19 (Top) Average daily discharge rate, in cubic feet per second, at the Crater Lake Spring, Oregon (data from U.S. Geological Survey, 2008). (Bottom) Monthly precipitation, in inches, at Crater Lake, Oregon (data from Oregon Climate Service, 2008). The natural periodic cycle is presented with the sixth-order polynomial line.

Frequency analysis can be graphical (empirical) and analytical (theoretical). As emphasized by the U.S. Army Corps of Engineers (1993), every set of frequency data should be plotted graphically, even though the frequency curves are obtained analytically. It is important to visually compare the observed data with the derived curve. The graphical method of frequency curve determination can be used for any type of frequency study, but analytical methods have certain advantages when they are applicable. The principal advantages of graphical methods are that they are generally applicable, the derived curve can be easily visualized, and the observed data can be readily compared with the computed results. However, graphical methods of frequency analysis are generally less consistent than analytical methods, as different individuals draw different curves. Also, graphical procedures do not provide means for evaluating the reliability of the estimates. Comparison of the adopted curve with plotted points is not an index of reliability, but it is often erroneously assumed to be, thus implying a much greater reliability than is actually attained. For these reasons, graphical methods should be limited to those data types where analytical methods are known not to be generally applicable; that is, where frequency curves are too irregular to compute analytically.

Fitting data by an analytical procedure consists of selecting a theoretical frequency distribution, estimating the parameters of the distribution from the data by some fitting technique, then evaluating the distribution function at various points of interest. Some theoretical distributions that have been used in hydrologic frequency analysis are the normal (Gaussian), lognormal, exponential, two-parameter gamma, three-parameter gamma, Pearson type III, log-Pearson type III, extreme value (Gumbel), and log Gumbel.

Determining the frequency distribution of data by the use of analytical techniques has several advantages. The use of an established procedure for fitting a selected distribution results in consistent frequency estimates from the same data set by different persons. Error distributions have been developed for some of the theoretical distributions that enable computing the degree of reliability of the frequency estimates. Another advantage is that it is possible to regionalize the parameter estimates, which allows making frequency estimates at ungauged locations.

The term *frequency* usually connotes a count (number) of events of a certain magnitude, such as the number of daily spring flows equal to or less than 2 $ft^3/s/a$ (year). In hydrologic studies, the probability of some magnitude being exceeded (or not exceeded) is usually of primary interest. Presentation of the data in this form is accomplished by accumulating the probability (area) under the probability density function. This curve is termed the *cumulative distribution function*. In most statistical texts, the area is accumulated from the smallest event to the largest. The accumulated area then represents nonexceedance probability or percentage. It is more common in hydrologic studies to accumulate the area from the largest event to the smallest. The area accumulated in this manner represents the exceedance probability or percentage. Figures 4–20 and 4–21 show the histogram and the cumulative probability distribution, respectively, for the average daily flows at the Crater Lake Spring recorded in the 1977–1985 time period.

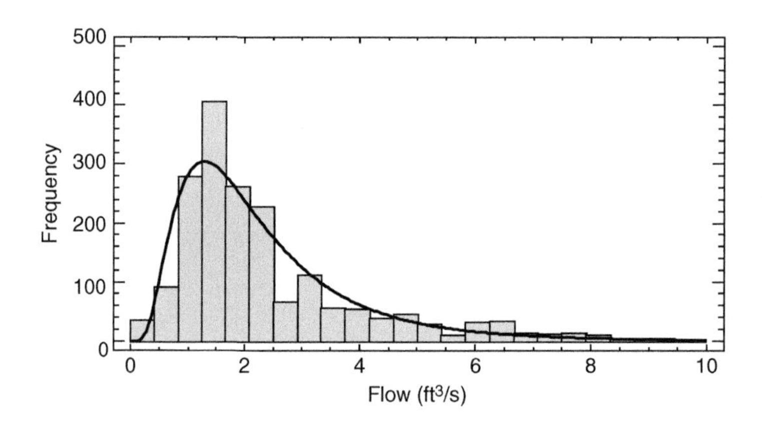

FIGURE 4–20 Histogram of average daily flows at the Crater Lake Spring, Oregon, recorded in 1977–1985. The most frequent flow is about 1.5 cfs, occurring 400 days out of 2920 (eight years).

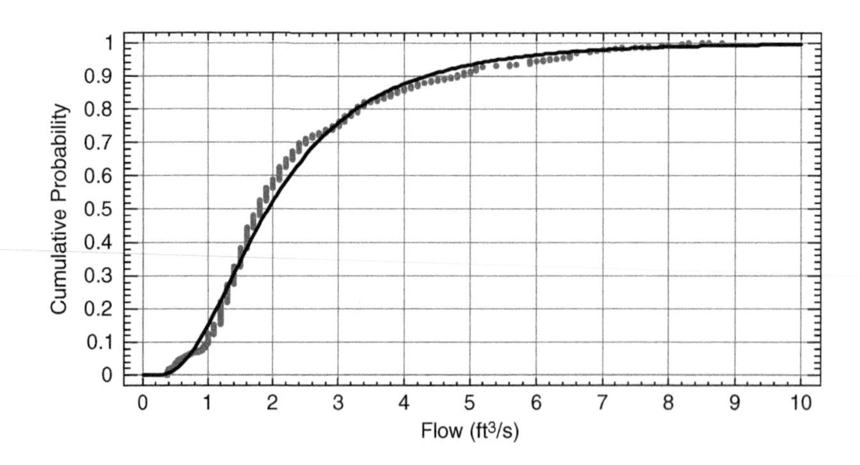

FIGURE 4–21 Cumulative empirical probability distribution (points) and fitted theoretical lognormal probability function of average daily flows at the Crater Lake Spring, Oregon, recorded in 1977–1985.

Table 4–1 Percentiles of the Crater Lake Spring Daily Flow Rates for 1977–1985

Percentile	Flow Rate (cfs)
1%	0.39
5%	0.62
10%	1.0
25%	1.3
50%	1.8
75%	3.0
90%	4.8
95%	6.2
99%	7.9

Two other terms are often used to describe spring flow probability: the *return period* (sometimes called *recurrence interval*) and the *percentiles* (or *quantiles*) of a probability distribution. The theoretical return period is the inverse of the probability that the event will be exceeded in any one year. For example, a 10 year maximum flow has a 1 in 10 = 0.1 or 10 percent chance of being exceeded in any one year and a 50 year maximum flow has a 0.02 or 2 percent chance of being exceeded in any one year. Conversely, a 10 year minimum flow has a 90 percent chance of being exceeded in any one year.

Percentiles are values below which specific percentages of the data are found. For example, for the nine year daily record of the Crater Lake the 25th percentile for the empirical data (measured flows) is 1.3 cubic feet per second (cfs), which means that 25 percent of the time the flow rate is lower and 75 percent of the time it is higher. Table 4–1 shows other characteristic percentiles.

Percentiles are often used to plot monthly or weekly average daily flows for the period of record and compare them with the currently measured flow at the spring (Figure 4–22). Combined with the similar plots of historic and current (recent) monthly precipitation, this practice can be very useful in spring management by anticipating likely flows in the near future. For example, if the currently measured flow is less than the 10th percentile (i.e., the flow is higher 90 percent of the time), the same is true for the recent precipitation, and if the spring discharge is significantly dependent on precipitation, it may be necessary to impose some type of use restriction before the hydrologic and meteorological conditions improve.

4.4.1 Probability of minimum and maximum flows

The procedure for identifying the probability of flow extremes is illustrated with examples of two large karst springs in Croatia. Table 4–2 presents the main characteristics of the two springs (name of spring and gauging station, elevation, estimated basin area, and period of available data). Both stations are located in the central part of the Dinaric karst of Croatia. Table 4–3 presents characteristic discharges measured at two analyzed karst springs for the period given in the last column of Table 4–2.

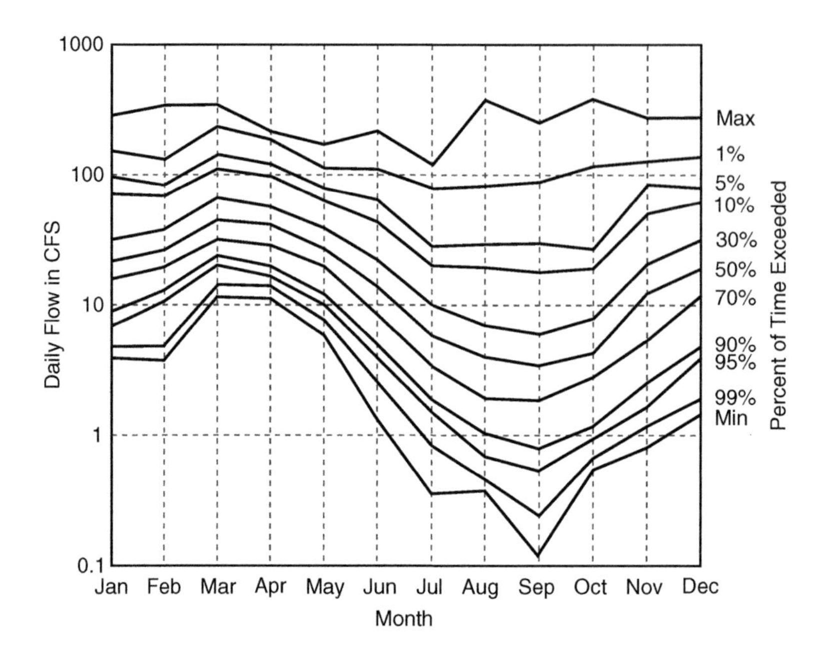

FIGURE 4–22 Daily flow duration curves for each month. (Modified from U.S. Army Corps of Engineers, 1993.)

Table 4–2 Main Characteristics of the Two Analyzed Karst Spring Discharge Gauging Stations

No.	River Spring Name	Gauging Station Name	Elevation, H (m asl)	Estimated Basin Area, *A* (km^2)	Period of Available Data
1	Kupa	Kupari	304.43	208	1951–2007
2	Ombla	Komolac	2.38	Between 800 and 900	1968–2007

Table 4–3 Characteristic Annual Discharges, in m^3/s, at the Two Analyzed Karst Springs for the Periods Given in the Last Column of Table 4–2

No.	Gauging Station and River Spring Name	Minimum, Q_{min}	Mean, Q_{av}	Maximum, Q_{max}
1	Kupari-Kupa	0.31	13.5	195
2	Komolac-Ombla	3.96	25.0	104

It should be noted that the recent awareness of possible global warming has emphasized the need for the frequency analysis of extreme discharges, as the question of whether global climate changes might increase their number is being often raised.

The large numbers of karst springs (Bonacci, 2001; Panagopoulos and Lambrakis, 2006; Fleury, Plagnes, and Bakalowicz, 2007; Herman et al., 2008) show that, under conditions of extremely intense precipitation, a maximum limiting value exists for the discharge of the main springs in a catchment, independent of the catchment size, watershed conditions (e.g., antecedent soil moisture content and groundwater levels), and amount of precipitation. For example, recent observations of the Arch Spring karst drainage system (United States) during exceptional runoff from Hurricanes Frances and Ivan revealed quantitatively the magnitude of discharge necessary to override the karst drainage system and spill excess discharge into the surface channel (Herman et al., 2008).

The phenomenon of limited maximum-discharge capacity of karst springs is often not included in rainfall-runoff process modeling, which is probably one of the main reasons for the present poor quality of predictive models of maximum spring flows. Possible reasons for the limitation on the maximum flow rate are (Bonacci, 2001) (1) limited dimensions of the karst conduit(s); (2) flow under pressure; (3) inter-catchment overflow; (4) overflow from a main spring to intermittent springs within the same or adjacent catchments; (5) available water-storage capacity in the epikarstic zone of the spring catchment (Williams, 1983); and (6) factors such as climate, soil and vegetation cover, catchment elevation, and geological conditions.

Similarly, in some cases, the minimum discharges from a karst springs have a limit that cannot be exceeded regardless of duration and severity of drought (period without precipitation). For example, the Jadro Spring and the Ombla Spring minimum discharge is about 4 m^3/s, and the entire Gacka River spring zone never had minimum discharge less than 2.3 m^3/s. All three karst springs are located in the Croatian part of the Dinaric karst region. The Fontaine de Vaucluse karst spring (France) has a minimum discharge of more than 4 m^3/s (Fleury et al., 2007). This phenomenon can be explained by their large, prolific aquifers, which provide constant water supply during the long-lasting dry periods.

The Ombla Spring represents a typical karst spring with limited maximum and minimum outflow capacity (Bonacci, 1995, 2001). The minimum and maximum annual discharge time series measured from 1968 to 2007 at the Komolac gauging station are presented in Figures 4–23 and 4–24, respectively. It can be seen that

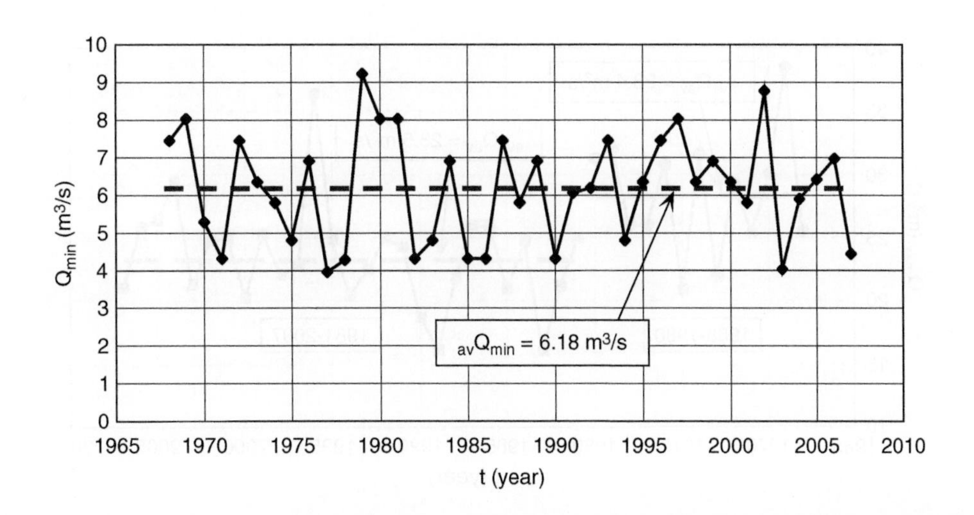

FIGURE 4–23 Time series of minimum annual discharges measured at the Komolac discharge gauging station (Ombla Spring) for the period 1968–2007.

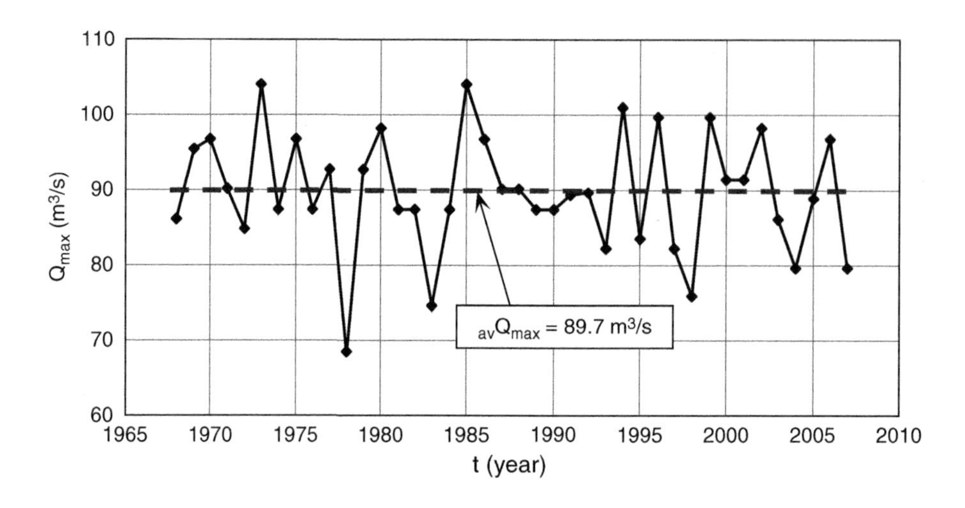

FIGURE 4–24 Time series of maximum annual discharges measured at the Komolac discharge gauging station (Ombla Spring) for the period 1968–2007.

minimum discharges never drop below about 4 m^3/s, while the maximum discharges never exceeded a value of 104 m^3/s, despite a large catchment and very intensive rainfall falling on it. Two time subseries of mean annual discharges measured from 1968 to 1980 and 1981 to 2007 at the Ombla karst spring at Komolac are presented in Figure 4–25. The sharp drop in mean annual discharge of about 5 m^3/s in 1981 was caused by large civil engineering works (canalization of the 60-km-long Trebišnjica River watercourse with spray concrete) in the upper part of the spring catchment (Bonacci, 1987, 1995). It is very interesting that these massive civil engineering works strongly influenced the mean annual discharge but not the extreme discharges (maximum and minimum) of the Ombla Spring.

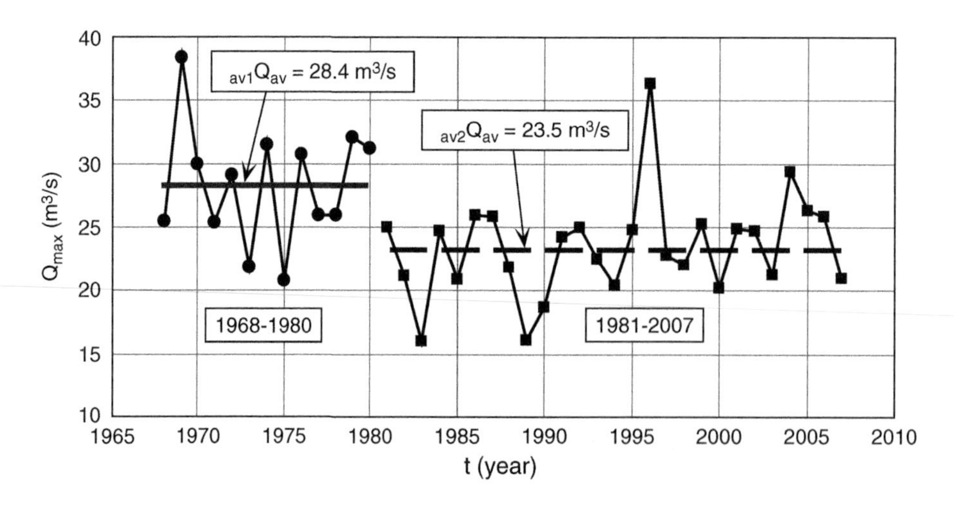

FIGURE 4–25 Two time subseries of mean annual discharges measured at the Komolac discharge gauging station (Ombla Spring) for the subperiods 1968–1980 and 1981–2007.

4.4.2 Time series analysis

Stochastic hydrology considers chronological sequences of hydrological events with the goal of explaining the irregularities of occurrence and, in particular, forecasting the incidence of important extremes (Shaw, 1994). Extreme discharges may be considered random events, since their future occurrences cannot be readily predicted. As explained in more detail in Chapter 5, time series analysis includes identification of the trend, periodic, and stochastic (random) components in the observed data. The effects of rare events cause occurrence of the so-called catastrophic component of a time series.

A period of longer than usually available historic measurements is often needed to determine whether the record has captured a normal fluctuation or an anomaly. The optimum requirement is long records of continuous homogeneous measurements. Fundamental assumptions for the most statistical analyses of discharge time series are (1) consistency, (2) freedom from trends, and (3) that it constitutes a stochastic process whose random component follows the appropriate probability distribution function (mostly normal distribution) (Adeloye and Montaseri, 2002).

Consistency implies that all the collected data belong to the same statistical population. A time series is stationary if the statistics of sample (e.g., mean, standard deviation, and variance) are not a function of the timing or the length of the sample. A trend in a data set exists if there is a significant correlation (positive or negative) between the observations and time. Occurrence of trend and periodicity in a time series means that it is not stationary. The modeling of a time series is much easier if it is stationary. Generally, randomness in a discharge time series means that the data arise from natural causes (Adeloye and Montaseri, 2002).

Different statistical tests are used to evaluate the aforementioned characteristics of time series. The most widely used method for evaluating the consistency of a time series is the double mass curve (Bras, 1990). The Spearman rank order correlation nonparametric test can be used to investigate the trend (McGhee, 1985; Adeloye and Montaseri, 2002). The nonparametric run test is appropriate for assessment of time series randomness (McGhee, 1985; Adeloye and Montaseri, 2002).

A trend or nonstationarity is very often introduced by human activities but can also be a consequence of natural changes. Levi (2008) stresses that scientists studying the time series of many climate-related variables had been noticing a rather sudden change, an inflection point, around the mid-1980s. They wondered whether the change was due to natural variability or to the greenhouse-gas impact.

Figure 4–26 shows a time series of maximum annual discharges ($Q_{\max}$) with the linear trend line for the period 1951–2007 measured at the discharge gauging station Kupari on the Kupa River (Croatia). This station controls discharge of the Kupa River karst spring with an estimated catchment area of 208 km^2. From the graphical presentation, it appears that there is a statistically insignificant decreasing linear trend.

A time series analysis can detect and quantify trends and fluctuations in records. The rescaled adjustment partial sums (RAPS) method (Garbrecht and Fernandez, 1994; Bonacci, Trninić, and Roje-Bonacci, 2008) can be used successfully for this purpose. A visualization approach based on RAPS overcomes small systematic changes in records and the variability of data values. The RAPS visualization highlights trends, shifts, data clustering, irregular fluctuations, and periodicities in the record. The visualization pattern identification is to be considered a complement to standard statistical tests (Garbrecht and Fernandez, 1994).

The values of RAPS are defined by the following equation:

$$\mathrm{RAPS}_k = \sum_{t=1}^{k} \frac{Y_t - \bar{Y}}{S_Y} \tag{4.34}$$

where $\bar{Y}$ is the sample mean, S_Y is the standard deviation, n is the number of values in the time series, and $(k = 1, 2, \ldots, n)$ is the counter limit of the current summation. The plot of the RAPS in time is the

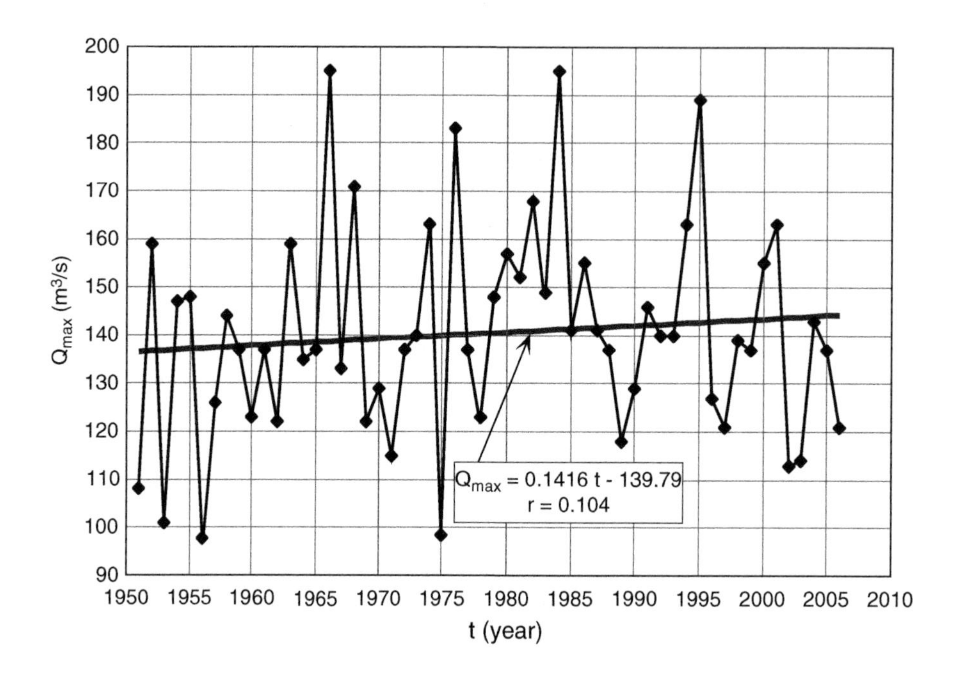

FIGURE 4–26 Time series of the maximum annual discharges (Q_{max}) with a linear trend line for the period 1951–2007 measured at the discharge gauging station Kupari on the Kupa River.

visualization of the trends and fluctuations of the analyzed variable Y_t. The rescaling and summation of the RAPS approach highlights small yet systematic changes over time that are often hidden in a standard time series plot by the comparatively large magnitude and variability of data itself (Garbrecht and Fernandez, 1994). The visualization is not a substitute for statistical evaluation. It allows a preliminary or exploratory inspection of the time series. The RAPS like all other methods is not without shortcomings.

Figure 4–27 represents time series of the RAPS for maximum annual Kupa River spring discharges in the period 1951–2007 shown in Figure 4–26. The total time series from Figure 4–26 was divided into three subsets: (1) 1951–1978, (2) 1979–1987, (3) 1988–2007. Three time subseries of the Kupa River karst spring maximum annual discharges (Q_{max}) with average lines for the three subperiods are shown in Figure 4–28. To investigate statistically significant differences between the averages of the three time subseries of maximum annual discharges, a t-test was used. The averages 137 m^3/s for the 1951–1978 subperiod, 156 m^3/s for the 1979–1987 subperiod, and 137 m^3/s for the 1988–2007 subperiod are statistically significant (beyond 5 percent).

Application of the RAPS method to a long-term (1878–2004) series of mean annual discharges of the karst spring Fontaine de Vaucluse (France) indicated existence of the following five statistically significant subseries: (1) 1878–1910, (2) 1911–1941, (3) 1942–1959, (4) 1960–1964, (5) 1965–2004 (Bonacci, 2007). Variations in the Fontaine de Vaucluse karst spring hydrological regime during a period of 127 years are very strong and cannot be ignored. Anthropogenic impacts are probably the main cause of such behavior of the mean annual spring discharge time series analyzed, although the natural cycles of dry and wet years are also a possible explanation. In any case, a strict division of natural and anthropogenic influences on the spring discharge regime is hardly possible.

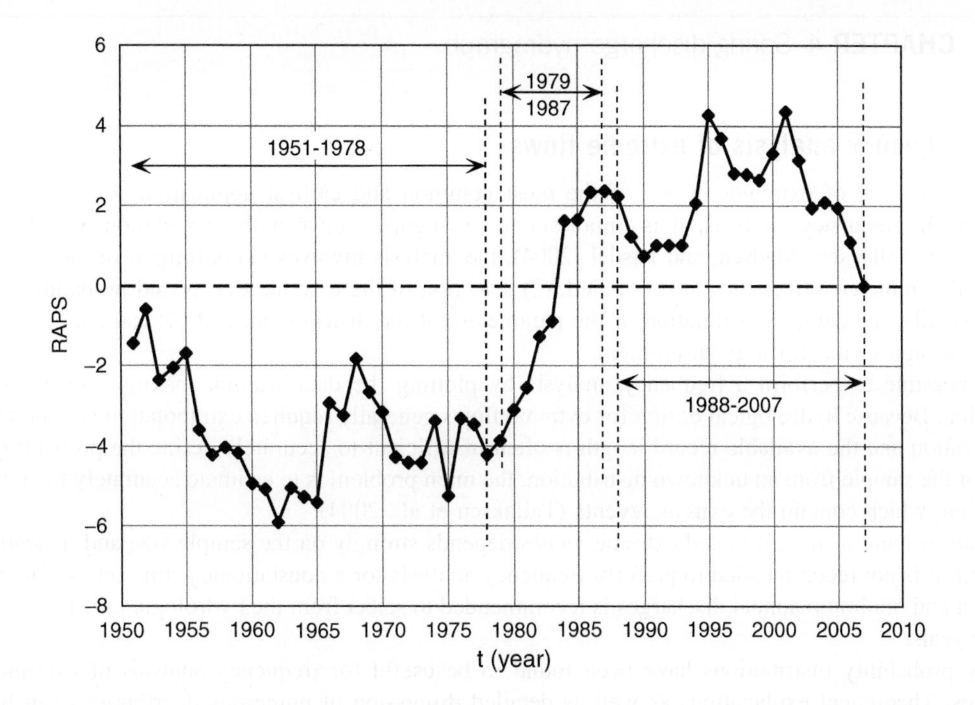

FIGURE 4–27 Time series of the RAPS for the maximum annual discharges, Q_{max} (Figure 4–24) for the period 1951–2007 measured at the discharge gauging station Kupari on the Kupa River with indication of three time subseries: (1) 1951–1978, (2) 1979–1987, (3) 1988–2007.

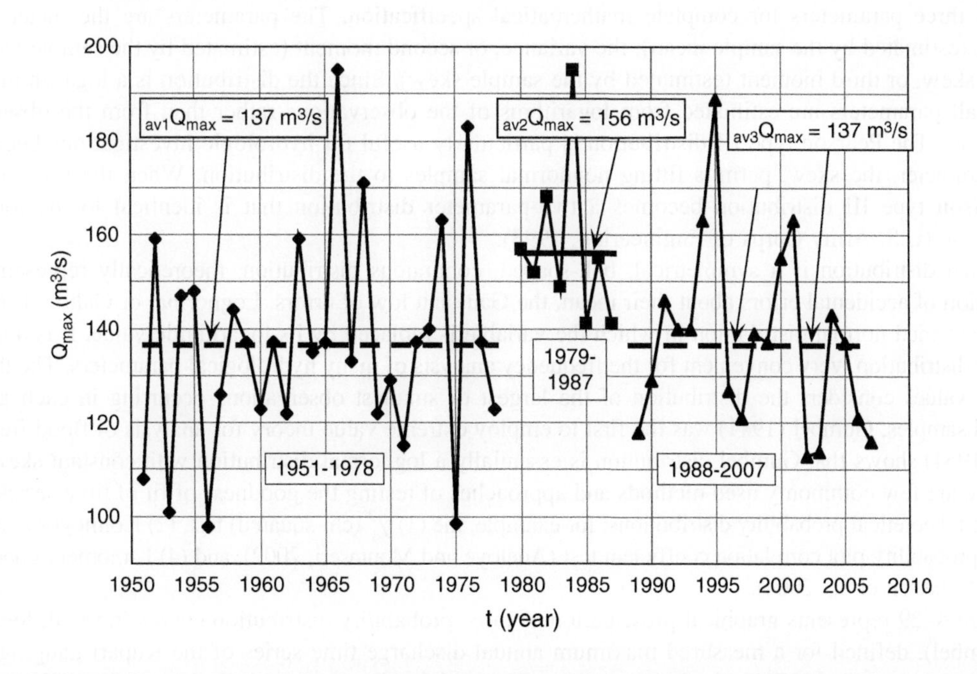

FIGURE 4–28 Three time subseries of the Kupa River karst spring maximum annual discharges, Q_{max}, with average lines for three defined subperiods: (1) 1951–1978, (2) 1979–1987, (3) 1988–2007.

4.4.3 Frequency analysis of extreme flows

Frequency analysis of extremes is one of the most common and earliest applications of statistics within hydrology. In frequency analysis, it is important to distinguish between the population and the sample. According to Tallaksen, Madsen, and Hisdal (2004), the analysis involves (1) definition of the hydrological event and extreme characteristics to be studied, (2) selection of the extreme events and probability distribution to describe the data, (3) estimation of the parameters of the distribution, and (4) estimation of extreme events or design values for a given problem.

It is possible to perform a frequency analysis by plotting the data without making any distributional assumption. Because hydrological design for extreme flows generally requires extrapolation beyond the range of observation and the available record length is often insufficient to accurately define the probability distribution for the sample from an unknown distribution, the main problem is to estimate accurately the tails of the distribution, which contain the extreme events (Tallaksen et al., 2004).

The uncertainty of the estimated extreme values depends strongly on the sample size and its stationarity. In general, it is not recommended to perform frequency analysis for a nonstationary time series. The value of maximum and minimum annual discharges is recommended to select from the hydrologic rather than from the calendar years.

Many probability distributions have been found to be useful for frequency analysis of extreme spring discharges. Theoretical explanations as well as detailed discussion of numerous distributions can be found in many standard textbooks on statistics. Here, we give a short explanation of three of them: (1) normal, (2) lognormal, and (3) Gumbel or extreme value distribution.

In the United States and quite a few other countries, the analytical frequency procedure recommended for annual maximum and minimum stream flows is the logarithmic Pearson type III distribution. This distribution requires three parameters for complete mathematical specification. The parameters are the mean, or first moment (estimated by the sample mean); the variance, or second moment (estimated by the sample variance); and the skew, or third moment (estimated by the sample skew). Since the distribution is a logarithmic distribution, all parameters are estimated from logarithms of the observations rather than from the observations themselves. The Pearson type III distribution is particularly useful for hydrologic investigations because the third parameter, the skew, permits fitting nonnormal samples to the distribution. When the skew is 0, the log-Pearson type III distribution becomes a two-parameter distribution that is identical to the lognormal distribution (U.S. Army Corps of Engineering, 1993).

Normal distribution is a symmetrical, bell-shaped, continuous distribution, theoretically representing the distribution of accidental errors about their mean, the Gaussian low of errors. Lognormal or Galton distribution is a transformed normal distribution in which the variable is replaced by its logarithmic value. It is a nonsymmetrical distribution, very convenient for the frequency analysis of many hydrological parameters. The theory of extreme values considers the distribution of the largest or smallest observations occurring in each group of repeated samples. Gumbel (1941) was the first to employ extreme value theory for analysis of flood frequency. Chow (1954) shows that Gumbel distribution is essentially a lognormal distribution with constant skewness.

There are few commonly used methods and approaches of testing the goodness of fit of time series data to postulated theoretical probability distributions; for example, the (1) χ^2 (chi-squared) test, (2) Kolmogorov-Smirnov test, (3) probability plot correlation coefficient test (Adeloye and Montaseri, 2002), and (4) L-moment goodness of fit test (Tallaksen et al., 2004).

Figure 4–29 represents graphical presentation of three probability distribution curves (normal, lognormal, and Gumbel), defined for a measured maximum annual discharge time series of the Kupari gauging station over the period 1951–2007. Table 4–4 gives the values of maximum annual discharges for different recurrence periods using the three previously mentioned distributions.

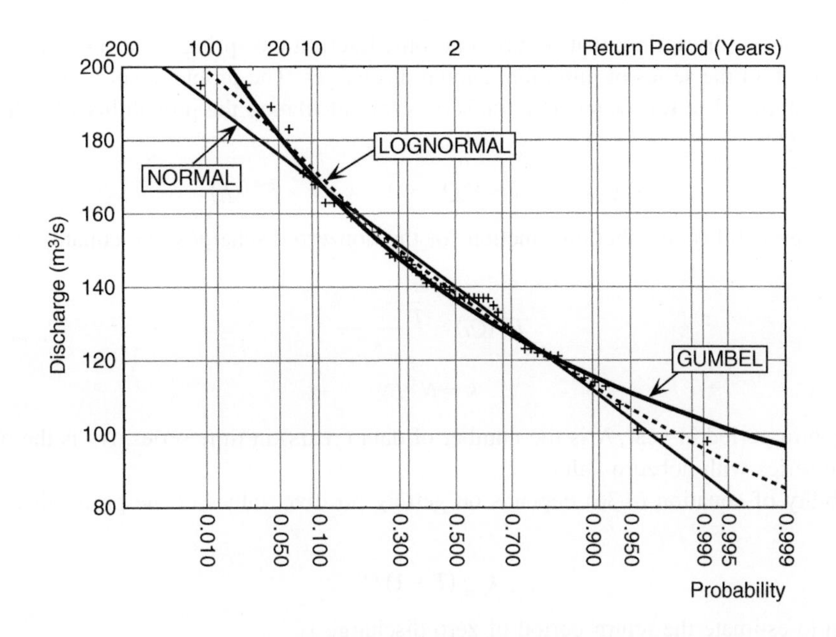

FIGURE 4–29 Graphical presentation of three probability distribution curves (normal, lognormal, and Gumbel), defined for measured maximum annual discharge time series of the Kupari gauging station over the period 1951–2007.

Table 4–4 Normal, Lognormal, and Gumbel Values, Defined for the Measured Maximum Annual Discharge Time Series of the Kupari Gauging Station over the Period 1951–2007

Frequency	Recurrence Period (year)	Normal (m^3/s)	Lognormal (m^3/s)	Gumbel (m^3/s)
0.5	2	140	138	136
0.2	5	159	158	156
0.1	10	168	169	169
0.05	20	176	179	181
0.04	25	178	182	185
0.02	50	185	191	197
0.01	100	192	199	209
0.005	200	197	207	221
0.002	500	204	217	237
0.001	1000	208	225	249

The zero flow values (dry spring) that occurred at some intermittent springs pose a special problem for the frequency analysis of a time series of minimum annual discharges. The problem can be solved using the theorem of total probability. The following equation serves for calculating the probability of a discharge less or equal to discharge Q_i:

$$P(Q_{\min} \leq Q_i) = F(Q_i) = 1 - k + k \times F^*(Q_i) \tag{4.35}$$

where $F^*(Q_i)$ is a probability distribution function for the nonzero discharges, the equation for which is

$$F^*(Q_i) = \frac{\frac{1}{T} - 1 + k}{k} \tag{4.36}$$

$$k = N^*/N \tag{4.37}$$

where T is the return period in year, N is the number of data (years) in time series, N^* is the number of data (years) in a time series with nonzero values.

The applicability of equation (4.36) depends on getting positive values of the probability $F^*(Q_i)$, which means that

$$k \geq (T - 1)/T \tag{4.38}$$

The equation to estimate the return period of zero discharge is

$$T(0) = N/(N - N^*) \tag{4.39}$$

If a karst spring has a limited maximum or minimum outflow capacity, a classical probability concept of the frequency analyses of extreme discharges should be applied carefully. In this case, the definition of discharges for the higher recurrence intervals (for example, 100 years and more) has no physical sense.

REFERENCES

Adeloye, A.J., Montaseri, M., 2002. Preliminary streamflow data analyses prior to water resources planning study. Hydrol. Sci. J. 47 (5), 679–691.

Amorocho, J., 1964. Nonlinear hydrologic analysis. Advances in Hydroscience, vol. 9. Academic Press, New York.

Atkinson, T.C., 1977. Diffuse flow and conduit flow in limestone terrain in Mendip Hills, Somerset (Great Britain). J. Hydrol. 35, 93–100.

Avdagic, I., 1990. Osnove hidrologije krsa [Fundamentals of Karst Hydrology; in Serbo-Croatian]. Zavod za hidrotehniku Gradjevinskog fakulteta u Sarajevu, radovi br. 27, Sarajevo.

Bonacci, O., 1987. Karst Hydrology with Special Reference to the Dinaric Karst. Springer-Verlag, Berlin.

Bonacci, O., 1993. Karst springs hydrographs as indicators of karst aquifers. Hydrological Sciences 38 (1), 51–62.

Bonacci, O., 1995. Ground water behaviour in karst: Example of the Ombla Spring (Croatia). J. Hydrol. 165 (1–4), 113–134.

Bonacci, O., 2001. Analysis of the maximum discharge of karst springs. Hydrogeol. J. 9 (4), 328–338.

Bonacci, O., 2004. Hazards caused by natural and anthropogenic changes of catchment area in karst. Nat. Haz. and Earth Sys. Sci. 4 (5–6), 655–661.

Bonacci, O., Trninić, D., Roje-Bonacci, T., 2008. Analyses of the water temperature regime of the Danube and its tributaries in Croatia. Hydrol. Processes 22 (7), 1014–1020.

Boussinesq, J., 1904. Recherches théoriques sur l'écoulement des nappes d'eau infiltrées dans le sol et sur les débits des sources. Journal de Mathématiques Pures et Appliquées, Paris 10, 5–78.

Boyer, D.G., Kuczynska, E., 2003. Storm and seasonal distributions of fecal coliforms and cryptosporidium in a spring. J. Am. Water Resour. Assoc. 39 (6), 1149–1156.

Bras, R.L., 1990. Hydrology—An introduction to hydrological science. Addison-Wesley, Reading, MA.

Castany, G., 1967. Traité Pratique des Eaux Souterraines. Dunod, Paris.

Chow, V.T., 1954. The log-probability and its engineering application. Proceedings of the ASCE 80. Paper No. 536, 1–25.

Desmarais, K., Rojstaczer, S., 2002. Inferring source waters from measurements of carbonate spring response to storms. J. Hydrol. 260, 118–134.

Dreiss, S.J., 1989. Regional scale transport in a karst aquifer. 1. Component separation of spring flow hydrographs. Water Resour. Res. 25 (1), 117–125.

Drogue, C., 1972. Analyse statistique des hydrogrammes de decrues des sources karstiques. J. Hydrol. 15, 49–68.

Fleury, P., Plagnes, V., Bakalowicz, M., 2007. Modelling of the functioning of karst aquifers with a reservoir model: Application to Fontaine de Vaucluse (South of France). J. Hydrol. 345 (1–2), 38–49.

Garbrecht, J., Fernandez, G.P., 1994. Visualization of trends and fluctuations in climatic records. Wat. Res. Bulletin 30 (2), 297–306.

Goeppert, N., Goldscheider, N., 2008. Solute and Colloid Transport in Karst Conduits under Low- and High-Flow Conditions. Ground Water 46 (1), 61–68.

Gumbel, E.J., 1941. The return period of flood flows. Ann. Math. Statist. 12 (2), 163–190.

Healy, R.W., Winter, T.C., LaBaugh, J.W., Franke, O.L., 2007. Water budgets: Foundations for effective water-resources and environmental management. U.S. Geological Survey Circular 1308, Reston, VA.

Herman, E.K., Toran, L., White, W., 2008. Threshold events in spring discharge: Evidence from sediment and continuous water level measurement. J. Hydrol. 351, 98–106.

Jevdjevic, V., 1956. Hidrologija, part 1 [Hydrology, volume 1; in Serbian]. Hidrotehnicki institut Ing. Jaroslav Cerni, Posebna izdanja, knjiga 4, Belgrade, Serbia.

Kresic, N., 1997. Quantitative Solutions in Hydrogeology and Groundwater Modeling. CRC Press/Lewis Publishers, Boca Raton, New York.

Kresic, N., 2007. Hydrogeology and Groundwater Modeling, Second ed. CRC Press/Taylor and Francis, Boca Raton, FL.

Kresic, N., 2009. Groundwater Resources: Sustainability, Management, and Restoration. McGraw Hill, New York.

Levi, B.G., 2008. Trends in the hydrology of the western US bear the imprint of manmade climate change. Physics Today 61 (4), 16–18.

Maillet, E., (Ed.) 1905. Essais díhydraulique souterraine et fluviale, Vol. 1. Herman et Cie, Paris.

McGhee, J.W., 1985. Introductory statistics. West Publishing Company, St. Paul, MN.

Oregon Climate Service, 2008. Climate Data. Available at: www.ocs.orst.edu/pub_ftp/climate_data/daily/prec/prec1946.lf (Accessed December 2008).

Panagopoulos, G., Lambrakis, N., 2006. The contribution of time series analysis to the study of the hydrodynamic characteristics of the karst systems: Application on two typical karst aquifers of Greece (Trifilia, Almyros Crete). J. Hydrol. 329 (3–4), 368–376.

Shaw, E.M., 1994. Hydrology in practice, third ed. Routledge, Abingdon, UK.

Soulios, G., 1991. Contribution à l'étude des courbes de récession des sources karstiques: Exemples du pays Hellénique. J. Hydrol. 127, 29–42.

Tallaksen, L.M., Madsen, H., Hisdal, H., 2004. Frequency analysis. In: Tallaksen, L.M., Van Lanen, H.A.J. (Eds.), Hydrological drought—Processes and estimation methods for streamflow and groundwater. Elsevier, Amsterdam, pp. 199–271.

U.S. Army Corps of Engineers, 1993. Hydrologic Frequency Analysis. Engineering manual 1110-2-1415, Washington, DC. Available at: http://140.194.76.129/publications/eng-manuals.

U.S. Geological Survey, 2008. USGS Ground-Water Data for the Nation. Available at: http://waterdata.usgs.gov/nwis/gw.

Vandike, J.E., 1996. The Hydrology of Maramec Spring. Water Resources Report Number 55. Missouri Department of Natural Resources, Division of Geology and Land Survey, Rola, MO.

Williams, P.W., 1983. The role of subcutaneous zone in karst hydrology. J. Hydrol. 61 (1–3), 45–67.

Yevjevich, V.M., (Ed.) 1981. Karst Water Research Needs. Water Resources Publications, Littleton, CO.

CHAPTER

5 Modeling

Neven Kresic
MACTEC Engineering and Consulting, Inc., Ashburn, Virginia

5.1 INTRODUCTION

What makes large springs particularly attractive as modeling targets also makes the task at hand more complex when compared to almost any other groundwater project. Namely, a large spring draining an extensive aquifer is the final result of *all* processes, both natural and anthropogenic, that generate the spring flow. As these processes act on the aquifer, groundwater continuously converges toward the spring from *all* directions and parts of the aquifer. Creating a quantitative, physically based, and reliable model of what "happens" and comes out of the spring is therefore very challenging but, at the same time, potentially very rewarding; having such a model would amount to having the ultimate tool for spring management and protection. Unfortunately, as discussed further, characterizing and mathematically describing the multifaceted complexity of aquifers feeding large springs present ongoing challenges.

The main explanation for these challenges is straightforward: In aquifers with intergranular porosity, the equations of groundwater flow are based mostly on the relatively simple Darcy's law. On the other hand, the nature of porosity in fractured-rock, karst, and pseudokarst aquifers requires the application of different sets of equations for distinct porous media: (1) rock matrix, (2) common rock discontinuities such as fractures (fissures) and bedding planes, and (3) solutionally enlarged voids, such as channels and conduits developed from the initial discontinuities. Any meaningful quantitative integration of various equations describing these distinct flow regimes is further complicated by the uncertainties associated with the field distribution and identification of different porosity types.

Unfortunately, the results of the related efforts have generally lagged behind the advances in hydrogeology of intergranular (nonfractured, nonkarstic) aquifers. As a consequence, the majority of spring modeling approaches so far are based on various applications of time series analyses as well as general statistical and probabilistic methods developed in surface water hydrology. Such methods have one common thread, the necessity for a relatively long time series of data on aquifer recharge and spring discharge, that is, various input-output relationships. This, however, is also the key limiting factor for many practical engineering projects with short execution times.

It is therefore not surprising that quite a few practicing hydrogeologists still elect to characterize a fractured-rock or karst aquifer as an "equivalent porous medium" then describe groundwater flow or even the fate and transport of contaminants with deterministic (physically based) models based on Darcian equations, thus arriving at conclusions that are often questionable (Kresic, 2009a).

Copyright © 2010, Elsevier Inc. All rights reserved.

In any case, with the continued improvement in field and laboratory methods for fractured rock and karst aquifer characterization, including the development of new analytical and numerical solutions for groundwater flow, it is likely that the successful management of large springs will, in many cases, require simultaneous application of both time series and deterministic models, thus requiring close cooperation between "traditional" surface water hydrologists and "traditional" hydrogeologists.

To be useful as an engineering or spring management tool, any model should be able to answer at least some of the following or similar questions:

- What will the spring flow rate be at the end of this month if there is no rain? What would it be if there is average rainfall this month?
- If there is future permitted well pumpage of 200 L/s 3 km from the spring, how will that affect the spring?
- What water would be available from that spring if there were a "catastrophic" drought of historic proportions?
- Can we regulate the spring and withdraw more water in August-September than what it naturally discharges? Can we count on being able to do that every year? How will that affect the stream?
- Can we count on good water quality from that spring? All the time? How often do we have to treat water?
- Can we use that spring for a trout hatchery?
- What capacity hydroelectric plant should we build at the spring to be profitable at current energy prices?
- How long would it take the contaminant to reach the spring? What would the contaminant concentration be once it reaches the spring?

It is obvious that the ability to quantitatively (i.e., using a model) answer these questions depends primarily on the available data and the funds required to obtain such data, including the time involved. However, regardless of the selected modeling approach and the available data, any answer would be more or less probable by default. For example, even though deterministic spring models are based on some established law(s) of groundwater flow in the subsurface (aquifer), every prediction about the future state of the spring to some degree depends on future precipitation, which is a random periodic process.

Models that use mathematical equations to describe the elements of groundwater flow are called *mathematical*. Depending on the nature of the equations involved, these models can be empirical (experimental), probabilistic, or deterministic. Empirical models are derived from experimental data that are fitted to some mathematical function. A good example is Darcy's law. (Note that Darcy's law was later found to be theoretically grounded and actually became a physical or deterministic law.) Although empirical models are limited in scope, they can be an important part of a more complex modeling effort. For example, tracer studies in a spring catchment (drainage area) can lead to development of mathematical expressions describing the tracer(s) transport toward the spring, including fast (conduit) and slow (diffuse) flow components where applicable; these functions can then be used in stochastic time series modeling or as calibration targets in numeric deterministic models.

Probabilistic models are based on laws of probability and statistics. They can have various forms and complexity, starting with a simple probability distribution of a spring discharge or precipitation and ending with complicated stochastic time series models. Such models do not attempt to describe the physics of how various water inputs are transformed into spring discharge. Their main goal is to find an appropriate statistical-mathematical expression that can with reasonable accuracy perform this transformation of one (or more) time series into spring discharge or some other component of discharge (e.g., water turbidity or coliform count). The main limitations for a wider use of time series (stochastic) models in hydrogeology and hydrology of springs are that they (1) require large data sets for parameter identification and (2) typically cannot be used to answer (predict) some common questions from hydrogeologic practice, such as the effects of planned new pumping from the aquifer.

Deterministic models, which can be analytical or numeric, assume that the stage or future reactions of the system (aquifer) are predetermined by physical laws governing groundwater flow. An example is the flow of groundwater toward a fully penetrating well in a confined aquifer, as described with the Theis equation (Theis, 1935). Most problems in traditional hydrogeology are solved using deterministic models, which are described in more detail in Section 5.7.

5.2 CORRELATION AND REGRESSION

Regression between different variables that are correlated in some manner and represented with a sufficient number of data points is probably the most common quantitative method in many scientific fields. In the hydrology of springs, this usually refers to finding a simple or multiple regression equation describing spring flow rate or some of its characteristics (the so-called dependent variable) using an observed time series of variables known to influence it. These independent variables may include various water budget components and parameters, such as rainfall, aquifer water levels, and soil moisture (see Figure 4–1).

As with all other models developed from time series data, regression is performed between variables observed with the same time interval and represented by a sufficient number of observations within samples of the same length: Daily flows are correlated with daily rainfall, monthly flows are correlated with monthly rainfall, and so on. It is always recommended that single and multiple regression analyses be performed with as many related variables as possible before more complex time series models are developed. Regression is included by default in all commercial statistical computer programs, most spreadsheet programs, and many public-domain programs available on the Internet. Accompanied by various plots (graphs), it can provide invaluable insight into the aquifer-spring system.

Figures 5–1 through 5–5 illustrate one such analysis of daily flows at Comal and San Marcos Springs draining Edwards Aquifer in Texas (see Chapter 10.9 for the locations of these springs). The daily pumpage from the aquifer in the San Antonio area, the aquifer levels at the Bexar County Index Well J17, and the precipitation in San Antonio are plotted against the daily flow rates at the two springs. As can be seen without any quantitative analysis, Comal Springs, which is closer to well J17 and San Antonio, shows a better correlation with the pumpage and the aquifer levels. In fact, the simple regression model of the Comal Springs discharge based on the J17 water levels (Figure 5–2) appears almost perfect, judging from the model correlation coefficient ($r = 0.978$). This value is very close to 1, which corresponds to a mathematical function. However, when looking at the graph of model residuals (Figure 5–3), it is apparent that the model cannot accurately describe likely but unknown cyclic components. In this particular case, adding more variables to the regression, including pumping from the aquifer, does not improve the model nor remove the obvious periodic component of the model residuals.

Since both springs are ascending artesian springs, there is no rapid reaction to daily precipitation at either one, except for the Comal Springs in early 2005. This is confirmed by the complete absence of correlation, as shown in Figure 5–4. Even for springs that more readily react to daily or weekly rainfall events, the simple linear regression between flows and total precipitation in the spring drainage area is always weak. In addition to many days with zero precipitation during a hydrologic year, there are other reasons for this: The same amount of rainfall produces different seasonal responses at the spring, depending on the antecedent rains, soil moisture, and evapotranspiration rates; and there usually is some delay in the spring response as the newly infiltrated water travels to the water table then to the spring.

The analysis also demonstrates that some correlations may be linear or close to linear (Figure 5–2), whereas others may show strong nonlinearity (Figure 5–5). One possible explanation for the "strange" zigzag pattern visible in Figure 5–5 is that some threshold aquifer levels (hydraulic heads) in certain portions of the

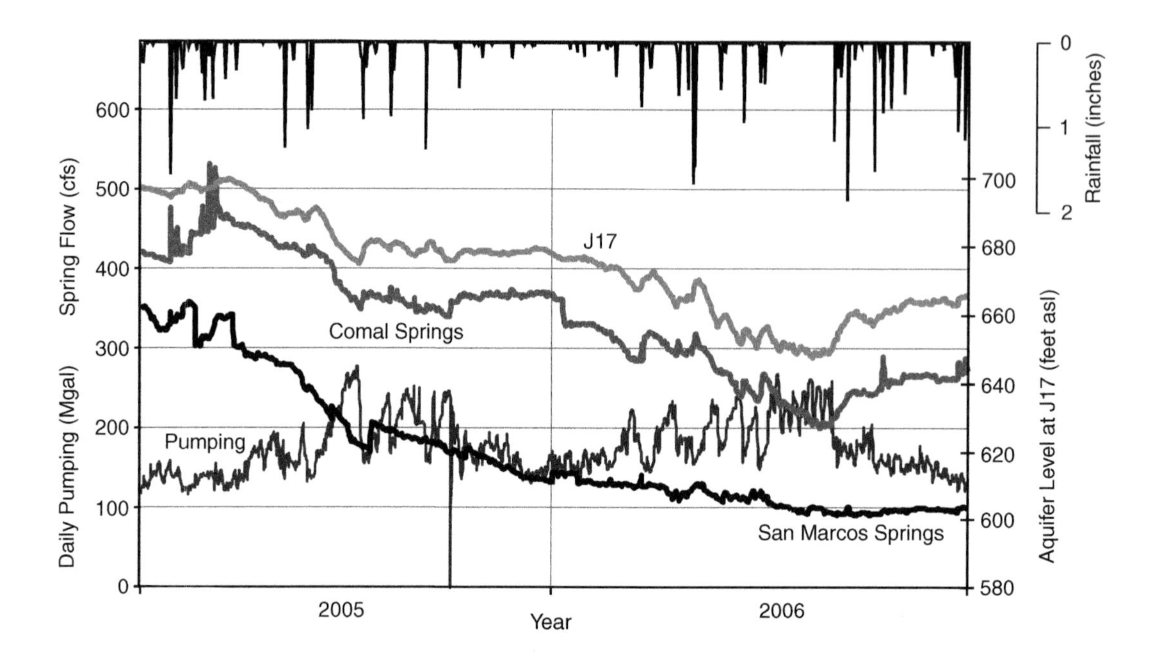

FIGURE 5–1 Daily flows at Comal and San Marcos Springs (in cubic feet per second) versus pumpage from the Edwards aquifer in the San Antonio area (in million gallons per day), daily aquifer level at the Bexar County Index Well J17, and daily precipitation in San Antonio. (Raw data from San Antonio Water System, 2008.)

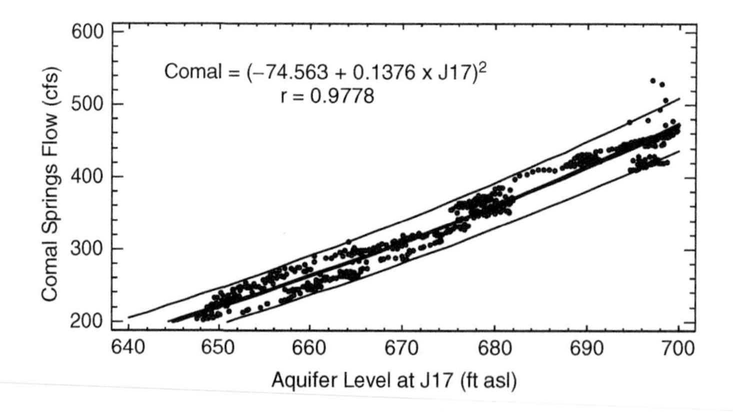

FIGURE 5–2 Simple regression model of the Comal Springs flow on aquifer level at the Bexar County Index Well J17, with the 95 percent prediction limits.

aquifer may be important for influence of the pumping on the spring discharge. As the pumping increases, the aquifer level and the spring discharge rate remain stable (indicated with arrows), then slowly decrease, only to drop significantly to the next lower level when the threshold is reached. This is common for karst aquifers, as illustrated in Figure 4–10.

The possible nonlinearity of the relationship between aquifer levels and spring discharge should be always considered when developing predictive models. Figure 5–6 shows one typical example and quite a few others

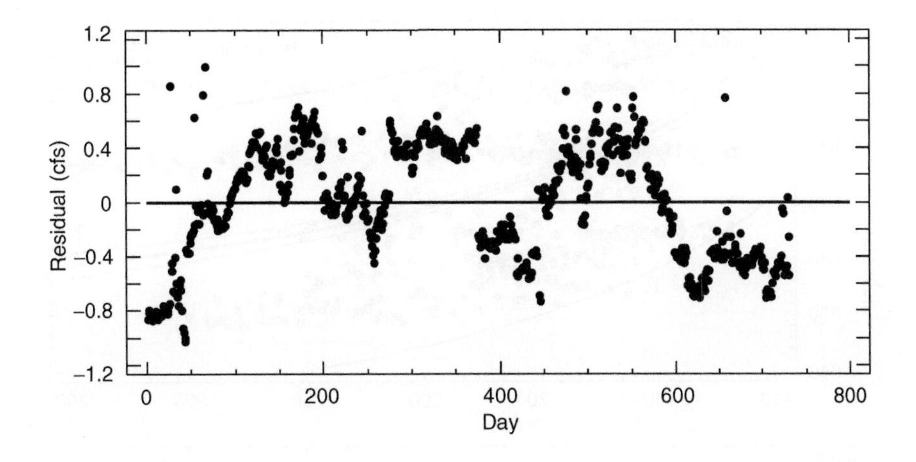

FIGURE 5–3 Residuals of the regression model shown in Figure 5–2.

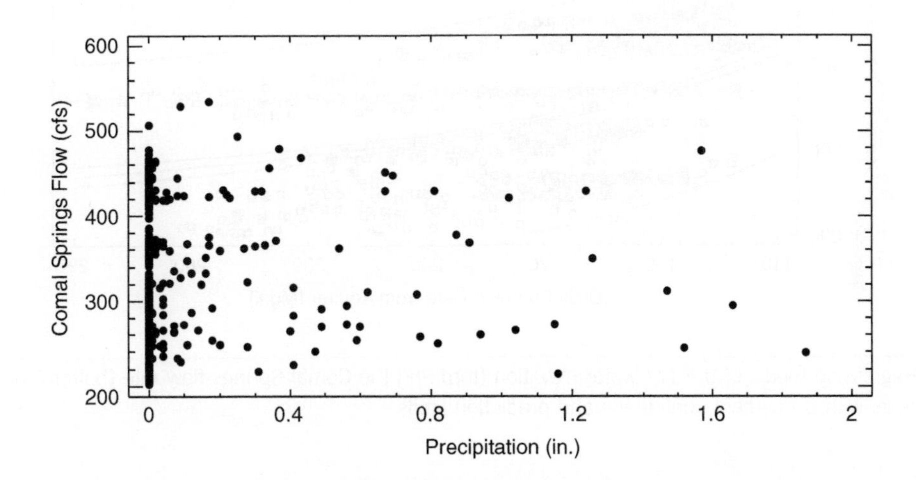

FIGURE 5–4 Comal Springs flow versus precipitation.

are explained in detail by Bonacci (1987, 1995). The relationship between the spring yield and the piezometer level is linear as long as the orifice is not flooded and the discharge is by overflow. Once the orifice is completely filled, the increase in discharge capacity is limited, while the piezometric level in the aquifer continues to rise at a much higher rate.

One convenient way to improve regression of spring discharge on precipitation is to use the so-called index of antecedent precipitation (IAP), which indirectly accounts for the influence of soil moisture and aquifer recharge due to precipitation that fell prior to the present day. The general form of IAP for springs with rapid response to rainfall is

$$\text{IAP} = \sum_{t=1}^{i} (C_t \times P_t) \tag{5.1}$$

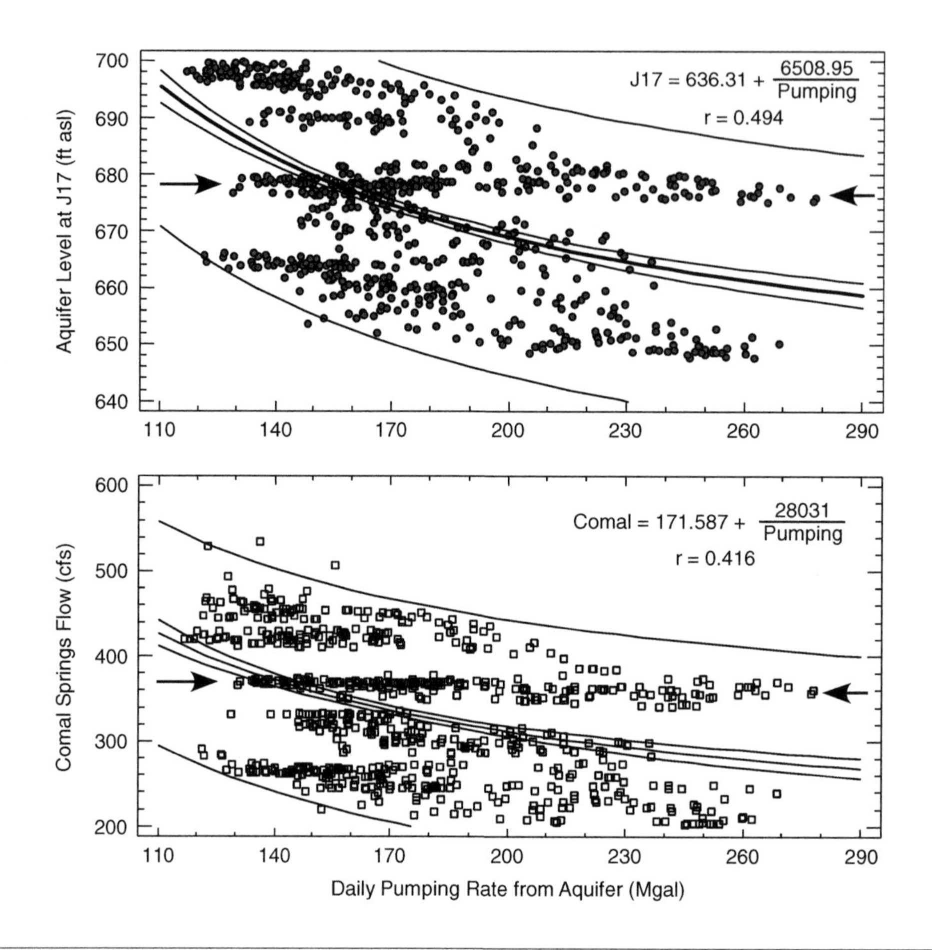

FIGURE 5–5 Regression model of the J17 water elevation (top) and the Comal Springs flow rate (bottom) on the aquifer pumpage, with 95 percent confidence and prediction limits.

where t is time interval (e.g., one day), i is total number of time intervals for the calculation of the IAP, C_t is the empirical coefficient, and P_t is precipitation at time t. Figure 5–7 shows some common expressions for C_t and the corresponding decrease in relative weights given to precipitation that fell over the 10 preceding days.

When the spring has a delayed response to rainfall (recharge), the IAP has a peak that is shifted d days to the past, so that the precipitation that fell on that day has the highest influence on the present day discharge (Figure 5–8). The delay (lag) time can be determined by trial based on the highest value of the regression model coefficient that incorporates the IAP. It can also be selected from the cross-correlogram of spring discharge and precipitation (see the next section).

5.3 AUTOCORRELATION AND CROSS-CORRELATION

The regression for the spring discharge can be significantly improved if one of the variables used to describe it is the discharge itself, the so-called autoregression. It is evident that the present day discharge is influenced by (related to) the discharge of the preceding day, the discharge two days ago, and so on. This is true for both

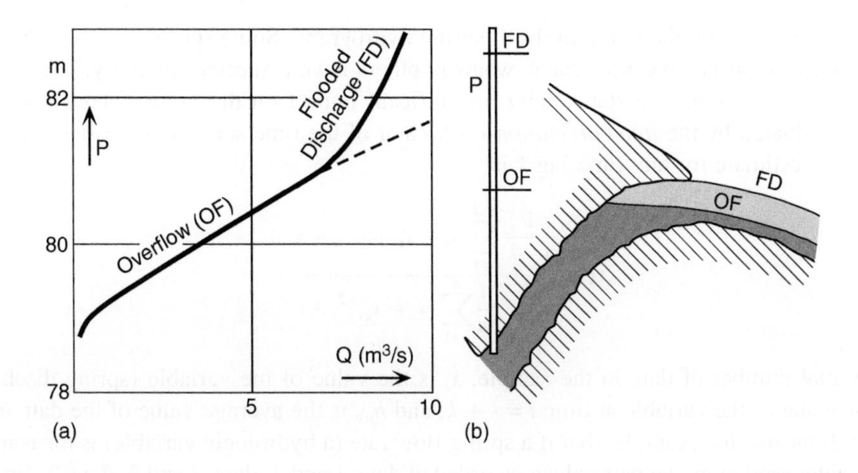

FIGURE 5–6 (a) Flow rate versus hydraulic head at piezometer P, Ljubovija intermittent spring near Mostar, Herzegovina. (b) Overflow, an increasing discharge causes increase in the flow cross section at the outlet and thus a slight increase in the hydraulic head at P; flooded discharge, the outlet cross section cannot increase further as the entire orifice is full; further increase in flow causes rapid increase in the hydraulic head at P. (From Hajdin, 1981.)

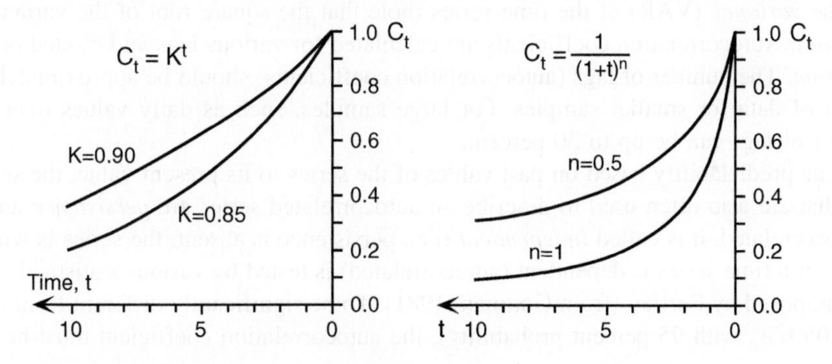

FIGURE 5–7 Common expressions for coefficient C_t used to define index of antecedent precipitation (IAP), when the spring shows a rapid response to rainfall.

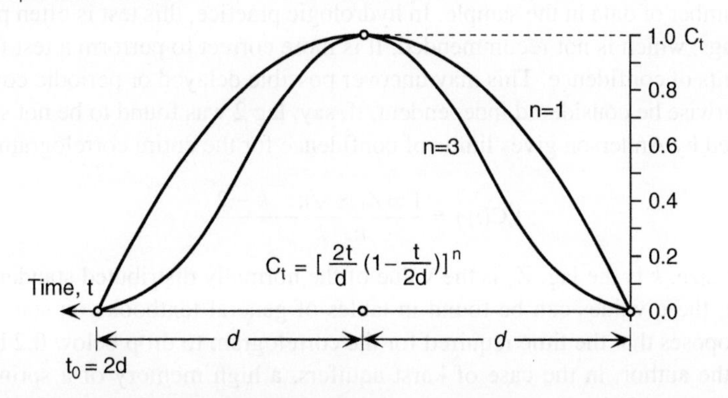

FIGURE 5–8 Coefficient C_t for springs with delayed response to rainfall.

the descending (recession) and the rising limbs of spring hydrograph. Some springs exhibit a higher degree of this interdependence of successive flow rates, whereas others have a shorter "memory." The flow rate, say, 10 and more days ago may have a statistically insignificant impact on the present day flow. This memory of the spring is evaluated by the *autocorrelation coefficient* of the time series (also called *serial correlation coefficient*), whose estimate for any time lag k is

$$r_k = \frac{\frac{1}{n-k}\sum_{i=1}^{n-k}(x_i - x_{av})(x_{i+k} - x_{av})}{\frac{1}{n}\sum_{i=1}^{n}(x_i - x_{av})^2} \tag{5.2}$$

where n is the total number of data in the sample, x_i is the value of the variable (spring discharge) at time $t = i$, x_{i+k} is the value of the variable at time $t = i + k$, and h_{av} is the average value of the data in the sample.

Equation (5.2) means, for example, that if a spring flow rate (a hydrologic variable) is measured on a daily basis, for lag 1 autocorrelation, we pair values recorded at days 1 and 2, days 2 and 3, days 3 and 4, and so on. The number of pairs in the autocorrelation is therefore $n - 1$, where n is the number of data points. For lag 2, we pair days 1 and 3, days 2 and 4, and so on. Consequently, the number of pairs in correlation decreases again and it is now $n - 2$.

The numerator in equation (5.2) is called the *autocovariance* (or just covariance, COV), and the denominator is called the *variance* (VAR) of the time series (note that the square root of the variance is called the *standard deviation*). Autocorrelation coefficients are calculated for various lags and plotted on a graph called an *autocorrelogram*. The number of lags (autocorrelation coefficients) should be approximately 10 percent of the total number of data for smaller samples. For large samples, such as daily values over one or several years, the number of lags can be up to 30 percent.

If there is some predictability based on past values of the series to its present value, the series is *autocorrelated*. Terms that are also often used to describe an autocorrelated series are *persistence* and *memory*. If a series is not autocorrelated, it is called *independent* (i.e., persistence is absent, the series is without memory). The hypothesis that a time series is dependent (autocorrelated) is tested by various statistical tests. One of the simpler tests is proposed by Bartlett (from Gottman, 1981). To be significantly different from zero at the level of confidence 0.05 (i.e., with 95 percent probability), the autocorrelation coefficient must be

$$r_k > \frac{2}{\sqrt{n}} \tag{5.3}$$

where n is the total number of data in the sample. In hydrologic practice, this test is often performed just for the first or the first two lags, which is not recommended. It is more correct to perform a test for the entire correlogram, introducing limits of confidence. This may uncover possible delayed or periodic components in the time series that would otherwise be considered independent, if, say, lag 2 was found to be not significantly different from 0. A test proposed by Anderson gives limits of confidence for the entire correlogram (Prohaska, 1981):

$$\text{LC}(r_k) = \frac{1 \pm Z_\alpha \times \sqrt{n-k-2}}{n-k-1} \tag{5.4}$$

where n is the sample size, k is the lag, Z_α is the value of the normally distributed standardized variable at the α level of confidence; these values can be found in tables of general textbooks on statistics.

Mangin (1984) proposes that the time required for the correlogram to drop below 0.2 be called the *memory effect*. According to the author, in the case of karst aquifers, a high memory of a spring indicates a poorly developed karst network with large groundwater flow reserves (storage). In contrast, a low memory is believed to reflect low storage in a highly karstified aquifer. However, Grasso and Jeannin (1994) analyzed the autocorrelograms of a synthetic, regular discharge time series and demonstrate that the increase in the frequency of flood events resulted in a steeper decreasing limb in the associated correlogram. They also point out that the sharper the peak of the flood event, the steeper a decreasing limb of the correlogram. Similarly, the decrease of the recession coefficient entails a steeper decreasing limb of the correlogram. Numerical forward simulation of spring hydrographs by Eisenlohr et al. (1997) confirms that the shape of the resulting correlogram strongly depends on the frequency of precipitation events. These authors also show that the spatial and temporal distribution of rainfall and the ratio between diffuse and concentrated infiltration have a strong influence on the shape of the hydrograph and subsequently on the correlogram. Consequently, the shape of the correlogram and the derived memory effect depend not only on the state of maturity of the karst system but also on the frequency and distribution of the precipitation events under consideration (Kresic, 1995; Kovács and Sauter, 2007).

Figures 5–9 and 5–10 show hydrographs of the same spring for two characteristic years (wet and dry—note the difference in the precipitation axis scale) and their autocorrelograms, respectively. As can be seen in Figure 5–10, the limits of confidence diverge slightly, since the number of pairs in correlation (i.e., the accuracy of correlation) decreases with the increasing lag. For smaller samples, this divergence is more pronounced. For the dry year, the autocorrelogram exceeds the confidence limits for approximately 60 days, indicating that the autocorrelation coefficients for this period are significantly different from zero and the process of spring discharge is not independent. In other words, the system has a long memory, and generally speaking, the flows for up to the 60 preceding days are important for the present day flow. A possible explanation is that the aquifer storage, including matrix porosity, is significant and it releases water gradually.

For the wet year, the picture is quite different. The system memory of about 20 days is considerably shorter, which could be explained by the fact that the inflow from precipitation is frequent enough to cause a dominant role of groundwater flow through large open fissures and fractures or karst conduits in the case of karst aquifers. These fast-draining pathways have high flow amplitudes, which can be seen from the spring hydrographs in Figure 5–9. The frequent flow changes in these pathways dampen the influence of the surrounding matrix seen during the dry year.

As discussed repeatedly in previous chapters, spring flow is more or less influenced by precipitation (or any other water input), and this influence may be delayed for a variety of reasons. In the cross-correlation analysis, the time-dependent relationship between an output or the dependent variable (e.g., daily spring flow) and an input or the independent variable (e.g., daily precipitation) is analyzed by computing the *coefficients of cross-correlation* for various time lags and plotting the corresponding *cross-correlogram.* The cross-correlation coefficient for any lag k is given as

$$r_k = \frac{\mathrm{COV}(x_i, y_{i+k})}{(\mathrm{VAR}x_i \times \mathrm{VAR}y_i)^{\frac{1}{2}}} \tag{5.5}$$

where COV is the covariance between the two series; x_i and y_i are the observed daily precipitation and flow, respectively; and VAR is the variance of each series. In practice, the coefficient of cross-correlation for lag k is estimated from the sample using the following equation:

$$r_k = \frac{\sum_{i=1}^{n-k} x_i \times y_{i+k} - \frac{1}{n-k}\left(\sum_{i=1}^{n-k} x_i\right)\left(\sum_{i=1}^{n-k} y_{i+k}\right)}{\left[\sum_{i=1}^{n-k} x_i^2 - \frac{1}{n-k}\left(\sum_{i=1}^{n-k} x_i\right)^2\right]^{\frac{1}{2}} \times \left[\sum_{i=1}^{n-k} y_{i+k}^2 - \frac{1}{n-k}\left(\sum_{i=1}^{n-k} y_{i+k}\right)^2\right]^{\frac{1}{2}}} \tag{5.6}$$

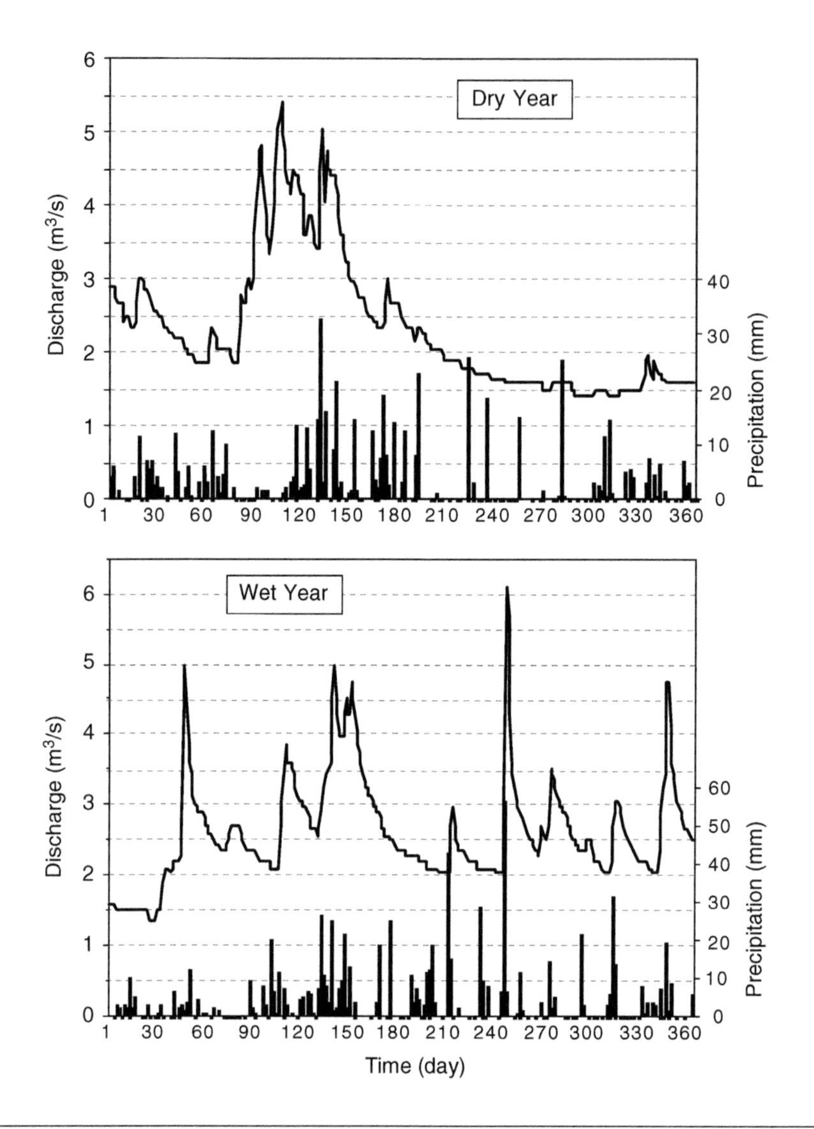

FIGURE 5–9 Characteristic discharge hydrographs of a large karst spring for typical dry and wet years. (From Kresic, 1997; copyright Taylor & Francis, reprinted with permission.)

Figure 5–11 shows cross-correlograms for the spring hydrographs in Figure 5–9 (typical dry and wet years). In both cases, the cross-correlation coefficients are statistically insignificant after about 12 days, which indicates that antecedent precipitation beyond this period has no direct influence on spring discharge. A distinct peak at lag 1 for the wet year shows an important, 1 day delayed response of large transmitters in the aquifer to significant rainfall events. In a typical dry year, this response is much less pronounced, but it has the same lag (1 day), which indicates the same discharge mechanism after major rainfall. Generally, low values of the cross-correlogram for both years show that the influence of the fast-flow component is significantly attenuated by the porous medium.

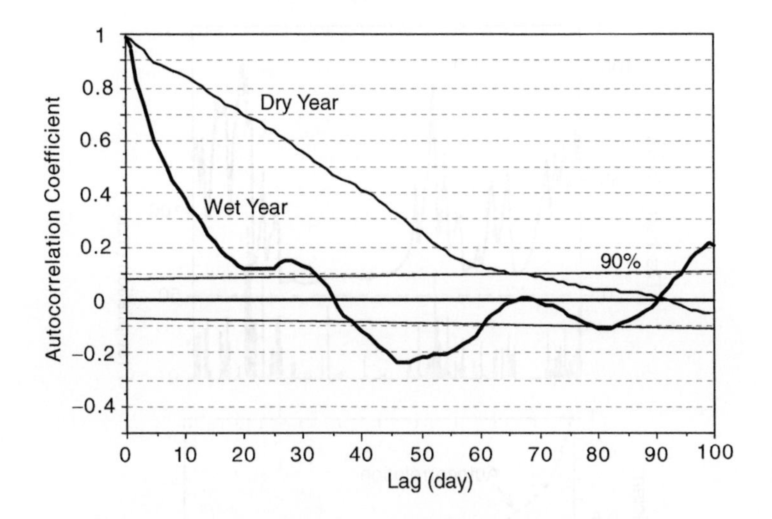

FIGURE 5–10 Autocorrelograms of the spring hydrographs shown in Figure 5–9 for typical wet and dry years. (From Kresic, 1997; copyright Taylor & Francis, reprinted with permission.)

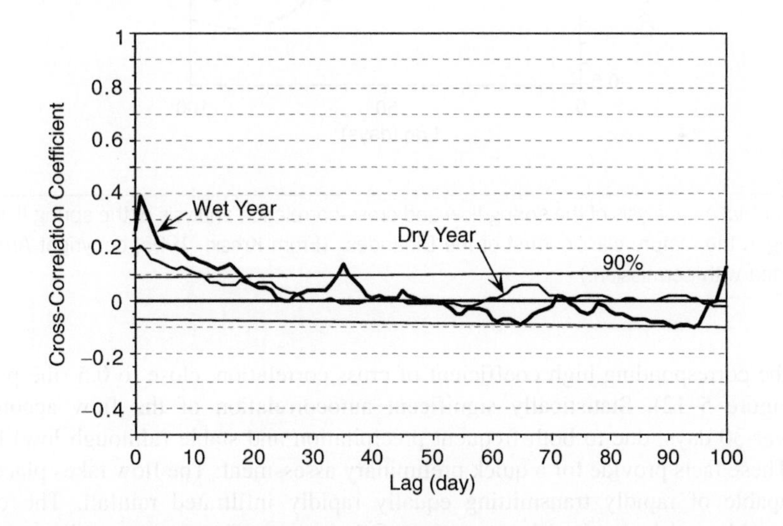

FIGURE 5–11 Cross-correlograms of the spring flow and precipitation in its drainage area for typical wet and dry years. (From Kresic, 1997; copyright Taylor & Francis, reprinted with permission.)

The following examples illustrate possible applications of the autocorrelation and cross-correlation analyses when developing conceptual models of spring discharge. Ombla Spring (Figure 5–12), tapped for the water supply of the Croatian coastal city of Dubrovnik, drains more than 600 km^2 of pure mature classic karst terrain of the Dinarides. The ratio of maximum to minimum flow (coefficient of spring nonuniformity) is more than 10 for most years, and a very rapid response to major rain events is evidenced by a short time lag

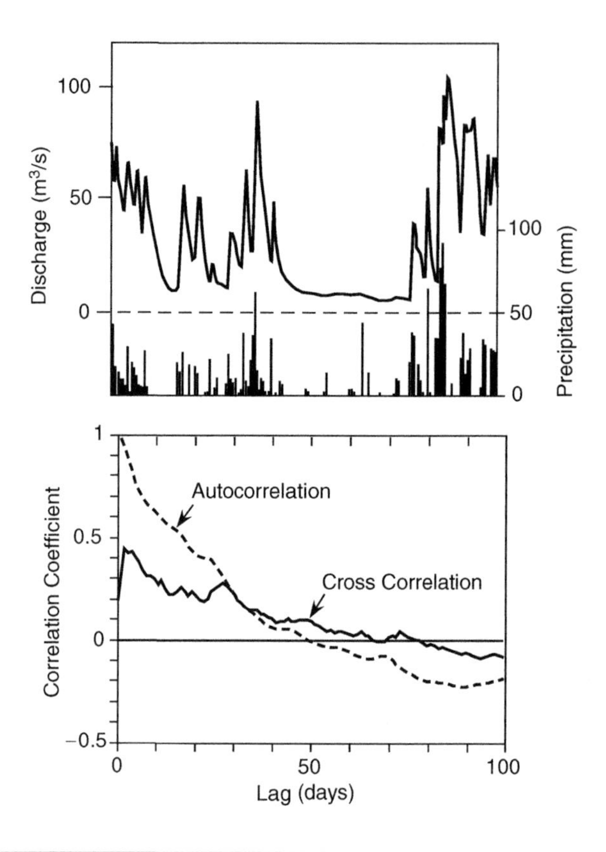

FIGURE 5–12 Autocorrelation analysis of the spring flow and cross-correlation analysis of the spring flow and precipitation for the Ombla Spring in the mature classic karst of the Dinarides. (From Kresic, 1995; copyright American Institute of Hydrology; reprinted with permission.)

of 2–3 days and the corresponding high coefficient of cross correlation, close to 0.5 (the peak on the cross-correlogram in Figure 5–12). Statistically significant autocorrelation of the flow according to Mangin ($r_k > 0.2$) lasts over 30 days, due to both frequent precipitation and stable (although low) base flow during summer months. These facts provide for a quick preliminary assessment: The flow takes place mainly through large conduits capable of rapidly transmitting equally rapidly infiltrated rainfall. The conduit network, however, drains quickly and lacks significant storage. Other types of porosity contribute to a very uniform regional base flow (between, 6 and 7 m^3/s) during long summer periods. However, knowing that the spring drainage area is probably more than 600 km^2, it appears that the effective matrix porosity of the aquifer is quite low.

Grza Spring (Figure 5–13), located in the semicovered karst of eastern Serbia, has a very high coefficient of nonuniformity of 22.5 but, at the same time, a significantly higher and longer autocorrelation. The cross correlation is statistically insignificant, although the precipitation in the drainage area is frequent and relatively uniformly distributed throughout the year. Preliminary assessment is that the infiltration is quite slow for a karst terrain. The conduit flow is not predominant, and other nonconduit types of effective porosity

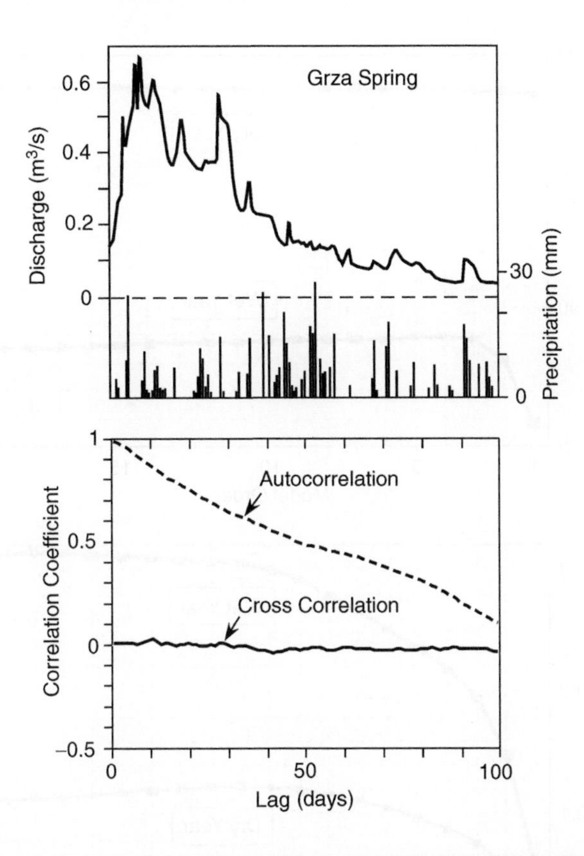

FIGURE 5–13 Autocorrelation analysis of the spring flow and cross-correlation analysis of the spring flow and precipitation for Grza Spring in the semi-covered karst of eastern Serbia. (From Kresic, 1995; copyright American Institute of Hydrology; reprinted with permission.)

and storage are more significant. It also helps to know that the drainage area is a mountainous terrain with a significant snowcap, which melts relatively quickly during spring. This snowmelt contributes to peak flows not directly related to the ongoing precipitation events.

5.4 AUTOREGRESSIVE–CROSS-REGRESSIVE MODELS (ARCR)

A separate analysis of simple autoregressive (AR) and cross-regressive (CR) models of spring discharge provides useful information on the aquifer structure. Figure 5–14 shows comparison of these models for typical wet and dry years (spring hydrographs are shown in Figure 5–9). The autoregressive model of the order p is (Kresic, 1997)

$$Q_t = a + b_1 Q_{t-1} + b_2 Q_{t-2} + \ldots + b_p Q_{t-p} \tag{5.7}$$

where Q_t is the predicted discharge at time t; $Q_{t-1}, \ldots, Q_{t-p}$ is the discharge for $1, \ldots, p$ preceding days; $a, b_1, \ldots, b_p$ are the model parameters.

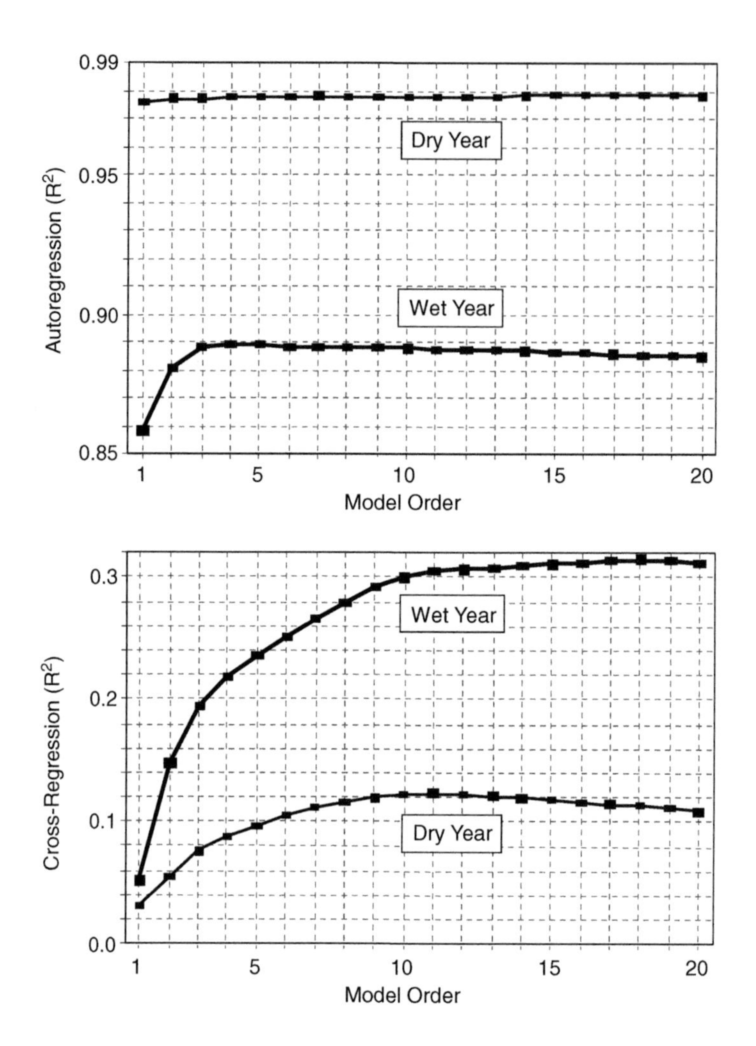

FIGURE 5–14 Comparison of simple autoregressive (top) and cross-regressive (bottom) models of the spring discharge shown in Figure 5–9. The goodness of the models is measured by the squared coefficient of multiple regression (R^2). (From Kresic, 1997; copyright Taylor & Francis, reprinted with permission.)

The cross-regressive model of order q is

$$Q_t = a + c_1 P_{t-1} + c_2 P_{t-2} + \ldots + c_q P_{t-q} \tag{5.8}$$

where Q_t is the predicted discharge at time t; $P_{t-1}, \ldots, P_{t-q}$ is the precipitation for $1, \ldots, q$ preceding days; $a, c_1, \ldots, c_q$ are the model parameters.

As can be seen in Figure 5–14, autoregressive flow models have much higher coefficients of multiple regression (which is the measure of the model goodness) than cross-regressive models. This is mainly because the system has a long internal memory, resulting from the gradual release of groundwater from storage. The same can be concluded from the shape of the autocorrelograms in Figure 5–10. In addition, the cross-regressive model uses only gross precipitation, and the magnitude and time distribution of the effective infiltration remain unknown. This introduces a lot of zero values into the multiple cross regression and significantly decreases its coefficient.

The autoregression model coefficients remain consistently high for both years. For the dry year, they are higher than for the wet year and change insignificantly for all tested model orders. This again can be explained by the fact that the flow generated by the slow draining of small fissures and filled cavities is always statistically predominant. In the wet year, however, groundwater flow through large fractures (or karst channels) for the preceding three to four days is more important (see the maximum on the autoregression graph in Figure 5–14). The cross-regression graph for the wet year changes its slope considerably for the order of 3 and, less abruptly, again for the order of 10, after which it remains steady. This also indicates that, in the wet year, discharge of the spring is statistically influenced by the precipitation of the preceding 10 days, with the first 3 days being the most important.

Because of their statistical insignificance, simple cross-regressive models cannot be used for the prediction of spring discharge: The coefficient of multiple regression for the wet and dry years for all orders is less than 0.32 and 0.13, respectively. On the other hand, the autoregressive models for both years have a very high coefficient of multiple regression, which for the dry year approaches almost a functional dependence (R^2 is close to 0.98). They are therefore good predictive models and can be used for forecasting when the spring discharge is not under the direct influence of precipitation, that is, in the periods of recession. Autoregressive models, however, cannot be used for forecasting based on some expected precipitation, because they cannot cause change in the direction of the hydrograph (e.g., from falling to rising). They also cannot be used for generating flow from the historic precipitation.

ARCR models belong to the category of multivariate time series models and are similar to ARMAX models, that is, ARMA models with exogenous variables (Salas, 1993). However, they are less complex, since the error term is not included in the parameter estimation. The order of the models is generally low—discharge and precipitation for up to three or four preceding days are included. Higher orders do increase multiple correlation coefficients (R) as a measure of the model goodness, but these increases are usually statistically insignificant. However, when a drainage area is large (50 km^2 or more) or there is a significant delay in the influence of precipitation on the spring discharge, an increase in its cross-regressive part may considerably improve the validity of an ARCR model.

In this case, the combination of autoregressive and cross-regressive models significantly improves validity of the new ARCR model for the wet year. For the dry year, this merging increases the coefficient of multiple regression insignificantly, since its value for the simple autoregressive model of all tested orders is already close to 0.98. The ARCR model is given as

$$Q_t = a + b_1 Q_{t-1} + \ldots + b_p Q_{t-p} + c_1 P_{t-1} + \ldots + c_q P_{t-q} \tag{5.9}$$

where the notation is the same as in equations (5.7) and (5.8).

The model order (p, q) can be chosen from various combinations of its autoregressive (p) and cross-regressive (q) parts, as shown in Figure 5–15. For the wet year, the model coefficient (R^2) has maximum for $p = 3$ and $q = 3$. A further increase in the orders of AR and CR parts results in slight negative changes of R^2, which remains close to 0.925. The model for the dry year has no single maximum of the multiple regression coefficient. The autoregressive part has the first abrupt positive change for the order of 2 then for 7, when it also reaches the first maximum. R^2 increases again for $p = 14$ and 15. However, these changes of the coefficient of multiple regression are with an insignificant magnitude, that is, less than 0.001. Therefore, introducing more variables into the autoregression by increasing the AR order does not increase the model validity. Generally speaking, when a time series is highly autocorrelated (as in this case), AR models of first and second orders are considered sufficient in hydrologic practice.

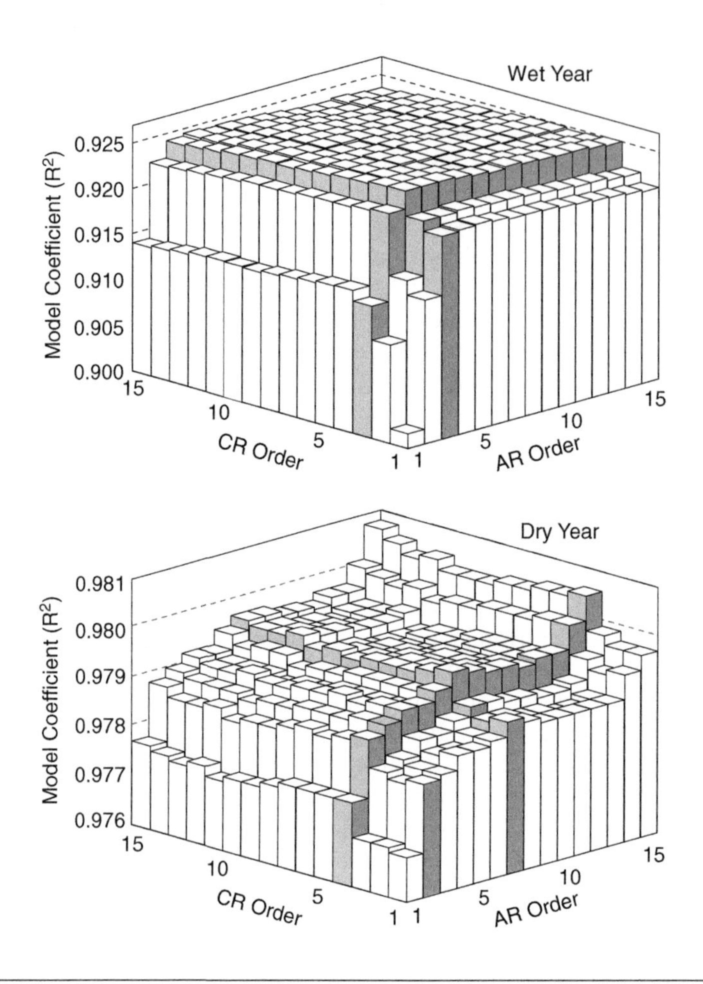

FIGURE 5–15 Dependence of the ARCR model coefficient (R^2) on the order of cross-regressive and autoregressive parts. (From Kresic, 1997; copyright Taylor & Francis, reprinted with permission.)

The model's cross-regressive part for the dry year is consistent and its order of 4 has the most significant increase throughout the analyzed domain (see Figure 5–15). The ARCR model order for the dry year can be chosen as $p = 2$, and $q = 4$, that is, (2, 4). The model-predicted versus recorded spring discharge for a part of the wet year is shown in Figure 5–16.

When the model uses actual gross precipitation, the predicted values differ significantly from the measured ones during the recession period (top hydrograph in Figure 5–16). This is because the model gives the same weight to all precipitation data, including isolated summer storms. It does not account for increased evapotranspiration and moisture deficit in the unsaturated zone or other factors during the summer months that greatly reduce effective precipitation.

Correct estimation of the effective (net) precipitation is a major difficulty in hydrologic modeling of the rainfall-runoff relationship. This is particularly true for the groundwater systems where the rainfall infiltration may be significantly attenuated by the porous media. The response of aquifers to recharge episodes varies

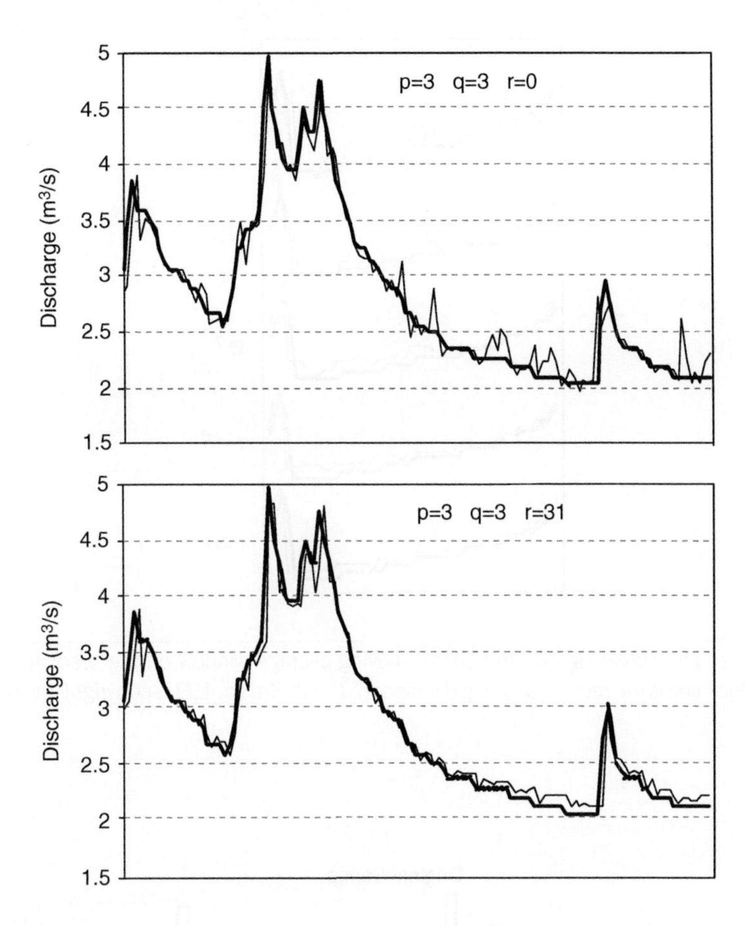

FIGURE 5–16 Recorded (bold line) versus predicted discharge of the spring for the summer period of a typical wet year. (Top) Result of the ARCR model without the moving average filter ($r = 0$). Both the autoregressive and the cross-regressive parts have an order of three. (Bottom) Result of the application of a 31-day moving average linear filter on the time series of total precipitation in the spring drainage area. The recessed section of the hydrograph is also shown in Figure 5–17. (From Kresic, 1997; copyright Taylor & Francis, reprinted with permission.)

throughout the year depending on the conditions and thickness of the unsaturated zone, water table levels, position of the discharge areas, and other factors.

A simple way to mathematically describe the attenuation of gross precipitation is to transform data using a moving average linear filter. Figure 5–17 shows the results of such transformation with different moving averages, while Figure 5–18 illustrates the principle of transformation with the moving average window of 5.

The ARCR model incorporating the modified precipitation input data is given as

$$Q_t = a + b_1 Q_{t-1} + \ldots + b_p Q_{t-p} + c_1 P_{r,t-1} + \ldots + c_q P_{r,t-q} \tag{5.10}$$

where r is the length of the linear filter window used to transform the precipitation series. The other notation is as in equations (5.7) and (5.8).

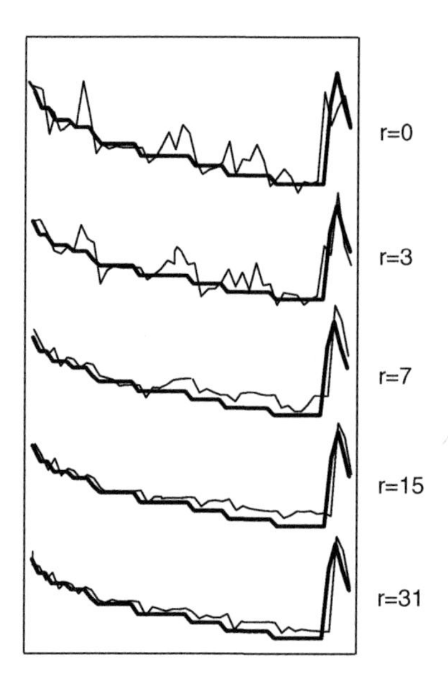

FIGURE 5–17 Influence of the increasing (top to bottom) moving average window on the predicted discharge of the spring (thin line). The thick line is the recorded spring discharge. (From Kresic, 1997; copyright Taylor & Francis, reprinted with permission.)

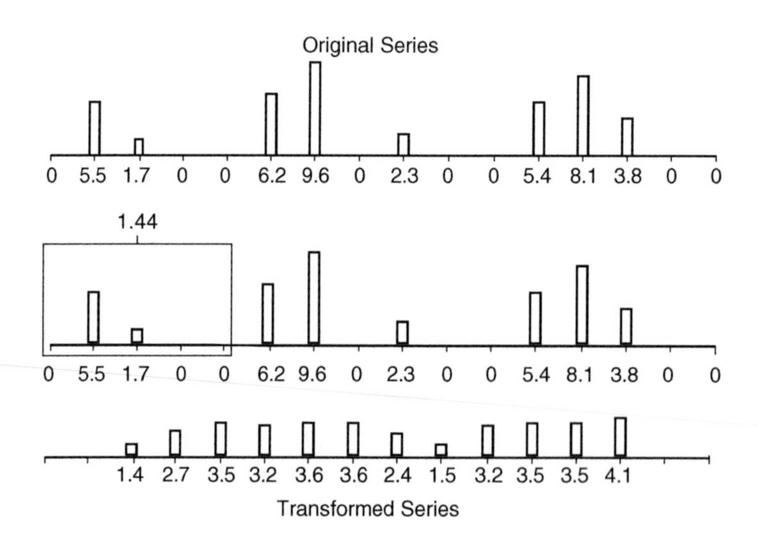

FIGURE 5–18 Application of a linear moving average filter for the transformation of precipitation series. Note that the total number of data in the transformed series is reduced by m-1 where m is the length of the filter window (m = 5 in this case). (From Kresic, 1997; copyright Taylor & Francis, reprinted with permission.)

The application of the moving average filter actually slightly decreases the model's coefficient of multiple regression: This is the reason why the visual determination of the filter length is preferred. It can be seen in Figure 5–16 that, for the periods of recharge, both gross and transformed precipitation give good predictions. The difference between the two is obvious in the case of recession when the transformed precipitation produces a much better result.

A simple ARCR model can be built by preparing data using any spreadsheet and multiple regression programs; the discharge and precipitation time series are arranged in a number of columns corresponding to the total number of ARCR orders, then shifted one lag at a time. For an ARCR(3, 3) model, the spring flow is predicted based on precipitation recorded that day, and flows and precipitation for the preceding three days:

$$Q_t = a + b_1Q_{t-1} + b_2Q_{t-2} + b_3Q_{t-3} + c_0P_t + c_1P_{t-1} + c_2P_{t-2} + c_3P_{t-3}$$

As mentioned earlier, the cross-regressive models generally have a low coefficient of multiple regression, which is one measure of their validity. This is because the continuous daily spring flow is regressed on daily precipitation, which is often zero. One way to transform the precipitation series is to apply a simple linear moving average filter, as shown in Figure 5–18. In this example, the filter window is 5, which means that five data points are summed at one time and their average is assigned to the central point in the window. The window then moves one position and the averaging is repeated. The resulting series is smoother: The high precipitation values decrease and the low values (including zeros) increase. The new transformed series is then included in the ARCR model instead of the total (gross) precipitation data. Various other mathematical and statistical filters can be tested and used to produce a better fit (e.g., see Figures 5–7 and 5–8).

ARCR models give valid future predictions for only one step at a time (e.g., next day), because for the second and following intervals, the true values of spring discharge are not known and the model confidence and prediction limits rapidly deteriorate.

5.5 SYSTEM ANALYSIS AND TRANSFER FUNCTIONS

Transformation of precipitation to flow in channels has been the primary focus of applied surface water hydrologists for decades. As mentioned before, because hydrographs of many large springs closely resemble hydrographs of surface streams, most if not all of the quantitative methods developed in surface water hydrology are applicable to hydrology of springs as well. The following discussion, which draws parallels between these two closely related systems (surface streams and large springs), is based on the groundbreaking publication by Dooge 1973, *Linear Theory of Hydrologic Systems*, published by the U.S. Department of Agriculture (Dooge, 1973). In subsequent years, the principles and mathematical models described by Dooge and other surface water hydrologists have been increasingly implemented by groundwater professionals, including those studying springs.

The distinction shown in Figure 5–19a between overland flow, interflow, and groundwater flow is not generally made in applied surface water hydrology, because it is virtually impossible to separate the three types. The analogy of this scheme in the hydrology of springs is shown in Figure 5–19b. Overland flow, the fastest component of stream-flow generation after rainfall, is equivalent to conduit type flow ("quick flow") in karst, pseudokarst, and fractured-rock aquifers, which exhibit behaviors closest to those of surface streams. Interflow is equivalent to fracture-fissure systems, and groundwater (base flow of surface streams) is equivalent to matrix storage (diffuse flow or the "spring base flow" component of spring discharge).

Applied hydrologists distinguish between surface flow and base flow and use the simplified model of the hydrologic cycle shown in Figure 5–20a. The analogy with springs is shown in Figure 5–20b. The precipitation is divided into (1) precipitation excess and (2) infiltration and other losses. The precipitation excess

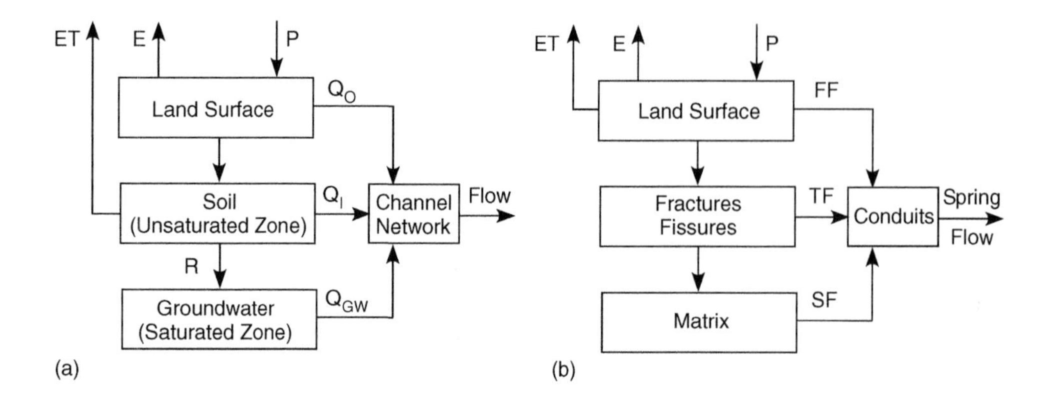

FIGURE 5–19 (a) Drainage area (catchment) of a surface stream as a hydrologic system: P = precipitation, E = evaporation, ET = evapotranspiration, R = recharge, Q_O = overland flow, Q_I = interflow, Q_{GW} = base flow from groundwater (modified from Dooge, 1973). (b) Aquifer drained by a spring as a hydrologic system: FF = fast flow (conduits), TF = transitional flow (fractures), SF = slow flow (matrix, small fissures).

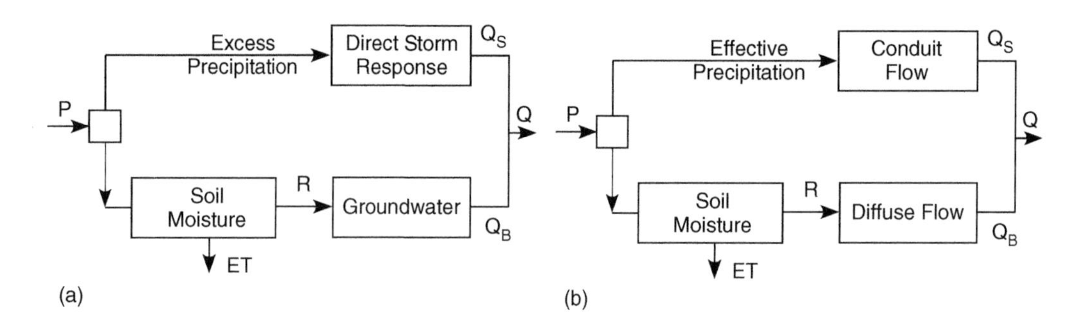

FIGURE 5–20 (a) The simplified surface stream catchment model: Q_S = storm response, Q_B = base flow, Q = total flow, P = precipitation, ET = evapotranspiration, R = recharge of groundwater. (Modified from Dooge, 1973). (b) Equivalent aquifer-spring model.

produces direct storm runoff. In the hydrology of springs, the precipitation excess is equivalent to rapid infiltration to the saturated zone, which produces the first response at the spring. The infiltration replenishes soil storage, which in turn is depleted by transpiration. Any excess infiltration after soil moisture is satisfied forms recharge to groundwater, which eventually emerges as base flow. In the hydrology of springs, "groundwater" is storage in the aquifer matrix, and base flow is flow of springs during recession periods.

The presence of the threshold in the soil storage phase of the system makes it impossible to treat the whole system as linear, even where the evaporation and transpiration are completely known. In surface water hydrology, the development of the unit hydrograph theory as a linear relationship between precipitation excess and storm runoff avoided this difficulty by the elimination of the base flow and the infiltration. It is the existence of this threshold, rather than the difference in response time between the surface response and the groundwater response, that necessitates the separation. In applied hydrology, the full model shown in Figure 5–20 is seldom used. Rather, the base flow is separated from the total hydrograph in some arbitrary fashion, and the precipitation excess is then taken to be equal in value (volume) to the storm runoff. If soil moisture accounting is included in the analysis, then the inherent threshold effect in soil moisture storage must be defined.

As emphasized by Dooge, if there is a desire to consider the whole system shown in either Figure 5–19 or 5–20, then one would have to deal with a nonlinear system, which brings all the difficulties of nonlinear mathematics. It is not surprising, therefore, that the concentration in both surface water and groundwater hydrology has been on the individual components shown in Figure 5–20. Various unit hydrograph techniques (impulse-response functions) have been developed for dealing with the direct response in runoff, and all are based on the assumption of linear behavior. Only recently has the nonlinear system modeling of spring response to recharge taken the lead (see Jukić and Denić-Jukić, 2006). The unsaturated phase involving soil moisture storage remains the most difficult part of the hydrologic cycle to handle. Not only does a threshold exist, but there is a feedback mechanism, because the state of the soil moisture determines the amount of infiltration.

The foundation of the most commonly applied methods of runoff and spring discharge modeling using a linear system approach are the following three principles: (1) superposition, (2) time invariance, and (3) lumping together the system inputs and outputs. In reality, none of the three principles is valid, but they greatly simplify the procedure and may produce satisfactory results (Dooge, 1973).

Superposition means that any number of inputs (x) can be added together so that the output (y) is the sum of the individual corresponding outputs (see also Figure 4–4). The system linearity defined by the principle of superposition must be distinguished from the existence of a general linear (that is, a straight line) functional relationship between input and output.

A system is said to be time invariant when its parameters do not change over time. For such a system, the form of the output depends only on the form of the input and not on the time at which the input is applied.

The problems of systems analysis and synthesis are also greatly simplified if the input and output of a system are assumed to be lumped together. In a lumped system with a single input and a single output, the behavior of the system would be described by a set of differential equations. If the inputs and outputs are not lumped together, then the system behavior must be described by partial differential equations, which are much more difficult to handle than ordinary differential equations.

The popularity of the linear system approach in hydrology can be appreciated when considering photographs in Figures 5–21 and 5–22. One needs to monitor only the rainfall (and maybe a few more meteorological parameters that play important roles in water budget calculations) and the spring discharge to mathematically describe the system (aquifer). Of course, many questions regarding the nature of the system remain unanswered, but there is a predictive tool as well as some useful quantitative parameters, which may help guide further characterization efforts.

The mathematical formulation for a lumped linear time-invariant system is based on the concept of an impulse (input) function and an impulse response (output) as follows (Dooge, 1973):

$$y(t) = \int_{-\infty}^{\infty} h(t-\tau)x(\tau)d\tau \tag{5.11a}$$

where $y(t)$ is the output in time t; $x(\tau)$ is the input in time τ represented by the orthogonal coefficients of the impulse or delta function; $h(t-\tau)$ is the impulse response function.

The right-hand side of equation (5.11a) represents the well-known mathematical operation of convolution, which is often expressed by an asterisk:

$$y(t) = h(t)^{*}x(t) \tag{5.11b}$$

FIGURE 5–21 Hydrometeorologic station equipped with instruments for measurements of direct evaporation (pan evapometer in the front), wind speed, precipitation, air temperature, and insolation (hours of sunshine).

FIGURE 5–22 The Cossaux spring used for water supply of the city of Yverdon-les-Bains in western Switzerland. A detailed description of the spring is given by Pronk et al., 2005. (Photograph courtesy of Nico Goldscheider.)

The impulse response function in equation (5.11) is called *instantaneous unit hydrograph* (IUH) in hydrology; in mathematics, it is called a *kernel function.*

For two discrete finite series that are related (e.g., rainfall and spring discharge), the convolution equation becomes (Dooge, 1973; Dreiss, 1982, 1989a, 1989b)

$$y_i = \Delta t \sum_{j=0}^{i} x_j h_{i-j} + \epsilon_i \qquad i = 0, 1, 2, \ldots, N \tag{5.12}$$

where N is the number of sampling intervals of equal length Δt; y_i is the mean value of the output during the interval i; x_j is the mean value of the input during the interval j; h_{i-j} is the kernel function during the interval i–j; ϵ_i is error due to inherent nonlinearities in the system and measurements errors.

If x_j and h_{i-j} are known, then y_i can be determined directly by convolution. If x_j and y_i can be identified, then h_{i-j} (the kernel function) can be determined through deconvolution (Dooge, 1973; Dreiss, 1982, 1989b). The solution of equation (5.12) is obtained by minimizing the sum of the square errors

$$\sum_{i=0}^{N} \epsilon_i^2 \Rightarrow \min \tag{5.13}$$

assuming that the discrete kernel function is nonnegative

$$h_k \geq 0 \quad k = 0, 1, 2, \ldots, M \tag{5.14}$$

where M is the system memory or the length of the period in which the impulse (input) has the influence on the output. The solution is obtained when the total volumes of the input and output series are equal to each other, that is, when the area under the kernel function is equal to 1:

$$\Delta t \sum_{k=0}^{M} h_k = 1 \tag{5.15}$$

System identification can also be performed using the correlation methods as described by Dooge (1973), who gives the following form of the convolution equation connecting two discrete time series and optimized in the least squares sense:

$$\varphi_{xy}(k) = \sum_{j=0}^{j=\infty} h_{\text{opt}}(j)\varphi_{xx}(k-j); \quad \text{when} \;\; k > 0 \tag{5.16}$$

where φ_{xy} is the cross correlation of the two series, which is defined as the limit

$$\varphi_{xy}(k) = \frac{1}{n} \sum_{i=-p}^{p} x(i)y(i+k) \tag{5.17}$$

as p tends to infinity. The term φ_{xx} is the autocorrelation of a time series defined as the limit

$$\varphi_{xx}(k) = \frac{1}{n} \sum_{i=-p}^{i=p} x(i)x(i+k) \tag{5.18}$$

where $n = 2p + 1$ is the number of data points as p tends toward infinity.

As illustrated in Figure 5–23, the impulse response function (kernel) describes the transfer of the fast discrete input, such as fast infiltration, into the output, such as spring discharge. The principle of superposition

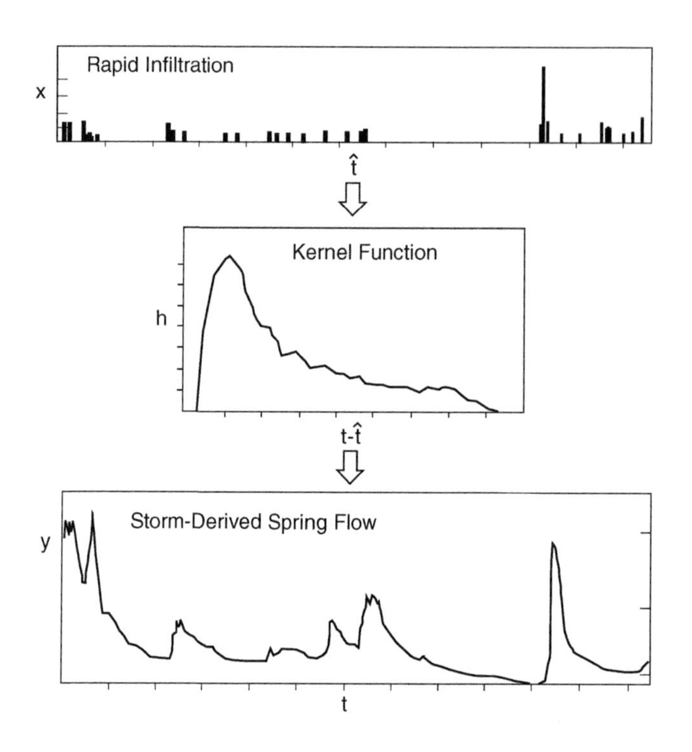

FIGURE 5–23 Schematic diagram of linear systems analysis of karst spring flow. (From Dreiss, 1989b; copyright American Geophysical Union.)

allows that the simulated unit hydrographs from isolated rainfall episodes (unit impulses) be combined into the "real" (complex) spring hydrograph. The kernel identification should be performed on as many isolated hydrographs as possible for which the precipitation episodes that caused them can be defined. It may be possible to apply one average kernel or try with a selective application of several kernels for different seasons if they are notably different.

One of the main limitations of convolution in simulating spring discharge is that long periods of base flow (without precipitation), longer than the base of the kernel(s), that is, longer than the system memory, cannot be simulated because the kernel calculates zero flows. In addition, most commonly applied numeric methods generally have problems in accurately describing the tails of recession hydrographs. In such cases, Snyder (1968) suggests that unit hydrographs be added to the previously separated base flow, so that the beginning of the output hydrograph does not start from zero. To avoid complicated algorithms, Canceill (1974) suggests adding a simple threshold with a decreasing trend into the convolution:

$$y(t) = \int_{-\infty}^{t} h(t-\tau)x(\tau)d\tau + C_k e^{-\infty(t-t_k)} \tag{5.19}$$

where t_k is the date of the last effective (nonzero) rainfall and C_k is the parameter to be optimized. The main idea behind equation (5.19) is to separate two types of discharge: the linear fast-flow component (impulse response) and the slow, nonlinear base flow during recession, as described by Maillet (1905; see Chapter 4).

Dreiss (1989a, 1989b) identifies kernels of different moments by combining the results of a tracer test performed at Maramec Spring, in Missouri, with a long-term, 12-month sequence of naturally occurring regional tracer events and three short-term subsets of the record.

5.5.1 Composite transfer functions

A new form of the transfer functions for karst aquifers, the so-called composite transfer function (CTF), was recently developed by Denić-Jukić and Jukić (2003). The CTF simulates discharges by two transfer functions adapted for the quick-flow and slow-flow hydrograph component modeling. A nonparametric transfer function (NTF) is used for the quick-flow component. The slow-flow component is modeled by a parametric transfer function that is an instantaneous unit hydrograph mathematically formulated and defined from a conceptual model. By using the CTF, the irregular shape of the tail of the identified transfer function can be avoided, and the simulation of long recession periods as well as the simulation of a complete hydrograph becomes more successful. The NTF, the Nash model, the Zoch model, and similar conceptual models can be considered separately as simplified forms of the CTF. The rainfall-runoff model based on the convolution between rainfall rates and the CTF was successfully tested on the Jadro Spring in Croatia. The results of the application were compared with the results obtained by applying NTFs independently.

Considering a karst aquifer as the linear, time-invariant, and casual system, the discharge from the karst aquifer at a time t may be represented by the superposition of three components (Figure 5–24a):

$$y(t) = y^D(t) + y^S(t) + y^Q(t) \tag{5.20}$$

where $y^Q(t)$ is the quick-flow component, $y^S(t)$ is the slow component, and $y^D(t)$ is the antecedent recession. The discharge $y^D(t)$ is the result of an initial storage, while $y^Q(t)$ and $y^S(t)$ are the result of rainfall $x(t-\tau)$, $t \in [0, t]$.

The relation between the rainfall and the resulting karst aquifer discharge can be represented by the linear convolution integral, so equation (5.20) becomes

$$y(t) = y^D(t) + \int_0^t h^S(\tau)x(t-\tau)d\tau + \int_0^t h^Q(\tau)x(t-\tau)d\tau \tag{5.21}$$

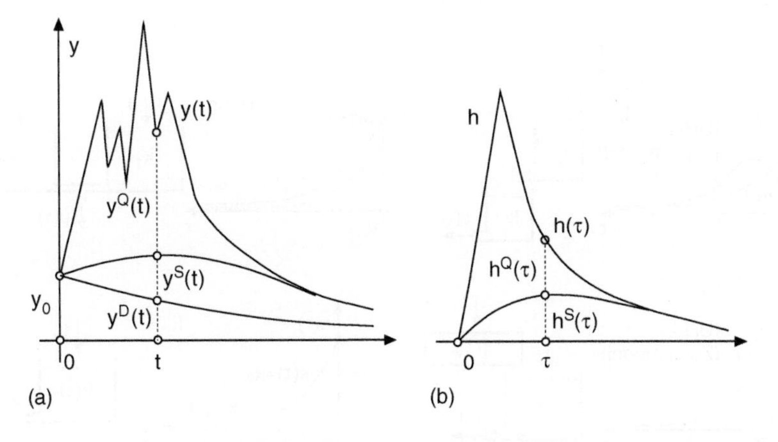

FIGURE 5–24 (a) Components of a karst spring hydrograph: $y^Q(t)$ = quick-flow component; $y^S(t)$ = slow component; $y^D(t)$ = antecedent recession. (b) Components of the CTF: $h^S(\tau)$ = slow flow transfer function; $h^Q(\tau)$ = quick flow transfer function. (From Denić-Jukić and Jukić, 2003; copyright 2003 Elsevier Science B.V., reprinted with permission.)

where $h^S(\tau)$ is the slow-flow transfer function and $h^Q(\tau)$ is the quick-flow transfer function (Figure 5–24b). The slow-flow transfer function may be obtained by $h^S(\tau) = \mu^S u(\tau)$, where the constant μ^S is the area under the slow-flow transfer function and $u(\tau)$ is the IUH of the slow-flow component.

Equation (5.21) can be written as

$$y(t) = y^D(t) + \mu^S \int_0^t u(\tau)x(t-\tau)d\tau + \int_0^t h^Q(\tau)x(t-\tau)d\tau \tag{5.22}$$

For discrete data points collected at intervals of equal length Δt, equation (5.22) becomes

$$y_i - y_i^D = a_i\mu^S + \sum_{j=0}^{j=\min(i,m)} b_{i,j}h_j^Q + \epsilon_i; \quad i = 0, 1, 2, \ldots, N$$

$$a_i = \Delta t \sum_{j=0}^{i} u_j x_{i-j}, \quad b_{i,j} = \Delta t x_{i-j} \tag{5.23}$$

where x_i are rainfall rates, y_i are discharge rates, h_j^Q are unknown values of the quick-flow transfer function at times $j\Delta t$, u_j are the values of $u(\tau)$ at times $j\Delta t$, y_i^D are the values of $y^D(t)$ at times $i\Delta t$, m is the memory length of the quick-flow component, and ϵ_i are residuals due to measurement errors or nonlinearities and temporal changes in the system. Equation (5.23) is solved using the constrained least squares procedure; that is, the values μ^S and h_j^Q can be found by minimizing the sum of squares of residuals.

IUH for the slow flow

The authors present four types of the IUH for the slow-flow component modeling derived from linear conceptual models (Figure 5–25). Detailed forms of IUH-1 and IUH-2 are presented in Singh (1988), while IUH-3 and IUH-4 are defined by assuming that the recession part of karst spring hydrographs has an exponential form (e.g., Bonacci, 1987; see also Chapter 4).

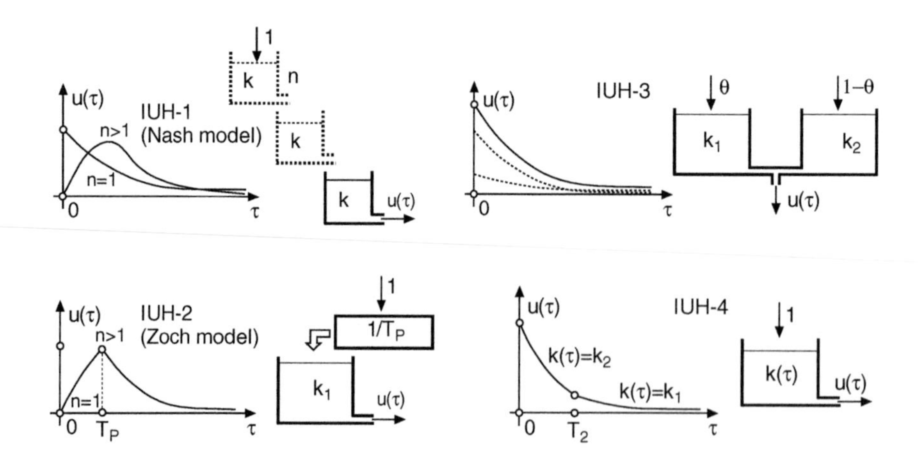

FIGURE 5–25 Conceptual models and their IUHs for the slow-flow component modeling. (From Denić-Jukić and Jukić, 2003; copyright 2003 Elsevier Science B.V., reprinted with permission.)

IUH-1 is the linear Nash reservoirs model, based on the concept of a cascade of n unequal linear reservoirs:

$$u(\tau) = \frac{1}{k\Gamma(n)}\left(\frac{\tau}{k}\right)^{n-1} e^{-\tau/k} \tag{5.24}$$

where $\Gamma(n)$ is the gamma function.

The second type, IUH-2, combines the concept of a linear channel and a linear reservoir (the Zoch model):

$$u(\tau) = \begin{cases} \dfrac{1}{T_p}(1 - e^{-\tau/k_1}) & \text{for } 0 \le \tau < T_p \\ \dfrac{1}{T_p}(1 - e^{-T_p/k_1})e^{-(\tau - T_p/k_1)} & \text{for } \tau \ge T_P \end{cases} \tag{5.25}$$

where T_P is the time of concentration (time to the peak of IUH).

IUH-3 uses the concept of two parallel unequal linear reservoirs with recession coefficients k_1 and k_2 ($k_1 > k_2$). Parameter θ presents the ratio of discharges from those two reservoirs. The expression is

$$u(\tau) = \frac{\theta}{k_1} e^{-\tau/k_1} + \frac{1-\theta}{k_2} e^{-\tau/k_2} \tag{5.26}$$

The last one, IUH-4, is derived from the concept of a time-variant linear reservoir, whose recession curve is described by the coefficient k as a function of time: $k = k_2$ for $k \le T_2$ and $k = k_1$ for $k > T_2$. The result is a IUH with two recession coefficients (k_1 and k_2) and a break point in T_2:

$$u(\tau) = \begin{cases} \dfrac{e^{T_2/k_2}}{k_2(e^{T_2/k_2} - 1) + k_1} e^{-\tau/k_2} & \text{for } 0 \le \tau < T_2 \\ \dfrac{e^{T_2/k_1}}{k_2(e^{T_2/k_2} - 1) + k_1} e^{-\tau/k_1} & \text{for } \tau \ge T_2 \end{cases} \tag{5.27}$$

which can be considered as time invariant at intervals $[0, T_2]$ and $[T_2; \infty]$.

When $t \to \infty$, it is evident that IUH-2, IUH-3, and IUH-4 have the exponential form of the master recession curve with the recession coefficient k_1. The master recession curve of IUH-1 has the form of the recession curve valid for the Nash model (for $n = 1$, the Nash model has the exponential form of the master recession curve also).

Antecedent recession

The function $y^D(t)$ represents the component of the discharge resulting from antecedent rainfalls. It can be defined by using the superposition law and the property of independence of discharge events (one discharge event does not affect another). Practically, it means that the component of discharge describing antecedent rainfalls is equal to the discharge resulting from a period without rainfalls; that is, the antecedent recession curve has the same form as the master recession curve. Consequently, the function $y^D(t)$ for IUH-2, IUH-3, and IUH-4 has the exponential form with the recession coefficient k_1:

$$y^D(t) = y_0 e^{-t/k_1} \tag{5.28}$$

where y_0 is the karst spring discharge at the beginning of the analyzed period (Figure 5–23). Equation (5.28) represents a well-known integral hydrograph separation method for groundwater recharge estimations (e.g., Ketchum, Donovan, and Avery, 2000; Bonacci, 2001). The antecedent recession curve for IUH-1 has the form of the Nash model recession curve. By applying $n = 1$, the antecedent recession curve of the Nash model becomes Equation (5.28) as well.

Effective rainfall

In the CTF model by Denić-Jukić and Jukić (2003), the effective rainfall is defined as a portion of total rainfall that ultimately reaches the observed spring, and it is identical to the groundwater recharge. It is calculated by Palmer's (1965) fluid-mass balance method. According to this approach, the effective rainfall available for groundwater recharge is the moisture left after evapotranspiration and the soil moisture-holding capacity is reached.

The moisture lost in the process of evapotranspiration is calculated by using the expression of Eagleman (1967):

$$\begin{aligned} ET_P &= Ce_{max}(100 - RH)^{1/2} \\ e_{max} &= 6.1e^{17.1T/(234.2+T)} \end{aligned} \tag{5.29}$$

where ET_P is the potential evapotranspiration in millimeters, T is the air temperature in degrees centigrade, and RH is the relative humidity in percent. The coefficient C depends on T: $C = 0.63$ for $T < 0°C$; $C = 0.63 + 0.024T$ for $0 < T < 21°C$; and $C = 1.13$ for $T > 21°C$.

All other catchment characteristics are included in calculations through the soil moisture-holding capacity, S_{max}, which therefore has no unique value. According to Soulios (1984), the values of S_{max} for the karst aquifers are in the range of 10–60 mm. The assumed value of S_{max} and the initial soil moisture conditions define the amount and temporal distribution of the calculated effective rainfall.

Parameter estimation

The values of parameters of IUH and the value of parameter k_1 of the antecedent recession curve may be defined using the methods of nonlinear least squares or a trial-and-error procedure based on changing values of parameters until a minimum error between simulated and observed discharges is reached. The trial-and-error procedure can be applied because the number of the IUH parameters is not significant (maximum three) and the majority of the proposed IUH parameters have physical meaning and relatively short intervals of possible values. If the karst spring catchment area is not accurately defined, the additional model parameter is the soil moisture-holding capacity, S_{max}, and the catchment area is defined as the area under the calculated CTF. If the karst spring catchment area is delineated, S_{max} needs to be defined as the value giving the volume of effective rainfall equal to the volume of discharge for the analyzed period. The memory length of the quick-flow method needs to be long enough to include the quick-flow component completely.

5.5.2 Application for water management

Fleury et al. (2008) present application of a rainfall-discharge model of the Lez Spring (Figure 5–26) used for the water supply of Montpellier, France. Groundwater is withdrawn by powerful pumps (Figure 5–27) from three wells placed in the main drain of this ascending spring, 400 m upstream from the main pool and 48 m below the overflow level. The average pumping rate for the 1997–2005 period was 1.1 m^3/s. When pumping begins to withdraw the reserves, the water levels in the drain and in the spring pool drop to the level of the pool overflow and the spring dries up. Pumping then causes a drawdown within the drain that can reach several tens of meters at the end of the low-water period. These reserves are recharged during the autumn and winter. The hydraulic water level in the drain rises above that of the pool, and the spring begins to flow again. During the active pumpage, when the Lez River flow is lower than the pumped flow, part of the pumped flow is channeled to the river. This return flow, set by the Declaration of Public Utility of June 5, 1981, is 160 L/s.

The Lez Spring model consists of three simulated reservoirs, all recharged by estimated effective precipitation: (1) saturated reservoir, (2) slow-flow infiltration zone reservoir, and (3) rapid infiltration zone

FIGURE 5–26 Natural overflow of the Lez Spring near Montpellier, France. (Photograph courtesy of Arnaud Vestier, Ville de Montpellier.)

FIGURE 5–27 Replacement of pumps at the Lez Spring. (Photograph courtesy of Jean Michel Valéry.)

reservoir. Effective infiltration is calculated based on the analysis of time series and the definition of transfer functions using TEMPO software by BRGM. Two modules are added to the saturated zone reservoir: a pumped volume and the returned flow. The model is used for estimating the volume of water (effective precipitation) needed to recharge the aquifer and it can accurately simulate the discharge at the spring and the water level in the drain.

5.6 TIME SERIES MODELS

As discussed by Gabric and Kresic (2009) and Kresic (1997), a time series is a series of time-dependent hydrologic variables, such as the flow rate in a surface stream or at a spring. When analyzing a time series, one deals with a limited amount of recorded data, a sample. This sample, regardless of its size, consists of a limited number of realizations of the same hydrologic process. All possible realizations of that process constitute a population. The goal of most hydrologic and hydrogeologic studies is to understand and quantitatively describe the population, as well as the process that generates it, based on a limited number of samples (actual field measurements of limited duration).

A time series can be continuous (such as the flow rate of a spring) or discrete (such as daily precipitation). For practical and computational purposes, most continuous time series are converted into discrete time series by introducing the recording (or modeling) time interval, such as one day, one week, or one month. When a time series is described with statistical and probabilistic parameters, it represents a probability of occurrence (realization) of one of its possible stages. A good example is a time series of monthly precipitation at a certain location in a moderate climate. Our long-term experience can tell us that, for example, April through June is the wet period and July through September is the dry period of the year. Accordingly, it can be expected, with a high probability, that in the near future (say, next year), these two periods will again last about the same time. However, no one can state with 100 percent accuracy that this will indeed happen (for example, June may be an unusually dry month next year), because it is impossible to accurately predict the annual or monthly amount of precipitation using some physical laws of nature. One can only apply tools of statistics and make predictions about the future using probabilistic models based on past data. A time series studied in this way is called a *stochastic time series*. In contrast, the stage of a deterministic process at time t is defined with certainty, knowing its stage at some earlier time t_0. In other words, a deterministic process is described with physical laws rather than laws of probability. An example is the flow of groundwater from point A to point B when described with equations such as the Dupuit equation, the Laplace equation, or the Theis equation, to name just a few. Quantitative hydrogeology is based on the physical laws of groundwater flow, as are the traditional numeric models presented in the next section.

Strictly speaking, most time series in hydrologic and hydrogeologic studies are stochastic, since they depend on at least one random variable, with precipitation often being the most important one. This also means that the result of a deterministic calculation, such as the drawdown in a well after one year of future pumping, although given explicitly, is actually just more or less probable.

In general, a time series has the following five components, all of which may or may not be present (adapted from McCuen and Snyder, 1986):

- **Trend**, which is a tendency to increase or decrease continuously for an extended period of time in a systematic manner. This component can often be described by fitting a functional form, such as a line or polynomial. The coefficients of the equation are commonly evaluated using regression analysis. The trend is also referred to as a *deterministic component*, even though its physical explanation may not always be clear.
- **Periodicity**, which is very common in a hydrologic time series: annual and seasonal periodicity of precipitation, temperature, and flow. The period(s) in a time series can be identified using a moving-average analysis, an autocorrelation analysis, or a spectral analysis, after which it is described by one or more trigonometric functions.
- **Cycle**, which occurs with an irregular period and is hard to detect (for example, hydrometeorological time series are thought to be influenced by sunspot activity, which has an irregular period).
- **Episodic variation**, which results from extremely rare or one-time events, such as hurricanes. Identification of this component requires supplementary information.

- **Random fluctuations**, which are often a dominant source of the variation in a time series and are the main target of a probabilistic identification.

Figure 5–28 illustrates these components with a small sample of the actual spring hydrographs created from long-term records kept by the U.S. Geological Survey.

Stochastic models describe time series formally and do not consider their physical nature. Simply stated, they statistically (mathematically) analyze the past of the time series, as system input, then predict the present or the future as the system output. They can also analyze the past of a one time series and use it to predict the present and future of some other time-dependent series proven to be correlated with the first one. Stochastic models can also combine several inputs and give one or several outputs. Examples would be a model that predicts water table elevation based on its position in the past, a model that predicts spring discharge based on its own past and the antecedent precipitation, or a model that includes past stages of a nearby river as well.

Two main applications of the time series models are the generation of synthetic samples and forecasting of hydrologic events. Generated time series, which are statistically indistinguishable from historic time series, serve as input to the analysis of complex water resources systems. They can also be used to provide a probabilistic framework for analyses and design. Generated series show many possible hydrologic conditions that do not explicitly appear in the historic record. Consequently, using synthetic time series, different designs and operational schemes can be tested under many different conditions contained in these time series. Forecasted data from known historic observations can help in evaluating options for a real-time system operation.

Time series modeling originated in different scientific fields, but it has subsequently become very important in stochastic hydrology, and the applications of generated time series are numerous. Development of stochastic modeling in hydrology began at the beginning of the 1960s, when time series analyses of hydrologic phenomena were extended to the synthetic generation of stream flow using a table of normal random numbers. Thomas and Fiering (1962) were the first to propose a first-order Markov model to generate stream-flow data. The classic book on time series analysis by Box and Jenkins (1976) presents the foundation of modern hydrologic stochastic modeling.

The general form of an input-output stochastic model of a discrete time series (or a continuous time series transformed into a discrete one) with the same recording time interval is

$$y_t = f(x_t, x_{t-1}, x_{t-2}, \ldots; y_{t-1}, y_{t-2}, \ldots; \theta_1, \theta_2, \ldots) + \epsilon_t \tag{5.30}$$

where f is the selected mathematical function; y_t is the predicted output at time t; $y_{t-1}, y_{t-2}, \ldots$, are the successive members of the output time series recorded at corresponding time intervals $t, t-1, t-2$; $x_t, x_{t-1}, t_{t-2}, \ldots$, are the successive members of the input time series recorded at time intervals $t, t-1, t-2$; $\theta_1, \theta_2, \ldots$, are the model parameters found by mathematically minimizing the differences between estimated (calculated) and observed y_t values; ϵ_t is the model error (residual) given as the difference between the calculated and the recorded value of the output series at time t.

Stochastic modeling generally follows the approach proposed by Box and Jenkins (1976), who introduced autoregressive moving average (ARMA) models. The mathematical formulation of ARMA models is

$$z_t = \sum_{j=1}^{p} \varphi_j z_{t-j} + \sum_{j=0}^{q} \theta_j \epsilon_{t-j} + \epsilon_t \tag{5.31}$$

where z_t represents the time dependent series with mean zero and variance one; $\theta_1, \ldots, \theta_p$ are time varying autoregressive coefficients; $\theta_0, \ldots, \theta_q$ are time varying moving average coefficients; ϵ_t is an independent normal variable.

Time series models used to generate synthetic time series can be classified into autoregressive models (AR(p)), moving average models (MA(q)), and their combination, autoregressive moving average (ARMA (p, q)) with variations, such as autoregressive integrated moving average (ARIMA) models (p, d, q) and

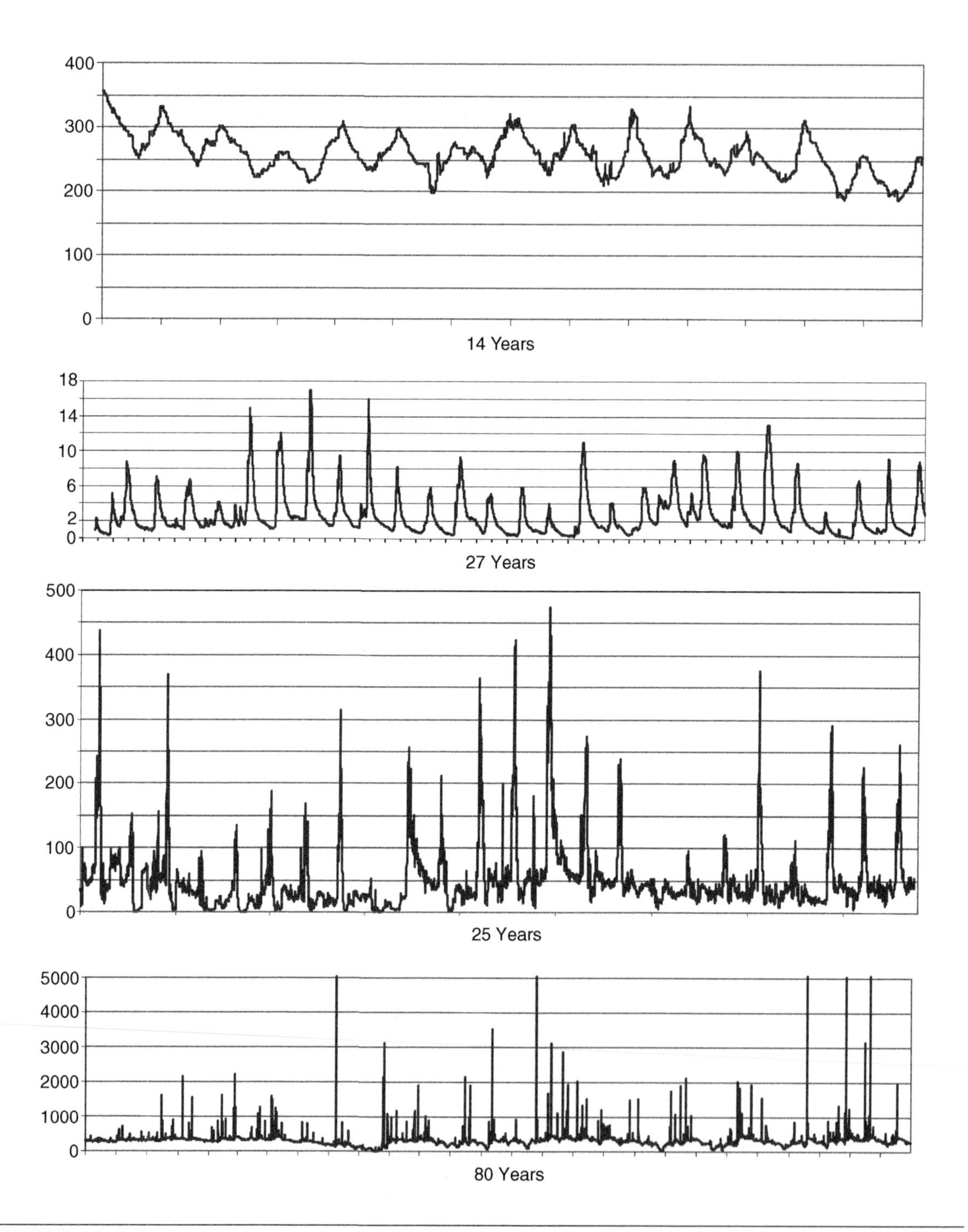

FIGURE 5–28 Actual hydrographs of springs showing various components of a time series: trends, periodicity, cycles, episodic variation, and random fluctuations. All flow rates are in cubic feet per second. (Source of data U.S. Geological Survey, 2008.)

others, where p and q are the orders of autoregressive and moving average terms, respectively, and d is the differentiation order. An autoregressive model estimates values for the dependent variable, Z_t, as a regression function of previous values $Z_{t-1}, Z_{t-2}, \ldots, Z_{t-n}$. A moving average model is conceptually a linear regression of the current value of the series against the white noise or random shocks of one or more prior values of the series. A pure autoregressive (AR) model, commonly called a *Thomas-Fiering model*, has been applied extensively in hydrology for modeling annual and periodic hydrologic time series. Because the parsimony ("the less the better") in the number of parameters is very desirable (since the parameters are estimated from data), the second order of these models is usually the highest lag necessary in representing hydrologic time series. A parsimonious model can be achieved using a mixed ARMA model as a combination of a moving average process and an autoregressive process rather than a pure AR or MA model. Therefore, low-order ARIMA models have been widely used in hydrological practice (Salas, Boes, and Smith, 1982; Weeks and Boughton, 1987; Padilla et al., 1996; Montanari, Rosso, and Taqqu, 2000).

An important aspect of stochastic modeling is the problem of nonstationarity in hydrologic time series. Stationarity is usually assumed when modeling annual time series. When dealing with a monthly or weekly time series, seasonal nonstationarity is present, and it may be necessary to use a model that has seasonally varying properties. Significant contributions in developing periodic models were made by Hirsch (1979) and Salas et al. (1985). For modeling seasonal time series, two approaches can be used. The first is a direct approach, in which a model with periodic parameters is fitted directly to the seasonal flows. This method requires a considerable number of years of data. The number of coefficients involved can be very large. If available historical data are limited, the parameters are poorly estimated. Consequently, the main problem of all seasonal models with time-varying coefficients is the lack of parsimony. The second approach is decomposition ("disaggregation"), in which the seasonal flows are generated at two or more levels. For instance, the first level involves modeling and generating annual flows, and the second level is their decomposition into seasonal flows based on a linear model. However, if the autocorrelation structure of a historical time series shows a significant periodicity, then seasonal models that explicitly incorporate a periodic structure must be used. If the seasonality of time series under consideration is in the mean and the variance, then such seasonality can be removed by simple seasonal standardization, and a stationary model can be applied. Another peculiarity of hydrologic processes is the skewed distribution functions observed in most cases. Therefore, attempts have been made to adapt standard models to enable treatment of skewness (Bras and Rodriguez-Iturbe, 1994).

A generalized framework for time series model development consists of three phases: (1) identification, (2) parameter estimation, and (3) verification and diagnostic checking. Model identification is not a standardized, automated procedure but rather heuristic. The usual approach is an iterative trial-and-error procedure. The first step is to investigate whether the time series data are stationary and if any significant seasonality needs to be modeled. A visual inspection of the time plot of the historic times series can help in deciding between seasonal and nonseasonal models, whether local differentiation is needed to produce stationarity, and to get a general feeling about the order of possible models. A further identification process is to examine the shapes of autocorrelation and the partial autocorrelation functions of the historic time series. To allow for possible identification errors, a set of several models with a close structure are considered. It is advisable to always select the simplest acceptable model.

When the model order is selected, the estimation of parameters follows. The parameters are estimated from recorded data by either a method of moments or methods of maximum likelihood. The final stage of modeling is verification as to what extent the selected historic statistics are reproduced by the model and proving the adequacy of the model. The verification involves a check of possible overfitting (the confidence limits of the parameters) and a check of randomness of the residuals (the autocorrelation function, for the residuals resulting from a good ARIMA model should have statistically insignificant autocorrelation coefficients). Sometimes, a sufficient objective for simulation purposes, and adequate for short-term forecasting, is to preserve the first- and second-order moments of the time series. When comparing several possible models, the one with the best goodness of fit is selected based on the minimum Akaike information criterion (Akaike, 1974).

When the selected mathematical function, f in equation (5.30), ignores the physical laws that govern the transformation of input(s) into output(s), the model is a pure stochastic one. If, in any form, the mathematical function incorporates physical laws, the model is called a *stochastic-conceptual model*. The knowledge of various physical processes and relationships related to the system of interest is invaluable and provides the physical background for stochastic modeling (Klemeš, 1978; Vecchia et al., 1983; Koch, 1985; Salas and Obeysekera, 1992; Knotters and Bierkens, 2000; Lee and Lee, 2000). It is always preferable to make detailed structural and physical analyses of the hydrologic processes involved before performing stochastic modeling.

As mentioned earlier, there are many possible uses of time series models. Typical applications of generated time series in surface water engineering are reservoir design, risk and reliability assessment, planning of hydropower production, and flood and drought hazard analysis. In groundwater studies, stochastic models can be used to analyze and forecast the hydraulic head fluctuations, fill in data gaps, and detect and quantify trends. (Houston, 1983; Padilla et al., 1996; Ahn, 2000; Knotteres and Bierkens, 2000; Bierkens, Knotters, and Hoogland, 2001; Kim, Hyun, and Lee, 2005). Time series models of spring discharge rates are becoming increasingly popular. They use observed flows and incorporate the factors that influence it, such as groundwater levels, precipitation, evapotranspiration, and anthropogenic hydrologic disturbances, such as pumpage from the aquifer.

A very important factor that limits wider use of stochastic models in hydrogeology is the lack of recorded data. Since these models are based on statistical and probabilistic calculations, very short time series do not allow for meaningful derivation of model parameters. Spring discharge rates measured for a couple of years on a quarterly basis are obviously not good candidates for stochastic modeling. On the other hand, if an appropriate amount of data is available, every attempt should be made to develop one or several stochastic models of the input-output type. This is mainly because the process of building even the simplest stochastic model reveals a great deal of information on the possible structure(s) of the system and connections between various hydrologic variables (Kresic, 1995, 1997).

More sophisticated stochastic models can include various linear and nonlinear transformations of input series, consideration of periodicity (seasons), and filtering (minimizing) of the model residuals. One widely used linear filter for the residuals is the Kalman filter, which usually significantly improves ARMA models (Kalman, 1960; Birkens et al., 2001). Note, however, that both ARMA models, like the previously explained ARCR models, give valid predictions for only one step at a time (say, the next day), because for the second and following intervals, the true values of the dependent variable (e.g., spring flow) are not known and the error rapidly increases.

The following example, courtesy of Ivana Gabric of the University of Split, illustrates the application of synthetic hydrologic time series in karst water resources management. Jadro Spring (see Figure 1–14), with an average discharge rate of 9.82 m^3/s, provides water supply for the city of Split and its 270,000 inhabitants. The main characteristic of the spring discharge is its large fluctuations in response to precipitation. During high discharge rates, there is more intense washing away of the soil and sediment, which accumulated in the subsurface, resulting in sudden, short-lived changes in water quality. Consequently, the spring water is characterized by the occasional high turbidity, exceeding allowable standards. Turbidity is the key problem in the management of the water quality of this spring, as well as many other karst springs. For management purposes, it is important to know the nature of the turbidity and predict its occurrence as early as possible, because elevated turbidity is often associated with high bacteria counts, indicating possible contamination. An accurate prediction of elevated turbidity can help optimize sampling strategies. In the case of Jadro Spring, the turbidity monitoring was not systematic, leaving insufficient information for reliable water supply system management. A stochastic time series model can therefore provide for a more comprehensive understanding of the turbidity, using available short-term measurements of the turbidity and long-term recording of the spring discharge (Margeta and Fistanić, 2004; Rubinić and Fistanić, 2005).

Time series analyses of discharge and turbidity showed that the turbidity is higher during the first large rainfall following a dry period. Similar discharge rates following the first rainy period produce generally lower turbidities. Consequently, a reliable prognosis of turbidity can be carried out using different regression functions of discharge and turbidity for different parts of the year. A stochastic model was built, based on 3 years of daily measurements of turbidity and discharge and 28 years of daily discharge measurements, as illustrated in Figure 5–29. Box A presents the identification, estimation, and verification phases of the Jadro Spring stochastic model. Synthetic time series of monthly spring discharge were generated using a Thomas-Fiering AR(2) model with constant coefficients, which provided the best preservation of historic statistical characteristics of the spring discharge: seasonal means, variances, and correlations of the processes. Box B describes the procedure of the turbidity time series generation. Based on the functional dependence between daily turbidity and discharge, the generated time series of the average monthly discharge (box A) is used to generate up to 100 time series of the average monthly turbidity. The time series of maximum daily turbidity is then generated using the time series of the average monthly turbidity.

Figure 5–30 shows the maximum daily values of the generated turbidity at Jadro Spring as a function of the number and length of the generated time series. The analysis of the occurrences of high turbidity indicates that their seasonal behavior is preserved in the model, as shown in Figure 5–31. The highest average, as well

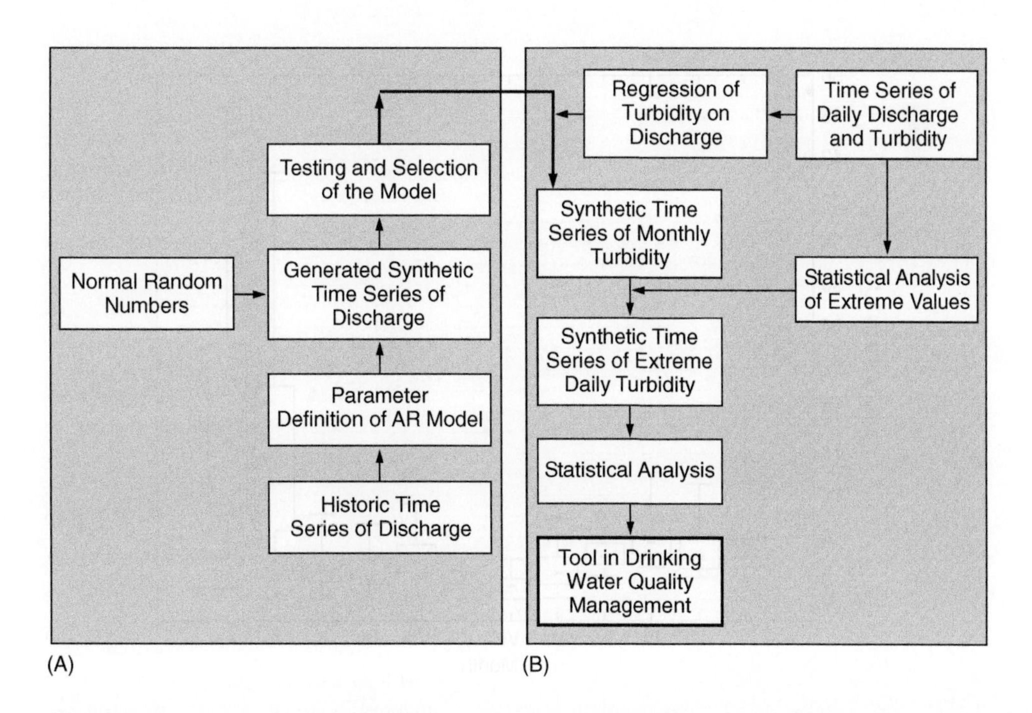

FIGURE 5–29 Building steps of the stochastic time series model of daily turbidity at Jadro karst spring. (Courtesy of Ivana Gabric, University of Split.)

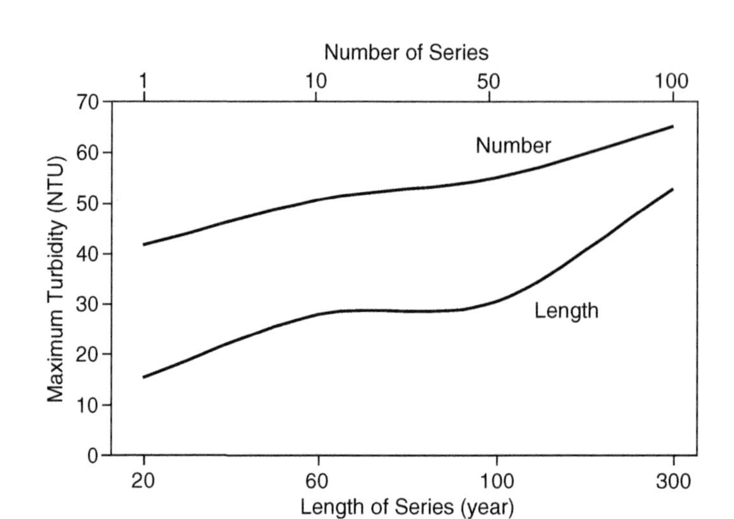

FIGURE 5–30 Maximum values of generated turbidity at Jadro Spring versus number and length of generated time series. (Courtesy of Ivana Gabric, University of Split.)

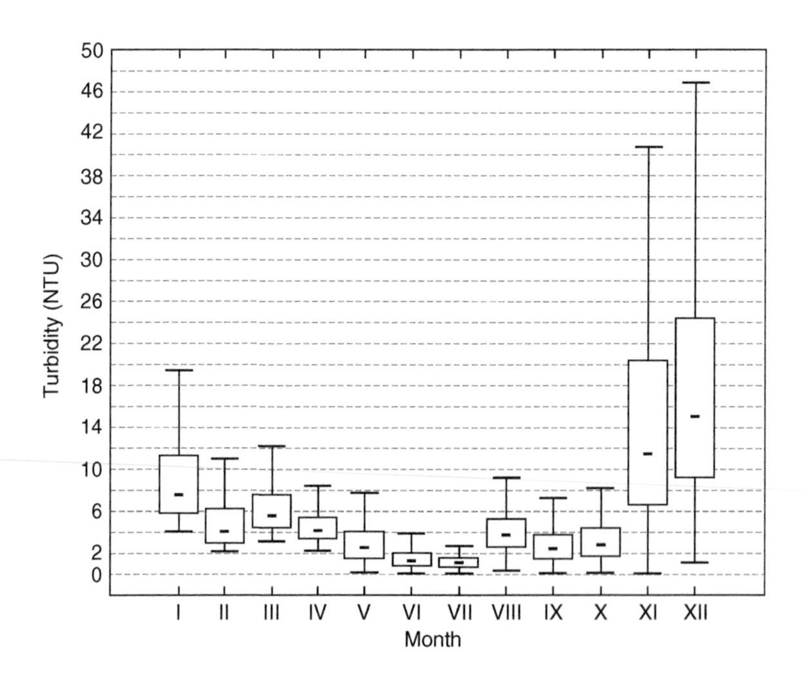

FIGURE 5–31 Box-whisker diagram of generated daily values of turbidity, without extremes and outliers. (Courtesy of Ivana Gabric, University of Split.)

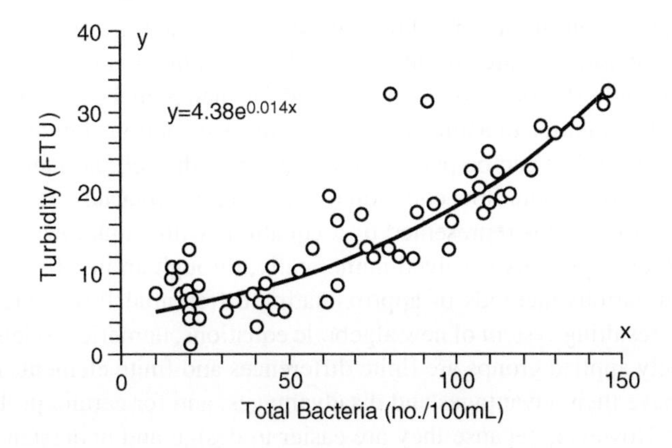

FIGURE 5–32 Total coliform bacteria (no./100 mL) versus turbidity of the Petnica karstic spring near Valjevo, western Serbia. (From Kresic et al., 1992; copyright Springer-Verlag New York Inc.)

as maximum turbidity values, usually appear during late autumn months, when rainfall after long summer droughts mobilizes deposited sediments and brings them into the karst underground.

One springwater quality problem often associated with elevated turbidity is bacteriological contamination, as shown in Figure 5–32. If such a relationship is established, then both water quality parameters can be modeled in the same manner or used simultaneously to improve the model predictions.

Once developed and verified, a time series model can be used for real-time spring management, based on monitoring various model inputs and outputs and a continuous updating of the model for the next-day prognosis. For example, rainfall above a certain threshold may trigger diversion of the spring flow to a clarification facility, additional disinfection, use of a standby (emergency) reservoir, or free outflow of the spring.

5.7 DETERMINISTIC MODELS

There are two large groups of deterministic models, depending on the type of solutions of the mathematical equations involved: analytical and numeric. Simply stated, *analytical models* solve one or several equation of groundwater flow at a time, and the result can be applied to one or several point locations in the analyzed flow field (aquifer). For example, if one wants to find (i.e., to "model") what the drawdown at 50 m from a pumping well would be after 24 hours of pumping, one would apply one of the commonly used equations describing flow toward a well. To find the drawdown at 1000 m from the well, one would have to solve the same equation (say, the Theis equation) for this new distance. If the aquifer is not homogeneous, these solutions would be applicable for just a limited radial distance of 50 or 1000 m within the same distribution of aquifer transmissivity. Obviously, if the aquifer is quite heterogeneous and one wants to know drawdown at "many" points, one might spend a rather long period of time solving the same equation (with slightly changed variables) again and again. If the situation gets really complicated, such as when there are several boundaries, more pumping wells, and several hydraulically connected aquifers, the feasible application of analytical models terminates (Kresic, 2007a).

Numeric models describe the entire flow field of interest at the same time, providing solutions for as many data points as specified by the user. The area of interest is subdivided into many small areas (referred to as *cells* or *elements*), and a basic groundwater flow equation is solved for each model cell, considering its water balance (water inputs and outputs). The solution of a numeric model is the distribution of hydraulic heads at points representing individual cells. These points can be placed in the center of the cell, at intersections between adjacent cells, or elsewhere. The basic differential flow equation for each cell is replaced (approximated) by an algebraic equation, so that the entire flow field is represented by x equations with x unknowns where x is the number of cells. This system of algebraic equations is solved numerically, through an iterative process, hence, the name *numeric models*. Based on various methods of approximating differential flow equations and methods used for numerically solving the resulting system of new algebraic equations, numeric models are divided into several groups. The two most widely applied groups are finite differences and finite elements numeric models.

Both types of models have their advantages and disadvantages, and for certain problems, one may be more appropriate than the other. However, because they are easier to design and understand and require less mathematical involvement, finite-difference models have prevailed in hydrogeologic practice. In addition, several excellent finite-difference modeling programs have been developed by the U.S. Geological Survey (USGS) and are in public domain, which ensures their widest possible use. One of these is Modflow, probably the most widely used, tested, and verified modeling program today; and it has become the industry standard thanks to its versatility and open structure: Independent subroutines, called *modules*, are grouped into "packages," which simulate specific hydrologic features. New modules and packages can be easily added to the program without modifying the existing packages or the main code. The USGS recently made public a significantly upgraded version of the finite element model SUTRA3D, now capable of simulating three-dimensional flow (Voss and Provost, 2002). This computer program can simulate both unsaturated and saturated flow, heat and contaminant transport, as well as variable density flow, which makes it a powerful tool for modeling just about any real-life condition. Unfortunately, unlike Modflow, SUTRA3D is not yet part of any of the most popular user-friendly commercial programs for processing model input and output data, which severely limits its greater application. In 2008, the USGS released a highly anticipated addition to Modflow-2005, Conduit Flow Process (Schoemaker et al., 2008), capable of simulating various flow types in karst aquifers explicitly using a physically based approach. For a thorough explanation of finite element and finite difference models and their various applications, the reader should consult the excellent work by Anderson and Woessner (1992).

Groundwater models can be used for three general purposes (Kresic, 2007a):

- To predict or forecast expected artificial or natural changes in the system (aquifer) studied. The term *predict* is more appropriately applied to deterministic (numeric) models, since it carries a higher degree of certainty, while *forecasting* is the term used with probabilistic (stochastic) models. Predictive models are by far the largest group of models built in hydrogeologic practice.
- To describe the system to analyze various assumptions about its nature and dynamics. Descriptive models help to better understand the system and plan future investigations. Although not originally planned as a predictive tool, they often grow to be full predictive models.
- To generate a hypothetical system that will be used to study principles of groundwater flow associated with various general or more specific problems. Generic models are used for training and are often created as part of a new computer code development.

Predictive numeric models are divided into two main groups: models of groundwater flow and models of contaminant fate and transport. The latter ones cannot be developed without first solving the groundwater flow field of the system studied; that is, they use the solution of the groundwater flow model as the base for fate and transport calculations.

Regardless of the intended use or type, it is essential that, for any groundwater model to be interpreted and used properly, its limitations be clearly understood. In addition to strictly "technical" limitations, such as accuracy of computations (hardware or software), the following is true for any model (Kresic, 2007a):

- It is based on various assumptions regarding the real natural system being modeled.
- Hydrogeologic and hydrologic parameters used by the model are always just an approximation of their actual field distribution, which can never be determined with 100 percent accuracy.
- Theoretical differential equations describing groundwater flow are replaced with systems of algebraic equations that are more or less accurate.

It is therefore obvious that models have varying degrees of reliability and cannot be "misused" as long as all the limitations involved are clearly stated, the modeling process follows industry-established procedures and standards, and the modeling documentation and any generated reports are transparent, also following the industry standards.

The following industry standards, created by the leading industry experts for the groundwater modeling community under the auspices of the American Society for Testing and Materials, cover all major aspects of groundwater modeling and should be followed when attempting to create a defensible groundwater model that can be used for predictive purposes (American Society for Testing Materials, 1999a, 1999b):

- Guide for application of ground-water flow model to a site-specific problem (D 5447-93).
- Guide for comparing ground-water flow model simulations to site-specific information (D 5490-93).
- Guide for defining boundary conditions in ground-water flow modeling (D 5609-94).
- Guide for defining initial conditions in ground-water flow modeling (D 5610-94).
- Guide for conducting a sensitivity analysis for a ground-water flow model application (D 5611-94).
- Guide for documenting a ground-water flow model application (D 5718-95).
- Guide for subsurface flow and transport modeling (D 5880-95).
- Guide for calibrating a ground-water flow model application (D 5981-96).
- Practice for evaluating mathematical models for the environmental fate of chemicals (E 978-92).
- Guide for developing conceptual site models for contaminated sites (E 1689-95).

The following language accompanies the U.S. EPA OSWER Directive 9029.00 (U.S. EPA, 1994):

The purpose of this guidance is to promote the appropriate use of ground-water models in EPA's waste management programs. More specifically, the objectives of the "Assessment Framework for Ground-Water Model Applications" are to:

- Support the use of ground-water models as tools for aiding decision-making under conditions of uncertainty;
- Guide current or future modeling;
- Assess modeling activities and thought processes; and
- Identify model application documentation needs.

"Guidelines for Evaluating Ground-Water Flow Models," published by the U.S. Geological Survey (Reilly and Harbaugh, 2004), states in its introduction: "Ground-water flow modeling is an important tool frequently used in studies of ground-water systems. Reviewers and users of these studies have a need to evaluate the accuracy or reasonableness of the ground-water flow model. This report provides some guidelines and discussion on how to evaluate complex ground-water flow models used in the investigation of ground-water systems. A consistent thread throughout these guidelines is that the objectives of the study must be specified to allow the adequacy of the model to be evaluated."

5.7.1 Analytic models (equations of groundwater flow)

Equations of groundwater flow, for all porous media alike (matrix, fissures or fractures, and conduits or channels) are based on the same principle of conservation of mass, also called the *principle of flow continuity*, as illustrated in Figure 5–33. This principle, derived from the general hydraulics of fluid flow, states that the flow rate (Q) through an elementary flow tube is directly proportional to the cross-sectional area of the tube (a) and the flow velocity (v):

$$dQ = vda \tag{5.32}$$

"Many" such elementary flow tubes can be combined into a realistic three-dimensional portion of an aquifer through which the groundwater flow is taking place; as long as there is no loss or gain of water inside this (combined) realistic flow tube, however, the principle is the same:

$$Q = \int_A dQ = \int_A vda \tag{5.33a}$$

$$Q = v_{av} A \tag{5.33b}$$

where A is the sum of all cross sections of the elementary flow tubes, and v_{av} is the average flow velocity within the realistic flow tube.

Two or more flow tubes can merge into one, and the resulting flow rate is additive (Figure 5–34). Consequently, one flow tube can split into several, where the sum of the new flows is equal to the initial flow. It should be noted that the term *tube* does not imply that the portion of the aquifer under consideration must look like (have the shape of) a tube; what it means is that the flow lines (streamlines) of groundwater particles inside the tube remain inside and no new particles enter the tube.

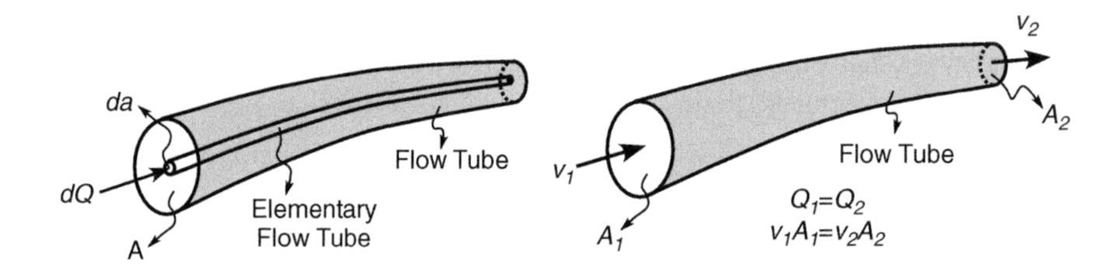

FIGURE 5–33 (Left) Flow tube in porous media: da = cross section of elementary flow tube, dQ = elementary flow rate. (Right) Principle of continuity of flow (conservation of mass) for incompressible fluids: A_1 and A_2 = cross sectional area of flow; v_1 and v_2 = average flow velocity.

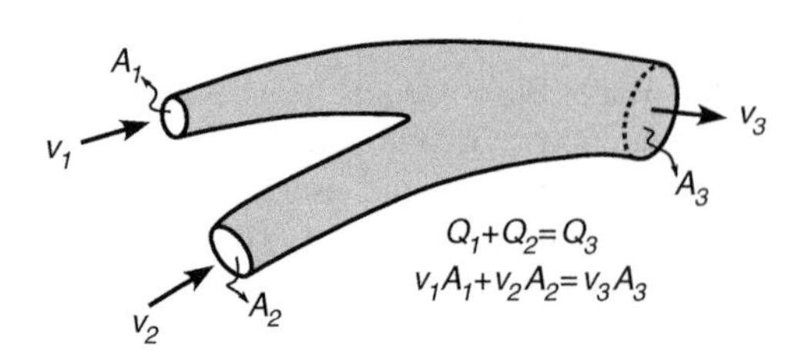

FIGURE 5–34 Principle of continuity of flow for three flow tubes.

When considering flow of groundwater toward a spring, regardless of the aquifer type drained by the spring and the spring size, the following additional principles should be always taken into account when attempting to quantitatively describe the flow at the spring using analytic equation(s):

- Groundwater flow is from the portions of the aquifer that have higher hydraulic heads than the elevation of the spring discharge; the flow of groundwater is always from the higher hydraulic head toward the lower hydraulic head.
- Groundwater flow converges toward the spring.
- The spring is always located at the end of a preferential flow path, in all types of aquifers, including aquifers developed in intergranular porous media (unconsolidated sediments).
- More preferential flow paths may merge into the final preferential flow path that feeds the spring.
- The preferential flow path has a higher permeability than the surrounding porous media.
- The spring emerges because there is a discontinuity in the three-dimensional groundwater flow field (this flow field is always three-dimensional).
- The flow at the spring is to a certain degree always time dependent (transient); that is, it varies over time.

Figure 5–35 illustrates some of these groundwater flow principles that lead to the emergence of springs. Unfortunately, no single equation or group of equations can be solved analytically in an efficient way and, at the same time, describe (honor) most of these principles. Virtually all commonly applied equations of groundwater flow, for any type of porous media, are based on various simplifying assumptions and are almost all for nontransient (steady-state) conditions describing one-dimensional or, at best, planar (two-dimensional) flow.

One of the main problems for a wider application of analytical equations in spring hydrogeology, in addition to the strongly transient nature of spring discharge, is that the transmissive properties of the porous media have to increase closer to the spring to accommodate an increasing flow rate through a decreasing cross-sectional area of the aquifer that ultimately results in a point discharge (Figure 5–36). This means that the equations have to be partial differential and integrated for the changing initial and boundary conditions, something that is beyond feasible analytical solutions. It is also the main reason why numeric groundwater models today are irreplaceable in quantitative hydrogeology.

Nevertheless, analytical equations are useful as part of the overall assessment and development of a conceptual spring model; they can help test various hypotheses quickly and provide some general ("rough") estimates of various aquifer (hydrogeologic) parameters including transmissivity and hydraulic conductivity, flow velocity (velocities), or effective porosity. They also are part of numeric models and, as such, should be familiar to practicing hydrologists and hydrogeologists.

There are three general types of groundwater flow equations, depending on the nature of the aquifer flow tube, shown in Figure 5–33, that is, the nature of the porous media. These are briefly described next. The reader interested in more detail should consult general hydrogeology textbooks (e.g., Freeze and Cherry, 1979; Kresic, 2007a), books on groundwater flow in fractured rock (e.g., Bear, Tsang, and de Marsily, 1993; Faybishenko, Witherspoon, and Benson, 2000), and books on karst hydrogeology (Ford and Williams, 2007; White, 1988).

Aquifer in unconsolidated sediments

The flow tube is filled with unconsolidated sediments, that is, grains of porous material. In this case, the flow of groundwater is mostly laminar (slow), and it can be described with Darcy's equation, the most widely used equation in hydrogeology, which has the following common forms:

$$Q = K\frac{\Delta h}{l}A \quad [\mathrm{m^3/s}] \tag{5.34a}$$

$$v = K\frac{\Delta h}{l} = Ki \quad [\mathrm{m/s}] \tag{5.34b}$$

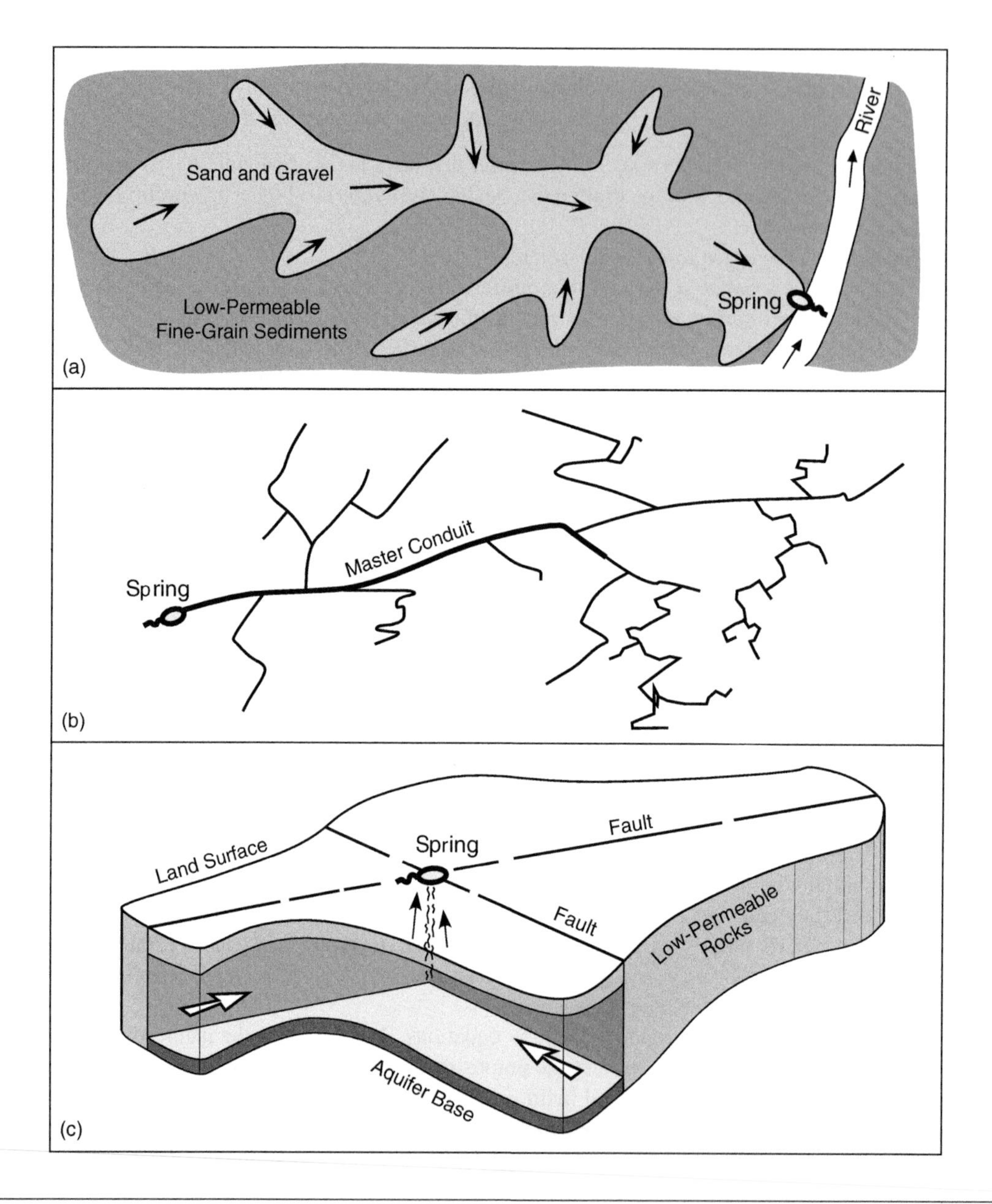

FIGURE 5–35 (a) Map view of emergence of a descending gravity spring in an intergranular alluvial aquifer incised by a river. The interconnected sand and gravel preferential pathways, surrounded by less permeable fine-grained sediments, collect and transfer groundwater to the spring. (b) Map view of karst conduits or cave passages collecting groundwater and transferring it to the master conduit or channel that feeds the spring. (c) Ascending spring that emerges at the intersection of regional faults by flowing through the enhanced permeability zone created in the overlying low-permeable rocks. The spring acts like a pumping well with a semi-radial flow.

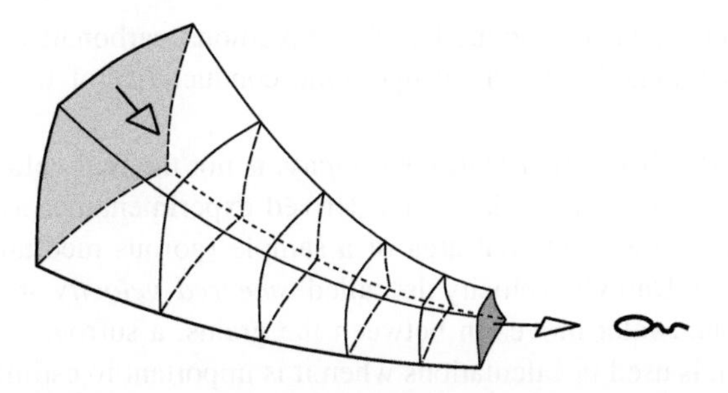

FIGURE 5–36 Decreasing cross-sectional area of groundwater flow toward a spring due to converging three-dimensional flow lines.

This linear law states that the rate of fluid flow (Q) through porous medium is directly proportional to the flow velocity (v) and the cross-sectional area of flow (A), where the flow velocity is given as the product of the constant, K, and the hydraulic gradient (i). The hydraulic gradient is defined as the loss of the hydraulic head (Δh) along the flow path of length l. K is the proportionality constant of the law called *hydraulic conductivity* and has units of velocity. This constant is the most important quantitative parameter characterizing the flow of groundwater through porous media. Figure 5–37 gives the range of hydraulic conductivity for

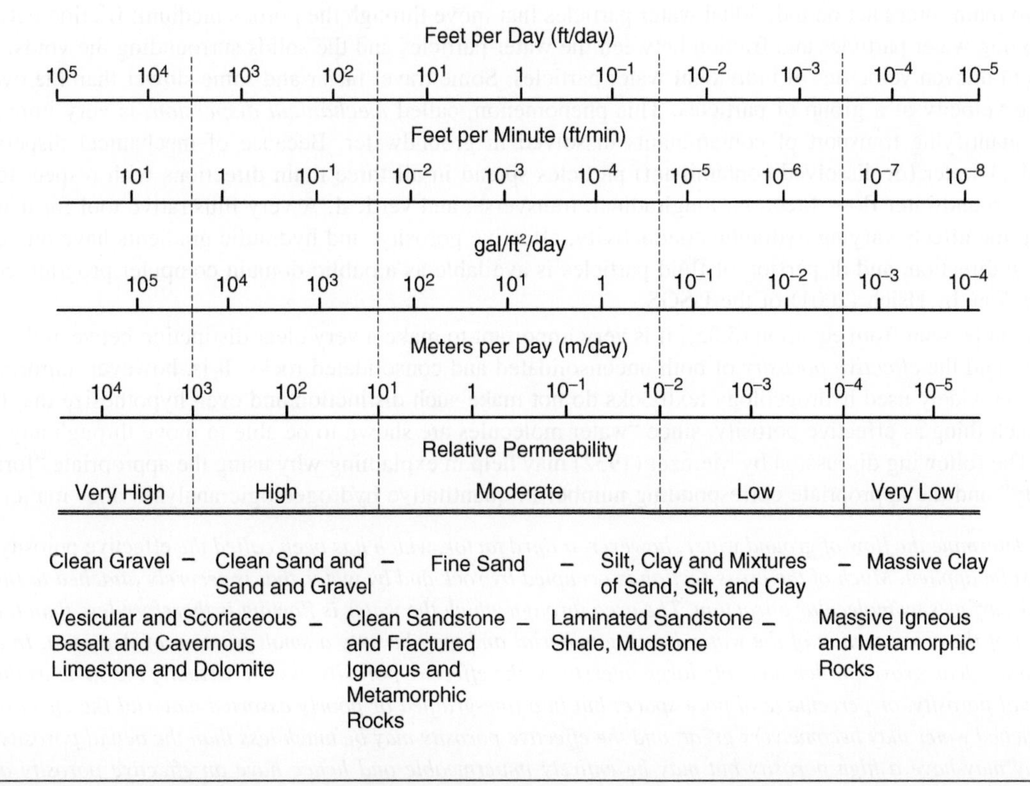

FIGURE 5–37 Range of hydraulic conductivites for different rock types. (From U.S. Bureau of Reclamation, 1977.)

different rock types. For consolidated fractured rocks, cavernous carbonates, and permeable basalts, the values given represent the combined, equivalent hydraulic conductivity of the entire rock mass, including discontinuities.

The velocity in equation (5.34b), called *Darcy's velocity*, is not the real velocity at which water particles move through the porous medium. Darcy's law, first derived experimentally, assumes that the groundwater flow occurs through the entire cross-sectional area of a sample (porous medium), including both the voids and the grains (appropriately, Darcy's velocity is called *smeared velocity* in the Russian literature). To account for the fact that groundwater moves in between the grains, a surrogate groundwater velocity, often called the *linear velocity* (v_L), is used in calculations when it is important to estimate the groundwater velocity in addition to the flow rate:

$$v_L = \frac{v}{n_{\text{ef}}} \tag{5.35}$$

where n_{ef} is the effective porosity. Since the actual cross-sectional area of flow is smaller than the total area (the flow takes place only through voids between the grains), it follows that the linear velocity must be greater than the Darcy's velocity: $v_L > v$.

The linear groundwater velocity is accurate when used to estimate the average travel time of groundwater, and Darcy's velocity is accurate for calculating flow rates. Neither, however, is the real groundwater velocity, which is the time of travel of a water particle along its actual path through the voids. It is obvious that, for practical purposes, the real velocity cannot be measured or calculated.

Two main forces act on individual water particles that move through the porous medium: friction between the moving water particles and friction between the water particles and the solids surrounding the voids. This results in uneven velocities of individual water particles: Some travel faster and some slower than the overall average velocity of a group of particles. This phenomenon, called *mechanical dispersion*, is very important when quantifying transport of contaminants dissolved in groundwater. Because of mechanical dispersion, individual water (or dissolved contaminant) particles spread in all three main directions with respect to the overall groundwater flow direction: longitudinal, transverse, and vertical. A very illustrative tool for demonstrating the effects varying hydraulic conductivity, effective porosity, and hydraulic gradients have on velocity, flow direction, and dispersion of fluid particles is available as a public-domain computer program called *Particleflow* by Hsieh (2001) of the USGS.

As can be seen from equation (5.35), it is very important to make a very clear distinction between the total porosity and the *effective porosity* of both unconsolidated and consolidated rocks. It is, however, unfortunate that some widely used hydrogeology textbooks do not make such distinction and even hypothesize that there is no such thing as effective porosity, since "water molecules are shown to be able to move through any pore size." The following discussion by Meinzer (1932) may help in explaining why using the appropriate "form of porosity" and the appropriate corresponding number in quantitative hydrogeologic analyses does matter:

> *To determine the flow of ground water, however, a third factor, which has been called the* effective porosity, *must be applied. Much of the cross section is occupied by rock and by water that is securely attached to the rock surfaces by molecular attraction. The area through which the water is flowing is therefore less than the area of the cross section of the water-bearing material and may be only a small fraction of that area. In a coarse, clean gravel, which has only large interstices, the effective porosity may be virtually the same as the actual porosity, or percentage of pore space; but in a fine-grained or poorly assorted material the effect of attached water may become very great, and the effective porosity may be much less than the actual porosity. Clay may have a high porosity but may be entirely impermeable and hence have an effective porosity of zero. The effective porosity of very fine grained materials is generally not of great consequence in*

determinations of total flow, because in these materials the velocity is so slow that the computed flow, with any assumed effective porosity, is likely to be relatively slight or entirely negligible. The problem of determining effective porosity, as distinguished from actual porosity, is, however, important in studying the general run of water-bearing materials, which are neither extremely fine nor extremely coarse and clean. Hitherto not much work has been done on this phase of the velocity methods of determining rate of flow. No distinction has generally been made between actual and effective porosity, and frequently a factor of 33⅓ per cent has been used, apparently without even making a test of the porosity. It is certain that the effective porosity of different water-bearing materials ranges between wide limits and that it must be at least roughly determined if reliable results as to rate of flow are to be obtained. It would seem that each field test of velocity should be supplemented by a laboratory test of effective porosity, for which the laboratory apparatus devised by Slichter (1905) could be used.

Effective porosity is often equated to the *specific yield* of the porous material or assumed that the volume of water in the pore space can be freely drained by gravity due to the change in the hydraulic head. *Effective porosity* is also defined as the volume of interconnected pore space that allows free gravity flow of groundwater. The volume of water retained by the porous media, which cannot be easily drained by gravity, is called *specific retention.* One important distinction between the specific yield and the effective porosity concepts is that the specific yield relates to *volume* of water that can be freely extracted from an aquifer, while the effective porosity relates to groundwater *velocity* and *flow* through the interconnected pore space.

Figures 5–38 and 5–39 are plots of average total porosity and porosity ranges for various rock types. A similar list of actually determined values of effective porosity, including a clear explanation of what exactly was tested, does not exist to the best of this author's knowledge. On the other hand, values of specific yield are readily found in literature and, as expected, vary widely due to inevitable heterogeneity of natural aquifers and different field testing methods (Kresic, 2007a). In general, the presence of fine-grained sediments, such as silt and clay, even in relatively small quantities, can greatly reduce the specific yield (effective porosity) of coarse-grained sediments. A very detailed discussion of the specific yield concept, various methods of measurements, and case studies is given by Johnson (1967).

Two simple cases of estimating steady-state groundwater flow rates under unconfined and confined conditions are shown in Figure 5–40. In both cases, the flow is planar, through a constant rectangular cross section of width a and over an impermeable horizontal base. The hydraulic conductivity is spatially constant (aquifers are homogeneous), and the hydraulic gradient is also constant. Equations (5.36) and (5.37) describe these simple conditions. For the confined flow case, the relationship is linear, since the saturated aquifer thickness (b) does not change along the flow path. Equation (5.37), which describes unconfined conditions and includes a possible recharge rate (w), is nonlinear, because the saturated thickness (position of the water table) does change between h_1 and h_2.

$$Q = a \times bK\frac{h_1 - h_2}{L} \tag{5.36}$$

$$Q = a \times K\frac{h_1^2 - h_2^2}{2L} + w\left(x - \frac{L}{2}\right) \quad \text{for} \quad x > 0 \tag{5.37a}$$

$$Q = a \times K\frac{h_1^2 - h_2^2}{2L} + w\frac{L}{2} \quad \text{for} \quad x = 0 \tag{5.37b}$$

In reality, flow conditions are almost always more complicated, including changing aquifer thickness, converging flow, possible transition between confined and unconfined flow, a nonhorizontal base, heterogeneous

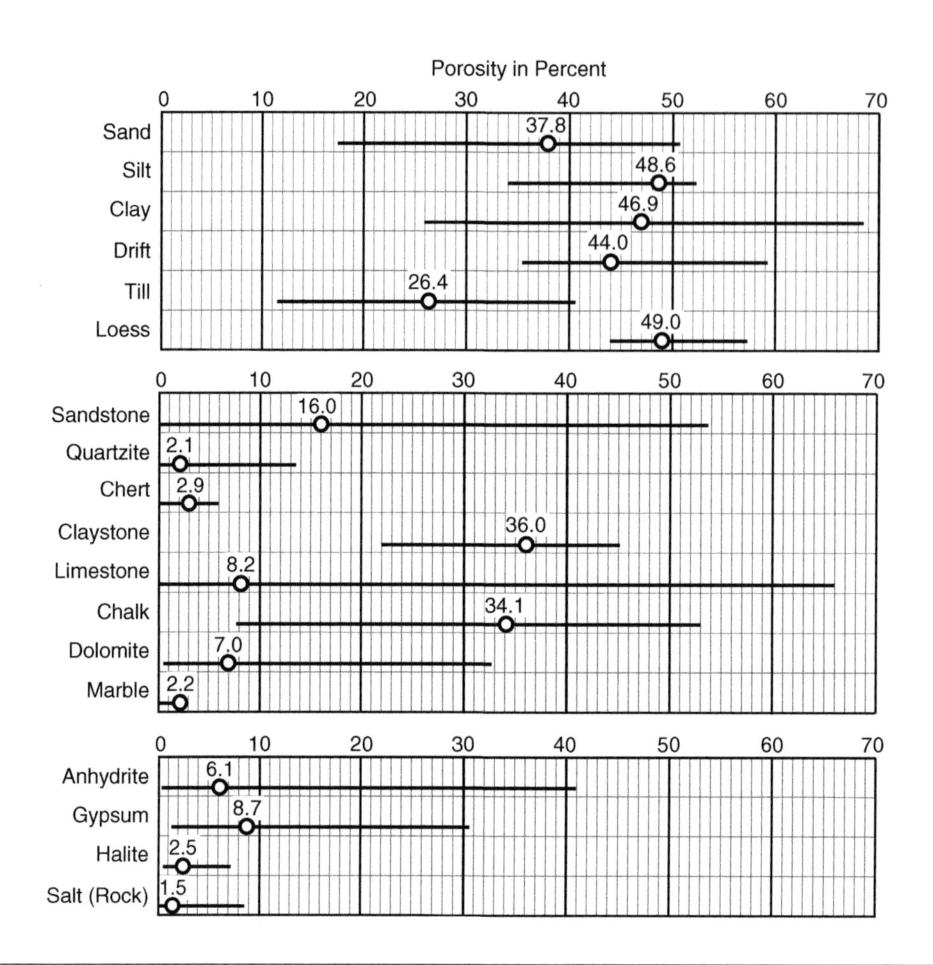

FIGURE 5–38 Porosity range (horizontal bars) and average porosities (circles) of unconsolidated and consolidated sedimentary rocks. (From Kresic, 2007a; copyright Francis & Taylor, printed with permission.)

porous media, and time-dependent (changing in time) recharge from different directions. Although various equations have been developed to describe some of these conditions (see, e.g., Freeze and Cherry, 1979; Kresic, 2007a), numeric models are now routinely used instead.

A key parameter for various calculations of groundwater flow rates is the *transmissivity* of porous media. For practical purposes, it is defined as the product of the aquifer thickness (b) and the hydraulic conductivity (K):

$$T = b \times K \quad (5.38)$$

It follows that an aquifer is more transmissive (more water can flow through it) when it has higher hydraulic conductivity and is thicker. Despite the many laboratory and field methods for determining hydraulic conductivity and the transmissivity of aquifers, the most reliable are long-term field pumping tests that register hydraulic response of all porous media present in the system. Aquifer testing is not a focus of this book, and the reader can consult various general and special publications on designing and analyzing aquifer tests, including pumping tests, such as guidance documents by the U.S. Geological Survey (Ferris et al., 1962;

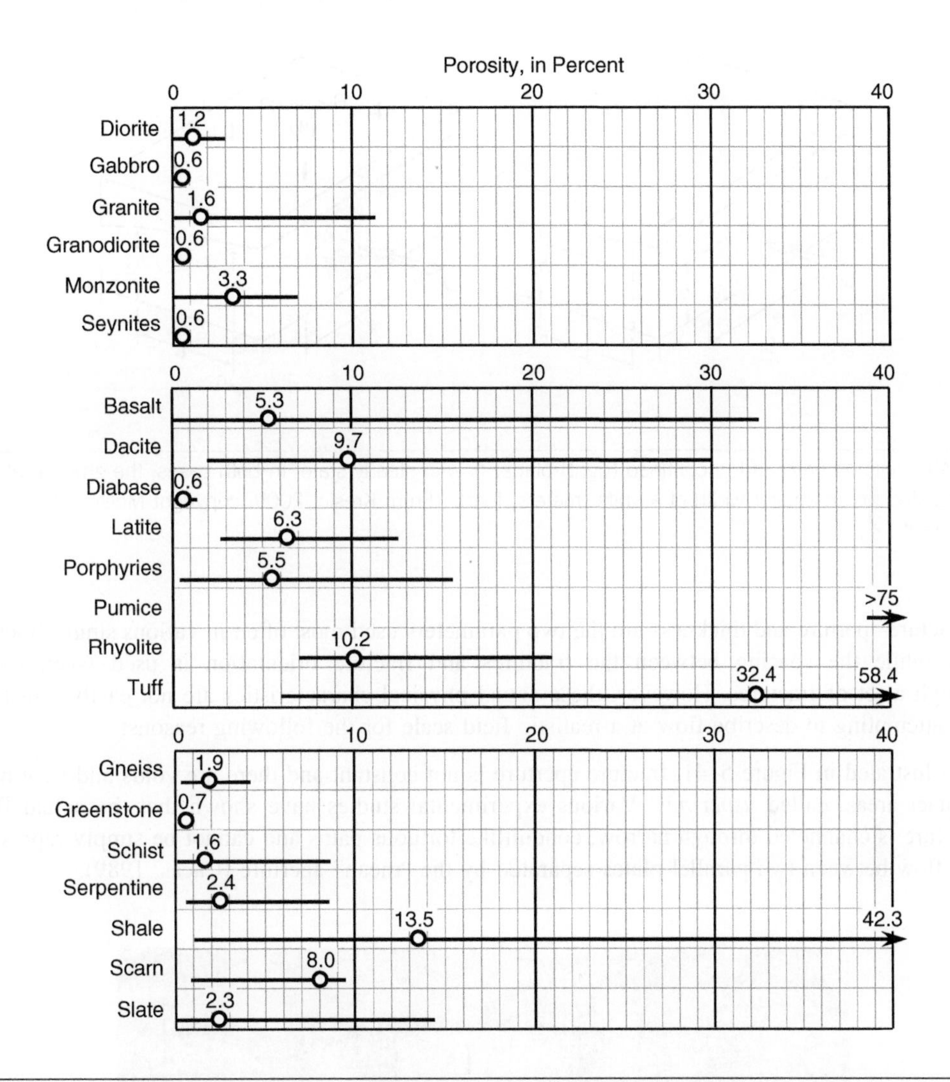

FIGURE 5–39 Porosity range (horizontal bars) and average porosities (circles) of magmatic and metamorphic rocks. (From Kresic, 2007a; copyright Francis & Taylor, reprinted with permission.)

Stallman, 1971; Lohman, 1972), the U.S. Environmental Protection Agency (Osborne, 1993), the U.S. Bureau of Reclamation (1977), the American Society for Testing and Materials (1999a, 1999b), and the books by Driscoll (1989); Walton (1987); Kruseman, de Ridder, and Verweij (1991); Dawson and Istok (1992); and Kresic (2007a).

Fractured rock aquifer

The dual nature of the aquifer is represented by two types of flow tubes (Figure 5–33): The first type is filled with grains and minerals of the rock matrix and it behaves in the same way as intergranular aquifers where Darcy's law is valid; the second flow tube type is a fracture of limited planar extent. There is fluid exchange between these two porous continua.

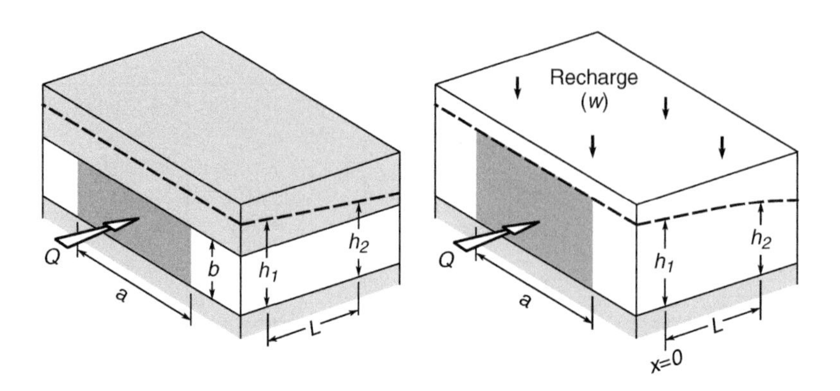

FIGURE 5–40 Confined (left) and unconfined (right) planar flow in steady state. In both cases, the groundwater flow rate is calculated for an aquifer width of *a* units (meters, feet). (From Kresic, 2009; copyright McGraw Hill, reprinted with permission.)

The fracture aperture and thickness are the two parameters used most often in various single-fracture flow equations, while the spacing between the fractures and fracture orientation is used when calculating flow through a set of fractures. However, these actual physical characteristics are not easily translated into equations attempting to describe flow at a realistic field scale for the following reasons:

1. As illustrated in Figure 5–41, fracture aperture is not constant and there are voids and very narrow or contact areas, called *asperities*. Various experimental studies have shown that the actual flow in a fracture is channeled through narrow, conduitlike tortuous paths and cannot be simply represented by the flow between two parallel plates separated by the "mean" aperture (Cacas, 1989).

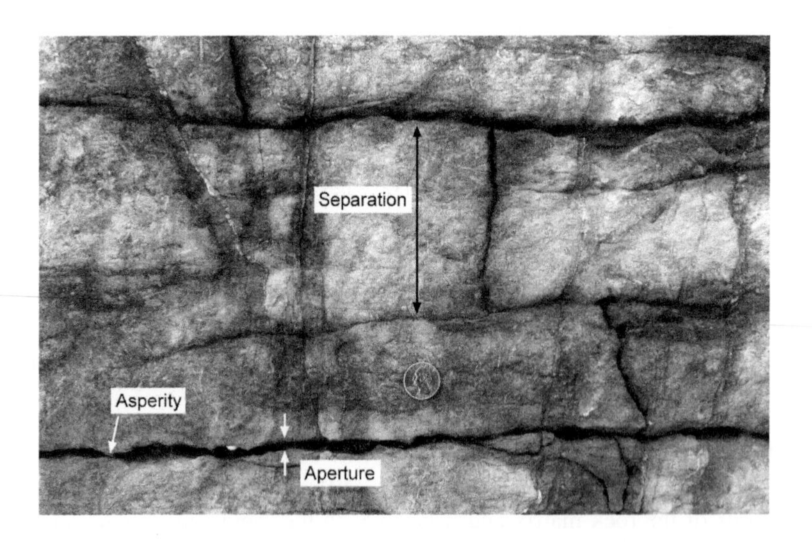

FIGURE 5–41 Fractures in dolomitic limestone, Saltville, Virginia, with varying aperture, separation, and asperities. Note 25-cent coin for scale in the middle bottom.

2. Because of stress release, the aperture measure at outcrops or inaccessible cave passages is not the same as an in situ aperture. An aperture measured on drill cores and in borings is also not a true one—the drilling process commonly causes bedrock adjacent to fractures to break out, thereby increasing the apparent width of fracture openings as viewed on borehole-wall images (Williams et al., 2002).
3. Fractures have limited length and width, which can also vary among individual fractures in the same fracture set. Spacing between individual fractures in the same set can also vary. Since all these variations take place in the three-dimensional space, they cannot be directly observed, except through continuous coring or logging of multiple closely spaced boreholes, which is the main cost-limiting factor.

Evolution of analytical equations and various approaches in quantifying fracture flow is given by Witherspoon (2000) and Faybishenko et al. (2000). In the simplest form, the hydraulic conductivity of a fracture with an aperture B (in Witherspoon's notation $B = 2b$) and represented by two parallel plates is

$$K = B^2 \frac{\rho g}{12\mu} \tag{5.39}$$

where ρ is fluid density, g is acceleration of gravity, and μ is fluid viscosity. The groundwater flow rate through a cross-sectional area $A = Ba$ (where B is the fracture aperture and a is the fracture width perpendicular to the flow direction) is

$$Q = Av = -a\left(\frac{\rho g}{12\mu}\right)\frac{dh}{dx}B^3 \tag{5.40}$$

where dh/dx is the change in the hydraulic head (h) along the flow direction (x). In the definite form, this change is denoted Δh and the minus sign disappears due to integration. The fracture flow approximation, called the *cubic law*, assumes that the representative aquifer volume acts as an equivalent porous medium (Darcian continuum). Witherspoon gives another form of the cubic law:

$$\frac{Q}{\Delta h} = \frac{CB^3}{f} \tag{5.41}$$

where C is a constant that depends on the geometry of the flow field and f is roughness that accounts for deviations from ideal conditions, which assume smooth fracture walls and laminar flow. The roughness f is related to the Reynolds number (Re; the indicator whether the flow is turbulent or laminar), and the friction factor (Ψ) through the following equation:

$$f = \frac{\Psi \mathrm{Re}}{96} \tag{5.42}$$

The Reynolds number is given as

$$\mathrm{Re} = \frac{Dv\rho}{\mu} \tag{5.43}$$

where D is the fracture hydraulic diameter, approximately equal to four times the hydraulic radius of the fracture. For a relatively smooth fracture, where the ratio between the fracture asperity and aperture is less than 0.1, the transition to turbulent flow is at a Reynolds number of about 2400. As this ratio, which is the indicator of fracture roughness, increases, the Reynolds number for transition to turbulent flow decreases significantly. The friction factor is given by

$$\Psi = \frac{D}{v^2/2g}\Delta h \tag{5.44}$$

where v is the flow velocity.

To simulate flow through a network of fractures, one has to decide on the spatial geometry of individual fractures and their interconnectivity within the entire aquifer volume of interest. This is done in a variety of ways, as illustrated in Figure 5–42 and discussed by Chilès (1989a, 1989b), Chilès and Marsily (1993), Long (1983), and Long, Gilmour, and Witherspoon (1985).

The hydraulic conductivity of a set of parallel fractures (K_f), with N fractures per unit distance, average fracture aperture B, and the fracture set porosity n_f (where $n_f = NB$), is given by (Snow, 1968)

$$K_f = \left(\frac{\rho g}{\mu}\right)\frac{NB^3}{12} \tag{5.45}$$

This approximation assumes that the representative aquifer volume acts as an equivalent porous medium (Darcian continuum). Snow concludes that the fracture porosity of a representative aquifer volume with an isotropic three-dimensional system of similar fractures is $n_f = 3NB$, while the hydraulic conductivity (K_{3f}) of such a three-dimensional system is double that of any one of its individual fracture sets:

$$K_{3f} = \left(\frac{\rho g}{\mu}\right)\frac{NB^3}{6} \tag{5.46}$$

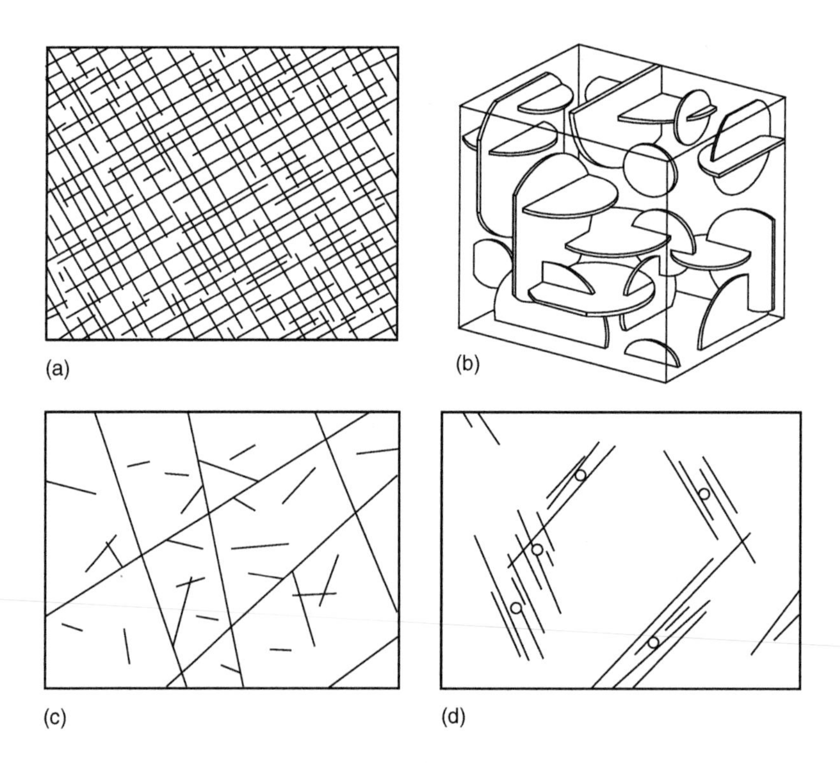

FIGURE 5–42 Several approaches of representing fracture networks: (a) Two-dimensional, nearly orthogonal random fractures, (b) three-dimensional orthogonal disks (from Long et al., 1985; copyright American Geophysical Institute), (c) two-dimensional hierarchical model of Conrad and Jacquin, and (d) disc clusters centered on seeds generated by a Poisson process. (From Chilès, 1989a; copyright Battelle Press.)

Equations of fluid flow in nonideal fractures with influences of various geometric irregularities and fracture network modeling approaches are discussed in detail by Bear et al. (1993), Zimmerman and Yeo (2000), and Faybishenko et al. (2000).

Karst and pseudokarst aquifers

There is a third type of flow tube (Figure 5–33) in addition to the types present in fractured rock aquifers (matrix and planar fracture flow tubes). In this case, the flow most closely resembles actual tubes and can be with either a free water surface (channel flow) or under pressure (flow in conduits and pipes).

Flow through a pipe of varying cross-sectional area is described by the Bernoulli equation for real viscous fluids, as illustrated in Figure 5–43. Since there is no gain or loss of water in the pipe, the flow rate remains the same while other hydraulic elements change from one cross section to another. The total energy line (*E*) of the flow can only decrease from the up-gradient cross section toward the down-gradient cross section of the same flow tube (pipe or conduit) due to energy losses. On the other hand, the hydraulic head line (*H*) may go "up" and "down" along the same flow tube as the cross-sectional area increases or decreases, respectively. The total energy surface, which includes the flow velocity component ($\alpha v^2/2g$), can be directly measured only by the Pitot device, whose installation is not feasible under most field conditions. Monitoring wells and piezometers, on the other hand, record only the hydraulic head, which does not include the flow velocity component. It is therefore conceivable that two piezometers in or near the same karst conduit with rapid flow may not provide useful information for calculation of the real flow velocity and flow rate between them and may even falsely indicate the opposite flow direction. In fact, as discussed by Bögli (1980), it has been shown that water rising through a tube in an enlargement passage can flow backward over the main flow conduit and into another tube that begins at a narrow passage in the same main conduit.

There are additional complicating factors when attempting to calculate flow through natural karst conduits using the pipe approach:

1. Flow through the same conduit may be both under pressure and with a free surface.
2. Since pipe and conduit walls are more or less irregular ("rough"), the related coefficient of roughness has to be estimated and inserted into the general flow equation.
3. The conduit cross section may vary significantly over short distances and in the same general area.

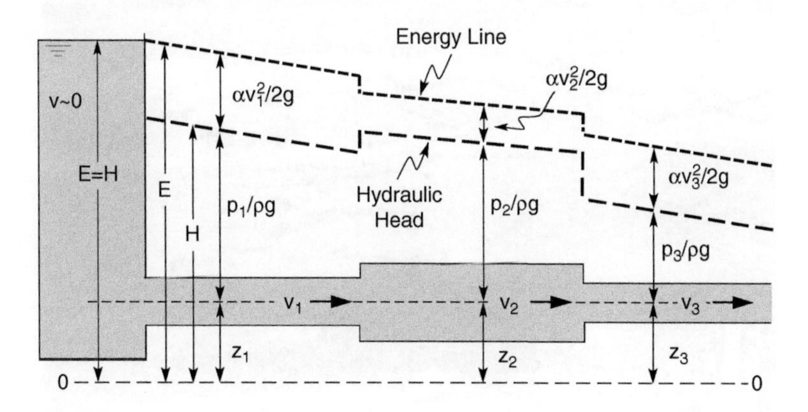

FIGURE 5–43 Illustration of the Bernoulli equation for the flow of real viscous fluids through a pipe with the varying cross-sectional area. The energy line (*E*) at any given cross section is the sum of the elevation head (*z*), the fluid pressure head ($p/\rho g$), and the velocity head ($\alpha v^2/2g$). The fluid pressure head and the elevation head give the hydraulic head (*H*).

4. The flow may be both laminar and turbulent in the same conduit, depending on the flow velocity, cross-sectional area, and wall roughness. The irregularities that cause turbulent flow are mathematically described through the Reynolds number and the friction factor.

According to Palmer (1991), flow in pre-solutional openings is almost exclusively laminar. As dissolution proceeds, turbulent flow develops in those openings that enlarge to a sufficient size, about w = 1–20 mm, depending on hydraulic gradient and temperature (Figure 5–44). Closed-conduit laminar flow is governed by the Hagen-Poiseuille equation:

$$Q = \frac{R^2 \gamma i A}{c\mu} \tag{5.47}$$

where R is the hydraulic radius ($r/2$ in tubes, $w/2$ in fissures), $c = 2$ in tubes and 3 in fissures, γ and μ are the specific weight and dynamic viscosity of water, and i is the hydraulic gradient expressed as a positive dimensionless number (head loss/flow length). Linear discharge, q ($= Q/b$ = velocity $\times$ width), is a convenient term used in fissures of unspecified b dimension.

Turbulent flow in any closed conduit or open channel is expressed by the Darcy-Weisbach equation:

$$Q = A\sqrt{\frac{8Rgi}{f}} \tag{5.48}$$

where g is the gravitational field strength and f is the friction factor.

In fully turbulent flow, f = 0.03–0.1 in a typical straight tube or fissure. Apparent friction factors, which include local head losses at bends and cross-sectional area changes, can be much higher. Turbulent flow develops at a Reynolds number (Re) > 500, where

$$\mathrm{Re} = \frac{\rho v R}{\mu} \tag{5.49}$$

where ρ and v are water density and velocity, respectively.

FIGURE 5–44 Small conduit, about 20 cm across, developed in fissured limestone.

Because of the widely varying effective porosity of karstified carbonates, even within the same aquifer system, the groundwater velocity in karst can vary over many orders of magnitude. One should therefore be very careful when making a (surprisingly common) statement such as "groundwater velocity in karst is generally very high." Although this may be true for turbulent flow taking place in karst conduits, the disproportionately larger volume of any karst aquifer has relatively low groundwater velocities (laminar flow) through small fissures and rock matrix (Kresic, 2007b). One common method for determining groundwater flow directions and apparent flow velocities in karst is dye tracing. However, most dye tracing tests in karst are designed to analyze possible connections between known (or suspected) locations of surface water sinking and groundwater discharge (springs). Because such connections involve some kind of preferential flow paths (sink-spring type), the apparent velocities calculated from the dye tracing data are usually biased toward the high end. For example, based on the results of 43 tracing tests in karst regions of West Virginia, the median groundwater velocity is 716 m/d, while 50 percent of the tests show values between 429 and 2655 m/d (the 25th and 75th percentile of the experimental distribution, respectively). It is interesting that, based on 281 dye tracing tests, the most frequent velocity (14 percent of all cases) in the classic Dinaric karst of Herzegovina, as reported by Milanovic (1979), is quite similar, between 864 and 1728 m/d. Twenty-five percent of the results show groundwater velocity greater than 2655 m/day in West Virginia and greater than 5184 m/day in Herzegovina.

Various approximate calculations of flow velocity have been based on the geometry of hydraulic features, such as scallops and flutes visible on walls, floors, and ceilings of accessible cave passages (Figure 5–45; see White, 1988, pp. 97–98, and Bögli, 1980, pp. 163–164). For example, the calculated flow velocity for a canyon passage in White Lady Cave, Little Neath Valley, United Kingdom, is 1.21 m/s and the flow rate is 9.14 m^3/s, for the cross-sectional area of flow of 7.6 m^2 and scallop length of 4.1 cm (White, 1988).

5.7.2 Representative hydraulic heads

The importance of field-measured hydraulic heads for any type of groundwater flow calculations cannot be emphasized enough. They are part of all groundwater flow equations presented in the preceding section,

FIGURE 5–45 Scallops on the walls of a gypsum cave near Carlsbad, New Mexico.

because they determine the hydraulic gradients, which are the main driving force for groundwater flow. They are also the main calibration target in numeric groundwater modeling, describe the initial conditions for the model, and are part of the majority of model boundary conditions. Without sufficient information on the time-dependent hydraulic head distribution within the area of interest, any calculation of the groundwater flow rates is highly speculative.

The hydraulic heads measured in monitoring wells, piezometers, surface streams and lakes (ponds), and at springs are combined into groundwater contour maps. At least several data sets collected in different hydrologic seasons should be used to draw groundwater contour maps for the area of interest. In addition, one should gather information about hydrometeorologic conditions in the area for the preceding days and weeks, paying special attention to the presence of extended wet or dry periods. All this information is essential for making a correct contour map.

A contour map is a two-dimensional representation of a three-dimensional flow field, and as such, it has limitations. If the area (aquifer) of interest is known to have significant vertical gradients and enough field information is available, it is always recommended to construct at least two contour maps: one for the shallow and one for the deeper aquifer depth. As with geologic and hydrogeologic maps in general, a contour map should be accompanied by several cross sections showing locations and vertical points of the hydraulic head measurements with posted data or ideally showing the entire cross-sectional flow net.

Probably the most incorrect and misleading case is when data from monitoring wells screened at different depths in a groundwater system with vertical gradients are lumped together and contoured as one data package. A perfect example would be a fractured rock or karst aquifer with thick residuum (regolith) deposits or a layer of unconsolidated sediments overlying it. If data from wells screened at different depths in the unconsolidated sediments and the bedrock were lumped and contoured together, it would be impossible to interpret where the groundwater is actually flowing for the following reasons: (1) The shallow sediments act primarily as an intergranular porous medium in unconfined conditions, and horizontal flow directions may be influenced by local (small) surface drainage features, and (2) the bedrock has discontinuous flow through fractures at different depths, which are often under pressure (confined conditions) and may be influenced by regional features, such as major rivers or springs. The flow in two distinct porous media (the residuum and the bedrock) may therefore be in two different general directions at a particular site, including strong vertical gradients from the residuum toward the underlying bedrock. Creating one "average" contour map for such a system would not make hydrogeologic sense.

In highly fractured rock and karst aquifers, where groundwater flow is discontinuous, (i.e., it takes place mainly along preferential flow paths), Darcy's law does not apply and flow nets are not an appropriate method for the flow characterization. However, contours maps in highly heterogeneous aquifers are routinely made by many professionals, who often find themselves excluding certain "anomalous" data points while trying to develop a "normal-looking" map. Figure 5–46 shows that, in karst, hardly any "unreasonable" hydraulic head measurement should be dismissed.

Contour maps showing a regional flow pattern in a fractured rock or karst aquifer may be justified, since the groundwater flow starts from the recharge areas and converges toward large springs, which should be apparent if there is a sufficient number of evenly dispersed monitoring wells.

The problems usually rise when interpreting local flow patterns, as schematically shown in Figure 5–47, which illustrates some key differences between a karst aquifer (left) and an intergranular aquifer (right). Both aquifers have the same general flow direction, from the north to the south, as shown on the map view. A triplet of wells, installed anywhere in the intergranular aquifer (provided they form a relatively "normal"-looking triangle) would reasonably accurately determine the general groundwater flow direction based on the measured hydraulic head. The same cannot be stated for the karst aquifer case, where the three-well principle may give very different results, depending on the position of individual wells relative to the preferential flow paths (karst conduits or large fracture zones). Moreover, a group of closely spaced wells (say, in the meter to

FIGURE 5–46 Waterfall in a limestone cave, West Virginia. (Photograph courtesy of William Jones, Karst Waters Institute.)

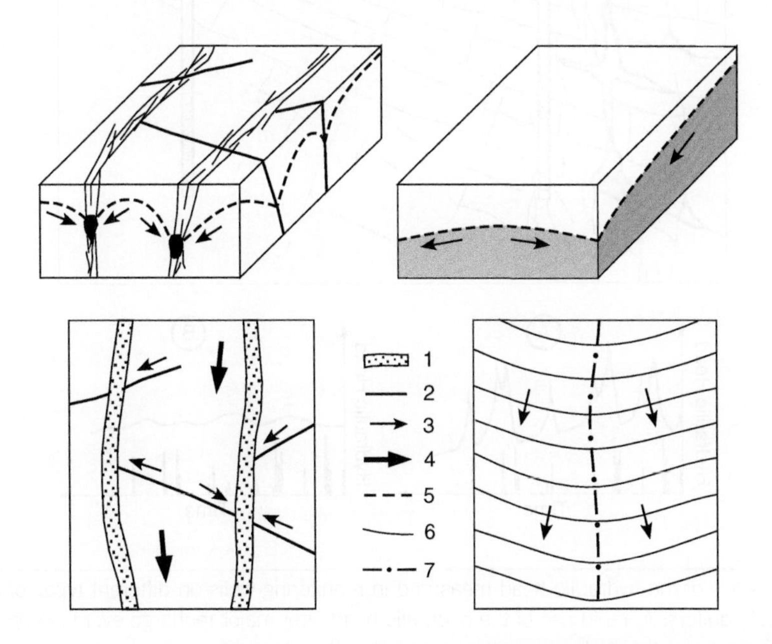

FIGURE 5–47 Groundwater flow in three-dimensions and its map presentation (top) for a fractured rock or karst (left) and an intergranular (right) aquifer: (1) Preferential flow path (e.g., fracture or fault zone or karst conduit or channel), (2) fracture or fault, (3) local flow direction, (4) general flow direction, (5) position of the hydraulic head (water table in the intergranular aquifer), (6) hydraulic head contour line, (7) groundwater divide. (From Kresic, 1991.)

dekameter scale) may show a completely "random" distribution of the measured hydraulic head; one well may be completed in a homogeneous rock block, with no significant fractures and low matrix porosity, and may even exhibit the so-called glass effect (no fluctuation of the water table, regardless of the precipitation-infiltration dynamics). A well 10 or so meters away, on the other hand, may show the hydraulic head fluctuation of several meters or more (Figure 5–48).

Another common "complication" in karst is the presence of preferential flow paths (conduits) at different depths. The same conduits may have different roles depending on seasonal recharge patterns, as illustrated in Figure 5–49 (Bonacci, 1987). Finally, drainage areas of large springs in karst usually extend beyond topographic divides, which should be always considered as a real possibility in any particular case. In other words, unless it is apparent from the data collected at numerous wells, the topographic divides should not be used as inferred groundwater divides when constructing contour maps.

In conclusion, measuring hydraulic heads and subsequently determining hydraulic gradients and groundwater flow directions is by no means a straightforward task and requires good planning by an experienced hydrogeologist. Ultimately, the number of monitoring wells and their depths, screen lengths, and frequency of water level recordings are based on the final goal of the study. As already mentioned, one common mistake is to apply the same approach of hydraulic head measurements in different types of aquifers; fractured

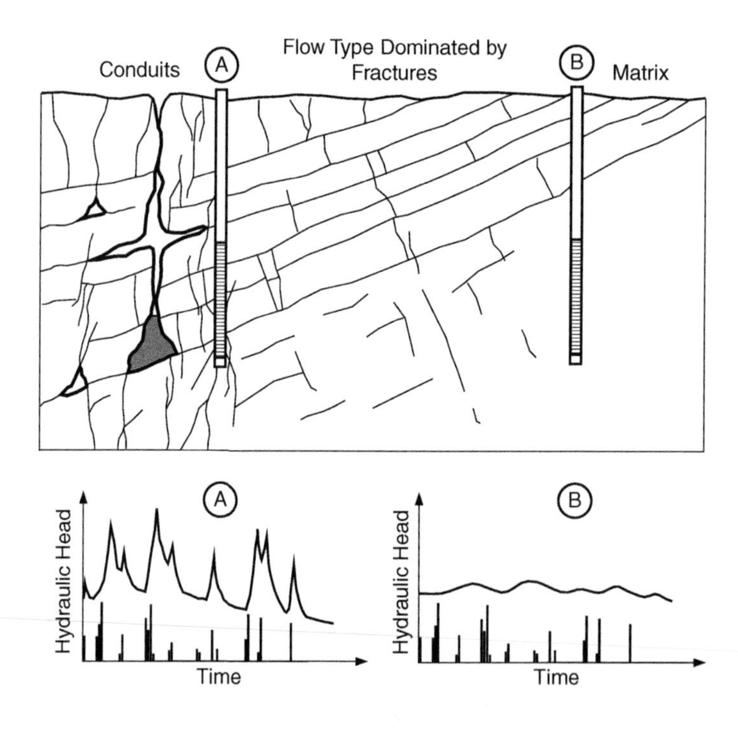

FIGURE 5–48 Dependence of the hydraulic head measured in monitoring wells on different types of effective porosity (specific yield) in karst aquifers: A, rapid rise of the hydraulic head after major recharge events in portions of the aquifer with large conduits and no significant storage in the matrix; B, delayed and dampened response of aquifer matrix. Flow dominated by fractures may include any combination of these two extremes. (From Kresic, 2007a; copyright Francis & Taylor, reprinted with permission.)

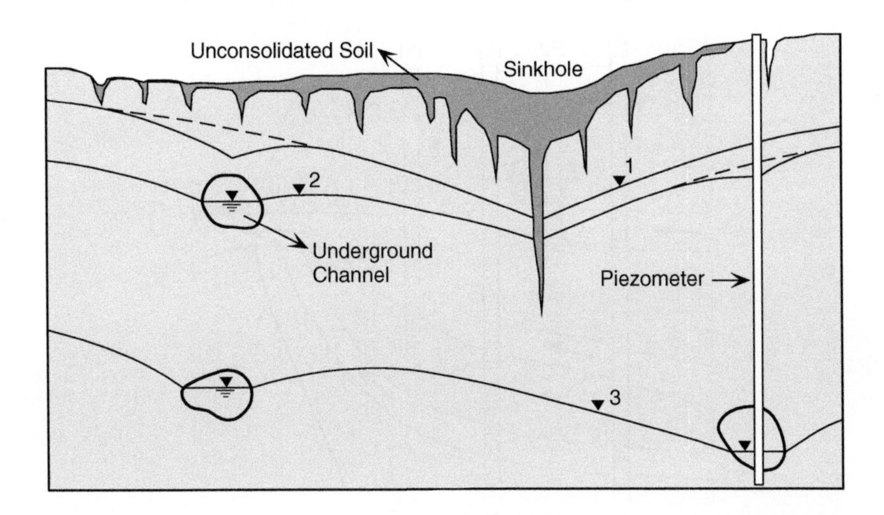

FIGURE 5–49 Influence exerted by karst landforms and phenomena on the change in groundwater levels. (From Bonacci, 1987; copyright Springer-Verlag, reprinted with permission.)

rock and karst aquifers present a special challenge, even to more experienced professionals. Because one portion of the groundwater flow takes place in fractures or conduits and the other part within the rock matrix porosity, measurements of the hydraulic heads do not provide a unique answer. Their interpretation should always be made in the overall hydrogeologic context. Using only the hydraulic head information to assess representative groundwater flow directions (hydraulic gradients) would usually not be sufficient. In general, the hydraulic head measurements should be combined with hydrogeologic mapping, possibly dye tracing, and certainly a thorough understanding of various hydraulic factors, such as flow through interconnected fractures and pipes. One extreme but real example is shown in Figure 5–50. By looking at the hydraulic heads measured in piezometers P4, P3, and P2, one could erroneously conclude that groundwater flows away from the spring.

When planning field measurements of the hydraulic head, the following facts should always be taken into consideration (Kresic, 2007a):

1. Hydraulic heads change in response to aquifer recharge, both seasonally and, especially in unconfined aquifers, after each recharge episode (rainfall). Measurements in multiple wells should therefore be performed within the shortest time interval feasible (so-called synoptic measurements). To accurately assess seasonal influences on the hydraulic head fluctuations, at least one round of synoptic measurements should be performed per season.
2. Hydraulic heads in confined aquifers change in response to barometric pressure fluctuations; this may also be true for unconfined aquifers, in some cases. The only reasonable method to accurately determine the magnitude and importance of such changes is to measure the hydraulic head and the barometric pressure continuously using pressure transducers and data loggers.
3. Hydraulic heads in coastal aquifers respond to harmonic tidal fluctuations. These changes can be accurately quantified only by performing continuous measurements.
4. Hydraulic heads may change in response to some local hydraulic stress on the aquifer, such as cyclic operation of extraction wells in the vicinity.

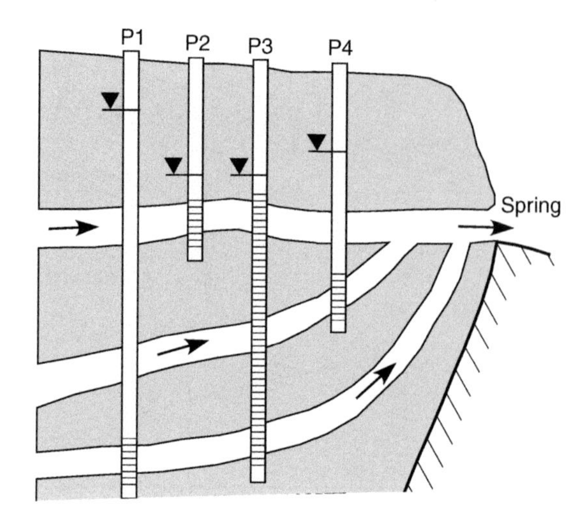

FIGURE 5–50 Example of how closely spaced monitoring wells in clusters may register very different hydraulic heads, depending on the depth and length of well screens. Well P3 with the longest screen, connecting all three channels, shows the same water level as the shallowest well, P2. Considering the measured water levels at P4 and either at P2 or P3, one could erroneously conclude that the groundwater flow is away from the spring. (Modified from Kupusović, 1989.)

5.7.3 Numeric Models

As explained in earlier chapters, large springs of sufficient magnitude to have been developed or considered to be developed for public water supply are in most cases draining carbonate karst aquifers. At the same time, most if not all widely used computer programs for numeric groundwater modeling lack the capability for a physically based simulation of the characteristic triple porosity of karst aquifers. Consequently, hydrogeologists and groundwater modelers either avoid numeric modeling of springs altogether or are engaged in searching for an equivalent porous media (EPM) approach that may work.

One such approach when using Modflow and similar codes based on Darcy's law (and therefore not suitable for modeling fractured rock and karst aquifers) includes assigning very high values of the hydraulic conductivity to those model cells known or suspected to contain highly transmissive conduits. Figure 5–51 illustrates how this procedure may result in a reasonable representation of the preferential pathways and the field-observed hydraulic head distribution.

Unfortunately, none of the EPM models can accurately reproduce the fast (conduit) flow component of the spring discharge after major rainfall events and the very important hydraulic interaction between the conduits and the surrounding matrix. That is, as the hydraulic head rapidly rises in the conduits, there may be a very significant transfer of water into the surrounding matrix and storage of water, provided that the matrix has some effective porosity. At the same time, the spring may react rapidly (sometimes in a matter of hours) to the propagation of pressure through the conduits caused by the newly infiltrated water. Finally, this newly infiltrated water may start discharging at the spring only several days after a localized rainfall in a remote portion of the spring drainage area.

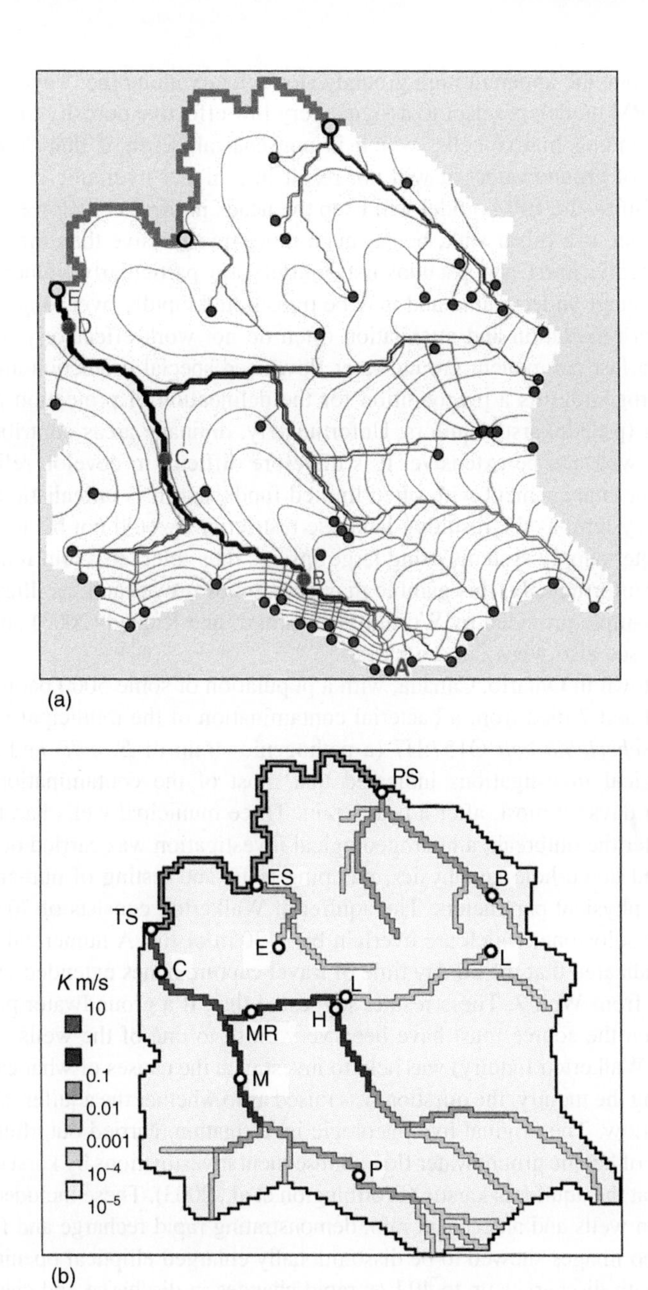

FIGURE 5–51 (a) Modeled hydraulic heads with an EPM numeric model of a karst aquifer showing good agreement between actual tracer paths and the hydraulic gradients. All 54 tracer paths go to the correct spring. (b) Hydraulic conductivity distribution that resulted in the map shown in (a). (After Worthington, 2003.)

In an attempt to simulate the apparent high groundwater velocity along the "conduit cells" and the related spring response, some EPM modelers select to assign a very low effective porosity to those model cells, much lower than in the surrounding matrix cells, which is nonsensical. Even if that "procedure" produces the desired effect (fast flow of groundwater), it will not result in a higher hydraulic head in the "conduit" cells at any given time by default—the EPM model will keep the heads in these cells lower than in the surrounding cells, since they always act as a linear sink, being much more transmissive than the rest of the model area.

As mentioned repeatedly, karst and pseudokarst aquifers are particularly vulnerable to contamination. Contaminants can easily enter underground and may be transported rapidly over long distances in the aquifer. Processes of contaminant retardation and attenuation often do not work effectively in such systems. Therefore, aquifers that have a fast-component groundwater flow need special protection and attention. A detailed knowledge of karst hydrogeology is a precondition for the delineation of protection areas and management of springs draining karst (pseudokarst) aquifers. Unfortunately, drainage areas contributing water to a single large spring or a supply well can be extensive. It is therefore difficult to develop reliable, physically based numeric models for spring management with often limited funds, and it is unrealistic to designate maximum protection for the entire system, as the resulting land-use restrictions would not be acceptable to some stakeholders. Many public water supply systems using large springs thus select to avoid related efforts and choose to rely on luck instead. Unfortunately, this gamble may sometimes prove fatal, as illustrated with the following widely publicized example provided by Worthington, Smart, and Ruland (2003) and Goldscheider, Drew, and Worthington (2007; see also www.iah.org/karst).

Walkerton is a rural town in Ontario, Canada, with a population of some 5000 people. In May 2000, about 2300 of them became ill and 7 died from a bacterial contamination of the municipal water supply. The principal pathogens were *Escherichia coli* O157:H7 (a pathogenic strain of *E. coli*) and *Campylobacter jejuni*. Subsequent epidemiological investigations indicated that most of the contamination of the water supply occurred within hours or days, at most, after a heavy rain. Three municipal wells had been in use at the time of the outbreak. Soon after the outbreak, a hydrogeological investigation was carried out that included drilling 38 boreholes, surface and down-hole geophysics, pumping tests, and testing of numerous water samples for both bacteriological and physical parameters. The aquifer at Walkerton consists of 70 m of thick flat-bedded Paleozoic limestone and dolostone, which are overlain by 3–30 m of till. A numerical model of groundwater flow (using Modflow) indicated that the 30 day time of travel capture zones extended 290 m from Well 5, 150 from Well 6, and 200 m from Well 7. These results suggested that, if a groundwater pathway was implicated in the contamination, then the source must have been very close to one of the wells.

A public inquiry (the Walkerton Inquiry) was held to investigate the causes of what came to be known as the *Walkerton tragedy*. During the inquiry, the question was raised as to whether the aquifer might be karstic and thus have rapid groundwater flow. The original hydrogeologic investigation, carried out after the outbreak, had not mentioned the possibility of karstic groundwater flow. Subsequent investigations by karst experts found that there were many indications that the aquifer is karstic (Worthington et al., 2003). These included a correlation between bacterial contamination in wells and antecedent rain, demonstrating rapid recharge and flow to wells; localized inflows to wells that video images showed to be dissolutionally enlarged elliptical openings on bedding planes; the presence of springs with discharges up to 40 L/s; rapid changes in discharge and chemistry at these springs following rain; and rapid, localized changes to electrical conductance in a well during a pumping test.

All these tests strongly suggested that the aquifer is karstic, but the most persuasive evidence was the results of aquifer tracing tests. Earlier numerical modeling had suggested that groundwater velocities were typically in the range of a few meters per day, but tracer tests demonstrated that actual velocities were some 100 times faster (Figure 5–52). In conclusion, the investigations by the karst experts demonstrated that the source for the pathogenic bacteria could have been much farther from the wells than the earlier investigations and modeling had indicated (Goldscheider et al., 2007).

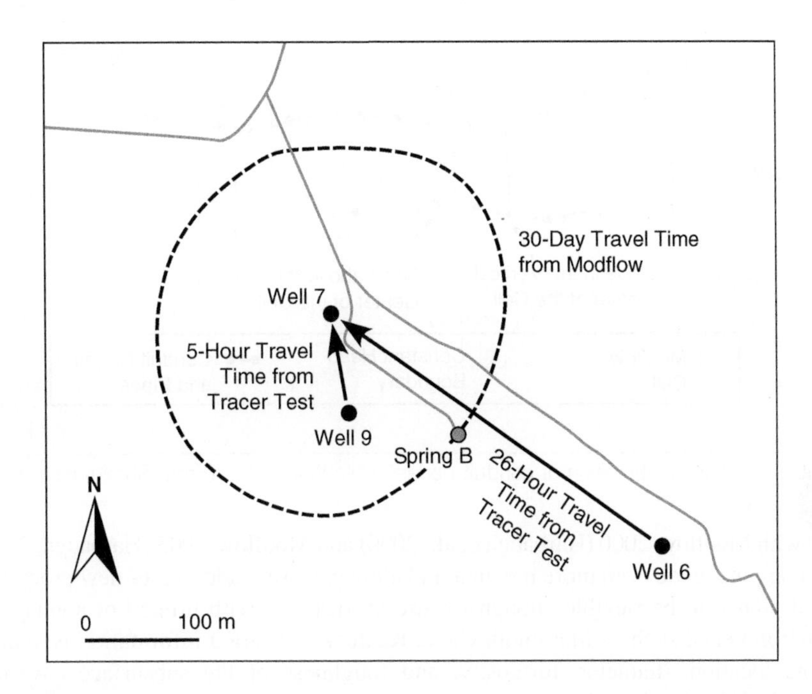

FIGURE 5–52 Trajectories and travel times for tracers injected in monitoring Wells 6 and 9 and recovered in pumping Well 7 showing velocities >300 m/d, compared to a 30-d capture zone for Well 7 predicted using Modflow. (Modified from Worthington et al., 2003.)

This example shows that care should be taken when using numeric models to simulate hydrogeology of large karst and pseudokarst springs, including design options for spring capture, estimation of spring drainage area, or simulation of fate and transport of contaminants that may affect the availability of springwater. With the introduction of the conduit flow process (CFP) for Modflow-2005 (Shoemaker et al., 2008), no practical reasons are left to justify the use of equivalent porous media models for modeling springs in dual-porosity aquifers with fast conduit flow.

The CFP has the ability to simulate turbulent groundwater flow conditions by (1) coupling the traditional groundwater flow equation with formulations for a discrete network of cylindrical pipes (Mode 1), (2) inserting a high-conductivity flow layer that can switch between laminar and turbulent flow (Mode 2), or (3) simultaneously coupling a discrete pipe network while inserting a high-conductivity flow layer that can switch between laminar and turbulent flow (Mode 3). Conduit flow pipes (Mode 1) may represent dissolution or biological burrowing features in carbonate aquifers, voids in fractured rock, or lava tubes in basaltic aquifers and can be fully or partially saturated under laminar or turbulent flow conditions. Preferential flow layers (Mode 2) may represent (1) a porous media where turbulent flow is suspected to occur under the observed hydraulic gradients; (2) a single secondary porosity subsurface feature, such as a well-defined laterally extensive underground cave; or (3) a horizontal preferential flow layer consisting of many interconnected voids.

The conduit flow process was developed in response to a need for a computer program that accounts for the dual-porosity nature of many aquifers. There also was a desire to provide compatibility with recent advancements to the USGS modular groundwater model (Modflow). Many research computer programs are available for simulating dual-porosity aquifers but have not been fully documented for wider use (for example, Clemens et al., 1996; Kiraly, 1998; Bauer, 2002; Birk, 2002). Additionally, the structure of

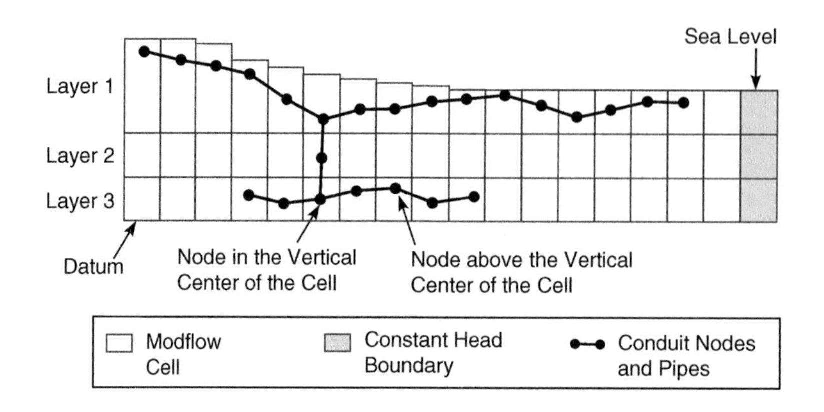

FIGURE 5–53 Possible variations in elevation of conduit nodes in Modflow cells. (From Shoemaker et al., 2008.)

Modflow changed with Modflow-2000 (Harbaugh et al., 2000) and Modflow-2005 (Harbaugh, 2005), making the groundwater flow computer code even more modular and allowing easier addition of new processes to the code.

The CFP was designed to be flexible enough for use in locations with limited or abundant field data. In some geologic environments, such as Mammoth Cave, Kentucky, detailed information is available (or could be derived) on the location, diameter, tortuosity, and roughness of the subsurface caverns. CFP Mode 1 (CFPM1) was designed with these locations in mind. In other locations, such as the Biscayne aquifer of southern Florida, void connections and distributions are so complicated within preferential flow layers that a complete characterization is not possible. CFP Mode 2 (CFPM2) was designed with these locations in mind; specifically, laminar and turbulent flow through complicated void connections is represented by a limited number of "effective" or "bulk" layer parameters.

A powerful option in the CFP is that, in cases with abundant field data on the void architecture and hydraulic behavior, complex two- or three-dimensional networks of conduit flow pipes and nodes can be designed to represent interconnected or dead-end voids in the subsurface. Flow calculations assume pipe nodes located in the center of Modflow cells. An exception is in the vertical direction, for which there are two options. First, pipe nodes can be assigned elevations above a datum and, therefore, are not restricted to center elevations of Modflow cells. Second, pipe nodes can be assigned a distance above or below the center of the Modflow cell (Figure 5–53). With this second option, if the distance is set to zero, pipe nodes are assumed to exist at the vertical center of the Modflow cell (Shoemaker et al., 2008).

REFERENCES

Ahn, H., 2000. Modeling of groundwater heads based on second-order difference time series models. J. Hydrol. 234, 82–94.

Akaike, H., 1974. A new look at the statistical model identification. IEEE Trans. Autom. Control AC-19 (6), 716–723.

Anderson, M.P., Woessner, W.W., 1992. Applied Ground Water Modeling; Simulation of Flow and Advective Transport. Academic Press, San Diego, CA.

American Society for Testing and Materials, 1999a. ASTM standards on determining subsurface hydraulic properties and ground water modeling, second ed. ASTM, West Conshohocken, PA.

American Society for Testing and Materials, 1990b. ASTM standards on ground water and vadose zone investigations; drilling, sampling, geophysical logging, well installation and decommissioning, second ed. ASTM, West Conshohocken, PA.

Bauer, S., 2002. Simulation of the genesis of karst aquifers in carbonate rocks. Vol. 62 of Tübinger Geowissenschaftliche Arbeiten: Tübingen, Germany. Reihe C. Institut und Museum für Geologie und Paläontologie der Universität Tübingen.

Bear, J., Tsang, C.F., de Marsily, G. (Eds.), 1993. Flow and contaminant transport in fractured rock. Academic Press, San Diego, CA.

Birk, S., 2002. Characterization of karst systems by simulating aquifer genesis and spring responses: Model development and application to gypsum karst: Vol. 60 of Tübinger Geowissenschaftliche Arbeiten: Tübingen, Germany. Reihe C. Institut und Museum für Geologie und Paläontologie der Universität Tübingen.

Birkens, M.F.P., Knotters, M., Hoogland, T., 2001. Space-time modeling of water table depth using a regionalized time series model and the Kalman filter. Water Resour. Res. 37 (5), 1277–1290.

Bögli, A., 1980. Karst hydrology and physical speleology. Springer-Verlag, Berlin.

Bonacci, O., 1987. Karst Hydrology with Special Reference to the Dinaric Karst. Springer-Verlag, Berlin.

Bonacci, O., 1995. Ground water behaviour in karst: Example of the Ombla Spring (Croatia). J. Hydrol. 165 (1–4), 113–134.

Bonacci, O., 2001. Monthly and annual effective infiltration coefficients in Dinaric karst: Example of the Gradole karst spring catchment. J. Hydrol. Sci. 46 (2), 287–299.

Box, G.E.P., Jenkins, G.M., 1976. Time Series Analysis. Forecasting and Control. Holden-Day, San Francisco.

Bras, R.L., Rodriguez-Iturbe, I., 1994. Random functions and hydrology. Dover Publications, New York.

Cacas, M.C., 1989. Développment d'un modèle tridimensionel stochastique discret por la simulation de l'écoulement et des transports de masse et de chaleur en milieu fracturé. Ph.D. thesis, Ecole des Mines de Paris, Fontainebleau, France.

Canceill, M.F., 1974. In: Convolution et tarissement dans l'etude des relations pluie-debit. Pub. Com. Fr. AIH, Congress de Montpellier, AIH Memoires, Tome X, Paris, pp. 181–184.

Chilès, J.P., 1989a. Three-dimensional geometric modeling of a fracture network. In: Buxton, B.E. (Ed.), Geostatistical Sensitivity, and Uncertainty Methods for Ground-Water Flow and Radionuclide Transport Modeling. Battelle Press, Columbus, OH, pp. 361–385.

Chilès, J.P., 1989b. Modélisation géostatistique de réseaux de fractures. In: Armstrong, M. (Ed.), Geostatistics, vol. 1. Kluwer Academic Publ., Dordrecht, the Netherlands, pp. 57–76.

Chilès, J.P., de Marsily, G., 1993. Stochastic models of fracture systems and their use in flow and transport modeling. In: Bear, J., Tsang, C.F., de Marsily, G. (Eds.), Flow and Contaminant Transport in Fractured Rock. Academic Press, San Diego, CA, pp. 169–236.

Clemens, T., Hückinghaus, D., Sauter, M., Liedl, R., Teutsch, G., 1996. A combined continuum and discrete network reactive transport model for the simulation of karst development. In: Calibration and reliability in groundwater modelling—Proceedings of the ModelCARE 96 Conference, Golden, Colorado, September 1996. International Association of Hydrological Sciences Publication 237, 309–318.

Dawson, K., Istok, J., 1992. Aquifer testing: Design and analysis. Lewis Publishers, Boca Raton, FL.

Denić-Jukić, V., Jukić, D., 2003. Composite transfer functions for karst aquifers. J. Hydrol. 274, 80–94.

Dooge, J.C.I., 1973. Linear Theory of Hydrologic Systems. Technical Bulletin No. 1468, U.S. Department of Agriculture, Washington, DC.

Dreiss, S.J., 1982. Linear kernels for karst aquifers. Water Resour. Res. 18 (4), 865–876.

Dreiss, S.J., 1989a. Regional scale transport in a karst aquifer. Part 1. Component separation of spring flow hydrographs. Water Resour. Res. 25 (1), 117–125.

Dreiss, S.J., 1989b. Regional scale transport in a karst aquifer. Part 2. Linear systems and time moment analysis. Water Resour. Res. 25 (1), 126–134.

Driscoll, F.G., 1989. Groundwater and Wells. Johnson Filtration Systems Inc., St. Paul, MN.

Eagleman, J.R., 1967. Pan evaporation, potential and actual evapotranspiration. J. Appl. Meteorol. 6, 482–488.

Eisenlohr, L., Kiraly, L., Bouzelboudjen, M., Rossier, I., 1997. Numerical versus statistical modeling of natural response of a karst hydrogeological system. J. Hydrol. 202, 244–262.

Faybishenko, B., Witherspoon, P.A., Benson, S.M. (Eds.), 2000. Dynamics of Fluids in Fractured Rock, Geophysical Monograph 122. American Geophysical Union, Washington, DC.

Ferris, J.G., Knowles, D.B., Brown, R.H., Stallman, R.W., 1962. Theory of aquifer tests. U.S. Geological Survey Water Supply Paper 1536-E, Washington, DC.

Fleury, P., Ladouche, B., Conroux, Y., Jourde, H., Dörfliger, N., 2008. Modelling the hydrologic functions of a karst aquifer under active water management—The Lez Spring. J. Hydrol. DOI: 10.1016/j.hydrol.2008.11.037.

Ford, D.C., Williams, P.W., 2007. Karst hydrogeology and geomorphology. Wiley, Chichester, UK.

Freeze, A.R., Cherry, J.A., 1979. Groundwater. Prentice-Hall, Englewood Cliffs, NJ.

Gabric, I., Kresic, N., 2009. Time Series Models. In: Kresic, N. (Ed.), Groundwater Resources: Sustainability, Management, and Restoration. McGraw Hill, New York, pp. 657–661.

Goldscheider, N., Drew, D., Worthington, S., 2007. Introduction. In: Goldscheider, N., Drew, D. (Eds.), Methods in karst hydrogeology. International Contributions to Hydrogeology 26. International Association of Hydrogeologists, Taylor & Francis, London, pp. 1–8.
Gottman, J.M., 1981. Time-series analysis. A comprehensive introduction for social scientists. Cambridge University Press, Cambridge, UK.
Grasso, D.A., Jeannin, P.Y., 1994. Etude critique des methodes d'analyse de la reponse globale des systemes karstiques. Application au site de Bure (JU, Suisse). Bulletin d'Hydrogéologie. [Neuchatel] 13, 87–113.
Hajdin, G., 1981. An example of the hydraulic explanation of the flow out of a karst spring and the piezometric levels in its hinterland. Nas Krs Bull. Speleological Society VI (10–11), 109–115.
Harbaugh, A.W., 2005. MODFLOW-2005, the U.S. Geological Survey modular ground-water model—The ground-water flow process. U.S. Geological Survey Techniques and Methods 6-A16, Reston, VA.
Harbaugh, A.W., Banta, E.R., Hill, M.C., McDonald, M.G., 2000. MODFLOW-2000, The U.S. Geological Survey modular ground-water model—User's guide to modularization concepts and the ground-water flow process. U.S. Geological Survey Open-File Report 00-92, Reston, VA.
Hirsch, R.M., 1979. Synthetic hydrology and water supply reliability. Water Resour. Res. 15 (6), 1603–1615.
Houston, J.F.T., 1983. Ground-water systems simulation by time-series techniques. Ground Water 21 (3), 301–310.
Hsieh, P.A., 2001. Topodrive and Particleflow—Two computer models for simulation and visualization of groundwater flow and transport of fluid particles in two dimensions. U.S. Geological Survey Open-File Report 01-286, Menlo Park, CA.
Johnson, A.I., 1967. Specific yield—Compilation of specific yields for various materials. U.S. Geological Survey Water-Supply Paper 1662-D, Washington, DC.
Jukić, D., Denić-Jukić, V., 2006. Nonlinear kernel functions for karst aquifers. J. Hydrol. 328, 360–374.
Kalman, R.E., 1960. A new approach to linear filtering and prediction problems. Trans. ASME. J. Basic Engineering 82, 35–43.
Ketchum, J.N., Donovan, J.J., Avery, W.H., 2000. Recharge characteristics of a phreatic aquifer as determined by storage accumulation. Hydrogeology Journal 8, 579–593.
Kim, S.J., Hyun, Y., Lee, K.K., 2005. Time series modeling for evaluation of groundwater discharge rates into an urban subway system. Geosciences Journal 9 (1), 15–22.
Kiraly, L., 1998. Modelling karst aquifers by the combined discrete channel and continuum approach. Bulletin d'Hydrogeologie 16, 77–98.
Klemeš, V., 1978. Physically based stochastic hydrologic analysis. In: Chow, V.T. (Ed.), Advances in hydroscience, vol. 11. Academic Press, New York, pp. 285–352.
Knotteres, M., Bierkens, M.F.P., 2000. Physical basis of time series models for water table depths. Water Resour. Res. 36 (1), 181–188.
Koch, R.W., 1985. A stochastic streamflow model based on physical principles. Water Resour. Res. 21 (4), 545–553.
Kovács, A., Sauter, M., 2007. Modelling karst hydrodynamics. In: Goldscheider, N., Drew, D. (Eds.), Methods in karst hydrogeology. International Contributions to Hydrogeology 26. International Association of Hydrogeologists, Taylor & Francis, London, pp. 201–222.
Kresic, N., 1991. Kvantitativna hidrogeologija karsta sa elementima zaštite podzemnih voda (in Serbo-Croatian; Quantitative karst hydrogeology with elements of groundwater protection). Naučna knjiga, Beograd.
Kresic, N., 1995. Stochastic properties of spring discharge. In: Dutton, A.R. (Ed.), Toxic substances and the hydrologic sciences. American Institute of Hydrology, Minneapolis, MN, pp. 582–590.
Kresic, N., 1997. Quantitative solutions in hydrogeology and groundwater modeling. Lewis Publishers/CRC Press, Boca Raton, FL.
Kresic, N., 2007a. Hydrogeology and groundwater modeling, second ed. CRC Press, Taylor & Francis Group, Boca Raton, FL.
Kresic, N., 2007b. Hydraulic methods. In: Goldscheider, N., Drew, D. (Eds.), Methods in karst hydrogeology. International Contributions to Hydrogeology 26. International Association of Hydrogeologists, Taylor & Francis, London, pp. 65–92.
Kresic, N. (guest editor), 2009a. Foreword: Ground Water in Karst. *Ground Water*, Theme Issue, vol. 47, no. 3, pp. 319–320.
Kresic, N., 2009b. Groundwater Resources: Sustainability, Management, and Restoration. McGraw Hill, New York.
Kresic, N., Papic, P., Golubovic, R., 1992. Elements of groundwater protection in a karst environment. Environ. Geol. Water Sci. 20 (3), 157–164.
Kruseman, G.P., de Ridder, N.A., Verweij, J.M., 1991. Analysis and Evaluation of Pumping Test Data, second ed. International Institute for Land Reclamation and Improvement (ILRI) Publication 47, Wageningen, the Netherlands.
Kupusovic, T., 1989. Measurements of piezometric pressures along deep boreholes in karst area and their assessment. Naš Krš XV (26–27), 21–30.
Lee, J.Y., Lee, K.K., 2000. Use of hydrologic time series date for identification of recharge mechanism in a fractured bedrock aquifer system. J. Hydrol. 229 (3–4), 190–201.
Lohman, S.W., 1972. Ground-water hydraulics. U.S. Geological Survey Professional Paper 708, Washington, DC.

Long, J.C.S., 1983. Investigation of equivalent porous medium permeability in networks of discontinuous fractures. Ph.D. dissertation, Univ. of California, Berkeley.

Long, J.C.S., Gilmour, P., Witherspoon, P.A., 1985. A model for steady fluid in random three-dimensional networks of disc-shaped fractures. Water Resour. Res. 21 (8), 1105–1115.

Maillet, E. (Ed.), 1905. Essais d'hydraulique souterraine et fluviale. Herman, Paris.

Mangin, A., 1984. Pour une meilleure connaissance des systemes hydrologiques a partir des analyses correlatoire et spectrale. Journal of Hydrology, v. 67, pp. 25–43.

Margeta, J., Fistanić, I., 2004. Water quality modelling of Jadro Spring. Water Sci. Technol. 50 (11), 59–66.

McCuen, R.H., Snyder, W.M., 1986. Hydrologic modeling. Statistical Methods and Applications. Prentice-Hall, Englewood Cliffs, NJ.

Meinzer, O.E., 1932. [reprinted 1959] Outline of methods for estimating ground-water supplies. Contributions to the hydrology of the United States, 1931. U.S. Geological Survey Water-Supply Paper 638-C, Washington, DC, pp. 99–144.

Milanović, P., 1979. Hidrogeologija karsta i metode istraživanja [Karst hydrogeology and methods of investigations; in Serbo-Croatian]. HE Trebišnjica, Institut za korištenje i zaštitu voda na kršu, Trebinje, Yugoslavia.

Montanari, A., Rosso, R., Taqqu, M.S., 2000. A seasonal fractional ARIMA model applied to the Nile River monthly flows at Aswan. Water Resour. Res. 36 (5), 1249–1259.

Osborne, P.S., 1993. Suggested operating procedures for aquifer pumping tests. Ground Water Issue, U.S. Environmental Protection Agency, EPA/540/S-93/503, Washington, DC.

Padilla, A., Pulido-Bosch, A., Calvache, M.L., Vallejos, A., 1996. The ARMA models applied to the flow of karstic springs. Water Resour. Res. 32 (5), 917–928.

Palmer, W.C., 1965. Meteorological Drought. Research Paper No. 45, U.S. Department of Commerce, Washington, DC.

Palmer, A.N., 1991. Origin and morphology of limestone caves. Geol. Soc. Am. Bull. 103, 1–21.

Prohaska, S., 1981. Stohastički model za dugoročno prognoziranje rečnog oticaja [Stochastic model for long-term prognosis of river flow; in Serbo-Croatian]. Vode Vojvodine, Special Edition 1981, Novi Sad, Serbia.

Pronk, M., Goldscheider, N., Zopfi, J., 2005. Dynamics and interaction of organic carbon, turbidity and bacteria in a karst aquifer system. Hydrogeology Journal 14, 473–484.

Reilly, T.E., Harbaugh, A.W., 2004. Guidelines for evaluating ground-water flow models. U.S. Geological Survey Scientific Investigations Report 2004-5038, Reston, VA.

Rubinić, J., Fistanić, I., 2005. Application of time series modeling in karst water management. In: Stevanovic, Z., Milanovic, P. (Eds.), Water resources and environmental problems in karst. Proceeding of the international conference and field seminar, Belgrade and Kotor, Institute of Hydrology, University of Belgrade, Serbia, pp. 417–422.

Salas, J.D., 1993. Analysis and modeling of hydrologic time series. In: Maidment, D.R. (Ed.), Handbook of Hydrology. McGraw-Hill, New York, pp. 19.1–19.72.

Salas, J.D., Boes, D.C., Smith, R.A., 1982. Estimation of ARMA models with seasonal parameters. Water Resources Research 18 (4), 1006–1010.

Salas, J.D., Obeysekera, J.T.B., 1992. Conceptual basis of seasonal streamflow time series models. Journal of Hydraulic Engineering 118 (8), 1186–1194.

Salas, J.D., Delleur, J.W., Yevjevich, V., Lane, W.L., 1985. Applied modeling of hydrologic time series. Water Resources Research Publication, Littleton, CO.

San Antonio Water System, 2008. Aquifer Levels and Stats. Available atwww.saws.org/our_water/aquifer (Accessed December 2008).

Shoemaker, W.B., Kuniansky, E.L., Birk, S., Bauer, S., Swain, E.D., 2008. Documentation of a Conduit Flow Process (CFP) for MODFLOW-2005. U.S. Geological Survey Techniques and Methods 6-A24, Reston, VA.

Singh, V.P., 1988. Rainfall-runoff Modeling, Hydrologic Systems, vol. 1. Prentice-Hall, Englewood Cliffs, NJ.

Snow, D.T., 1968. Rock fracture spacings, openings, and porosities. J. Soil Mech. Found. Div., Amer. Soc. Civil Engineers 94, 73–91.

Snyder, W.M., 1968. Subsurface implications from surface hydrograph analysis. In: Proceedings of Second Seepage Symposium, Phoenix, AZ. U.S. Department of Agriculture, Washington, DC, pp. 35–45.

Soulios, G., 1984. Effective infiltration into Greek karst. J. Hydrol. 75, 343–356.

Stallman, R.W., 1971. Aquifer-test, design, observation and data-analysis: U.S. Geological Survey Techniques of Water-Resources Investigations, book 3, chap. B1. U.S. Geological Survey, Washington, DC.

Theis, C.V., 1935. The lowering of the piezometric surface and the rate and discharge of a well using ground-water storage. Transactions, American Geophysical Union 16, 519–524.

U.S. Bureau of Reclamation, 1977. Ground water manual. U.S. Department of the Interior, Bureau of Reclamation, Washington, DC.

Thomas, H.A., Fiering, M.B., 1962. Mathematical synthesis of streamflow sequences for the analysis of river basin by simulation. In: Maas, A. et al. (Ed.), Design of Water Resources Systems, Chapter 12. Harvard University Press, Cambridge, MA, pp. 459–493.

U.S. EPA, 1994. Assessment framework for ground-water model applications. OSWER Directive 9029.00, U.S. Environmental Protection Agency, Office of Solid Waste and Emergency Response, Washington, DC.

Vecchia, A.V., Obeysekera, J.T.B., Salas, J.D., Boes, D.C., 1983. Aggregation and estimation for low-order periodic ARMA models. Water Resour. Res. 19 (5), 1297–1306.

Voss, C.I., Provost, A.M., 2002. SUTRA: A model for saturated-unsaturated, variable-density ground-water flow with solute or energy transport. U.S. Geological Survey Water-Resources Investigations Report 02-4231, Reston, VA.

Walton, W.C., 1987. Groundwater Pumping Tests, Design and Analysis. Lewis Publishers, Chelsea, MI.

Weeks, W.D., Boughton, W.C., 1987. Tests of ARMA model forms for rainfall-runoff modelling. J. Hydrol. 91, 29–47.

White, B.W., 1988. Geomorphology and hydrology of karst terrains. Oxford University Press, New York.

Williams, J.H., Lane Jr., J.W., Singha, K., Haeni, F.P., 2002. Application of advanced geophysical logging methods in the characterization of a fractured-sedimentary bedrock aquifer, Ventura County, California. U.S. Geological Survey Water-Resources Investigations Report 00-4083, Troy, NY.

Witherspoon, P.A., 2000. Investigations at Berkeley on fracture flow in rocks: From the parallel plate model to chaotic systems. In: Faybishenko, B., Witherspoon, P.A., Benson, S.M. (Eds.), Dynamics of Fluids in Fractured Rock, Geophysical Monograph 122. American Geophysical Union, Washington, DC, pp. 1–58.

Worthington, S., 2003. Characterization of the Mammoth Cave aquifer. In: Significance of Caves in Watershed Management and Protection in Florida, Workshop Proceedings, April 16–17, 2003. Florida Geological Survey Special Publication No. 53. Florida Geological Survey, Ocala, FL.

Worthington, S.R.H., Smart, C.C., Ruland, W.W., 2003. Assessment of groundwater velocities to the municipal wells at Walkerton. In: Proceeding of the 2002 joint annual conference of the Canadian Geotechnical Society and the Canadian chapter of the IAH. Niagara Falls, Ontario, pp. 1081–1086.

Zimmerman, R.W., Yeo, I.W., 2000. Fluid flow in rock fractures: From the Navier-Stokes equations to cubic law. In: Faybishenko, P.A., Witherspoon, P.A., Benson, S.M. (Eds.), Dynamics of Fluids in Fractured Rock, Geophysical Monograph 122. American Geophysical Union, Washington, DC, pp. 213–224.

CHAPTER

Springwater geochemistry

6

William B. White

Materials Research Institute and Department of Geosciences,
The Pennsylvania State University, University Park

6.1 PHYSICAL CHEMISTRY OF NATURAL WATERS

6.1.1 Introduction

Springs result from a concentration of groundwater flow paths, so that water issues from a single location (or a small number of nearby locations) instead of diffuse flow into surface streams. As such, the chemistry of springwaters reflects the interaction of groundwater with the aquifer host rock as well as any chemical constituents that may be introduced from surface sources. Springwater chemistry is, therefore, not intrinsically different from groundwater chemistry and the same principles apply.

Aqueous chemistry in general, and springwater chemistry in particular, can be approached from two points of view. One can begin with the minerals that make up the host rock and their thermodynamic properties and calculate mineral solubilities and interactions. From this, in principle, the water chemistry can be predicted. The other point of view is to begin with hard data—chemical analyses of the springwater—then try to rationalize the data with theoretical interpretations. Both viewpoints are used in the discussion that follows.

Most springwaters are dilute aqueous solutions with concentrations of dissolved species in the range of tens to hundreds of milligrams per liter. For dilute solutions, where the density of the solution is not significantly different from the density of pure water, the volume concentration (mg/L) and the mass concentration (parts per million, ppm) are numerically equivalent. In a few cases, such as brine waters, the density of the solution is high enough that a distinction must be made.

The common cations in springwaters are Ca^{2+}, Mg^{2+}, Na^{+}, and K^{+}; while the most common anions are HCO_3^-, SO_4^{2-}, and Cl^-. Together, these form the basis for the classification of springwater (also groundwater and surface water). The three main groups are bicarbonate waters, where Ca and Mg tend to be the dominant cations; sulfate waters, where Mg may be greater than Ca and alkali ions are also present; and chloride waters, where alkali ions tend to be dominant. The dominant chemistry usually depends on the host rock of the aquifer from which the spring emerges.

There exist tens of thousands of springs in a great variety of geologic settings. There also exist tens of thousands of chemical analyses of springwater. All that can be accomplished in a short review is to sort these waters into categories and give some indication of the chemical compositions and chemical reactions to be expected. To this end, tables are provided that are intended to illustrate the range in springwater chemistry. The tables are informative but certainly not comprehensive.

Copyright © 2010, Elsevier Inc. All rights reserved.

6.1.2 Chemical equilibrium and mineral saturation

Low-temperature aqueous geochemistry has become a well-developed science over the past several decades. Excellent detailed information on the chemistry of natural waters may be found in such books as Hem (1985), Morel and Hering (1993), Stumm and Morgan (1996), Drever (1997), and Langmuir (1997).

Some minerals simply dissolve in water, dissociating into their component ions. Other minerals react chemically with water, often forming some soluble ions and some insoluble residues. For those that dissolve by ionic dissociation, the range in solubilities spans many orders of magnitude. For three common minerals at ambient temperatures,

Halite	Nacl	350,000 mg/L
Gypsum	$CaSO_4 \bullet 2H_2O$	2500 mg/L
Calcite	$CaCO_3$(in pure water)	7 mg/L

For any of these minerals, there is a certain equilibrium concentration, defined thermodynamically by the concentration at which the free energies of the species on both sides of the dissolution reaction balance exactly or kinetically where the rate of dissolution is equal to the rate of precipitation. In the real world, the concentrations of dissolved species are often not at equilibrium with the minerals of the host rock. This is particularly true in carbonate rocks, where the nonequilibrium is a useful interpretive tool.

However, calculations of mineral solubilities provide a useful reference for the interpretation of springwater chemistry. The water chemistry predicted from equilibrium calculations can be compared with real-world water chemistry determined by chemical analysis. Because of the large number of interactions, such calculations are best made by computer programs such as PHREEQC, developed by the U.S. Geological Survey (Parkhurst and Appelo, 2008).

6.2 SPRINGWATER FROM SILICATE ROCKS

6.2.1 The dissolution of silica and silicates

The dominant mineral in most silicate rocks is quartz, SiO_2. SiO_2 dissolves by chemical reaction with water:

$$SiO_2 + 2H_2O \leftrightarrow H_4SiO_4$$

$$a_{H_4SiO_4} = K_{qtz} = 1.06 \times 10^{-4} \tag{6.1}$$

As a dilute neutral species, the activity of orthosilicic acid, H_4SiO_4, is essentially equal to the concentration, which is 6.39 mg/L at 25°C. Some silica in springwater is derived from the breakdown of other silicate minerals, not from the dissolution of highly crystalline quartz. Amorphous silica has a higher solubility, 117 mg/L based on the thermodynamic data given in Drever (1997). The dissolution reaction for quartz is not reversible; crystalline quartz does not precipitate from cold solutions. Any silica taken into solution by any reaction tends to remain in solution. As a result, the concentrations of SiO_2 found in silicate rock springs is often higher than the lower limit imposed by crystalline quartz.

The solubility of quartz is independent of pH through the acid regime up to pH values in the range of 9. The ionization of silicic acid and the formation of oligomeric species cause the solubility to increase strongly at high pH, conditions rarely found in natural water (Figure 6–1).

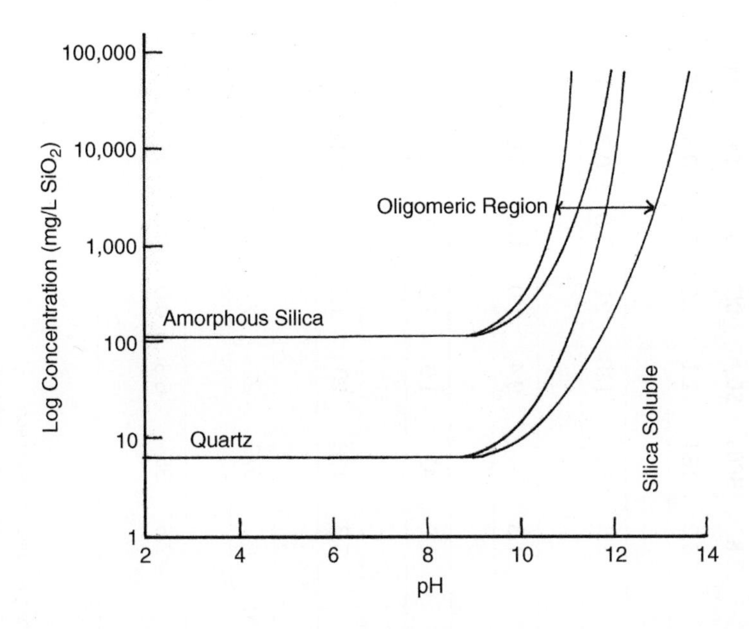

FIGURE 6–1 Speciation diagram for the SiO_2-H_2O system at 25°C. The solubilities of quartz and amorphous silica in the low pH region were calculated from thermodynamic data. The solubility curves in the high pH oligomeric region are schematic because of the complex silica species that exist in this region.

Most complex silicate minerals also react with water rather than dissolve by dissociation. For example, in granitic aquifers, the breakdown of feldspars produces clay minerals and releases alkali ions into solution:

$$2KAlSi_3O_8 + 2H^+ + 7H_2O \leftrightarrow Al_2Si_2O_5(OH)_4 + 2K^+ + 4H_4SiO_4$$

Aluminum tends to be immobile in near-neutral water, so that aluminum-containing species are rarely found in low-temperature springwater. In the example of orthoclase, the Si/Al ratio changes, the aluminum remains as kaolinite, while both potassium and some of the silica are released into solution. Under more extreme weathering conditions, clay minerals such as kaolinite also break down, with aluminum converting to gibbsite and the remaining silica going into solution:

$$Al_2Si_2O_5(OH)_4 + 5H_2O \leftrightarrow 2Al(OH)_3 + 2H_4SiO_4$$

6.2.2 Springs in shales, sandstones, and granites

There are thousands of springs in shales, siltstones, and sandstones as well as in granites and metamorphic rocks, but these usually have a low discharge and are located mostly on fracture zones or related structural weaknesses. Many older farmsteads were located close to such springs because they provided a reliable source of high-quality water. Silicate rock springs are not high magnitude but they do provide water supplies for many homes.

Some representative analyses of silicate rock springwater are given in Table 6–1. These are selected mostly from the compilation of D. White, Hem, and Waring (1963). Springwater from silicate rocks tends

Table 6–1 Selected Analyses of Springwaters from Silicate Rocks

Location	T (°C)	pH	SiO_2	Ca^{2+}	Mg^{2+}	Na^+	K^+	HCO_3^-	SO_4^{2-}	Cl^-	Al	Fe	Mn
Melbourne Spring, AK	16.7	7.4	12	50	6.0	2.4	3.0	184	2.1	1.8	0.2	0.6	0.0
Ordovician. St. Peter sandstone													
Park Lake Spring, KY	10.0	4.9	22	15	7.5	36	3.5	3.2	128	21	1.5	0.39	0.22
Devonian Ohio shale													
Weathertop Spring, PA	10.0	—	2.0	3.3	2.7	0.4	1.3	—	9.4	1.0	0.0	0.0	0.0
Silurian Clinton shale													
East Fork Spring, NM	12.8	7.2	55	4.4	1.4	11	1.2	42	1.9	2.0	0.1	0.08	0.0
Tertiary rhyolite													
Snyder's Spring, ID	7.8	7.5	27	34	7.3	8.5	3.3	136	20	1.2	0.1	0.05	0.0
Mesozoic quartz monzonite													
Buell Park Spring, AZ	12.2	8.2	31	20	42	19 (Na + K)		279	22	7	—	—	—
Olivine basalt													
Bear Springs, ID	—	7.7	8.9	12	0.5	1.8	2.6	38	6.3	0.0	0.0	0.0	0.0
Permian volcanics													

Notes:
All concentrations in milligrams per liter.
All data extracted from tables in D. White et al. (1963) except Weathertop Spring, which are unpublished data of the author.

to have very low total dissolved solids. Silica is almost always the dominant species with concentrations of a few tens of milligrams per liter, considerably higher than the expected equilibrium concentration from quartz. Concentrations of iron, aluminum, and manganese range from very low to undetectable. If not tied up in other minerals, these elements form oxides and hydroxides that are extremely insoluble in the near-neutral pH regime. One shale spring has a lower pH, 4.9, and it has a measurable concentration of aluminum of 1.5 mg/L.

Most of the springwater contains substantial concentrations of calcium, magnesium, and bicarbonate. Although these species are inconsistent with pure silicate, most shales and sandstones contain at least minor quantities of carbonate minerals. These minor minerals tend to be the main contributors to the dissolved species.

6.2.3 Cold water springs in volcanic rocks

Volcanic rocks in the form of lava flows tend to be highly porous. Fracture systems and lava tube caves provide the same conduit permeability found in carbonate rocks. Pyroclastic flows and pumice deposits are extremely permeable. Other volcanic rocks are tight, glassy, and have almost no permeability. As an aquifer, a volcanic terrain can be highly heterogeneous. As a result, some of the largest springs in the world are found in volcanic terrain, comparable to those found in karstic terrain. Examples are the springs in the Snake River Gorge in southern Idaho and the comparable large spring draining lava flows in Iceland (Figure 6–2).

Volcanic rocks—basalts, andesites, and their more coarsely crystalline counterparts, such as gabbros—contain minerals that have lower concentrations of silica and higher concentrations of magnesium and iron. These are igneous minerals, formed at high temperatures, and are generally not stable in the presence of water.

FIGURE 6–2 Hraunfoss, large springs emerging from a lava flow in Iceland. (Photo by the author.)

Table 6–2 Average Compositions of Spring- and Groundwaters from the Snake River Plain, Idaho

Species	Mean	Standard Deviation	Coefficient of Variation
Ca^{2+}	44.0	19.7	45%
Mg^{2+}	14.1	8.4	60%
Na^{+}	30.2	24.3	80%
K^{+}	4.6	3.8	83%
HCO_3^{-}	204.8	78.9	39%
Cl^{-}	20.6	19.3	94%
SO_4^{2-}	40.0	33.4	84%
SiO_2	36.4	15.4	42%

Notes:
Mean concentrations are in units of milligrams per liter.
Table obtained by averaging the 230 analyses given in Table 20A of Wood and Low (1988).

The igneous minerals react with groundwater, releasing alkali ions as well as Mg, Ca, and silica. A few examples are given in Table 6–1. Note that Mg tends to exceed Ca, but otherwise the compositions of cold volcanic springwater are similar to the compositions of other springwater from silicate rocks. Because of the oxidizing environment, iron and manganese concentrations are very low or not detected in spite of the high concentrations of these elements in the host rock.

Mean values for the major ions from the Snake River Plain aquifer in southern Idaho have been extracted from the report of Wood and Low (1988) (Table 6–2). It is immediately apparent from the concentrations of Ca, Mg, and bicarbonate ions that carbonate minerals, although likely present in minor amounts, dominate the chemistry of this extensive volcanic terrain. Chloride concentrations are highly variable and may result in part from contamination of the aquifer by salt used for road deicing in the winter.

6.3 SPRINGWATER FROM CARBONATE ROCKS

Most of the largest springs in the world discharge from carbonate rocks. Groundwater in carbonate aquifers is often organized into systems of conduits, fragments of which are sometimes accessible as explorable caves. Carbonate springs vary widely in their physical appearance. Some appear as rise pools in river valleys or along carbonate coast lines (Figure 6–3). Some emerge from rubble piles that block access to any feeder system. Some emerge from zones of concentrated fractures that have undergone only modest dissolutional modification. But the most impressive are the springs that emerge from open cave mouths (Figure 6–4). During base flow, the feeder conduits to these springs can often be explored for long distances. More intense exploration by divers reveals that many of the feeder conduits undulate in the vertical plane, so that at low flow, there are air-filled segments with free surface streams and segments that are completely water filled but only to shallow depths. During flood flow, the entire conduit may be submerged.

Carbonate rock (karstic) springs that discharge from conduit systems often obtain rapid recharge from sinking streams and open sinkholes. Such recharge has a very short residence time and is likely not in

FIGURE 6–3 Echo River Spring, Mammoth Cave National Park, a rise pool spring with the spring feeder-conduit 6–9 m below the level of the nearby Green River. (Photo by the author.)

FIGURE 6–4 Mammoth Spring, a large spring emerging from a master conduit system, Mifflin County, Pennsylvania. (Photo courtesy of Joe Kearns.)

equilibrium with the limestone or dolomite wall rock. Further, direct injection of surface water is a ready source of contaminants, which may make their way to the spring with little dilution or adsorption.

Interpretation of the chemistry of carbonate springwater requires a consideration of the aqueous chemistry of calcite and dolomite.

6.3.1 The dissolution of limestone and dolomite

The dominant minerals in carbonate rocks are calcite, $CaCO_3$, and dolomite, $CaMg(CO_3)_2$. These minerals may dissolve by simple dissociation, but the solubilities in pure water are both very low, about 7 mg/L, comparable to quartz:

$$CaCO_3 \leftrightarrow Ca^{2+} + CO_3^{2-}$$

$$a_{Ca^{2+}} a_{CO_3^{2-}} = K_C \tag{6.2}$$

$$CaMg(CO_3)_2 \leftrightarrow Ca^{2+} + Mg^{2+} + 2CO_3^{2-}$$

$$a_{Ca^{2+}} a_{Mg^{2+}} a^2_{CO_3^{2-}} = K_D \tag{6.3}$$

However, unlike the silicate minerals, carbonate minerals are very susceptible to attack by acids. Carbonate minerals can react with any organic or inorganic acid, but the most important one in most carbonate aquifers is carbonic acid.

Carbonic acid is derived from CO_2 in the atmosphere or in the soil. When CO_2 is dissolved in water, most of it remains aqueous CO_2 but a small amount, about 0.17 percent, reacts chemically with the water to produce carbonic acid. Because it is difficult to separate aqueous CO_2 from carbonic acid, the two reactions are usually taken together:

$$CO_2(\text{gas}) \leftrightarrow CO_2(\text{aqueous})$$

$$CO_2(\text{aqueous}) + H_2O \leftrightarrow H_2CO_3$$

$$\frac{a_{H_2CO_3}}{P_{CO_2}} = K_{CO_2} \tag{6.4}$$

Carbonic acid is a weak acid that only partially ionizes and does so in two steps:

$$H_2CO_3 \leftrightarrow H^+ + HCO_3^-$$

$$\frac{a_{H^+} a_{HCO_3^-}}{a_{H_2CO_3}} = K_1 \tag{6.5}$$

$$HCO_3^- \leftrightarrow H^+ + CO_3^{2-}$$

$$\frac{a_{H^+} a_{CO_3^{2-}}}{a_{HCO_3^-}} = K_2 \tag{6.6}$$

In these equations, the *a* terms are thermodynamic activities for the designated species. The equilibrium constants, *K*, are temperature dependent. Numerical values at various temperatures are tabulated in Drever (1997).

The interpretation of karstic springwater under equilibrium conditions requires solution of all equilibrium equations simultaneously. For the case of limestone, there is a three-phase system: solid limestone, assumed to be pure $CaCO_3$; a gas phase with CO_2 as the only active component, described by a partial pressure, P_{CO2}; and an aqueous phase containing varying concentrations of the species Ca^{2+}, H_2CO_3, HCO_3^-, CO_3^{2-}, H^+, and OH^-. Counting the CO_2 pressure and the concentrations of the six aqueous species, there are seven unknowns, requiring seven independent relationships among them. Equation (6.2) connects the solid limestone to the aqueous phase, and equation (6.4) connects the gas phase to the aqueous phase. Equations (6.5) and (6.6) provide two more relationships. A fifth relationship is provided by the ionization of water, which connects the species H^+ and OH^-:

$$H_2O \leftrightarrow H^+ + OH^-$$

$$a_{H^+} a_{OH^-} = K_W \tag{6.7}$$

The sixth relationship is provided by the necessity of electrical neutrality; the sum of all positive charges must be equal to the sum of all negative charges:

$$2m_{Ca^{2+}} + m_{H^+} = m_{HCO_3^-} + 2m_{CO_3^{2-}} + m_{OH^-} \tag{6.8}$$

The charge balance uses the actual molal concentrations, m, whereas the equilibrium expressions are written in terms of activities, a. These are connected through the activity coefficients, γ:

$$a_i = \gamma_i m_i \tag{6.9}$$

Both the concentration of dissolved carbonate and the pH of the solution are fixed by these reactions, providing a seventh relationship can be specified. There are three possibilities:

1. **Closed system**. The excess carbon dioxide in the system is fixed before the reactions are "turned on." The seventh relation is provided by mass conservation of the carbon-bearing species.
2. **System open to CO_2**. The seventh relation is provided by specifying the CO_2 partial pressure.
3. **System open to H^+**. The seventh relation is provided by fixing the pH.

Possibility 2 is the most useful for karst aquifers. With a considerable amount of algebraic manipulation (developed in detail by Drever (1997) and Langmuir (1997)), the set of equations can be solved to give both the equilibrium concentration of dissolved carbonate and the pH for any specified temperature and CO_2 partial pressure:

$$m_{Ca^{2+}} = (P_{CO_2})^{\frac{1}{3}} \left(\frac{K_1 K_C K_{CO_2}}{4K_2 \gamma_{Ca^{2+}} \gamma^2_{HCO_3^-}} \right)^{\frac{1}{3}} \tag{6.10}$$

$$pH = -\frac{2}{3} \log P_{CO_2} - \frac{1}{3} \log \left(\frac{K_1^2 K_2 K_{CO_2}^2 \gamma_{Ca^{2+}}}{2K_C \gamma_{HCO_3^-}} \right) \tag{6.11}$$

These equations are the simplest approach. The activity coefficients must be estimated and this requires information about the other constituents in the water. Also complexation reactions should be taken into account. Similar equations can be constructed for dolomite:

$$m_{Ca^{2+}} = (P_{CO_2})^{\frac{1}{3}} \left(\frac{K_1 K_D^{\frac{1}{2}} K_{CO_2}}{4K_2 \gamma^2_{HCO_3^-} (\gamma_{Ca^{2+}} \gamma_{Mg^{2+}})^{\frac{1}{2}}} \right)^{\frac{1}{3}} \tag{6.12}$$

The equation for the concentration for Mg^{2+} is exactly the same. If the composition of dolomite is written equimolar with calcite, $Ca_{1/2}Mg_{1/2}CO_3$, the solubilities of the two minerals can be compared (Figure 6–5).

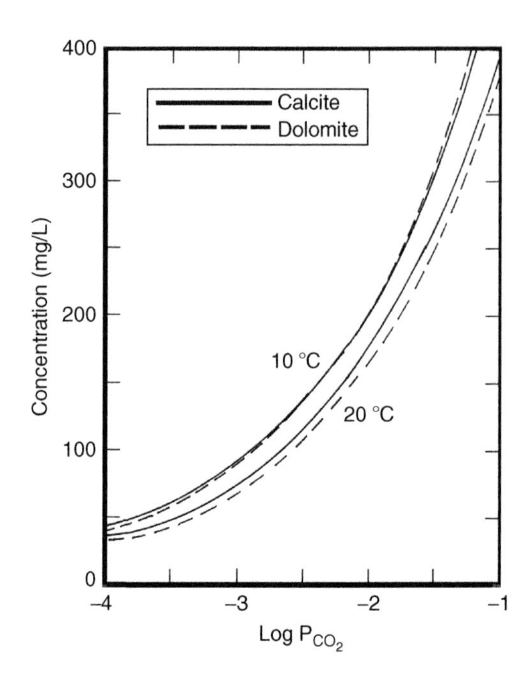

FIGURE 6–5 Solubility curves for calcite and dolomite at 10 and 20°C. Note that these curves are plotted on an equimolar basis with dolomite written as $Ca_{1/2}Mg_{1/2}CO_3$.

At 20°C, calcite is slightly more soluble than dolomite. At 10°C, the solubilities are almost identical. In fact, the curves cross, with calcite slightly more soluble at low CO_2 pressure while dolomite is slightly more soluble at high CO_2 pressure. One should not confuse equilibrium solubility with kinetics. Although the equilibrium solubilities of calcite and dolomite are very similar, the rates at which the dissolution reactions achieve equilibrium are quite different.

The solubilities of both minerals increase rapidly with increasing CO_2 partial pressure, with almost an order of magnitude increase between the solubility in water in equilibrium with atmospheric CO_2 and water in equilibrium with typical soil CO_2 pressures. The curves continue to flatten at lower CO_2 pressures and, in the limit of extremely low CO_2 pressures, approach the mineral solubilities calculated for pure water.

6.3.2 Chemical kinetics and nonequilibrium

Karst aquifers differ from most others because of the short travel time for water to move from recharge areas to the springs. The various dissolution reactions described earlier apply to equilibrium conditions. The time required for aqueous solutions to achieve equilibrium varies greatly among the reactions. The literature on calcite dissolution kinetics is huge (Morse and Arvidson, 2002) with rate equations proposed by Plummer, Wigley, and Parkhurst (1978) and many others. Rate equations for the dissolution of dolomite have also been proposed (Busenberg and Plummer, 1982; J. Herman and White, 1985; Pokrovsky and Schott, 2001; Pokrovsky, Golubev, and Schott, 2005). Dolomite appears to have two dissolution regimes. At extremely high undersaturation, dolomite dissolves at rates in the same range as calcite. However, the dissolution curves flatten at about 1 percent saturation, after which dolomite dissolves much more slowly than calcite, requiring times of months to years to achieve complete equilibrium.

Formal dissolution kinetics has been used with good success in the interpretation of the development of conduit permeability. Some quite elaborate computer models have been constructed (Dreybrodt, Gabrovšek, and Romanov, 2005). For an interpretation of the chemistry of karst spring waters, however, it is sufficient to understand the timescale for the various reactions:

1. The time constant for hydration of CO_2 is about 30 seconds.
2. The ionization of carbonic acid is in milliseconds, essentially instantaneous.
3. The time required for the dissolution of calcite is several days.
4. The time required for the dissolution of dolomite is several days followed by weeks.

As a result of the sluggish reaction kinetics of limestone and dolomite dissolution, springwater that drains aquifers with well-developed conduit systems or substantial allogenic recharge is not likely to be at equilibrium.

6.3.3 Chemical characterization of carbonate springwater

Carbonate rock springs can be characterized by tables of chemical analyses but these are not as instructive as secondary parameters that can be calculated from the primary analytical data. The four most useful of these parameters are described next.

Hardness

Hardness is one of the oldest water quality parameters and was originally a measure of the capacity of the water to precipitate soap. Indeed, the old hardness tests used a standard soap solution. Hardness is a measure of the combined concentrations of Ca and Mg in the water. The convention has been to define hardness as milligrams per liter of calcium carbonate:

$$\mathrm{Hd} = 100.09\left(\frac{\mathrm{Ca(mg/L)}}{40.08} + \frac{\mathrm{Mg(mg/L)}}{24.305}\right) \tag{6.13}$$

Hardness is of limited scientific value because of treating all dissolved alkaline earth ions as if they were calcium, but it does give a good sense of the quantity of dissolved carbonate and, as such, is a useful water quality parameter. As a practical matter, hardness is proportional to specific conductance in most karstic waters. Figure 6–6 gives a selection of these relationships. Because the specific conductance also depends on other ions present in the water, it is important to calibrate each study site with a set of chemical analyses. Conductance is one of the least complicated parameters to be instrumented for continuous monitoring.

Ca/Mg ratio

The molar ratio of calcium ion concentration to magnesium ion concentration provides information on the rock type from which a spring is emerging. Dolomite springs have a Ca/Mg ratio near unity. Because most limestones contain some magnesium, Ca/Mg ratios for limestone springs are typically in the range of 6–8. Intermediate values would indicate both limestone and dolomite are present along the flow path leading to the spring.

Calculated CO_2 partial pressure

Using the basic equations for carbonate chemistry, it is possible to back-calculate the CO_2 partial pressure with which a springwater would be in equilibrium:

$$P_{CO_2} = \frac{\gamma_{HCO_3^-} m_{HCO_3^-} 10^{-pH}}{K_1 K_{CO_2}} \tag{6.14}$$

Using the equilibrium constants listed by Drever (1997) and the concentration of bicarbonate ion in molal units, the CO_2 pressure has units of atmospheres.

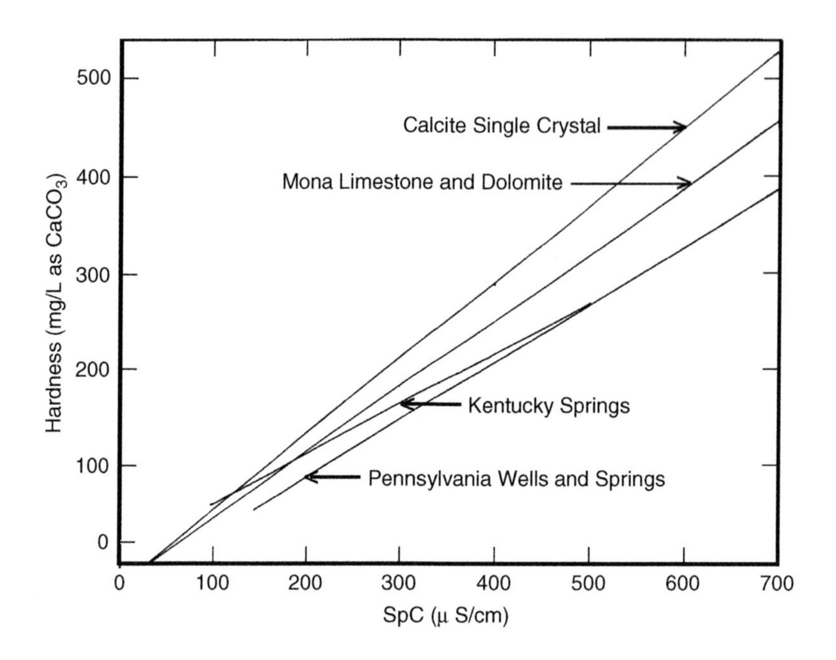

FIGURE 6–6 Relationship between specific conductance and hardness. The lines shown are based on many data points. The fitted regression lines and the goodness of fit follow.

Calcite crystal: Dissolution of single crystal calcite in CO_2-saturated water, unpublished data from J. Herman and White.
Hd = 0.778 SpC − 22.2 $r^2 = 0.997$

Lirio limestone and Mona dolomite from Isla de Mona, Puerto Rico. Discs dissolved in CO_2-saturated water in the laboratory. Data from Martinez and White (1999).
Hd = 0.679 SpC − 22.4 $r^2 = 0.967$

Analyses of 53 springs and wells in limestone and dolomite, central Pennsylvania, data from Langmuir (1971).
Hd = 0.589 SpC − 29.6 $r^2 = 0.93$

Analyses of nearly 100 spring- and cave waters, Turnhole Spring basin, Mammoth Cave National Park, Kentucky, data from Hess (1974).
Hd = 0.515 SpC + 8.886 $r^2 = 0.815$.

The saturation index

Because the waters from karst springs are not likely to be at equilibrium, the saturation index provides a numerical value for the disequilibrium, either undersaturation or supersaturation. The starting point is equation (6.2), the solubility product for calcite. If the waters were exactly at equilibrium, this equation would hold. If the activities of Ca^{2+} and CO_3^{2-} are determined experimentally, the product can be compared with the equilibrium one:

$$\frac{a_{Ca^{2+}}a_{CO_3^{2-}}(\text{meas.})}{a_{Ca^{2+}}a_{CO_3^{2-}}(\text{theor.})} = \frac{a_{Ca^{2+}}a_{CO_3^{2-}}(\text{meas.})}{K_C} \tag{6.15}$$

Two further adjustments are needed. The carbonate ion, CO_3^{2-}, is a minority species in waters at near-neutral pH. The activity of CO_3^{2-} must be calculated from the measured concentration of bicarbonate ion and pH, as in equation (6.6). It is also necessary to estimate the activity coefficients for Ca^{2+} and HCO_3^-, so that activities can be calculated from measured concentrations. If the measured ion activity product is greater than K_C, the water is supersaturated; if it is less than K_C, the water is undersaturated. To maintain equal scale intervals on both sides of equilibrium, it is convenient to define the saturation index as a logarithm. The final equation for the saturation index for calcite is

$$SI_C = \log \frac{\gamma_{Ca^{2+}} m_{Ca^{2+}} \gamma_{HCO_3^-} m_{HCO_3^-} K_2}{10^{-pH} K_C} \tag{6.16}$$

The saturation index for dolomite is given by the expression

$$SI_D = \log \left(\frac{\gamma_{Ca^{2+}} m_{Ca^{2+}} \gamma_{Mg^{2+}} m_{Mg^{2+}} \gamma^2_{HCO_3^-} m^2_{HCO_3^-} K_2^2}{10^{-2pH} K_D} \right)^{\frac{1}{2}} \tag{6.17}$$

Similar equations can be derived to describe the state of equilibrium or disequilibrium of other minerals.

6.3.4 The chemistry of karst springs

To give at least a flavor of the range of values of the parameters that have been measured on karst springs from many locations and geologic settings, a few examples have been chosen from a very large literature. The chosen papers presented both detailed chemical analyses of the springwaters and the calculated parameters of CO_2 pressure and saturation index. Data for 53 samples of springwater are plotted in Figures 6–7 and 6–8; 6 to 12 data

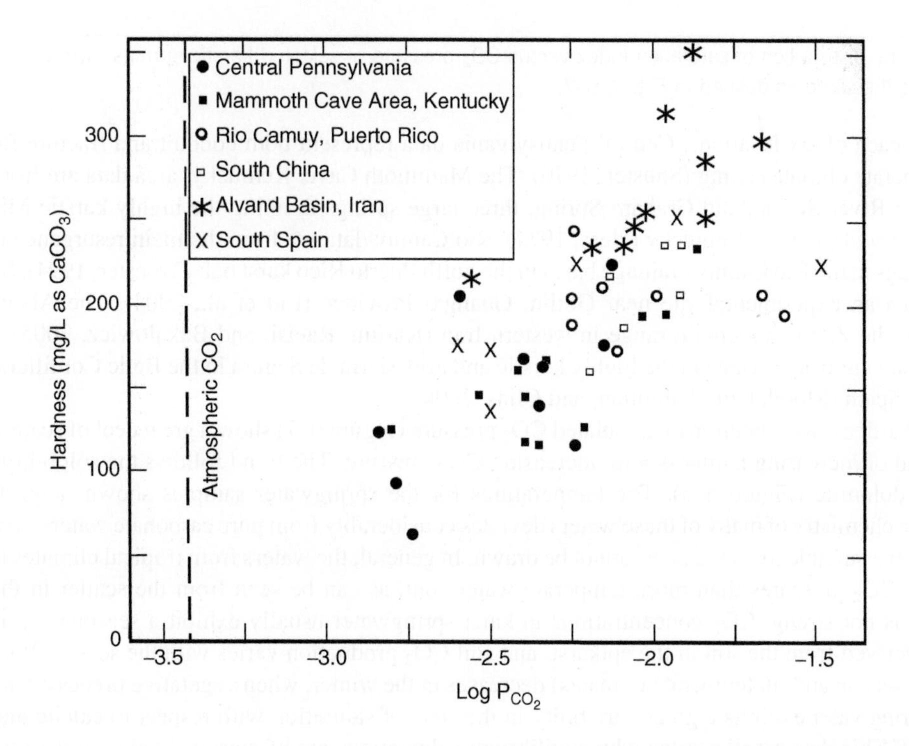

FIGURE 6–7 The distribution of hardness versus CO_2 pressure for selected karstic springs. Symbols are for specific sets of data. See the text for a description of the data sets.

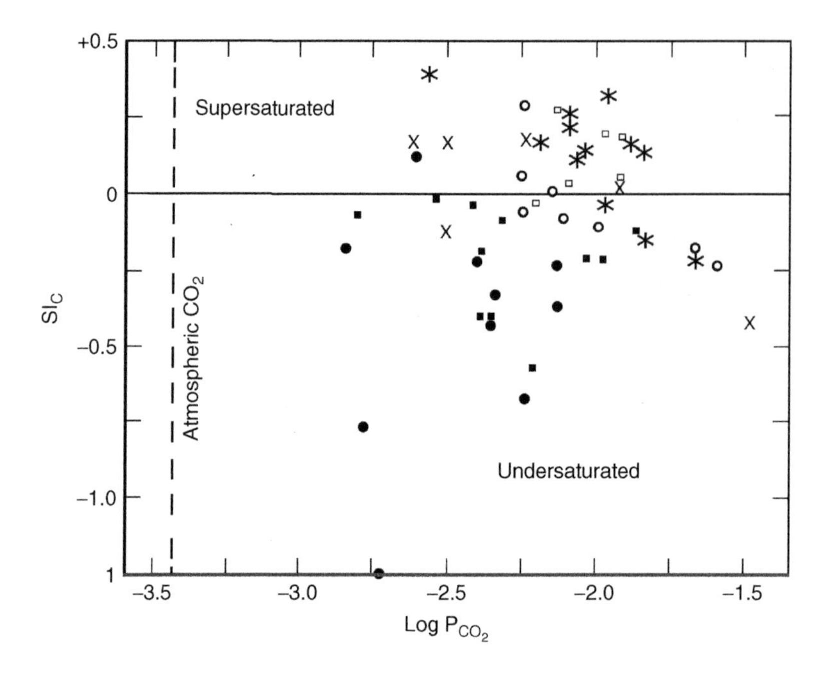

FIGURE 6–8 The distribution of saturation index versus CO_2 pressure of selected karstic springs. Symbol meanings are the same as defined in Figure 6–7.

points from each of six locations. Central Pennsylvania data represent both conduit and fracture flow springs from a temperate climate setting (Shuster, 1970). The Mammoth Cave, Kentucky, area data are from Turnhole Spring, Echo River Spring, and Graham Spring, three large springs draining the highly karstic Mississippian limestones of south-central Kentucky (Hess, 1974). Rio Camuy data are from the main resurgence and several smaller springs in the Rio Camuy drainage basin in the north Puerto Rico karst belt (Troester, 1994). South China data are from an experimental site near Guilin, Guangxi Province (Liu et al., 2004). The Alvand basin is northwest of the Zagros mountain range in western Iran (Karimi, Raeisi, and Bakalowicz, 2005). Data from southern Spain are from springs in the highly karstic and arid Sierra de Segura in the Betic Cordillera, northeast of Granada, Spain (Moral, Cruz-Sanjulián, and Olias, 2008).

Plots of hardness as a function of calculated CO_2 pressure (Figure 6–7) show a great deal of scatter but with a general trend of increasing hardness with increasing CO_2 pressure. The trend follows the solubility curves for calcite and dolomite (Figure 6–5). The temperatures for the springwater samples shown range from 10 to 20°C, but the chemistry of many of these waters deviates considerably from pure carbonate waters, so a reference solubility curve suitable to all the data cannot be drawn. In general, the waters from tropical climates have somewhat higher CO_2 pressures than more temperate waters but, as can be seen from the scatter in the data, the relationship is not strong. CO_2 concentrations in karst springwater usually exhibit a seasonal cycle. Most of the CO_2 is derived from the soil in the epikarst, and soil CO_2 production varies with the season. It is highest in the growing season and (in temperate climates) decreases in the winter, when vegetative processes are dormant.

Karst springwater exhibits a great variability in the state of saturation with respect to calcite and dolomite (Figure 6–8). This data set illustrates why equilibrium calculations are of marginal value in the interpretation of karst water chemistry. The temperate climate springs, Pennsylvania and Kentucky, are mostly undersaturated. The springs with highly undersaturated water drain open conduit systems. However, the state of saturation is not, in itself, an indication of the presence of conduit drainage. Because of CO_2 degassing along conduits, springs discharging from conduits may be supersaturated. The large springs of Florida that receive much of their recharge from overlying soils are fed by large conduits and discharge waters that range near saturation or are undersaturated (Katz et al., 1999).

6.3.5 Time-dependent spring chemistry: Chemographs, turbidographs, and storm flow

The chemistry of springs discharging from noncarbonate rocks and fracture systems in carbonate rocks tends to be time independent. Although spring discharge may vary somewhat between wet and dry seasons, the concentrations of dissolved species vary very little. Chemical reactions between the water and the aquifer minerals tend to be close to equilibrium.

The same is not true for springs draining from karstic carbonate aquifers. As noted in several reviews of karst hydrogeology (e.g., W. White, 2002), karst springs fall into three broad categories: (1) those with response times that are short with respect to the average spacing between storms, (2) those with response times much longer than the average spacing between storms, and (3) intermediate cases where the response times are comparable to the spacing between storms. Because of the rapid throughput of categories (1) and (3), storm waters reach the spring before the water has time to come into chemical equilibrium with carbonate wall rock. As a result, the base flow water moving to the spring is diluted with relatively fresh storm water. The arrival of dilute storm water at the spring is revealed by an abrupt drop in hardness and other chemical parameters.

An illustration is given in Figure 6–9. In this example, the storm pulses are well separated, giving sharp peaks with very rapid recessions on the spring hydrograph. Continuous monitoring of specific conductance

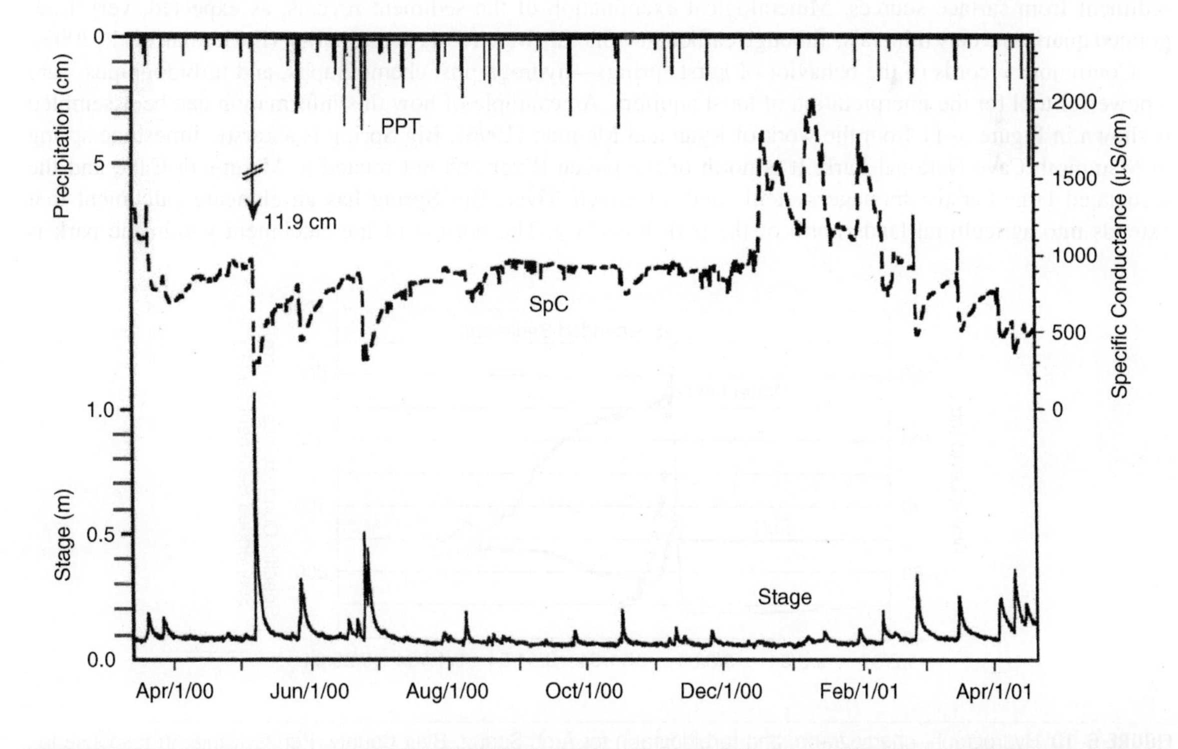

FIGURE 6–9 Precipitation record, discharge hydrograph, and specific conductance chemograph for Bluegrass Spring near St. Louis, Missouri. (Adapted from Winston and Criss, 2004.)

provides a proxy for the chemical composition of the water as a function of time. The chemical composition is plotted as a chemograph. Base flow on the hydrograph is matched by an approximately constant and relatively high specific conductance on the chemograph. Storm peaks on the hydrograph have their counterparts as abrupt dips on the chemograph. However, the recession limb of the hydrograph typically has a shorter response time than the recession limb of the chemograph.

Karst springs draining integrated conduit systems have the potential for transmitting clastic particles: clays, silt, and sometimes coarser material. These clastic sediments are often observed in conduits accessible by cave exploration. The materials are derived from sinking streams and injection of surface soils and regolith through open fractures and sinkhole drains. Particle sizes range from colloidal and clay-sized particles to cobbles and occasionally boulders. Deposits seen in caves can be organized into facies depending on particle size and degree of sorting (Bosch and White, 2004). Materials in the small particle size range are easily mobilized and carried by storm flow and are responsible for karst springs becoming turbid or muddy during storms. Measurement of suspended particles in spring discharge, by either frequent sampling or continuous turbidity measurements, gives yet another probe into the aquifer system: the turbidograph. The flushing of the coarse particles requires extreme storms, so few quantitative data have been collected.

In the example shown in Figure 6–10, an intense storm caused an abrupt rising limb of the spring hydrograph, which because of the rainfall distribution, had a long and unusually curved recession limb. The turbidograph was sharply peaked with sediment movement only during the rising limb and initial crest of the hydrograph. One interpretation is that the loose fine-grained sediment that had accumulated on conduit walls and floors was mobilized by the first flush of the storm flow. Once the reservoir of loose sediment had been depleted, the turbidity at the spring disappeared. In other instances, springs remains turbid because of the input of fresh sediment from surface sources. Mineralogical examination of the sediment reveals, as expected, very fine-grained quartz and clay minerals, although carbonate minerals were found in one spring (E. Herman et al., 2007).

Continuous records of the behavior of karst springs—hydrographs, chemographs, and turbidographs—are a powerful tool for the interpretation of karst aquifers. An example of how this information can be assembled is shown in Figure 6–11 from the work of Ryan and Meiman (1996). Big Spring is a karstic limestone spring in Mammoth Cave National Park. It is north of the Green River and not related to Mammoth Cave and the associated large karstic drainage system south of Green River. Big Spring has an elongate catchment that extends into agricultural lands north of the park boundary. The portion of the catchment within the park is

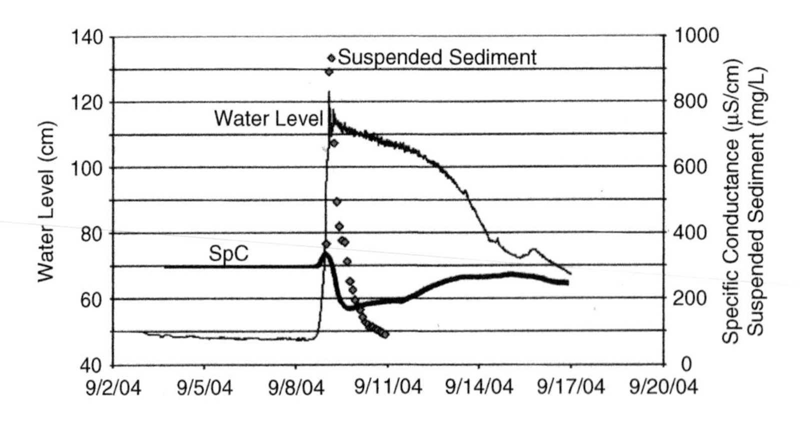

FIGURE 6–10 Hydrograph, chemograph, and turbidograph for Arch Spring, Blair County, Pennsylvania, in response to Hurricane Francis. (From E. Herman, Toran, and White, 2008.)

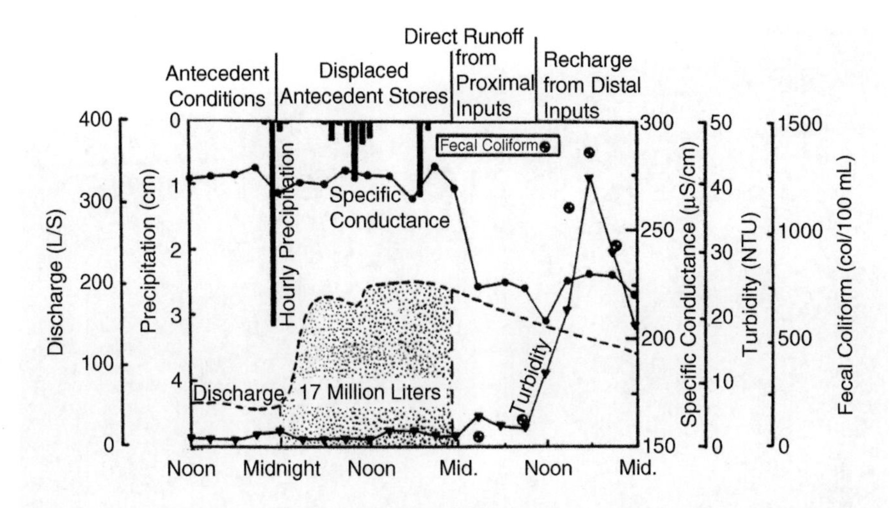

FIGURE 6–11 Storm response of Big Spring, Mammoth Cave National Park, Kentucky, showing arrival times for storm water and for sediment load. (From Ryan and Meiman, 1996.)

forested and undeveloped. The hydrograph of Big Spring rose very quickly following a large storm, but the chemograph remained constant and fell only about a day later than the rise of the hydrograph. The turbidograph remained flat and low for another 18 hours then a peak in turbidity occurred. The turbidity peak coincided with a rise in coliform bacteria in the springwater. The interpretation is that the discharge between the initial rise of the hydrograph and the drop in the chemograph represents water in storage in flooded conduits. Increased discharge began immediately following the storm because of a rising head in the upstream reaches of the system but only 24 hours later did the storm water itself arrive at the spring. A further 18 hours later, storm water from the upstream agricultural land, carrying both sediment and coliform bacteria, arrived at the spring. The latter interpretation was supported by a dye trace, which showed that the travel time from the upstream end of the system to the spring was the same as the travel time indicated by the turbidograph.

6.3.6 Travertine-depositing springs

Most carbonate springwater has CO_2 pressures well in excess of the CO_2 pressure of the atmosphere. As a result CO_2 is degassed at the spring mouth and along the spring run as the water flows away from the spring. The water approaches or exceeds saturation and the pH rises. Some spring mouths contain extensive deposits of travertine. Others do not. Tropical karst springs often have travertine deposits while temperate climate karst springs do not. In the case of thermal springs, cooling also initiates precipitation.

There is some nomenclatural confusion concerning these deposits. The term *travertine* is widely used in the United States for all freshwater carbonate deposits, including cave travertines, usually dense and coarsely crystalline; hot spring deposits, also dense and coarsely crystalline; and cold-water surface stream deposits, usually porous, fine grained, and with considerable incorporated plant and microbial material. Elsewhere, the last deposits are called *tufa*. Travertine and tufa deposits occur widely throughout the world (Herman and Hubbard, 1990; Ford and Pedley, 1996). Some of these deposits are very massive. An example is the Plitvice Lakes in Croatia, fed by karst springs and held back by travertine dams (Emeis, Richenow, and Kempe 1997; Figure 6–12). An example of a large carbonate travertine deposited from a hot spring is Mammoth Hot Springs in Yellowstone National Park, Wyoming (Barger, 1978; Figure 6–13).

FIGURE 6–12 Massive travertine deposits, Plitvice Lakes, Croatia. (Photo by the author.)

FIGURE 6–13 Travertine (tufa) deposit at Mammoth Hot Springs, Yellowstone National Park. (Photo by the author.)

If the deposition of travertine is considered a purely chemical process, the presence of travertines in some springs and its absence in others can be explained by the chemical pathway taken by the springwater as CO_2 is lost. Temperate climate springs, particularly those draining conduit systems, have CO_2 pressures from 10–20 times the atmospheric background and many such springs are undersaturated with respect to calcite. Tropical springs and springs fed by water derived from organic-rich soils have higher CO_2 pressures when the degassing process begins and may also be closer to equilibrium. When CO_2 is degassed with no calcite precipitation, the saturation index increases, while the hardness remains constant. A saturation index of about +0.5 or greater is required to drive the nucleation of calcite in the absence of catalysts or templates. As shown in Figure 6–14, typical temperate climate springs reach the CO_2 pressure of the atmosphere and thus come to a new equilibrium before calcite deposition begins. Hence, there is usually no travertine in these springs. Tropical springs and springs with high CO_2 loading cross the $SI_C = +0.5$ boundary, after which further CO_2 loss drives travertine deposition.

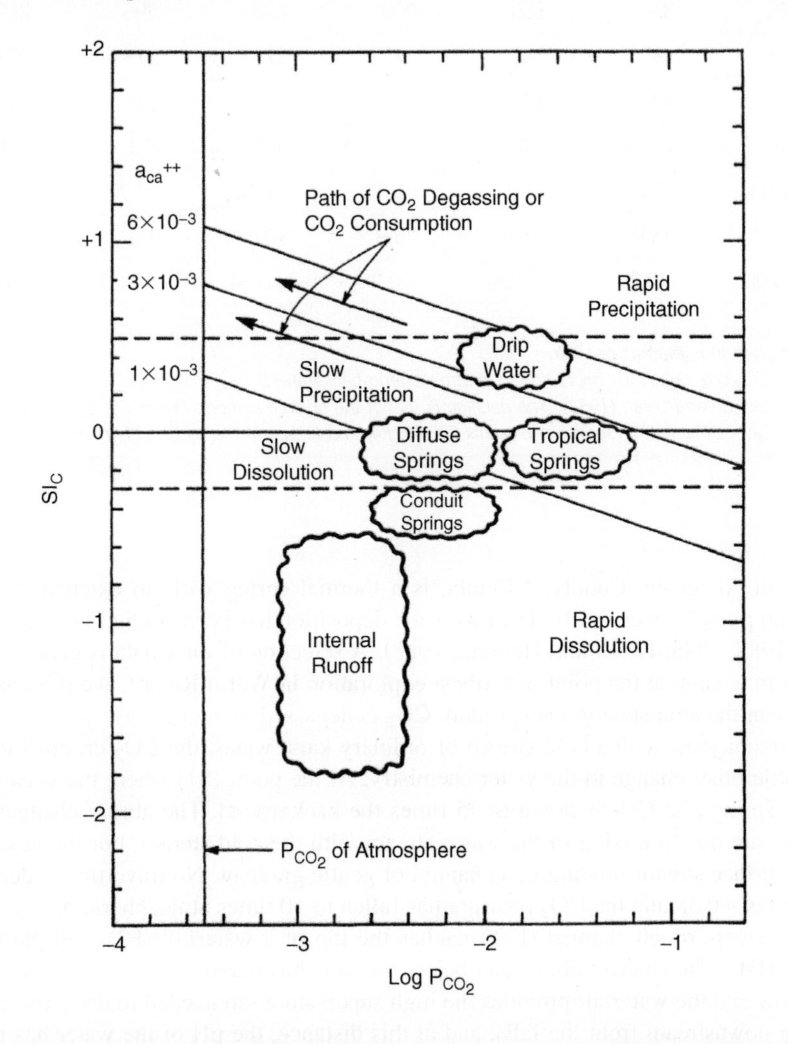

FIGURE 6–14 Evolutionary pathways for waters degassing carbon dioxide. (From W. White, 1997.)

Table 6–3 Chemical Composition of Falling Spring, Virginia, along the Flow Path

	C9	C7	C3	S1	F1	F3	F4	F2
T (°C)	35.0	32.0	27.0	25.0	23.8	23.0	21.5	21.0
pH	6.65	7.38	7.00	7.25	7.88	8.17	8.18	8.43
Ca^{2+}	232	217	204	164	165	163	163	151
Mg^{2+}	47.3	45.4	36.2	27.9	27.9	27.3	29.1	28.8
Na^+	6.9	6.8	6.1	4.4	5.3	4.4	4.5	5.1
K^+	26.2	26.1	19.7	13.7	15.5	15.1	15.1	15.3
HCO_3^-	392	387	359	306	298	296	269	246
SO_4^{2-}	431	435	327	279	275	277	277	277
F^-	1.6	1.6	1.2	1.0	1.0	1.0	1.0	1.0
Cl^-	4.2	4.2	3.4	3.3	3.3	3.3	3.3	3.3
Log P_{CO2}	−1.04	−1.80	−1.47	−1.80	−2.46	−2.77	−2.83	−3.14
Rel. CO_2	264	45.9	98.1	45.9	10.0	4.9	4.3	2.1
SI_C	+0.08	+0.73	+0.26	+0.35	+0.94	+1.20	+1.15	+1.30

Notes:
All concentrations are given in milligrams per liter.
Rel. CO_2 = (calculated CO_2 pressure)/(CO_2 pressure of the atmospheric background).
Selected data from J. Herman and Lorah (1987). The notation is that of the original authors. Points C9, C7, and C3 are in the warm stream in Warm River Cave; S1 is Falling Spring; and points F1, F3, F4, and F2 are in the surface channel.

Falling Spring in Alleghany County, Virginia, is a thermal spring with an extensive travertine deposit where the spring run plunges over a cliff. The travertine deposition has been studied in great detail (J. Herman and Lorah, 1986, 1987, 1988; Lorah and Herman, 1988). A selection of their data is given in Table 6–3. The water emerging from a sump at the point of farthest exploration in Warm River Cave (C9) has a CO_2 content 264 times higher than the atmospheric background. CO_2 is degassed along the cave passage. At the point C3, where the warm stream joins with a cold stream of ordinary karst water, the CO_2 dropped to 98 times background but with little other change in the water chemistry. At the point (S1) where the stream emerges to the surface as Falling Spring, the CO_2 is down to 45 times the background. The abrupt changes in composition between C3 and S1 are due to mixing of the warm stream with the cold stream. For the next 0.8 km, Falling Spring Creek is a surface stream flowing in a channel of gentle gradient. No travertine is deposited, although the saturation index is +0.35 and the CO_2 pressure has fallen to 10 times atmospheric background. At 0.8 km, the stream enters a steep, rough channel (F1), reaches the top of a waterfall (F3), and plunges 20 m to the bottom of the falls (F4). The channel above and below the falls has massive deposits of travertine. Degassing due to turbulent flow and the waterfall provides the high supersaturation needed to drive travertine deposition. Point F2 is 0.5 km downstream from the falls, and at this distance, the pH of the water has risen to 8.43 and the CO_2 pressure dropped to only twice the atmospheric background.

The processes are much more complicated in surface tufas, where microbial processes become important (Pentecost, 2005). *Cyanobacteria*, in particular, extract CO_2 from the water as part of their metabolism. This lowers the CO_2 pressure and drives the precipitation of travertine even in the absence of CO_2 degassing.

6.3.7 Contaminant transport in carbonate springs

Carbonate aquifers with their sinking stream and sinkhole inputs and rapid-throughput conduit systems are particularly susceptible to contamination. First, of course, there is little or no filtration. Contaminants are swept immediately into the groundwater. Second, contaminants in karst aquifers do not disperse. Instead of spreading as a slow-moving plume, as is typical for contaminant movement in porous media aquifers, the contaminants remain in the conduit system and move directly to the spring. Further, contaminants may be adsorbed onto clastic sediments, where they remain in storage for long periods of time until flushed by exceptional storms.

The transport and storage of contaminants in karst aquifers depend on the chemical nature of the contaminant. In broad-brush terms, contaminants may be classified as follows.

Water-soluble compounds

Water-soluble compounds are taken completely into solution at all concentrations likely to be found in natural environments. They include inorganic ions such as nitrate, nitrite, and ammonia; low molecular-weight polar organic compounds such as alcohols, carboxylic acids, and phenols; many agricultural chemicals; and pharmaceuticals.

The nitrate ion, NO_3^-, is a widespread contaminant in karst springwater. It is formed by the oxidation of ammonia released from animal and human waste, often with the nitrite ion, NO_2^-, as an intermediate species. Once formed, nitrate is a conservative ion, so that the concentration in springwater depends only on concentration at the source and dilution along the flow path. Many studies have been made of nitrate contamination, such as Katz, Chelette, and Pratt (2004) and Panno and Kelly (2004). The nitrate ion is found in most carbonate springwater; ammonia and nitrite are less common. Ammonia and nitrite easily oxidize to nitrate, so their occurrence in springwater is an indication of a nearby source of contamination.

Agricultural chemicals—fertilizers, insecticides, and herbicides—accumulate in the epikarst and are flushed into carbonate aquifers during storms. Pulses of agricultural chemicals appear at karst springs in response to storm flow (Mahler and Massei, 2007). Another class of organic compounds that have become of concern is pharmaceuticals. A significant fraction of the medication taken by the human population is passed through the body and into the waste disposal system. In karst areas, drainage from septic tanks and related sources carries these compounds into the groundwater and to karst springs. The natural estrogen, 17 β-estradiol, has been found in Maramec Spring, Missouri (Wicks, Kelley, and Peterson, 2004).

Light, nonaqueous phase liquids

Light, nonaqueous phase liquids (LNAPLs) are slightly soluble organic compounds that float on water. Gasoline, diesel fuel, home heating oil, and related petroleum hydrocarbons are the most common examples. LNAPLs float on underground streams and pond where the streams percolate through obstructions, such as piles of breakdown. Rising water levels may force the ponded LNAPL through the obstruction to produce a pulse of contaminant at the spring or it may force the LNAPL upward into fractures, where it may appear at the surface in basements as toxic or explosive fumes. LNAPLs tend to be trapped in pockets in the ceilings of conduits, so flushing the system may take a long time (Ewers et al., 1991).

Dense, nonaqueous phase liquids

Dense, nonaqueous phase liquids (DNAPLs) are slightly soluble organic compounds that are denser than water. Mostly, they are chlorinated (or brominated) compounds. They include such low weight, relatively volatile compounds as methylene chloride, CH_2Cl_2; trichloroethylene, C_2HCl_3 (TCE); and perchloroethylene, C_2Cl_4 (PCE), which are widely used as solvents, degreasers, and dry cleaning agents. These materials are transported in tanker car quantities and are often stored in underground tanks. A truck wreck, tanker car derailment, or leaking tank can inject these materials directly into karst aquifers. Other DNAPLs include the polychlorinated biphenyls (PCBs). PCBs are nonvolatile, oily liquids that were once used extensively in electrical transformers and have been injected into karst aquifers by salvage operations intended to recover copper from scrap transformers. For some reason, old limestone quarries have frequently been the site of such operations.

Depending on quantity, DNAPLs can accumulate in pools beneath the water in the beds of cave streams and also sink into the pores of clastic sediments, displacing water. DNAPL flow in karst aquifers may not follow the hydraulic gradient of the water but instead follow fractures and bedding plane partings along different pathways. DNAPLs in storage in cave sediments may remain sequestered until extreme storms flush the sediment pile. Overall, the possible transport of DNAPLs through a karst aquifer can follow many pathways and take a very long time to reach the discharging spring (Loop and White, 2001).

Metals

About two thirds of the elements on the periodic table are metals, but most of these do not create environmental problems, because they are rare in nature and seldom used in commercial products. Two metals, iron and manganese, make up most of the natural background. These metals occur widely in sedimentary rocks, and their oxides and hydrated oxides are common in cave deposits. Nickel and chromium appear in waste from chrome-plating and other nonferrous metals industry activities. Zinc and the chemically similar but more toxic cadmium occur widely as "galvanized" coatings on utensils, building materials, and other objects likely to end up in trash dumps. Lead, arsenic, vanadium, and other metals occur in commercial products that, when discarded as waste materials, can be leached into groundwater systems.

Most metals form highly insoluble oxides, hydroxides, and carbonates, which limit their solubility in groundwater. However, solubility limits are often higher than concentrations specified as drinking water standards. Metals can be adsorbed onto clays and other particulates and carried through karst aquifers as suspended particles during storm flow. Evidence for this mechanism is provided by a series of chemographs showing peaks in various metal concentrations matching the peak in aluminum concentration, aluminum in the analysis of unfiltered samples being a proxy for suspended clay minerals (Figure 6–15).

Pathogens

Viruses, bacteria, protozoa, and larger organisms are easily transported into karst aquifers, because of the absence of filtering from the soil. The most widespread of these are fecal coliform and fecal streptococci bacteria. The presence of these organisms is the most common indicator of pollution from sewage or animal waste. Of most concern among protozoa is *Giardia lamblia*, which is released in a cyst form in animal feces and is present in many surface waters. Sinking streams carry the stable cysts to the subsurface, where they reappear in karst springs. Pathogens move freely through karst aquifers, either as free organisms or, more likely, adsorbed onto particles of clastic sediment (Pronk et al., 2008).

Trash

Rural residents and even entire communities have, from long tradition, used sinkholes as waste disposal sites. Farmers routinely use sinkholes to dispose of dead animals and empty containers of their agricultural chemicals. Unique among aquifers, sinking streams and sinkhole drains provide routes along which bulk trash can be carried for long distances inside the aquifer. The deposition of trash as "clastic sediment" provides a source term for the leaching and release of contaminants over long periods of time. Transported trash is also located where it is difficult or impossible to clean up.

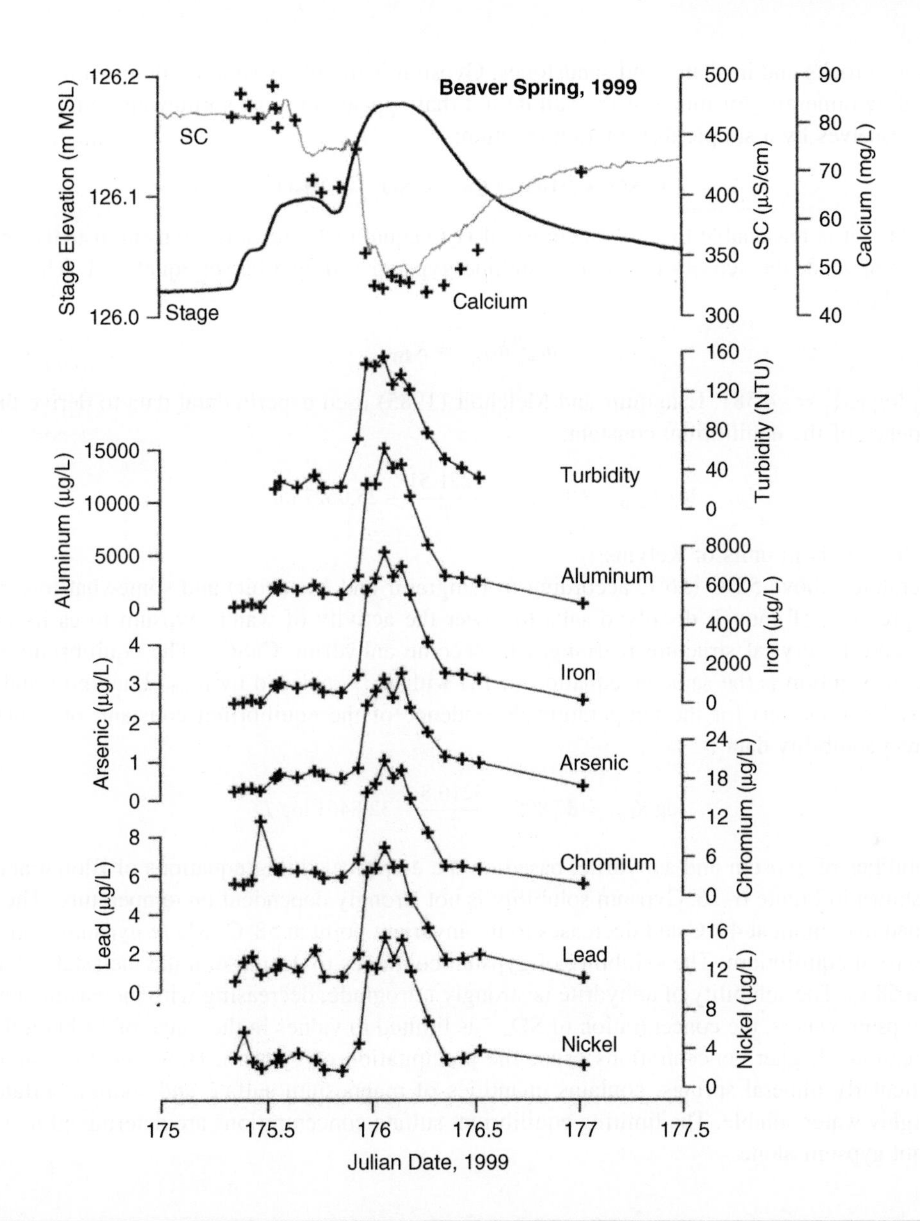

FIGURE 6–15 Storm chemograph for various heavy metals in comparison with the storm hydrograph. Beaver Spring, Fort Campbell Army Base, western Kentucky. (From Vesper and White, 2003.)

6.4 GYPSUM SPRINGS

Karstic drainage systems are most commonly found in carbonate rocks, but karst also develops in gypsum. Although gypsum karst is found mainly in arid and semiarid regions, gypsum karst occurs widely throughout the world (Klimchouk et al., 1996). With respect to controls on springwater chemistry, gypsum has a much wider influence and is not restricted to regions of gypsum karst. Gypsum occurs widely as minor interbeds within carbonate rocks and in shales and sandstones. Gypsum is the likely source (along with oxidized pyrite and other sulfide minerals) for much of the sulfate ion that appears in most springwater analyses.

Gypsum dissolves by a simple dissociation reaction:

$$CaSO_4 \bullet 2H_2O \leftrightarrow Ca^{2+} + SO_4^{2-} + 2H_2O$$

In freshwater, it is reasonable to set the activity of H_2O equal to 1, and since gypsum usually appears as a nearly pure compound, the activity of solid crystalline gypsum can also be set equal to 1. The equilibrium expression is then

$$a_{Ca^{2+}} a_{SO_4^{2-}} = K_{gyp} \tag{6.18}$$

At 25°C, log K_{gyp} = −4.581. Langmuir and Melchoir (1985) used experimental data to derive the temperature dependence of the equilibrium constant:

$$\log K_{gyp} = 68.2451 - \frac{3221.51}{T} - 25.0627 \log T \tag{6.19}$$

The temperature, T, is in units of Kelvins.

At temperatures above 58°C (56°C according to Langmuir and Melchoir) and somewhat lower temperatures in the presence of enough dissolved salts to lower the activity of water, gypsum loses its two waters of hydration and the crystal structure rearranges to become anhydrite, $CaSO_4$. The equilibrium expression for anhydrite dissolution is the same as equation (6.18) with K_{gyp} replaced by K_{anh}. Langmuir and Melchoir derived several expressions for the temperature dependence of the equilibrium constant, of which the one based on direct solubility data is

$$\log K_{anh} = 87.805 - \frac{3210.8}{T} - 32.8461 \log T \tag{6.20}$$

The solubilities of gypsum and anhydrite, based on the empirical fitting equations of Blount and Dickson (1973), are shown in Figure 6–16. Gypsum solubility is not strongly dependent on temperature. The solubility rises to a broad maximum at 40°C and decreases to the invariant point at 58°C, where gypsum, anhydrite, and solution coexist at equilibrium. The solubility of gypsum continues to decrease in the metastable temperature region above 58°C. The solubility of anhydrite is strongly retrograde, decreasing with increasing temperature.

In pure gypsum waters, the concentration of SO_4^{2-} is limited to values in the range of 1400 mg/L, depending on temperature. Higher concentrations cause the precipitation of gypsum. However, the water of many springs, particularly mineral springs, contains quantities of magnesium sulfate and sodium sulfate, both of which are highly water soluble. The limiting equilibrium sulfate concentrations are determined by total water chemistry, not gypsum alone.

6.5 MINERAL SPRINGS AND THERMAL SPRINGS

Mineral springs have a high concentration of dissolved salts. Many of these are also thermal springs, in that the temperature of the springwater is higher than the mean annual temperature of the region. Thermal spring temperatures vary from a few degrees above ambient to the boiling point of water (or above the boiling point of water in the case of some volcanic springs). Spas and health resorts have often been built around thermal and mineral springs because of the supposed curative properties of the waters (Back et al., 1995; LaMoreaux and Tanner, 2001). More than a thousand thermal springs have been cataloged in the United States and thousands more in other parts of the world (Waring, 1965). The distribution of thermal springs (Figure 6–17) shows that

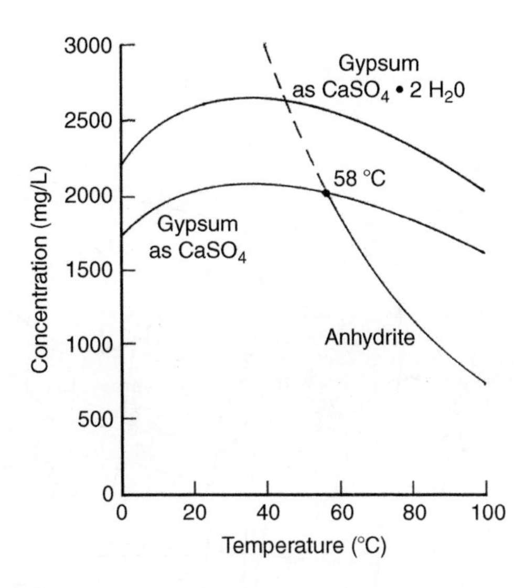

FIGURE 6–16 Solubility of gypsum and anhydrite in the system $CaSO_4$-H_2O. Solubility curves calculated with the empirical equations of Blount and Dickson (1973).

most are limited to the western portion of the United States. There is a concentration of thermal springs along the southern Rocky Mountains and a rather curious concentration of thermal springs along the southern Appalachians. The latter are mainly in sedimentary rocks in the border region between the folded Appalachians and the Appalachian plateaus.

Table 6–4 lists analyses for a small selection of mineral springs. These springs are (or have been) the sites of resorts, spas, and have been used as medicinal water. Warm water is desirable for spas. The medicinally useful component is the high sulfate concentration, along with the magnesium concentration. The Bedford Spring has one of the highest magnesium concentrations of the mineral springs. Bedford Spring issues from limestone at ambient temperature and, except for the high content of dissolved solids, is not intrinsically different from the many karstic springs in the folded Appalachians.

The heat source for the Appalachian thermal springs has been the subject of considerable investigation because there is no known volcanic source. According to Hobba et al. (1979), the Appalachian thermal springs are the result of geothermal heating. Boiling Spring (Table 6–4) has the highest temperature and according to calculations requires a minimum depth of groundwater circulation of 1600 m. The springs also require flow paths to be sufficiently open so that water can rise from these depths without cooling due to contact with rock near the surface.

6.5.1 Sulfur springs

Sulfur springs discharge reduced sulfur. The waters characteristically have the pronounced "rotten egg" odor of H_2S. The source of the H_2S is not known in most cases but is thought to be derived from nearby petroleum reservoirs. When H_2S from depth reaches the oxidizing meteoric waters near the surface, oxidation produces sulfuric acid and sometimes native sulfur as an intermediate step. Microbial processes are important in the oxidation reactions; indeed, the microbiology of sulfur caves and springs is a rich field for research (Engel, 2007). Waters from sulfur springs have been prized as medicinal waters.

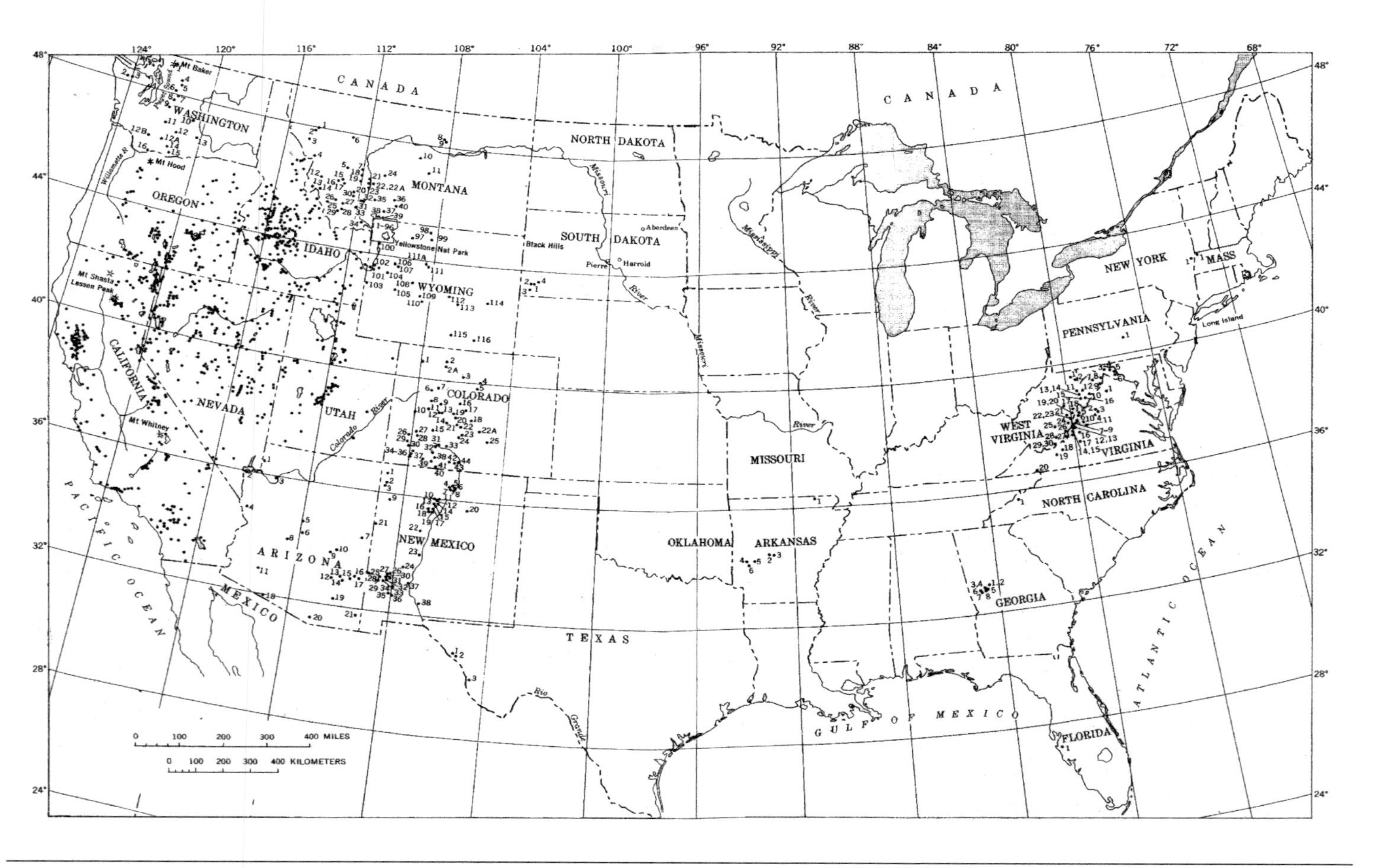

FIGURE 6–17 Distribution of thermal springs in the United States. (From Waring, 1965.)

Table 6–4 Some Typical Analyses of Mineral Springs

	Telese	DBT	BC	Old Spa	New Spa	Bedford	Boiler
T (°C)	20.1	31.0	53.3	40	40	10	39.9
pH	6.23	7.30	6.53	7.40	7.46	—	6.65
SiO_2	—	40.3	54.9	—	—	—	21
Ca^{2+}	423	162.5	488	68.13	96.7	558	132
Mg^{2+}	68	25.4	67.9	31.6	16	146	40
Na^+	100	133	325	—	—	<5	7.0
K^+	16	12.6	24.7	—	—	<5	13
HCO_3^-	1617	195	107	189	164	183	454
SO_4^{2-}	35.8	532	1765	144	140	1764	130
Cl^-	152	68	244	56.7	78	6.0	2.6

Notes:
All concentrations in units of milligrams per liter.
Telese Spring, southern Apennines, Italy, mean of set of analyses of waters collected at different times. Data from Paternoster and Mongelli (2000).
DBT = Trastullina Bassa Spring; BC = Doccione Spring, both part of the Bagni di Lucca medicinal spring group, Tuscany, Italy. Data from Boschetti et al. (2005).
Old Spa (Balneario Viejo) and New Spa (Balneario Nuevo) at the Alhama de Granada baths, Granada, Spain. Data from Torija Isasa et al. (2002).
Bedford Spring at Bedford Springs Resort, Bedford County, Pennsylvania. Data from Lohman (1938).
Boiler Spring is the highest temperature spring of the thermal spring group at Hot Springs, Virginia. Data from Hobba et al. (1977).

Springs with particularly active bacterially induced sulfur chemistry include the Frasassi Caves in Italy (Macalady et al., 2006), the Kane Caves in Wyoming (Engel, Stern, and Bennett, 2004), and Cueva de Villa Luz, Tabasco, Mexico (Hose and Pisarowicz, 1999; Hose et al., 2000). All these caves are in limestone. A chemical analysis of the waters from Cueva de Villa Luz (Table 6–5) reveals few surprises. The temperature is only slightly above ambient. The spring discharges a mixed bicarbonate-sulfate-chloride water that is supersaturated with respect to calcite and undersaturated with respect to gypsum. The only anomaly is the concentration of dissolved H_2S in some of the infeeders. The pH of the infeeders and the spring is close to neutral. However, some waters dripping from bacterial mats have pH values as low as 0.

Although active sulfur springs are relatively rare, relict solution openings are common. It is becoming recognized that some of the largest caves in the country, for example, Carlsbad Caverns and Lechuguilla Cave, were excavated by sulfuric acid generated by oxidation of H_2S (Hill, 1994). These belong to the set of hypogene caves, now widely recognized (Klimchouk, 2007).

6.5.2 Brine and brackish springs

Sodium chloride and most other chloride salts have high solubilities. NaCl has a solubility limit of about 35.5 weight percent depending on temperature. Many springs have significant chloride and alkali metal concentrations as demonstrated by the analyses given in the tables. When concentrations reach values comparable to seawater (3.5 weight percent or higher), they are referred to as *brines*. Springs with intermediate

Table 6–5 Example of a Sulfur Spring: Waters from Cueva Villa Luz, Tobasco, Mexico

	1	2	3	4	5	6	7	8
T (°C)	27.5	28.3	28.1	28.1	28.1	28.1	28.0	28.0
pH	6.61	7.23	7.01	7.14	7.00	7.16	7.14	6.31
H_2S	500	<0.1	150	375	100	25	17.5	750
SiO_2	5	6	6	7	7	8	7	15
Ca^{2+}	396	383	394	417	424	407	393	285
Mg^{2+}	81	97	80	69	61	78	88	65
Na^+	484	477	—	—	—	—	—	—
HCO_3^-	498	451	479	475	494	506	477	636
SO_4^{2-}	940	980	980	940	920	940	910	520
Cl^-	814	792	767	782	807	792	803	735
P_{CO2} atm	0.105	0.023	0.041	0.030	0.043	0.030	0.030	0.292
SI_{cal}	+0.13	+0.71	+0.52	+0.67	+0.56	+0.71	+0.65	−0.14
SI_{gyp}	−0.36	−0.35	−0.34	−0.33	−0.34	−0.35	−0.37	−0.54

Notes:
All concentrations in milligrams per liter.
Samples 1–6 are various internal "springs" of water draining into the cave. Sample 7 is the actual spring where the combined water exits to the surface. Sample 8 is the El Azufre Spring at Teapa.
Data compiled from Hose et al. (2000).

concentrations but with enough NaCl to produce a salty taste are "brackish." The following mechanisms may produce springs with high salt content:

1. Springs containing salt derived from evaporite beds within the spring drainage basin.
2. Springs contaminated by anthropogenic sources, particularly salt used for highway deicing.
3. Coastal springs affected by seawater intrusion.
4. Springs discharging oil-field brines.

Carbonate aquifers in coastal regions developed conduit systems that extend below present day sea level during Pleistocene low sea stands. Rising sea levels flooded some of these conduits, giving rise to off-shore springs. Other springs lie some distance inland but may discharge brackish water. A review of these types of springs is given by Fleury, Bakalowicz, and de Marsily (2007). Increased discharge of freshwater during storm flow may dilute the brackish water to acceptable levels (Figure 6–18). Water quality in coastal brackish springs is thus a matter of mixing and not chemistry. The process is amenable to modeling, and several models have been constructed (e.g., Lambrakis et al., 2000; Arfib and de Marsily, 2004; Maramathas, Pergialiotis, and Gialamas, 2006).

Table 6–6 lists the compositions of the waters from five springs that appear to be derived from oil-field brines. The first three of these are chloride-bicarbonate waters with high concentrations of both sodium chloride and sodium bicarbonate. Mercey Spring is only slightly brackish and, except for being a thermal spring, is not greatly different from many other springs. Stinking Spring has a concentration of NaCl of 3.4 weight percent, bringing it very close to the composition of seawater. Except for Mercey Spring, the others have high

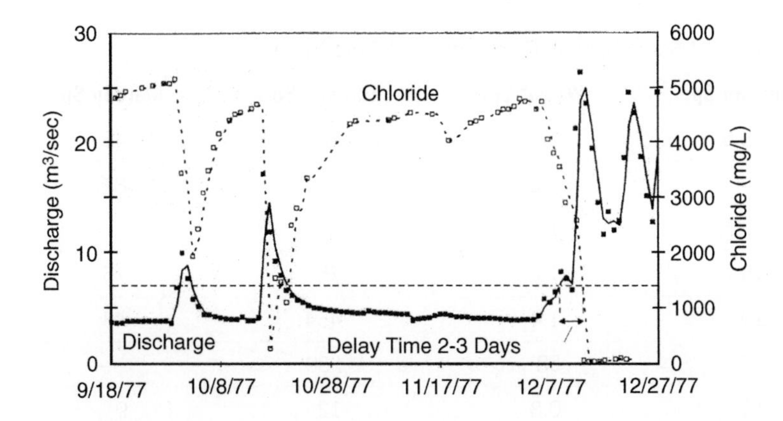

FIGURE 6–18 Hydrograph and chemograph for the Almyros Spring in Crete showing the dilution of brackish water by storm flow. (Adapted from Lambrakis et al., 2000.)

concentrations of hydrogen sulfide, suggesting the same petroleum environment as described in the previous section for sulfur springs.

6.5.3 Carbonated springs from deep sources

Unusual among the mineral springs are those that are naturally carbonated. Analytical data for a set of the springs at Saratoga, New York, and a set of the springs at Manitou Springs, Colorado, are given in Table 6–7. The two sets of springs have in common their high pressures of carbon dioxide, much above the surface ambient. Both are bicarbonate waters. The Saratoga springs are also brine springs with a high sodium and chloride content. The sulfate concentration in the Saratoga springs is extremely low. Manitou Springs are mainly bicarbonate springs with a nominal sulfate concentration.

The source of the high concentration of carbon dioxide has been the subject of considerable discussion. Both the concentration and carbon isotope consideration rule out atmospheric and organic sources. Both sets of springs are located on major fault systems, suggesting a deep source, lower crust or upper mantle, for the CO_2. Much of the bicarbonate content could then result from reaction of CO_2 with carbonate rocks along the flow path. The temperatures of the springs are at most only slightly above ambient. If there is a deep source, the migration of water (or gases) up the faults must be slow enough for the water to come into thermal equilibrium. There is little evidence for heating along the geothermal gradient.

6.5.4 Water chemistry at high temperatures

All the equilibrium constants for any water-rock interactions in spring feeder systems are temperature dependent. Further, some minerals undergo phase changes or dehydration reactions as the temperature is raised, as was the case of gypsum. Mostly, however, the effect of increasing temperature is to change the mix of aqueous species that result from equilibrium reactions. Many minerals become more soluble at high temperatures, but because CO_2 is less soluble at high temperature, the solubilities of calcite and dolomite decrease at the same CO_2 pressure.

In contrast to equilibrium considerations, the rates of reaction usually increase dramatically at higher temperature. Reaction rates usually follow the Arrhenius equation:

$$\text{Rate} = Ae^{-E_a/RT} \tag{6.21}$$

In this equation, A is a constant, E_a is the activation energy for reaction, and R is the gas constant. The temperature, T, is in units of Kelvins. Rates increase exponentially with temperature, which explains why hot

Table 6–6 Compositions of Some Springs Discharging Oil-Field Brines

	Tuscan Sp.	Wilbur Sp.	Tolenas Sp.	Mercey Sp.	Stinking Sp.
T (°C)	28.5	57	20	46	48
pH	8.4	7.2	6.7	8.6	6.7
H_2S	172	178	0	—	60
SiO_2	15	190	75	75	48
Ca^{2+}	19	1.4	454	43	946
Mg^{2+}	17	58	239	—	297
Sr^{2+}	13	0.8	12	9	31
Ba^{2+}	72	1	1.3	0	4.1
Na^+	7900	9146	6100	830	12,600
K^+	59	460	180	7.1	571
Li^+	2.0	14	9.0	0.1	6.9
HCO_3^-	1060	7390	6340	13	324
SO_4^{2-}	0	23	0.3	5	111
F^-	5	1.1	2	0.4	1.9
Cl^-	11,800	11,000	7510	1300	21,600
Br^-	5.3	15	20	0	15
I^-	1.3	16	3	20	1.3
Mn	0.3	0.3	—	—	0.00
Fe	0.2	0.1	0.1	—	0.03

Notes:
All concentrations in milligrams per liter.
Data extracted from Table 15 of D. White et al. (1963).
Tuscan Spring, Tehama County, California.
Wilbur Spring, Colusa County, California.
Tolenas Spring, near ridge crest, 6 km north of Fairfield, Solano County, California.
Mercey Spring, Fresno County, California.
Stinking Spring, immediately north of Great Salt Lake, Box Elder County, Utah.

springwater generally shows much more evidence of a reaction between the water and the host rock than cold springwater.

The most dramatic result of high temperature on springwater chemistry is on the solubility of silica. According to Blatt, Middleton, and Murray (1980), the temperature dependence of the solubility of quartz is described by

$$\log C_{qtz} = 4.83 - \frac{1132}{T} \tag{6.22}$$

Table 6–7 Chemistry of Some Deep-Seated Carbonate-Rich Springs

	Hathorn	Orenda	Big Red	Peerless	Shoshone	Cheyenne	Soda	Navajo
T (°C)	*	*	*	*	14.0	14.2	15.3	15.3
pH	6.0	5.9	6.0	5.9	6.50	6.18	6.32	6.30
P_{CO2} atm	4.3	3.25	2.65	2.0				
CO_2 (aq)					2.20	2.38	2.43	—
SiO_2	12	10	55	10	51	47	48	47
Ca^{2+}	950	700	340	420	500.9	527.3	474.3	518.6
Mg^{2+}	410	270	260	100	81.3	94.0	84.8	89.5
Sr^{2+}	17	13	12	4.2	—	—	—	—
Ba^{2+}	21	15	6.3	3.1	—	—	—	—
Na^+	4400	2420	1500	800	490.8	529.2	459.3	475.0
K^+	320	240	100	93	75.1	78.9	67.9	70.6
Li^+	10	7.8	4.8	2.4	0.76	0.88	0.66	0.71
HCO_3^-	5000	3680	3050	2350	2681	2562	2495	2508
SO_4^{2-}	4	20	7	11	212.0	197.0	196.0	187.0
F^-	0.45	0.78	0.20	0.7	7.7	7.0	6.8	6.6
Cl^-	6900	3800	2300	1000	242.0	229.0	220.0	211.0
Br^-	110	61	40	20	1.03	0.92	1.10	0.88
I^-	3.6	2.0	0.7	0.4	—	—	—	—
Mn	—	—	—	—	3.80	2.50	2.40	2.80
Fe	2.2	2.7	11	1.9	0.04	0.05	0.23	0.02

Notes:
All concentrations in milligrams per liter except CO_2 (aq), which is in units of grams per liter.
Hathorn, Orenda, Big Red, and Peerless Springs are in Saratoga, New York. Data from Putnam and Young (1985). Shoshone, Cheyenne, Soda, and Navajo Springs are in Manitou Springs, Colorado. Data from Luiszer (1994).
**Temperatures are in the range of 9–13° C.*

The concentration, C, is in units of milligrams per liter, and the temperature, T, is in Kelvins. The solubility of amorphous silica varies linearly with temperature:

$$C_{amor} = 3.82T - 1000 \quad (6.23)$$

In general, the concentration of dissolved silica increases with increasing temperature but with a considerable scatter (Figure 6–19). The solubility curves for quartz and amorphous silica form lower and upper bounds, although two of the geyser waters fall above the top of the diagram.

The increased solubility, particularly of silica, at high temperatures means that most hot springs deposit extensive quantities of mineral matter (Figure 6–20) when the water cools. Many of these deposits consist of reprecipitated silica although some contain carbonates. Iron and manganese oxides also occur.

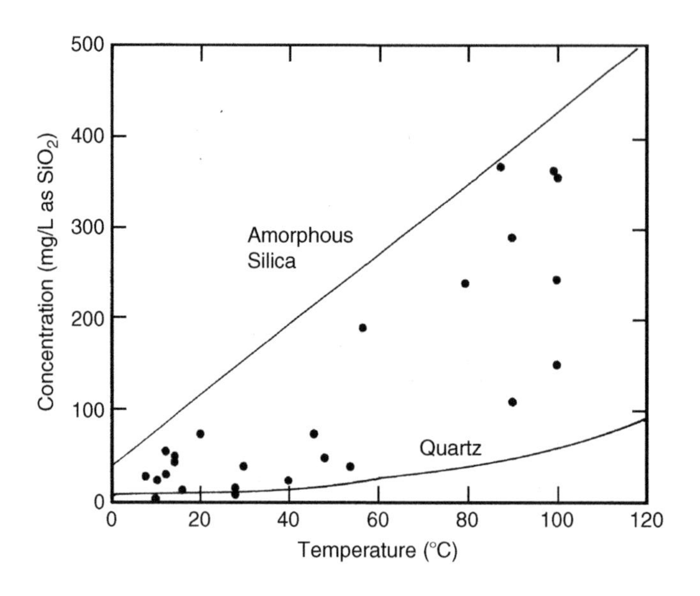

FIGURE 6–19 Concentrations of SiO_2 taken from Tables 6–1, 6–4, and 6–5 through 6–9 as a function of temperature. The solubility curves are plots of equations (6.22) and (6.23).

FIGURE 6–20 A hot spring deposit north of Durango, Colorado. (Photo by the author.)

6.5.5 Volcanic hot springs

Meteoric waters in active volcanic areas infiltrating through the deep porous zones can be superheated and ejected through orifices to form geysers (Figure 6–21). The chemical compositions of some springs in geyser areas are shown in Table 6–8. The temperatures are near the atmospheric boiling point of water, as expected for geyser areas. The silica concentration is high, but the concentrations of other dissolved species are

FIGURE 6–21 The rising burst of water from a geyser. Near Geysir, Iceland. (Photo by the author.)

Table 6–8 Compositions of Some Geyser-Field Springs

	Upper Basin	Norris	Steamboat	Geyser	Haukadular	Hveravellir
T (°C)	94	84.5	89.2	101.5	100	90.5
pH	9.6	7.5	7.9	6.9	9.7	8.7
H_2S	2.6	0.0	4.7	—	2.7	2.5
SiO_2	363	529	293	150	359	609
Ca^{2+}	0.8	5.8	5.0	40	0.4	2.0
Mg^{2+}	0.0	0.2	0.8	0.2	0.5	0.5
Sr^{2+}	0.0	0.4	0.5	0.3	0.02	0.003
Na^+	352	439	653	350	233	156
K^+	24	74	71	18	11	15
Li^+	5.2	8.4	7.6	2	0.2	0.2
HCO_3^-	0	27	305	29	0.0	112
SO_4^{2-}	23	38	100	130	102	178
F^-	25	4.9	1.8	1.2	12	3.3
Cl^-	405	744	865	482	126	63
Br^-	1.5	0.1	0.2	—	0.2	0.0
I^-	0.3	0.0	0.1	—	0.0	0.0
Al	0.2	—	0.5	—	0.89	0.55
Mn	0.00	—	0.05	0.01	0.00	0.00
Fe	0.06	—	0.05	0.1	0.02	0.21

Notes:
All concentrations in milligrams per liter.
Data extracted from Table 17 of D. White et al. (1963).
Upper Basin: Spring located in the Upper Geyser Basin, Yellowstone National Park, Wyoming.
Norris: Small spring located in the Norris Basin, Yellowstone National Park, Wyoming.
Steamboat: Spring near the east edge of the main terrace, Steamboat Springs 16 km southeast of Reno, Nevada.
Geyser: Located 7 km southeast of Geyser Bight, Umnak Island, Alaska.
Haukadalur: Sisjodandi Spring near Geysir, northeast of Reykjavik, Iceland.
Hveravellir: Blahver Spring in Hveravellir thermal area, west-central Iceland.

relatively low, suggesting only a short contact time with the hot rock. The pH range is from near neutral to slightly alkaline.

Other waters in active volcanic areas react strongly with the wall rock and discharge highly acid water (Table 6–9). Such waters are responsible for boiling "mud pots" and similar features. The chemistry is extremely variable. The waters tend to be reducing, some contain H_2S, so that high concentrations of iron appear. Ferric iron is essentially insoluble in near-neutral water but becomes soluble at the low pH values

Table 6–9 Compositions of Some Acid Volcanic Springs

	Green Dragon	White Island	Devils Kitchen	Bumpass Hell	Locomotive
T (°C)	87	"Hot"	Boiling	79.0	90
pH	2.47	"Acid"	1.8	—	1.97
NH_4^+	3.4	17	1400	14	30
SiO_2	369	164	225	240	109
Ca^{2+}	6.5	2370	47	6.5	2.2
Mg^{2+}	0.0	6770	281	5.3	0
Na^+	243	7100	12	32	2
K^+	61	926	5	13	3
HCO_3^-	0	—	0	—	0
SO_4^{2-}	454	2170	5710	718	758
Cl^-	408	57,300	0.5	1.1	15
Al	1.5	1880	14	31	2.4
Mn	—	24	1.4	—	—
Fe^{2+}	—	10,500	63	18	—
Fe^{3+}	0.8	130	0	5.5	0.8

Notes:
All concentrations in milligrams per liter.
Data drawn from Tables 19 and 20 of D. White et al. (1963).
Green Dragon Spring, southern part of Norris Basin, Yellowstone National Park, Wyoming.
Hot spring pool in crater of White Island, Bay of Plenty, northeast coast of North Island, New Zealand.
Devils Kitchen Spring, "The Geysers," Sonoma County, California.
Spring in thermal area, Bumpass Hell, Shasta County, California.
Locomotive Spring, Norris Basin, Yellowstone National Park, California.

found in these springs. Reducing conditions produce ferrous iron, which is more soluble than ferric iron. Curiously, the ammonium ion, NH_4^+, also appears.

6.6 CONCLUSIONS

Springs are found in many geologic environments but their chemistry is surprisingly similar. The same sets of cations and anions are found in most springwater. What distinguishes one springwater from another is

1. The relative concentrations of the dominant anions, HCO_3^-, SO_4^{2-}, and Cl^-.
2. The relative mix of alkali ions, Na^+ and K^+, and alkaline earth ions, Ca^{2+} and Mg^{2+}.
3. The total concentration of dissolved species.
4. Temperature.
5. The presence of dissolved gases, especially H_2S and excessive CO_2.

ACKNOWLEDGMENTS

I thank Elizabeth L. White for her assistance with the illustrations. James J. Van Gundy and William K. Jones are thanked for providing some needed reference material. Joe Kearns provided the photograph of Mammoth Spring.

REFERENCES

Arfib, B., de Marsily, G., 2004. Modeling the salinity of an inland coastal brackish karstic spring with a conduit-matrix model. Water Resour. Res. 40, W11506 doi:10.1029/2004WR003147.

Back, W., Landa, E.R., Meeks, L., 1995. Bottled water, spas, and early years of water chemistry. Ground Water 33, 605–615.

Barger, K.E., 1978. Geology and thermal history of Mammoth Hot Springs. Yellowstone National Park, Wyoming U.S. Geological Survey Bulletin 1444.

Blatt, H., Middleton, G., Murray, R., 1980. Origin of Sedimentary Rocks. Prentice-Hall, Englewood Cliffs, NJ.

Blount, C.W., Dickson, F.W., 1973. Gypsum-anhydrite equilibria in systems $CaSO_4$-H_2O and $CaSO_4$-NaCl-H_2O. Am. Mineral. 58, 323–331.

Boschetti, T., Venturelli, G., Toscani, L., Barbieri, M., Mucchino, C., 2005. The Bagni di Lucca thermal waters (Tuscany, Italy): An example of Ca-SO_4 waters with high Na/Cl and low Ca/SO_4 ratios. J. Hydrol. 307, 270–293.

Bosch, R.F., White, W.B., 2004. Lithofacies and transport of clastic sediments in karst aquifers. In: Sasowsky, I.D., Mylroie, J. (Eds.), Studies of Cave Sediments. Kluwer Academic, New York, pp. 1–22.

Busenberg, E., Plummer, L.N., 1982. The kinetics of dissolution of dolomite in CO_2–H_2O systems at 1.5 to 65°C and 0 to 1 atm P_{CO2}. Am. J. Sci. 282, 45–78.

Drever, J.I., 1997. The Geochemistry of Natural Waters, third ed. Prentice Hall, Upper Saddle River, NJ.

Dreybrodt, W., Gabrovšek, F., Romanov, D., 2005. Processes of speleogenesis: A modeling approach. Carsologica 4, ZRC Publishing, Karst Research Institute at ZRC SAZU, Postojna, Slovenia.

Emeis, K.C., Richnow, H.H., Kempe, S., 1987. Travertine formation in Plitvice National Park, Yugoslavia: Chemical versus biological control. Sedimentology 34, 595–609.

Engel, A.S., 2007. Observations on the biodiversity of sulfidic karst habitats. Journal of Cave and Karst Studies 69, 187–206.

Engel, A.S., Stern, L.A., Bennett, P.C., 2004. Microbial contributions to cave formation: New insights into sulfuric acid speleogenesis. Geology 32, 369–372.

Ewers, R.O., Duda, A.J., Estes, E.K., Idstein, P.J., Johnson, K.M., 1991. The transmission of light hydrocarbon contaminants. In: Proceedings of the Third Conference on Hydrogeology, Ecology, Monitoring and Management of Ground Water in Karst Terranes. National Ground Water Association, Nashville, TN, pp. 287–305.

Fleury, P., Bakalowicz, M., de Marsily, G., 2007. Submarine springs and coastal karst aquifers: A review. J. Hydrol. 339, 79–92.

Ford, T.D., Pedley, H.M., 1996. A review of tufa and travertine deposits of the world. Earth-Sci. Rev. 41, 117–175.

Hem, J.D., 1985. Study and Interpretation of the Chemical Characteristics of Natural Waters. U.S. Geological Survey Water-Supply Paper 2254.

Herman, E.K., Tancredi, J.H., Toran, L., White, W.B., 2007. Mineralogy of suspended sediment in three karst springs. Hydrogeology Journal 15, 255–266.

Herman, E.K., Toran, L., White, W.B., 2008. Threshold events in spring discharge: Evidence from sediment and continuous water level measurement. J. Hydrol. 351, 98–106.

Herman, J.S., Lorah, M.M., 1986. Groundwater geochemistry in Warm River Cave, Virginia. National Speleological Society Bulletin 48, 54–61.

Herman, J.S., Lorah, M.M., 1987. CO_2 outgassing and calcite precipitation in Falling Spring Creek, Virginia, U.S.A. Chem. Geol. 62, 251–262.

Herman, J.S., Lorah, M.M., 1988. Calcite precipitation rates in the field: Measurement and prediction for a travertine-depositing stream. Geochim. Cosmochim. Acta 52, 2347–2355.

Herman, J.S., White, W.B., 1985. Dissolution kinetics of dolomite: Effects of lithology and fluid flow velocity. Geochim. Cosmochim. Acta 49, 2017–2026.

Herman, J.S., Hubbard Jr., D.A., 1990. Travertine-marl: Stream deposits in Virginia. Virginia Division of Mineral Resources Publication 101, Charlottesville, VA.

Hess, J.W., 1974. Hydrochemical investigations of the Central Kentucky Karst aquifer system. Ph.D. thesis, The Pennsylvania State University.

Hill, C.A., 1994. Sulfuric acid speleogenesis of Carlsbad Cavern and its relationship to hydrocarbons, Delaware Basin, New Mexico and Texas. AAPG Bull. 76, 1685–1694.

Hobba, W.A., Chemerys, J.C., Fisher, D.W., Pearson Jr., F.J., 1977. Geochemical and hydrologic data for wells and springs in thermal-spring areas of the Appalachians. U.S. Geological Survey Water-Resources Investigations 77–25.

Hobba, W.A., Fisher, D.W., Pearson Jr., F.J., Chemerys, J.C., 1979. Hydrology and geochemistry of thermal springs of the Appalachians. U.S. Geological Survey Professional Paper 1044-E.

Hose, L.D., Pisarowicz, J.A., 1999. Cueva de Villa Luz, Tabasco, Mexico: Reconnaissance study of an active sulfur spring cave and ecosystem. Journal of Cave and Karst Studies 61, 13–21.

Hose, L.D., Palmer, A.N., Palmer, M.V., Northup, D.E., Boston, P.J., DuChene, H.R., 2000. Microbiology and geochemistry in a hydrogen-sulphide-rich karst environment. Chem. Geol. 169, 399–423.

Karimi, H., Raeisi, E., Bakalowicz, M., 2005. Characterizing the main karst aquifers of the Alvand basin, northwest of Zagros, Iran, by a hydrochemical approach. Hydrogeology Journal 13, 787–799.

Katz, B.G., Hornsby, H.D., Bohlke, J.F., Mokray, M.F., 1999. Sources and chronology of nitrate contamination in spring waters, Suwannee River Basin, Florida. U.S. Geological Survey Water-Resources Investigations Report 99–4252.

Katz, B.G., Chelette, A.R., Pratt, T.R., 2004. Use of chemical and isotopic tracers to assess nitrate contamination and groundwater age, Woodville Karst Plain, USA. J. Hydrol. 289, 36–61.

Klimchouk, A., 2007. Hypogene speleogenesis: Hydrogeological and morphogenetic perspective. National Cave and Karst Research Institute Special Paper No. 1.

Klimchouk, A., Lowe, D., Cooper, A., Sauro, U., 1996. Gypsum karst of the world. International Journal of Speleology 25.

Lambrakis, N., Andreou, A.S., Polydoropoulos, P., Georgopoulos, E., Bountis, T., 2000. Nonlinear analysis and forecasting of a brackish karstic spring. Water Resour. Res. 36, 875–884.

LaMoreaux, P.E., Tanner, J.T., 2001. Springs and Bottled Waters of the World. Springer, Berlin.

Langmuir, D., 1971. The geochemistry of some carbonate ground waters in central Pennsylvania. Geochim. Cosmochim. Acta 35, 1023–1045.

Langmuir, D., 1997. Aqueous Environmental Geochemistry. Prentice Hall, Upper Saddle River, NJ.

Langmuir, D., Melchior, D., 1985. The geochemistry of Ca, Sr, Ba and Ra sulfates in some deep brines from the Palo Duro Basin, Texas. Geochim. Cosmochim. Acta 49, 2423–2432.

Liu Zaihua, Groves, C., Yuan Daoxian, Meiman, J., Jiang Guanghui, He Shiyi and Li Qiang, 2004. Hydrochemical variations during flood pulses in the south-west China peak cluster karst: Impacts of $CaCO_3$-H_2O-CO_2 interactions. Hydrologic Processes, v. 18, pp. 2423–2437.

Lohman, S.W., 1938. Ground water in south-central Pennsylvania. Pennsylvania Topographic and Geologic Survey Bulletin W-5, Harrisburg, PA.

Loop, C.M., White, W.B., 2001. A conceptual model for DNAPL transport in karst ground water basins. Ground Water 39, 119–127.

Lorah, M.M., Herman, J.S., 1988. The chemical evolution of a travertine-depositing stream: Geochemical processes and mass transfer reactions. Water Resour. Res. 24, 1541–1552.

Luiszer, F.G., 1994. Speleogenesis of Cave of the Winds, Manitou Springs, Colorado. Karst Waters Institute Special Publication 1. Karst Waters Institute, Charles Town, WV, pp. 91–106.

Macalady, J.L.L., Lyon, E.H., Koffman, B., Albertson, L.K., Meyer, K., Galdenzi, S., et al., 2006. Dominant microbial populations in limestone-corroding stream biofilms, Frasassi cave system, Italy. Appl. Environ. Microbiol. 72, 5596–5609.

Mahler, B., Massei, N., 2007. Anthropogenic contaminants as tracers in an urbanizing karst aquifer. J. Contam. Hydrol. 91, 81–106.

Maramathas, A., Pergialiotis, P., Gialamas, I., 2006. Contribution to the identification of the sea intrusion mechanism of brackish karst springs. Hydrogeology Journal 14, 657–662.

Martinez, M.I., White, W.B., 1999. A laboratory investigation of the relative dissolution rates of the Lirio Limestone and the Isla de Mona Dolomite and implications for cave and karst development on Isla de Mona. Journal of Cave and Karst Studies 61, 7–12.

Moral, F., Cruz-Sanjulián, J.J., Olias, M., 2008. Geochemical evolution of groundwater in the carbonate aquifers of Sierra de Segura (Betic Cordillera, southern Spain). J. Hydrol. 360, 281–296.

Morel, F.M.M., Hering, J.G., 1993. Principles and Applications of Aquatic Chemistry. John Wiley, New York.

Morse, J.W., Arvidson, R.S., 2002. The dissolution kinetics of major sedimentary carbonate minerals. Earth-Sci. Rev. 58, 51–84.

Panno, S.V., Kelly, W.R., 2004. Nitrate and herbicide loading in two groundwater basin of Illinois' sinkhole plain. J. Hydrol. 290, 229–242.

Parkhurst, D.L., Appelo, C.A.J., 2008. User's guide to PHREEQC (version 2)—A computer program for speciation, batch-reaction, one-dimensional transport, and inverse geochemical calculations. Available at:wwwbrr.cr.usgs.gov/projects/GWC_coupled/phreeqc/html/final/html.

Paternoster, M., Mongelli, G., 2000. Hydrochemistry of the Telese spring, southern Apennines, Italy. Mineral. Petrogr. Acta 43, 167–178.

Pentecost, A., 2005. Travertine. Springer-Verlag, Berlin.

Plummer, L.N., Wigley, T.M.L., Parkhurst, D.L., 1978. The kinetics of calcite dissolution in CO_2–water systems at 5 to 60°C and 0.0 to 1.0 atm CO_2. Am. J. Sci. 278, 179–216.

Pokrovsky, O.S., Schott, J., 2001. Kinetics and mechanism of dolomite dissolution in neutral to alkaline solutions revisited. Am. J. Sci. 301, 597–626.

Pokrovsky, O.S., Golubev, S.V., Schott, J., 2005. Dissolution kinetics of calcite, dolomite and magnesite at 25°C and 0 to 50 atm p_{CO2}. Chemical Geology 217, 239–255.

Pronk, M., Goldscheider, N., Zopfi, J., Zwahlen, F., 2008. Percolation and particle transport in the unsaturated zone of a karst aquifer. Ground Water. doi: 10.1111/j.1745-6584.2008.00509.x.

Putnam, G.W., Young, J.R., 1985. The bubbles revisited: The geology and geochemisty of "Saratoga" mineral waters. Northeast. Geol. 7, 53–77.

Ryan, M., Meiman, J., 1996. An examination of short-term variations in water quality at a karst spring in Kentucky. Ground Water 34, 23–30.

Shuster, E.T., 1970. Seasonal variations in carbonate spring water chemistry related to ground water flow. M.S. thesis, The Pennsylvania State University.

Stumm, W., Morgan, J.J., 1996. Aquatic Chemistry, third ed. John Wiley, New York.

Torija Isasa, E., Orzáez Villa-Nueva, T., García Mata, M., Tenorio Sanz, D., de Pradena Lobón, J.M., 2002. Análisis físico-químico de las aquas del Balneario de "Alhama de Granada" (Granada). Anales de la Real Academia Nacional de Farmacia 68, 359–371.

Troester, J.W., 1994. The geochemistry, hydrogeology, and geomorphology of the Rio Camuy drainage basin, Puerto Rico: A humid tropical karst. Ph.D. thesis, The Pennsylvania State University.

Vesper, D.J., White, W.B., 2003. Metal transport to karst springs during storm flow: An example from Fort Campbell, Kentucky/Tennessee, USA. J. Hydrol. 276, 20–36.

Waring, G.A., 1965. Thermal springs of the United States and other countries of the world—A summary. U.S. Geological Survey Professional Paper 492.

White, D.E., Hem, J.D., Waring, G.A., 1963. Chemical composition of subsurface waters. U.S. Geological Survey Professional Paper 440-F.

White, W.B., 1997. Thermodynamic equilibrium, kinetics, activation barriers, and reaction mechanisms for chemical reactions in karst terrains. Env. Geol. 30, 46–58.

White, W.B., 2002. Karst hydrology: recent developments and open questions. Eng. Geol. 65, 85–105.

Wicks, C., Kelley, C., Peterson, E., 2004. Estrogen in a karstic aquifer. Ground Water 42, 384–389.

Winston, W.E., Criss, R.E., 2004. Dynamic hydrologic and geochemical response in a perennial karst spring. Water Resour. Res. 40, W05106.

Wood, W.W., Low, W.H., 1988. Solute geochemistry of the Snake River Plain regional aquifer system, Idaho and eastern Oregon. U.S. Geological Survey Professional Paper 1408-D.

CHAPTER

Springwater treatment

7

Farsad Fotouhi[1] **and Neven Kresic**[2]
[1]Pall Corporation, Ann Arbor, Michigan:
[2]Mactec, Inc., Ashburn, Virginia

7.1 INTRODUCTION

Water sustains life. It drives natural processes. It is a primary component of or is used in virtually every human endeavor, whether domestic, agricultural, or industrial. Its abundance is a boon. Its scarcity is a curse, triggering disputes and battles, even ending empires. Its importance, past and future, cannot be overstated. While natural water supplies are essentially constant, the demand for them increases as the human population does. Therefore, every source of water must be utilized to the fullest.

As discussed in Chapter 6, springs issuing from different types of rocks have different chemical compositions and may have highly variable water quality, in both the long and short terms. For example, springs arising from igneous rock typically have lower dissolved inorganic solid content than those from limestone, sedimentary rock, or soil. Springs emerging from soils with high organic-matter levels typically have higher dissolved organic compound concentrations and notable water color. In general, however, the chemical and physical nature of the water from many springs remains relatively stable over time. The notable exceptions are springs arising from karst formations, as they are influenced more quickly by the aboveground occurrences, such as storms and human activities, including farming and improper waste disposal practices. This is because their primary flow paths are often fissures, fractures, and channels within and between rock layers. The variability in the flow and storage volume of a karst spring, as well as travel distances and land use in its recharge area, can make the spring particularly susceptible to anthropogenic contamination. This vulnerability increases water quality variability and the complexity of treatment that may be required before springwater can be used.

Anthropogenic contaminants, which are typically organic compounds, include industrial chemicals, fuel components, agricultural supplies such as pesticides and fertilizers, and most recently, pharmaceuticals. These compounds are typically stable in the dissolved phase. For this reason, water treatment methods must be chemically or mechanically energetic, which increases their capital and operating costs relative to more traditional techniques.

Most springs considered for water supply must initially be viewed as potentially influenced by surface water (or, as designated in the United States, groundwater under direct influence, GWUDI), and therefore subject to regulations applicable to surface water treatment. Groundwater and springwater are classified as being GWUDI if they exhibit either of the following:

Copyright © 2010, Elsevier Inc. All rights reserved.

1. Significant occurrence of particulate matter, insects or other macroorganisms, algae, or large-diameter pathogens, such as *Giardia lamblia* or *Cryptosporidium.*
2. Significant and relatively rapid shifts in water characteristics, such as turbidity, temperature, conductivity, or pH, which closely correlate to climatological or surface water conditions.

The purpose of the GWUDI regulations in the United States is to protect public health. *Giardia lamblia* and *Cryptosporidium* are parasites commonly found in surface water. They enter the environment through fecal contamination, such as sewage or animal waste. When these parasites make their way into springwater supplies, they can cause severe outbreaks of gastrointestinal illness, including diarrhea, nausea, or stomach cramps. Although they do not pose a grave danger for most people, they can be fatal to immunosuppressed individuals and infants. *Giardia lamblia* and *Cryptosporidium* are extremely hard to kill with conventional amounts of chlorine sufficient to destroy bacteria. GWUDI must therefore be treated more thoroughly than most groundwater extracted by deep or artesian wells.

The U.S. Environmental Protection Agency (EPA) developed a procedure for determining if groundwater is under the direct influence of surface water using microscopic particulate analysis (MPA). MPA identifies organisms that occur in surface water whose presence in groundwater would clearly indicate mixing of the two (U.S. Environmental Protection Agency, 1991, 1992, 1998).

As discussed in Chapters 6 and 8, turbidity and the associated microbiological contamination are the main concerns regarding the drinking water quality of many springs and most karstic springs, small and large. Consequently, they are also the main target of springwater treatment for potable use in most cases. More on the nature and origin of turbidity and microbiological contamination of groundwater in general can be found in textbooks on environmental water chemistry (e.g., Matthess, 1982; Drever, 1988; Appelo and Postma, 2005).

Some of the key elements of an initial assessment of springwater quality for drinking water purposes, using portable field instruments and simple laboratory analyses, are briefly discussed next. They are sometimes used interchangeably or inconsistently, which may cause confusion or even erroneous assessments. The same elements should also be frequently (in cases of small public systems or individual spring users) or continuously (in case of large public supply systems) monitored in both raw water (influent) and treated water (effluent).

- **Total dissolved solids**. The total concentration of dissolved material in groundwater is called *total dissolved solids* (TDS). It is commonly determined by weighing the dry residue after heating the water sample usually to 103°C or 180°C (the higher temperature is used to eliminate more of the crystallization water). TDS can also be calculated if the concentrations of major ions are known. However, for some water types, a rather extensive list of analytes may be needed to accurately obtain the total. During evaporation, approximately one half of the hydrogen carbonate ions are precipitated as carbonates and the other half escapes as water and carbon dioxide. This loss is taken into account by adding half of the HCO_3^- content to the evaporation (dry) residue. Some other losses, such as precipitation of sulfate as gypsum and partial volatilization of acids, nitrogen, boron, and organic substances, may contribute to a discrepancy between the calculated and the measured total dissolved solids (Kresic, 2009).
- **Specific conductance**. Solids and liquids that dissolve in water can be divided into electrolytes and nonelectrolytes. Electrolytes, such as salts, bases, and acids, dissociate into ionic forms (positively and negatively charged ions) and conduct electrical current. Nonelectrolytes, such as sugar, alcohols, and many organic substances, occur in aqueous solution as uncharged molecules and do not conduct electrical current. The ability of 1 cm^3 of water to conduct electrical current is called *specific conductance* (or sometimes simply *conductance*, although the units are different). Conductance is the reciprocal of resistance and is measured in units called *Siemen* (International System) or *mho* (1 *Siemen* equals 1 *mho*; the name *mho* is derived from the unit for resistance, ohm, by spelling it in reverse). Specific conductance is expressed as

Siemen per centimeter or mho per centimeter. Since the mho is usually too large for most groundwater types, the specific conductance is reported in micromhos per centimeter or micro-Siemens per centimeter (μS/cm), with instrument readings adjusted to 25°C, so that variations in conductance are a function of only the concentration and type of dissolved constituents present (water temperature also has a significant influence on conductance). Measurements of specific conductance can be made rapidly in the field with a portable instrument, which provides for a convenient method to quickly estimate total dissolved solids and compare general types of water quality. For a preliminary (rough) estimate of total dissolved solids, in milligrams per liter, in fresh potable water, the specific conductance in micromhos per centimeter can be multiplied by 0.7. Pure water has a conductance of 0.055 micromhos at 25°C, laboratory distilled water between 0.5 and 5 micromhos, rainwater usually between 5 and 30 micromhos, potable groundwater ranges from 30 to 2000 micromhos, seawater from 45,000 to 55,000 micromhos, and oil-field brines have commonly more than 100,000 micromhos (Davis and DeWiest, 1991).

- **Salinity**. The term *salinity* is often used for total dissolved salts (ionic species) in groundwater, in the context of water quality for agricultural uses or human and livestock consumption. Various salinity classifications, based on certain salts and their ratios, have been proposed (see Matthess, 1982). One problem with the term *salinity* is that a salty taste may be already noticeable at somewhat higher concentrations of sodium chloride, NaCl (e.g., 300–400 mg/L), even though the overall concentration of all dissolved salts may not "qualify" a particular groundwater to be considered saline. In practice, it is common to call water with less than 1000 milligrams per liter (1 g/L) dissolved solids *fresh* and water with more than 10,000 mg/L *saline*. Brackish water, which is becoming more and more interesting for water supply development worldwide, has TDS between 2000 mg/L and 10,000 mg/L.

7.2 DRINKING WATER STANDARDS

7.2.1 Primary drinking water standards

The U.S. National Primary Drinking Water Regulations (primary standards) are legally enforceable standards that apply to public water systems. Primary standards protect drinking water quality by limiting the levels of specific contaminants that can adversely affect public health and are known or anticipated to occur in water. They take the form of maximum contaminant levels (MCLs) or treatment techniques. These standards must be met at the point of delivery to any user of a public system (i.e., point of use or point of discharge from the water distribution system) or, in some cases, at various points throughout the distribution system (U.S. Environmental Protection Agency, 2003).

Once the U.S. EPA has selected a contaminant for regulation, it examines the contaminant's health effects and sets a maximum contaminant level goal (MCLG). This is the maximum level of a contaminant in drinking water at which no known or anticipated adverse health effects would occur and which allows an adequate margin of safety. MCLGs do not take cost and technologies into consideration. MCLGs are nonenforceable public health goals. In setting the MCLG, the U.S. EPA examines the size and nature of the population exposed to the contaminant and the length of time and concentration of the exposure. Since MCLGs consider only public health and not the limits of detection and treatment technology, they are sometimes set at a level that water systems cannot meet. For most carcinogens (contaminants that cause cancer) and microbiological contaminants, MCLGs are set at zero because a safe level often cannot be determined (U.S. Environmental Protection Agency, 2003, 2006a).

Maximum contaminant levels, which are enforceable limits that finished drinking water must meet, are set as close to the MCLG as feasible. The Safe Drinking Water Act defines *feasible* as the level that may be achieved with the use of the best available technology, treatment technique, or other means specified by U.S. EPA, after examination for efficacy under field conditions (that is, not solely under laboratory conditions) and taking cost into consideration (U.S. Environmental Protection Agency, 2003).

For some contaminants, especially microbiological contaminants, no reliable method is economically and technically feasible to measure a contaminant at particularly low concentrations. In these cases, the U.S. EPA establishes treatment techniques. A treatment technique is an enforceable procedure or level of technological performance that public water systems must follow to ensure control of a contaminant. Examples of rules with treatment techniques are the surface water treatment rule (aimed primarily at biological contaminants and water disinfection) and the lead and copper rule.

As of January 2009, the U.S. EPA set MCLs or treatment techniques for 87 contaminants included in the National Primary Drinking Water Standards list (Table 7–1).

7.2.2 Secondary drinking water standards

The U.S. National Secondary Drinking Water Regulations (NSDWRs, or secondary standards) are non-enforceable guidelines regarding contaminants that may cause cosmetic effects (such as skin or tooth discoloration) or have aesthetic effects (such as affecting the taste, odor, or color of drinking water). The U.S. EPA recommends secondary standards to water systems but does not require systems to comply. However, states may choose to adopt them as enforceable standards. NSDWRs are intended to protect "public welfare" (U.S. Environmental Protection Agency, 2003). Fifteen constituents are included in the National Secondary Drinking Water Standards list (Table 7–2).

7.3 TREATMENT TECHNOLOGIES

Dangerous and objectionable constituents of water must be removed, destroyed, or neutralized before the water can be used or consumed. One constituent is usually addressed first, due to the risk or nuisance it presents relative to all others, even though combinations of effects are possible when two or more contaminants are present or the major contaminant present has multiple effects.

Many methods to address these risks have been developed over the centuries. And research to reduce costs, improve performance, and address emerging contaminants continues. The techniques potentially applicable to treatment of springwater are summarized next. The methods and equipment are discussed in greater detail in numerous documents, including *Recommended Standards for Water Works* (Health Research Inc., 2007), and various U.S. EPA reports (1999a, 1999b, 1999c, 2005a, 2005b, and 2006b).

The choice of technology for a particular situation is determined by the water flow rate to be treated and its highest use; the regional and local geology, hydrology, and meteorology; the cost of materials, equipment, labor, and supplies; and other factors whose importance is more limited overall but critical under certain circumstances.

The threat posed by groundwater constituents is primarily chemical, biological, or physical in nature. However, the impact of each treatment method is not as defined. For example, filtration can change turbidity (physical impact) and remove bacteria (biological impact), while aeration to oxidize and precipitate iron (chemical impact) can reduce phosphate and arsenic concentrations through coprecipitation (physical impact). Therefore, groundwater constituents and the treatment technologies used to control the risks and nuisances they pose are presented here approximately in order of increasing complexity of the problem addressed and increasing complexity of the treatment.

Table 7–1 National Primary Drinking Water Standards (From U.S. EPA, accessed January 2009; available at: www.epa.gov/safewater/contaminants/index.html#listmcl)

	Contaminant	MCL or TT[1] (mg/L)[2]	Potential Health Effects from Exposure above the MCL	Common Sources of Contaminant in Drinking Water	Public Health Goal
OC	Acrylamide	TT[8]	Nervous system or blood problems	Added to water during sewage/ wastewater increased risk of cancer treatment	0
OC	Alachlor	0.002	Eye, liver, kidney, or spleen problems; anemia; increased risk of cancer	Runoff from herbicide used on row crops	0
R	Alpha particles	15 picocuries per liter (pCi/L)	Increased risk of cancer	Erosion of natural deposits of certain minerals that are radioactive and may emit a form of radiation known as alpha radiation	0
IOC	Antimony	0.006	Increase in blood cholesterol; decrease in blood sugar	Discharge from petroleum refineries; fire retardants; ceramics; electronics; solder	0.006
IOC	Arsenic	0.01	Skin damage or problems with circulatory systems, and may have increased risk of getting cancer	Erosion of natural deposits; runoff from orchards, runoff from glass and electronics production wastes	0
IOC	Asbestos (fibers >10 mm)	7 million fibers per liter (MFL)	Increased risk of developing benign intestinal polyps	Decay of asbestos cement in water mains; erosion of natural deposits	7 MFL
OC	Atrazine	0.003	Cardiovascular system or reproductive problems	Runoff from herbicide used on row crops	0.003
IOC	Barium	2	Increase in blood pressure	Discharge of drilling wastes; discharge from metal refineries; erosion of natural deposits	2
OC	Benzene	0.005	Anemia; decrease in blood platelets; increased risk of cancer	Discharge from factories; leaching from gas storage tanks and landfills	0

(Continued)

Table 7–1 National Primary Drinking Water Standards (From U.S. EPA, accessed January 2009; available at: www.epa.gov/safewater/contaminants/index.html#listmcl)—Cont'd

	Contaminant	MCL or TT[1] (mg/L)[2]	Potential Health Effects from Exposure above the MCL	Common Sources of Contaminant in Drinking Water	Public Health Goal
OC	Benzo(a)pyrene (PAHs)	0.0002	Reproductive difficulties; increased risk of cancer	Leaching from linings of water	0
IOC	Beryllium	0.004	Intestinal lesions	Discharge from metal refineries and coal-burning factories; discharge from electrical, aerospace, and defense industries	0.004
R	Beta particles and photon emitters	4 millirems per year	Increased risk of cancer	Decay of natural and human-made deposits of certain minerals that are radioactive and may emit forms of radiation known as photons and beta radiation	0
DBP	Bromate	0.010	Increased risk of cancer	By-product of drinking water disinfection	0
IOC	Cadmium	0.005	Kidney damage	Corrosion of galvanized pipes; erosion of natural deposits; discharge from metal refineries; runoff from waste batteries and paints	0.005
OC	Carbofuran	0.04	Problems with blood, nervous system, or reproductive system	Leaching of soil fumigant used on rice and alfalfa	0.04
OC	Carbon tetrachloride	0.005	Liver problems; increased risk of cancer	Discharge from chemical plants and other industrial activities	0
D	Chloramines (as Cl_2)	MRDL = 4.01	Eye/nose irritation; stomach discomfort; anemia	Water additive used to control microbes	MRDLG = 41
OC	Chlordane	0.002	Liver or nervous system problems; increased risk of cancer	Residue of banned termiticide	0

D	Chlorine (as Cl_2)	MRDL = 4.01	Eye/nose irritation; stomach discomfort	Water additive used to control microbes	MRDLG = 41
D	Chlorine dioxide (as ClO_2)	MRDL = 0.81	Anemia; infants & young children: nervous system effects	Water additive used to control microbes	MRDLG = 0.81
DBP	Chlorite	1.0	Anemia; infants and young children: nervous system effects	By-product of drinking water disinfection	0.8
OC	Chlorobenzene	0.1	Liver or kidney problems	Discharge from chemical and agricultural chemical factories	0.1
IOC	Chromium (total)	0.1	Allergic dermatitis	Discharge from steel and pulp mills; erosion of natural deposits	0.1
IOC	Copper	TT[7]; action level = 1.3	Short-term exposure: Gastrointestinal distress. Long-term exposure: Liver or kidney damage. People with Wilson's disease should consult their doctor if the amount of copper in their water exceeds the action level	Corrosion of household plumbing systems; erosion of natural deposits	1.3
M	*Cryptosporidium*	TT[3]	Gastrointestinal illness (e.g., diarrhea, vomiting, cramps)	Human and animal fecal waste	0
IOC	Cyanide (as free cyanide)	0.2	Nerve damage or thyroid problems	Discharge from steel/metal factories; discharge from plastic and fertilizer factories	0.2
OC	2,4-D	0.07	Kidney, liver, or adrenal gland problems	Runoff from herbicide used on row crops	0.07
OC	Dalapon	0.2	Minor kidney changes	Runoff from herbicide used on rights of way	0.2
OC	1,2-Dibromo-3-chloropropane (DBCP)	0.0002	Reproductive difficulties; increased risk of cancer	Runoff/leaching from soil fumigant used on soybeans, cotton, pineapples, and orchards	0
OC	o-Dichlorobenzene	0.6	Liver, kidney, or circulatory system problems	Discharge from industrial chemical factories	0.6

(*Continued*)

Table 7–1 National Primary Drinking Water Standards (From U.S. EPA, accessed January 2009; available at: www.epa.gov/safewater/contaminants/index.html#listmcl)—Cont'd

	Contaminant	MCL or TT[1] (mg/L)[2]	Potential Health Effects from Exposure above the MCL	Common Sources of Contaminant in Drinking Water	Public Health Goal
OC	p-Dichlorobenzene	0.075	Anemia; liver, kidney, or spleen damage; changes in blood	Discharge from industrial chemical factories	0.075
OC	1,2-Dichloroethane	0.005	Increased risk of cancer	Discharge from industrial chemical factories	0
OC	1,1-Dichloroethylene	0.007	Liver problems	Discharge from industrial chemical factories	0.007
OC	cis-1,2-Dichloroethylene	0.07	Liver problems	Discharge from industrial chemical factories	0.07
OC	trans-1,2-Dichloroethylene	0.1	Liver problems	Discharge from industrial chemical factories	0.1
OC	Dichloromethane	0.005	Liver problems; increased risk of cancer	Discharge from drug and chemical factories	0
OC	1,2-Dichloropropane	0.005	Increased risk of cancer	Discharge from industrial chemical factories	0
OC	Di(2-ethylhexyl) adipate	0.4	Weight loss; liver problems; possible reproductive difficulties	Discharge from chemical factories	0.4
OC	Di(2-ethylhexyl) phthalate	0.006	Reproductive difficulties; liver problems; increased risk of cancer	Discharge from rubber and chemical factories	0
OC	Dinoseb	0.007	Reproductive difficulties	Runoff from herbicide used on soybeans and vegetables	0.007
OC	Dioxin (2,3,7,8-TCDD)	0.00000003	Reproductive difficulties; increased risk of cancer	Emissions from waste incineration and other combustion; discharge from chemical factories	0
OC	Diquat	0.02	Cataracts	Runoff from herbicide use	0.02

OC	Endothall	0.1	Stomach and intestinal problems	Runoff from herbicide use	0.1
OC	Endrin	0.002	Liver problems	Residue of banned insecticide	0.002
OC	Epichlorohydrin	TT[8]	Increased cancer risk; over a long period of time, stomach problems	Discharge from industrial chemical factories; an impurity of some water treatment chemicals	0
OC	Ethylbenzene	0.7	Liver or kidney problems	Discharge from petroleum refineries	0.7
OC	Ethylene dibromide	0.00005	Problems with liver, stomach, reproductive system, or kidneys; increased risk of cancer	Discharge from petroleum refineries	0
IOC	Fluoride	4	Bone disease (pain and tenderness of the bones); children may get mottled teeth	Water additive that promotes strong teeth; erosion of natural deposits; discharge from fertilizer and aluminum factories	4
M	*Giardia lamblia*	TT[3]	Gastrointestinal illness (e.g., diarrhea, vomiting, cramps)	Human and animal fecal waste	0
OC	Glyphosate	0.7	Kidney problems; reproductive difficulties	Runoff from herbicide use	0.7
DBP	Haloacetic acids (HAA[5])	0.06	Increased risk of cancer	By-product of drinking water disinfection	N/A[6]
OC	Heptachlor	0.0004	Liver damage; increased risk of cancer	Residue of banned termiticide	0
OC	Heptachlor epoxide	0.0002	Liver damage; increased risk of cancer	Breakdown of heptachlor	0
M	Heterotrophic plate count (HPC)	TT[3]	No health effects; it is an analytic method used to measure the variety of bacteria common in water. The lower the concentration of bacteria in drinking water, the better maintained the water system is	HPC measures a range of bacteria that are naturally present in the environment	N/A
OC	Hexachloro-cyclopentadiene	0.05	Kidney or stomach problems	Discharge from chemical factories	0.05

(*Continued*)

Table 7–1 National Primary Drinking Water Standards (From U.S. EPA, accessed January 2009; available at: www.epa.gov/safewater/contaminants/index.html#listmcl)—Cont'd

	Contaminant	MCL or TT[1] (mg/L)[2]	Potential Health Effects from Exposure above the MCL	Common Sources of Contaminant in Drinking Water	Public Health Goal
IOC	Lead	TT[7]; action level = 0.015	Infants and children: Delays in physical or mental development; children could show slight deficits in attention span and learning abilities. Adults: Kidney problems; high blood pressure	Corrosion of household plumbing systems; erosion of natural deposits	0
M	*Legionella*	TT[3]	Legionnaire's disease, a type of pneumonia	Found naturally in water; multiplies in heating systems	0
OC	Lindane	0.0002	Liver or kidney problems	Runoff/leaching from insecticide used on cattle, lumber, gardens	0.0002
IOC	Mercury (inorganic)	0.002	Kidney damage	Erosion of natural deposits; discharge from refineries and factories; runoff from landfills and croplands	0.002
OC	Methoxychlor	0.04	Reproductive difficulties	Runoff/leaching from insecticides used on fruits, vegetables, alfalfa, livestock	0.04
IOC	Nitrate (measured as nitrogen)	10	Infants below the age of six months who drink water containing nitrate in excess of the MCL could become seriously ill and, if untreated, may die. Symptoms include shortness of breath and blue-baby syndrome	Runoff from fertilizer use; leaching from septic tanks, sewage; erosion of natural deposits	10
IOC	Nitrite (measured as nitrogen)	1	Infants below the age of six months who drink water containing nitrite in excess of the MCL could become seriously ill and, if untreated, may die. Symptoms include shortness of breath and blue-baby syndrome	Runoff from fertilizer use; leaching from septic tanks, sewage; erosion of natural deposits	1

OC	Oxamyl (Vydate)	0.2	Slight nervous system effects	Runoff/leaching from insecticide used on apples, potatoes, and tomatoes	0.2
OC	Pentachlorophenol	0.001	Liver or kidney problems; increased cancer risk	Discharge from wood preserving factories	0
OC	Picloram	0.5	Liver problems	Herbicide runoff	0.5
OC	Polychlorinated biphenyls (PCBs)	0.0005	Skin changes; thymus gland problems; immune deficiencies; reproductive or nervous system difficulties; increased risk of cancer	Runoff from landfills; discharge of waste chemicals	0
R	Radium 226 and Radium 228 (combined)	5 pCi/L	Increased risk of cancer	Erosion of natural deposits	0
IOC	Selenium	0.05	Hair or fingernail loss; numbness in fingers or toes; circulatory problems	Discharge from petroleum refineries; erosion of natural deposits; discharge from mines	0.05
OC	Simazine	0.004	Problems with blood	Herbicide runoff	0.004
OC	Styrene	0.1	Liver, kidney, or circulatory system problems	Discharge from rubber and plastic factories; leaching from landfills	0.1
OC	Tetrachloroethylene (PCE)	0.005	Liver problems; increased risk of cancer	Discharge from factories and dry cleaners	0
IOC	Thallium	0.002	Hair loss; changes in blood; kidney, intestine, or liver problems	Leaching from ore-processing sites; discharge from electronics, glass, and drug factories	0.0005
OC	Toluene	1	Nervous system, kidney, or liver problems	Discharge from petroleum factories	1
M	Total coliforms (including fecal coliform and *E. coli*)	5.0%[4]	Not a health threat in itself; it is used to indicate whether other potentially harmful bacteria may be present[5]	Coliforms are naturally present in the environment as well as feces; fecal coliforms and *E. coli* come only from human and animal fecal waste	0
DBP	Total trihalomethanes (TTHMs)	0.08	Liver, kidney, or central nervous system problems; increased risk of cancer	By-product of drinking water disinfection	N/A[6]

(Continued)

Table 7–1 National Primary Drinking Water Standards (From U.S. EPA, accessed January 2009; available at: www.epa.gov/safewater/contaminants/index.html#listmcl)—Cont'd

	Contaminant	MCL or TT[1] (mg/L)[2]	Potential Health Effects from Exposure above the MCL	Common Sources of Contaminant in Drinking Water	Public Health Goal
OC	Toxaphene	0.003	Kidney, liver, or thyroid problems; increased risk of cancer	Runoff/leaching from insecticide used on cotton and cattle	0
OC	2,4,5-TP (Silvex)	0.05	Liver problems	Residue of banned herbicide	0.05
OC	1,2,4-Trichlorobenzene	0.07	Changes in adrenal glands	Discharge from textile finishing factories	0.07
OC	1,1,1-Trichloroethane	0.2	Liver, nervous system, or circulatory problems	Discharge from metal degreasing sites and other factories	0.2
OC	1,1,2-Trichloroethane	0.005	Liver, kidney, or immune system problems	Discharge from industrial chemical factories	0.003
OC	Trichloroethylene (TCE)	0.005	Liver problems; increased risk of cancer	Discharge from metal degreasing sites and other factories	0
M	Turbidity	TT[3]	A measure of the cloudiness of water, it is used to indicate water quality and filtration effectiveness (e.g., presence of disease-causing organisms). Higher turbidity levels often associated with higher levels of disease-causing microorganisms, such as viruses, parasites, and some bacteria, which can cause symptoms such as nausea, cramps, diarrhea, and associated headaches	Soil runoff	N/A
R	Uranium	30 μg/L	Increased risk of cancer, kidney toxicity	Erosion of natural deposits	0
OC	Vinyl chloride	0.002	Increased risk of cancer	Leaching from PVC pipes; discharge from plastic factories	0

M	Viruses (enteric)	TT^3	Gastrointestinal illness (e.g., diarrhea, vomiting, cramps)	Human and animal fecal waste	0
OC	Xylenes (total)	10	Nervous system damage	Discharge from petroleum factories; discharge from chemical factories	10

1. *Definitions:*
   Maximum contaminant level goal (MCLG)
   Maximum contaminant level (MCL)
   Maximum residual disinfectant level goal (MRDLG)
   Maximum residual disinfectant level (MRDL)
   Treatment technique (TT)
2. *Units are in milligrams per liter (mg/L) unless otherwise noted. Milligrams per liter are equivalent to parts per million (ppm).*
3. *The EPA's surface water treatment rules require systems using surface water or groundwater under the direct influence of surface water to (1) disinfect their water and (2) filter their water or meet criteria for avoiding filtration so the following contaminants are controlled at these levels:* Cryptosporidium, *99 percent removal;* Giardia lamblia, *99.9 percent removal/inactivation; viruses, 99.99 percent removal/inactivation.Turbidity, may never exceed 1 NTU and must not exceed 0.3 NTU in 95 percent of daily samples in any month.HPC, no more than 500 bacterial colonies per milliliter.*
4. *No more than 5.0 percent of samples total coliform positive in a month. (For water systems that collect fewer than 40 routine samples per month, no more than 1 sample can be total coliform positive per month.) Every sample that has total coliform must be analyzed for either fecal coliforms or* E. coli. *If two consecutive samples are TC-positive and one is also positive for* E. coli *fecal coliforms, system has an acute MCL violation.*
5. *Fecal coliform and* E. coli *are bacteria whose presence indicates that the water may be contaminated with human or animal wastes.*
6. *Although there is no collective MCLG for this contaminant group, there are individual MCLGs for some of the individual contaminants: Haloacetic acids, dichloroacetic acid (0), trichloroacetic acid (0.3 mg/L), trihalomethanes, bromodichloromethane (0), bromoform (0), dibromochloromethane (0.06 mg/L). If more than 10 percent of tap water samples exceed the action level, water systems must take additional steps. For copper, the action level is 1.3 mg/L; and for lead, it is 0.015 mg/L.*
7. *Lead and copper are regulated by a treatment technique that requires systems to control the corrosiveness of their water.*
8. *Each water system must certify, in writing, to the state (using third-party or manufacturers certification) that, when it uses acrylamide and/or epichlorohydrin to treat water, the combination (or product) of dose and monomer level does not exceed the levels specified, as follows: Acrylamide = 0.05 percent dosed at 1 mg/L (or equivalent); Epichlorohydrin = 0.01 percent dosed at 20 mg/L (or equivalent).*

Table 7–2 National Secondary Drinking Water Standards (From U.S. EPA, accessed January 2008; available at: www.epa.gov/safewater/contaminants/index.html#listsec)

Contaminant	Secondary Standard
Aluminum	0.05 to 0.2 mg/L
Chloride	250 mg/L
Color	15 (color units)
Copper	1.0 mg/L
Corrosivity	Noncorrosive
Fluoride	2.0 mg/L
Foaming agents	0.5 mg/L
Iron	0.3 mg/L
Manganese	0.05 mg/L
Odor	3 threshold odor number
pH	6.5–8.5
Silver	0.10 mg/L
Sulfate	250 mg/L
Total dissolved solids	500 mg/L
Zinc	5 mg/L

7.3.1 Turbidity

Water is naturally colorless. Physical problems associated with water supplies are often visible to the unaided eye and, for this reason, may have been the first addressed groundwater characteristic treated.

Springwater, and in fact water from any natural source, may be cloudy, tinted, or turbid. The cause may be suspended inorganic (silt or clay) or organic (plant matter or algae) particles or dissolved organic compounds such as humic, fulvic, or tannic acids. True solutes are in the state of separated molecules and ions, which all have very small dimensions (commonly between 10^{-6} and 10^{-8} cm), thus making a water solution transparent to light. Colloidal solutions have solid particles and groups of molecules larger than the ions and molecules of the solvent (water). When colloidal particles are present in large enough quantities, they give water an opalescent appearance by scattering light. Although there is no one agreed-on definition of what exactly colloidal sizes are, a common range cited is between 10^{-6} and 10^{-4} cm (Matthess, 1982). The amount of a solute in water is expressed in terms of its concentration, usually in milligrams per liter (mg/L) or parts per million (ppm) and micrograms per liter (parts per billion, ppb). It is sometimes difficult to distinguish between certain true solutes and colloidal solutions that may carry particles of the same source substance. Filtering or precipitating colloidal particles before determining the true dissolved concentration of a solute may be necessary in some cases. This is especially true for drinking water standards, because amounts for most substances are based on dissolved concentrations. Laboratory analytical procedures are commonly designed to determine total concentrations of a substance and do not necessarily provide an indication of all the individual species (chemical forms) of it. If needed, however, such speciation can be requested.

Clarification without pretreatment

Clarification can be a single or multistep process. In the simplest mode, water is passed through a large tank, sized so that water moves slowly. The reduction in speed and turbulence allows particles with a small surface-to-mass ratio to settle before the water exits the tank, typically through an overflow structure. This process emulates natural clarification in some large karst aquifers that receive turbid water via sinking streams at locations distant from major ascending springs (Figure 7–1). The aquifer provides treatment by acting as a large settling tank. Rapidly moving water, such as the water racing down recharge streams during storm events, carries sediment and debris. When recharge water enters the aquifer it slows down, and the sediment load and organic materials it carries begin to settle out (as in the jar in Figure 7–1). Divers in the Edwards Aquifer reported that underwater cave floors are covered with a thick layer of silt that is easily stirred up, causing a blinding "silt-out" in the normally crystal clear water (Eckhardt, 2009; www.edwardsaquifer.net/geology.html#movement).

Surface water reservoirs constructed up-gradient from major sinks and located in nonkarstic rocks can become part of the springwater treatment: Turbid water flowing into the reservoir after major rainfall events is allowed to settle before it is released and enters the karst aquifer via sinks (Figure 7–2).

Filtration without pretreatment

Filtration is conceptually a simple treatment method. Raw water is passed through a tank or vessel containing filter media. The relative sizes of the targeted material and the media, the depth (thickness) of the media, and the water flow rate through the media determine system performance. Simply put, the interconnected spaces

FIGURE 7–1 (Left) Sample of storm water runoff collected from Helotes Creek in the recharge zone of Comal Springs, Edwards Aquifer, Texas. (Right) Sample taken from Comal Springs on the same day; the springs are known for their clarity. (Photograph courtesy of Gregg Eckhardt.)

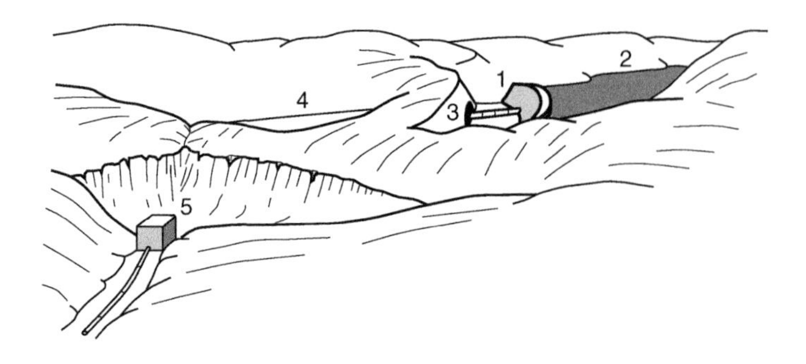

FIGURE 7–2 Schematic of the Bucje aquifer regulation, Serbia and Montenegro. (1) Dam; (2) reservoir; (3) Tmusa River cave sink; (4) dry valley; (5) Bucje karstic spring. (From Kresic, 1988.)

between the media particles must be small enough to block the passage of the suspended soil or organic matter particles. Many variations on the theme exist.

Slow sand Slow sand filters may be the oldest water treatment method after clarification. The filter media is placed in a vault or tank. Fine media is placed on top of coarse media. The coarser media itself lies on the underdrain system. Water passes through a slow sand filter, as the name implies, slowly. This means the water flow rate through the filter can be very slow before it is unacceptable. The extra time allows a thicker layer of captured particles to accumulate between rejuvenation events. This allows a bacterial community to develop in the layer of captured particles. The physical nature of the bacterial layer and the metabolism of the bacteria increase the ability of the filter to capture fine suspended particles. When the water treatment rate becomes unacceptable, the upper media layer is removed and a new cycle started.

Slow sand filters can remove suspended solids but do not capture clay particles or reduce color levels. They are simple to build and operate. (Media removal can be done by unskilled labor.) Because they are simple to operate and maintain, slow sand filters are often used by low-income communities or in emergencies. Typical loading rates are 0.2 to 1.0 gpm/ft^2 (0.1–0.4 m^3/m^2/h). Typical filter cycles are two to six months. Longer cycles are associated with better-quality raw water.

Rapid sand The rapid sand technique is commonly used in modern municipal water treatment. Vaults or tanks are filled with media, typically sand, as the name implies. Other materials, such as granular activated carbon or anthracite, can be added to adsorb dissolved chemicals, particularly organic compounds. Water is applied from above and flows down through the media to an underdrain collection system under the force of gravity. Particles captured by the filter reduce the pore spaces and, as a consequence, the rate at which water passes through the filter. When the rate becomes unacceptably low, the filter is backwashed.

To backwash, the raw water supply is shut off. Water or water and air (to increase agitation) are pumped into the vault below the filter media. The pumping rate is selected to lift and agitate the media, causing media particles to scour one another, dislodging and resuspending previously captured particles. The backwash water is then removed through collection points or troughs and disposed of. The media is allowed to settle in the tank. Raw water is again applied to its upper surface of the filter and the process repeated. Slow sand filters obviously pre-dated rapid sand filters, because the latter technique was not possible before the invention of pumps large and powerful enough to pump enough water to suspend the media and backwash the filter.

Typical loading rates are 2 to 5 gpm/ft^2 (5 to 12 m^3/m^2/h) of sand filter surface. Operation, monitoring, and maintenance of a rapid sand filter requires extensive experience and training. Nonetheless, their improved performance compared to slow sand justifies the added filter management and expense.

Precoated media These filters are used to remove fine solids from water. A slurry of very fine media is applied to a permeable rigid element. It is allowed or induced to settle onto the element with minimal disturbance or mixing, which promotes an even, uniform coating on the element.

Very fine media, such as diatomaceous earth or ground perlite or vermiculite, are used so that fine to very fine particles can be captured. When the combination of media and captured fines begins to exert an unacceptable head loss, the deposited media and accumulated solids are flushed from the rigid element and disposed of. A new layer of fine media is then applied to the permeable rigid element.

Loading rates are 0.5 to 2 gpm/ft^2 (1.2 to 5 m^3/m^2/h) of filter surface. Process capital costs are relatively low. However, disadvantages include cracks in the fine media layer when the fine media is not applied properly, which may allow only partially treated water to pass, and a tendency for filters to plug quickly when high-turbidity water is treated.

Clarification or filtration after coagulation

Steps can be taken to improve clarification or filtration. The raw water pH can be raised or a coagulant or flocculant (e.g., aluminum or iron salts or proprietary organic compounds) can be added. Both processes reduce electrostatic forces, which cause suspended particles to repel one another, preventing clumping (agglomeration). Clumps of particles have a lower surface area to mass ratio and settle more quickly. Destruction of the electrostatic forces promotes clumping. Coagulation or the technology described in the next section, oxidation, increases the efficacy of either clarification or filtration. This is because bacteria and dissolved constituents of groundwater also can be physically or chemically altered by the coagulant or oxidant. The bacteria and chemicals can be trapped in or attracted to clumps or the filter media or even precipitated directly.

For clarification, treated water is placed in a settling tank where the agglomerated particles settle more quickly than from untreated water. Treatment of the water allows the use of smaller tanks or shorter retention times.

For filtration, treated water is applied to any of the filters described earlier. For high-volume filters targeting relatively large particles, such as slow sand and rapid sand, particle clumps that have a larger average size tend to stay on the surface of the filter, potentially making their regeneration easier. For filters targeting relatively small particles (silts and clays), addition of a coagulant or flocculant may enable an upstream filter to capture larger particles and clumps of fine particles, a cost-effective part of the treatment train.

A basic water treatment plant, incorporating flocculation, sedimentation, filtration, pH control, and disinfection, is shown in Figure 7–3.

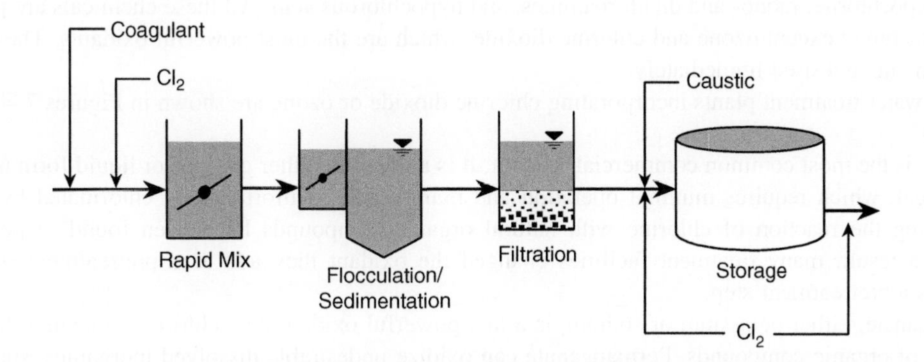

FIGURE 7–3 Basic drinking water treatment plant. (U.S. Environmental Protection Agency, 2005b.)

A new clarification process was recently developed. Small air bubbles are released near the bottom of the tank. The rising bubbles accumulate flocs, carry them to the surface, where they overflow into a gutter or are skimmed off and disposed of.

7.3.2 Oxidation

Occasionally, dissolved organic compounds, inorganic salts, or elements are present at levels that create objectionable tastes, odors, or colors. Many of these can be destroyed or precipitated through oxidation. The treated water is then clarified or filtered. Oxidation is also efficient in destroying many microorganisms and is often used for that sole purpose.

A variety of oxidants are available. Factors to be considered during selection of the oxidant include

- Oxidant strength (half-cell potential) required to control impurity.
- Impurity concentration.
- Oxidant form (liquid, gas, or solid).
- Complexity of the treatment system.
- Personnel needs.
- Oxidant cost.
- Safety.

The simplest form of oxidation is aeration. Contact between the atmosphere and the raw water is maximized during passage through one of several types of structures. The simplest structures are cascade aerators. Water tumbles down a series of steps or a pile of coarse media as air is blown in the opposite direction. As oxygen diffuses into the water, metals, such as iron, manganese, and arsenic, are oxidized to a solid form and precipitated. Gases, such as hydrogen sulfide and carbon dioxide, and volatile organic compounds move from the dissolved phase to the gas state in the air.

Other common aerators are packed columns, tray diffusers, sprayers, and bubble diffusers (compressed air is released at the bottom of a tank of water). Operating and maintenance costs vary, depending on the volume of water treated, concentration of impurity, and the mass of solids produced, which determines the required capacity and performance of any downstream filters. Finally, the nature and mass of gases and vapors transferred to the atmosphere determine whether these must be controlled.

All other oxidation requires the purchase or production of an oxidizing chemical or compound. The most common chemical oxidants are chlorine, chlorine dioxide, ozone, and permanganate. Others are hydrogen peroxide, hypochlorite, mono- and dichloroamines, and hypochlorous acid. All these chemicals are purchased from manufacturers except ozone and chlorine dioxide, which are the most powerful oxidants. They must be produced on site and used immediately.

Typical water treatment plants incorporating chlorine dioxide or ozone are shown in Figures 7–4 and 7–5, respectively.

Chlorine is the most common commercial oxidant. It is applied in either gaseous or liquid form using simple equipment, which requires minimal operation and maintenance. Unfortunately, chlorinated by-products formed during the reaction of chlorine with natural organic compounds have been found to pose health threats. As a result, many treatment facilities changed the oxidant they add as a pretreatment or stopped oxidation as a pretreatment step.

Permanganate, either potassium or sodium, is a less powerful oxidant than chlorine. It cannot destroy the same range of organic compounds. Permanganate can oxidize undesirable dissolved inorganics, such as iron and manganese. Permanganate ions, which are pink in aqueous solution, are visible to the naked eye at concentrations as low as 1 mg/L (ppm). If permanganate concentrations are not below its visible threshold before the

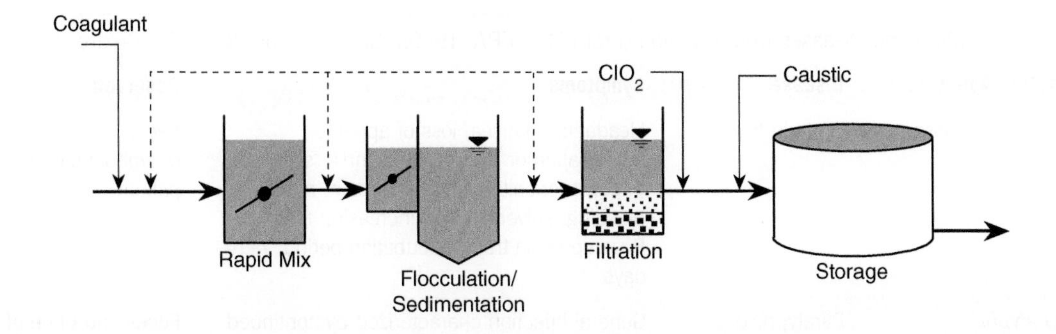

FIGURE 7–4 P Oxidation with chlorine dioxide plant. Point of chlorine dioxide addition may be (1) prior to rapid mix, (2) prior to flocculation, (3) prior to filtration, or (4) postfiltration. (U.S. Environmental Protection Agency, 2005b.)

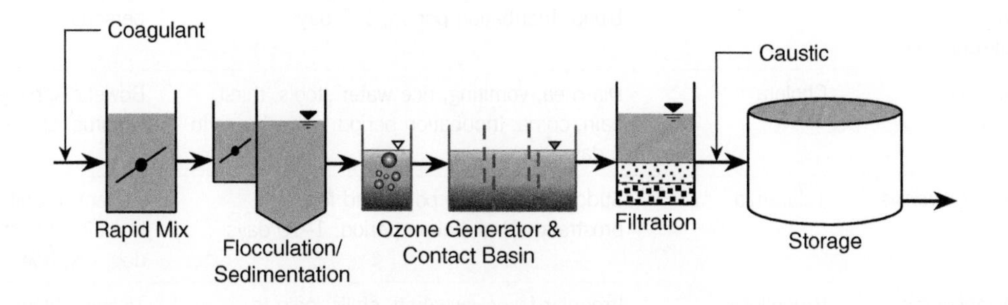

FIGURE 7–5 Ozone oxidation plant. (U.S. Environmental Protection Agency, 2005b.)

end users see the water, they may become concerned, even though permanganate levels below 1 mg/L pose no health risk to healthy individuals. For this reason, the permanganate dose should be carefully calculated and its effects and residual carefully observed. In unusual cases, a permanganate neutralizer can be added.

Ozone is the most powerful oxidant used in water treatment. It is manufactured on site by passing air or oxygen through a powerful electrical field. Due to the capital, operational, and maintenance costs, ozone is only a part of large treatment systems. The ozone is used immediately. Because ozone is so powerful, it can oxidize more organic and inorganic constituents of the raw water than any other oxidant. Unfortunately, because of ozone's power, its use and dose must be carefully managed to avoid production of oxidized forms of naturally occurring compounds or elements that may pose risks to the public. The most common inorganics of concern are bromate and hexavalent chromium.

Chlorine dioxide has become a popular oxidant because it does not form chlorinated by-products. It is produced on site and used immediately. Its capital, operational, and maintenance costs are lower than ozone. Chlorine dioxide must not be overapplied because residual concentrations as low as less than 1 mg/L can cause odor and taste problems.

7.3.3 Disinfection

Pathogens, such as protozoan parasites (e.g., *Giardia* and *Cryptosporidium*), bacteria, and viruses in springwater can cause debilitating diseases and even death (Table 7–3). The basic methods and procedures for

Table 7–3 Waterborne Diseases from Bacteria (From U.S. EPA, 1999a. Sources: Salvato, 1972; Geldreich, 1972)

Causative Agent	Disease	Symptoms	Reservoir
Salmonella typhosa	Typhoid fever	Headache, nausea, loss of appetite, constipation or diarrhea, insomnia, sore throat, bronchitis, abdominal pain, nose bleeding, shivering, and increasing fever. Rose spots on trunk. Incubation period, 7–14 days	Feces and urine of typhoid carrier or patient
S. paratyphi *S. schottimuelleri* *S. hirschfeldii*	Paratyphoid fever	General infection characterized by continued fever, diarrhea disturbances, sometimes rose spots on trunk. Incubation period, 1–10 days	Feces and urine of carrier or patient
Shigella flexneri *S. dysenteriae* *S. sonnei* *S. paradysenteriae*	Bacillary dysentery	Acute onset with diarrhea, fever, tenesmus, and stool frequently containing mucus and blood. Incubation period, 1–7 days	Bowel discharges of carriers and infected persons
Vibrio comma *V. cholerae*	Cholera	Diarrhea, vomiting, rice water stools, thirst, pain, coma. Incubation period, a few hours to 5 days	Bowel discharges, vomitus, carriers
Pasteurella tularensis	Tularemia	Sudden onset with pains and fever; prostration. Incubation period, 1–10 days	Rodent, rabbit, horsefly, woodtick, dog, fox, hog
Brucella melitensis	Brucellosis (undulant fever)	Irregular fever, sweating, chills, pain in muscles	Tissues, blood, mold, urine, infected animal
Pseudomonas pseudomallei	Melioidosis	Acute diarrhea, vomiting, high fever, delirium, mania	Rats, guinea pigs, cats, rabbits, dogs, horses
Leptospira icterohaemorrhagiae (Spirochaetales)	Leptospirosis (Well's disease)	Fevers, rigors, headaches, nausea, muscular pains, vomiting, thirst, prostration, and jaundice may occur	Urine and feces of rats, swine, dogs, cats, mice, foxes, sheep
Enteropathogenic *E. coli*	Gastroenteritis	Water diarrhea, nausea, prostration, and dehydration	Feces of carrier

disinfection are the same as described for oxidation. Because water quality deteriorates as soon as it enters the distribution system, many utilities apply a secondary disinfectant to maintain the microbiological quality of water. Water in the distribution system can be contaminated by a variety of pollution sources, such as backflow, pipe leaks and intrusion, and bacterial regrowth in the distribution pipes. In the United States, the EPA mandates that treated water contain sufficient excess disinfection chemical to maintain a residual in the distribution system and ensure no microbial regrowth or recontamination in the water as it is distributed (U.S. EPA, 1999a, 2006b). However, the idea of the need for a residual disinfectant is not universally accepted. For example, some European municipalities do not implement secondary disinfection (Franchi, 2009).

The amount of required additional treatment for a spring under direct influence of surface water is dependent on the results of source water *Cryptosporidium* monitoring and the existing level of treatment. Systems can treat for *Cryptosporidium* by (1) removing *Cryptosporidium* through filtration processes, like granular media filtration, cartridge filters, or membranes, or (2) using disinfectants that are effective against *Cryptosporidium*, such as chlorine dioxide, ultraviolet radiation, and ozone. Chlorine and chloramines are largely ineffective at inactivating *Cryptosporidium.*

The advantages and disadvantages of various disinfection methods are presented in Table 7–4, and the effectiveness of various disinfection methods for various pathogens is listed in Table 7–5.

UV light is an effective disinfectant for bacteria, viruses, *Giardia*, and *Cryptosporidium* and does not form disinfection by-products, such as trihalomethanes or haloacetic acids. Because particulate matter may affect the performance of ultraviolet (UV) systems, they are normally installed downstream from a filter for turbidity removal. A UV dose of 40 mJ/cm^2 has been shown to be sufficient for 3 log inactivation of *Cryptosporidium* and *Giardia* and 1 to 2 log inactivation of viruses. Studies have shown that a UV dose of 200 mJ/cm^2 is adequate for 4 log inactivation of viruses. It is, however, not possible to validate a UV reactor for 4 log virus inactivation (U.S. Environmental Protection Agency, 2005b).

Low-pressure (LP) UV lamp–based systems have been used for small treatment plants but are not typically installed at larger facilities due to the high number of lamps that would be required. Medium-pressure lamp systems are not typically used for smaller utilities due to higher capital costs in comparison to LP systems at low flow rates.

All UV systems are designed with an equipment redundancy of one extra UV reactor ($n + 1$) or 15 percent capacity above design flow, whichever is greater. UV disinfection systems are sensitive to power

Table 7–4 Summary of Advantages and Disadvantages of Disinfection Techniques (From Franchi, 2009. Source: Earth Tech, Canada, 2005)

Consideration	Chlorine	Chloramines	Ozone	Chlorine Dioxide	Ultraviolet
Equipment reliability	Good	Good	Good	Good	Medium
Relative complexity of technology	Less	Less	More	Medium	Medium
Safety concerns	Low to high*	Medium	Medium	High	Low
Bactericidal	Good	Good	Good	Good	Good
Virucidal	Good	Medium	Good	Good	Medium
Efficacy against protozoa	Medium	Poor	Good	Medium	Good
By-products of possible health concern	High	Medium	Medium	Medium	None
Persistent residual	High	High	None	Medium	None
pH dependency		High	Medium	Low	Low
Process control	Well developed	Well developed	Developing	Developing	Developing
Intensiveness of operations and maintenance	Low	Moderate	High	Moderate	Moderate

**Safety concern is high for gaseous chlorine but low for hypochlorites.*

Table 7–5 Effectiveness of Disinfectants on Selected Pathogens (Franchi, 2009, modified from U.S. EPA, 1999a)

	Microorganism Reduction Ability			
Disinfectant	***E. Coli***	***Giardia***	***Cryptosporidium***	**Viruses**
Chlorine	Very effective	Moderately effective	Not effective	Very effective
Ozone	Very effective	Very effective	Very effective	Very effective
Chloramines	Very effective	Moderately effective	Not effective	Moderately effective
Chlorine dioxide	Very effective	Moderately effective	Moderately effective	Very effective
Ultraviolet radiation	Very effective	Very effective	Very effective	Moderately effective

Note: The reduction levels in the table are for normal dose and contact time conditions, and they are only for general comparison purposes. The effectiveness of different disinfectants depends on the dose, contact time, and water characteristics.

interruptions and fluctuations. When a UV reactor goes down, it can take 4–10 minutes for the UV lamps to regain full power. A utility with poor power quality might have problems with the UV system going down too frequently. One way to prevent this problem is to install an uninterruptible power supply, which is essentially a battery that smoothes out the power interruptions and fluctuations (U.S. Environmental Protection Agency, 2005b).

7.3.4 Disinfection by filtration

The advanced filter technologies discussed here are able to remove particles of all sizes, including silt and clay, and capture microorganisms as small as viruses (Figure 7–6).

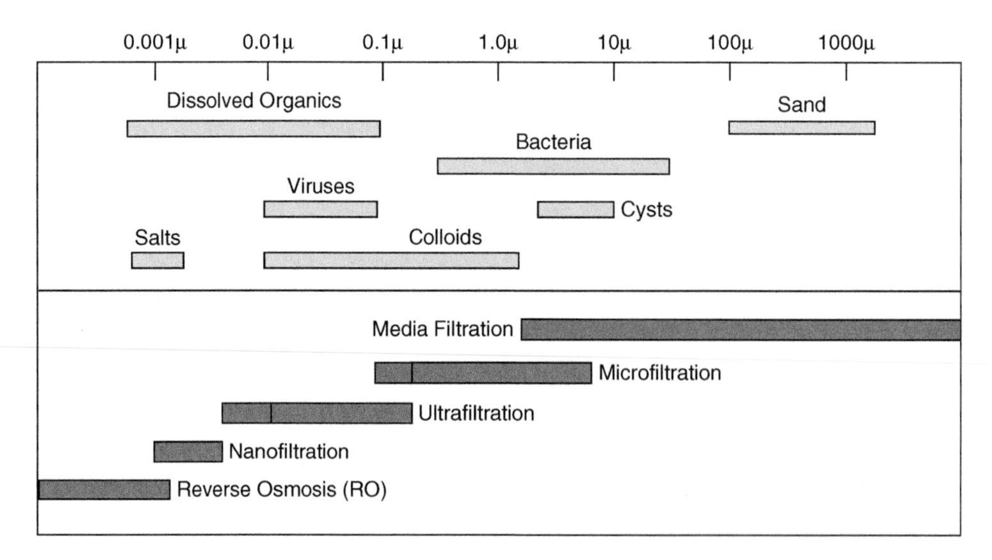

FIGURE 7–6 Pressure-driven membrane separation spectrum. (U.S. Environmental Protection Agency, 2005b.)

Media and replaceable element filters

Media filters, which are typically used to remove both turbidity and bacterial contamination of springwater, are susceptible to the following two problems:

- High-turbidity water can quickly plug the filter. Because media filters are capable of capturing very small particles, which can significantly reduce their throughput, a prefiltering to remove medium to fine particles before the final filter removes the smallest targets may be advisable.
- A disinfectant may have to be added to the water, occasionally or continuously, to assure that bacterial populations, metabolizing captured organic matter, do not taint or discolor the water or colonize the filter media, reducing its efficiency.

Replaceable element filters are used to remove very small to microscopic particles (silt turbidity to microorganisms). A replaceable element, either cartridge or bag, is placed in a metal or plastic housing. Water is passed through the unit until the differential between pressure meters upstream and downstream of the element becomes unacceptable. The valves to and from the housing are turned off. The water is drained from the element and discarded. The housing is opened, the used element is replaced and discarded, and the housing is closed. The valves are opened, trapped air is vented, and the filter unit is returned to use. Though this is the typical scenario, some filter elements in some applications can be backwashed one or more times before they are replaced.

Filter elements are available in a variety of sizes and shapes. They may be used independently or as a component of a treatment train, where they constitute the final polishing step. Filter elements are also used to protect other units targeting dissolved organics or inorganics or color. They come in a variety of sizes and element capabilities based on required flow rates.

Ceramic filters

Ceramic filter elements target very fine particles and bacteria. Ceramic elements are predominately clay. Consequently, the pore spaces between particles are very small and ideal for capturing very fine and microscopic particles. However, the small average pore sizes and tortuous flow path make these filters highly susceptible to physical plugging. Therefore, another filter must be placed upstream of a ceramic filter to remove the majority of fine to medium particles, or short cycles of filter cleaning or replacement must be implemented.

Ceramic filter throughput is very low. This limits their economical use to producing relatively small volumes of very clean water for drinking or cooking at single-family residences. If significant amounts of organic particulates in the water can be attacked by bacteria, it may be necessary to add a disinfectant to the filter or water stream to prevent taste and odor concerns related to bacterial metabolism of the organic matter.

Membrane filters

The cost to remove pathogens from drinking water is well spent in terms of both human well-being and productivity, which increase the health of the individual and the society. Specialized filter material can remove the finest of suspended particles, whether silt, clay, or pathogens. Due to the variety of filter materials and filter housing sizes and shapes available, a membrane filter can be found for virtually any application.

Membrane filters are easy to operate and perform well for a variety of purposes. Raw water is placed on one side of a permeable membrane and one of the following forces is applied:

- The raw water is placed under positive pressure.
- The treated water is placed under negative pressure.
- The raw water is pressurized while the treated water is placed under vacuum.

Membrane pore sizes are selected to prevent movement of undesirable solids, colloids and pathogens, and dissolved materials, including elemental salts and ions, through the membrane. Water and acceptable suspended or dissolved materials flow through the membrane because the system pressure is higher on the supply (raw water) side. The treated water is pumped directly into a supply system or is treated further before this step. The reject water, which now has a higher concentration of materials that could not pass through the membrane than the raw water, is typically treated before or after discharge under permit.

The size of particles controlled or removed by a specific membrane generally ranges from 1 to 1.5 orders of magnitude. Treatment or filter classifications approximate the range of particle sizes controlled. From largest to smallest particles removed, the classifications are media filtration, microfiltration, ultrafiltration, nanofiltration, and reverse osmosis. Media filtration can remove fine dusts and large bacteria; reverse osmosis can remove metal ions and salts and simple organic molecules, such as sugars. The latter is used for seawater desalinization.

A typical microfiltration or ultrafiltration treatment plant is shown in Figure 7–7. A nanofilter could replace the micro- or ultrafiltration unit in the treatment train, if needed.

Membrane materials are generally synthetic, typically polymers. The materials available for filters and treatment vessel or housing materials allow the use of membrane filtration for purification of industrial and commercial fluids as well as water supplies. The costs to install, operate, and maintain membrane filters have fallen significantly in the last 10–20 years, leading to increased use in raw water treatment as well as many more industrial and commercial processes.

Membranes can be configured as small tubes, large cartridges, or sheets. Examples of a hollow tube and a spiral wound cartridge (module) are shown in Figures 7–8 and 7–9, respectively.

Large or numerous particles (clay turbidity) may quickly obstruct a membrane. While some membranes can be backwashed to restore this function, those intended to control the smallest particles typically cannot be and the membrane must be changed. For this reason, the nature of the raw water and the potential benefits of pretreatment to remove the majority of suspended particles should be considered if the planned processing rate or liquid volume is large.

Pressure

Positive or negative pressure can be used to improve the throughput rate of most filters. However, the use of very high pressures or vacuums exaggerates the pressure differential across the media. The filter may fail, depending on the physical characteristics of the media. The increased production typically associated with pressurized or vacuum filtration makes the additional subsystems required to remove turbidity and control this risk acceptable, in many cases.

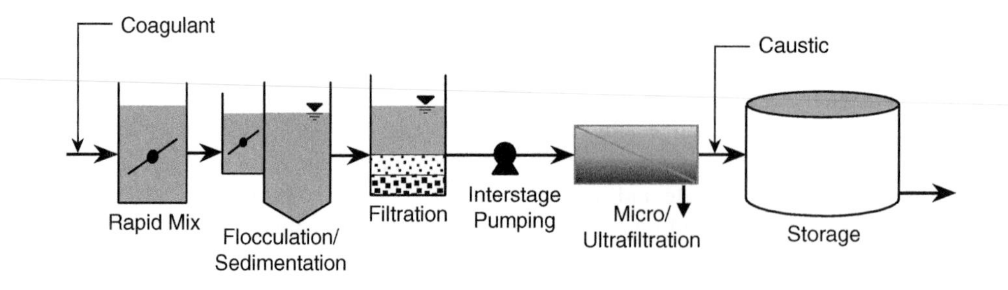

FIGURE 7–7 Microfiltration and ultrafiltration plant. (U.S. Environmental Protection Agency, 2005b.)

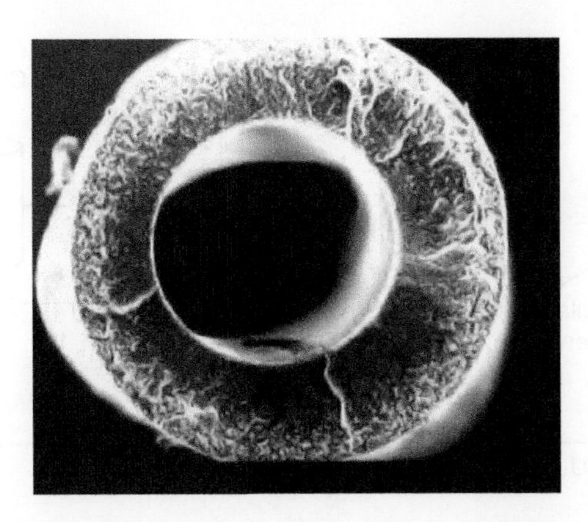

FIGURE 7–8 Hollow fiber cross-section photomicrograph. (U.S. Environmental Protection Agency, 2005a.)

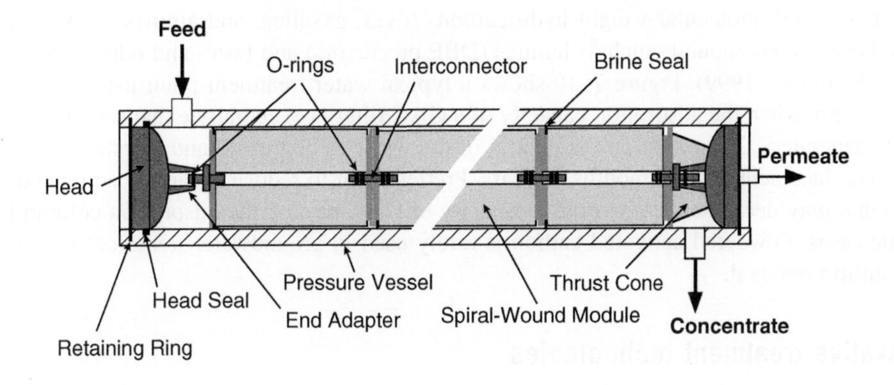

FIGURE 7–9 Typical spiral-wound (NF/RO) module pressure vessel. (U.S. Environmental Protection Agency, 2005a.)

An increased pressure differential between points upstream and downstream of a membrane filter is a common indicator that the filter element is clogged or obstructed and must be replaced or cleaned (backwashed). The same is true for ceramic filters and granular activated carbon vessels (see the next section).

Carbon adsorption

Granular activated carbon (GAC), made from a number of original source materials, such as bituminous coal, coconut shell, petroleum coke, wood, and peat, is used to adsorb undesirable components of raw water. Although GAC particles have a high surface area, which is physically very irregular, and is available in a variety of sizes, the media is not typically used for removal or capture of suspended particles, such as organic matter or clay. This is because a significant portion of the adsorptive capacity of the media, which typically relies on weak electrical charges, would be wasted if physical contact between the raw water and GAC were prevented by previously accumulated organic and inorganic particles.

As discussed by Franchi (2009), the main reason for using GAC is its capability for removing organic compounds. GAC can effectively remove synthetic organic compounds, such as aromatic solvents (benzene,

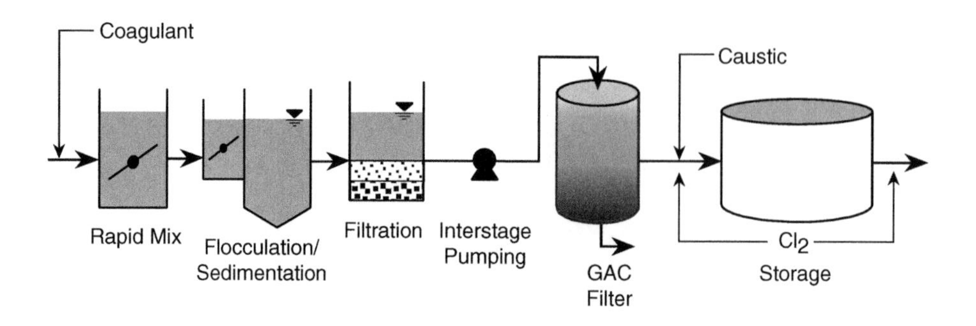

FIGURE 7–10 GAC filtration plant. (U.S. Environmental Protection Agency, 2005b.)

toluene, nitrobenzenes), chlorinated aromatics (PCBs, chlorobenzenes, and chloronaphthalene), phenol and chlorophenols, polynuclear aromatics (e.g., acenaphthene and benzopyrenes), pesticides and herbicides (e.g., DDT, aldrin, chlordane, and heptachlor), chlorinated aliphatics (e.g., carbon tetrachloride, and chloroalkyl ethers), and high molecular-weight hydrocarbons (dyes, gasoline, and amines). GAC can be used to remove natural organic compounds such as humics (DBP precursors) and taste- and odor-causing compounds (Snoeyink and Summers, 1999). Figure 7–10 shows a typical water treatment plant using GAC.

During operation, when the adsorption capacity of a GAC filter is eventually exhausted, the media must be replaced and regenerated. Depending on the quality of the water to be treated and filtration rates, the interval between media replacement may be months or years. Pretreatment to reduce organic loading and remove suspended solids that may decrease the adsorptive capacity of GAC or clog the adsorption column is a valuable option in some cases. Powdered activated carbon is rarely used in groundwater treatment with the exception of hydrogen sulfide removal.

7.3.5 Innovative treatment technologies

Other, sophisticated drinking water treatment methods include adsorption of inorganic ions on resins, which may be regenerated or replaced, and distillation. Both require careful management, significant energy, supplies, and labor and generate waste streams with high concentrations of impurities. The waste streams themselves may require treatment or disposal. Treatment systems of this sophistication and cost are rarely justified for water to be used for residential or high volume commercial or industrial purposes.

Biological treatment relies on development of a biofilm that metabolizes unwanted organic matter and organic compounds, typically those containing sulfur or nitrogen, which cause unacceptable taste and odors. The most efficient bacteria are aerobic. The optimum conditions for biological treatment are therefore high in dissolved oxygen, which allows specialized bacteria to oxidize and precipitate manganese, iron, and arsenic. A properly managed biological filter can achieve all realistic objectives. However, all biological systems require careful oversight. Therefore, although these systems are becoming more popular, they are not as common in the United States as in Europe.

7.3.6 Reverse osmosis technologies

Reverse osmosis (RO) technology can be used to extract potable water from seawater (approximately 3.5 percent total dissolved solids). The use of RO for desalination is similar to other applications where

RO is used. Using pumping pressure to overcome the osmotic pressure across a semipermeable membrane, these systems force water through the membrane to produce a high-quality permeate stream. A reject stream containing the salts and impurities from the feedwater at significantly higher concentrations must be properly disposed of.

In general, the membranes used for desalination are thin-film composite (TFC) membranes, consisting of a polysulfone support layer and a thin polyamide layer. Cellulose acetate membranes have been used for treatment of brackish water but are not suitable for desalination. TFC desalination membranes are typically two layers, but three-layer configurations are sometimes used where increased durability is required.

TFC membranes perform well over a wide range of pH and temperatures. The primary weakness of these membranes is their incompatibility with chemical oxidants, which may be used for pretreatment.

RO technology is also used for the treatment of brackish groundwater in areas where aquifers have been affected by saltwater intrusion, as well as in inland areas due to upconing of naturally occurring, highly mineralized waters. Treatment of these sources is similar to desalination of seawater, with lower operating pressures and less demanding pretreatment.

Desalination of brackish and submarine springs

Growing human populations throughout the world, and particularly in coastal and desert regions, continually increase the demand for potable water. The most logical solution in such areas is the use of reverse osmosis for desalination of seawater, brackish (onshore and submerged) springs, and mineralized groundwater. Power for the process can be obtained from conventional power plants or, in tropical latitudes, solar arrays.

Although the majority of the existing desalination plants use RO technology, there are a few viable alternatives, such as

- Electrodialysis and electrodialysis reversal.
- Multiple-stage flash distillation.
- Multiple effect distillation.
- Vapor compression (mechanical or thermal).

Despite continuing challenges associated with successfully applying the technology, RO typically has advantages over competing desalination technologies. The major advantage is that the life cycle cost (i.e., annualized capital plus operating and maintenance costs) is typically lower than for other technologies. For this reason, approximately two thirds of over 9500 desalination plants worldwide are RO facilities.

As with the other technologies, a major design and location challenge for an RO facility is brine (i.e., reject stream) disposal. This is governed by state and national regulations and is addressed on a case-by-case basis.

Pretreatment Effective pretreatment is crucial to design and operation of an RO facility for desalination. Pretreatment is required for control of several potential problems:

- Mineral scale (i.e., calcium carbonate, calcium sulfate).
- Biological fouling.
- Fouling by iron, manganese, and other metal oxides.
- Silica precipitation.
- Particulate and colloidal fouling.

Mineral scale, oxidation, and precipitation can be controlled or minimized using scale inhibitors, such as chelating agents and surfactants.

Biofouling can be difficult to control if there is consistent growth potential. In general, pilot testing is recommended to identify potential problems, including biofouling. Anaerobic system conditions are desirable, because anaerobic bacteria grow much more slowly than aerobic bacteria. In addition, a good mechanical design must ensure effective cleaning can be performed (i.e., adequate cross-flow velocities etc.). Good operator training is crucial to the integrity and operation of the membranes. A well-designed system will fail rapidly if operators are not well trained and attentive. Operational problems can quickly change the economics of RO treatment, for example, if original planning assumed a membrane life of three to five years but they have to be replaced every one to two years under actual operation.

Various separation technologies are suitable for control of suspended solids, precipitates, and colloidal material. These include all the techniques discussed previously, from clarification (possibly with chemical coagulation) to granular filtration to straining (i.e., cartridge filters, self-cleaning screens, microfiltration, ultrafiltration, and nanofiltration). Operators trained to monitor and maintain these systems are vital to particulate and colloid control and, thereby, performance of the RO system.

Capital cost Capital costs for desalination plants are variable, based on the pretreatment required, posttreatment residuals, raw water sources and intakes, method of brine disposal, and the like. For plants over 10 mgd in capacity, unit costs range from approximately $3 to $5 per gpd of plant capacity. Unit costs tend to decrease with increasing capacity.

Based on the empirical cost guidelines, a budgetary cost for a 25 mgd facility is approximately $100 million. This agrees well with the projected cost of the 25 mgd desalination plant in Tampa, Florida.

Operation and maintenance cost The principal operation and maintenance (O&M) cost components are energy, membrane replacement, labor, and chemicals. The unit O&M cost, like capital, is related to the plant size. Above 10 mgd, the total water cost (O&M plus debt service) ranges from approximately $1.50 to $2.50 per 1000 gallons. For the hypothetical 25 mgd plant, the cost of water would be approximately $2/kgal.

Energy recovery Various methods are used to recover the energy in the high-pressure reject stream from the RO system, to reduce energy costs: Pelton wheel, work exchanger, pressure exchanger, hydraulic turbocharger, and isobaric chambers. The first four technologies are relatively well established. The Pelton wheel is the device used most commonly in larger plants. Pressure exchangers are more common in smaller facilities (under 2.5 mgd).

Isobaric chambers are used to transfer pressure energy from the brine to the feed directly without using turbines or pumps (which introduce another layer of shaft inefficiency). One recent improvement to this technology uses a spinning ducted rotor to transfer the energy from the brine to the feed stream. Due to its configuration, this device also functions as a high-pressure pump (which reduces the required capacity for high-pressure feed pumps). Literature sources claim 95 percent efficiency for this device, compared with up to 85 percent (ignoring shaft inefficiency introduced by driving a high-pressure pump).

Relationship to power plants Several advantages accrue to locating desalination facilities adjacent to electric power generating plants. These advantages, which tend to reduce the capital and O&M costs associated with desalination, include

- Sharing common seawater intake.
- Combining RO brine discharge with power plant cooling water discharge (reducing the cooling water thermal plume and decreasing the salinity of the brine plume).
- Ability to consider a hybrid desalination plant consisting of RO and thermal processes (electricity for RO plant and low-pressure waste steam for thermal desalination).

7.4 CASE STUDIES

7.4.1 Small-scale springwater treatment system, Iraq

Groundwater and springs are water sources for many villages in northern Iraq. Groundwater from springs or wells is used in traditional farming (irrigation and livestock for local food supply) and for human consumption. The water is often of such poor quality from a health and aesthetic standpoint that it is ill-suited for many of these uses. Sulfide is one of the main naturally occurring contaminants. Local populations are often forced to migrate to locate better water resources. An estimated 120,000 people in this region are forced to depend on substandard water supplies (Stevanovic and Iurkiewicz, 2004).

The existing water sources include deep wells (100–200 ft), shallow wells (30–50 ft), and springs. A typical irrigation system includes a well pump (except for systems fed by springs), gravity concrete channels, small basins, and perforated galvanized pipe for drip irrigation. These systems have no water treatment equipment, but the channels can provide some sulfide reduction, if there is sufficient time for the release of dissolved hydrogen sulfide to the atmosphere.

Sampling and analysis of approximately 140 water sources is summarized in Table 7–6. For each water source type, the best and worst water qualities are presented. Water quality guidelines (i.e., World Health Organization guidelines and U.S. EPA regulatory levels) are also shown in the table. Values above the respective limits or guidelines are in bold. Also shown in bold are other values of concern, even if there are no applicable guidelines, such as a total hardness of 3552 mg/L, which is not suitable for domestic use.

These data show significant variability between the best and worst sources within each source type. It is assumed that the columns of minimum data reflect the sources with the best water quality within each category, as opposed to the minimum values pulled from numerous sources within that category. This seems reasonable, since a number of the parameters are linked from an analytical perspective (i.e., total dissolved solids comprises hardness, potassium, sodium, chloride).

It is assumed that turbidity was measured in the lab, rather than in the field. Therefore, it is possible that the turbidity was produced by the precipitation of dissolved constituents during the trip to the lab. Iron, if present, will precipitate on exposure to oxygen. In addition, calcium carbonate can precipitate as the sample warms up or dissolved carbon dioxide is released, depending on the level of dissolved calcium in the sample and other factors. Therefore, elevated turbidity levels do not necessarily serve as an indicator of microbial contamination in this case (but still reflect an aesthetic problem, at a minimum). It should be noted, however, that the elevated level of nitrate in some sources does suggest pollution from animal or human sources. Therefore, treatment for turbidity removal and disinfection is recommended.

Although not shown in Table 7–6, most of the sources sampled have significant sulfide concentrations. The sampling results also suggest that there may be sources that would be good quality for human consumption after the sulfide is removed, with little or no further treatment. Other supplies would require a high degree of treatment to be considered acceptable for human consumption.

Project scope

Small-scale, modular, water treatment systems have been initially proposed by Stevanovic and Lurkiewicz (2004). Such systems improve the integrity and quality of village water supplies for humans, livestock, and irrigation and human, animal, and plant health. Such benefits reduce the community cost of a system. Each village system consists of two treatment modules (Figure 7–11):

1. A hydrogen sulfide removal system would treat approximately 8000 gpd for livestock and irrigation uses, and supply treated water to the drinking water treatment module for further treatment. It would be located near an existing well or spring.

Table 7–6 Source Water Quality (From Stevanovic and Lurkiewicz, 2004)

Parameter	Deep Wells		Shallow Wells		Springs		World Health Org.		EPA	
	Min.	Max.	Min.	Max.	Min.	Max.	Health	Aesthetic	MCL	SMCL
Temperature (°C)	16	26	15.5	25	13	31.5			—	—
pH	7.1	8.2	7.1	8.1	6.3	8		6.5 to 8	—	6.5 to 8.5
Total dissolved solids	422	**18,269**	65	**1691**	104	**1385**		1200	—	500
Turbidity (NTU)	0.01	**166**	0.3	**10.4**	0.01	**13.1**	None[1]		TT[2]	—
Total hardness (as $CaCO_3$)	150	**3552**	138	**1336**	170	**1521**			—	—
Potassium	0.1	9.9	0.1	74	0.2	26			—	—
Sodium	2.8	**9660**	8	**483**	0.7	**397**			—	—
Calcium	24.8	886	38.5	305	41.7	223			—	—
Magnesium	9.7	326	7.3	156	7.3	94			—	—
Chloride	2.1	**10,100**	9.9	**810**	BDL	**225**		250	—	250
Sulfate	BDL	**9370**	12.5	**518**	1.9	**1354**		500	—	250
Bicarbonate	85.4	222	207	586	220	512			—	—
Nitrate	0.2	**181**	4.5	**1284**	0.1	**26**	50		44[3]	—

Note: Units are in milligrams per liter unless otherwise noted.
1. Recommend median <0.1 NTU for effective disinfection.
2. <0.3 in 95th percentile; 1 NTU max.
3. U.S. EPA MCL of 10 mg/L NO_3–N, expressed as mg/L NO_3.
BDL = below detection level.
MCL = maximum contaminant level.
NTU = nephelometric turbidity units.

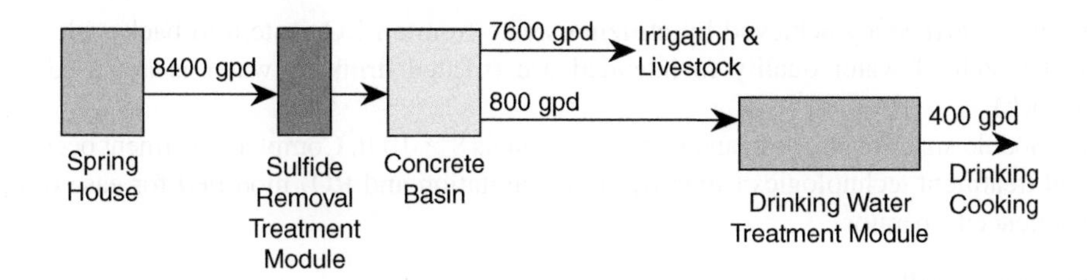

FIGURE 7–11 Small modular water treatment system configuration.

2. A drinking water treatment system would treat approximately 400 gpd for human consumption. The exact system components in the module would be based on the raw and pretreated water quality. The module would be located in the village, with a station for filling household water vessels or attachment to a pressurized piping system.

Dissolved sulfides in water are present as hydrogen sulfide gas and HS^- ion. The proportion of each is determined by the pH of the water. Sulfide removal can be accomplished by stripping hydrogen sulfide from solution, chemical oxidation, or biological oxidation.

The high sulfide concentrations and rural locations make chemical oxidation impractical for these systems. Biological oxidation is also not practical, due to the operator attention required. Stripping can cause nuisance odors, but the rural nature of the area makes it less likely that there will be homes close to the water treatment system. Therefore, a system would incorporate one or more of the following stripping technologies:

1. A stripping tower with forced draft aeration (pump to top of tower using well pumps).
2. Diffused air aeration in a concrete channel (suitable for both springs and wells).
3. Induced channel turbulence—engineer cascade sections as part of gravity channel systems or placement of obstructions in gravity channels to increase the stripping that already takes place.

Following sulfide removal, some sources may be suitable for human consumption with no further treatment. These sources would be sources with the characteristics shown in the Minimum columns in Table 7–6. For these sources, which are low in turbidity, hardness, and TDS, only disinfection and storage would be required.

Disinfection would consist of a passive chlorination device, which uses calcium hypochlorite tablets instead of liquid hypochlorite (i.e., bleach) or gaseous chlorine, followed by a small vessel to provide contact time prior to storage. With adequate disinfection contact prior to storage, short-circuiting in the storage tank is not a concern.

For all but the best-quality sources, treatment to remove a variety of constituents is necessary to produce water suitable for human consumption. These sources would require treatment for removal of turbidity, hardness, and a number of other dissolved constituents.

Treatment would consist of softening, turbidity removal, and total dissolved solids reduction. Softening would be accomplished using a VRTX unit, which would convert dissolved calcium and magnesium to insoluble forms. These would be filtered from the water using a sand filter prior to the TDS removal step, along with any turbidity in the water. TDS removal would consist of a small reverse osmosis unit. The concentrate from the RO system would be stored in a tank for daily backwashing of the sand filter. Disinfection would be as discussed previously, followed by storage.

The advantages of the described approach, which addresses constraints to service delivery, that is, low-cost treatment of very poor-quality water, are

- Application of technologies that do not require treatment chemicals, which results in very low operating cost.

- High water-use efficiency achieved by utilizing waste from an RO system to backwash the sand filter.
- Treatment match of water quality to intended use (treated drinking water is not used for irrigation and livestock).
- Compact module size. Disinfection and storage module is 8 × 10 ft. Complete treatment occupies 8 × 20 ft.
- Advanced treatment technologies (hydrodynamic cavitation and RO) modified for ease of operation by an inexperienced operator.

Outcome This system will

- Provide water suitable for irrigation and livestock, which will improve the ability of local populations to provide food locally and regionally.
- Provide water suitable for human consumption, which will improve the health and quality of life of local populations. This will eliminate the need for regional migration to locate potable water.
- Such improvements would be expected to improve local economies, which could also give rise to other forms of economic activity. Diversification would be beneficial for these areas.

Financial viability It is expected that villages will provide the labor for construction of water systems and a site for the environmentally responsible disposal of the treatment residuals and spent materials. This will give them a stake in the project outcome and build local support. Despite the challenging economic conditions in this region, it is important that the consumers pay for the water, even if it does not initially cover the cost of production. This would encourage responsible water use and conservation (based on experience of international aid agencies). This system would have lower operation and maintenance costs than alternative systems. The only major inputs are electricity (for the well pump and RO feed pump) and periodic membrane replacement (frequency to be determined). Therefore, it is anticipated that villages would be able pay the true cost for water production as the local economies develop.

Sustainability All the challenges posed by this project are manageable with the proper testing and attention to design. The challenges and their respective solutions are

- **Treatment systems have to treat a range of water qualities**. The system component would depend on local source water quality.
- **Systems need to operate without highly trained or educated operators**. The proposed treatment system requires no adjustment by operators.
- **Repair parts and treatment chemical deliveries are limited in the area**. Very few moving parts and no liquid treatment chemicals are used by the treatment system.
- **Village electrical supplies are limited**. Energy is conserved by matching the treatment capacity to the intended use (i.e., treat 400 gpd for human consumption, instead of treating the entire 8000 gpd). Solar power can be used to power the drinking water treatment module.

Replicability and scaling-up Design simplicity would allow the system to be placed in any part of the world with similar water quality, such as the Middle East or northern and sub-Saharan Africa. For regions with different water quality characteristics, the treatment system could be modified to meet other water quality demands. The system could be scaled up to produce larger flow rates.

7.4.2 Ozone technology, Ann Arbor, Michigan

An extensive 1,4-dioxane plume, extending 3–4 miles and over 200 ft deep, was created during permitted lagoon disposal of manufacturing process water. Remediation of the contaminated groundwater consists of recovery of 1300 gallons a minute from numerous production wells, transmission to a central treatment location through force mains, and advanced oxidation to destroy the organic contaminant. A particular challenge is the presence in the groundwater of naturally occurring bromine, which could be converted to bromate by overly aggressive treatment.

1,2-Dioxane is a very stable molecule. It is not readily degraded by permanganate, hydrogen peroxide, chlorine, or bacteria. Ozone is a powerful oxidant that can degrade contaminants via two mechanisms. The first, commonly referred to as the *direct mechanism*, involves the reaction of molecular ozone with the contaminant. The second mechanism, commonly known as the *indirect mechanism*, involves the oxidation of the contaminants by secondary oxidants, particularly the hydroxyl radical, $^{\bullet}OH$, which is produced when ozone is degraded. The importance of the direct and indirect pathways for contaminant degradation depends on the relative reactivity of the contaminant with ozone and secondary radicals.

Overview

The Pall ozone (O_3)/hydrogen peroxide (H_2O_2) system is a continuous flow reactor into which ozone (in gaseous form) and hydrogen peroxide are continuously added. The system is very effective in the destruction of various organics found in contaminated groundwaters and can purify and disinfect drinking and process waters.

The two major effects of mixing ozone with hydrogen peroxide are (1) increasing the oxidation efficiency by increased conversion of ozone molecules to hydroxyl radicals and (2) improvement of the ozone transfer from the gas phase to liquid due to an increase in ozone reaction rates. Hydroxyl radicals are created in water during the decomposition of ozone. By adding hydrogen peroxide, the decomposition rate is raised, which raises the concentration of hydroxyl radicals:

$$2O_3 + H_2O_2 \rightarrow 2^{\bullet}OH + 3O_3$$

Pall O_3/H_2O_2 treatment technology utilizes multiple inline injection points (Figure 7–12). The proper contact time between successive injection points was carefully calculated to maximize the organic(s) destruction rate. The system is unique, compared to commercially available similar systems, in that it does not require high-pressure ozone generation (>15 psi). The system can use either liquid oxygen or high-purity oxygen gas from an on-site oxygen generator as the feed for O_3 gas generation. The system is controlled by a networked programmable logic controller. An autodialer provides notification of alarm conditions to off-site operating personnel. The computer software allows operator interaction with the controller system. This provides a visual representation of the system operating conditions and allows the operator to start and stop equipment and adjust system set points as desired. The system is equipped with advanced interface panels to allow monitoring and operation from remote locations. A computer processor continuously monitors the ambient levels of oxygen and ozone. The system is automatically shut down and an exhaust fan or damper is started if a high level condition is experienced.

Ozone reacts directly with the bromide ion to form bromate. Ozone converts the natural bromide ion (Br^-) to a hypobromite ion (BrO^-) to a bromite ion (BrO_2^-) and subsequently to a bromate ion (BrO_3^-). Currently, the maximum contaminate level of bromate in drinking water is 10 μg/L. Pall O_3/H_2O_2 is capable of treating organics while maintaining bromate well below the required MCL concentration.

Advantages of the O_3/H_2O_2 treatment technology are

- No residual taste or odor left by the process.
- On-site total destruction of organics without the generation of residues.
- Ozone generated on site.
- Ability to be designed as a mobile unit to be placed in operation for residential or commercial settings (Figure 7–13).
- Utilization of fewer chemicals than other treatment systems, thus reducing the need for on-site storage with the associated hazards and costs.

FIGURE 7–12 Full-scale advanced oxidation and ozone treatment system for flow rate of 1300 gpm. Design and engineering Farsad Fotouhi, Pall Corporation.

FIGURE 7–13 Mobile field unit for flow rate of 200 gpm. Design and engineering by Farsad Fotouhi, Pall Corporation.

- Ability of ozone to partially oxidize nontarget organics in the water to biodegradable compounds that can be removed by biological filtration.
- Inability of microorganisms to become resistant to ozone.

7.5 SUMMARY

The eventual use and natural condition of springwater determine the treatment applied. The demand for potable water and livestock water continues to grow as the human population grows. Therefore, lower-quality water, virtually the only water not fully utilized at this time, will eventually require treatment before use.

The constituents of the water requiring reduction or removal are similar to those found in groundwater, although dissolved gases, such as hydrogen sulfide, may be more common in certain parts of the world, depending on the geology from which the springwater and groundwater arise.

The technologies used for springwater conditioning are those used to treat groundwater and surface water. Their engineering and operation are well understood and readily available. The most significant challenge is adequate monitoring of springs that are quickly affected by surface events, such as rain events, and adjustment of treatment plant operations to remove new constituents or continue control of increased levels of those already present.

REFERENCES

Appelo, C.A.J., Postma, D., 2005. Geochemistry, groundwater and pollution, second ed. Taylor & Francis/Balkema, Leiden, the Netherlands.

Davis, S.N., DeWiest, J.M., 1991. Hydrogeology. Krieger Publishing Company, Malabar, FL.

Drever, J.I., 1988. The Geochemistry of Natural Waters. Prentice-Hall, Englewood Cliffs, NJ.

Earth Tech (Canada), 2005. Chlorine and Alternative Disinfectants Guidance Manual. Prepared for: Province of Manitoba (Canada) Water Stewardship. Canada Office of Drinking Water, Winnipeg, Manitoba.

Eckhardt, G., 2009. Hydrogeology of the Edwards Aquifer. Available at: www.edwardsaquifer.net/geology.html#movement.

Franchi, A., 2009. Groundwater Treatment. In: Kresic, N. (Ed.), Groundwater Resources: Sustainability, Management and Restoration. McGraw-Hill, New York, pp. 437–482.

Geldreich, E.E., 1972. Water-borne pathogens. In: Mitchell, R. (Ed.), Water Pollution Microbiology. John Wiley & Sons, New York, pp. 207–241.

Health Research Inc., 2007. Recommended Standards for Water Works, 2007 Edition. Health Education Services Division, Albany, NY. Available at: www.leafocean.com/test/10statepreface7.html.

Kresic, N., 1988. Karst aquifers of the Lim catchment (Serbia). In: Des comptes rendus des séances de la Sociéte Serbe de Geologie pour l'année 1985–1986. Belgrade, pp. 217–223.

Kresic, N., 2009. Groundwater Resources: Sustainability, Management, and Restoration. McGraw-Hill, New York.

Matthess, G., 1982. The Properties of Groundwater. John Willey & Sons, New York.

Salvato Jr., J.A., 1972. Environmental Engineering and Sanitation, second ed. John Wiley & Sons, New York.

Snoeyink, V.L., Summers, R.S., 1999. Adsorption of Organic Compounds. In: Letterman, R.D (Ed.), Water Quality and Treatment, fourth ed. American Water Works Association. McGraw Hill, New York, pp. 13.1–13.76.

Stevanovic, Z., Iurkiewicz, A., 2004. Hydrogeology of northern Iraq, Vol. 2, Regional hydrogeology and aquifer systems. Spec. Ed. FAO (Spec. Emerg. Prog. Serv.), Rome.

U.S. Environmental Protection Agency, 1991. Guidance manual for compliance with the filtration and disinfection requirements for public water systems using surface water sources. Office of Drinking Water, Washington, DC, various paging.

U.S. Environmental Protection Agency, 1992. Consensus method for determining groundwater under the direct influence of surface water using microscopic particulate analysis (MPA). Office of Drinking Water, Washington, DC, EPA 910/9-92-029.

U.S. Environmental Protection Agency, 1998. National Primary Drinking Water Regulations: Interim Enhanced Surface Water Treatment; Final Rule. Fed. Regist. 63 (241), Rules and Regulations, 69477–69521.

U.S. Environmental Protection Agency, 1999a. Alternative Disinfectants and Oxidants Guidance Manual. Office of Water, EPA 815-R-99-014 Available at: www.epa.gov/safewater/mdbp/alternative_disinfectants.guidance.pdf (Accessed January 2009).

U.S. Environmental Protection Agency, 1999b. Enhanced Coagulation and Enhanced Precipitative Softening Guidance Manual. Office of Water. EPA 815-R-99-012. Available at: www.epa.gov/safewater/mbdp/coaguide.pdf (Accessed January 2009).

U.S. Environmental Protection Agency, 1999c. Guidance Manual for Conducting Sanitary Surveys of Public Water Systems; Surface Water and Ground Water under the Direct Influence (GWUDI). Office of Water, EPA 815-R-99-016.

U.S. Environmental Protection Agency, 2003. Overview of the Clean Water Act and the Safe Drinking Water Act. Available at: www.epa.gov/OGWDW/dwa/electronic/ematerials (Accessed in September 2008).

U.S. Environmental Protection Agency, 2005a. Membrane Filtration Guidance Manual. Office of Water. EPA 815-R-06-009. Available at: www.epa.gov/safewater/smallsys/pdfs/guide-smallsystems-sdwa.pdf (Accessed January 2009).

U.S. Environmental Protection Agency, 2005b. Technologies and Costs Document for the Final Long Term 2 Enhanced Surface Water Treatment Rule and Final Stage 2 Disinfectants and Disinfection Byproducts Rule. Office of Water. EPA 815-R-05-013 Available at: www.epa.gov/safewater/disinfection/lt2/regulations.html (Accessed January 2009).

U.S. Environmental Protection Agency, 2006a. Setting standards for safe drinking water. Office of Water, Office of Ground Water and Drinking Water. Available at: www.epa.gov/safewater/standard/setting.html (updated November 28, 2006; accessed December 2008).

U.S. Environmental Protection Agency, 2006b. Point-of-Use or Point-of-Entry Treatment Options for Small Drinking Water Systems. Office of Water. EPA 815-R-06-010 Available at: www.epa.gov/safewater.smallsys/ssinfo.htm (Accessed January 2009).

CHAPTER

Delineation of spring protection zones 8

Nico Goldscheider
Centre of Hydrogeology, University of Neuchatel, Neuchatel, Switzerland

8.1 INTRODUCTION

Freshwater from springs is the prime example of a renewable natural resource: It can be used and the infiltration of rainwater or meltwater in the catchment will renew it—as long as the climate does not change dramatically, which it may do, however. One of the most obvious ways to waste this resource is groundwater contamination. Therefore, the delineation of spring protection zones and the implementation of proper land-use practices in these zones, resulting in a reduction of polluting activities, are keys to the sustainable use of these valuable drinking water resources (Adams and Foster, 1992; Ravbar, 2007). Other important elements include springwater quality monitoring and drinking water treatment, typically by filtration and disinfection (Figure 8–1). All these activities are required to ensure safe drinking water from springs as well as from other freshwater resources.

Most human activities cause contamination. Reasonable land-use planning consists of finding a balance between groundwater protection, on the one hand, and human activities, such as agriculture and industry, on the other. Groundwater protection must be a high priority in zones where contamination can easily enter

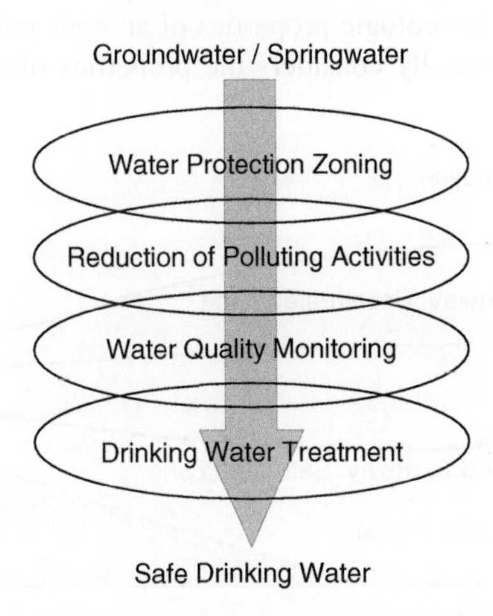

FIGURE 8–1 The security chain required to ensure safe drinking water from springs.

Copyright © 2010, Elsevier Inc. All rights reserved.

the aquifer and affect a spring or pumping well used for the drinking water supply of a large population. The challenge of protection zoning is to identify these zones and assign them high priority. However, even in less sensitive zones, polluting activities should be minimized as much as possible. Groundwater resources that are not used at present should be protected for possible future use; furthermore, clean groundwater is also crucial for groundwater-dependent ecosystems (Eamus and Froend, 2006) and the "hidden biodiversity" in aquifers (Danielopol and Pospisil, 2001).

There are two general approaches to groundwater protection: resource protection and source protection. The term *resource* is often used to describe an entire groundwater body. A "source" can be a tapped spring, a pumping well, or another type of groundwater abstraction point (Daly et al., 2002). The concepts of resource and source protection are indivisibly connected: It is impossible to protect a particular source without protecting the resource. However, from a conceptual point of view, it is useful to differentiate between the two approaches.

This leads to another basic concept, the origin-pathway-target conceptual model of environmental management: The "origin" is the place of potential contaminant release, usually the land surface; the "pathway" includes all components of the natural system through which a potential contaminant has to travel from the origin to the target; and the "target" is the water that is potentially affected by contamination and has to be protected. For resource protection, the target is the groundwater table, and the pathway essentially consists of the unsaturated zone, where water and contaminants move downward by percolation. For source protection, the target is the spring or well, and the pathway additionally includes the saturated zone of the aquifer, or aquifer system, as illustrated in Figure 8–2.

The first and most important question for source protection is, Where is the catchment area of the spring? Catchment areas are sometimes also referred to as *drainage areas*, *recharge areas*, or *watersheds*. Different hydrogeologic methods for the delineation of catchment areas are available, such as artificial tracer tests. The second question is, How easily can contaminants enter the aquifer and be transported to the spring? This second question leads to another important concept, the vulnerability of groundwater to contamination, which can be defined as the ease at which contaminants could enter the groundwater.

The terms *vulnerability to contamination* and *natural protection against contamination* can be used alternatively. High vulnerability means low natural protection and vice versa. Two main types of groundwater vulnerability can be distinguished: intrinsic vulnerability and specific vulnerability. Intrinsic vulnerability depends only on the geologic and hydrogeologic properties of an area and is independent of the type of contaminant. Specific vulnerability additionally considers the properties of a specific contaminant or group of

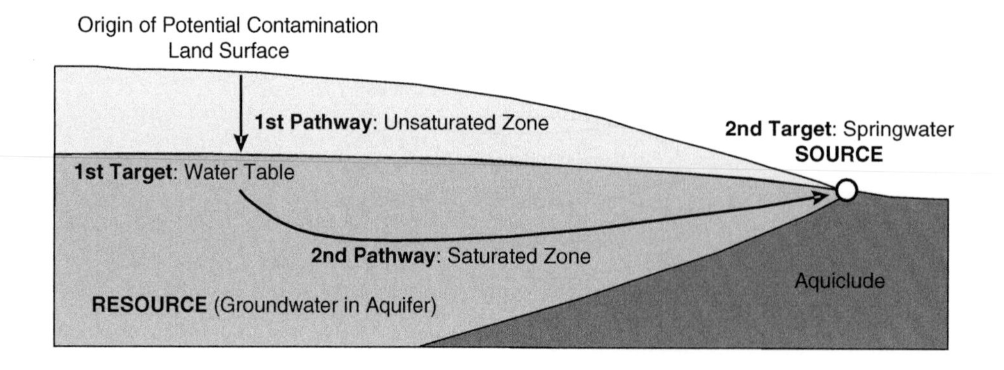

FIGURE 8–2 The origin-pathway-target model for groundwater resource and source protection.

contaminants and their interaction with the geologic environment. Vulnerability maps can be used as a basis for the delineation of groundwater resource and source protection zones.

Vulnerability maps are not stand-alone documents but can be combined with other types of information, depending on the purpose. A hazard map shows the location and type of actually or potentially polluting activities, such as agriculture, industry, and waste disposal sites. A risk map combines the information of a vulnerability map and a hazard map and makes it possible to identify zones where groundwater is actually at risk of being contaminated. A risk map consequently shows the necessity to act, ideally by removing existing hazards and changing land-use practices or otherwise by implementing early-warning systems. Another element of proper groundwater protection schemes is the evaluation of the importance of the groundwater or springwater for use as drinking water, for connected surface water ecosystems, or for other purposes. A large spring that is a source of drinking water for a large population will always be classified as very important, and its protection needs to be given a high priority (Ravbar and Goldscheider, 2007).

A large number of national and international legislations and guidelines deal with the delineation of groundwater source and resource protection zones. It is beyond the scope of this chapter to review, discuss, and compare all these guidelines. Its goals are to provide a short overview of springwater contamination problems and natural attenuation processes, discuss the hydrogeologic basis for the delineation of spring protection zones, and present the relevant methods and techniques. As most large springs of the world are karst springs, there is a special focus on this type of hydrogeologic environment.

8.2 SUMMARY OF CONTAMINATION PROBLEMS IN SPRINGWATER

8.2.1 Introduction

Contaminants in springwater can originate from various sources, such as agricultural land use, waste disposal sites, leakage of storage tanks or pipelines, release of untreated domestic wastewater (Figure 8–3), industrial activities, and transportation accidents in the catchment of the spring. Salinization caused by improper irrigation can also affect springwater quality (Scanlon et al., 2007).

Contaminants can be grouped into inorganic chemical compounds, organic chemical compounds, and microbial pathogens. Some contaminants, such as nitrate (NO_3^-), are not acutely toxic, but high concentration levels in drinking water over long periods of time are unfavorable to human health. Nitrate is a common groundwater contaminant in agricultural areas (Boyer and Pasquarell, 1999). Other contaminants are dangerous even at very low levels. For example, a single Norwalk virus in drinking water can cause infection (Szewzyk et al., 2000). Hexavalent chromium and benzene are examples of highly toxic inorganic and organic contaminants, respectively.

Not only the toxicity, but also the fate and transport of contaminants in aquifers are crucial for their actual risk to human health. For example, lead is a highly problematic toxic metal in soils, plants, and aerosols but generally not of concern in groundwater from springs, due to its limited mobility (Shokes and Moller, 1999). In a simplified way, contaminants can be grouped into conservative and reactive ones. Conservative contaminant transport depends only on the hydraulic properties of the aquifer, particularly on the flow velocity and heterogeneities causing dispersion. Reactive contaminants are additionally influenced by various processes, such as adsorption, precipitation, chemical transformation, biodegradation, and for particulate and particle-bound contaminants, filtration and sedimentation (Fetter, 1999).

The persistence of contaminants in groundwater with respect to the transit time to the spring is the key issue in delineation of source protection zones. Groundwater transit times can range from hours in karst conduits to thousands of years or more in deep circulation systems (Toth, 1999). Toxic metals have infinite

FIGURE 8–3 Former injection well near a restaurant at a skiing station in the northern Austrian Alps, where untreated wastewaters were released into a karst aquifer (Photo: N. Goldscheider). A tracer injected into this shaft reappeared at five karst springs at 4.1–5.7 km distance after 14 to 27 h. For details see Goldscheider (2005a).

lifetimes but can be transformed into less toxic species by redox and other processes, such as the conversion of highly toxic hexavalent chromate into less toxic trivalent chromium. Some organic compounds are biodegradable, while others are recalcitrant, depending on their chemical structure and biogeochemical conditions. The persistence of microbial pathogens in groundwater ranges from nearly absent to nearly infinite.

Pathogens found in springwater generally originate from fecal contamination, often caused by agricultural activities in the catchment area or untreated domestic wastewater release (Figure 8–3). Most fecal pathogens cannot grow or reproduce inside the aquifer and persist for only limited periods of time. Therefore, most groundwater protection legislation uses transit time as a key criterion for the delineation of "inner source protection zones" (see Section 8.4). For example, the German legislation uses the 50-day line of travel time, assuming that most pathogens will be filtered or inactivated by then, an approach that has proven feasible in most cases (Deutscher Verein des Gas- und Wasserfaches, 2006). Fecal indicator bacteria, such as *E. coli*, are used to control the hygienic safety of water. However, some pathogens can persist much longer, such as *Cryptosporidium* cysts (Bonadonna et al., 2002) and specific viruses (Schaub and Oshiro, 2000), some of which can also persist despite water treatment by chlorination (Lisle and Rose, 1995; Mackenzie et al., 1994). Therefore, the absence of *E. coli* provides no absolute safety (Brookes et al., 2005).

Contaminant levels in springwater are often highly variable, particularly in response to intense precipitation. Therefore, continuous water quality monitoring is important, complementary to detailed analyses of water samples taken at regular time intervals. However, microbial pathogens and many other contaminants cannot be monitored continuously, so that surrogate parameters need to be monitored as contamination indicators. For example, the simultaneous increase of total organic carbon and turbidity in karst springwater was demonstrated to correlate well with *E. coli* levels (Figure 8–4). Therefore, continuous monitoring of these two parameters can be used as an "early-warning system" for microbial contamination (Pronk, Goldscheider, and Zopfi, 2006), which can be further improved by measurements of particle-size distribution (Pronk, Goldscheider, and Zopfi, 2007).

8.2.2 Sources of contamination

Contaminant sources can be grouped according to their spatial extension (areal sources, line sources, point sources), their temporal characteristics (permanent, periodic, intermittent, only one time), the general origin

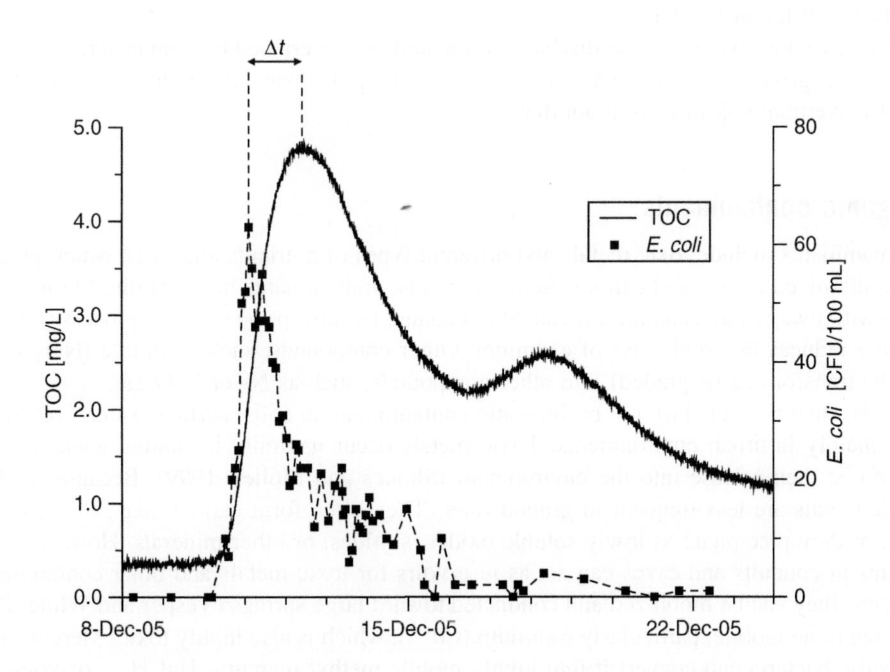

FIGURE 8–4 Evolution of total organic carbon (TOC) and fecal bacteria (*E. coli*, CFU = colony forming units) observed at a karst spring in Switzerland following intense rainfall. Both TOC and bacteria originate from a sinking stream draining agricultural land, which explains the high correlation. High TOC levels consequently indicate poor drinking water quality. Earlier peak times for bacteria (Δ*t*) can be explained by higher transport velocities and a time-dependent die-off function for bacteria. The high temporal variability of fecal contamination also illustrates the necessity of event-based sampling approaches and continuous monitoring techniques. For details see Pronk et al. (2006, 2007).

of the contaminants (agricultural, industrial, domestic, urban, military, natural), and according to the type and quantity of contaminants released. Fetter (1999) proposed a detailed classification, which is reproduced here in a condensed way:

- Contaminant sources designed to discharge substances include septic tanks and cesspools, different types of injection wells, and land applications, such as the application of sludge from wastewater treatment plants to the soil as a fertilizer.
- Sources designed to store, treat, or dispose of substances, such as different types of landfills and waste dumps, mine wastes, material stockpiles, graveyards, animal burials, storage tanks above and below ground, as well as radioactive waste disposal sites.
- Sources designed to retain substances during transport, including pipelines for oil, gas, or other liquids as well as wastewater sewers; the transport of hazardous substances by truck or train also represents potential contamination sources.
- Sources discharging substances as a consequence of other planned activities; this group includes many agricultural activities, such as irrigation causing salinization, the application of pesticides, synthetic fertilizers and farm animal wastes, but also the application of road salt.
- Sources providing a conduit for contaminated water to enter aquifers, such as production and monitoring wells, excavation work, and geothermal energy installations. Although these are not contamination sources in the proper sense, such installations can introduce contaminants or allow for cross contamination between different aquifers.
- Naturally occurring sources whose discharge is created or exacerbated by human activities; this includes surface water–groundwater interactions induced by pumping (bank filtration) and seawater intrusions caused by overpumping of coastal aquifers.

8.2.3 Inorganic contaminants

Inorganic contaminants include toxic metals and different types of nutrients and salts, which generally occur in the form of dissolved anions and cations. Some inorganic contaminants have infinite lifetimes: For example, cadmium will always stay cadmium; it can be attenuated by adsorption or other processes, but no natural process actually reduces the total mass of cadmium. Other compounds, such as nitrate (NO_3^-) or ammonia (NH_4^+), can be transformed (degraded) into other compounds, such as N_2 or N_2O gas.

Toxic metals, such as lead (Pb), are problematic contaminants in soils, surface waters, recent sediments, and aerosols, mainly in urban environments. Toxic metals occur naturally in mining areas, but the mining activities reinforce their release into the environment (Shokes and Moller, 1999). Because of their limited mobility, toxic metals are less frequent in groundwater. They often form cations (e.g., Pb^{2+}) that are prone to adsorption, or they precipitate as lowly soluble oxides, sulfides, or other minerals. However, in karst systems, sediments in conduits and caves can act as reservoirs for toxic metals and other contaminants; during high-flow events, they can be mobilized and conducted toward large springs (Vesper and White, 2004). Some metal cations are more mobile, particularly cadmium (Cd^{2+}), which is also highly toxic. Mercury is extremely toxic and specific bacteria can convert it into highly mobile methyl mercury, $HgCH_3^+$, or volatile dimethyl mercury, $Hg(CH_3)_2$ (Ehrlich, 1997).

Arsenic (As) is a toxic metal that causes large-scale geogenic groundwater contamination and diseases in several regions of the world, for example, in China, Bangladesh, and West Bengal (India), where tens or hundreds of millions of people are affected (Nickson et al., 2000). Natural As contamination occurs in arid to semiarid inland basins and in reducing alluvial aquifers, that is, in flat, low-lying areas, where groundwater is sluggish and often exploited by deep pumping wells (Smedley and Kinniburgh, 2002). Therefore, large springs are usually not affected by this type of contamination. However, arsenic-rich water is also found in mining areas; in the past, it was also used for insecticides, some of which might still be in use in some regions.

Nitrate (NO_3^-) is a common groundwater contaminant, mainly in agricultural areas, where different N-compounds are used as fertilizers (Boyer and Pasquarell, 1999). Nitrate is not very harmful, but high concentrations over long periods of time are unfavorable to human health. High nitrate levels often go along with other type of agricultural contamination, such as fecal bacteria or pesticides, and are therefore a bad sign for the general water quality. As nitrate is a nutrient, increased concentrations also influence the ecosystem. Under oxidizing conditions, nitrate behaves conservatively; under reducing conditions, it is microbiologically transformed into nitrogen gas (denitrification). Effluent waters contain high levels of ammonia (NH_4^+), which is often converted to nitrate under oxidizing conditions (Chapelle, 2000).

Dissolved salts are natural water constituents, but increased levels caused by human activities can render the water unusable. The application of road salt locally affects soils and vegetation but rarely causes large-scale groundwater contamination (Lundmark and Olofsson, 2007). Salt mining is another contamination source (Siefert, Buchel, and Lebkucher-Neugebauer, 2006). However, the severest problems are due to improper agricultural and water management practices: salinization by irrigation, and seawater intrusion caused by overpumping of coastal aquifers, often also for irrigation purposes (Milnes and Renard, 2004; Scanlon et al., 2007).

8.2.4 Organic contaminants

Organic compounds in groundwater can be grouped into natural and anthropogenic substances; the latter include synthetic chemical products but also petroleum, which is of natural origin but typically released into the environment by human activity.

Natural organic carbon (OC) can be present in dissolved or particulate form, the sum of which is referred to as *total organic carbon*. Natural OC in groundwater originates mostly from soils and surface waters infiltrating the aquifer and from all types of organic waste materials, such as liquid manure. High OC levels in drinking water are unfavorable, because (1) the presence of OC indicates the possible presence of fecal bacteria and pathogens, which originate from similar sources (Pronk et al., 2006; see the section that follows); (2) organic compounds can mobilize toxic metals, including radioisotopes, by complexation (Ranville et al., 2007); and (3) during water treatment, chlorine reacts with natural organic matter to produce carcinogenic disinfection by-products (Bull et al., 1995).

Synthetic organic contaminants are generally of greater environmental concern. They can be grouped according to their chemical structure and composition (e.g., aromatic compounds, halogenated hydrocarbons), according to their use and purpose (e.g., fuels, solvents, pesticides), according to their physical and chemical properties (e.g., solubility in water, density with respect to water, volatility, biodegradability), or according to their toxicity. For more details on organic contaminants, the reader is referred to related papers and textbooks (e.g., Fetter, 1999; Appelo and Postma, 2005; Kresic, 2009).

8.2.5 Microbial pathogens

Microbial pathogens in drinking water cause several billions of infections and several millions of deaths per year, mainly of children in developing countries (Montgomery and Elimelech, 2007). However, waterborne diseases also occur in rich countries (Herwaldt et al., 1992). Pathogens in groundwater include bacteria (e.g., *Vibrio cholerae*, *Campylobacter jejuni*, *Shigella* spp., *Salmonella* spp.), protozoans and their cysts (e.g., *Cryptosporidium parvum, Giardia lamblia, Entamoeba histolytica*), and viruses (e.g., Norwalk virus, polio, hepatitis, rotavirus) (Auckenthaler and Huggenberger, 2003; Szewzyk et al., 2000).

The most important transmission pathway is the fecal-oral route, that is, wastewater that contaminates drinking water. Feces of human origin are generally more dangerous than animal feces, because some pathogens, particularly viruses, are highly host specific (Smith, 2001). Therefore, human feces are more likely to include human pathogens. The general health status of the population is also important: Feces from a diseased population include more pathogens. Therefore, poor sanitary conditions and insufficient groundwater protection are often starting points of a vicious circle: Contamination of drinking water causes disease, which generates more pathogens, which in turn causes more water contamination.

Pathogens in groundwater are generally allochthonous; that is, they originate from outside and enter the aquifers on different pathways. Most pathogens cannot grow or reproduce inside the aquifer and persist for only limited periods of time. Therefore, most groundwater protection legislations use transit time as a key criterion for the delineation of an inner source protection zone. For example, the German legislation uses the 50-day line of travel time, assuming that most pathogens are filtered or inactivated by then, an approach that has proven feasible in most cases. However, some pathogens can persist much longer, particularly protozoan cysts (Bonadonna et al., 2002) and some types of viruses (Schaub and Oshiro, 2000).

Many possible pathogens may be in water, most of which are difficult to detect. Therefore, fecal indicator bacteria (FIB) are commonly used to assess the hygienic safety of water. The most important FIB species is *Escherichia coli*, which is always present in very large numbers in the feces of warm-blooded animals and humans and only occurs there. Therefore, the presence of *E. coli* or other FIB in a water sample indicates fecal contamination and therefore the possible presence of various pathogens (Figure 8–4). Although this is a well-established and generally safe approach, it has one major drawback: Some viruses and protozoan cysts, particularly *Cryptosporidium*, persist in the environment much longer than all commonly used FIB and can also survive water treatment (Lisle and Rose, 1995; Mackenzie et al., 1994). Therefore, the absence of *E. coli* provides no absolute safety (Brookes et al., 2005).

8.3 METHODS FOR DELINEATION OF CATCHMENT AREAS

8.3.1 Introduction

As mentioned earlier, the first and most fundamental question for source protection zoning is, Where is the catchment area of the spring? The delineation of catchment areas requires a combination of topographic, geologic, hydrogeologic, and hydrologic considerations, using a variety of investigation methods, including water balance, natural tracers, and artificial tracer tests (Goldscheider and Drew, 2007).

In karst hydrogeologic systems, spring catchments can often be subdivided into autogenic and allogenic recharge areas. The autogenic recharge area is where rainwater or meltwater infiltrates diffusely through the soil (if present) and unsaturated zone into the aquifer. The allogenic recharge area comprises adjacent areas consisting of other geologic formations that drain toward the aquifer, often via surface streams that sink into swallow holes (sinks) near the geologic contact. The Danube-Aach system in southwestern Germany is a well-known example: During low-flow conditions, the entire Danube River sinks into the karst aquifer via several swallow holes and flows toward the Aach spring, Germany's largest spring, 12 km away (Hötzl, 1996). Under such conditions, the entire surface catchment of the Danube upstream of this point represents the allogenic catchment area of the Aach spring.

Another characteristic of karst aquifer systems is that groundwater divides, and hence, spring catchments may change as a function of the hydrologic conditions. In some cases, swallow holes transform into springs during high-flow conditions, so that the catchment of the surface stream is no longer part of the spring catchment. Cavities that alternately act as swallow holes and springs, depending on the hydrologic conditions, are referred to as *estavelles* or *inversacs* (Figure 8–5). Hydrologic variability must be considered both for the delineation of spring catchment areas and, consequently, for groundwater source protection zoning.

FIGURE 8–5 A cave entrance in an alpine karst valley during different hydrologic conditions: (a) During low-flow conditions, a stream sinks into the cave and an injected tracer (uranine) reappears at several karst springs farther downstream, confirming that the stream catchment belongs to the allogenic recharge area of the springs; (b) during high-flow conditions, the cave acts as a spring, and the upstream catchment is no longer connected to the karst aquifer. (Photos: N. Goldscheider.)

8.3.2 Topography, geology, and groundwater flow

Topography is a major influence on groundwater flow patterns at different scales (Toth, 1963, 1999), but the relation between topography and spring catchment divides is often not that simple. Topographic limits at the land surface, such as mountain ridges, are not necessarily significant limits for groundwater flow, particularly in karst systems, where there are many examples of flow crossing below mountain ridges and valleys (Bianchetti et al., 1992; Goldscheider, 2005a). In such cases, it is incorrect to delineate spring catchments only on the basis of topography. Nevertheless, topography needs to be considered: It is clear that the entire recharge area of a spring is always situated above the level of the spring, although flow may also occur below this level, with flow lines rising up from the deeper aquifer portions toward springs (Figure 8–6) or streams (Ophori and Toth, 1990). Thermal springs outside volcanic areas are the most obvious manifestation of deep, warm groundwater rising up to the surface, but the recharge area is always located at higher altitudes (with the rare exception of flow induced by tectonic stress). Topography is also useful in delineating allogenic recharge areas, such as the catchments of losing streams or sinking streams.

An aquifer draining toward a spring is a geologic body, characterized by its geometry and internal structure, hydrogeologic properties, and hydraulic boundary conditions (Burkhard et al., 1998). The geometry of an aquifer is usually defined by stratigraphic, tectonic, and topographic elements, such as an aquifer consisting of sand and gravel in a valley surrounded by impermeable schist or an aquifer consisting of karstified limestone occupying a low topographic position, separated by normal faults from crystalline rocks that drain by surface flow toward the karst aquifer. The orientation and hydraulic characteristics of fractures are crucial for groundwater flow; faults and thrusts can limit an aquifer or create hydraulic connections between different aquifers; anticlines often act as water divides, while synclines tend to form the main pathways for groundwater flow in karst aquifer systems (Figure 8–6) (Goldscheider and Andreo, 2007). Detailed geologic and hydrogeologic characterization of a region makes it possible to better define and delineate the catchments of springs.

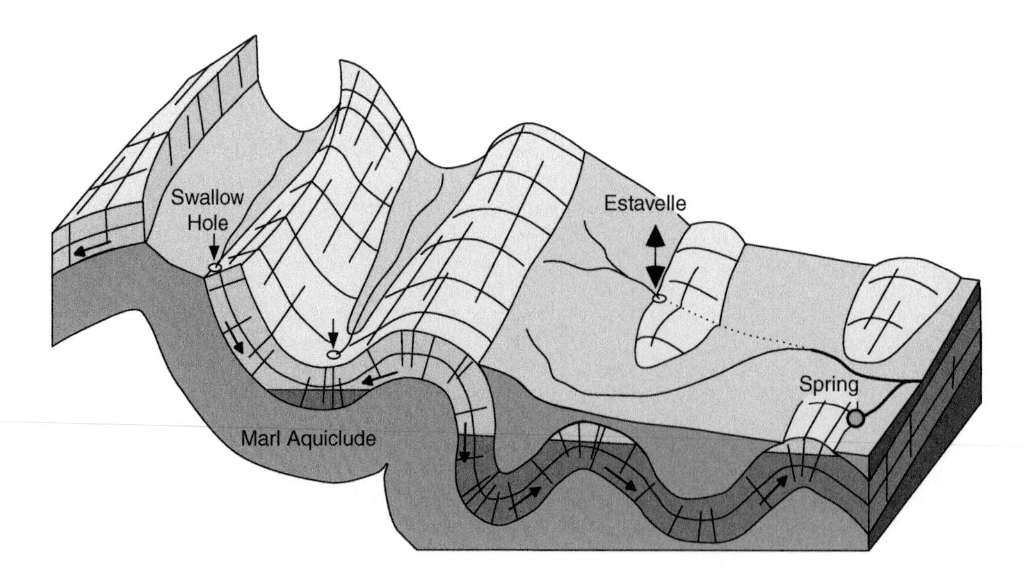

FIGURE 8–6 Schematic block diagram illustrating the importance of topography, stratigraphy, and geological structure for groundwater flow in a karst aquifer system. Light and dark areas with arrows indicate the unsaturated and saturated zone of the karstified limestone and the flow directions, respectively. (Modified after Goldscheider, 2005a.)

8.3.3 Water balance

Water balance determination helps to quantify the size of the spring catchment area. In general, all water balance expressions have the following form:

$$\text{Outflow} = \text{Inflow} + \text{Change in storage}$$

In the simplest case, the long-term water balance of a spring and its catchment, without surface runoff and changes in storage can be written as

$$Q_S = A(P - \text{ET})$$

where Q_S is the mean annual spring discharge, A is the autogenic recharge area, P is the annual precipitation, and ET is annual evapotranspiration ($P - \text{ET} = R$, where R is recharge).

Although the concept of water balance is simple in theory, there are many complications when it comes to its practical applications. The most obvious difficulty is a precise determination (measurement) of the different elements of the water balance: precipitation, evapotranspiration, and the different forms of runoff; regionalization of hydrologic data is also problematic. Furthermore, all elements of the water balance are usually variable in time, particularly in the case of karst systems, and there is also storage of water and release of stored water. In many cases, the spring under consideration is not the only outlet of the aquifer: There might be other outlets, some of which are accessible (such as at other springs), while others might be hidden (such as direct discharge into streams, rivers, wetlands, lakes, or the sea). If unknown discharge locations are possible, a water balance delivers only the minimum size of the spring catchment. Inflow from other aquifers or upward flow of deep groundwater from large-scale regional flow systems can further complicate or limit the application of water balances.

8.3.4 Natural and artificial tracers

Natural and artificial tracers (explained in more detail in Chapter 3) can help to delineate spring catchment areas and source protection zones. Natural tracers include water temperature, water chemistry, stable isotopes, and other parameters.

Springwater temperatures and their variability are easy to measure, even continuously, and deliver much information. For example, constantly high temperatures indicate an inflow of thermal water from greater depth; very low temperatures can indicate rapid inflow of snow or glacier meltwaters from higher altitudes; and highly variable springwater temperatures often indicate inflow from surface waters (Vaneverdingen, 1991).

Stable isotopes of the water molecule (^{18}O and D) are often used to determine the mean altitude of the recharge area, which is particularly useful in mountainous regions. Stable isotopes also help to conclude on the origin of the water: rainfall, snowmelt, glacier melt, or infiltration of surface water (Clark and Fritz, 1997; Criss et al., 2007).

The hydrochemical water composition provides information about the aquifer(s) contributing to spring discharge, inflow from surface waters, contamination problems, and other aspects (Hunkeler and Mudry, 2007). For example, calcium-bicarbonate dominated water chemistry is characteristic of limestone aquifers. Increased levels of magnesium point to the presence of dolomite rocks in the spring catchment; sulfate can originate from gypsum or anhydrite layers or from the oxidation of sulfide minerals. The weathering of granite and other silicate rocks usually generates potassium and other cations; with increasing transit times, potassium tends to be exchanged for sodium (Krauskopf and Bird, 1995). Natural tracers rarely deliver unambiguous evidence; for example, sodium can originate from the solution of rock salt, plagioclase (albite) weathering, cation exchange, seawater intrusions, or road salt. Therefore, indications obtained from natural tracers should always be combined with other types of information.

Artificial tracers are a powerful and reliable method for the delineation of spring catchment areas (see Chapter 3). Tracer tests are commonly applied in karst aquifer systems but can also be used in other hydrologic environments. Fluorescent dyes, such as uranine, also referred to as sodium fluorescein or, for short, fluorescein (Figure 8–5), are the most favorable groundwater tracers in many cases (Benischke, Goldscheider, and Smart, 2007; Käss, 1998). Positive tracer detection at a spring is strong evidence that there is a hydraulic connection between the injection site and the spring; that is, the injection site is located within the spring catchment.

Fully quantitative tracer tests, which deliver detailed concentration-time-discharge data series, so-called tracer breakthrough curves, provide much insight into flow toward springs. In the context of protection zoning, the tracer recovery, that is, the absolute or relative quantity of the tracer reaching the spring, is a highly valuable but often underexploited result. A recovery of less than 100 percent can be due to many different causes, which are often difficult to quantify, such as imperfect injection conditions or tracer loss by adsorption or degradation. However, when a conservative tracer is used (e.g., uranine behaves conservatively in most hydrogeologic environments) and the experimental conditions are favorable (i.e., direct injection into flowing groundwater), tracer recoveries, along with discharge data, make it possible to quantify underground flow rates and point to unknown discharge locations. This approach can best be illustrated by means of a theoretical example. The sketch in Figure 8–7 shows a simple karst aquifer system, consisting of a swallow hole with a flow rate of 5 L/s and a karst spring with a discharge of 100 L/s. A conservative tracer was injected into the swallow hole, and a complete breakthrough curve was recorded at the spring; the calculated tracer recovery is 17 percent. These readily obtained data, together with a simple conceptual model of the conduit network, make it possible to quantify the underground flow rates in the system using simple water and tracer balances (equations and solutions in Figure 8–7).

These types of considerations can substantially improve the water balance concept described earlier. In this particular example, the total discharge of the system (Q_1) is 5.88 times larger than the observed spring discharge. Correspondingly, the catchment area must be 5.88 times higher than would be expected by applying simple water balance equations.

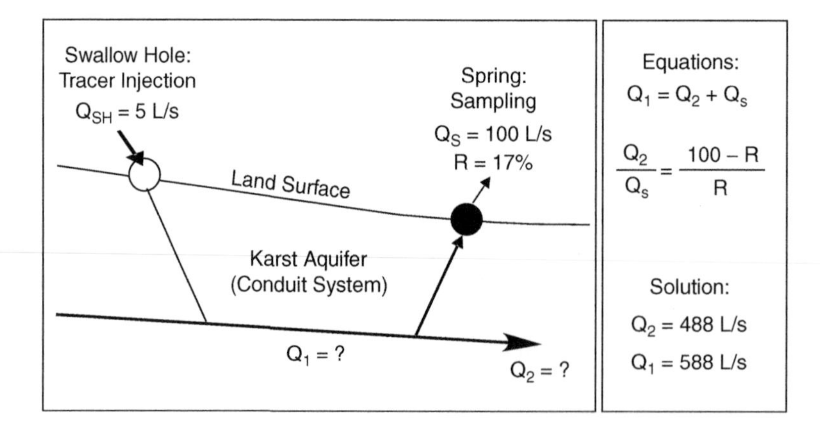

FIGURE 8–7 Calculation of underground flow rates by means of tracer recovery and flow measurements in a simple karst aquifer system: Q = discharge, R = tracer recovery. Further explanations in the text.

8.4 GROUNDWATER SOURCE PROTECTION ZONES AND LAND-USE RESTRICTIONS

Many national groundwater protection schemes differentiate at least three type of source protection zones, often referred to as the *wellhead protection zone* (zone I), *inner protection zone* (zone II), and *outer protection zone* (zone III); zones II and III are sometimes further subdivided. In the most typical and simplest case, these zones are arranged in succession around a groundwater pumping well or up-gradient from a tapped spring (Figure 8–8).

Potentially polluting land-use activities are prohibited or restricted in source protection zones, with decreasing restrictions from zone I to zone III. Outside these zones, polluting activities are also often restricted, or should be restricted, for reasons of water resources protection (both groundwater and surface water) or general environmental and human health protection.

The wellhead or springhead protection zone (zone I) comprises the area immediately surrounding a pumping well or a tapped spring, respectively. The purpose of this zone is to protect the well or spring from any type of direct impact, mechanical damage, or contamination. Depending on the respective national guidelines, zone I typically extends 10–20 m around the well or spring; it is usually fenced and all activities not directly related to drinking water supply are forbidden. Artificial recharge facilities and swallow holes in karst areas are also often included into this zone.

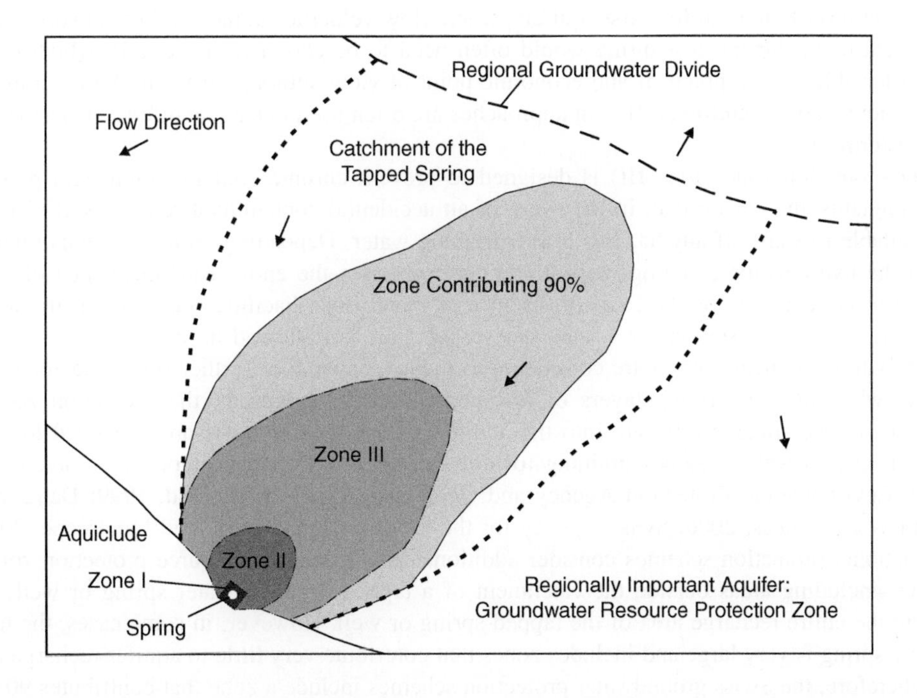

FIGURE 8–8 Typical arrangements of groundwater source protection zones for a large spring situated at a geologic contact between an aquifer and an aquiclude. Zone I is the springhead protection zone; II and III are the inner and outer protection zones, respectively. Several national legislations include additional or alternative protection zones, such as the entire spring catchment or a zone contributing 90 percent to the water source. Regionally important aquifers outside the catchment are often classified as groundwater resource protection zones. The figure is not to scale, but the dimensions of the map are on the order of magnitude of kilometers.

The inner protection zone (zone II) is primarily designed to prevent drinking water contamination with fecal and pathogenic microorganisms. Therefore, the travel time in the aquifer is often used as the primary criterion for the delineation of zone II. Different national legislations take different travel times as a limit, for example, 10 days in Switzerland, 50 days in Germany, and 100 days in Ireland, assuming that most microbial pathogens are inactivated or filtered after this period of time (Department of Environment and Local Government, Environmental Protection Agency and Geological Survey of Ireland, 1999; Deutscher Verein des Gas- und Wasserfaches, 2006; Swiss Agency for the Environment, Forests and Landscape, 2004). Activities releasing microbial contaminants, such as the application of liquid manure, are prohibited in zone II; other types of polluting land-use practices and construction works that might obstruct groundwater flow are also restricted.

In sand and gravel aquifers, where effective porosities are high and flow velocities are often around 1 m/d, this transit time approach often results in a zone II extending several tens to several hundreds of meters around the pumping well; the higher the groundwater flow velocity in the natural flow field, the more zone II is elongated up-gradient from the well. Two types of methods are typically applied to delineate zone II around a pumping well: (1) hydraulic methods, ranging from relatively simple calculations to more sophisticated numerical groundwater flow and transport models, and (2) artificial tracer tests, which make it possible to directly measure the transit time from a particular injection site, such as a monitoring well, to the pumping well. Ideally, the two methods are combined (Goldscheider et al., 2008).

If the same criteria were used for karst aquifers, where flow velocities in the conduit network often exceed 100 m/h, the entire catchment of a spring would often need to be classified as zone II, which is, however, generally not feasible from a practical and economic point of view, although it would be favorable in terms of drinking water safety. Therefore, different approaches are often used in karst aquifers, such as vulnerability mapping (see further).

The outer protection zone (zone III) is designed to prevent chronic contamination with persistent and mobile contaminants and ensure that, in the event of an accidental contaminant release, sufficient time and space are available to ward off any hazards to the drinking water. Depending on the respective national regulations and the hydrogeologic setting, zone III either comprises the entire catchment area of a spring or pumping well or is delineated on the basis of distance or travel time. Facilities that pose a substantial threat to groundwater, such as gas stations or wastewater seepage, are not allowed in this zone.

Different hydrogeological criteria are considered to reduce, expand, or further subdivide zones II and III. For example, when thick overlying layers of low permeability are present, the protection zones can be reduced; when surface streams infiltrate into the aquifer, the protection zones are expanded to include the stream catchments, as well as slopes draining toward the aquifer (Department of Environment and Local Government, Environmental Protection Agency and Geological Survey of Ireland, 1999; Deutscher Verein des Gas- und Wasserfaches, 2006; Swiss Agency for the Environment, Forests and Landscape, 2004).

Several national protection schemes consider additional zones, such as resource protection zones for the entire aquifer (including areas outside the catchment of a tapped drinking water spring or well) or a zone encompassing the entire recharge area of the tapped spring or well. However, in some cases, the total catchment area of a spring is very large and includes zones that contribute very little to aquifer recharge and spring discharge. Therefore, the Swiss groundwater protection schemes include a zone that contributes 90 percent to the recharge of the water source under consideration. This zone is typically larger than zone III but sometimes significantly smaller than the entire catchment area (Figure 8–8) (Bussard et al., 2006; Swiss Agency for the Environment, Forests and Landscape, 2004). The land-use restrictions in these wider zones are less stringent than in zone III, but they provide additional safety against chronic and persistent contamination, for example, due to nitrate, which behaves conservatively in oxygen-rich groundwater (Chapelle, 2001; Johnston et al., 1998).

The Irish groundwater protection schemes differentiate between an "inner zone" and an "outer zone" around a well or spring, generally based on a travel time of 100 days. These zones are then combined with a simple vulnerability assessment (see further), taking into account overlying layer thickness and permeability to obtain source protection zones and land-use responses (Department of Environment and Local Government, Environmental Protection Agency and Geological Survey of Ireland, 1999). For source protection zoning in Slovene karst areas, Ravbar and Goldscheider (2007) proposed a similar concept without using the 90 percent criterion or a 100 day line of travel time: The Slovene approach distinguishes between zones that always and directly contribute to the spring and zones that are only indirectly or marginally connected to the spring or only during extreme hydrologic events. The spring catchment is then further subdivided on the basis of a vulnerability map to obtain source protection zones.

For drinking water safety, source protection zones should be delineated on the basis of hydrogeologic criteria, and the corresponding land-use restrictions should be enforced. However, as land-use restrictions also have socioeconomic implications, such as equalization payments to farmers or restrictions on industrial development, the inverse approach is often applied, and the protection zones are bent, twisted, and fragmentized to minimize the short-term costs (Figure 8–9). Ironically, this approach often results in higher costs for the society in the long term, because the remediation of contaminated aquifers and the treatment of polluted drinking water are generally more expensive than the implementation of appropriate protection zones. Although socioeconomic interests other than groundwater protection are legitimate and must be

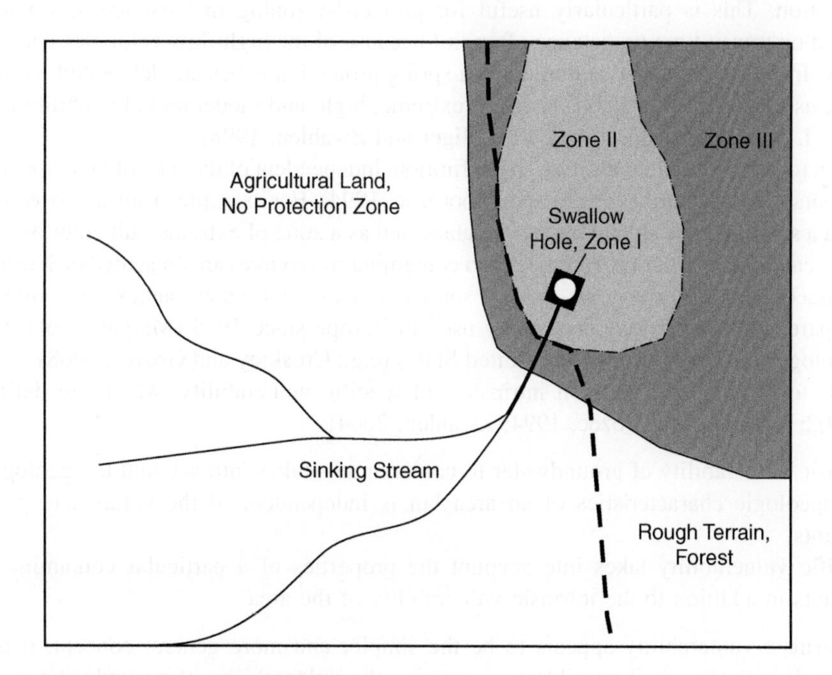

FIGURE 8–9 Example of obviously inappropriate protection zones violating basic hydrogeologic principles but minimizing the short-range costs: The catchment of a stream draining agricultural land and sinking into a swallow hole, which is connected to a karst spring used for a drinking water supply, is not included in any protection zone, while the nearby forested rough terrain is classified as protection zones, minimizing equalization payments to farmers (based on a real case but generalized).

considered in comprehensive land-use management schemes, hydrogeologists and engineers should not distort hydraulic laws to achieve "convenient" protection zones.

8.5 GROUNDWATER VULNERABILITY ASSESSMENT AND MAPPING

8.5.1 Generalities and terminology

The term *vulnerability of groundwater to contamination* was first introduced by the French hydrogeologist Jean Margat (1968). The concept of groundwater vulnerability is based on the assumption that the physical environment provides some degree of natural protection against contaminants entering the subsurface (Vrba and Zaporozec, 1994). Vulnerability assessment therefore consists of quantifying the factors that determine the degree of natural protection against contamination, such as the thickness and properties of the overlying layers and the infiltration conditions. Vulnerability mapping means subdividing a given area, such as the catchment of a spring, into zones of different degrees of vulnerability. Red colors are typically used to indicate high degrees of vulnerability (low natural protection), while blue stands for lower vulnerability (high natural protection).

Vulnerability maps can help to find a balance between groundwater protection and other economic interests, such as agriculture and industry. The maps can also be used as a basis for the delineation of source protection zones, alternatively or complementary to the conventional approach to protection zoning described in the previous section. This is particularly useful for protection zoning in karst areas, where conventional approaches based on transit time or distance often fail because of the high flow velocities and the high degree of heterogeneity. In Switzerland, for example, karst spring protection zones are delineated on the basis of vulnerability maps, using the EPIK method. Areas of extreme, high, and moderate vulnerability are classified as protection zones I, II, and III, respectively (Doerfliger and Zwahlen, 1998).

It is important to note that vulnerability is, by definition, independent of the type of land use and the presence or absence of contamination sources (Vrba and Zaporozec, 1994). For example, a surface stream sinking into a karst aquifer via a swallow hole should always be classified as a zone of extreme vulnerability, independent of the land use in its catchment. Land-use practices and contamination sources are presented on hazard or risk maps, which are discussed later. Unfortunately, there is some confusion and inconsistency concerning terminology. Table 8–1 compares the terminology commonly used in Europe since 1968 (Margat, 1968; Zwahlen, 2004) with the terminology sometimes used in the United States (e.g., Croskrey and Groves, 2008).

It is possible to differentiate between intrinsic and specific vulnerability, which are defined as follows (Daly et al., 2002; Vrba and Zaporozec, 1994; Zwahlen, 2004):

- The intrinsic vulnerability of groundwater to contaminants takes into account the geologic, hydrologic, and hydrogeologic characteristics of an area but is independent of the nature and properties of the contaminants.
- The specific vulnerability takes into account the properties of a particular contaminant or group of contaminants in addition to the intrinsic vulnerability of the area.

Although intrinsic vulnerability appears to be the simpler and more general concept, it is scientifically more difficult to define: How is it possible to determine the vulnerability of groundwater to contamination without considering any specific contaminant properties? One possibility is to assume a "general" contaminant that behaves as completely conservative, that is, shows no specific reactions or interactions with the aquifer material, such as retardation or biodegradation. For the purpose of protection zoning, intrinsic vulnerability maps are generally more useful and applicable than specific vulnerability maps, because protection zones aim at preventing all types of contamination. However, when one particular contaminant or group of

Table 8–1 Use of Terminology in Europe Compared to the Terminology Found in Some American Publications

Term used in the European literature and in this chapter	Vulnerability	Risk
Term used in several U.S. publications	Sensitivity	Vulnerability
Meaning	The ease at which contaminants can enter the groundwater, dependent only on the physical properties of the hydrogeologic environment, and independent of the land use and contamination hazards	The likelihood of groundwater contamination to occur, dependent on both the physical properties of the hydrogeologic environment and the presence of contamination hazards
Example	A swallow hole in a karst area is a zone of high vulnerability (Europe) or sensitivity (USA), respectively	A waste dump near a swallow hole is a zone of high risk (Europe) or vulnerability (USA)
Comments	Both terms are correct; they have the same meaning and can be used as synonyms	The term *risk* is more consistent in this context, while the term *vulnerability* is misleading and should be avoided

Note: This chapter uses European terminology, as the concept of vulnerability mapping was first proposed by a French hydrogeologist (Margat, 1968) and is practiced more in Europe.

contaminants is of special concern, specific vulnerability maps might be more useful, such as for pesticides or chlorinated solvents.

As already mentioned in the introduction, vulnerability assessment is based on an origin-pathway-target conceptual model (Figure 8–2), which also illustrates the difference between resource and source vulnerability mapping. A resource vulnerability map shows the vulnerability of an entire aquifer to contamination; the water table is consequently taken as the target. A source vulnerability map shows the vulnerability of a specific water source, such as a pumping well or a spring, which is consequently specified as the target. Therefore, resource vulnerability mapping considers only the mostly vertical percolation of water (and contaminants) from the land surface through the unsaturated zone toward the groundwater table, while source vulnerability mapping additionally considers the contaminant migration in the saturated zone of the aquifer toward the spring or well.

Unfortunately, there are a large number of methods of vulnerability mapping, which are often based on different definitions, terminology, and assumptions. Some methods are part of national legislation or groundwater protection guidelines, such as EPIK in Switzerland, while others were developed for scientific purposes. Vrba and Zaporozec (1994) and Gogu, Hallet, and Dassargues (2003) provide a review of various methods of vulnerability mapping. However, experience has shown that the application of different methods to the same test site, using the same database, often leads to different and sometimes contradictory results (e.g., Gogu et al., 2003). Therefore, the European COST Action 620 on "vulnerability and risk mapping for the protection of carbonate (karst) aquifers" gathered about 50 scientists, practitioners, and decision makers from 15 European countries to propose a common conceptual framework for vulnerability, hazard, and risk mapping (Daly et al., 2002; Zwahlen, 2004). This chapter mostly uses the terminology and methodology proposed by this group, but also briefly discusses some previously existing methods, such as DRASTIC (Aller et al., 1987).

8.5.2 Quantitative approach to vulnerability assessment

The term *vulnerability of groundwater to contamination* is often intuitively understood by decision makers in the planning process and by average citizens. However, there are also disadvantages in using a qualitative approach alone. A property that is not precisely defined cannot be derived unambiguously from measurable quantities and cannot be validated. Therefore, COST Action 620 proposed a quantitative basis for vulnerability assessment (Brouyère et al., 2001; Daly et al., 2002; Goldscheider et al., 2001; Zwahlen, 2004).

When a contamination occurs somewhere in the catchment of a spring, the water supplier and the water users may ask several basic questions (Figure 8–10): (1) When will the contamination first arrive at the spring or well and when will it reach its maximum level? (2) What will be the maximum concentration level of contamination? (3) How long will the contamination remain above a critical threshold, such as the legal limit for drinking water? (4) Which proportion of the released contaminant quantity will actually arrive at the spring?

Therefore, the basic aspects for groundwater vulnerability assessment are transit time, maximum concentration, expected duration of a contamination, and relative quantity of a contaminant that can actually reach the target—a proportion of the contaminant may never reach the target but leave the catchment by surface runoff or go to another spring (Daly et al., 2002). Among these criteria, transit time is probably the most relevant, particularly in the context of microbial contamination, because time-dependent die-off and inactivation are important attenuation processes for pathogens (Figure 8–4). Most other processes of contaminant attenuation are also directly or indirectly related to transit time—more time generally means more possibilities for ion exchange, degradation, filtration, and other processes to operate. Furthermore, longer transit times also

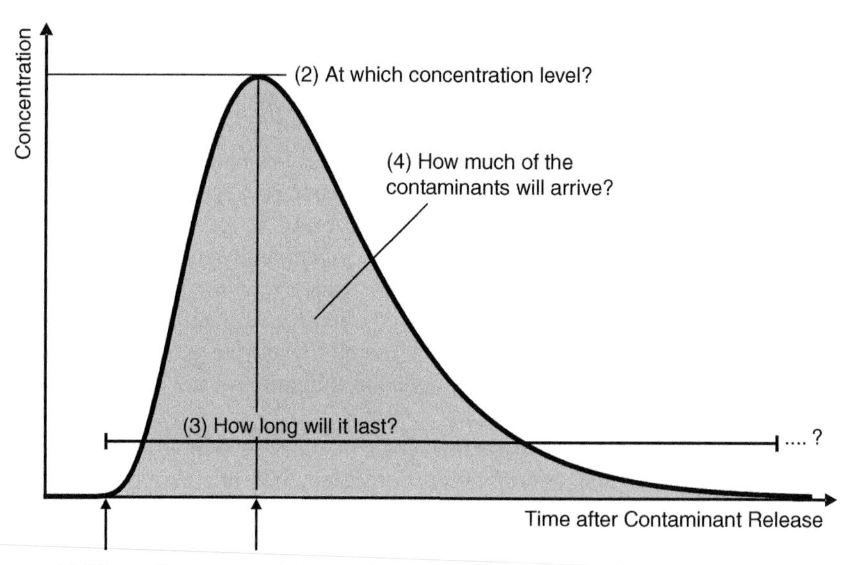

FIGURE 8–10 Basic questions that a water user may ask when a contamination occurs in the catchment of a drinking water spring. These questions can also serve as a quantitative basis for vulnerability assessment and source protection zoning, such as a high degree of vulnerability means that a contaminant would arrive at the spring after a short time and at high concentrations. Stringent protection measures would consequently be required. (Modified after Brouyère et al., 2001, and Zwahlen, 2004.)

mean more time to react to accidental contamination events. Transit time and the other three criteria are interdependent; for example, the maximum concentration levels systematically decrease with increasing transit time and travel distance. The duration of a possible contamination, on the other hand, is quite difficult to determine.

Arguably, transit times and other aspects of contamination listed earlier can never be determined accurately, directly assessed, or mapped in the field. Therefore, vulnerability mapping consists of identifying, quantifying, and mapping those factors that control transit time (the most important aspect), concentration decline, duration, and relative quantity of contaminants that can reach the target. For example, the transit time depends on the thickness and hydraulic properties of the overlying layers, which should consequently be a key factor for vulnerability assessment.

8.5.3 Importance of hydrologic variability for source protection zoning

While groundwater flow velocities and potentiometric levels are often relatively stable in alluvial aquifers and deep, confined hydrogeologic systems, significant hydrologic variability can be observed in many karst aquifer systems. Water table fluctuations in karst aquifers often exceed 10 m and sometimes 100 m; flow velocities during high-flow conditions are often 5–10 or even more times higher than during low-flow conditions (Göppert and Goldscheider, 2008; Pronk et al., 2007; Figure 8–11); spring discharge and the hydrologic state of swallow holes can be subject to significant variability (Figure 8–5).

This raises the question of how to include hydrologic variability into vulnerability mapping. One possibility would be to prepare different vulnerability maps for different hydrologic conditions. However, since vulnerability maps are often used as a basis for the delineation of groundwater protection zones, this approach is not feasible from a practical point of view. Another possibility is to take average hydrologic conditions as a basis for vulnerability mapping. However, the highest contamination levels in karst

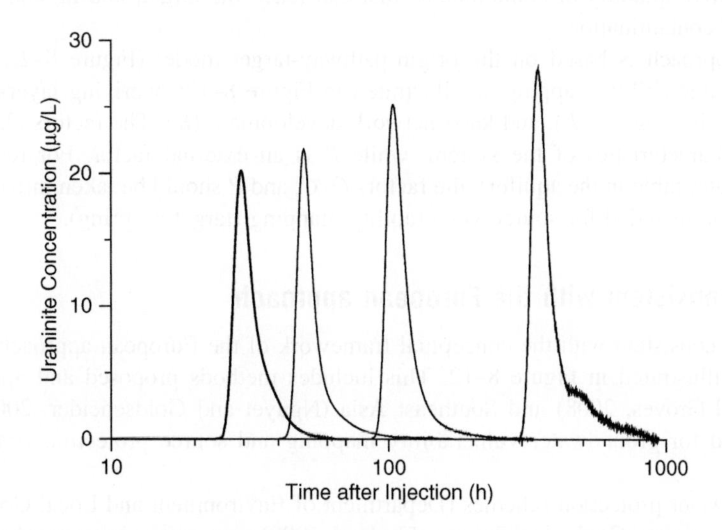

FIGURE 8–11 Results of four tracer tests with each 1 kg of uranine carried out between a swallow hole and a karst spring during different hydrologic conditions, ranging from very low- to very high-flow conditions. The peak times range between less than 30 and more than 300 hours, while the maximum concentrations are less variable (Pronk et al., 2007). Source protection zoning based on transit times requires specification of the hydrologic conditions.

springwater often occur during high-flow conditions, caused by extreme hydrologic events (Pronk et al., 2007; Figure 8–4); therefore, average conditions have little relevance to questions of groundwater vulnerability. Accordingly, Vias et al. (2006) proposed to use extreme hydrologic conditions as a reference point for vulnerability assessment, whereas Goldscheider et al. (2000) consider high-flow conditions that occur at least once per year. The most detailed assessment scheme for considering hydrologic variability was proposed for Slovene karst areas, which are characterized by particularly high vertical fluctuations of the water table and lateral fluctuations of groundwater divides (Ravbar and Goldscheider, 2007).

As illustrated in Figure 8–1, protection zoning is only one element of a security chain. The other three elements can also help to address the issue of hydrologic variability: reduction of polluting activities during high-flow and high-recharge periods; continuous water quality monitoring and event-based sampling approaches; and last but not least, increased drinking water treatment during high-flow periods, when higher contamination risk has to be expected.

In conclusion, hydrologic variability should be considered for groundwater source protection, particularly in the case of karst springs characterized by very high variability.

8.5.4 Factors for vulnerability assessment

Based on earlier works, the European COST Action 620 proposed an approach to intrinsic groundwater vulnerability mapping that is a general and flexible conceptual framework rather than a prescriptive method (Daly et al., 2002; Zwahlen, 2004). Although this approach was essentially developed for application in karst areas, it can also be used for other types of hydrogeologic environments, which requires some modest adaptations. Several existing or newly developed methods can be used within this framework. According to the quantitative basis described earlier, the approach aims at assessing those properties that influence the travel time of a potential contaminant from the origin to the target (most important aspect), as well as its concentration decline, the relative quantity of contaminants that can reach the target, and in some specific cases, the duration of potential contamination.

The European approach is based on the origin-pathway-target model (Figure 8–2). Up to four factors are considered for vulnerability mapping, as illustrated in Figure 8–12: overlying layers (O), concentration of flow (C), precipitation regime (P), and karst network development (K). The factors O, C, and K represent the hydrogeologic characteristics of the system, while P is an external factor. For resource vulnerability mapping (target = water table in the aquifer), the factors O, C, and P should be taken into consideration, while the factor K should be included for source vulnerability mapping (target = spring).

8.5.5 Methods consistent with the European approach

Several methods are consistent with the conceptual framework of the European approach (Daly et al., 2002; Zwahlen, 2004), as illustrated in Figure 8–12. This includes methods proposed and applied in the United States (Croskrey and Groves, 2008) and Southeast Asia (Nguyet and Goldscheider, 2006). These methods can be recommended for groundwater vulnerability mapping and source protection zoning, particularly in karst areas.

The Irish groundwater protection schemes (Department of Environment and Local Government, Environmental Protection Agency and Geological Survey of Ireland, 1999) represent an important basis for the European approach, as they are based on a clear origin-pathways-target conceptual model, as illustrated in Figure 8–2. Groundwater source and resource protection zones are obtained by combining different types of maps. A resource vulnerability map takes into account mainly the thickness and properties of the overlying layers, but also specific infiltration conditions in karst areas (corresponding to the O and C factors, respectively).

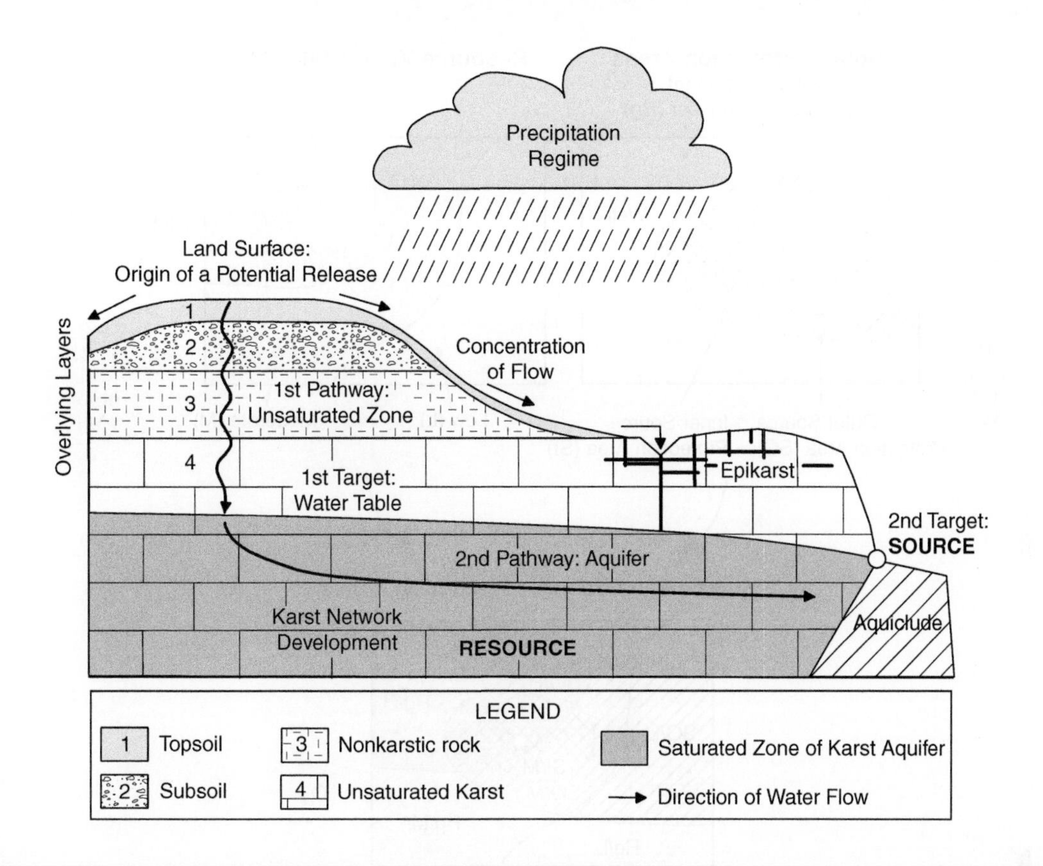

FIGURE 8–12 The European approach (EA) to groundwater vulnerability mapping is based on an origin-pathway-target conceptual model and can be used for resource and source vulnerability mapping. The main factors include the precipitation regime, the overlying layers, the concentration of flow, and the karst network development. (Modified after Goldscheider et al., 2000; Daly et al., 2002; Zwahlen, 2004.)

Another map shows the drinking water abstraction points (springs or wells) with the "inner source protection areas" (SI) and "outer source protection areas" (SO), which are delineated by means of a 100 day line of travel time. Source protection zones are then obtained by combining the resource vulnerability map with the source protection areas. The remaining land surface is subdivided into different resource protection zones, taking into account the value or importance of the resource and its vulnerability (Figure 8–13).

The EPIK method (Doerfliger et al., 1999) is the first method specifically developed for source vulnerability mapping in karst areas. The method considers four factors: epikarst development (*E*), protective cover (*P*), infiltration conditions (*I*), and karst network development (*K*). In Switzerland, EPIK is the official method used for the delineation of source protection zones in karst areas (Doerfliger and Zwahlen, 1998). The different degrees of vulnerability are directly translated into protection zones: very high vulnerability means zone I, high vulnerability means zone II, moderate vulnerability means zone III, and areas of low vulnerability are not included in the protection zones.

The PI method (Goldscheider et al., 2000) first proposed the conceptual model that later served as a basis for the European approach (Figure 8–12). The method includes two factors also included in the European

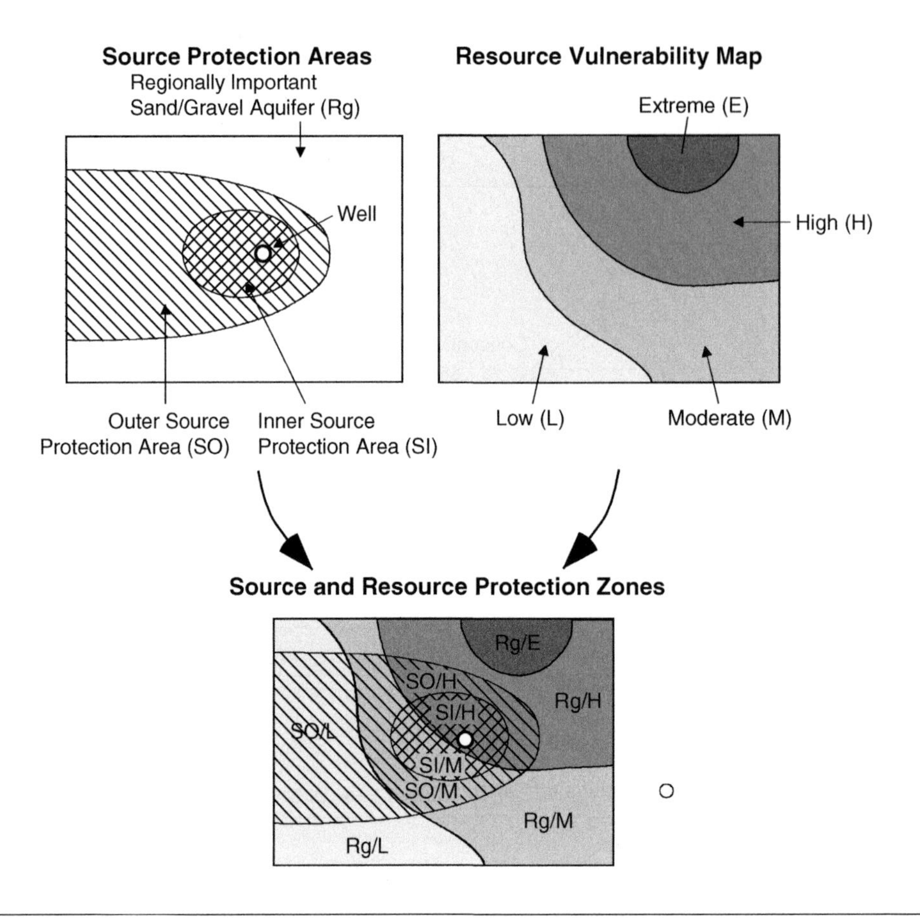

FIGURE 8–13 The Irish groundwater protection schemes. Source protection zones and the corresponding land-use restrictions are obtained by combining "source protection areas," which are delineated on the basis of travel time in the aquifer, and a resource vulnerability map. (Modified after Department of Environment and Local Government, Environmental Protection Agency and Geological Survey of Ireland, 1999.)

approach, although the terminology is different: the protective cover (*P* factor) and the infiltration conditions (*I* factor) correspond to the *O* and *C* factors of the European approach, respectively. The PI method can be used for groundwater resource vulnerability mapping in all types of hydrogeologic environments, but it includes specific tools for karst aquifers. The method requires a relatively detailed database.

The COP method (Vias et al., 2006) was directly derived from the European approach and includes three of the four factors: concentration of flow, overlying layers, and precipitation regime. An assessment scheme for the karst saturated zone factor (*K*) is in preparation, so that the method can be used for karst groundwater resource and source vulnerability mapping. The method has been applied in Spanish karst aquifer systems (Andreo et al., 2006) and elsewhere.

For application in developing countries or other areas with limited data availability and technical resources, Nguyet and Goldscheider (2006) proposed a simplified method for groundwater resource vulnerability mapping, hazard assessment, and risk mapping, which was recently expanded for source vulnerability mapping. Just like the PI method, it can be used for all types of aquifers but includes specific tools for karst aquifers. Due to its simplicity, the method nicely illustrates the idea of vulnerability by overlaying several preliminary maps, which is facilitated by geographic information systems (Figure 8–14).

Croskrey and Groves (2008) also proposed a relatively simple but consistent method for vulnerability mapping (or "sensitivity mapping"; see Table 8–1) in Kentucky, where karst areas are widespread and important for freshwater supply. Very similar to the concept illustrated in Figure 8–14, the method also considers so-called high-risk runoff zones, where contaminants could be transported in runoff from less sensitive to higher sensitivity areas.

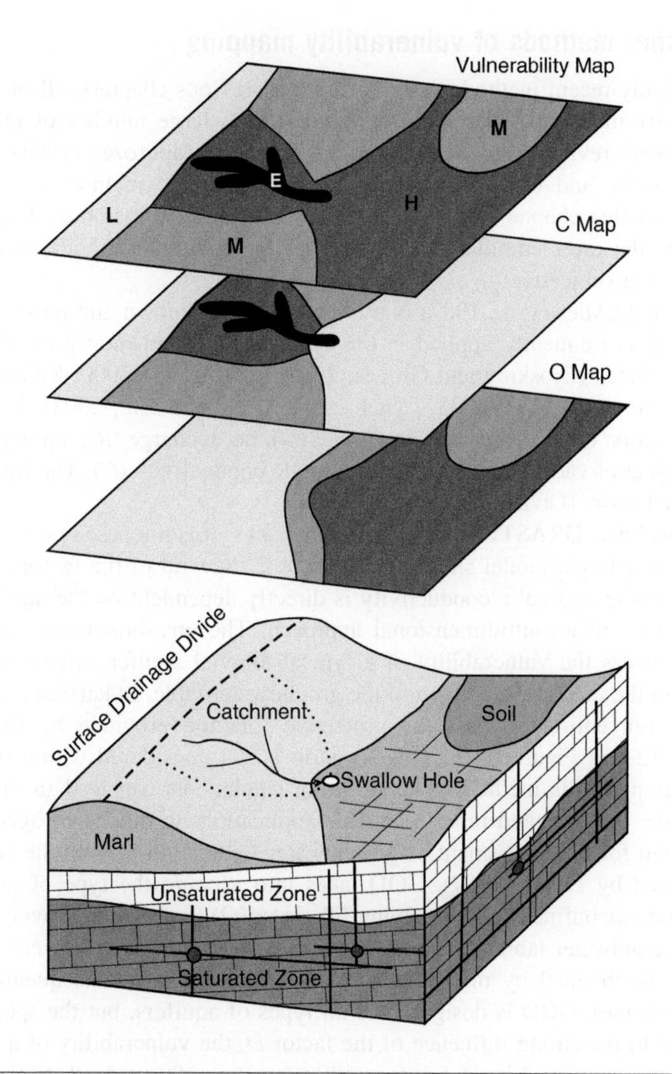

FIGURE 8–14 The simplified method of groundwater vulnerability mapping uses two factors: overlying layers (*O*) and concentration of flow (*C*). *L*, *M*, *H*, and *E* stand for low, medium, high, and extreme vulnerability, respectively. It is important to note that surface runoff toward a swallow hole can completely bypass the protective function of the overlying layers. In such a setting, the *C* factor simply overrides the *O* factor. (Modified after Nguyet and Goldscheider, 2006.)

A comprehensive approach to groundwater protection for karst aquifers and springs in Slovenia has been proposed by Ravbar and Goldscheider (2007). This Slovene approach represents the most complete interpretation of the European approach. It includes detailed methods of groundwater resource and source vulnerability mapping (Figure 8–15), the assessment of contamination hazards, the evaluation of the value or importance of a groundwater source or resource, and risk mapping. Figures 8–16 and 8–17 show a resource vulnerability map and a source vulnerability map, respectively, prepared for a Slovene karst spring catchment. The source vulnerability map can be translated into spring protection zones.

8.5.6 Review of other methods of vulnerability mapping

In addition to the relatively recent methods described in the previous chapters, all of which can be used for groundwater vulnerability mapping in karstic catchments, are a large number of other methods, some of which have been compiled, reviewed, and compared by Vrba and Zaporozec (1994), Gogu and Dassargues (2000), Gogu et al. (2003), and others. Magiera (2000) counted 69 methods of vulnerability mapping. It is beyond the scope of this chapter to present and discuss all these methods. The following paragraphs briefly discuss some of the most commonly used and most important methods, acknowledging that their selection is to some degree subjective.

The DRASTIC method (Aller et al., 1987) is probably the best-known and most widely used method of vulnerability mapping. It is frequently applied in the context of groundwater protection and contamination problems in the United States (Croskrey and Groves, 2008; Fritch et al., 2000; Kalinski et al., 1994) and in many other regions of the world (Lee, 2003; Lynch et al., 1997; Rahman, 2008). Vulnerability is assessed on the basis of seven factors: depth to groundwater table (D), net recharge (R), aquifer media (A), soil media (S), topography (T), impact of vadose zone (I), and hydraulic conductivity (C). The Italian SINTACS method uses seven very similar factors (Civita and De Maio, 2000).

Despite its widespread use, DRASTIC has several drawbacks: It is not based on a clear conceptual model, such as the origin-pathway-target model shown in Figure 8–2. Several of the factors are redundant, such as the factors A and C, because hydraulic conductivity is directly dependent on the aquifer medium. The most severe problem is the lack of a multidimensional approach. The one-dimensional approach of DRASTIC might be sufficient to assess the vulnerability of a typical alluvial aquifer, where water and contaminants percolate vertically from the land surface down to the groundwater table. In karst areas, however, lateral flow toward swallow holes can entirely bypass the protective function provided by the overlying layers, as illustrated in Figure 8–12 and Figure 8–13. This scenario is not considered within the DRASTIC method. Therefore, the application of this method to karst spring catchments can lead to drastically inappropriate results, while most of the authors cited report plausible applications to other hydrogeologic environments.

Another rating system for the assessment of groundwater vulnerability, with the slightly immodest acronym GOD, was proposed by Foster (1987). GOD takes into account the type of groundwater occurrence (G) (e.g., none, confined, unconfined), the overlying lithology (O) (e.g., loam, gravel, sandstone, limestone), and the depth to the groundwater table (D). Each factor is assigned a value between 0 and 1. The numeric value for vulnerability is obtained by multiplying the three factors and consequently ranges between 0.0 (negligible) and 1.0 (extreme). GOD is designed for all types of aquifers, but the special properties of karst are not considered. Due to the strong influence of the factor D, the vulnerability of a karst is often underestimated; for example, an unconfined karst aquifer with more than 100 m depth to the groundwater table is assigned a moderate vulnerability (0.4), although the protective function of karstified limestone is often very limited, due to the presence of epikarst and vertical shafts.

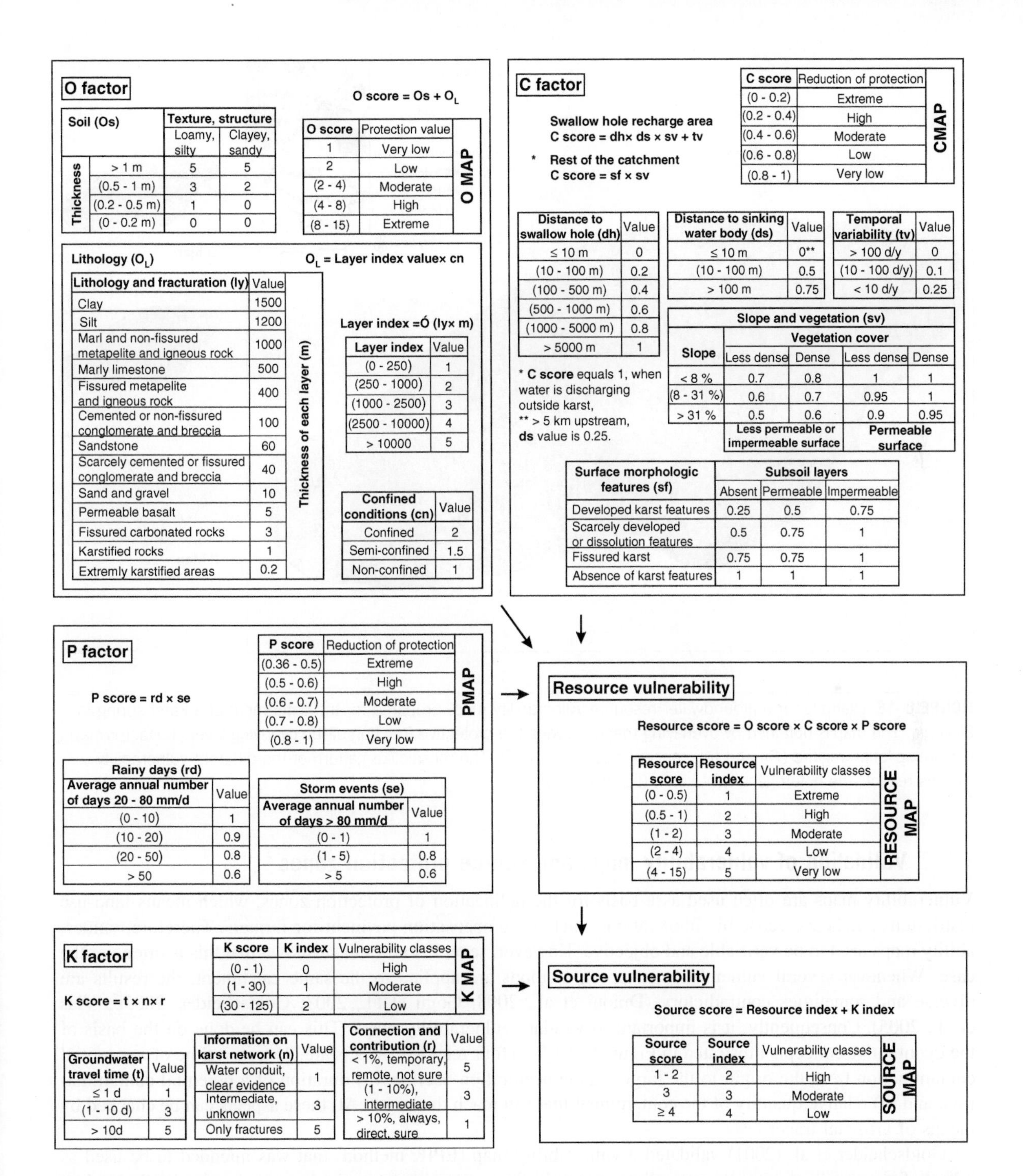

O factor

O score = Os + O_L

Soil (Os)		Texture, structure: Loamy, silty	Clayey, sandy
Thickness	> 1 m	5	5
	(0.5 - 1 m)	3	2
	(0.2 - 0.5 m)	1	0
	(0 - 0.2 m)	0	0

O score	Protection value	
1	Very low	O MAP
2	Low	
(2 - 4)	Moderate	
(4 - 8)	High	
(8 - 15)	Extreme	

Lithology (O_L)

O_L = Layer index value× cn

Lithology and fracturation (ly)	Value	
Clay	1500	Thickness of each layer (m)
Silt	1200	
Marl and non-fissured metapelite and igneous rock	1000	
Marly limestone	500	
Fissured metapelite and igneous rock	400	
Cemented or non-fissured conglomerate and breccia	100	
Sandstone	60	
Scarcely cemented or fissured conglomerate and breccia	40	
Sand and gravel	10	
Permeable basalt	5	
Fissured carbonated rocks	3	
Karstified rocks	1	
Extremly karstified areas	0.2	

Layer index =Ó (ly× m)

Layer index	Value
(0 - 250)	1
(250 - 1000)	2
(1000 - 2500)	3
(2500 - 10000)	4
> 10000	5

Confined conditions (cn)	Value
Confined	2
Semi-confined	1.5
Non-confined	1

C factor

Swallow hole recharge area
C score = dh× ds × sv + tv

* Rest of the catchment
C score = sf × sv

C score	Reduction of protection	
(0 - 0.2)	Extreme	CMAP
(0.2 - 0.4)	High	
(0.4 - 0.6)	Moderate	
(0.6 - 0.8)	Low	
(0.8 - 1)	Very low	

Distance to swallow hole (dh)	Value
≤ 10 m	0
(10 - 100 m)	0.2
(100 - 500 m)	0.4
(500 - 1000 m)	0.6
(1000 - 5000 m)	0.8
> 5000 m	1

Distance to sinking water body (ds)	Value
≤ 10 m	0**
(10 - 100 m)	0.5
> 100 m	0.75

Temporal variability (tv)	Value
> 100 d/y	0
(10 - 100 d/y)	0.1
< 10 d/y	0.25

Slope and vegetation (sv)

Slope	Vegetation cover: Less dense	Dense	Less dense	Dense
< 8 %	0.7	0.8	1	1
(8 - 31 %)	0.6	0.7	0.95	1
> 31 %	0.5	0.6	0.9	0.95
	Less permeable or impermeable surface		Permeable surface	

* **C score** equals 1, when water is discharging outside karst,
** > 5 km upstream, **ds** value is 0.25.

Surface morphologic features (sf)	Subsoil layers: Absent	Permeable	Impermeable
Developed karst features	0.25	0.5	0.75
Scarcely developed or dissolution features	0.5	0.75	1
Fissured karst	0.75	0.75	1
Absence of karst features	1	1	1

P factor

P score = rd × se

P score	Reduction of protection	
(0.36 - 0.5)	Extreme	PMAP
(0.5 - 0.6)	High	
(0.6 - 0.7)	Moderate	
(0.7 - 0.8)	Low	
(0.8 - 1)	Very low	

Rainy days (rd)

Average annual number of days 20 - 80 mm/d	Value
(0 - 10)	1
(10 - 20)	0.9
(20 - 50)	0.8
> 50	0.6

Storm events (se)

Average annual number of days > 80 mm/d	Value
(0 - 1)	1
(1 - 5)	0.8
> 5	0.6

Resource vulnerability

Resource score = O score × C score × P score

Resource score	Resource index	Vulnerability classes	
(0 - 0.5)	1	Extreme	RESOURCE MAP
(0.5 - 1)	2	High	
(1 - 2)	3	Moderate	
(2 - 4)	4	Low	
(4 - 15)	5	Very low	

K factor

K score = t × n× r

K score	K index	Vulnerability classes	
(0 - 1)	0	High	K MAP
(1 - 30)	1	Moderate	
(30 - 125)	2	Low	

Groundwater travel time (t)	Value
≤ 1 d	1
(1 - 10 d)	3
> 10d	5

Information on karst network (n)	Value
Water conduit, clear evidence	1
Intermediate, unknown	3
Only fractures	5

Connection and contribution (r)	Value
< 1%, temporary, remote, not sure	5
(1 - 10%), intermediate	3
> 10%, always, direct, sure	1

Source vulnerability

Source score = Resource index + K index

Source score	Source index	Vulnerability classes	
1 - 2	2	High	SOURCE MAP
3	3	Moderate	
≥ 4	4	Low	

FIGURE 8–15 Assessment schemes of the Slovene approach to (karst) groundwater resource and source vulnerability mapping. The method uses four factors: overlying layers (*O*), concentration of flow (*C*), precipitation regime (*P*), and karst network development (*K*). (Ravbar, 2007.)

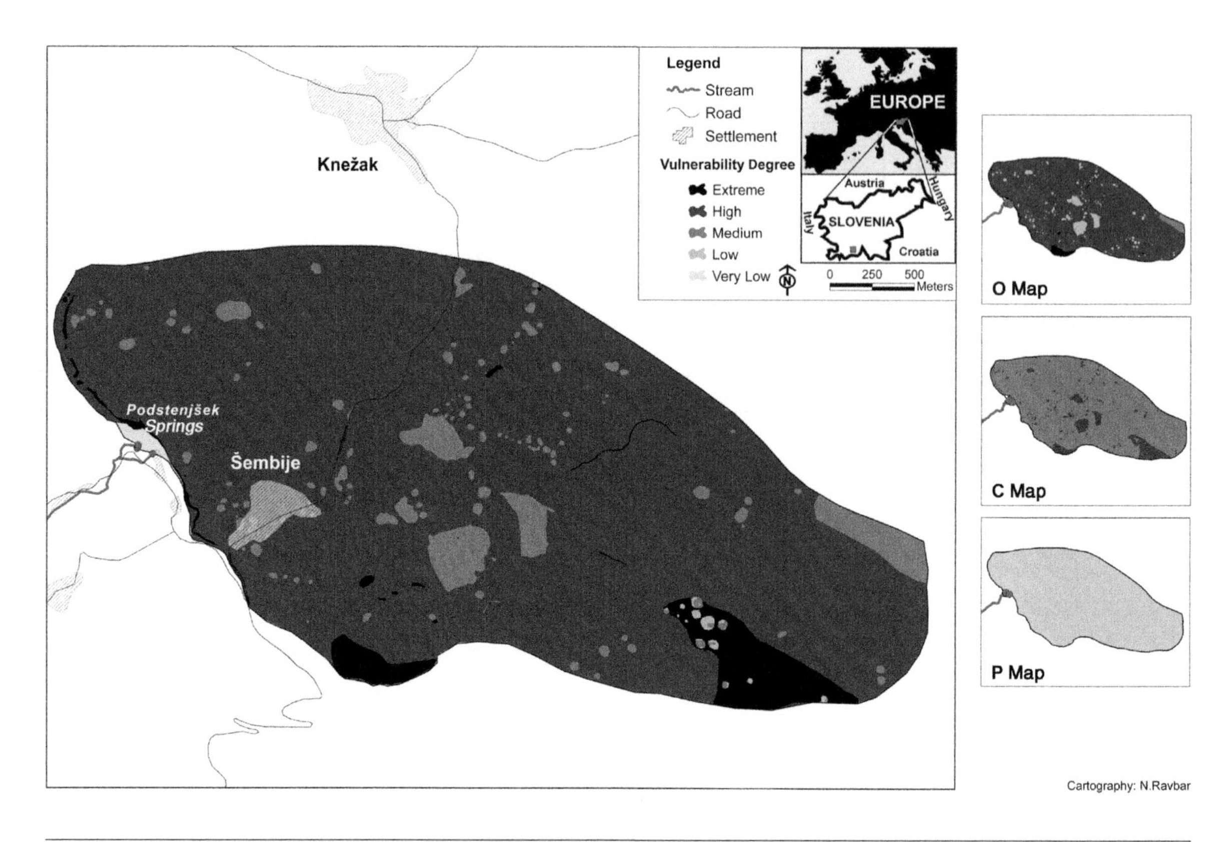

FIGURE 8–16 Example for a groundwater resource vulnerability map, prepared for the catchment of a karst spring in Slovenia. The map is obtained by overlaying maps showing the protective function of the overlying layers (*O* factor/map), the precipitation regime (*P*), and the concentration of flow (*C*). The mosaiclike pattern of the different vulnerability classes reflects the heterogeneity of karst. (Ravbar, 2007.)

8.5.7 Validation of vulnerability maps and source protection zones

Vulnerability maps are often used as a basis for the delineation of protection zones, which means land-use restrictions and hence economic implications, such as compensation payment for farmers. Therefore, vulnerability maps need to be verifiable and objective. However, several studies have shown that this is often not the case. Whenever several vulnerability mapping methods are applied to the same catchment, the results are diverse and sometimes contradictory (Draoui et al., 2008; Gogu et al., 2003; Goldscheider, 2005b; Vias et al., 2005). Consequently, it is important to validate vulnerability maps. This can be done on the basis of the quantitative concepts illustrated in Figure 8–10. The four relevant aspects are (1) transit time of a potential contamination from the origin to the target, (2) concentration decline, (3) duration of a potential contamination, and (4) relative quantity of the contaminant that can reach the target. All these aspects can be checked by means of artificial tracer tests.

Goldscheider et al. (2001) validated a vulnerability map (EPIK method) that was intended to be used as a basis for groundwater source protection zoning in the catchment area of a karst spring in the Swiss Jura

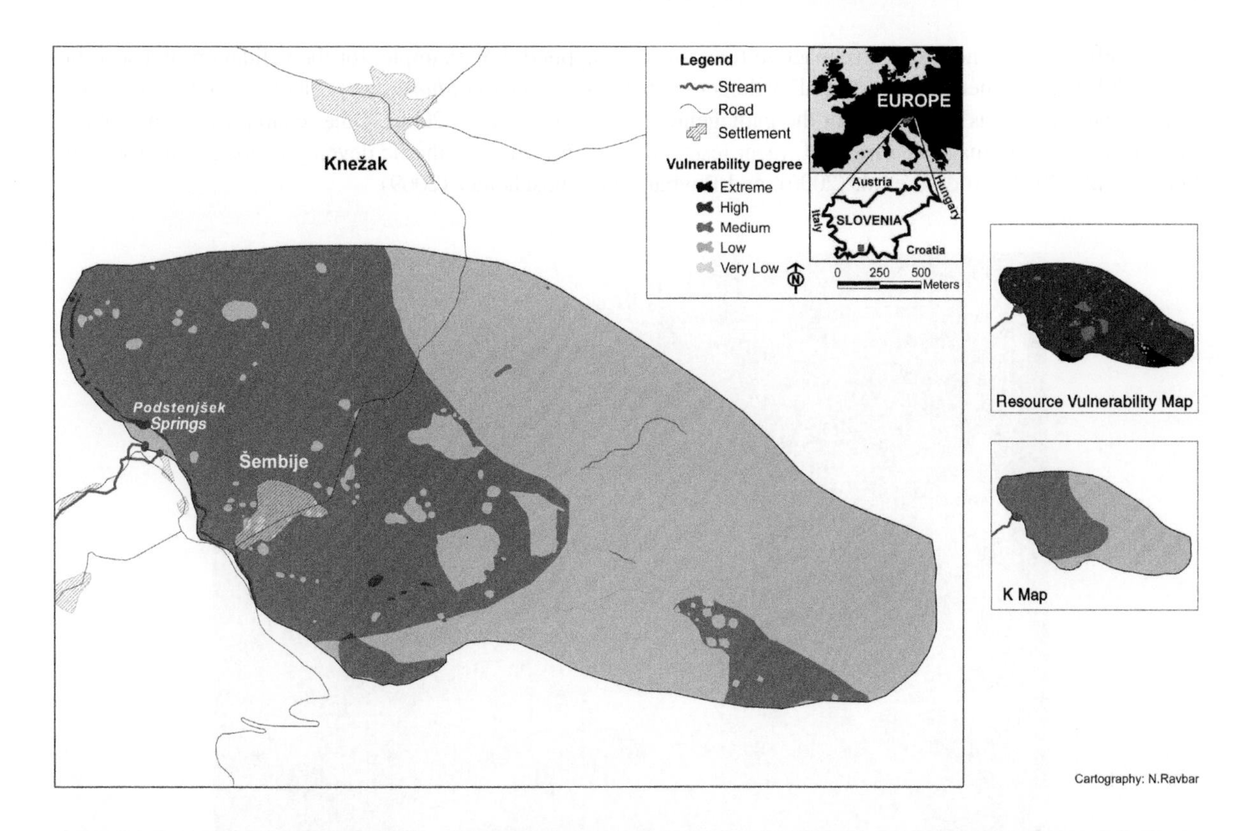

FIGURE 8–17 Example for a groundwater source vulnerability map, which can be used as a basis for the delineation of source protection zones for the Podstenjsek karst spring in Slovenia. The map is obtained by overlaying the resource vulnerability map (Figure 8–16) and the K map, which shows the pathway in the saturated zone of the karst aquifer, i.e., flow towards the spring. The source vulnerability map is a combination of a mosaic pattern and a concentric arrangement of protection zones. (Ravbar, 2007.)

Mountains. Different tracers were released at the land surface using a watering can, followed by 20 mm of artificial rainfall (irrigation). The tracer breakthrough curves were recorded at the spring. The following criteria were used for validation: the time of maximum concentration (peak time), the normalized maximum concentration (i.e., maximum concentration divided by injection quantity), and the tracer recovery (Figure 8–18). The results largely confirmed the vulnerability assessment but also allowed some optimizations.

In some cases, the duration of a potential contamination may also be an important aspect. Tracer tests in karst spring catchments sometimes deliver breakthrough curves with short peak times and high maximum concentrations but very long tails, although the injection was done instantaneously and directly into flowing groundwater. Different processes can explain this behavior, such as conduit-matrix interactions (double porosity effects) or intermediate storage in immobile fluid regions (Field and Pinsky, 2000; Goldscheider, 2005a, 2008). A released contaminant would consequently reach the target very rapidly and at high concentration levels, but the contamination would also last a long time (Figure 8–19).

For the validation of an intrinsic vulnerability assessment, conservative tracers should be used, such as uranine or several other fluorescent dyes or anionic salt tracers, such as chloride. For the validation of specific vulnerability (see the next section), reactive tracers can be applied. For example, for the validation of a specific vulnerability assessment for microbial pathogens, such as *Cryptosporidium* cysts, fluorescent tracer microspheres can be used as surrogates for the pathogenic cysts (Harvey et al., 2008). Other examples for the validation of vulnerability maps by means of tracer tests, along with further methodic developments, can be found in Perrin et al. (2004), Andreo et al. (2006), and Ravbar and Goldscheider (2009).

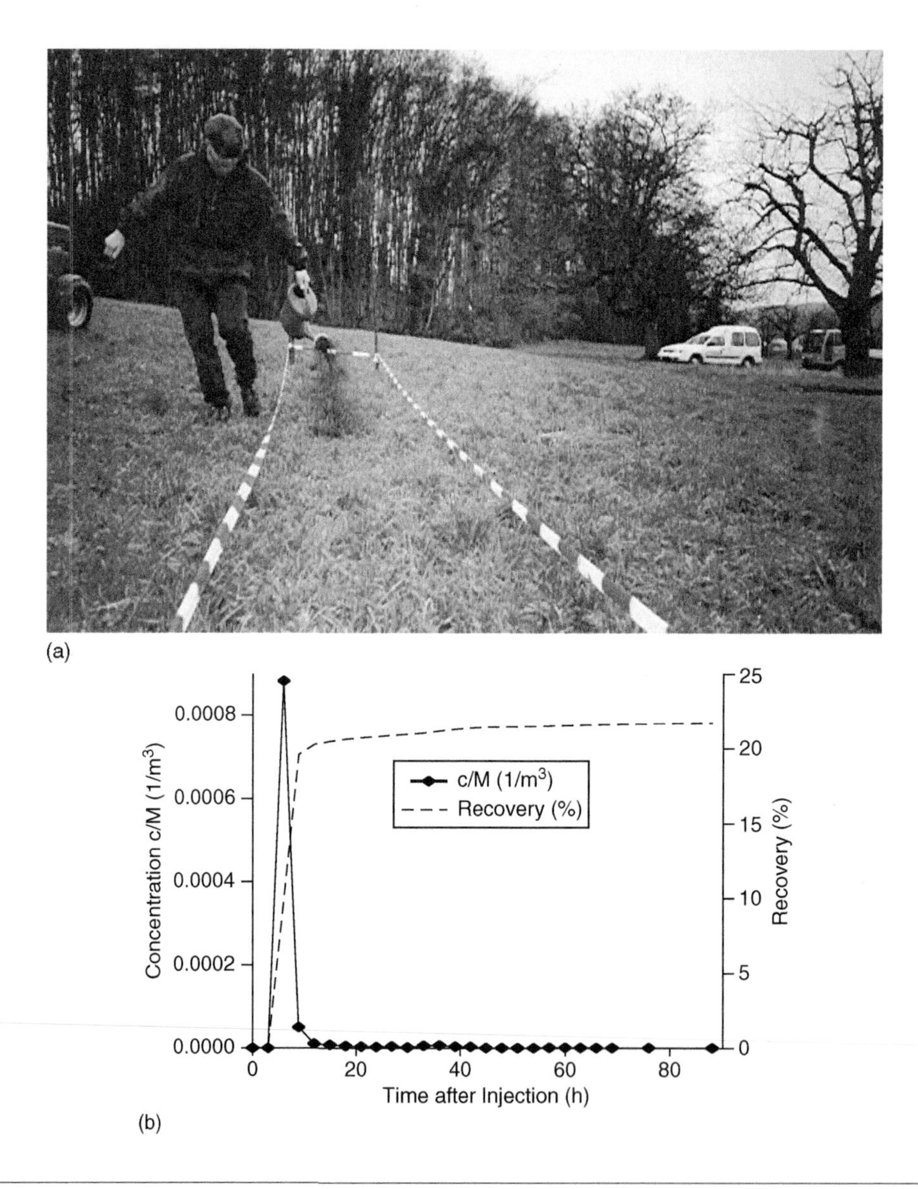

FIGURE 8–18 First example of the validation of a vulnerability map (EPIK) by means of multitracer tests for a karst spring catchment in the Swiss Jura Mountains: (a) spreading of a tracer (eosin) at the land surface; (b) eosin breakthrough curve observed at the spring, normalized by injection quantity;

(continued)

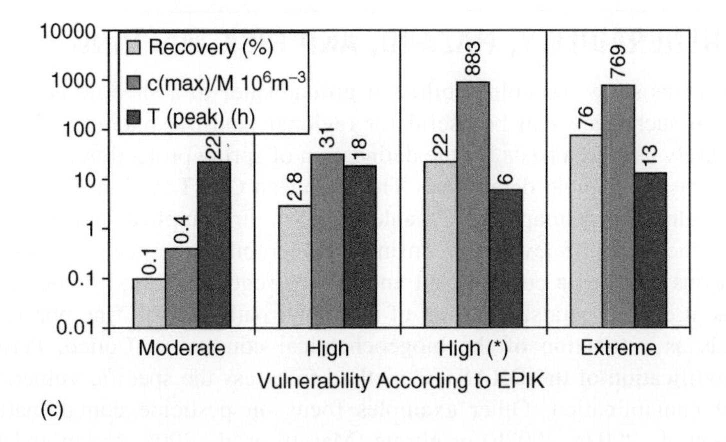

FIGURE 8–18, CONT'D (c) summary of the results from four injection sites. Peak time (*T*), normalized maximum concentration (*c/M*), and tracer recovery were used as validation criteria, largely confirming the vulnerability assessment of the EPIK method. Results obtained by the eosin injection are marked with an asterisk. Although eosin was released on the soil surface, it peaked at the spring only 6 hours after injection. (Modified after Goldscheider et al., 2001.)

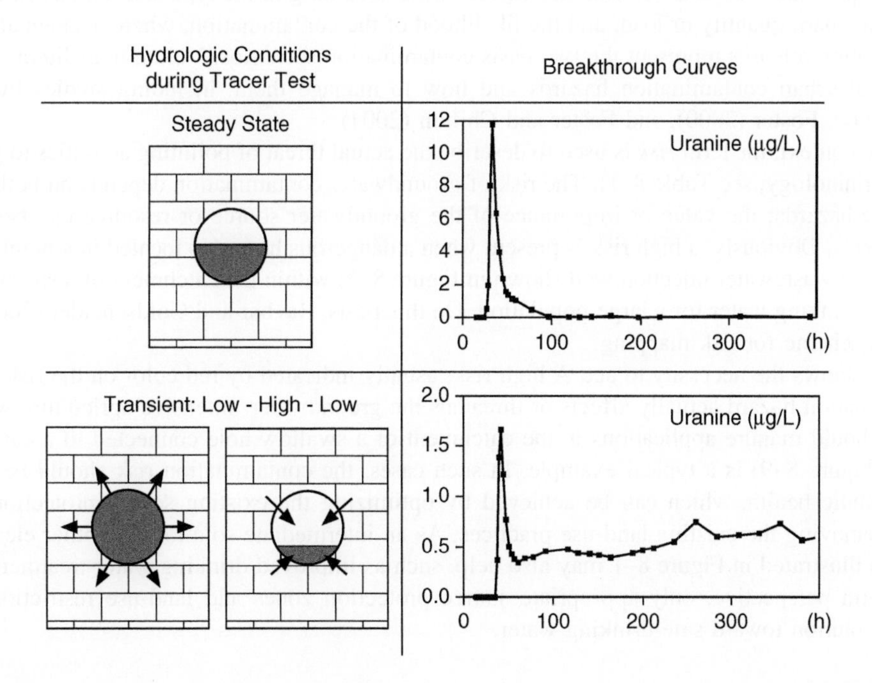

FIGURE 8–19 Breakthrough curves of two tracer tests with uranine, monitored at the same karst spring but done during different hydrologic conditions. During transient flow conditions, the tracer breakthrough curve displays a short peak time but a very long tail, indicating that karst aquifers are not always fast-flushing systems. Due to intermediate storage, even a short-term (accidental) contaminant release can cause long-term springwater contamination. (Modified after Goldscheider, 2005a.)

8.6 SPECIFIC VULNERABILITY, HAZARD, AND RISK MAPPING

Specific vulnerability maps show the vulnerability of groundwater to a specific contaminant or group of contaminants. Although such maps can be useful for regional land-use planning and environmental risk assessment, they are rarely used as a basis for the delineation of spring protection zones. Therefore, specific vulnerability is not discussed in great detail here. The European COST Action 620 also proposed a methodology for specific vulnerability mapping (Zwahlen, 2004), first applied and tested by Andreo et al. (2006). In addition to the factors relevant for intrinsic vulnerability, a specific vulnerability assessment considers the interactions between a contaminant and the hydrogeologic environment, such as adsorption of toxic metal cations at clay minerals, filtration of microbial pathogens in fine pores, or the degradation of organic compounds as a function of the biogeochemical conditions. Celico, Petrella, and Naclerio (2007) proposed a modification of the DRASTIC method to assess the specific vulnerability of carbonate aquifers to microbial contamination. Other examples focus on pesticide contamination (Arias-Estevez et al., 2008; Stenemo et al., 2007a, 2007b) or nitrate (Masetti et al., 2008; Nolan and Hitt, 2006; Parker, Booth, and Foster, 1987).

In the context of groundwater protection, hazards are defined as actually or potentially polluting activities or infrastructure. Hazards can be classified on the basis of their spatial extension into point hazards (e.g., septic tanks, gasoline station), linear hazards (e.g., roads, pipelines), and areal hazards (e.g., spreading of pesticides or liquid manure). Hazards can also be classified according to the type and toxicity of the contaminant, the contaminant quantity or load, and the likelihood of the contamination, where accidental and permanent contamination releases represent the two basis contamination scenarios. Voluminous literature discusses agricultural and urban contamination hazards and how to manage them, including studies by Boyer and Pasquarell (1999), Foster (2000), and Foster and Chilton (2004).

In the same context, the term *risk* is used to describe the actual threat of polluting activities to groundwater quality (for terminology, see Table 8–1). The risk of groundwater contamination depends on both the vulnerability and the hazards; the value or importance of the groundwater source or resource can be included as additional aspects. Obviously, a high risk is present when a dangerous hazard is located in a highly vulnerable zone (such as the wastewater injection well shown in Figure 8–3) within the catchment of a spring used as the only source of drinking water for a large population. On this basis, Ravbar and Goldscheider (2007) proposed an assessment scheme for risk mapping.

A risk map shows the necessity to act. A high risk, usually indicated by red color on the risk map, means that a contamination hazard actually affects or threatens the groundwater. Intensive agriculture with frequent pesticide and liquid manure applications in the catchment of a swallow hole connected to a karst spring (as illustrated in Figure 8–9) is a typical example. In such cases, the contamination risk should be reduced for the sake of public health, which can be achieved by optimizing the existing source protection zones and changing or removing the existing land-use practices. As an intermediate solution, the other elements of the security chain illustrated in Figure 8–1 may also help, such as improved drinking water treatment. However, in the long-term perspective, only appropriate source protection zones and land-use restrictions represent a sustainable solution toward safe drinking water.

ACKNOWLEDGMENTS

I am grateful to Michiel Pronk (Neuchâtel) and Natasa Ravbar (Postojna) for their valuable contributions in the form of several figures and maps and to David Drew for proofreading.

REFERENCES

Adams, B., Foster, S.D., 1992. Land-surface zoning for groundwater protection. Journal of the Institution of Water and Environmental Management 6 (3), 312–320.

Aller, L., Bennett, T., Lehr, J.H., Petty, R.J., 1987. DRASTIC: A standardized system for evaluating ground water pollution potential using hydrogeological settings. U.S. Environmental Protection Agency, Ada, Oklahoma.

Andreo, B., Goldscheider, N., Vadillo, I., Vias, J.M., Neukum, C., Sinreich, M., et al., 2006. Karst groundwater protection: First application of a Pan-European Approach to vulnerability, hazard and risk mapping in the Sierra de Libar (Southern Spain). Sci. Total Environ. 357 (1–3), 54–73.

Appelo, C.A.J., Postma, D., 2005. Geochemistry, groundwater and pollution, second ed. Balkema, Leiden, the Netherlands.

Arias-Estevez, M., Lopez-Periago, E., Martinez-Carballo, E., Simal-Gandara, J., Mejuto, J.C., Garcia-Rio, L., 2008. The mobility and degradation of pesticides in soils and the pollution of groundwater resources. Agriculture Ecosystems and Environment 123 (4), 247–260.

Auckenthaler, A., Huggenberger, P. (Eds.), 2003. Pathogene Mikroorganismen im Grund- und Trinkwasser. Birkhäuser Verlag, Basel, Switzerland.

Benischke, R., Goldscheider, N., Smart, C.C., 2007. Tracer techniques. In: Goldscheider, N., Drew, D. (Eds.), Methods in karst hydrogeology. International Contributions to Hydrogeology. Taylor & Francis/Balkema, London, pp. 147–170.

Bianchetti, G., Roth, P., Vuataz, F.D., Vergain, J., 1992. Deep groundwater circulation in the Alps—Relations between water infiltration, induced seismicity and thermal springs—The case of Val-d'Illiez, Wallis, Switzerland. Eclogae Geol. Helv. 85 (2), 291–305.

Bonadonna, L., Briancesco, R., Ottaviani, M., Veschetti, E., 2002. Occurrence of Cryptosporidium oocysts in sewage effluents and correlation with microbial, chemical and physical water variables. Environ. Monit. Assess. 75 (3), 241–252.

Boyer, D.G., Pasquarell, G.C., 1999. Agricultural land use impacts on bacterial water quality in a karst groundwater aquifer. J. Am. Water Resour. Assoc. 35 (2), 291–300.

Brookes, J.D., Hipsey, M.R., Burch, M.D., Regel, R.H., Linden, L.G., Ferguson, C.M., et al., 2005. Relative value of surrogate indicators for detecting pathogens in lakes and reservoirs. Environ. Sci. Technol. 39 (22), 8614–8621.

Brouyère, S., Jeannin, P.Y., Dassargues, A., Goldscheider, N., Popescu, I.C., Sauter, M., et al., 2001. Evaluation and validation of vulnerability concepts using a physically based approach. In: Seventh Conference on Limestone Hydrology and Fissured Media. Sci. Tech. Envir., Mém. H. S. no. 13, Besançon, France, 67–72.

Bull, R.J., Birnbaum, L.S., Cantor, K.P., Rose, J.B., Butterworth, B.E., Pegram, R., et al., 1995. Water chlorination: Essential process or cancer hazard? Fundam. Appl. Toxicol. 28 (2), 155–166.

Burkhard, M., Atteia, O., Sommaruga, A., Gogniat, S., Evard, D., 1998. Tectonics and hydrogeology of the Neuchatel Jura. Eclogae Geologicae Helvetiae 91 (1), 177–183.

Bussard, T., Tacher, L., Parriaux, A., Maitre, V., 2006. Methodology for delineating groundwater protection areas against persistent contaminants. Quarterly Journal of Engineering Geology and Hydrogeology 39, 97–109.

Celico, F., Petrella, E., Naclerio, G., 2007. Updating of a DRASTIC-based method for specific vulnerability assessment in carbonate aquifers. Water International 32 (3), 475–482.

Chapelle, F.H., 2000. The significance of microbial processes in hydrogeology and geochemistry. Hydrogeology Journal 8 (1), 41–46.

Chapelle, F.H., 2001. Ground-Water Microbiology and Geochemistry. John Wiley & Sons, New York.

Civita, M., De Maio, M., 2000. Valutazione e cartografia automatica della vulnerabilità degli acquiferi all'inquinamento con il systema parametrico SINTACS R5 [Assessment and mapping of aquifer vulnerability using the parametric system SINTACS R5]. Pitagora Editrice, Bologna, Italy.

Clark, I.D., Fritz, P., 1997. Environmental isotopes in hydrogeology. Lewis Publishers, Boca Raton, FL.

Criss, R., Davisson, L., Surbeck, H., Winston, W., 2007. Isotopic methods. In: Goldscheider, N., Drew, D. (Eds.), Methods in karst hydrogeology. International Contributions to Hydrogeology. Taylor & Francis/Balkema, London, pp. 123–145.

Croskrey, A., Groves, C., 2008. Groundwater sensitivity mapping in Kentucky using GIS and digitally vectorized geologic quadrangles. Env. Geol. 54 (5), 913–920.

Daly, D., Dassargues, A., Drew, D., Dunne, S., Goldscheider, N., Neale, S., et al., 2002. Main concepts of the "European approach" to karst-groundwater-vulnerability assessment and mapping. Hydrogeology Journal 10 (2), 340–345.

Danielopol, D.L., Pospisil, P., 2001. Hidden biodiversity in the groundwater of the Danube Flood Plain National Park (Austria). Biodivers. Conserv. 10 (10), 1711–1721.

Department of Environment and Local Government, Environmental Protection Agency and Geological Survey of Ireland, 1999. Groundwater protection schemes. Department of Environment and Local Government, Environmental Protection Agency and Geological Survey of Ireland, Dublin.

Deutscher Verein des Gas- und Wasserfaches, 2006. Richtlinien für Trinkwasserschutzgebiete, Part 1: Schutzgebiete für Grundwasser. Deutscher Verein des Gas- und Wasserfaches, DVGW-Regelwerk. Working Paper W 101, Eschborn.

Doerfliger, N., Jeannin, P.Y., Zwahlen, F., 1999. Water vulnerability assessment in karst environments: a new method of defining protection areas using a multi-attribute approach and GIS tools (EPIK method). Env. Geol. 39 (2), 165–176.

Doerfliger, N., Zwahlen, F., 1998. Practical guide: Groundwater vulnerability mapping in karstic regions (EPIK). Swiss Agency for the Environment, Forests and Landscape (SAEFL), Bern.

Draoui, M., Vias, J., Andreo, B., Targuisti, K., El Messari, J.S., 2008. A comparative study of four vulnerability mapping methods in a detritic aquifer under Mediterranean climatic conditions. Env. Geol. 54 (3), 455–463.

Eamus, D., Froend, R., 2006. Groundwater-dependent ecosystems: The where, what and why of GDEs. Aust. J. Bot. 54 (2), 91–96.

Ehrlich, H.L., 1997. Microbes and metals. Appl. Microbiol. Biotechnol. 48 (6), 687–692.

Fetter, C.W., 1999. Contaminant Hydrogeology, Prentice Hall, Upper Saddle River, NJ.

Field, M.S., Pinsky, P.F., 2000. A two-region nonequilibrium model for solute transport in solution conduits in karstic aquifers. J. Contam. Hydrol. 44 (3–4), 329–351.

Foster, S.S.D., 1987. Fundamental concepts in aquifer vulnerability, pollution risk and protection strategy. In: Van Duijevenboden, W., Van Waegeningh, W. (Eds.), Vulnerability of Soil and Groundwater to Pollutants, TNO Committee on Hydrogeological Research, Proceedings and Information, 38, The Hague, pp. 69–86.

Foster, S.S.D., 2000. Assessing and controlling the impacts of agriculture on groundwater—From barley barons to beef bans. Quarterly Journal of Engineering Geology and Hydrogeology 33 (4), 263–280.

Foster, S.S.D., Chilton, P.J., 2004. Downstream of downtown: urban wastewater as groundwater recharge. Hydrogeology Journal 12 (1), 115–120.

Fritch, T.G., McKnight, C.L., Yelderman, J.C., Arnold, J.G., 2000. An aquifer vulnerability assessment of the Paluxy aquifer, central Texas, USA, using GIS and a modified DRASTIC approach. Environ. Manage. 25 (3), 337–345.

Göppert, N., Goldscheider, N., 2008. Solute and colloid transport in karst conduits under low- and high-flow conditions. Ground Water 46 (1), 61–68.

Gogu, R.C., Dassargues, A., 2000. Current trends and future challenges in groundwater vulnerability assessment using overlay and index methods. Env. Geol. 39 (6), 549–559.

Gogu, R.C., Hallet, V., Dassargues, A., 2003. Comparison of aquifer vulnerability assessment techniques. Application to the Neblon river basin (Belgium). Env. Geol. 44 (8), 881–892.

Goldscheider, N., 2005a. Fold structure and underground drainage pattern in the alpine karst system Hochifen-Gottesacker. Eclogae Geol. Helv. 98 (1), 1–17.

Goldscheider, N., 2005b. Karst groundwater vulnerability mapping: Application of a new method in the Swabian Alb, Germany. Hydrogeology Journal 13 (4), 555–564.

Goldscheider, N., 2008. A new quantitative interpretation of the long-tail and plateau-like breakthrough curves from tracer tests in the artesian karst aquifer of Stuttgart, Germany. Hydrogeology Journal 16 (7), 1311–1317.

Goldscheider, N., Andreo, B., 2007. The geological and geomorphological framework. In: Goldscheider, N., Drew, D. (Eds.), Methods in Karst Hydrogeology. International Contributions to Hydrogeology. Taylor & Francis, London, pp. 9–23.

Goldscheider, N., Drew, D., 2007. Methods in karst hydrogeology. International Contributions to Hydrogeology 26. Taylor & Francis, London.

Goldscheider, N., Hötzl, H., Fries, W., Jordan, P., 2001. Validation of a vulnerability map (EPIK) with tracer tests. Seventh Conference on Limestone Hydrology and Fissured Media, Besançon, France, pp. 167–170.

Goldscheider, N., Klute, M., Sturm, S., Hötzl, H., 2000. The PI method—A GIS-based approach to mapping groundwater vulnerability with special consideration of karst aquifers. Z. Angew. Geol. 46 (3), 157–166.

Goldscheider, N., Milnes, E., Fries, W., Joppen, M., 2008. Markierungsversuche und Modellierung zur Bewertung der Gefährdung eines Trinkwasserbrunnens [Tracer experiments and modelling for risk assessment of a drinking water well]. Grundwasser, published online.

Harvey, R.W., Metge, D.W., Shapiro, A.M., Renken, R.A., Osborn, C.L., Ryan, J.N., et al., 2008. Pathogen and chemical transport in the karst limestone of the Biscayne aquifer: 3. Use of microspheres to estimate the transport potential of Cryptosporidium parvum oocysts. Water Resour. Res. 44 (8).

Herwaldt, B.L., Craun, G.F., Stokes, S.L., Juranek, D.D., 1992. Outbreaks of waterborne disease in the United States, 1989–1990. Journal American Water Works Association 84 (4), 129–135.

Hötzl, H., 1996. Origin of the Danube-Aach system. Env. Geol. 27 (2), 87–96.

Hunkeler, D., Mudry, J., 2007. Hydrochemical tracers. In: Goldscheider, N., Drew, D. (Eds.), Methods in karst hydrogeology. International Contributions to Hydrogeology. Taylor & Francis/Balkema, London, pp. 93–121.

Johnston, C.T., Cook, P.G., Frape, S.K., Plummer, L.N., Busenberg, E., Blackport, R.J., 1998. Ground water age and nitrate distribution within a glacial aquifer beneath a thick unsaturated zone. Ground Water 36 (1), 171–180.

Kalinski, R.J., Kelly, W.E., Bogardi, I., Ehrman, R.L., Yamamoto, P.D., 1994. Correlation between DRASTIC vulnerabilities and incidents of VOC contamination of municipal wells in Nebraska. Ground Water 32 (1), 31–34.

Käss, W., 1998. Tracing Technique in Geohydrology. Balkema, Rotterdam, the Netherlands.

Krauskopf, K.B., Bird, D.K., 1995. Introduction to geochemistry. McGraw-Hill, New York.

Kresic, N., 2009. Groundwater Resources, Sustainability, Management and Restoration. McGraw-Hill, New York.

Lee, S., 2003. Evaluation of waste disposal site using the DRASTIC system in Southern Korea. Env. Geol. 44 (6), 654–664.

Lisle, J.T., Rose, J.B., 1995. Cryptosporidium contamination of water in the USA and UK—A minireview. Journal of Water Supply Research and Technology-Aqua 44 (3), 103–117.

Lundmark, A., Olofsson, B., 2007. Chloride deposition and distribution in soils along a deiced highway—Assessment using different methods of measurement. Water Air Soil Pollut. 182 (1–4), 173–185.

Lynch, S.D., Reynders, A.G., Schulze, R.E., 1997. A DRASTIC approach to groundwater vulnerability in South Africa. S. Afr. J. Sci. 93 (2), 59–60.

Mackenzie, W.R., Hoxie, N.J., Proctor, M.E., Gradus, M.S., Blair, K.A., Peterson, D.E., et al., 1994. A massive outbreak in Milwaukee of Cryptosporidium infection transmitted through the public water-supply. N. Engl. J. Med. 331 (3), 161–167.

Magiera, P., 2000. Methoden zur Abschätzung der Verschmutzungsempfindlichkeit des Grundwassers [Methods for the estimation of groundwater vulnerability to contamination]. Grundwasser 5 (3), 103–114.

Margat, J., 1968. Vulnérabilité des nappes d'eau souterraine à la pollution. BRGM Publication 68 SGL 198 HYD, Orléans, France.

Masetti, M., Poli, S., Sterlacchini, S., Beretta, G.P., Facchi, A., 2008. Spatial and statistical assessment of factors influencing nitrate contamination in groundwater. J. Environ. Manage. 86 (1), 272–281.

Milnes, E., Renard, P., 2004. The problem of salt recycling and seawater intrusion in coastal irrigated plains: an example from the Kiti aquifer (Southern Cyprus). Journal of Hydrology 288 (3–4), 327–343.

Montgomery, M.A., Elimelech, M., 2007. Water and sanitation in developing countries: Including health in the equation. Environ. Sci. Technol. 41 (1), 17–24.

Nguyet, V.T.M., Goldscheider, N., 2006. A simplified methodology for mapping groundwater vulnerability and contamination risk, and its first application in a tropical karst area, Vietnam. Hydrogeology Journal 14 (8), 1666–1675.

Nickson, R.T., McArthur, J.M., Ravenscroft, P., Burgess, W.G., Ahmed, K.M., 2000. Mechanism of arsenic release to groundwater, Bangladesh and West Bengal. Appl. Geochem. 15 (4), 403–413.

Nolan, B.T., Hitt, K.J., 2006. Vulnerability of shallow groundwater and drinking-water wells to nitrate in the United States. Environ. Sci. Technol. 40 (24), 7834–7840.

Ophori, D., Toth, J., 1990. Relationships in regional groundwater discharge to streams—An analysis by numerical-simulation. J. Hydrol. 119 (1–4), 215–244.

Parker, J.M., Booth, S.K., Foster, S.S.D., 1987. Penetration of Nitrate from Agricultural Soils into the Groundwater of the Norfolk Chalk. Proceedings of the Institution of Civil Engineers Part 2—Research and Theory 83, 15–32.

Perrin, J., Pochon, A., Jeannin, P.Y., Zwahlen, F., 2004. Vulnerability assessment in karstic areas: validation by field experiments. Env. Geol. 46 (2), 237–245.

Pronk, M., Goldscheider, N., Zopfi, J., 2006. Dynamics and interaction of organic carbon, turbidity and bacteria in a karst aquifer system. Hydrogeology Journal 14 (4), 473–484.

Pronk, M., Goldscheider, N., Zopfi, J., 2007. Particle-size distribution as indicator for fecal bacteria contamination of drinking water from karst springs. Environ. Sci. Technol. 41 (24), 8400–8405.

Rahman, A., 2008. A GIS based DRASTIC model for assessing groundwater vulnerability in shallow aquifer in Aligarh, India. Applied Geography 28 (1), 32–53.

Ranville, J.F., Hendry, M.J., Reszat, T.N., Xie, Q.L., Honeyman, B.D., 2007. Quantifying uranium complexation by groundwater dissolved organic carbon using asymmetrical flow field-flow fractionation. J. Contam. Hydrol. 91 (3–4), 233–246.

Ravbar, N., 2007. The protection of karst waters. Carsologica, Postojna, Ljubljana, Slovenia.

Ravbar, N., Goldscheider, N., 2007. Proposed methodology of vulnerability and contamination risk mapping for the protection of karst aquifers in Slovenia. Acta Carsologica 36 (3), 397–411.

Ravbar, N., Goldscheider, N., 2009. Comparative application of four methods of groundwater vulnerability mapping in a Slovene karst catchment. Hydrogeology Journal. 17 (3), 725–733.

Swiss Agency for the Environment, Forests and Landscape, 2004. Wegleitung Grundwasserschutz [Practical guide: groundwater protection], Swiss Agency for the Environment, Forests and Landscape, Bern.

Scanlon, B.R., Jolly, I., Sophocleous, M., Zhang, L., 2007. Global impacts of conversions from natural to agricultural ecosystems on water resources: Quantity versus quality. Water Resour. Res. 43 (3).

Schaub, S.A., Oshiro, R.K., 2000. Public health concerns about caliciviruses as waterborne contaminants. J. Infect. Dis. 181, 374–380.
Shokes, T.E., Moller, G., 1999. Removal of dissolved heavy metals from acid rock drainage using iron metal. Environ. Sci. Technol. 33 (2), 282–287.
Siefert, B., Buchel, G., Lebkuchner-Neugebauer, J., 2006. Potash mining waste pile Sollstedt (Thuringia): Investigations of the spreading of waste solutes in the Roethian Karst. Grundwasser 11 (2), 99–110.
Smedley, P.L., Kinniburgh, D.G., 2002. A review of the source, behaviour and distribution of arsenic in natural waters. Appl. Geochem. 17 (5), 517–568.
Smith, J.L., 2001. A review of hepatitis E virus. J. Food Prot. 64 (4), 572–586.
Stenemo, F., Lindahl, A.M.L., Gardenos, A., Jarvis, N., 2007a. Meta-modeling of the pesticide fate model MACRO for groundwater exposure assessments using artificial neural networks. J. Contam. Hydrol. 93 (1–4), 270–283.
Stenemo, F., Ray, C., Yost, R., Matsuda, S., 2007b. A screening tool for vulnerability assessment of pesticide leaching to groundwater for the islands of Hawaii, USA. Pest Manag. Sci. 63 (4), 404–411.
Szewzyk, U., Szewzyk, R., Manz, W., Schleifer, K.H., 2000. Microbiological safety of drinking water. Annu. Rev. Microbiol. 54, 81–127.
Toth, J., 1963. A theoretical analysis of groundwater flow in small drainage basins. J. Geophys. Res. 68 (16), 4795–4812.
Toth, J., 1999. Groundwater as a geologic agent: An overview of the causes, processes, and manifestations. Hydrogeology Journal 7 (1), 1–14.
Vaneverdingen, R.O., 1991. Physical, chemical, and distributional aspects of Canadian springs. Memoirs of the Entomological Society of Canada 155, 7–28.
Vesper, D.J., White, W.B., 2004. Spring and conduit sediments as storage reservoirs for heavy metals in karst aquifers. Env. Geol. 45 (4), 481–493.
Vias, J.M., Andreo, B., Perles, M.J., Carrasco, F., 2005. A comparative study of four schemes for groundwater vulnerability mapping in a diffuse flow carbonate aquifer under Mediterranean climatic conditions. Env. Geol. 47 (4), 586–595.
Vias, J.M., Andreo, B., Perles, M.J., Carrasco, F., Vadillo, I., Jimenez, P., 2006. Proposed method for groundwater vulnerability mapping in carbonate (karstic) aquifers: The COP method. Hydrogeology Journal 14 (6), 912–925.
Vrba, J., Zaporozec, A. (Eds.), 1994. Guidebook on Mapping Groundwater Vulnerability. International Contributions to Hydrogeology, 6, Hannover, Germany.
Zwahlen, F., 2004. Vulnerability and risk mapping for the protection of carbonate (karst) aquifers, final report COST action 620. European Commission, Brussels.

CHAPTER

Utilization and regulation of springs

9

Zoran Stevanovic

Department of Hydrogeology, School of Mining and Geology, University of Belgrade, Serbia

9.1 INTRODUCTION

The utilization and tapping of springwater is an ancient art. Historically, to have easy access to water, cities were often situated near large springs, while those cities without a reliable water supply were destroyed or abandoned because they could not survive long sieges. As a rule, the cities with plentiful water drawn from successfully constructed spring intakes and reservoirs provided a base for prosperous development and a safe haven for their citizens. For example, in the narrow historical center of Rome, there were 23 springs, which initially supplied water throughout the city, while at the height of the Roman Empire 11 long aqueducts delivered more than 13 m^3/s of water to the city from distances ranging from 16 to 91 km (Lombardi and Corazza, 2008).

The history of capturing groundwater predates Roman times. In ancient China, Mesopotamia, and Egypt, systems were developed for tapping springs or digging deep wells and delivering water sometimes to very distant points. Even in old Babylon, securing water was not a problem. To ensure a water supply to the ancient Mesopotamian city Nineveh, the Assyrian king Sanherib, son of Sargon II, constructed an intake on the Khanis spring systems and one of the first aqueducts to deliver water to the city walls. Even then, springwater was found to be a much better solution than the nearby surface waters of the Tigris (Figure 9–1).

Especially difficult was the search for water along the Mediterranean coast, where there were many developed civilizations. Greek philosophers and mathematicians, such as Euclid, Pythagoras, and Archimedes, established the principles for the design and operation of hydraulic structures, which would be further developed during the Roman golden age of water supply.

"The culture of water" was emblematic of Roman times. Apart from drinking and sanitary purposes, water was important for physical medicine, balneotherapy, and aesthetics. The Roman architect Vitruvius was the first to leave a written record showing that springs on mountain slopes might be recharged by atmospheric precipitation, finding their interaction in the rapid propagation of infiltrated water. Many beautiful fountains were constructed throughout the Empire. In short, the Romans' supremacy and dominance were demonstrated through their knowledge of water, including the art of tapping and delivering springwater (Figure 9–2).

After the collapse of the Roman Empire, there was, in terms of the utilization of springs, a long period of engineering silence. Much later, the first centralized water supply system was constructed in Paris in the 12th century, in London in the 13th century, then in many other European cities. Apart from digging shallow wells in alluvial fans and primitive drilling of deeper wells in large artesian basins (for example, in Paris and London), many cities tried to solve their water problems by capturing different types of springs and generally by delivering springwater to the city centers by gravity pipelines made of wood, lead, iron, or ceramic (Figure 9–3).

Copyright © 2010, Elsevier Inc. All rights reserved.

(a)

(b)

FIGURE 9–1 (a) A reconstructed aqueduct supplying water to ancient Nineveh; (b) human-made caves above the ruins of the Khanis spring intake (Atrush, northern Iraq).

However, as discussed in Chapter 1, with the rapid advancement of drilling technologies during the 19th and especially the 20th centuries and thanks to affordable energy, the users of springwater have experienced an enormous pressure from the users of groundwater in general. The following is arguably the most dramatic such example on the global scale.

The Ras el Ain Spring in Syria near the Turkish border was very large and helped sustain the flow of the Euphrates River via its tributary Khabour. Some references even declare it to have been the largest spring in the world. According to Burdon and Safadi (1963), its discharge was between 34.5 and 107.8 m^3/s. They stated that groundwater issued from 13 springs that drained one basin of over 8000 km^2 consisting of Eocene limestone, Miocene evaporates and limestone, and basaltic rocks.

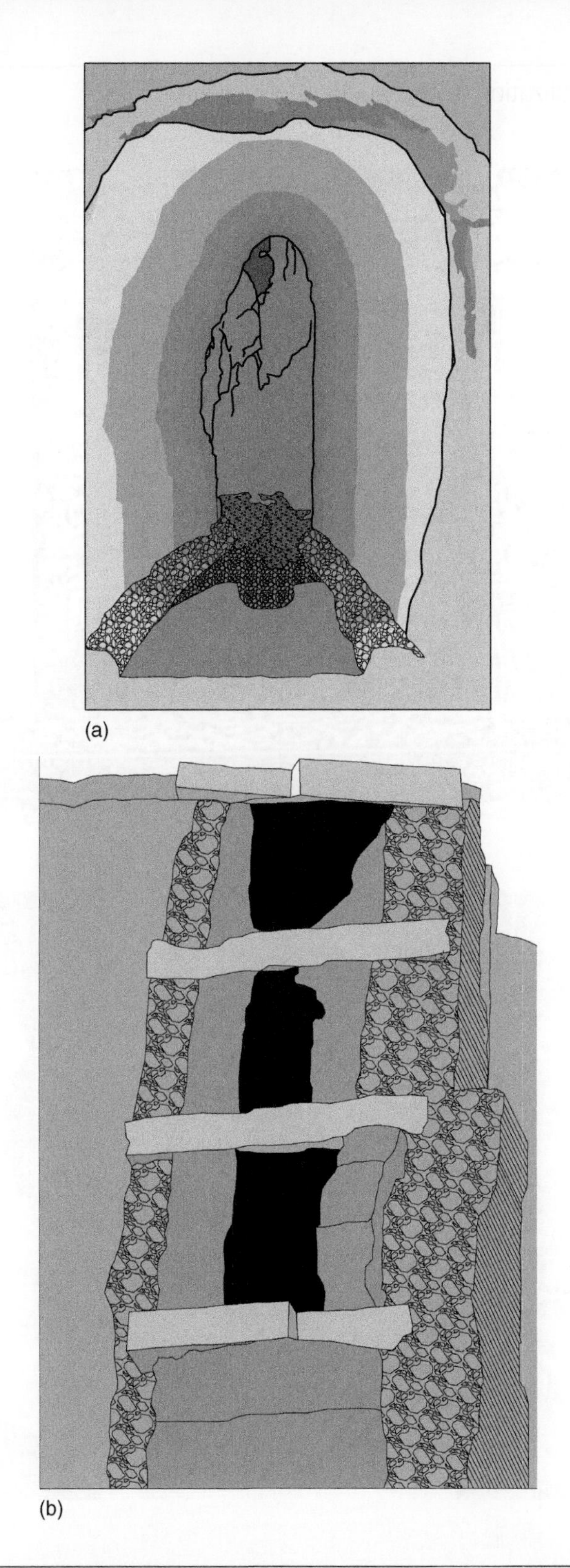

FIGURE 9–2 (a) Cistern tunnels at the Pincian hill (Rome, Cunicoli del Pincio); (b) section of three levels of channels delivering water, aqueduct on Porta Maggiore, Rome. (Modified from Lombardi and Corazza, 2008.)

(a)

(b)

FIGURE 9–3 (a) Sketch of the famous Trevi Fountain, Rome, from the 15th century; (b) the fountain today; the water is diverted from Salone spring 10 km away via ancient Aqueduct Vergine.

Unfortunately, this spring no longer flows. Overexploitation of the water resources by numerous drilled wells in the area caused serious depletion of the water table. Hole and Smith (2004) discussed the environmental and landscape changes in northeastern Syria over the last 100 years and its transformation from an open rangeland to an intensely cultivated landscape. As in most arid or semiarid zones, successful agriculture requires either supplemental or full irrigation. But very ambitious plans to develop the water resources for summer crops and decisions by individual farmers to install wells led to a fundamental alteration of the natural drainage in favor of groundwater extraction, storage reservoirs, and irrigation canals. Although this is today one of the most fertile and intensively cultivated regions in the Near East, it remains to be seen if such intense groundwater extraction is sustainable. As to the Ras el Ain Spring, it is virtually certain that it will never flow again, regardless of any future water management decisions in this troubled part of the world.

9.2 UTILIZATION OF SPRINGS

9.2.1 Springwater in the drinking water supply

Natural drainage of aquifers through springs can meet water demands on a wide scale: from the supply of large cities at the regional level, to the supply of just one or several households. Although the latter is not a big problem in terms of amount of water, for the big consumer, a very large aquifer and spring discharge are required. For example, a long tradition of setting priories on springwater exists in central and southeastern Europe, where as many as five capital cities obtain water from karstic springs (see Chapter 10.1). Details on one of the most famous European capital's springs, the Kläffer Spring utilized by the Vienna Waterworks, are provided in Chapter 10.2 (see also Figure 9–4). The increased demands have, however, caused many cities to substitute or enhance their primary water system, based on springwater, with surface water or groundwater from other aquifers, most commonly alluvial ones.

In the developing world, it is very difficult, sometimes even impossible, to meet increased demands and provide sufficient water to the population. In fact, population growth is the main reason why many previously utilized small springs are now abandoned. Orientation toward more abundant water resources, such as alluvial aquifers, river water, or water from newly constructed reservoirs, is a common trend.

If the spring's catchment is effectively protected from pollution, the quality of discharged groundwater is usually high. Therefore, from an environmental point of view, to watch precious drops of springwater used to wash streets, for example, or irrigate city greenery is distressing. Even if an expensive water treatment is applied, it is never easy to separate pipelines and supply the specific consumers with technical waters. The problem is the same everywhere.

Compared to water from other sources, springwater has many advantages in terms of water quality. However, as explained in Chapter 6, natural conditions do not always favor water quality: Examples of water deterioration caused by geology are known from areas where salty rocks, such as gypsum, anhydrite, or halite, are widely present as well as from areas that consist mostly of magmatic rocks (heavy metals or radioactivity could be major problems). In such conditions, water treatment is unavoidable and should be planned at the source or the point of use and must be undertaken before the water reaches the end users (see Chapter 7). Although drinking water standards are quite restrictive in most countries, national legislatures allow tolerance of some specific components, if they are the result of local natural conditions. For this reason, the World Health Organization (WHO) prescribes only "recommendations" regarding maximum allowed levels of chemical and organic components in water. Furthermore, other than the limits on nitrates and pesticides, no agreement has yet been reached by the Water Frame Directive of the European Union (WFD EU 2000/60) concerning the regulation of other water constituents. Therefore, it remains the responsibility of the member states to determine levels of As, Pb, Hg, Cd, SO_4, Cl, NH_3, as well as trichlorethene and tetrachlorethene.

FIGURE 9–4 Detail from Kaiserbrunn spring (Rax Mt. near Vienna), a masterpiece of spring capture design and quality from the 19th century.

9.2.2 Springwater in power generation

In hilly and mountainous regions, springs are often located at higher elevations than settlements, enabling gravity water transport with no expenditure of energy. An additional advantage in such cases is that the high hydraulic head of springs can be utilized for both water supply and hydropower generation. One example is a project completed in Austria to supply water and energy to Innsbruck, the famous Olympic ski resort and capital of Tyrol (Figure 9–5).

The first intake of Mühlau springs for Innsbruck was built in 1887 and reconstructed in the 1950s. Graziadei and Zötl (1984) describe the conditions and difficulties of construction of the new Rumerstollen tunnel through the karstic massif of Triassic limestone overlain by younger sediments. The tunnel opening is at 1140 m above mean sea level (amsl), while the head is at about 560 m amsl. The collecting galleries are 564 m long, with additional 1159 m of branches. Two pipelines 600 mm in diameter convey the water to two turbines in the power station. About 25 percent of the inhabitants of Innsbruck obtain energy from this plant. In addition, the water, which receives no treatment, is of excellent quality. The water temperature is constant at 5°C. The annual discharge ratio is around 1:2. The minimum, during early spring months, is 560 L/s, while the maximum, during summer, is over 1600 L/s. The relatively steady discharge is due to delayed infiltration of rainfall and snowmelt. There is also a longer residence time within a fissured limestone aquifer, which consists more of narrow joints than large caverns (Graziadei and Zötl, 1984).

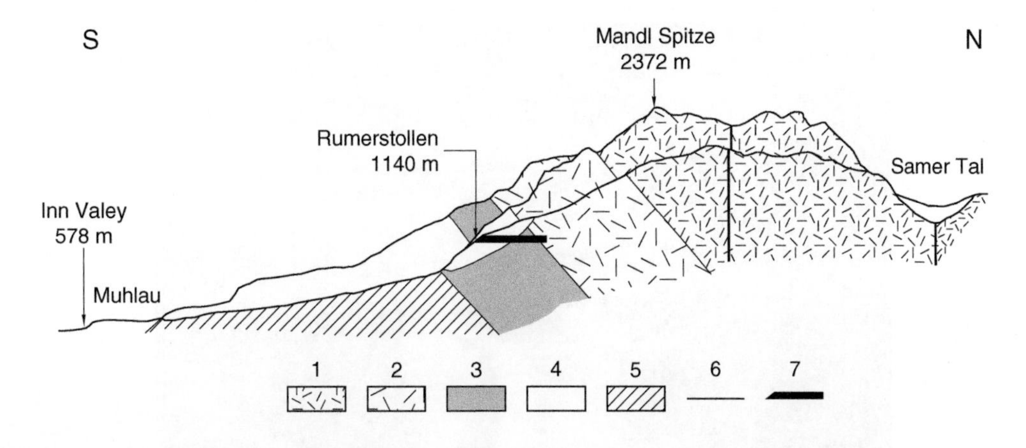

FIGURE 9–5 Cross section of the Mühlau gallery: (1) Wettersteinkalk limestone; (2) karstic aquifer of Muschelkalk limestone; (3) Bundsandstein sandstone and grauwacke; (4) river and talus deposits, breccia; (5) marl, sandstone, limestone, and dolomite; (6) fault; (7) gallery. (Modified from Grazidai and Zötl, 1984.)

Water from numerous springs has been stored in artificial surface reservoirs. Accordingly, in the case of multipurpose dams, the water from impounded springs can directly contribute to the generation of electricity. Some hydropower systems strongly depend on the spring base flow. For example, one of the largest springs in the world, the Dumanli Spring in Turkey, with an average discharge of 50 m^3/s (Karanjac and Günay, 1980), is now impounded by an artificial reservoir.

9.2.3 Springwater in agriculture

Irrigation by springwater has a very long tradition in arid and semiarid regions, as does the utilization of springs for drinking purposes. For example, Mesopotamia, the cradle of civilization, owed its economic growth and prosperity to its irrigation-based food production. While diversion of the channels from the main streams of the Tigris and the Euphrates was common in the plains, small traditional gravity irrigation systems, which diverted water from springs or perennial streams through permanent or temporary structures, provided the main source for irrigation in hilly and mountainous areas (Figure 9–6). In many places worldwide, such small irrigation schemes provided food for the local communities, alleviated poverty, and enabled surplus production for commerce at local markets.

A typical infrastructure of traditional irrigation systems consists of semipermanent rubble or masonry weirs and intakes and dugout canals that convey water to the irrigated area. Traditional infrastructure requires substantial maintenance, and water losses through seepage can be very high. The improved intake structures and conveyance systems regularly consist of concrete channels or installed pipes. The latter solution is very important for increasing system efficiency by reducing not only leakage but also losses by evaporation from the open channels. The piped water conveyance also helps in preventing sedimentation problems.

Water is a limited resource in most arid countries, and it will most probably become more limited in the future. While the water losses for an average field application using traditional surface gravity systems are regularly between 40 and 60 percent, by other irrigation techniques, such as pressurized irrigation or drip irrigation (water-saving technology), the efficiency can increase and losses can be as low as 25 percent. Improving the efficiency of water use, particularly if coming from the springs, is an important aspect of the water resource and irrigation strategy elsewhere. One of the most efficient ways of keeping and rationally utilizing water in arid areas is to construct subsurface dams. In fact, this type of structure, very commonly located near

FIGURE 9–6 Typical concrete channel diverting springwater from the mountains toward the plain (Botas, Zakho, northern Iraq).

the springs where the perennial or temporary streams originate, is regularly utilized to store groundwater in alluvial deposits. The objective is to place an impermeable barrier across the riverbed from the land surface down to the bedrock and collect the water from temporary streams, wadis (Figure 9–7).

Kresic (2009) lists several major advantages these structures have over surface reservoirs: very limited or negligible evaporation, no danger of dam failure, and impact on the environment of much lower magnitude. The main disadvantage is that none of the cost-effective construction methods can guarantee complete dam impermeability.

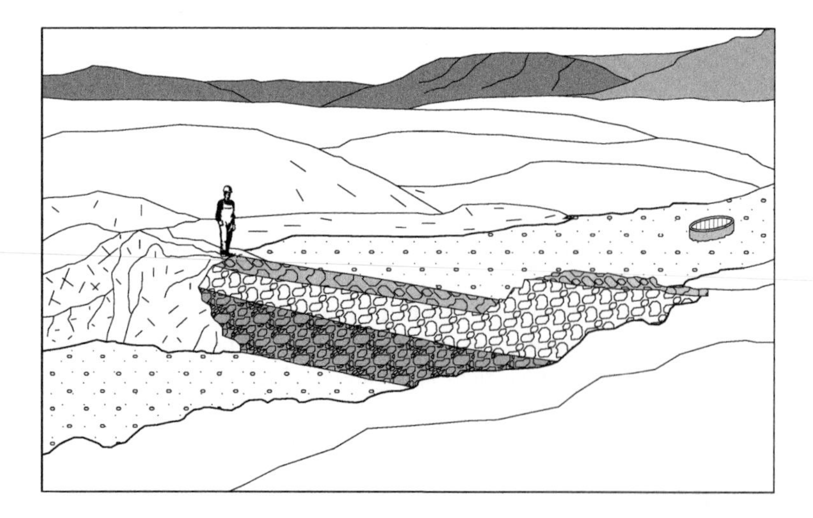

FIGURE 9–7 Typical masonry subsurface dam (built of stone and mortar) placed in alluvium.

Subsurface dams have been found to be an appropriate solution for arid areas and are widely applied in Brazil, India, Algeria, Ethiopia, Kenya, and many other countries. In cases where bedrock and riverbanks are fully impermeable, the benefit is a filled alluvial reservoir in the upstream zone (retention effect during the floods). In cases of permeable rocks in contact with alluvium, such a structure is multifunctional: It not only stores the water in an upstream area in alluvial and riverbed deposits, but it also regularly recharges underlying the bedrock aquifer throughout the year, as in the case of the FAO project implemented in 2001 in Gali Basera near the city of Dohuk in northern Iraq (Stevanovic, 2001; Figure 9–8).

FIGURE 9–8 Subsurface dam Gali Basera in northern Iraq. The goal is to store water coming from numerous smaller springs along the alluvial valley. The stored water is utilized throughout the year and contributes to the recharge of the underlying karstic aquifer. (Top) View from the upstream side, concrete barrier under construction; (bottom) view from downstream, after the construction was completed.

Finally, a “classical” use of the springwater is for watering animals. Good-quality springwater provides security for animal health and growth. Therefore, in a rural environment, it is common to see large numbers of animals occupying the springs or the ponds and swamps formed nearby.

9.2.4 Thermal springs in recreation and balneotherapy

Springs with a water temperature higher than the annual average air temperature are commonly considered thermal, although there are classifications that use a threshold of 20°C or 36°C (homeotherm) for thermal waters (see Chapter 2). The word *mineral* is often associated with the word *thermal* to indicate a higher content of dissolved mineral components (over 500 or 1000 ppm, depending on classifications) commonly found in thermal springs. Humans have always been attracted to thermal springs, and over time, many spas, baths, cities, and cultural heritages developed in their vicinity. Many countries still generate a substantial income through the exploitation of natural thermal waters. The discharge and water quality of several important thermal spring occurrences, all from central Europe but of different origins, are briefly presented here.

Karlovy Vary (Czech Republic), formerly Carlsbad, is one of the most famous European spas. Vrba (1996) stated that, “in the early 18th century, Karlovy Vary was already a social center popular with European rulers, nobility, and aristocracy.” Its value is based on 79 mineral springs with varying yields, of which 13 are controlled and used for drinking cures. The first chemical analyses of water date from the mid-19th century; the first recorded measurement of mineral water discharge of 35 L/s dates from the same period (Vrba, 1996).

The Karlovy Vary springs originate in Carboniferous-Permian granitic rocks intruded during the Hercynian orogenesis. The intensively fissured granites and granodiorites of Karlovy Vary Pluton cover an area of 1000 km^2. The rocks are weathered and hydrothermally altered. The oldest known spring was very close to the Tepla (which means “hot” in Slavic languages) River and consisted of shallow caverns developed in aragonite layers. To prevent uncontrolled seepage toward the riverbed, sealing by clay and cement was regularly applied. Eventually, in the 1980s, after complex geological investigations and exploratory drilling on both sides of one of the major faults, the four inclined wells were drilled to different depths (44–88 m) and they enabled tapping of some 30 L/s. The capture of the springs in this way resulted in an increase and stabilization of the yield of thermal waters and gases (Vrba, 1996).

The water temperature varies from one spring to another, from 30 to 72°C. The total dissolved solids (TDS) are regularly over 6000 mg/L, while the dominant ions are Na among the cations and SO_4, Cl, and HCO_3 among the anions. Free CO_2 is in the range of 500–1000 mg/L. The different temperatures and contents of CO_2 provide different balneological effects. The colder springs (Figure 9–9) have slightly laxative effects, whereas the warmer ones suppress and inhibit the secretion of bile and gastric juices.

Baden-Baden (southwestern Germany, Black Forest edge) is another famous European spa linked to the sedimentary and volcanic rocks of Hercynian orogenesis (Carboniferous-Permian). Twelve thermal springs discharge from an intensively fissured aquifer in a relatively narrow zone some 30 m above the valley floor, indicating that a major fault could be clogged and water laterally diverted upward (Wohnlich, 1996). The lowest temperature is 32°C, while the hottest spring water is 69°C. The TDS is around 3000 ppm, while Na and Cl ions are dominant in the chemical composition. The salty mineral waters probably originated in evaporitic rocks of the Rhine graben then mixed with groundwater from granites surrounding the Baden-Baden depression. Wohnlich (1996) states that the aquifer regime is very stable; temperature and mineralization have been almost constant since the beginning of continuous recording in 1894.

The total discharge of the springs is around 9 L/s. At the end of the 19th century, two tunnels were dug to divert water to the baths. In the 1960s, to tap more water, two wells were drilled to 300 and 500 m depths, and they increased the total groundwater discharge by some 30 percent. The main curative effect of these famous waters comes from their radioactivity, the ions Na, Li, F, Bo, Cl, as well as traces of cobalt, zinc, and copper.

FIGURE 9–9 Svoboda (Liberty) Spring in Karlovy Vary. (Photo courtesy of Zarko Veljkovic.)

The healing effects relate particularly to cardiovascular problems, metabolism upsets, respiratory complaints, rheumatism, and arthritis. The oldest bath in Baden-Baden is the famous Friedrichsbad, opened in 1877 (Figure 9–10).

The oldest spa in Hungary, a country with a long tradition of thermal and mineral spring development, is Heviz near the southern shores of the Balaton Lake. Its active use dates back to Roman times. Legend has it that the Holy Virgin caused a spring to appear after a prayer of appeal by a Christian nurse who wanted to heal an invalid child; hot water and mud started to flow and the steaming mud cured the weak child completely. That child later became the East Roman emperor Flavius Theodosius.

The origin of the Heviz waters is very different from Karlovy Vary and Baden-Baden. It is a large karstic ascending spring of the vauclusian type. This large lake of some 4.4 hectares formed above two main thermal springs (Figure 9–11). Exploratory diving at the bottom depth of 38 and 41 m indicated small caves that discharge water with temperatures of around 40°C. However, the hot waters from those springs mix with the cold water springs (17°C), providing an average lake water temperature of 31°C. The total discharge is estimated to be around 400 L/s with no significant variation throughout the year. The water is actively used for medical treatment and has a curative character due to its carbonic and sulfuric components and light radon emanation. A thick mud layer covers the lake bed. It is also exploited for medical purposes, due to its rich organic and inorganic constituents, as well as to the presence of radium salts.

FIGURE 9–10 (Left) The tap of potable Baden-Baden mineral water; (right) Friedrichsbad bath.

FIGURE 9–11 Thermal Heviz Lake.

In Romania, the most important and famous occurrence of thermal waters is Baile Herculane. Several natural springs occur along a narrow canyon of the Cerna River. The mechanism of formation and discharge of these waters is very complex and linked to the presence of granites, Permian and Jurassic clastic rocks, and Cretaceous limestone, while the discharge points are predisposed by regional faults and folds. Therefore, it can be said that water from these springs represents a mixture of carbonate and volcanic rocks, that is, karstic and fissured aquifers. This assertion is confirmed by the high content of Ca and of H_2S.

The long linear discharge zone contains 16 springs with a total spring flow varying between 55 and 164 L/s. The water temperatures are between 17 and 61°C. The water origin and discharge mechanism, which have been studied by numerous researchers (Povara and Marin, 1984) are shown in Figure 9–12.

9.2.5 Springs as geothermal sources

Thermal springs are primary indicators of the presence of geothermal fluids. Sometimes a very small spring or diffuse ascending seepage at the surface indicates an underground geothermal reservoir. Usually such natural drainage is possible through faults, fractures, or joints. The spring yield then directly depends on the size and aperture of the discontinuities allowing the circulation of the fluid (water, steam, or gas). Dislocation lines also enable meteoric water to seep to greater depths and collect the heat (Figure 9–13; see also Chapter 2).

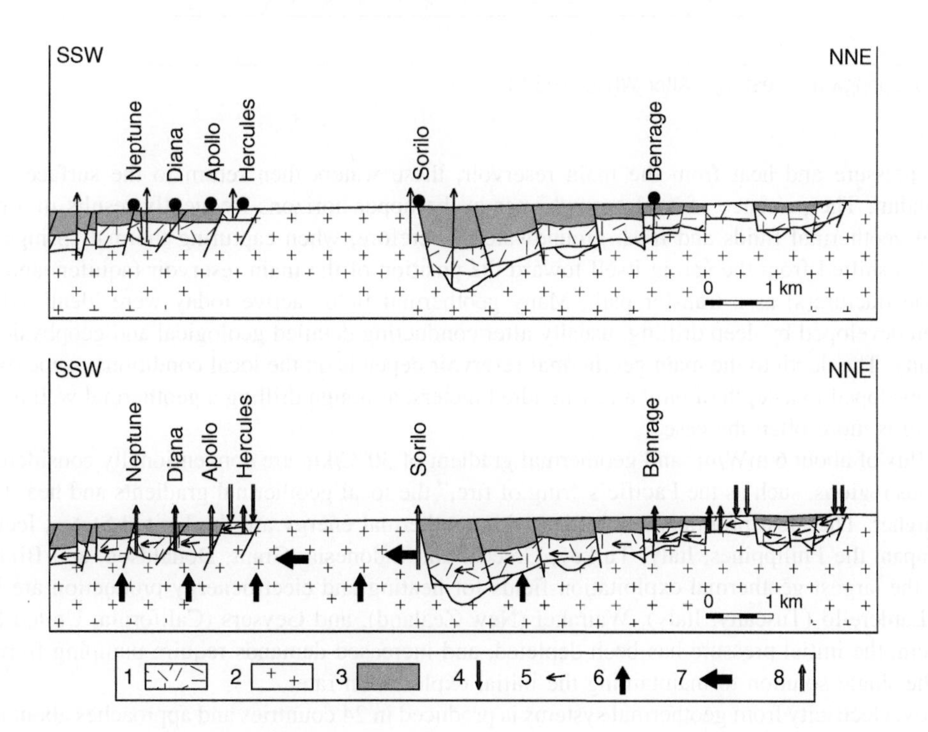

FIGURE 9–12 Baile Herculane. Geologic cross section along the major springs in the valley (top) and scheme showing the origin of thermal waters: (1) Karstic limestone aquifer, (2) fractured granite aquifer, (3) marls and flysch (impermeable rocks), (4) descending flow (recharge), (5) groundwater flow within karst aquifer, (6) ascending flow, (7) horizontal circulation within fractured aquifer, (8) drainage through faults. (From Povara and Marin, 1984.)

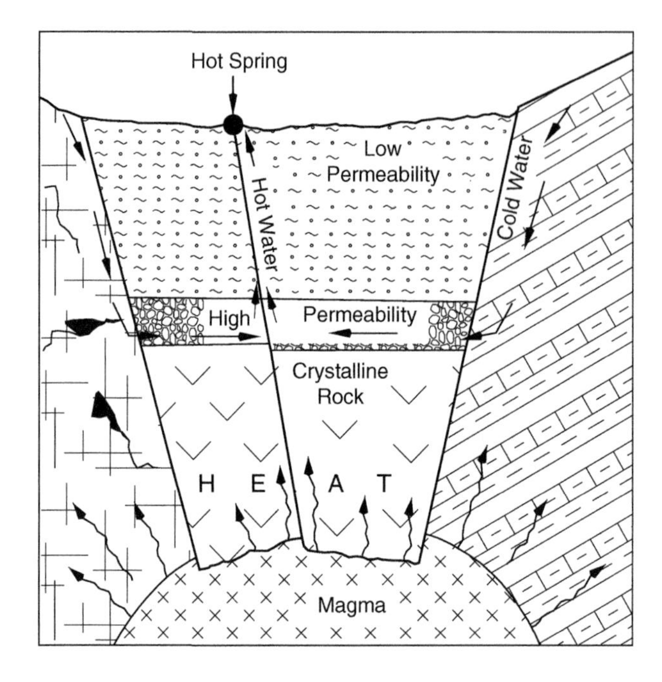

FIGURE 9–13 Springwater heating. (After White, 1967.)

Driven by pressure and heat from the main reservoir, these waters then return to the surface, usually by another conduit. The presence of cool groundwater in the upper horizons frequently results in a mixture of the original geothermal fluids and fresh groundwater. Therefore, when capturing thermal springs, the main focus is often shifted from the spring itself toward the position of the main reservoir (aquifer) and the orientation of the fracture(s) as a transfer path. Many geothermal fields active today were identified based on springs then developed by deep drilling, usually after conducting detailed geological and geophysical surveys in the vicinity. The depth to the main geothermal reservoir depends on the local conditions; some exploitation fields are developed to a depth of only a few hundred meters, although drilling a geothermal well to a depth of over 1000 m is more often the case.

A heat flux of about 6 mW/m^2 and geothermal gradient of 30°C/km are conventionally considered normal. In anomalous regions, such as the Pacific's "ring of fire," the local geothermal gradients and heat fluxes may be much higher. The most prominent countries with geothermal energy are the United States, Iceland, New Zealand, Japan, the Philippines, Italy, Turkey, Greece, and Indonesia. Grant, Donaldson, and Bixley (1982) stated that the largest geothermal exploitation fields for heating and electroenergy production are Reykjavik (Iceland), Larderello (Tuscany, Italy), Wairakei (New Zealand), and Geysers (California, United States). In most of them, the initial pressure has been depleted, and increased demands require pumping from the deep wells, as the single solution to maintaining the initial exploitation rate.

Currently, electricity from geothermal systems is produced in 24 countries and approaches about 60 TWh/a, in addition to the direct use of geothermal energy, which is close to 80 TWh/a worldwide. Some estimates show that it is possible to increase the capacity of the installed world geothermal electricity from the current 10 GW to 70 GW with present technology and to 140 GW with enhanced technology (Fridleifsson and Albertsson, 2008).

FIGURE 9–14 Great Geyser, a famous tourist attraction of Iceland. (Photo courtesy of M. Martinovic.)

A special group of hot springs are geysers, which are characterized by an intermittent or pulsation flow (Chapter 2). The term *geyser* is in fact derived from the famous Geyser in Iceland (Figure 9–14). Iceland is famous for its very rich geothermal system, with water present at some 200°C at a depth lower than 1000 m. Numerous geysers exist in the United States and New Zealand, as well as in Iceland, all of them connected to postvolcanic activity.

The interest in substitutions for nonrenewable energy sources has significantly increased during the last decade. Along with geothermal energy and the exploitation of hot springs and groundwaters, there is great interest in utilizing cold to subthermal or relatively warm springwater. The utilization of these low-enthalpy springs (and groundwater in general) for space heating and cooling is enabled by heat pump technology and has gained increasing acceptance, particularly within the European Union. Groundwater heat pumps typically use a compressor, which extracts renewable heat from the groundwater. The heat effect delivered by those pumps is typically 3–4 times the electrical power input. This means that heat pumps deliver heat 3–4 times more cheaply, with a significantly lower CO_2 emission.

9.2.6 Bottled (spring) water industry

Bottling and selling water have a very long tradition, particularly in Europe and North America. In the 16th century, waters from the Belgium spa were distributed in some major European towns, while Italian Acqua dei Navigatori was widely used on the boats sailing to the New World.

In 1767, Joseph Priestley discovered the technique of adding CO_2 to water to make it sparkle. This opened a whole new prospect and enabled further commercial development for the refreshing waters and for soft drinks, which were derived from sparkling water.

During the 19th century, many French, British, Italian, and German brands were established. The exportation of bottled water started in their numerous colonies. The German Apollinaris, the carbonated water from volcanic rocks, was introduced in England and served in western American railroad systems.

Today the annual revenue of the bottled water industry worldwide is estimated at about $13 billion. Among the water sources used in the bottling industry, water from natural springs is the favorite and the most widely consumed. The types of water for bottling are commonly classified as (1) low-mineralized (table) water, (2) mineral water, and (3) sparkling water. In Europe and the United States, the mineral content of water is determined according to different limits. In the European Union, 500 mg/L or less dissolved solids is the standard for low-mineralized water, while other European countries set that limit at 1000 mg/L. In the United States, mineral water contains no less than 250 mg/L dissolved solids, while the labels on waters with more than 1500 mg/L must declare a "high mineral content." Sparkling water contains natural or added CO_2 equal to the amount in the water when it emerged from the source. However, to get a more refreshing taste, it is also common to add an amount of CO_2 considerably higher than the natural level. Such processes are not considered artificial treatments as, for example, distillation or reverse osmosis, which can also be applied before bottling. It is therefore very important to clearly distinguish "natural water" from water that undergoes any technological process.

The bottling industry often does not differentiate between types of intake structure and may declare both water from springs and water tapped from wells as "springwater."

Probably the most famous among the thousands of important commercialized springs worldwide are Perrier and Evian, both of French origin. The Evian Spring in the Haute-Savoie has been in use since 1826. The water originates from rainfall and snowmelt recharging a karstic aquifer in the Alps. The spring discharges mostly as subterranean outflow into the overlying glacial sediments, mainly sands, where additional natural filtering and purification take place. The water has low mineralization and is of the calcium-magnesium-bicarbonate type.

While Evian is a typical karstic spring with atypical discharge, Perrier issues from volcanic rocks. Legend has it that, after crossing the Pyrenean Alps on their way to conquer Rome, Hannibal's army decided to set up camp near a spring with a high content of CO_2 gas. The spring, which became known as Les Bouillens and later by its current name, Perrier, looked like a cool, bubbling pool. These springs have been actively used for commercial purposes since 1863, when Napoleon III acknowledged in writing that the spring yielded natural mineral water. In 1908, the water bottling plant produced about 5 million bottles annually and by 1933, the annual production had risen to 19 million bottles (www.en.wikipedia.org).

Perrier waters are low mineralized; TDS is less than 500 mg/L. These are naturally carbonated (CO_2), typical calcium-bicarbonate waters, with a pH of 5.5. Today, spring capture is via wells in a specific way: The water and natural carbonic gas are captured independently in different isolated layers. The appropriate balance of minerals and carbonation is reestablished in the bottling process.

9.3 CAPTURE OF SPRINGS

When we think about a natural spring, what might come to mind first? Its beautiful waterfalls or the murmur of its waters? Its cold, fresh water, the ultimate drinking water? As appealing as these images are, it is well known that not all springs are beautiful nor is their water always drinkable. They do, though, represent one of the essential sources of life on our planet: A spring is in fact the place where groundwater, an invisible and precious resource, reaches the surface and creates a precious, continuous flow.

Now, again envisage a spring and this time ask, How many people come to it to capture the water for common use? On a regular basis, just a very few. In addition to engineers, local inhabitants who are not always skilled or experienced can also perform this task. And all those who have the chance to perform this task should feel honored—and even beyond that, they should feel responsible. Why? First, because a chance to change nature is not given to everybody; and second, because to do this in an environmentally friendly way is not an easy task. And the latter mandate has become more and more imperative during recent decades. Inexperienced people might not understand many important environmental issues of spring capture, such as allowing some water to flow downstream for other users, including animals and stream biota, or preventing garbage from being left at the spring site. The list could be very long, emphasizing that capturing a spring in the "right way" requires many considerations and steps.

9.3.1 Natural characteristics of springs and their impact on capture design

In Chapter 2 of this book, the classification of springs is discussed in detail. We learned that, to distinguish among springs, several important factors such as hydrogeological and hydraulic characteristics, discharge pattern (magnitude, fluctuation, and seasonality), and the physical and chemical properties of their waters should be considered.

The preference, of course, is to capture water of a perennial spring. Only this approach provides a reliable solution and results in a stable water supply. However, if there are no other options, even control of seasonal flow is reasonable. Such control can be achieved by tapping deeper parts of the aquifer by wells or using subsurface dams to store water for later use.

The volume of an engineered reservoir at a spring very much depends on the amount of water that needs to be controlled. For small springs, sometimes a very simple structure, such as a spring box, is sufficient. If properly developed, even a very small spring flow during the dry season can supply desperately needed water. In other cases, large and complicated structures, such as reinforced concrete basins or even dams, are required (Kresic, 2009). But, whatever the size, the main task of the design is to try to control as much additional water as possible. Only then are we in a position to manage the water, use the amount needed, and enable surplus water to flow out freely downstream.

To make the discussion about "small spring" or "low-discharge spring" more accurate, when using these relative terms, we propose considering a yield lower than 0.1 L/s as the base value. For that yield, a collection chamber of less than 10 m^3 (e.g., 2.5 × 2 × 2 m) would be big enough to store the entire flow of one day (24 h). For a yield 10 times lower (0.01 L/s), the construction of a large chamber is unnecessary, since a very small spring box of just 1 m^3 could alone accommodate the daily amount of water discharged. To allow free flow-through, it is even better that such a box be fully filled with gravel and have no storage space. By contrast, for a yield of 1 L/s or higher, it is recommended that the collection chamber be expanded to 10 m^3 or a larger tank or reservoir be constructed next to it so that, in terms of timing, delivery of the water can be dealt with efficiently.

9.3.2 Principles of spring capture

Apart from the spring type and flow regime, the following aspects should be carefully evaluated before deciding to tap springwater:

1. Vicinity of the consumers or the point of use.
2. Topography of the spring site, the potential pipeline route, and the usage area. An assessment and survey of the spring site is important before construction of all necessary facilities, such as reservoirs, water treatment units, electrical installations, and similar facilities.
3. Geology of the spring site and the entire catchment area (topographic and hydrogeologic basins).
4. Water demands of different consumer groups (drinking, animal watering, irrigation, small and large industries, recreation). The current and future demands should be evaluated as well as the expected seasonal and daily variations.
5. Cost of the project. The financial component is essential for most projects: Comparing the alternatives, optimizing solutions, and assessing the environmental impact are all prerequisite to the implementation of the project.

From a practical point of view, not all projects require the same technical level and details. A modern engineering approach recommends the following actions:

1. Ensure that access to the spring site is possible.
2. Assess the general topography and conditions of the site and the possibility to modify and adapt it.
3. Compute the distance and hydraulic gradient between the spring and points of use.
4. Consider additional benefits such as power generation or energy saving.
5. Conduct a (hydro)geological reconnaissance.
6. Collect information about the rainfall regime and roughly assess the water budget.
7. Estimate minimum spring discharge or measure (recommended) and obtain data about the spring flow regime, particularly of the low-water season.
8. Check water quality—physical properties in situ, take samples and perform chemical analysis in the laboratory. It is highly recommended to observe seasonal variations in water quality and, for karst springs, the impact of storm events.
9. Presume water protection in the catchment—observe possible pollutants, evaluate preventive measures to be taken against pollution, roughly delineate sanitary protection zones, and approximate where the fencing zone around the spring should be located.
10. Estimate water demands—type of consumers, their number, future growth.
11. Consider environmental requirements—check general and local water policies and practices, evaluate possible conflicts from water utilization, minimize the impact on downstream consumers, wildlife, and surface water biota.
12. Think about alternatives—consider other sources, tapping groundwater in a different way, regulating spring regime, artificial recharge.
13. Optimize the capture design.
14. Select the type of design and its creation, fully respecting the environmental and aesthetic conditions of the site.
15. Calculate the construction, and operation and maintenance costs.

When a decision is needed about a small spring and a relatively simple structure, fewer of these actions can be undertaken. Finally, success is greater when the simplest intake structure is introduced and the same results are obtained. "Why use an elephant when a donkey will do?"

Optimization involves a compromise between wishes and opportunities. In general, the quality of the chosen option can be assessed based on the responses to the following questions:

1. How are the water demands met (fully satisfied or not)?
2. How are the minimal capacity and stability of the flow ensured?
3. How are the water quality and protection maintained and monitored?
4. How is the negative environmental impact mitigated?
5. How is the cost of the construction reduced?

9.3.3 Structures for spring capture

There are two main groups of structures for spring development:

1. Simple capture structures without large artificial interventions. The aim is to collect and protect water and secure the delivery of a stable flow to the users in the simplest possible way.

2. Complex capture structures for regulation of water discharge. These include a variety of means to maintain or increase the flow (installed pumps, drilled wells, galleries, shafts, trenches, perforated collector pipes, and dams).

The basic, but not mandatory, elements of a spring capture are

- Spring box (collection chamber).
- Storage (reservoir).
- Pipes (for distribution, usage, overflow, cleanout).
- Pump.
- Water treatment.
- Other (such as backfill, sediment box, cutoff walls, retention walls, drainage ditch, cover and seal, vent, valves, fence, monitoring equipment, electrical facilities, maintenance room).

The *spring box* or collection chamber is the main part of the simple capture design. In the case of descending (gravity) springs, the simple design comprises just a box laterally tied directly to the rock formation yielding water (Figure 9–15). The wall coupled with the aquifer is perforated or fully open (in case of a consolidated-rock aquifer) to allow the inflow. It is very common for the capture structure of a gravity spring to be located at the foot of the cliffs.

When water is issuing under pressure (ascending springs), the box should be directly over the inflow and open at the bottom. In the case of gas-lift mineral springs, it is common to have an additional hole drilled or dug below the spring box that reaches the main fault or fissure conveying the water up. A small-diameter pipe is often used to bring water to the surface by keeping the gases at the maximum possible level (Filipovic and Dimitrijevic, 1991).

The spring box should be designed with an overflow, allowing surplus water to flow out freely and not to flood the intake. Some additional space between the maximum water level in the box and the ceiling is always

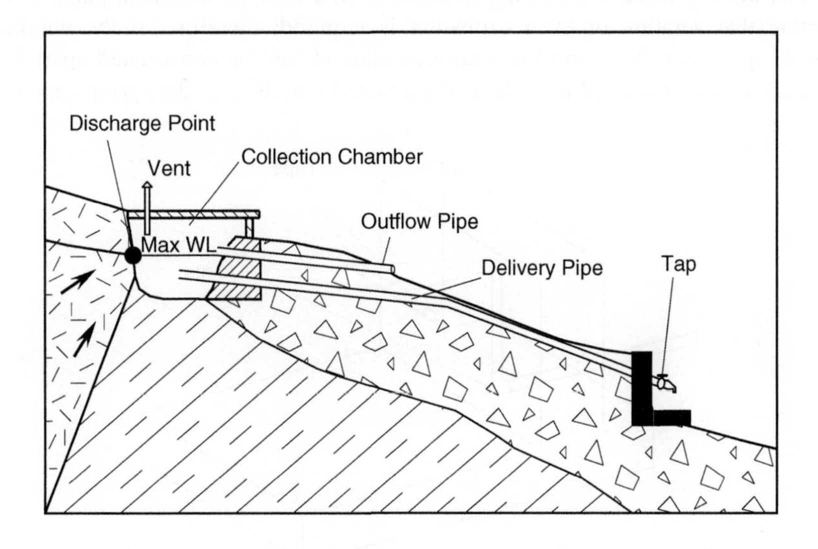

FIGURE 9–15 Simple spring box (collection chamber) and nearby point of use.

required. The collection chamber could be filled with gravel or stone. In this case, an empty space ("pocket"), usually in the center, should be left for the inlet of the delivery pipe.

The box acts as *storage* and should ideally accommodate daily variation in water consumption. If the captured groundwater has increased turbidity, an additional reservoir of clear water can be constructed along with the collection chamber. The water enters the first chamber, called the *sediment box*, where the fines (particles) are settled; it then overflows into the clear water reservoir.

Pipes are integral parts of any spring capture (Figures 9–15 through 9–17). Apart from the delivery of water to the points of use or the distribution network, they are also needed for the removal of any undesirable excess water inside the intake (overflow pipe). If the delivery pipe is blocked for any reason, the water will rise in the box but only to the height of the overflow pipe. In addition, the pipes ensure drainage of all stored water when cleaning the box, which is periodically required (bottom outlet pipe or cleanout drain). All pipes should have screens on both ends to prevent clogging or entrance of small rodents and insects.

The inlet of the delivery pipe should be placed at least 0.3 m below the level of the lowest-positioned point where water issues from the rocks. Ideally, the pipe should be laid on a uniform downward grade from the inlet to the outlet. Furthermore, there should be no high space or circuit where air or sediment could be trapped.

Pumps are installed inside the tapping structure only if water pumping is necessary. The relationship between the elevation of the intake and usage point dictates such conditions. The pumps can also be part of the regulation structure, which enables access to more water than provided by the natural discharge. The separate pump chamber can be constructed next to the main tapping structure (Figure 9–18). However, due to operation costs and other concerns, water lifting should be avoided whenever possible.

The design of *water treatment* facilities depends on the quality of raw water. A water treatment plant can be very large and the applied treatment technologies very complex if the water quality is unsatisfactory or variable throughout the year. If turbidity is only slightly elevated, the backfill and sediment box should be sufficient to clarify the water. However, in the case of significant turbidity, basins with fast or slow filters are constructed either directly inside the tapping structure or in a separate treatment plant. Deteriorated water quality could dictate that aeration or even ozonation is required; therefore, if the number of processes increases, then more space is needed and the treatment plant should be constructed apart from the tapping structure. Such plants are often located near the end users and not close to the spring capture itself.

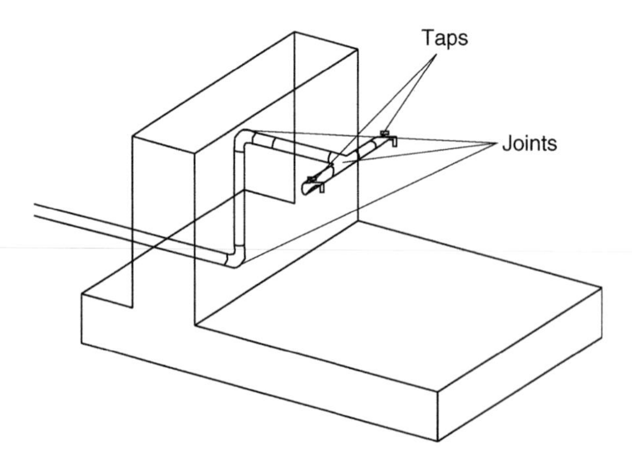

FIGURE 9–16 Taps, joints, and delivery pipes at the point of use.

FIGURE 9–17 Typical spring capture with several taps.

Many springs have water that needs no treatment or treatment is limited to elementary disinfection (elimination of bacteria), such that chlorination or ultraviolet radiation treatment may be sufficient. Small automated disinfection devices are regularly installed inside many tapping structures used for centralized water supply (Figure 9–19).

Backfill is the gravelly or sandy material ("gravel pack") that fills the space between the aquifer and the spring box (or is placed inside the box). The grain size of the filling material depends on the water quality and velocity. When longer filtration is required, the backfill material could consist of grain sizes from rock fragments to fine sands, so that water first flows through more porous and rough material and then reaches finer sediments. These layers are either lateral, as in the case of descending springs or vertically superpositioned if the spring is of an ascending type.

Cutoff walls or collector trenches are aimed at bringing water to the maximum amount and channeling it into the spring box. They are very important for the diffuse type of springs and can have different shapes,

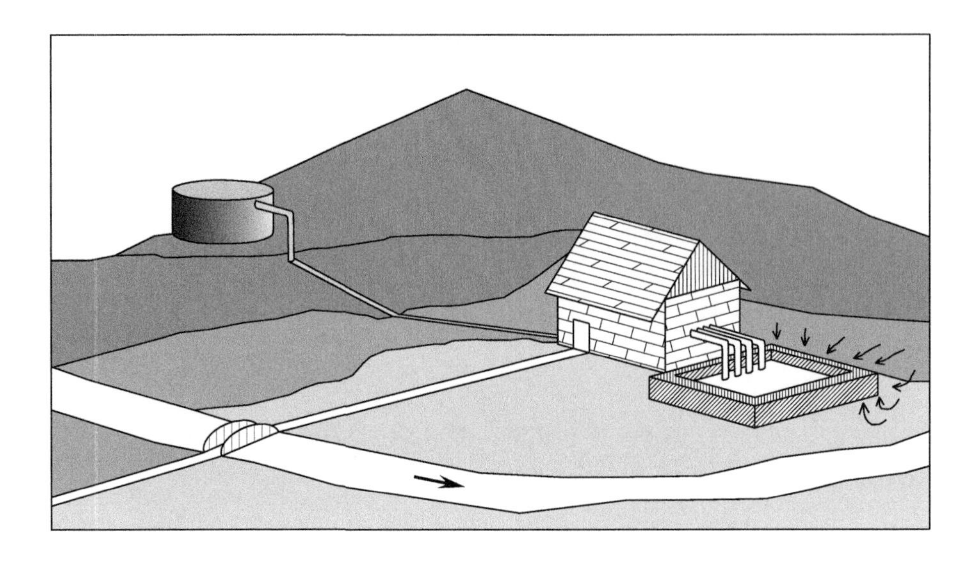

FIGURE 9–18 Spring capture with collection basin and pump room. Water is pumped to the cylindrical reservoir for further gravity distribution.

FIGURE 9–19 Automated chlorination device.

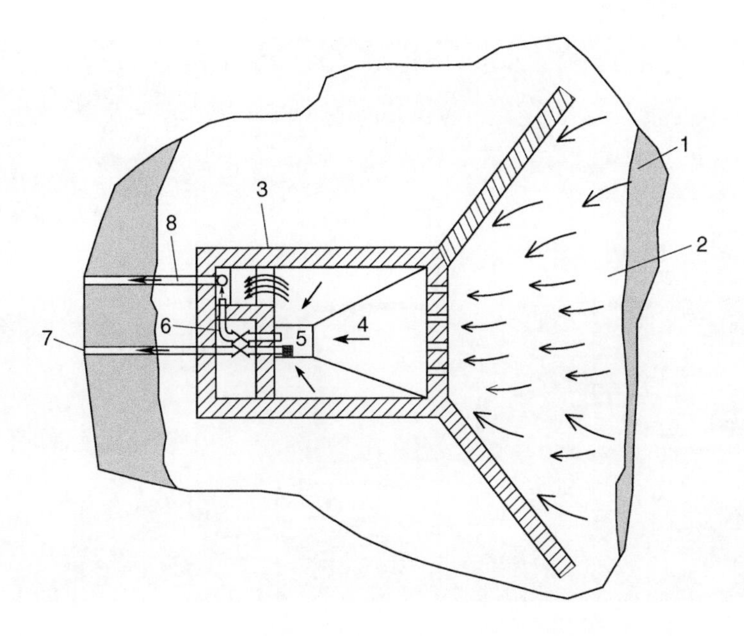

FIGURE 9–20 Schematic of spring capture with cutoff walls: (1) Aquifer, (2) backfill surrounding the spring, (3) spring box and cutoff walls from reinforced concrete, (4) sediment box (the first chamber), (5) intake, (6) clean water distribution pipes, (7) overflow pipe, (8) pipe to storage, treatment or directly to consumers. (Modified from Jahic, 1988.)

such as Y or V. To prevent bottom seepage and water losses, the walls should be carefully coupled with the underlain rocks and sealed with impermeable cement. An example with very long cutoff walls is shown in Figure 9–20.

Retention walls are very common around gravity springs. When the relief is rough, it protects the tapping structure from debris or landslides, but it could also support the collection and channeling of the water. Depending on the pressure of native materials (rocks), the retention wall can be made of reinforced concrete but can also be very simple and cheap to construct using stone (Figure 9–21).

A *drainage ditch* (*trench*) is primarily intended to collect and divert all rainfall or surface runoff water and prevent its infiltration into the spring box. The ditch should always be sized to divert the maximum flow. Trenches can also be constructed to collect the water and divert it to the common collection point.

A *cover* and a *seal* should ensure the watertightness of the capture. The sealing takes place after the cleaning and excavation, construction of the spring box, and finally the backfilling. The seal materials can be artificial, such as various types of plastic, or natural, such as a compacted clay layer, often grassed above (Figure 9–22). A concrete or metal cover is common for closing the entrances of the structures constructed for tapping ascending springs or reservoirs dug out below the surface. Both covers and seals should completely enclose the spring capture and prevent any surficial contamination and sunlight from entering the spring box.

Vents enable minimal airing of the interior of the tapping structure. The ventilation can be managed by placing windows on the reservoir walls, but a lattice or wire mesh is then obligatory to prevent access by undesirable visitors or birds. For larger reservoirs, more openings ensure intensive ventilation (Figure 9–23).

A *fence* usually surrounds the main intake and other adjacent facilities. It should correspond to the first sanitary protection zone. In fact, as a precautionary measure, the fence should prevent access to all persons

FIGURE 9–21 Long retention wall behind the spring capture, Topcider Spring, Belgrade, Serbia.

FIGURE 9–22 Kaiserbrunn spring used for the water supply of Vienna since the 19th century. The covered, grassed intake and reservoir are shown together with a large diameter overflow drain below the main entrance.

FIGURE 9–23 Maintenance room and reservoir covered by grass with one vent and metal cover entrance, Opicvet Spring near Sofia, Bulgaria.

and animals. In the case of good sealing over and around the spring box or ascending inflow, less restrictive measures and fencing may be applied (Figure 9–24).

Valves are part of almost every tapping structure (Figure 9–25). They can be installed on intake, overflow, distribution, and cleanout pipes. Valves on distribution pipes enable adjustment of the yields.

Monitoring equipment is regularly installed at larger springs and those tapped for a centralized water supply. It includes various instruments for measuring the flow, as well as the physical and chemical

FIGURE 9–24 Spring capture surrounded by wooden fence. The sign on the pole indicates the first sanitary protection zone (Haiduk Spring in Kosutnjak Belgrade, Serbia).

FIGURE 9–25 Large diameter pipes and valves used to divert water in different directions.

parameters of the tapped water, such as turbidity, temperature, pH, and conductance (Figure 9–26). Modern technologies enable the installation of different sensors (probes) at collection points or in reservoirs and automated recording of data. Data loggers can store large amounts of information and facilitate the monitoring process. Collection and analysis of monitoring data are particularly important for spring management and planning of future exploitation.

Larger spring captures are overseen by a *control* (operations) *room*. The system can include sophisticated computerized and automated operation over processes such as controlling the yield, observing the water

FIGURE 9–26 Nonstationary equipment for monitoring groundwater quality.

quality, or treating raw water. Many large water utilities are implementing this approach using technology commonly referred to as SCADA (supervisory control and data acquisition).

In conclusion, the capture of a spring and the installation of the facilities is not a routine process. Every spring has its own specificities, and no rule covers all cases. However, common engineering practice is to identify one of the following major spring types and adapt a corresponding basic tapping structure to the actual conditions at the location:

- Descending spring.
- Ascending spring.
- Spring with concentric flow.
- Spring with diffuse discharge (seepage flow).
- Spring with gas-lift flow.
- Impounded spring.

Figures 9–27 through 9–30 show some basic types of spring capture.

9.3.4 Construction work

Construction work for spring capture often does not strictly follow what was designed and planned, simply because many factors are seen only after the excavation and cleaning of the site. Therefore, the project should be flexible enough to allow the design details or volume of work to be adapted to the site-specific conditions. For this reason, continuous oversight by the design engineer is required.

During the spring site cleaning, all debris, loose and separated rock fragments, and barriers should be removed. Once the discharge points are completely cleaned and rocks removed, the flow origin and surrounding environment should be carefully observed. If water issues from one single opening, the capture task is much easier. If there are more openings close to each other, they should be tested by attempting to temporarily plug the smallest ones. If discharge of the major openings increases, then an intervention should be made to centralize the flow by permanently closing the smaller outlets.

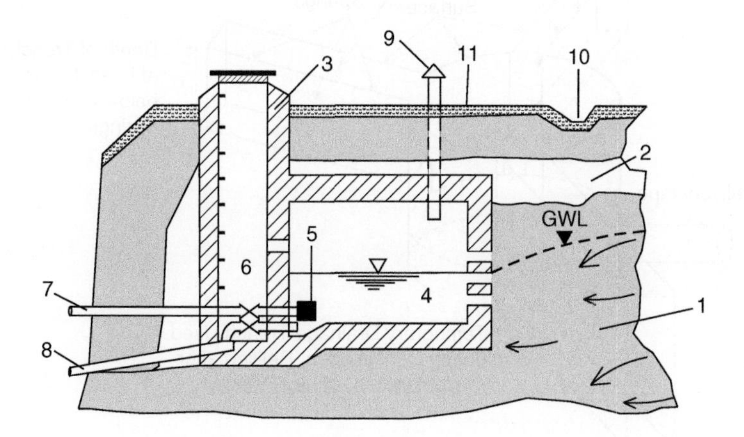

FIGURE 9–27 Capture of a descending spring: (1) Aquifer, (2) seal covering the spring box, (3) reinforced concrete spring box with openings at the back wall, (4) main collection chamber, (5) intake with filter, (6) maintenance room, (7) delivery pipe, (8) cleanout pipe, (9) vent, (10) drainage ditch, (11) top soil. (Modified from Jahic, 1988.)

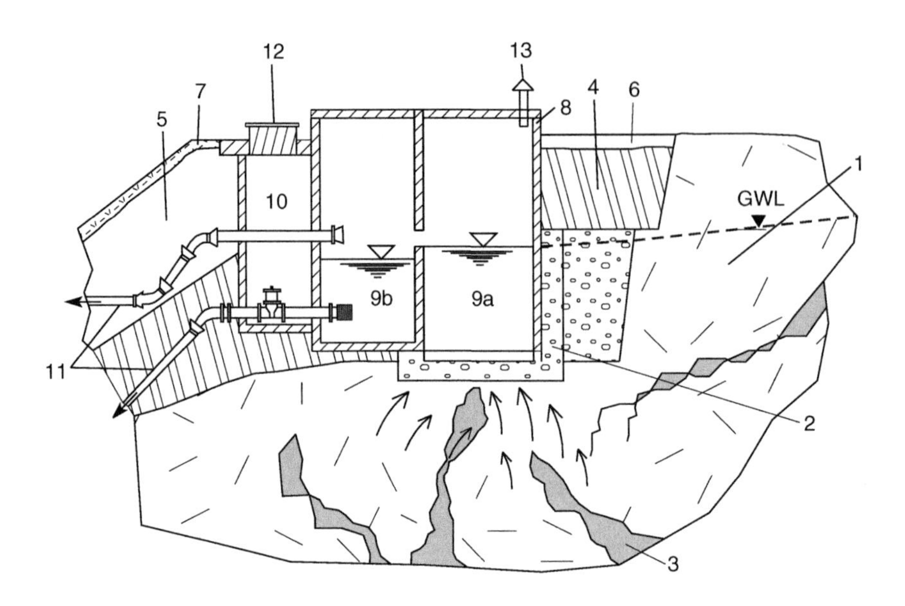

FIGURE 9–28 Capture of an ascending spring with concentric flow in a karstic aquifer: (1) Karst aquifer, (2) backfill, (3) caverns, (4) clay seal, (5) silty sand cover, (6) top soil, (7) grassed cover, (8) reinforced concrete for two chambers and access room, (9a) sediment box, (9b) reservoir of clean water, (10) maintenance room with valves, (11) overflow and delivery pipes, (12) entrance, (13) vent.

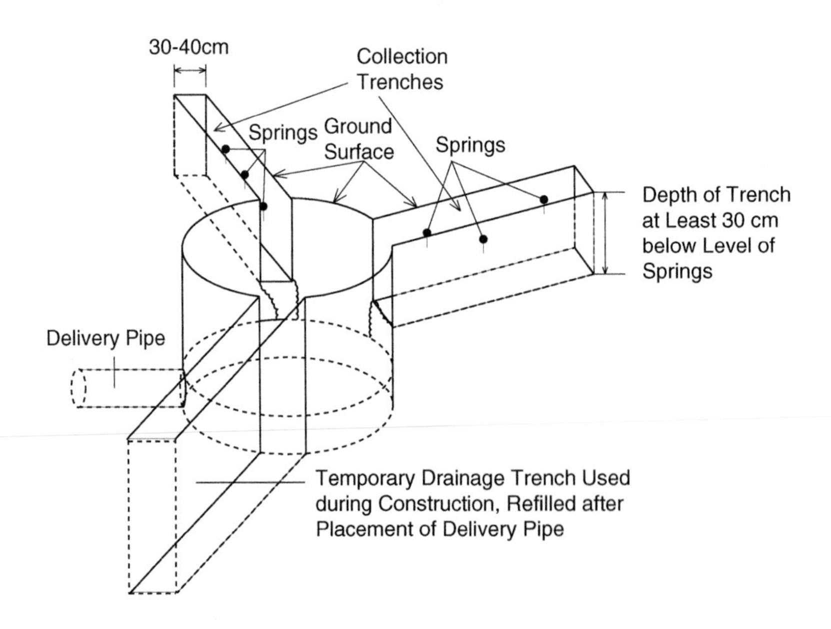

FIGURE 9–29 Capture of a diffuse discharge spring. Water collects in trenches and flows into a cylindrical reservoir (shaft). (Modified from Coffman and John, 1984.)

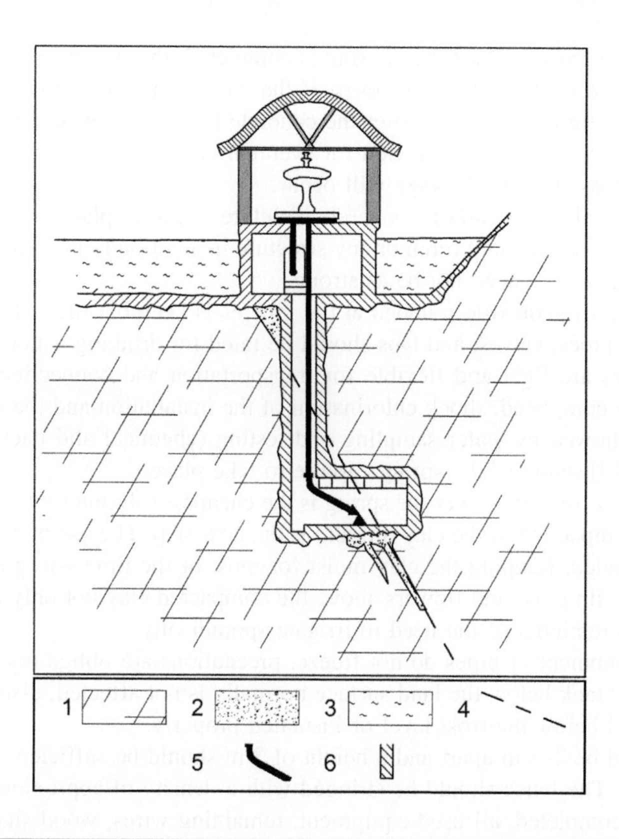

FIGURE 9–30 Capture of an ascending mineral spring with gas-lift flow: (1) Fissured rock, (2) fault with breccia, (3) clay plug, (4) probable fault, (5) small pipe inserted directly into fissure preventing escape of carbon dioxide, (6) reinforced concrete. (Modified from Filipovic and Dimitrijevic, 1991.)

In cases where the separate openings are not close to each other, perforated drainage pipes, cutoff walls, or collection trenches should be installed between them to collect and divert springwater. Deeper digging of soil or rock excavation often is required, but to avoid damaging the underlying impermeable layer, which directs the flow, it should always be done very carefully; otherwise, water could "escape." The impermeable layer may be used for the foundation of the spring box. If the search for the main discharge point and the fissure pathway leads into consolidated (hard) rock, the orientation and dimension of fissures should be closely analyzed. Digging out the fill of terra rossa or clay or mechanically widening fractures in karstified rocks could increase the flow. However, experience teaches that the use of explosives is not a suitable solution for getting more water (in addition to being dangerous for workers and the environment).

The collection chamber, tank, or box should be made of reinforced concrete. Consequently, carpenters' tools, boards, plywood, mixing bins, reinforced rods, rebar, wire mesh, and similar equipment and devices should be transported to the site and used for such construction. Wooden building materials should be cut into pieces of appropriate sizes and installed by skilled carpenters prior to pouring the concrete. Panels should be properly secured (braced) and fixed to the bottom to avoid or minimize leakage. The concrete should consist of cement, sand, and gravel mixed in the proportion 1:2:3 or similar. An adequate amount of water should be added; too much water makes the concrete weak. The concrete is best poured by a mixer. Good tamping should prevent air pockets or voids. Once the pouring is completed, the structure should be carefully covered with plastic to prevent additional wetting. However, if the weather is too sunny and dry, some very small water spraying of the concrete is beneficial. After the concrete has cured for several days, the wooden forms should be removed. At the end, careful inspection for eventual cracks or any similar concrete incoherence is mandatory. If open cracks exist, water leakage will occur.

When the springs are smaller, instead of a concrete structure, metal or plastic containers (always of a non-corrosive material) can be used. A pump room or any structure in addition to the tapping structure can be built from bricks and mortar because it need not be as strong.

Filter socks from nylon or geotextile rounded at the pipe ends keep out the sediment and organic matter. The material used for the pipes, valves, and taps should be rated for drinking water use, clean, and above all noncorrosive. Plastic pipes are light and flexible for transportation and connection, respectively. After the construction work is fully completed, shock chlorination of the installation and the water is mandatory. This disinfection should be followed by water sampling and testing (chemical and bacteriological) after several days, allowing for natural flushing of the spring capture to take place.

Sealing with puddle clay around or over the spring is the cheapest solution to the problem of infiltration of surficial contaminants. Compaction of the clay mixture is the next step. The use of a rolling machine to flatten the top layer is recommended. Keeping the clay moist for most of the time will prevent cracking. An additional soil (humus) layer with grass and flowers above the compacted clay not only makes the spring capture site pleasing but also is a reminder of the need to irrigate sporadically.

To ensure that the equipment or pipes do not freeze, precautions are obligatory. While water stored and covered in the chamber or tank below the land surface normally is not affected, distribution pipes and valves should always be installed below the frost level or insulated properly.

The fence posts should be 2–3 m apart and a height of 3 m should be sufficient. Their foundation should be at a depth of 0.5–1 m. The fence should be stringed with a density of approximately 0.2 m.

After construction is completed, all used equipment, remaining wires, wood shavings or similar, oil and lubricant cans, cement bags, and all garbage should be removed from the location and safely disposed of.

The maintenance of the collection chamber, spring box, pipes, and installed equipment is required. If the spring capture is used for a centralized water supply of a large number of consumers, the maintenance should be a well-organized activity, timed precisely. Clogging or silting of the collection chamber(s) is a frequent and major problem during exploitation. If clogging occurs, substantially less water will reach the outlet pipes. The best remedy is to temporarily drain the water (through the bottom outlet), remove the sediment, and wash the pipes and walls. From time to time, checking the water quality, installed equipment, fence, covers, seals, and window lattices is also recommended. The tapping structure, particularly if it has been recently constructed, requires careful inspection to detect any traces of leakage. If leakage is confirmed, additional sealing, plugging, and similar measures should be implemented.

9.3.5 Case studies

Capture of a seepage spring used for the water supply of the village of Mihajlovac in eastern Serbia is shown in Figures 9–31 and 9–32. The first step was to locate the primary discharge zone by partially excavating sediments in the narrow thin alluvial valley extending down-gradient from the seeps. The observed groundwater discharge of 8 L/s was from three constant small springs (all at the same level) along a linear drainage zone 5 m long. After removing large rock fragments and boulders, two parallel cutoff walls were built from reinforced concrete. After the walls were completed, alluvial material was washed, leveled, and covered by an impermeable clay plug. After compaction of the clay, a soil layer was placed on top and partially covered by rock fragments to prevent erosion. The channeled water, now fully captured and closed, was directed into a

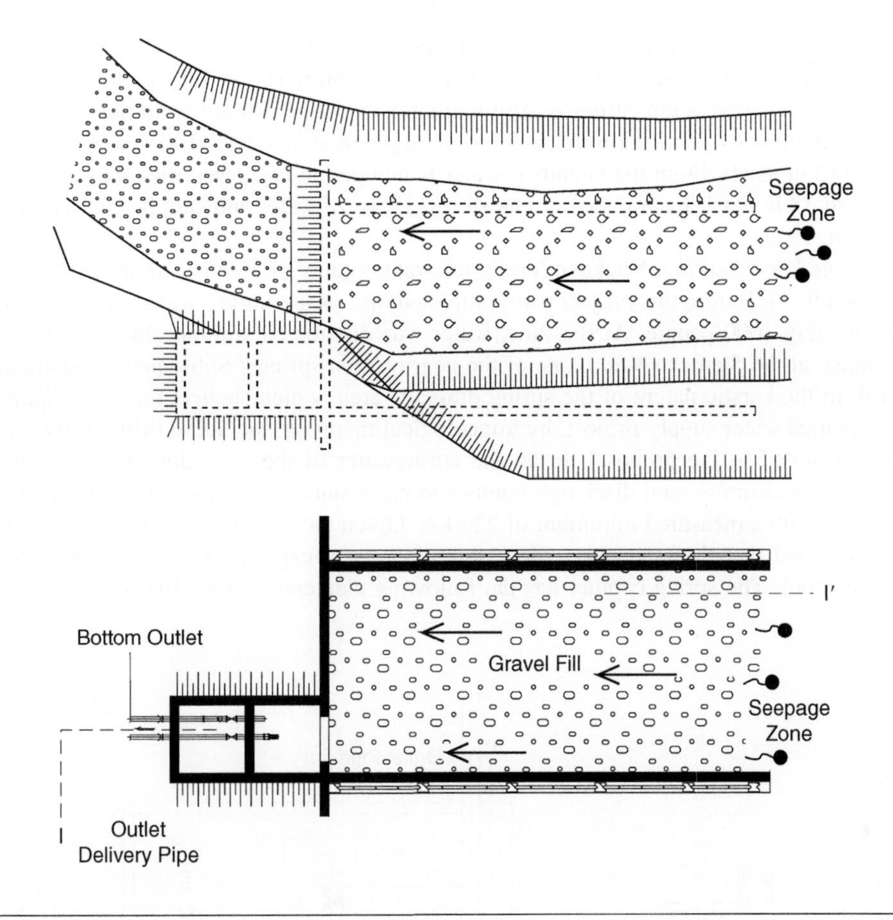

FIGURE 9–31 Design of Mihajlovac spring capture. (Courtesy of WIGA Co., Belgrade.)

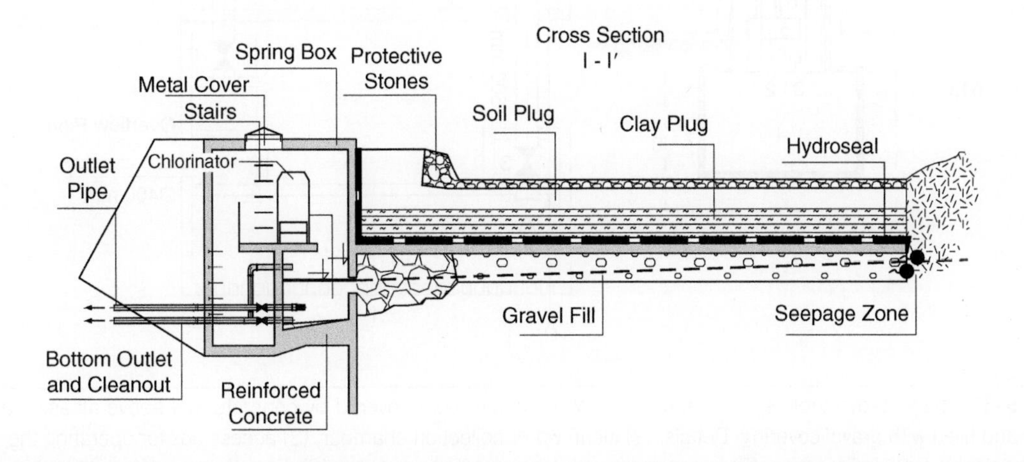

FIGURE 9–32 Cross section I–I′ of the Mihajlovac spring capture. (Courtesy of WIGA Co., Belgrade.)

storage reservoir, also built from reinforced concrete (Figure 9–32). Next to the storage reservoir is a maintenance room, housing the control valves and an automated chlorinator. The entire structure is covered by soil except for a concrete plate with a top entrance. Approximately 5–6 L/s of water from the spring capture (altitude 62 m amsl) is diverted by gravity through a 700 m long pipeline (diameter 100 mm) to a pumping chamber at the village (42 m amsl). From the chamber, water is pumped to two reservoirs above the village at the elevations 75 m and 78 m amsl. Each reservoir can store 125 m^3 of water and supplies the village's distribution network (Gaon, 1975).

Capture of a large karstic spring, Vuckovo Vrelo, for water supply of Sjenica, southwest Serbia, is shown in Figures 9–33 and 9–34. Before this project was completed, use of unsanitary water resulted in an epidemic affecting approximately 1000 people. At the time, water was distributed without chlorination by a pipeline from a simple intake at the Sjenicko Vrelo, one of the nearby karst springs. Subsequent investigations attributed the outbreak to the karstic nature of the spring drainage area, which shelters numerous animal farms.

For the new Sjenica water supply project, hydrogeological mapping, including drilling of several investigation boreholes, resulted in the proposal to tap the springwater of the ascending Vuckovo Vrelo (Gaon, 1987). The spring issues from several discharge points and has a stable yield throughout the year (coefficient of uniformity is 1.5) with a measured minimum of 220 L/s. Given the proximity of discharge locations at the spring, it was concluded that the most convenient solution was to capture them all together and put them under one common roof. The spring capture has the following elements (Gaon, 1987):

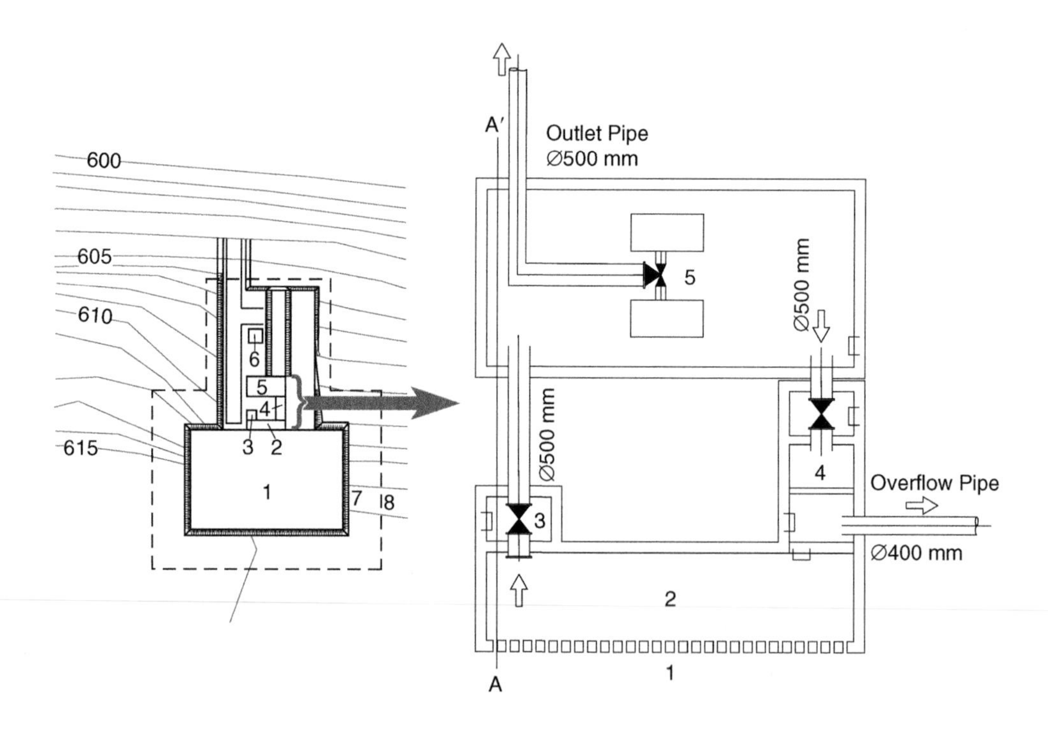

FIGURE 9–33 The Vuckovo Vrelo spring capture: (1) Main intake room covered by concrete roof above all ascending springs and filled with gravel covering; Details: (2) clean water collection chamber, (3) access box for operating the valve, (4) access room for overflow pipe, (5) pump room with two centrifugal pumps, (6) operations and maintenance room, (7) concrete walls, (8) fence. (Courtesy of WIGA Co., Belgrade.)

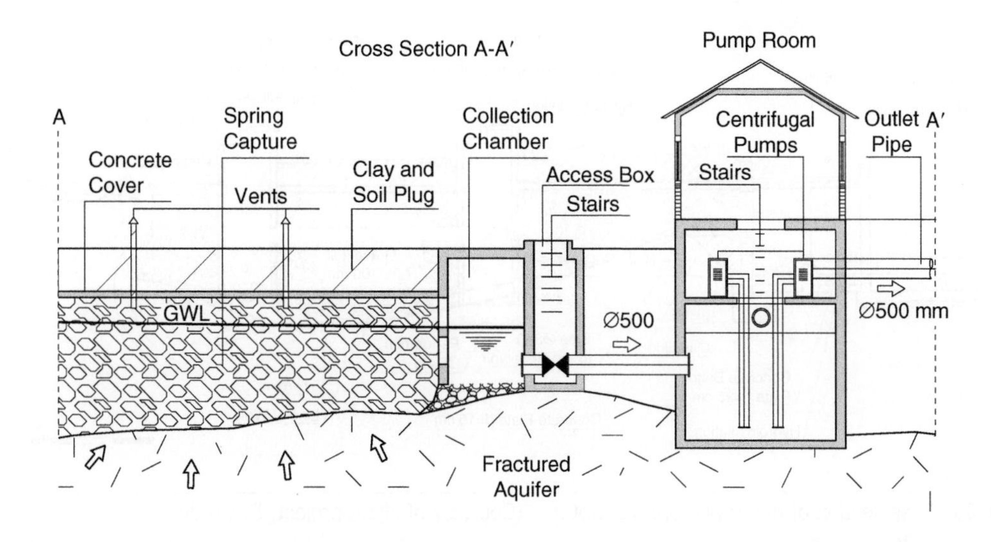

FIGURE 9–34 Cross section A–A′ of the Sjenicko Vrelo spring capture. (Courtesy of WIGA Co., Belgrade.)

- The main intake room covers the surface, where all the ascending springs discharge.
- The reinforced concrete walls are founded in unweathered limestone.
- The concrete ceiling is 1 m above the maximum water level inside the intake.
- The vents provide additional fresh air inside.
- The intake is filled with gravel and coarse sand materials.
- Water is channeled toward the clean water collection chamber with 100 m^3 capacity.
- From the chamber, water is diverted through a 500 mm pipe toward the pump room, where the chlorination takes place.
- Disinfection by chlorine is the only water treatment.
- Two centrifugal pumps lift the water to a 1400 m^3 reservoir above the city.

Figure 9–35 shows capture of Toplik Spring used for the water supply of the Holcim cement factory and the nearby village of Popovac, central Serbia. The spring got its name from a slightly higher water temperature, of about 15°C, which remains constant during cold winter months. Such springs are usually ascending, as is the case with the Toplik. The aquifer is karstic and consists of Jurassic well-karstified limestone. Minimum recorded discharge is 30 L/s. After the excavation was completed and rock fragments removed, it was decided to fill the depression with gravel and cover it with a large flat concrete roof. The gravel fill provides for filtration, while the roof prevents external impacts. The main characteristics of the spring capture are

- Three filter layers are installed in the main intake reservoir. They are, from bottom to top, large rock fragments, gravel, and sand.
- Several shafts 1 m in diameter are installed inside the structure to enable ventilation as well as access to the stored water.
- The connection drain pipe between the shafts diverts water by gravity to the clean water tank and the pump room.
- The reinforced concrete roof is partially covered by soil.
- The diversion ditch collects and diverts all surface runoff water.
- The overflow pipe drains excess groundwater inflow during periods of high water.

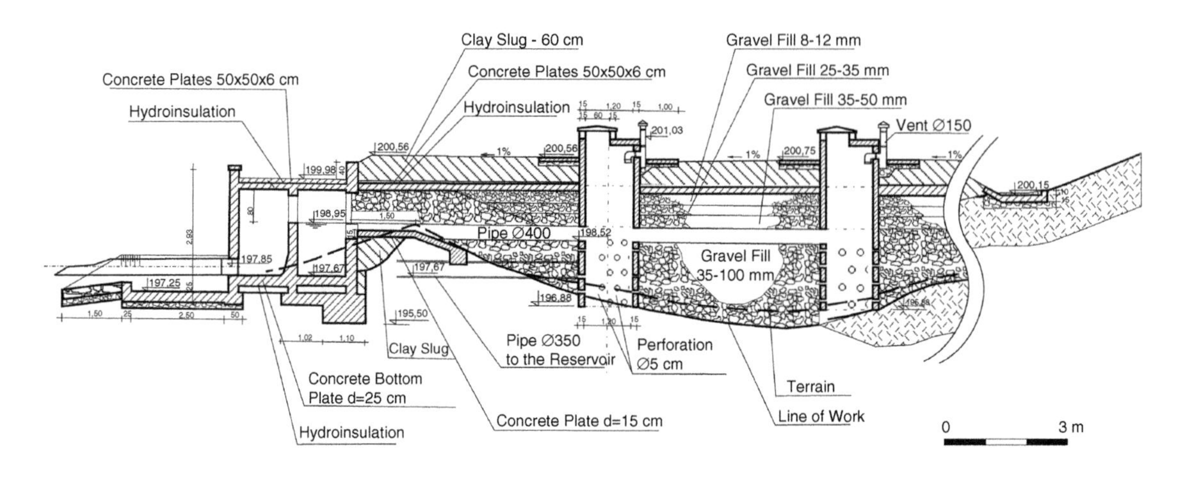

FIGURE 9–35 Cross section of the Toplik Spring capture. (Courtesy of Hydroproject, Belgrade.)

9.4 SPRING REGULATION

Regulation of a spring implies different engineering interventions aimed at controlling its flow and managing its water distribution. In general, controlling the discharge of a spring is analogous to managing a surface reservoir. Three main requirements for spring regulation are that is should be

1. Physically possible.
2. Environmentally sound.
3. Economically feasible.

Physical condition

Understandably, not all springs can be regulated. In addition, without prior comprehensive research, any direct attempt at their control is bound to fail. The knowledge of aquifer characteristics (the discharge regime and the position of the hydraulic head in the aquifer), the thickness of the saturated zone, and the aquifer transmissivity and storage are of key importance.

From the applicability standpoint, there are two main cases of stored groundwater in an aquifer (Figures 9–36 and 9–37). In the first, no significant storage below the elevation of spring discharge exists, leaving no opportunities for additional groundwater extraction beyond what naturally flows out. In the case of a thicker saturated zone and aquifer storage below the spring elevation, it is physically possible to extract more water than what discharges freely from the aquifer.

The majority of successfully implemented spring regulation projects are related to karst. Knowledge of karst aquifer characteristics was fundamental for these projects, in particular the discharge regime, the position of the karstification base, and the karstification intensity. Apart from the permeability of rock matrix, the dimensions and position of major karst channels (conduits) providing water to a spring are key elements of karst aquifer permeability. Investigations for spring regulation therefore often include speleological and cave-diving exploration. For example, in the Dinaric and Carpathian karst, several large ascending springs with explored channels to depths of 70–80 m have been regulated or are waiting for a new design of discharge control (Toulomudjian, 2005; S. Milanovic, 2005; Stevanovic et al., 2005).

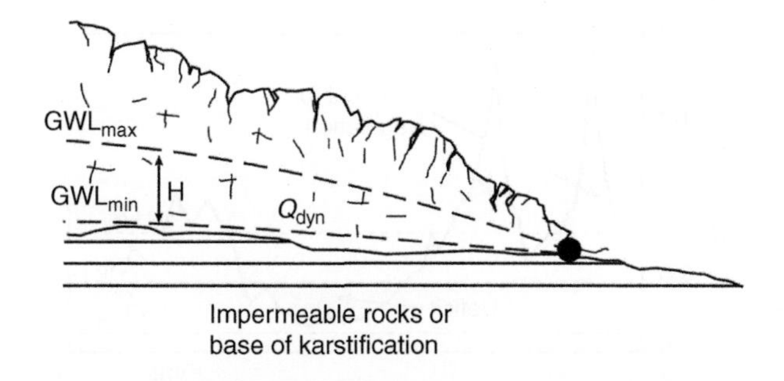

FIGURE 9–36 Gravity spring with shallow impermeable bedrock. Only dynamic groundwater reserves (Q_{dyn}) of karst aquifer are available.

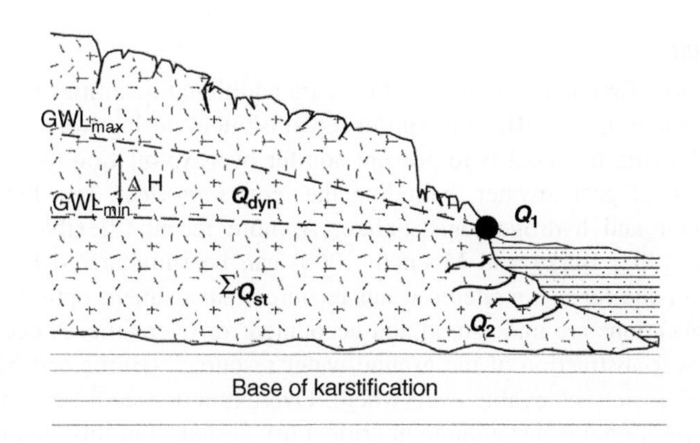

FIGURE 9–37 Karst aquifer with deep karstification base drained by an ascending spring (Q_1) and subsurface outflow (Q_2). Static (Q_{st}) groundwater reserves are available for additional withdrawal.

Fast recharge and fast discharge are not good signs if the aim is to regulate a spring. If an observed spring discharge hydrograph or the hydraulic head fluctuations show a quick response to rainfall in the drainage area and short travel times, then the aquifer likely has limited storage. For example, in karstic and fractured rocks, near-surface layers may contain fractures, joints, and cavities that cannot always receive all the percolating water and transfer it deeper into the aquifer. If this is the case, surface runoff increases, and after the saturation is completed, many temporary springs are quickly activated. Such situations with water flowing "everywhere" after storm episodes are common in karst and indicate low storage capacity.

In conclusion, for spring regulation, it is preferable that the spring be of an ascending type, discharging at a complete lateral barrier (Figure 9–37). If the barrier is not impermeable or deep, allowing for some subsurface discharge from the primary aquifer, the regulation structures beside the spring can focus on the capture of the deeper underground flow or on tapping the secondary aquifer.

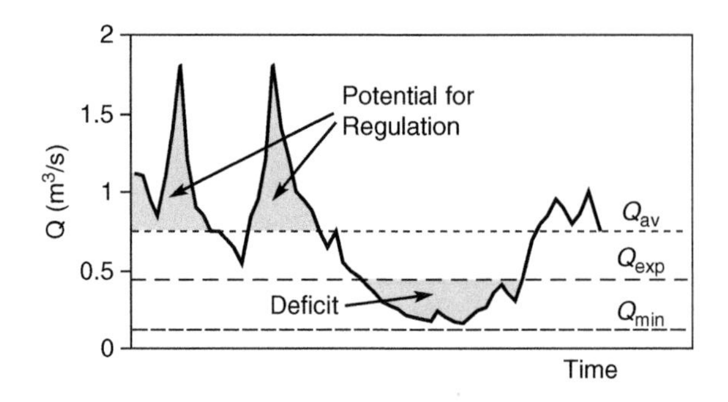

FIGURE 9–38 Typical hydrograph of a karstic spring with a potential for regulation. Exploitation capacity (Q_{exp}) is larger than the natural minimum discharge (Q_{min}) and still lower than dynamic groundwater reserves represented by the average annual discharge (Q_{av}).

Environmental condition

The aim of spring regulation, first and foremost, is to secure additional quantities of water during periods of increased demand, while counting on sufficient aquifer replenishment during wet seasons (Figure 9–38). The main balancing act in achieving this goal is to prevent aquifer overexploitation (aquifer mining).

Sustainable exploitation of groundwater, including the term *safe yield*, is a hot topic among decision makers in the water sector and hydrogeologists alike. A more recent "flexible" approach to this topic (Custodio, 1992; Margat, 1992; Burke and Moench, 2000) can be summarized by the following citation: "The planned mining of an aquifer is a strategic water resource management option where the full physical, social and economic implications are understood and accounted for over time. A declining water table does not necessarily indicate over-abstraction of the ground water resource" (Burke and Moench, 2000).

This "planned mining" is not achievable everywhere. Overextraction could be applied when it is physically possible to provide additional water volume in critical dry periods, but this "loan" of long-term reserves is meant to be returned. Two extreme scenarios are possible:

1. The replenishment potential is sufficient to cover the "loan" over a short period of time (within the same or the next hydrologic cycle).
2. The deficit cannot be compensated and the water table decline is guaranteed.

Although the first scenario is not too problematic, the second can be compared to a reckless, unsustainable credit-financed spending spree. Furthermore, water management can be a very delicate task, given that a large portion of the world's groundwater is stored in areas with arid or semiarid climates with little or no present-day aquifer recharge.

Economic condition

The implementation of large groundwater control projects, such as the construction of underground reservoirs, can be a very expensive and delicate task. Although numerous analyses show that benefits often surpass the negative consequences, the fear of uncertain (and invisible) results and large investment have put many such projects on hold.

Smaller projects, such as drilling wells near the springs and temporary pumping of a required amount of water, are less costly and can be easily implemented. However, where extraction rates provided by this "tool" are substantial, groundwater monitoring should be mandatory. Otherwise, an unlimited pumping not only results in negative impacts on the environment but also causes economic losses. The consequences might not be seen immediately; but if significant long-term depletion of the water table occurs, this can be taken as a clear sign that the "tool" is not sustainable.

9.4.1 Types of regulation structures

There are different types of regulation structures. In the case of deep vauclusian springs in karst, it is possible to install pumps in submerged channels below spring elevation and overpump the spring during periods of high water demand (Avias 1984; Stevanović and Filipović, 1994). Deep wells or horizontal galleries are also frequently used for the same purpose. Large-diameter vertical tube wells have so far yielded the best results, because drilling technology is widely practiced and continuously improved. Primarily because of their cost and uncertain results and despite the many good ideas and designs, more complicated structures, such as subsurface (underground) dams, are still rarely implemented. Several common means for obtaining more groundwater than discharges naturally at the spring are discussed briefly next.

Searching for more water in karstic or in fractured rock aquifers drained by a spring is always delicate. Anisotropic and nonhomogeneous media result in the formation of preferential flow pathways, such as along fractures, faults, conduits, and large cave channels. Finding such pathways and reaching them with wells and pumps is always uncertain but necessary for obtaining additional flow. When the saturated zone is under pressure (caused by overlying low-permeable rocks) and the topographic conditions are favorable, it is possible to install horizontal or inclined pipes (wells) for free gravity drainage of the aquifer (Figure 9–39).

In the case of barrier (contact) springs, horizontal or slightly inclined galleries are often constructed below discharge to reach the confined or trapped part of the aquifer and enable free gravity discharge. They are

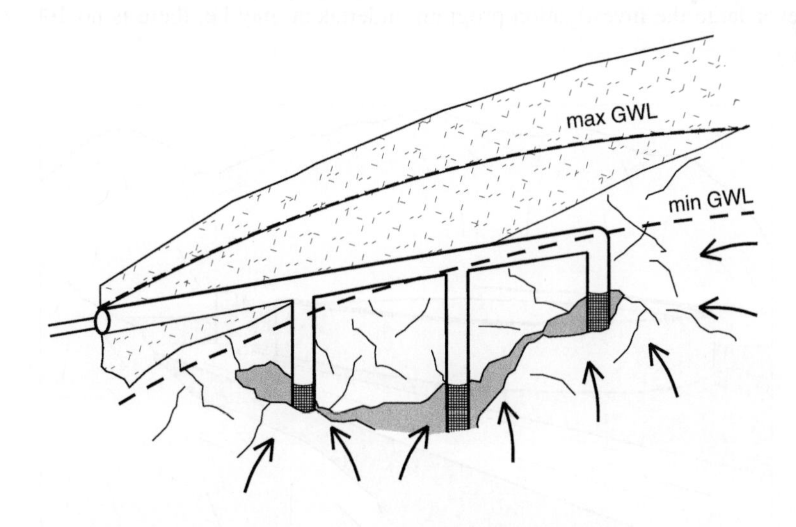

FIGURE 9–39 Capture of a confined karst spring with a drainage gallery. Three small shafts are drilled from the main gallery, controlling the ascending (artesian) flow during maximum groundwater levels (max GWL). The lowering of the artesian pressure during the dry season (min GWL) may necessitate pumping. The gallery provides water to the point of use by gravity.

sometimes called *bottom outlets*, due to their function, which is similar to cleanout drainage pipes from the spring boxes. By piping and installing valves, it is possible to have the full control over the flow from the gallery. This is important if water utilization is seasonal and water can be stored until it is needed. For blocking the inflow from the aquifer to the gallery, a barrage door is constructed, similar to ones used inside many underground mines.

Increasing the water level at the discharge point may sometimes result in an increase in the hydraulic head inside the aquifer. This is a classic retention effect (Figure 9–40). Therefore, at many springs worldwide, retention walls, small dams, weirs, and basins are constructed to maintain the water level at the highest position possible, enabling larger underground storage (see Chapter 10.7). The same principle is applied in case of underground dams.

An underground dam can significantly raise the hydraulic head in the aquifer, if it fully or mostly controls and blocks the discharge point. Such structures are not very different from the subsurface dams described earlier; the idea is the same, to store more groundwater inside the rocks. The difference is in the type of rocks and the position of the dam. While subsurface dams are relatively shallow structures for collecting water in the alluvial deposits downstream of the spring, underground dams are located at the spring and create water storage inside the primary aquifer.

There are many advantages to underground reservoirs over the surface ones. A few are as follows:

- No problems with flooding of infrastructure, fertile land, monuments, or compensation to relocate a population.
- No threat of dam collapse and major destruction downstream.
- Minimal loss of water due to evaporation.
- No negative impact on water quality as in surface reservoirs (eutrophication, sedimentation).

Over and above the expensive investigations and construction, an underground reservoir has two main problems: possible leakage from the reservoir and a not always predictable amount of storage space within the aquifer. However large the investigation program undertaken may be, there is no 100 percent guarantee

FIGURE 9–40 Small semielliptical dam constructed around a spring at the foothill and tied to the rock. Overflow is measured by a V-notch weir.

against leakage. Therefore, for such a project, it is best to identify the existence of a concentric flow and preferential pathways, preferably one main discharge channel. In addition, the same main drainage channel should be well developed inside the aquifer and extended into the network of secondary channels. Otherwise, the storage capacity would likely be limited. Karst aquifers are favorable from this point of view.

The key prerequisite for a successful underground dam is that the spring is in contact with impermeable rocks and, ideally, fully surrounded by them laterally. Given this condition, two main engineering options make creation of the reservoir inside the aquifer and above the natural water table possible:

1. Constructing a dam at the land surface directly in front of the spring. The dam should be properly founded at the bottom and coupled laterally with impermeable rocks to prevent leakage (Figure 9–41).
2. Sealing the spring channel with a concrete plug, a watertight barrage, or a carpet (usually built from concrete, as any other material can be easily washed out).

In both cases grouting is regularly implemented to seal against bottom and lateral leakage. However, sometimes, even a very dense curtain (boreholes at 1 m intervals) cannot stop the undesirable leakage, particularly when very karstified rocks are present. When the saturated zone is not very deep, installation of cutoff walls or a watertight diaphragm instead may be effective for regulating and directing the flow.

An option for a dam proposed at the Beli Drim Spring is described in Chapter 10.1 (Peric, Simic, and Milivojevic, 1980). For the same spring, an alternative regulation structure with a watertight barrage is shown in Figure 9–42.

Storing additional groundwater could also have some deficiencies. P. Milanovic (2000) emphasizes rock instability and local seismically induced collapse failure as main problems and describes several such cases from the engineering practice in karst regions of the Dinarides (the Balkans) and China. Very illustrative is the example of an experimental plugging of the Obod estavelle (Fatnicko polje, eastern Herzegovina). During the dry season, a concrete plug was installed in the main channel to prevent the flooding of the polje

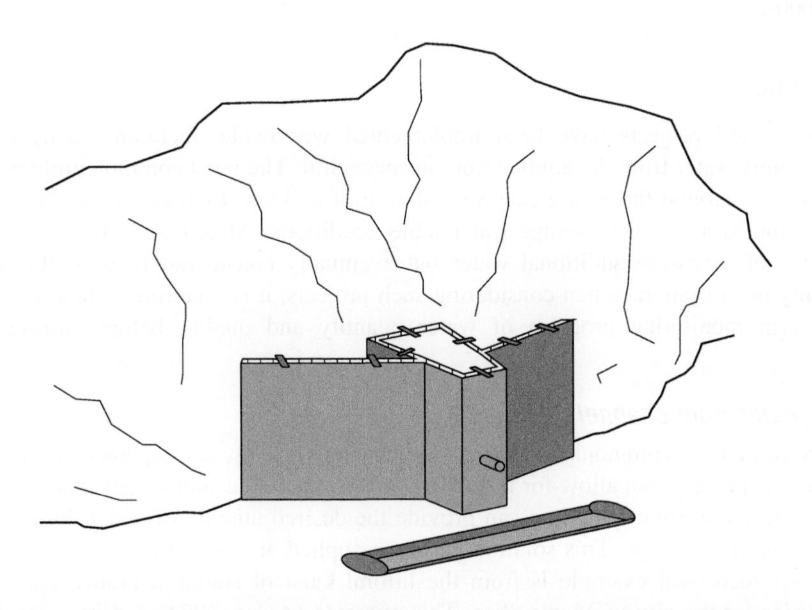

FIGURE 9–41 Capture of a small spring coupled with cutoff walls. The principle of building a large dam is very similar.

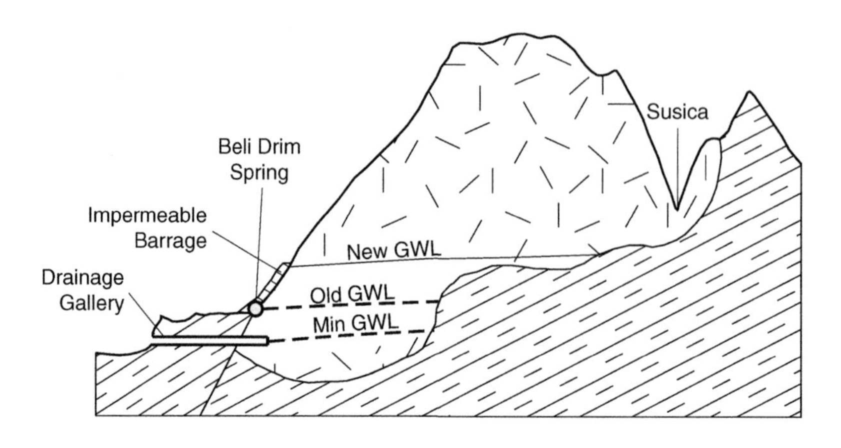

FIGURE 9–42 Alternative design for an underground reservoir at the Beli Drim Spring, Metohija, Serbia. A watertight barrage creates a new artificial groundwater level (GWL), while discharge is controlled by a horizontal gallery below the old groundwater level. (From Peric et al., 1980.)

(karst depression), which regularly happens every year. After heavy rains (200 mm in 24 h), the water pressure in the plugged channel increased to 10.6 bars and many new springs started to discharge at various elevations, as high as 100 m above the estavelle. The total yield of new springs was estimated at over 10 m^3/s. The emergence of new springs was followed by explosions of captured air and strong local seismic waves. After some houses in the vicinity were damaged, an urgent blasting of the plug was carried out. Water started to flow out and the pressure soon decreased. The stored water was completely discharged within approximately 6 hours, indicating a very good transmissivity but a very low retardation capacity of the karstic aquifer (P. Milanovic, 2000).

9.4.2 Case studies

Numerous spring control projects have been implemented worldwide, including many examples where attempts to obtain more water from the aquifer were unsuccessful. The most common problems resulted from water leakage below or around the spring capture, collapse of rocks and intakes, unproductive drilled wells, and galleries constructed above the average water table conditions. Also, many projects were successfully completed in terms of accessing additional water but eventually considered failures due to inappropriate use or water quality deterioration. When considering such projects, it is therefore critical to implement comprehensive long-term monitoring program of water quantity and quality before, during, and after the construction.

Pumping springwater from channels and siphons

Drainage of karst aquifers is commonly through large channels, caves, and siphons. As mentioned earlier, ascending vauclusian springs often allow for installation of pumps in the submerged channels located significantly below the spring elevation; pumping can provide the desired amount of water during periods of high demand and low natural discharge. This solution has been applied at many springs worldwide.

One well-known successful example is from the littoral karst of southern France and the Lez Spring, which supplies water for the city of Montpellier. This example (Avias, 1984) and its wide promotion have inspired many similar attempts worldwide. With regard to the Lez project, mentioned in Chapter 2, a few additional remarks follow.

The water demands of Montpellier increased more than 70-fold over the last 150 years, from just 25 L/s to over 1800 L/s. It is common in water practice to abandon supply based on springwater due to large expansion of the cities and rapid industrialization. This, however, was not the case in Montpellier, thanks to the engineering idea to tap additional water deep inside the aquifer.

Exploration of deep submerged cave channels and their precise mapping allowed for the construction of an extraction shaft, a gallery, and the placement of high-capacity pumps. Diving into channels to more than 500 m from the submerged cave entrance and to depths of over 100 m was a delicate task, but now serves as a proof of the concept worldwide.

A similar solution was recently applied at the vauclusian spring Modro Oko, one of the sources for the water supply of Nis, the second largest city in Serbia. Based on the initial pumping test conducted in the 1970s, which confirmed the presence of a deeper, large underground reservoir that drains a karstic basin of over 70 km^2, an investigation program for spring regulation was implemented in the early 2000s. It included exploratory diving to a depth of 86 m and a new pumping test by two pumps installed at a depth of about 30 m (Figure 4–43). For a pumping rate of 150–350 L/s, the maximum drawdown at the spring was 21 m. Based on the pumping test results, it was concluded that the spring could provide additional 260–280 L/s

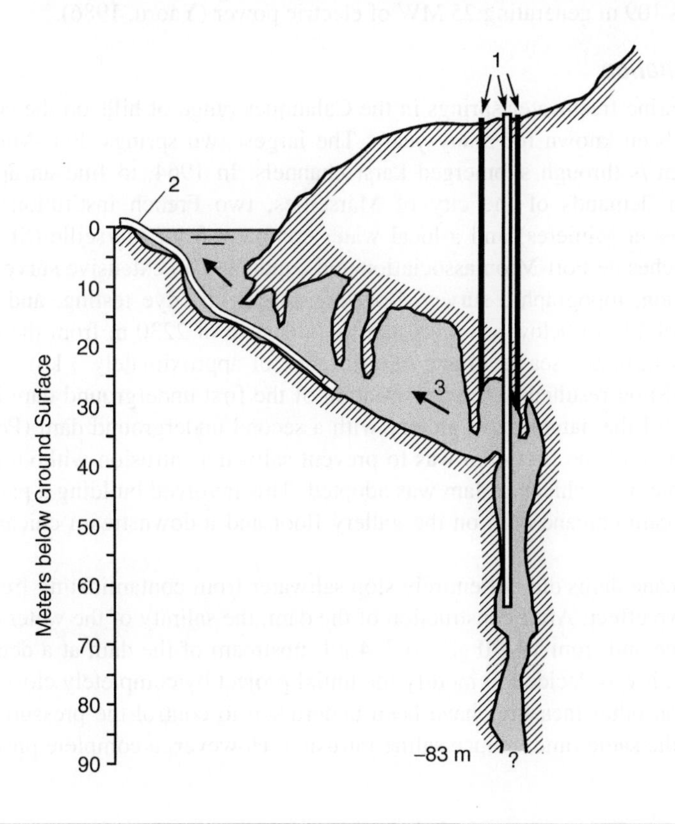

FIGURE 9–43 Modro Oko ("azure eye") vauclusian spring in Krupac, Nis, Serbia: (1) Well and exploratory boreholes, (2) pump and pipeline, (3) groundwater flow direction. (After Jevtic et al., 2005.)

of water constantly for two months, with drawdown in the karst channel of 50–60 m. If the average discharge of 50–150 L/s during the dry season is taken as a reference, the additional amount of water represents a significant increase in meeting the water supply needs (Jevtic et al., 2005). Two new 720 mm diameter wells, each 80 m deep, were drilled above the spring and completed in the vertical channel. Continuous monitoring has since confirmed that water abstracted from the karst aquifer during dry seasons (static reserves) is regularly replenished during periods of intensive aquifer recharge in winter and early spring.

Spring regulation with underground dams

Although several projects and experiments with underground dams have been conducted in Europe and in Northern America (Peric et al., 1980; P. Milanovic, 2000, 2004), the main practical experience has been obtained from projects implemented in China (Yaoru, Jie, and Zhang, 1973; Yaoru, 1986) and Japan (Ishida et al., 2005). In China alone, some 20 underground reservoirs have been completed for different purposes (water supply, irrigation, and hydropower) and enable water storage ranging from 1×10^5 m^3 to 1×10^7 m^3 in individual reservoirs. Although these projects vary in design, the solutions described earlier with a dam in front of the spring and plugging the discharge channel were among the most frequently implemented (Figure 9–44). Quibei in Yunnan Province is one of the largest such reservoirs, formed behind a 25 m high underground dam; the hydraulic head difference between the underground reservoir (dam) and a down-gradient surface power station is 109 m generating 25 MW of electric power (Yaoru, 1986).

Brackish spring regulation

The existence of submarine freshwater springs in the Calanques range of hills on the coastline between Marseilles and Cassis has been known for many years. The largest two springs, Port-Miou and Bestouan, discharge a combined 3 m^3/s through submerged karst channels. In 1964, to find an appropriate response to the fast-growing water demands of the city of Marseilles, two French institutions, BRGM (Bureau de Recherches Géologiques et Minières) and a local water company from Marseille (SEM), united to become the Syndicat de Recherches de Port-Miou association and conducted an extensive survey: geological and geophysical field exploration, topographic survey, flow measurements, dye testing, and diving exploration in Port-Miou (to a depth of 147 m below sea level and to a distance of 2230 m from the entrance), and in Bestouan (to a depth of 31 m below sea level and to a distance of approximately 3 km from the entrance).

The survey of Port-Miou resulted in the construction of the first underground dam in 1972 (Figure 9–45) and a complete closure of the natural cave gallery with a second underground dam (Potier, Ricour, and Tardieu, 2005). The objective of the first dam was to prevent saltwater intrusion without modifying pressure in the aquifer. The principle of a "chicane" dam was adopted. This involved building a pair of dams 530 m from the sea outlet: an upstream chicane dam on the gallery floor and a downstream chicane dam on the gallery ceiling (Figure 9–45).

Construction of chicane dams did not entirely stop saltwater from contaminating fresh groundwater, but it had a significant positive effect: After construction of the dam, the salinity of the water dropped from 4–5 g/L to 2–3 g/L at the surface and from 18–20 g/L to 3–4 g/L upstream of the dam at a depth of 20 m below sea level. A few years later, it was decided to modify the initial project by completely closing the gallery and creating a spillway. Several other measures have been undertaken to control the pressure in the aquifer during flood episodes and, at the same time, reduce saline intrusion. However, a complete prevention of salinization was not possible.

Monitoring the effects of spring regulation

Hydrogeological surveys undertaken during the last two decades of the 20th century made possible the construction of several successful systems for artificial control of karst aquifers in Serbia (Stevanovic

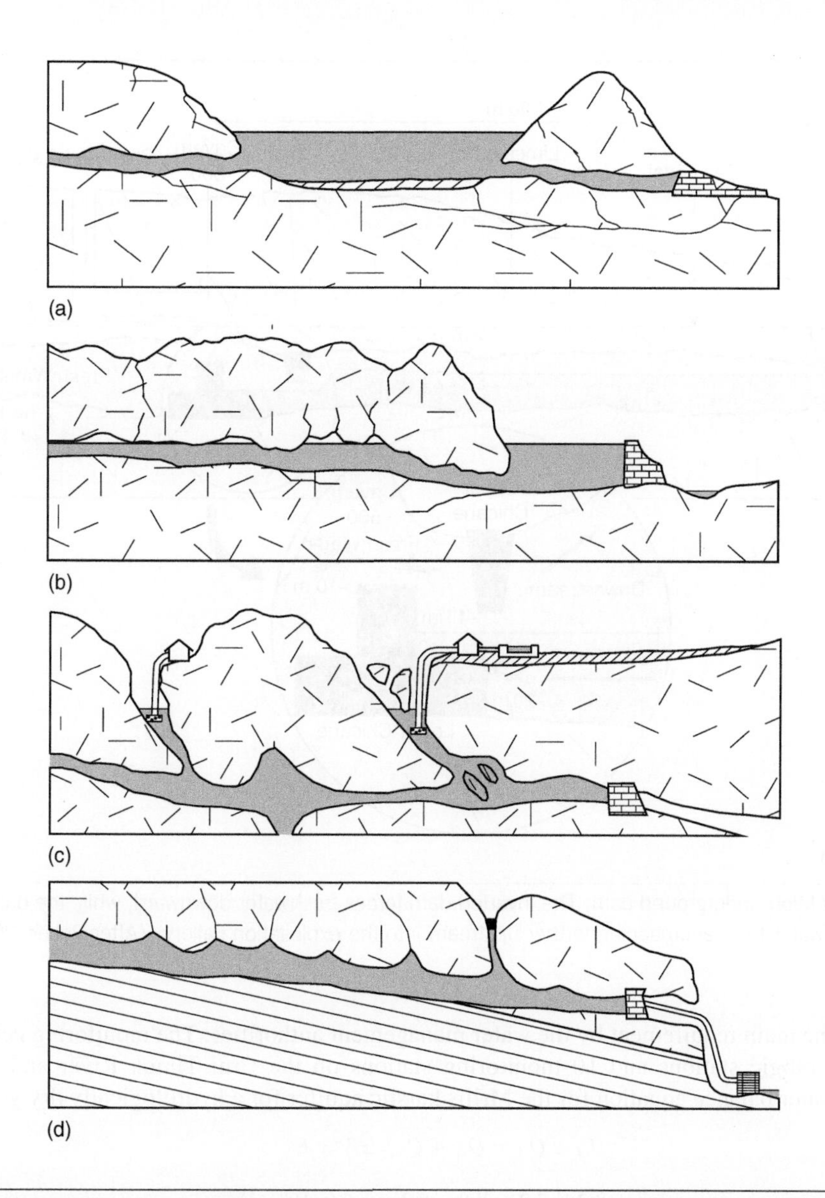

FIGURE 9–44 Implemented spring regulation projects in China, (a and b) with underground and surface reservoirs, (c) with underground dam and pumpage from deep sinkholes, (d) with underground dam and gravity transport of water to water tank. (Modified from Yaoru, 1986.)

et al., 2007). The largest system for aquifer control was constructed at the karstic Mrljis Spring (flow rate 80–1000 L/s) as part of the regional water system, "Bogovina," which supplies several towns of the Timok region in eastern Serbia. This project is a transitional solution until the surface dam is constructed on the nearby Crni Timok River. An important part of the project was establishment of a monitoring and warning system to prevent any disturbance of the historically observed river flow during base flow conditions

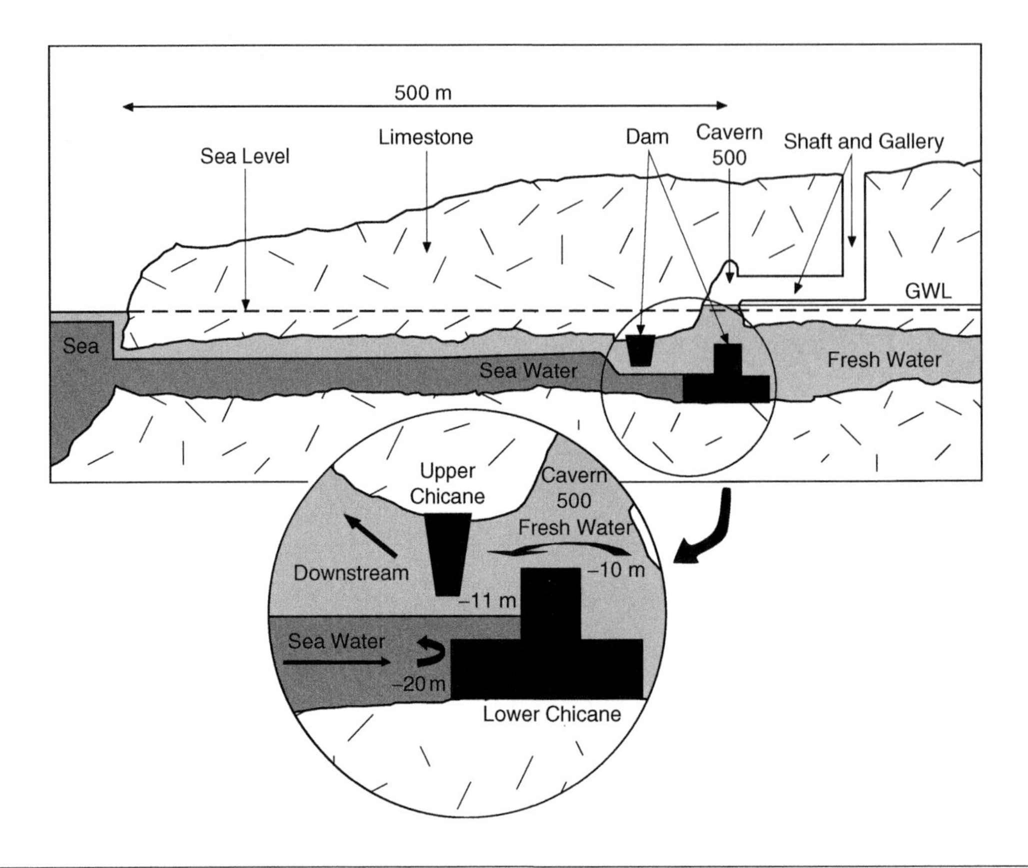

FIGURE 9–45 Port Miou underground dam. The inverted dam forces freshwater downward, while the dam on the cavern floor prevents saltwater from encroaching farther upstream into the exploitation gallery. (After Potier, 2005.)

(Figure 9–46), the main requirement by the water management authorities. The monitoring network consists of 4 new climatologic stations and 10 monitoring stations on the Crni Timok River and its tributaries. The estimated water balance equation for the Mrljis karstic aquifer for a hydrologically dry year is

$$I_{ef} + Q_{sf} = Q_{sp} + Q_{und} \pm V \pm E$$

$$8.2 + 7.2 = 8.8 + 8.2 - 1.6\,(\times 10^6\ \mathrm{m}^3)$$

where I_{ef} is the effective infiltration (in this case 20 percent of rainfall); Q_{sf} is the surface water infiltration from several sinking streams in the upper catchment; Q_{sp} is the discharge through the spring; Q_{und} is the underground drainage into the Crni Timok riverbed; V is the annual variation of groundwater reserves; and E is a possible calculation error.

To estimate the physical conditions for the aquifer control, a pumping test lasting several days was conducted at Mrljis Spring. Prior to pumping, the natural spring flow had been 172 L/s, while during the test, 325 L/s was constantly pumped out (Stevanovic and Dragisic, 1995). The drawdown at the discharging

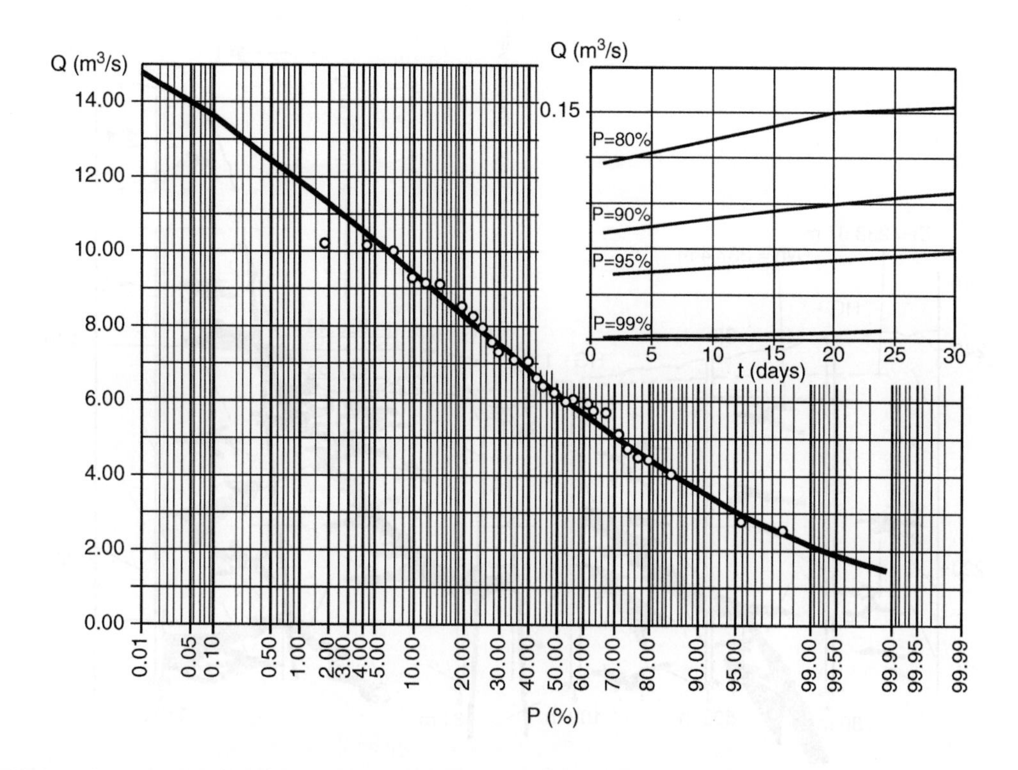

FIGURE 9–46 Probability and duration of minimal average monthly flow of Crni Timok.

siphon was 2 m, while in the nearest observation wells the level decreased only by 1 m (Figures 9–47 and 9–48). Figure 9–49 shows the calculated drawdown at the spring versus different pumping rates for an average aquifer transmissivity of 0.2 m^2/s determined from the pumping test.

During the pumping test, a continuous tracing of the Crni Timok River by fluorescein dye resulted in the identification of no tracers in the spring. This confirmed deep groundwater circulation toward the spring and no river flow loss for the given drawdown at the spring.

The final design of the Mrljis water supply system included four exploitation wells (Figure 9–50) and a monitoring network on the Crni Timok River. The pumping rates of the individual wells range from 50 to 110 L/s. The combined extraction rate of the new system, as compared to the natural minimal spring flow of Mrljis, increased almost 400 percent.

Based on the results of investigations, the exploitation permit for the Mrljis source was issued by the regulators for an unlimited pumping from the wells for all Crni Timok River flows over 850 L/s. A proportional restriction is placed for lower flows. However, since the new system became fully operational, neither significant drawdown nor decrease of the natural river flow during base flow conditions has been registered (Stevanovic et al., 2007). During the two analyzed years (2005 and 2006), the total extracted groundwater volume of 6×10^6 m^3 caused a flow reduction in the Crni Timok of only 7.5 percent (Figure 9–51).

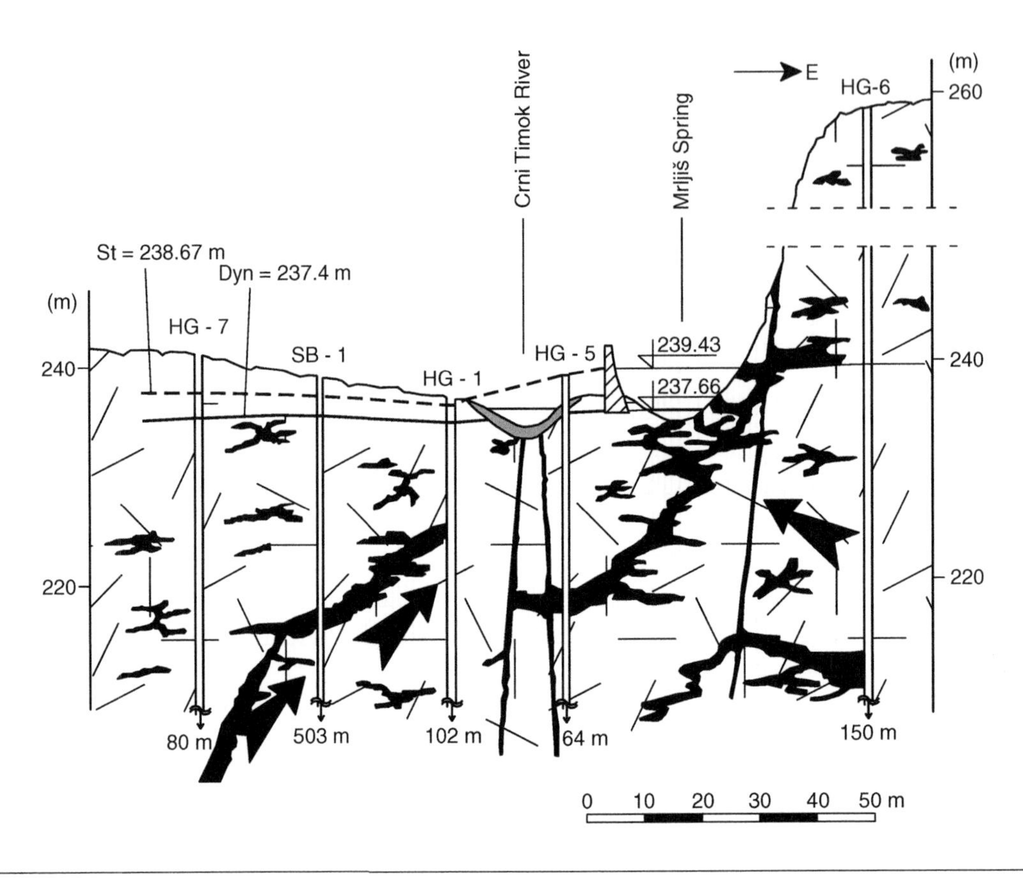

FIGURE 9–47 Pumping test at Mrljis Spring: St = static water table (before pumping); Dyn = drawdown (dynamic level).

The comparison of the annual precipitation and annual flow of the Crni Timok River before and after the regulation has also confirmed only a minor influence by the new intake on the river flow regime (Figure 9–52). However, this type of spring control has other implications for the environment, both positive and negative. As a consequence of intensive pumping, Mrljis Spring dries out every year during the recession period, while spring flows recover regularly in the late autumn or during the wintertime.

Comparison of groundwater quality before and after the construction of the wells for spring overpumping shows a generally positive impact from the spring regulation. Some previously unfavorable water quality indicators, such as slightly increased mineralization and content of the Mg and SO_4 ions in groundwater, which were the result of slower circulation within deeper parts of the aquifer, were improved by pumping. Moreover, the water quality problems common for karst aquifers, such as spring turbidity and bacteriological contamination, are now all but gone. In general, the extracted groundwater is of low mineralization and good quality (electrical conductance is 400–450 μS/cm), and no treatment except chlorination is required and applied before its consumption.

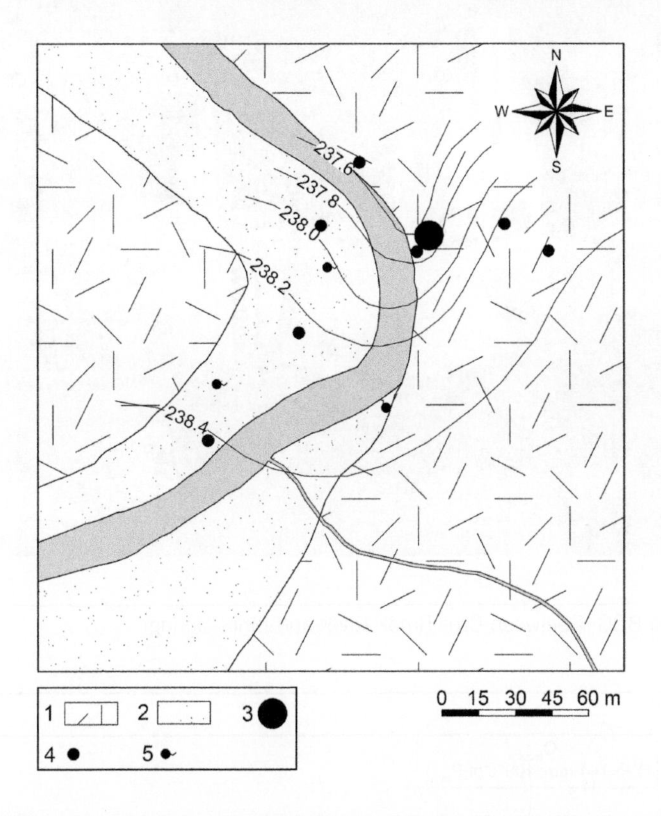

FIGURE 9–48 Map of Mrljis Spring and contour lines of water table during pumping of the spring: (1) Karstic aquifer, (2) alluvium, (3) Mrljis Spring, (4) well, (5) small spring.

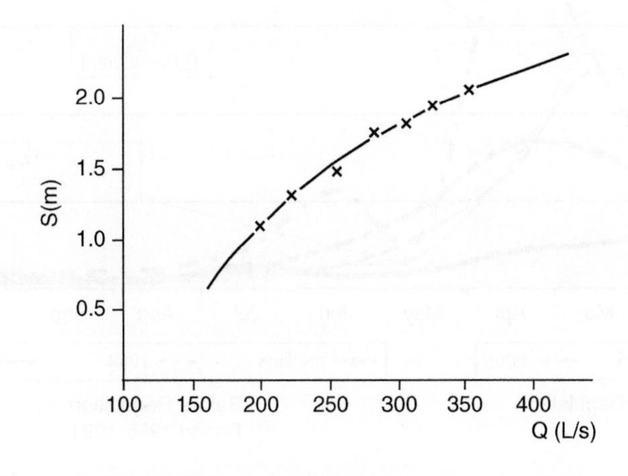

FIGURE 9–49 Drawdown (S) versus pumping rate (Q) at Mrljis Spring.

FIGURE 9–50 Pumping well BOG 4 between Crni Timok River and Mrljis Spring.

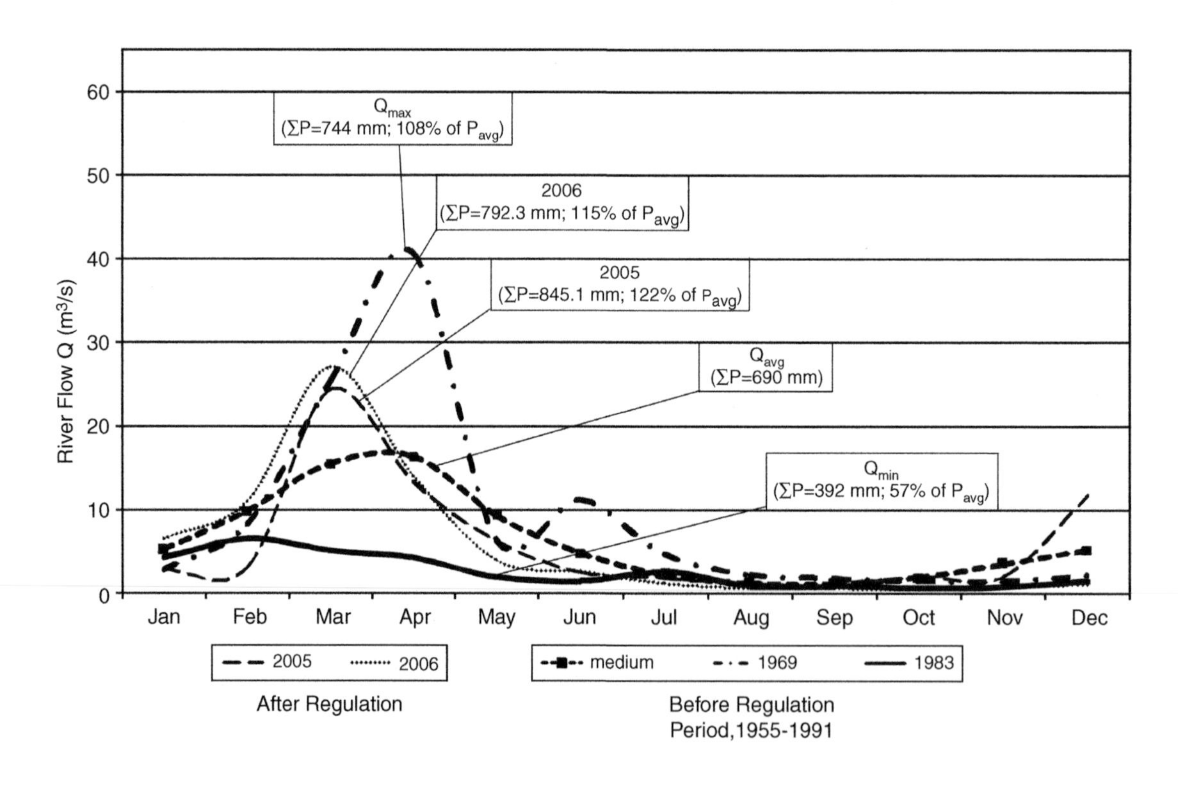

FIGURE 9–51 Flow rate of the Crni Timok before (1955–1991) and after (2005–2006) the Mrljis Spring regulation: P = precipitation; Q = flow rate.

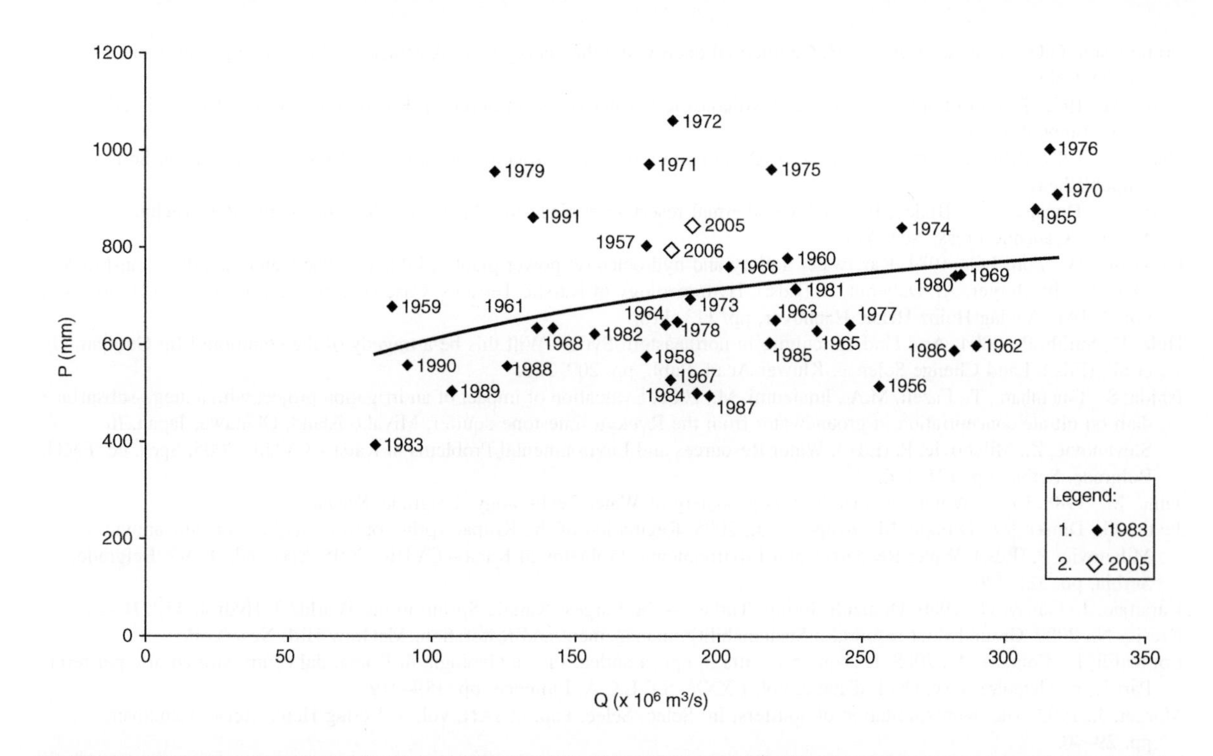

FIGURE 9–52 Annual precipitation in the drainage area and the Crni Timok River flow: (1) Before regulation, 1955–1991; (2) after regulation, 2005 and 2006; *Q* = Annual flow rate; *P* = annual precipitation.

ACKNOWLEDGMENTS

I gratefully acknowledge the technical support and dedication provided by my friends Beverly Lynch and Branislav Petrovic.

REFERENCES

Avias, J., 1984. Captage des sources karstiques avec pompage en periode d'etiage. L'example de la source du Lez. In: Burger, A., Dubertet, L. (Eds.), Hydrogeology of Karstic Terrains, Case Histories, Int. Contrib. to Hydrogeol., vol. 1. IAH, Verlag Heinz Heise, Hannover, pp. 117–119.

Burdon, D., Safadi, C., 1963. Ras-El-Ain: The great karst spring of Mesopotamia. An hydrogeological study. J. Hydrol. 1, 58–95.

Burke, J.J., Moench, H.M., 2000. Groundwater and society: Resources, tensions and opportunities. Spec ed. of DESA and ISET, UN public, ST/ESA/265, New York.

Coffman, C.R., John, B.C., 1984. Spring water development. Assistance to soil and water conservation programme—phase II. Field Doc. no. 7 ETH/81/003. Soil and water conservation. Dept. of Min. of Agri. Ethiopia and FAO, Addis Ababa.

Custodio, E., 1992. Hydrogeological and hydrochemical aspects of aquifer overexploitation. In: Selec. Pap. of IAH, vol. 3. Verlag Heinz Heise, Hannover, pp. 3–27.

Filipovic, B., Dimitrijevic, N., 1991. Mineral waters [in Serbian]. Spec. ed. Fac. Min. Geol, (FMG), University of Belgrade, Belgrade, Serbia.

Fridleifsson, G.O., Albertsson, A., 2008. Geothermal energy and the energy race. Abstracts of 33rd Intern, Geol. Congress, on CD, Oslo.
Gaon, M., 1975. Design of spring capture and Mihajlovac's waterworks [in Serbian]. Hidrosanitas, WIGA Co., Belgrade, Serbia (unpublished).
Gaon, M., 1987. Sjenica water supply project. Design and report [in Serbian]. Hidrosanitas, WIGA Co., Belgrade, Serbia (unpublished).
Grant, M., Donaldson, I., Bixley, P., 1982. Geothermal reservoir engineering. In: Energ. Sci. and Engin. Res. Techn. Manag. Academic Press, New York.
Graziadei, W., Zötl, J.G., 1984. Karstwater gallery and hydroelectric power plant "Mühlau"—The water supply of Innsbruck (Austria). In: Burger, A., Dubertet, L. (Eds.), Hydrogeology of Karstic Terrains, Case Histories, Int. Contrib. to Hydrogeol., vol. 1. IAH, Verlag Heinz Heise, Hannover, pp. 113–116.
Hole, F., Smith, R., 2004. Arid land agriculture in northeastern Syria—Will this be a tragedy of the commons? In: Gutman, G., et al., (Eds.), Land Change Science. Kluwer Acad. Publ., pp. 209–222.
Ishida, S., Tsuchihara, T., Fazeli, M.A., Imaizumi, M., 2005. Evaluation of impact of an irrigation project with a mega-subsurface dam on nitrate concentration in groundwater from the Ryukyu limestone aquifer, Miyako island, Okinawa, Japan. In: Stevanovic, Z., Milanovic, P. (Eds.), Water Resources and Environmental Problems in Karst—CVIJIĆ 2005, Spec. ed. FMG, Belgrade, Serbia, pp. 121–126.
Jahic, M., 1988. Urban Waterworks [in Serbian]. Society of Water Technology, Belgrade, Serbia.
Jevtic, G., Dimkic, D., Dimkic, M., Josipovic, J., 2005. Regulation of the Krupac spring outflow regime. In: Stevanovic, Z., Milanovic, P. (Eds.), Water Resources and Environmental Problems in Karst—CVIJIĆ 2005, Spec. ed., FMG, Belgrade, Serbia, pp. 321–326.
Karanjac, J., Günay, G., 1980. Dumanli Spring, Turkey—The Largest Karstic Spring in the World? J. Hydrol. 45, 219–231.
Kresic, N., 2009. Groundwater resources: Sustainability, management and restoration. McGraw-Hill, New York.
Lombardi, L., Corazza, A., 2008. L'acqua e la città in epoca antica. In: La Geologia di Roma, dal centro storico alla periferia, Part I. In: Memoire Serv. Geol. d'Italia, vol. LXXX. S.E.L.C.A, Florence, pp. 189–219.
Margat, J., 1992. The overexploitation of aquifers. In: Selec. Selec. Pap. of IAH, vol. 3. Verlag Heinz Heise, Hannover, pp. 29–40.
Milanovic, P., 2000. Geological engineering in karst. Dams, reservoirs, grouting, groundwater protection, water tapping, tunneling. Zebra Publ. Ltd, Belgrade, Serbia.
Milanovic, P., 2004. Water resources engineering in karst. CRC Press, Boca Raton, FL.
Milanovic, S., 2005. Hydrogeological characteristcs of some deep siphonal springs in Serbia and Montenegro karst. In: Stevanovic, P., Milanovic, P. (Eds.), Water Resources and Environmental Problems in Karst—CVIJIĆ 2005, Spec. ed. FMG, Belgrade, Serbia, pp. 451–458.
Peric, J., Simic, M., Milivojevic, M., 1980. Feasibility study of storing Beli Drim water within underground reservoirs for water supply and irrigation of Metohija [in Serbian]. In: Transactions of FMG, vol. 22. Belgrade, Serbia.
Potier, L., Ricour, J., Tardieu, B., 2005. Port-Miou and Bestouan freshwater submarine springs (Cassis–France) investigations and works (1964–1978). In: Stevanovic, Z., Milanovic, P. (Eds.), Water Resources and Environmental Problems in Karst—CVIJIĆ 2005, Spec. ed. FMG, Belgrade, Serbia, pp. 267–274.
Povara, I., Marin, C., 1984. Hercule thermomineral spring. Hydrogeological and hydrochemical considerations. In: Theoretical and Applied Karstology. vol. 1. Bucharest, pp. 183–193.
Stevanovic, Z., 2001. Subsurface dams—Efficient groundwater regulation scheme. Brayatti Press, 18 Erbil, Iraq, pp. 277–290.
Stevanovic, Z., Dragisic, V., 1995. An example of regulation of karst aquifer. In: Günay, G., Johnson, I. (Eds.), Karst waters and environmental impacts. Balkema, Rotterdam, pp. 19–26.
Stevanovic, Z., Filipović, B., 1994. Hydrogeology of carbonate rocks of Carpatho–Balkanides. In: Stevanovic, Z., Filipovic, B. (Eds.), Ground waters in carbonate rocks of the Carpathian–Balkan mountain range, Spec. ed. CBGA, Alston, Jersey, UK, pp. 35–112.
Stevanovic, Z., Jemcov, I., Milanovic, S., 2007. Management of karst aquifers in Serbia for water supply. Environ. Geol. 51 (5), 743–748.
Toulomudjian, C., 2005. The springs of Montenegro and Dinaric karst. In: Stevanovic, Z., Milanovic, P. (Eds.), Water Resources and Environmental Problems in Karst—CVIJIĆ 2005, Spec. ed. FMG, Belgrade, Serbia, pp. 443–450.
Vrba, J., 1996. Thermal mineral water springs in Karlovy Vary. Environ. Geol. 27 (2), 120–125.
White, D.E., 1967. Some principles of geyser activity, mainly from Steamboat Springs, Nevada. Amer. J. Sci. 265, 641–684.
Wohnlich, S., 1996. The spa of Baden-Baden, Germany. Environ. Geol. 27 (2), 108–109.
Yaoru, L., 1986. Karst in China, Landscapes, types, rules [in Chinese]. Spec. ed. Geol. Publ. House, Beijing, China.
Yaoru, L., Jie, X.A., Zhang, S.H., 1973. The development of karst in China and some of its hydrogeological and engineering geological conditions. Acta Geol. Sinica (1), 121–136.

CHAPTER

Case Study: Major springs of southeastern Europe and their utilization

10.1

Zoran Stevanovic
Department of Hydrogeology, Faculty of Mining and Geology, University of Belgrade, Serbia

10.1.1 INTRODUCTION

It is not that easy to determine where southeastern Europe begins and ends and where its real borders are. Recent European Union nomenclature includes the territories or part of the territories of 17 countries. However, it is most common to include the area south of the Alps, the eastern Apennines, the whole Balkan peninsula, and the region around the western and northern margins of the Black Sea (Figure 10.1–1). The largest river basin is the Danube. The catchment area of some 800,000 km^2 is inhabited by about 80 million people.

This is one of the most water-rich regions of the world. It is also characterized by numerous large springs, which are widely spread throughout the region. The majority of the large springs are connected to the karstic aquifers of the Alpine orogeny. Karst, as a specific type of topography and landscape, has always attracted and provoked the interest of humans, ultimately leading to scientific specializations embodied by hydrogeologists and karstologists. As it is such a specific environment, the various environmental problems found in karst ask for a nonstandard approach. Water management and control of karst aquifers, protection measures against the potential pollution of karstic surfaces and groundwater, difficult construction of human-made structures in karst and their influence on the environment, and protection of geo- and bioheritage objects (e.g., springs, caves, sinkholes, rare endemic species) are all important aspects that require careful study to ensure sustainable development within and around the karst environment (Stevanovic and Mijatovic, 2005).

Exploration and utilization of large springs discharging fresh or mineral water is an ancient art in the area. Two major ancient civilizations, the Greeks and the Romans, provided the foundation for modern water utilization and introduced advanced water intake and transport structures for the first time.

10.1.2 GEOLOGICAL STRUCTURES AND HYDROGEOLOGY

The geology of the region is complex. For most of the Mesozoic period, the Thetis Ocean covered this area, whereas during the Tertiary period, its central part was exposed to the Alpine orogenesis, when the majority of today's mountains were uplifted and folded. In flat areas in the south (Puglia) and in marginal parts in the northern (Pannonian basin) and eastern (Black Sea basin) regions, the sedimentation continued until the Quaternary period or is still ongoing.

Copyright © 2010, Elsevier Inc. All rights reserved.

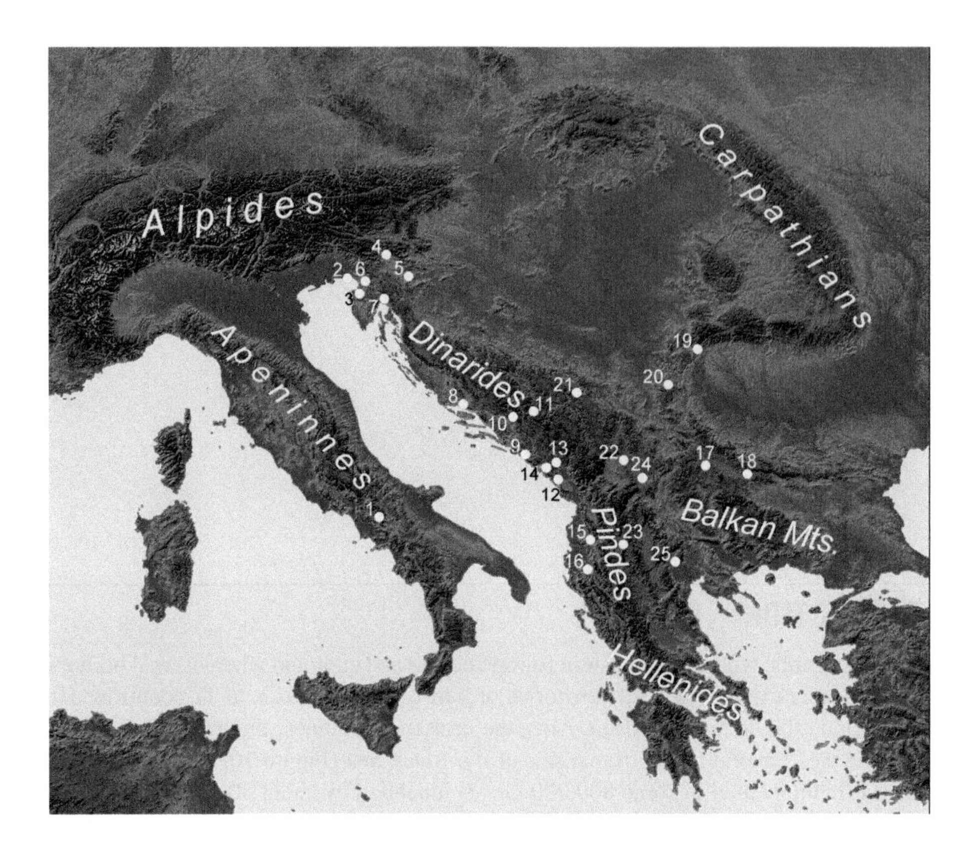

FIGURE 10.1–1 Southeastern Europe with main mountain chains and locations of springs described in the text.

Mesozoic carbonate rocks are widespread in the central part of the region, in the Alpides, Dinarides, Pindes, and Hellenides mountains. In the Apennines and in the Carpathian-Balkan in the north (including its terminal parts, the west and south Carpathians), carbonate rocks also prevail but with a greater extent of low-permeable volcanic rocks or sedimentary (flysch) formations (Figure 10.1–1).

Most of the carbonate rocks (limestones and dolomites) are well karstified and contain very large groundwater reserves. Along with alluvial groundwater and surface water from the reservoirs, the water from the karstic springs is the main source of the water supply in the region. Generally, the exploration of karst groundwater is more costly and less predictable than surface water use in terms of final outcomes, but the extraction is regularly much cheaper. Due to the unstable regime of the karstic springs, the main problem for most of the waterworks is to ensure water supply during low-water periods (summer and autumn). Several large cities with populations of over a million depend on karst aquifers and their discharge regimes. Among them are the capitals Rome, Vienna, Sarajevo, Tirana, Skopje, and Podgorica. In cases of insufficient discharge, some other big cities combine the utilization of karst waters with water from other sources. The Adriatic and Ionian coastal tourist areas strongly depend on groundwater from the karst. There, in addition to the problem of water shortage during the recession period (summer-autumn), water pollution or saltwater intrusion often means that local water and economy sectors cannot be expanded and properly developed.

Despite producing less water, the nonkarstic aquifers are also important mineral and thermal water sources (there are many developed spas), produce geothermal energy, support the bottled water industry, and provide water for irrigation or small industry.

10.1.3 REGIONAL DISTRIBUTION OF SPRINGS AND THEIR USE

10.1.3.1 Italy

Italy has numerous large springs that flow from various geological formations. The biggest, however, are those draining karstic aquifers of the Apennines mountain range. Surrounding Trieste in northern Italy is the carbonate rock area called Crasso, the name used at the end of the 19th century to describe all similar types of morphological and hydrological features worldwide (or, in German, *karst*; Cvijic, 1895). Therefore, the region is well known as the homeland of "classical karst."

Tapping springs and conveying water by aqueduct is an ancient art invented during the Roman Empire. Constructed in the 2nd century BC, Aqua Marcia, with a length of 90 km, was one of the first aqueducts and supplied Rome with water from karst springs from the Aniene catchment. The tapped amount of water (Q) was over 2 m^3/s on average (Bono and Boni, 1996). Other aqueducts constructed later also conveyed water captured from the large springs in the vicinity of Rome (Salone, Tepula, and Claudia).

A more recent system that has been supplying water to Rome since 1935 in the amount of 14 m^3/s is based on the group of springs of Peschiera and Capore (no. 1 in Figure 10.1–1). The springs of Peschiera in the Velino River catchment discharge an average of 18 m^3/s, half of which is tapped for Rome. The springs' altitude is 410 m above sea level (asl) and they drain an area of about 1000 km^2. The infiltration rate is assumed to be some two thirds of the rainfall, which is usually above 1000 mm/a (Boni and Bono, 1984). The discharge of the springs is very constant.

A very similar, stable discharge regime characterizes the group of Capore springs. The altitude of this discharge zone is much lower than in the case of Peschiera springs (246 m asl); the catchment area is also much smaller (280 km^2) and the rainfall rate lower (570 mm/a). All these result in a smaller average discharge of 5 m^3/s (Boni and Bono, 1984) for these springs.

Water from the springs of Peschiera and Capore is of excellent quality and requires no treatment except preventive chlorination.

Perhaps the most famous and the largest spring on the northern Italian coast is Timavo (Sorgenti del Timavo; no. 2 in Figure 10.1–1), which used to supply water to Trieste. This is a group of springs discharging at or below sea level near San Giovanni di Duino in the Trieste Gulf. These springs have a very large catchment in the Crasso area and their average discharge rate is 30 m^3/s (Bensi et al., 2005). At the end of the 19th century, one of the first tracing experiments in the world proved the connection with the sinking Reka River, which is today on the Slovenian side of the border.

An interesting way of tapping springwater by wells is applied on the Slovenian part of Crasso. Near the border city of Sezana, several highly productive wells were drilled in the 1980s to tap the water from deeper karstic channels oriented toward Timavo. For example, the pumping well VB-4 discharged 0.05 m^3/s for a drawdown of only 0.45 m.

During Roman times, the use of thermal water or heating springwater and developing bath centers around them was simply a standard way of life. Consequently, many traces and archaeological facts from such baths ("terme") are preserved at many places within the former Roman Empire. In Italy, the most famous spas are the Montecatini terme near Florence, the Abano terme near Padua, and the Acqui terme near Alessandria.

Bottling water from springs also has a long tradition in Italy. There is a large variety in chemical and physical characteristics, but the two main groups of water are those from karstic aquifers (sometimes covered by younger deposits) and volcanic rocks (very low-mineralized water). Over 200 brands are registered and many of them are well known on the world market (e.g., San Pellegrino, San Benedetto, Farrarelle).

10.1.3.2 Slovenia

As one of the "classic karst" countries, Slovenia (Figure 10.1–2) is very rich in groundwater. The minimum groundwater flow is assumed to be about 38 m^3/s, of which some 20 percent is used in the water supply. Among the more than 10,000 registered springs, the biggest are the karstic springs, and they supply water to more than half the inhabitants.

The Slovene coastal zone is supplied by water from the Rizana spring (no. 3) near Koper. The spring is situated at the contact of Upper Cretaceous and Paleocene limestones and Eocene flysch at an altitude of 70 m asl. Its discharge ranges from 0.03 to 91 m^3/s, and the mean discharge for the period 1961–1990 is

FIGURE 10.1–2 Entrance to Skocjanka Jama cave in the Slovenian part of Crasso. The cave is one of the largest speleological objects worldwide in terms of the distance between the floor and the ceiling of the cave; the underground canyon is some 100 m high.

4.3 m^3/s (Kolbezen and Pristov, 1998). The recharge area of Podgorski kras and Slavnik of 247 km^2 also includes the sinking streams of the Brkini flysch area (Krivic, Bricelj, and Zupan, 1989).

Water from the Rizana River has been used since the early 19th century, while the coastal water supply system was constructed in 1935. Today, during the summer season, the Rizana spring supplies water for more than 120,000 people. Water treatment includes a very sophisticated system of ultrafiltration to reduce turbidity and eliminate bacterial contamination. The average annual quantity of water pumped in the last few years is 0.2 m^3/s, which cannot satisfy increased demands, so that other water supply solutions are currently under evaluation.

One of the largest Danube tributaries and biggest domicile river basins of the former Yugoslavia is the Sava. One of its main sources is at the edge of Bohinj Lake in Slovenia. This is a typical gravity spring discharging a minimum of some 0.5 m^3/s (Figure 10.1–3; no. 4 in Figure 10.1–1).

FIGURE 10.1–3 Sava Spring, Bohinj, Slovenia.

The Ljubljanica group of springs in the southwestern part of the Dinaric karst in Slovenia is located at the contact between limestone and the very low-permeable sediments of the Ljubljana Moor (no. 5). The catchment of the Ljubljanica springs covers about 1100–1200 km^2 and consists mostly of carbonate rocks of the Cretaceous and Jurassic age (Gospodaric and Habic, 1976). Water fluctuations between the superficial and underground parts, sinking streams, temporarily flooded karst poljes on several levels, large karst springs, and numerous karst caves are the characteristics of that area. The Ljubljanica springs discharge ranges from 4.25 to 132 m^3/s, with a mean value of 39 m^3/s. Despite the large discharge, the springs are not used for the water supply. It has been concluded that bank filtration is a more secure solution regarding water quality, and the alluvial porous aquifer along the Sava River is tapped for the capital Ljubljana.

The Vipava group of seven permanent springs (no. 6) at the western foot of the Nanos karst plateau marks the contact of the Upper Cretaceous limestone with the impermeable Eocene flysch (Habic, 1980; Kranjc, 1997). The extent of the recharge area of the Vipava Spring is about 149 km^2 (Petric, 2002). On the southeastern, nonkarstic part of the drainage area, there are numerous small streams, all of which sink into the karst aquifer and flow toward the Vipava Springs. Groundwater flow below the surficial flysch deposits was also proven by conducting a tracing test. The Vipava discharge regime is characterized by two peaks in April and November and two minima in August and January. The average discharge is 6.78 m^3/s. During extreme droughts, it decreases to 0.73 m^3/s, but after heavy rains, the discharge can reach up to 70 m^3/s (Kolbezen and Pristov, 1998).

10.1.3.3 Croatia

Croatian karst springs are numerous, and those in coastal areas are the largest and essential for the maintenance of Croatia's intensive tourism industry and its national economy. Many large cities, such as Rijeka, Split, and Dubrovnik, as well as hundreds of small towns and islands depend on their water supply from the karstic springs, which are often affected by seawater intrusion.

The Zvir group of springs (no. 7) discharges from Upper Cretaceous limestone almost at sea level in the Rijecina gorge. They were tapped at the end of the 19th century for the water supply of the city of Rijeka, the largest Croatian port. The discharge varies between 0.6 and 3.0 m^3/s. Due to difficult access to and fast development of a nearby industrial zone, one gallery 400 m long has been excavated and five shallow wells drilled on its floor, each enabling the extraction of over 0.03 m^3/s during the recession period (Figure 10.1–4). It is

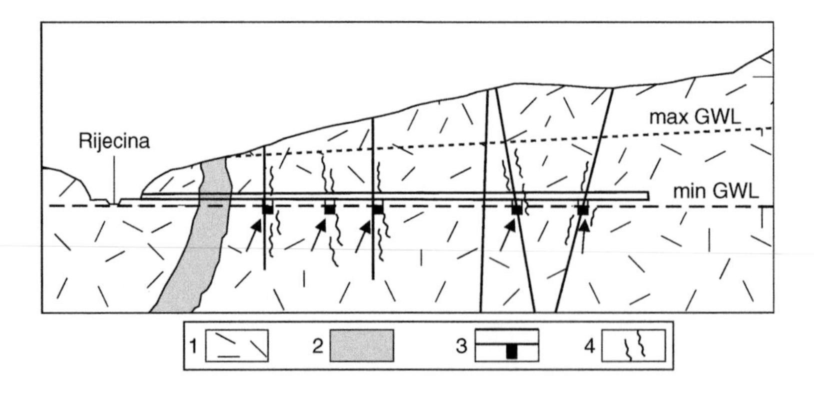

FIGURE 10.1–4 Zvir spring intake gallery: (1) Karstified Cretaceous limestone, (2) impermeable Eocene flysch, (3) shallow well, (4) fault; min GWL and max GWL are minimum and maximum groundwater levels, respectively. (After Biondic and Goatti, 1984.)

notable that, although the pumping drawdown is below sea level by about 1 m, no saltwater intrusion is registered (Biondic and Goatti, 1984a).

Jadro Spring is the main source for the water supply of Split (no. 8). It drains the thick Cretaceous limestone of Mosor Mountain. The main spring outlet is at 35 m asl at the contact with Eocene flysch sediments. This is the case for many springs along the Adriatic seashore. The average minimum discharges of Jadro during the recession period are 3–5 m^3/s, while maximum discharges are often over 50 m^3/s. Mijatovic (1968) identified the existence of one turbulent and two laminar subregimes (the determined recession coefficient α for the year 1962 is in the range of 0.016–0.0011). Denic-Jukic and Jukic (2003) developed a rainfall discharge model of this spring as a composite transfer function that simulates discharges by two transfer functions adapted for the quick-flow and slow-flow hydrograph components (Figure 10.1–5). By applying the groundwater hydrograph method, Bonacci (1987, 1993) estimated the Jadro catchment area to be between 225 and 260 km^2 (Figure 10.1–6). He also stated that, during the period of high groundwater levels, the catchment increased by approximately 10 percent. Tracing tests conducted in 1962 confirmed the connection with the Cetina River, which highly influences discharge regime. About 96 percent of the tracer appeared at the spring 23 days after the tracer had been injected into a ponor (swallow hole) in the riverbed.

Ombla is the biggest spring in southern Croatia and supplies the city of Dubrovnik (no. 9). The three outlets, one major and two smaller laterals, at the contact of Cretaceous limestone and Eocene flysch, form the Dubrovacka Rijeka River, which after a short course, flows into the Adriatic Sea (Figure 10.1–7). At 2.5 m asl, the direct contact of the flysch and karst aquifer is close to sea level, but there is no mixture of fresh- and saltwater even during the long drought periods (Milanovic, 2006).

The drainage area of the Ombla Spring is some 600 km^2, out of which a large portion is in the neighboring country Bosnia and Herzegovina. An additional amount of water percolated from numerous swallow holes (ponors) in the riverbed of the biggest European sinking river, the Trebisnjica River, before it was regulated (Figure 10.1–8). Since the completion of the Trebisnjica hydropower system, with one large and three smaller reservoirs, and seal treatment of the Trebisnjica riverbed, the average outflow of Ombla has been reduced from 34 m^3/s to 24 m^3/s. However, the minimum discharge (the lowest recorded is 2.3 m^3/s) is not affected by the applied measures (Milanovic, 2006).

Based on extensive exploration, which included different geophysical methods, speleology (more than 3 km of explored channels), diving, and drilling of boreholes and galleries, the construction of a large Ombla underground dam was proposed (Milanovic, 1996; Elektroprojekt, 1999). Three-level grout curtains supported by an existing natural barrier of Eocene flysch should maintain watertightness of the underground reservoir to the maximum possible level. Plugging the main channels is also envisaged. The underground hydropower plant is proposed to be installed in a large cavern downstream of the curtains (Milanovic, 1996).

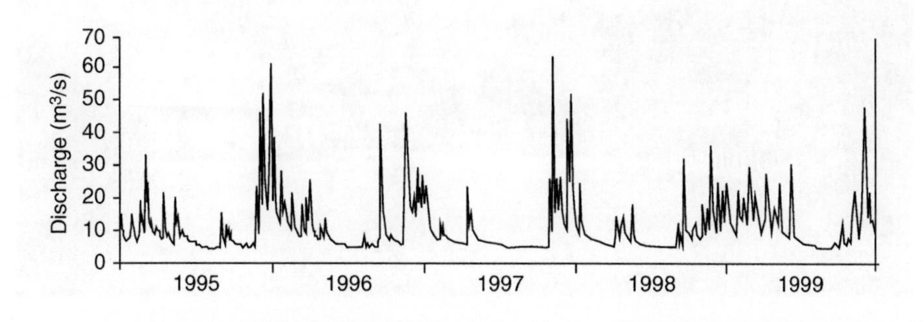

FIGURE 10.1–5 Hydrograph of Jadro Spring in the period 1995–1999. (Denic-Jukic and Jukic, 2003.)

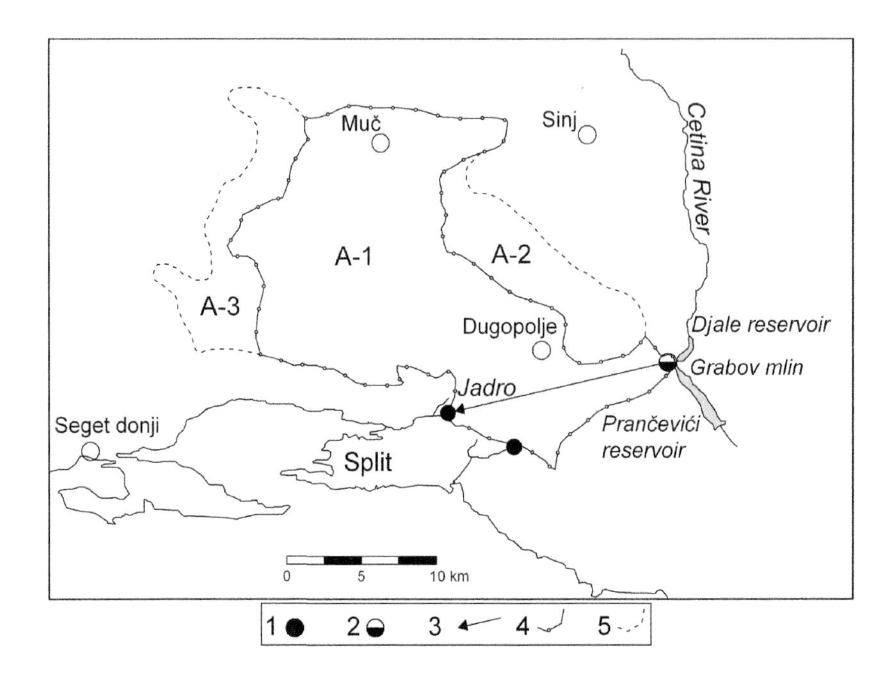

FIGURE 10.1–6 Direct (A-1) and indirect (A-2, A-3) catchments of Jadro Spring: (1) Spring, (2) ponor (sink), (3) direction of groundwater flow confirmed by tracing test, (4) dividing line of the Jadro catchment. (Denic-Jukic and Jukic, 2003.)

FIGURE 10.1–7 Ombla Spring. (Photo courtesy of P. Milanovic.)

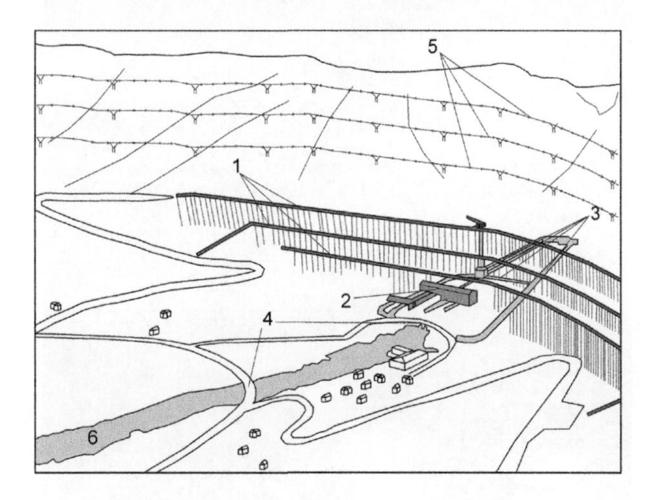

FIGURE 10.1–8 Ombla underground reservoir: (1) Grout curtains; (2) hydropower plant; (3) pipelines; (4) bridge; (5) electrical transmission lines; (6) Rijeka Dubrovacka River, formed from Ombla Spring. (Modified from original technical documentation of Electroproject, Zagreb 1999.)

10.1.3.4 Bosnia and Herzegovina

Bosnia and Herzegovina is very rich in karst groundwater, particularly in its southern and eastern parts. The system of successively lower-positioned poljes, such as Gacko, Fatnicko, Bilecko, and Trebinjsko, conveys the groundwater from the upper to the lower horizons and to the Adriatic Sea, which serves as the regional erosion base for groundwater discharge (Milanovic, 1979). Therefore, most of the very large springs are located near the coast. Buna and Bunica are two such springs (no. 10) in the Neretva River basin, near Blagaj and Mostar.

Buna Spring is a very attractive site for tourists and visitors (Figure 10.1–9). Water discharges from the cave underneath the high cliff. The total length of deep siphons explored by divers is 520 m, while the vertical distance between the deepest explored and the discharge point is 72 m (Touloumdjian, 2005). Minimum discharge is about 3 m^3/s, while the recorded maximum is 123 m^3/s. The spring is not captured.

The neighboring Bunica Spring is fed mostly by the percolating waters of the sinking Zalomka River. It discharges 0.7 m^3/s at a minimum, while during an extreme high period, water flow over 200 m^3/s is registered (Milanovic, 2006). The spring's siphons have also been explored by divers. Their total length is some 160 m.

Vrelo Trebisnjice near Bileca (no. 11) is the biggest karstic source of the former Yugoslavia and, arguably, of the entire Mediterranean basin, given its maximum recorded flow of over 850 m^3/s. It drains Upper Cretaceous limestone at the edge of a large karstic depression of the Bilecko Polje. The two main aquifer outlets are at a distance of some 400 m. Minimum discharge is about 2 m^3/s. Since the construction of the large Grncarevo reservoir as a part of the Trebisnjica hydropower system, the springs are completely submerged. The water column is now 75 m over the springs.

10.1.3.5 Montenegro

Regarding the territory and available water resources, Montenegro can be proclaimed the most water-rich country in the region. Skadar (Skutari) Lake is the largest lake on the Balkan peninsula. Around 60 percent of the lake is in Montenegro, while 40 percent is in Albania. The average surface of the lake is 475 km^2,

(a)

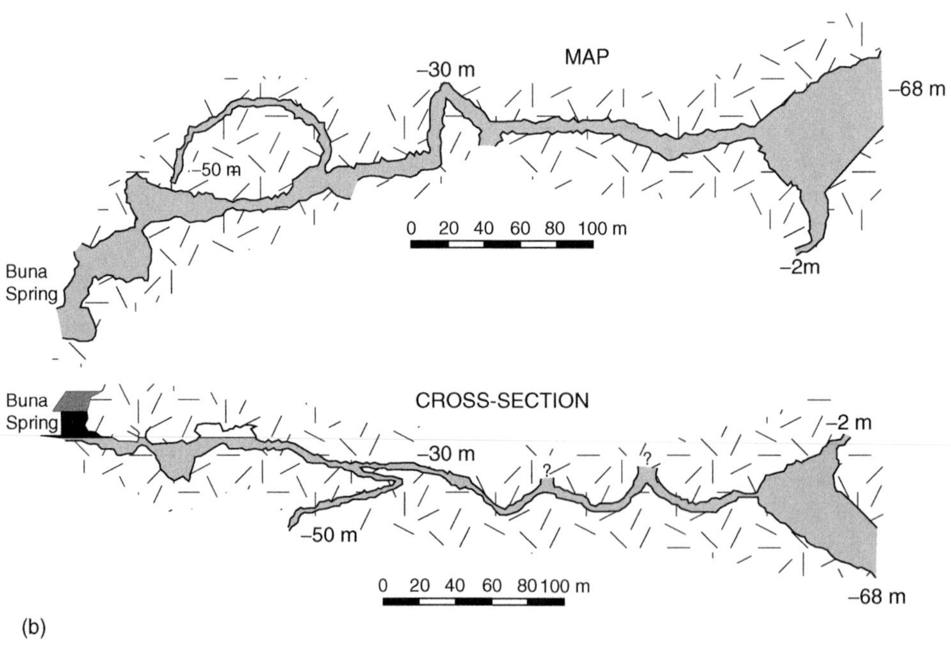

(b)

FIGURE 10.1–9 (a) Buna Spring in Blagaj near Mostar, Herzegovina (photo courtesy of S. Milanovic); (b) map and cross section of the submerged Buna Spring siphons. (After Touloumdjian, 2005.)

whereas the catchment area occupies around 5000 km^2. The strong prevalence of karstified Mesozoic rocks and the average annual amount of precipitation of 2000 mm make this basin one of the largest European groundwater reservoirs. Numerous sublacustrine springs ("eyes") are registered along the rim of Lake Skadar. The largest are Bazagur, Karuc, and Volac at the northwestern rim and Kaludjerovo Oko and Radusko Oko at the southwestern rim. The Radusko Oko sublacustrine spring is the best-known one (no. 12), with its bottom at about 60 m below the lake's floor or about 53 m below sea level. It is assumed that the total average discharge of all Montenegrin submerged springs of the lake exceeds 40 m^3/s.

The westernmost part of the Skadar basin belongs to Nikšićko Polje, the largest polje in the Montenegrin part of the Dinaric karst. Many strong karst springs exist along its margins, some of which are tapped for the city and industry water supply (e.g., Vidrovan, Vukova Springs, and Rastovac). Specific karstic occurrences, such as estavelles (Figure 10.1–10) or intermittent springs, are also present. These waters form the Zeta River, which sinks in the southern edge of Nikšićko Polje and appears again at the large springs Glava Zete (3–30 m^3/s).

Before the construction of three artificial reservoirs, Nikšićko Polje was periodically flooded in autumn and spring. The water from the polje was sinking through numerous ponors along the southern edge. Despite the construction of these reservoirs and undertaken consolidation works and grouting treatment of karst features, large water losses through highly karstified rocks still existed, and these waters continue to recharge Glava Zete and other springs at the foothills (no. 13).

The capital city of Podgorica obtains most of its drinking water from the Mareza source, also situated within the Skadar Lake catchment. This is a group of typical ascending springs at the 2 km long contact of Cretaceous limestone and limnoglacial sediments of the Podgoricko Polje. The springs' discharge is in the range of 2.0–10.0 m^3/s (Radulovic, 2000).

FIGURE 10.1–10 Estavelle Gornjepoljski Vir at Nikšićko Polje.

FIGURE 10.1–11 (Left) Ljuta Spring during no-flow conditions (photo courtesy of S. Milanovic); (right) Ljuta Spring discharging over 100 m^3/s.

In the littoral karst of Montenegro, the most important springs are those along the edges of the Boka Kotorska Bay (no. 14): Gurdic and Skurda Spring near Kotor, Ljuta Spring at Orahovac (Figure 10.1–11), Spila Spring at Risan, Morinj Springs and Opacica at Herceg Novi, and Plavda at Tivat. All these springs are characterized by high variations in discharge due to intensively karstified rocks in the catchment and an extremely fast propagation of rainfall. Some of those springs even dry out completely during summer (e.g., Sopot, Spila Risan), while after intensive rainfall or at the end of winter, some of them can discharge over 100 m^3/s. Therefore, despite the water abundance all along the edges of the Boka Kotorska Bay, it is a well-known fact that the population of this area suffers from an inadequate supply of freshwater, which affects the economy of this region (Radulovic et al., 2005). This is a paradox, however, if one considers the size of the catchment area of the Boka Kotorska Bay and the quantity of precipitation in it, which, according to the long-term average, is the highest in Europe—over 5000 mm per annum (Milanovic, 2005a).

The hydraulics of the Sopot Spring discharge are particularly interesting. The spring cave has been explored by speleologists and divers to a depth of 35 m and a length of about 380 m of both dry and flooded channels (Touloumdjian, 2005; Milanovic, 2005b). During the hydrological maximum, when the karst channel of the Sopot submerged spring is unable to accept all inflowing water, a huge amount of water erupts from the cave, and the groundwater level rises to over 40 m above sea level. The maximum recorded discharge of this periodic spring is over 50 m^3/s (measurements are extremely difficult due to local topography). The lower-positioned caverns continuously discharge groundwater to the sea bottom; this is a typical submarine spring. The conducted chemical analyses determined that, during the low-water period, even at the most remote inland point, some 500 m away from the sea, there is an extensive intrusion of seawater and mixture with the groundwater.

Such extremely complex hydrogeology attracts many experts to assess and evaluate technical conditions under which it would be possible to tap the discharging fresh groundwater and avoid salt intrusion. Underground dams, horizontal galleries, long wall barriers along the shoreline, big tank balloons to be filled with freshwater (and progressively emptied throughout the summer months) are just a few of the discussed alternatives. However, no such proposal is yet to be implemented.

10.1.3.6 Albania

Albania is another country very rich in springwater, originating mostly from karst aquifers.

At the northern and eastern margins of Skadar Lake are numerous springs; most are ascending and some are impounded by the lake water.

The state capital, Tirana, obtains its water in part from the source that drains the karstic aquifer of a nearby Mali me Gropa plateau. The limestones are of Upper Triassic and Lower Jurassic age (no. 15). The discharges of karst springs Selita (Figure 10.1–12) range from 0.24 to 0.86 m^3/s (Eftimi, 1971), while spring Shemria discharges at between 0.45 and 1.50 m^3/s. Both have been in use since the 1950s. The third important spring, Buvilla, issues from the Upper Cretaceous dolomite formation of the Dajti Mountain massif and flows into a human-made reservoir. Groundwater is of low mineralization and good quality (Eftimi, 1998). Although the natural conditions and unpopulated catchments are favorable in terms of the resource protection, the captured amount of water cannot meet the increasing demands of the population and industry, and many boreholes were drilled in an attempt to fill this gap.

One of the largest cities in central Albania, Gyrokastr, obtains water from karst springs. Two large springs are located 10 and 5 km from the city. The farthest, Shpella Skhoteni, is a vauclusian spring with a large siphon where pumps are installed. Touloumdjian with his team (2005) discovered 70 m deep galleries at the siphon's bottom. Syri i Zeze is also a siphonal spring, explored by divers to a depth of 83 m.

The Bistrica spring system, the largest in Albania (Eftimi, personal communication), drains the 440 km^2 Gjere Mountain karst massif located in the southeastern part of Albania near the border with Greece (Figure 10.1–13; no. 16 in Figure 10.1–1). The Gjere Mountain is a natural water divide between the Vjosa River

FIGURE 10.1–12 Mali me Gropa karstic plateau and Selita Spring (marked as the white dot). (Photo courtesy of R. Eftimi.)

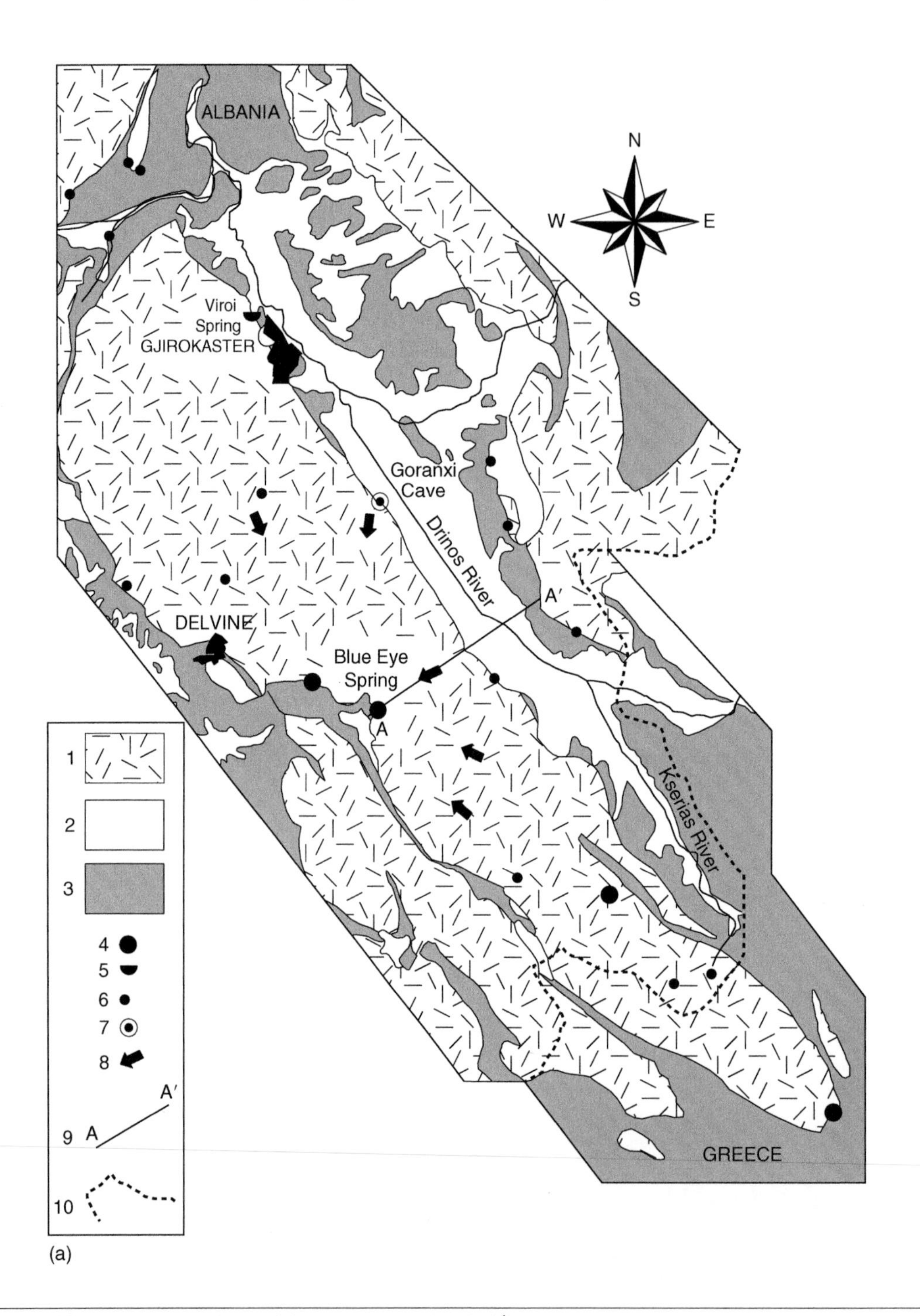

FIGURE 10.1–13 Hydrogeological map (a) and cross section A–A′

(continued)

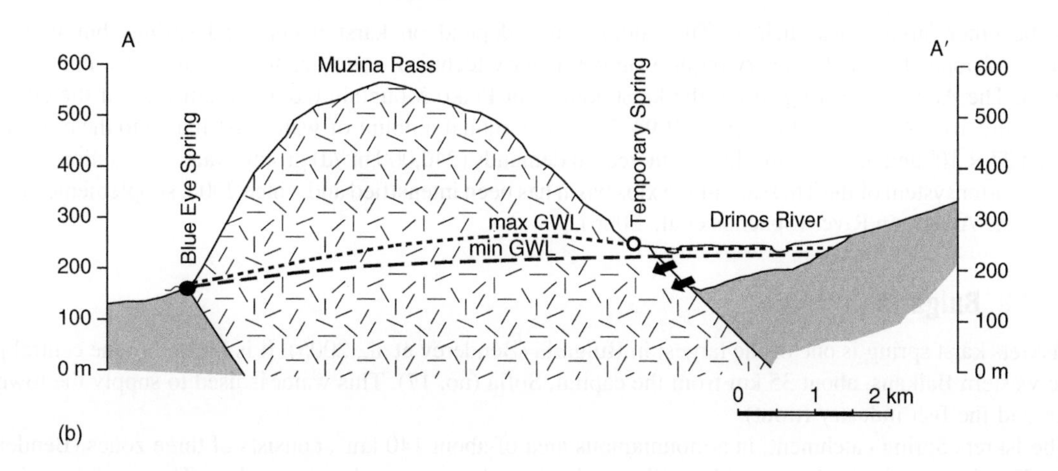

FIGURE 10.1–13, CONT'D (b) of Mali Gjere karst massif: (1) Karstic aquifer; (2) alluvium; (3) flysch, impervious rocks; (4) large spring, discharge >1 m^3/s; (5) temporary spring; (6) small spring; (7) cave with groundwater; (8) groundwater flow direction; (9) line of cross section; (10) national boundary. (Modified from Eftimi et al., 2007.)

basin located on its eastern side and the Bistrica River basin located on its western side. With its surface, the Gjere Mountain is one of the biggest and most abundant karstic basins of Albania. The total groundwater resources of this basin are calculated to be 743×10^6 m^3/a. The karst massif on the eastern side is in contact with the prolific gravel aquifer of the Drinos River valley. On the western side of the karst massif, at an elevation about 45 m lower than the Drinos River valley, Bistrica Springs discharge an average of 18.4 m^3/s. Based on isotopic investigations ($\delta^{18}O$, δD) during 1989–1990 and again in 1996, it is confirmed that about 70 percent of the groundwater resources of the massif are recharged by precipitation, while about 34 percent represent the Drinos valley groundwater seepage into the massif (Eftimi, Amataj, and Zoto, 2007; Figure 10.1–13).

The Blue Eye Spring is the largest within the Bistrica group of the springs. The Blue Eye drainage siphon has been also explored by divers, first by an Italian team in 1992 then by French divers that reached a depth of 70 m (Touloumdjian, 2005). It was extremely difficult to dive against very strong ascending flow. After several attempts, this was achieved with support of 20 kg weight belts and special ropes. The Bistrica springwater is planned for overseas supply of the Italian south Adriatic coast.

10.1.3.7 Greece

In Greece, karstic aquifers and large springs are widely distributed in the southern part, along the Ionian and Aegean coasts. The Peloponnesus peninsula and the islands are particularly rich in karst groundwater, but the problem of saline intrusions regularly disturbs drinking water supplies.

Almiros Spring near Heraklion on the island of Crete, one of the well-explored springs, is a typical example of saltwater intrusion. Groundwater drains from the karstic aquifer some 1 km inland at an elevation of 5 m asl, but the deeper part of the aquifer is under the strong influence of saltwater percolating from the estavelle identified at a depth of 25 m on the sea floor, some 500 m from the shore. The spring discharge varies from 3.3 to 30 m^3/s (Monopolis and Mastopis, 1969). The problem of salinity starts with discharge lower than about 15 m^3/s, and during the minimum flows, the concentration of the Cl ion can reach 6 g/L. The concrete dam constructed at the discharge location to increase the freshwater level is only partly mitigating the saltwater encroachment (Mijatovic, 2005).

Some other large cities such as Thessaloniki also depend on karst waters and springs but increased demands of population and industry require supplementary technical solutions to alleviate the water shortage problem. The Aravissos Springs drain the karst aquifer of Paiko Mountain (some 50 km west of the city; no. 25 in Figure 10.1–1). During the period 1994–2007, the annual amount of water distributed to the city varied between 27×10^6 and 55×10^6 m^3. Due to increased demands (240,000 m^3/d) groundwater abstraction from the porous aquifer system of the Thessaloniki-Axios basin has been intensified and, since 2003, supplemented by the water of the Aliakmon River (Spachos et al., 2006).

10.1.3.8 Bulgaria

The Iskrets karst spring is one of the largest in Bulgaria (Benderev et al., 2005). It is located in the central part of the western Balkans, about 35 km from the capital, Sofia (no. 17). This water is used to supply the town of Svoge and the fish industry (trout).

The Iskrets Spring catchment, in a mountainous area of about 140 km^2, consists of three zones (Benderev, 1989). The first one, built of exposed karstified rocks, is without any surface river flow. The second one is at a higher elevation, made of nonkarstic rocks, and has numerous small streams, which all sink at the contact with karst. The third zone encompasses areas where karstified Triassic rocks are covered by Lower-Middle Jurassic rocks.

The minimum spring yield is in the range of 0.09–0.28 m^3/s, while the maximum values vary from 3.6 to more than 15 m^3/s. The absolute maximum flow rate of the spring was measured in 1966, when it reached 56 m^3/s for a short period (Benderev, 1989). The peaks of the spring hydrographs are the result of the fast infiltration of rainfall or snowmelt at the end of winter. Tracing tests confirmed the travel time from a distance of 10 km to be regularly 1–2 days during periods of high water and 7–8 days during dry periods (Figure 10.1–14). During the high-water season, the upper cave outlet becomes active, while in dry periods, water seeps from the blocks that cover the primary discharge zone. The seasonal variation in the groundwater table is 15 m (Figure 10.1–15).

The Iskrets Spring has a very interesting discharge mechanism, which occasionally causes rare flow interruptions. It was noticed, during the 19th century, that the spring dried up several times (Benderev, personal communication). After the strong Vranchea earthquake in 1977 (which caused great damage to Bucharest), the discharge rate dropped from 5.5 to 0.5 m^3/s in 7.5 hours. After that, the discharge rate raised abruptly

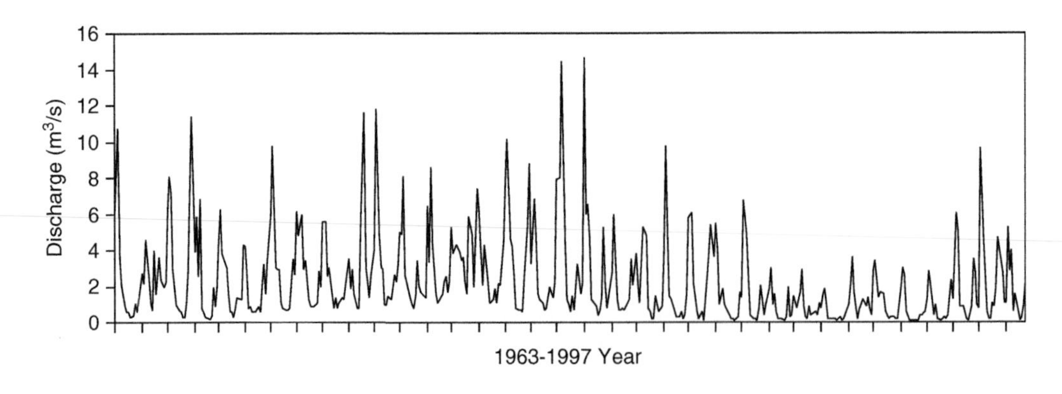

FIGURE 10.1–14 Variations of the Iskrets Spring discharge rate. (Data from the Bulgarian National Institute of Hydrology and Meteorology.)

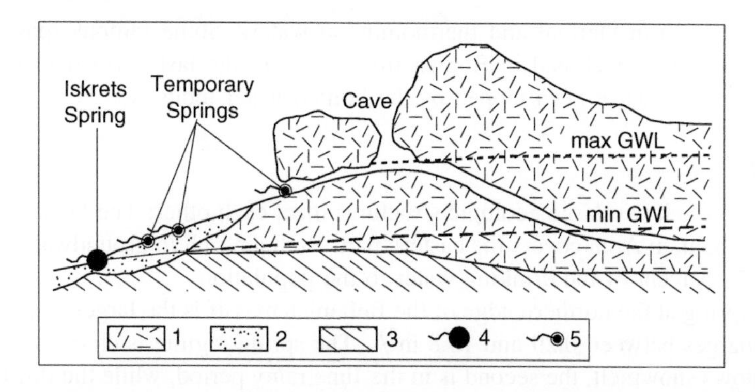

FIGURE 10.1–15 Iskrets Spring captured for water supply of Svoge: (1) Karstic aquifer, middle Triassic limestone; (2) alluvium; (3) nonkarstic rocks, Lower Triassic sandstone, and conglomerates; (4) main Iskrets Spring; (5) temporary springs.

to 13.5 m^3/s, then again began to decrease gradually. A similar event was recorded during the local Svoge earthquake in 1979, but the same thing happened on April 11, 1982, when no significant seismic activity was recorded nearby nor in the whole Balkans region. Similarly, accidental and very short (several hours) yield disruption is recorded at some other springs in the Carpathian karst, as in the case of the Serbian springs Vrelo Mlave or Jelovicko Vrelo on Stara Planina Mountain.

Glava Panega is the largest karstic spring of the Pleven region (no. 18). It emerges from Upper Jurassic limestone. The spring discharge is in the range of 0.58–35.7 m^3/s (Beron et al., 2006). One tracing test indicated relatively slow drainage from the aquifer, probably caused by the existing large underground reservoir: The tracer (uranine) traveled a linear distance of 6.5 km from the ponor for 250 hours (with a velocity of 7.2×10^{-3} m/s).

A relatively stable drainage regime characterizes the Devnenski Springs near Devnya. These springs discharge between 2.3 and 4.2 m^3/s from the Upper Jurassic and Lower Cretaceous karstic aquifer.

10.1.3.9 Romania

The very complex geology of Romania results in a variety of springs draining different types of aquifers and geological units. Despite limited areas underlain by karstified rocks (some 4 percent of the territory), the springs from karstic aquifers have the largest discharges. Oraseanu (1993) classified four hydrogeologic types of karst: Carpathian Orogen karst, North Dobrogea karst, platform karst, and Carpathians posttectonic cover karst. The first one is considered the most developed and possesses the largest water reserves. Iurkiewicz and Oraseanu (1995) describe the Banat, Padurea Craiului, Valcan, Bihor, and Codru Moma Mountains as the main karstic systems, often used for water supply of major Romanian cities.

The largest karst springs in the Carpathian Orogen have a median annual discharge of 1–2 m^3/s. The catchment area of the important Barza spring in the Mehedinti Mountains and the sources from Cheile de Jos ale Dambovitei in Piatra Craiului Mountains consist primarily of carbonate rocks, while the Izvarna Spring from the Valcan Mountains (see Chapter 10.4) and the Cerna Spring from the Cerna-Jiu trench, relatively large areas in their upper catchments, are nonkarstic terrains. The Cerna is the largest karstic source of Romania (no. 19), with an average discharge of 2 m^3/s (min/max = 0.5/10 m^3/s) from the drainage area of 85 km^2 (Iurkiewicz and Oraseanu, 1997).

Romania is also very rich in thermal and thermomineral waters. Some famous spas, such as Herculane Springs (see Chapter 9), were developed in Roman times. During the last two decades, numerous mineral and oligomineral springwaters have been tapped by the bottled water industry.

10.1.3.10 Serbia

Serbia is the country in which two large geologic structures meet each other: The Dinaric Belt is in the west, while the Carpathian Mountain Arch covers the eastern part. Both are rich in groundwater and karstic springs, which provide about 25 percent of the drinking water to the population.

The Vrelo Mlave Spring at the northern edge of the Beljanica massif is the largest spring of the Carpathian karst (no. 20). It discharges between 0.29 and 14.8 m^3/s. The spring regime is characterized by three annual maxima: The first follows snowmelt, the second is in the June rainy period, while the third is at the beginning of winter. The recession period is usually no longer than 60 days during late summer. The correlation of rainfall and discharge peaks shows an average response time of about five days (Stevanovic, 1994). One hundred years after the first bathymetrical exploration done by Jovan Cvijic (Stevanovic and Mijatovic, 2005), this vauclusian spring and its funnel-shaped siphon were explored again, this time by a specialized diving team (Milanovic, 2005b). After the knee passage, at a depth of about 30 m, the deeper vertical part that divers explored to a depth of 73 m expands to a large chamber (Figure 10.1–16). This exploration reaffirmed an old idea of utilizing Mlava Spring and nearby Krupaja Spring for the water supply of the capital city of Belgrade, some 100 km away (Stevanovic, 1997).

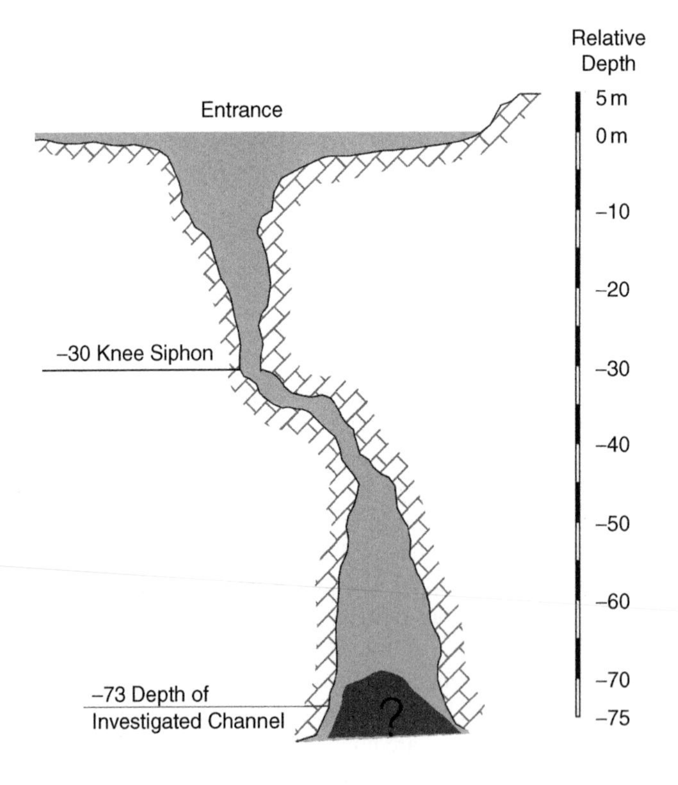

FIGURE 10.1–16 The Mlava Spring siphon. (Courtesy of S. Milanovic.)

The largest karstic spring in western Serbia is Vrelo Perucac Spring (no. 21). It drains Tara Mountain, the large karstic massif built of Triassic limestone (65 km^2). It is a gravitational spring, discharging between 1.3 and 6.2 m^3/s (Kresic, 1984). In wintertime, the Vrelo Perucac discharges a large amount of unutilized water, which flows to the Drina River, while during dry periods, the minimum discharge is insufficient for the water supply of a nearby town. In an attempt to equalize the yearly inbalance, the construction of an underground dam for groundwater storage has been proposed (Kresic, 1991; Milanovic, 2005c).

The biggest karstic spring in the southern part of the Metohija basin is Vrelo Beli Drim (Radavac). It drains the karstic massifs of Zljeb and Maja Rusolija Mountains (no. 22), with a surface area of at least 100 km^2. The discharge from the Triassic limestone occurs at the contact with the impermeable Jurassic diabase-chert formation. The minimum discharge is around 1 m^3/s, the average flow some 4 m^3/s, while the registered maximum is over 21 m^3/s. Similarly, along the contact of the Triassic karstic aquifer with Jurassic rocks, are two other big karstic springs in the Metohija basin. The Vrelo Spring discharges between 0.27 and 1.6 m^3/s, and the Istok Spring between 2.4 and 4.0 m^3/s. Both have been evaluated and proposed for regulation utilizing underground dams (Peric, Simic, and Milivojevic, 1980).

In Serbia, more than 200 mineral water sources are registered, and most of them have been explored to a certain extent. The hottest natural spring has a temperature of 78°C (Josanicka spa).

10.1.3.11 Former Yugoslav Republic of Macedonia

The former Yugoslav Republic of Macedonia possesses a large diversity of springs in terms of geological origin and chemical and physical composition. Many thermal springs are connected with Quaternary or Tertiary volcanic activity, but with respect to yield, the largest are still those draining karst aquifers. St. Naum Spring at the Ohrid lakeshore near the border with Albania drains a Triassic limestone aquifer. It has an average discharge of 5.5 m^3/s. Eftimi and Zoto (1997) confirmed that most water to the St. Naum Spring (no. 23) and the neighboring Tushemishit Spring on the Albanian side (average discharge 2.5 m^3/s) comes from the topographically higher Prespa Lake and Zaver swallow hole on the western margin of the lake.

The Paleozoic marble is very rich in groundwater, and this karstic-fissured type of aquifer ensures the water supply to several big towns in the western part of the country (Gostivar, Kicevo). The capital, Skopje, also obtains most of its drinking water from that aquifer. Rasce spring is situated on the Vardar River bank upstream of the city (no. 24). The spring appears from the Zeden massif, consisting of marble, but the tracing tests and isotopic analyses confirmed that only 30 percent of the water comes from those rocks. The majority of Rasce water is from a porous aquifer in the topographically higher Polosko Polje and the percolated Vardar River water lost in the upstream river sections (Kekic, 1982). The discharge regime is relatively stable; the mean flow is 2.5 m^3/s.

10.1.4 SPRINGWATER AND SOME ENVIRONMENTAL PROBLEMS

The karst is particularly vulnerable to pollution due to openings of karstic channels and fast groundwater circulation. Many projects have been executed aiming to reduce the negative anthropogenic impact. However, much untreated wastewater still discharges directly into aquifers or uncultivated solid waste disposals. Fast-growing tourist areas along the Montenegro and Albania coast are particularly in danger, and only during the last few years has a systematic action to construct a modern regional municipal landfill started.

The problems of fast urbanization and uncontrolled pollution of groundwater discharging through large karst springs in the vicinity of Rijeka (Croatia) in the late 1970s have been studied in detail by Biondic and Goatti (1984b). Based on their proposal, a solution with four sanitary protection zones has been implemented to mitigate negative consequences.

Similarly, the four protection zones in the recharge area of the Rizana Spring were defined based on detailed geological and hydrogeological research, as well as three combined tracer tests (Krivic et al., 1989). There are different sources of pollution within the recharge area of the Rizana Spring. The railway passes over the first protection zone in the immediate vicinity of the spring. Some other local, regional, and main roads and trails pass the second protection zone, and an eventual traffic accident could cause a catastrophic contamination of the spring.

Karst aquifers are highly vulnerable to human activities that result in contamination problems. For example, the main sources of pollution of the Ljubljanica springs are untreated wastewater, landfills, roads, a military training area, industry, and agriculture. The temporary appearance of organic pollution and heavy metals in spring sediments is also registered. A specific problem is the poor water quality of the sinking streams, which has a direct impact on karst aquifers.

Bensi et al. (2005) stated that the water supply of the Trieste region from the Timavo Springs is compromised by anthropogenic and industrial development and today represents only an emergency water resource (an alluvial aquifer is used as the main source of water supply).

ACKNOWLEDGMENTS

I gratefully acknowledge the data and information provided by my colleagues and friends from the region: Romeo Eftimi, Aleksey Benderev, Metka Petric, Ognjen Bonacci, Thomas Spachos, Adrian Iurkiewicz, Vesna Denic-Jukic, Borivoje Mijatovic, and Petar and Sasa Milanovic.

REFERENCES

Benderev, A., 1989. Karst and karst waters of Ponor Mountain. Ph.D. thesis, Scientific Institute for Mineral Researches, Sofia, Bulgaria.

Benderev, A., Spasov, V., Shanov, S., Mihaylova, B., 2005. Hydrogeological karst features of the Western Balkan (Bulgaria) and the anthropological impact. In: Stevanovic, Z., Milanovic, P. (Eds.), Water Resources and Environmental Problems in Karst—CVIJIĆ 2005. Spec. ed. FMG, Belgrade, Serbia, pp. 37–42.

Bensi, S., Casagrande, G., Cucchi, F., Zini, L., 2005. Vulnerability of karst aquifers related to the construction of a railway tunnel—Applied case study in the karst of Trieste (NE Italy). In: Stevanovic, Z., Milanovic, P. (Eds.), Water Resources and Environmental Problems in Karst—CVIJIĆ 2005. Spec. ed. FMG, Belgrade, Serbia, pp. 691–695.

Beron, P., Daaliev, T., Jalov, A., 2006. Caves and speleology in Bulgaria. Pensoft. Publ. Bulg. Fed. of Spel. and Nat. Mus. of Nat. Hist., Sofia, Bulgaria.

Biondic, B., Goatti, V., 1984a. La galerie souterraine "Zvir II" a Rijeka (Yougoslavie). In: Burger, A., Dubertret, L. (Eds.), Hydrogeology of karstic terrains: Case histories, Intern. Contrib. to Hydrogeology, vol. 1. IAH, Verlag Heinz Heise, Hannover, pp. 150–151.

Biondic, B., Goatti, V., 1984b. Sanitary protection zones of fresh water springs in the community of Rijeka. Proceedings of VIII Yugoslav. Symp. of Hydrogeol. and eng. geol, vol. 1, Budva, pp. 281–289.

Bonacci, O., 1987. Karst hydrology. Springer Verlag, Berlin.

Bonacci, O., 1993. Karst spring hydrographs as indicators of karst aquifers. J. Hydrol. Sci. 38 (1–2), 51–62.

Boni, C., Bono, P., 1984. Essai de bilan hidrogéologique dans une région karstique de l'Italie centrale. In: Burger, A., Dubertret, L. (Eds.), Hydrogeology of karstic terrains: Case histories. Intern. Contrib. to Hydrogeology, vol. 1. IAH, Verlag Heinz Heise, Hannover, pp. 27–31.

Bono, P., Boni, C., 1996. Water supply of Rome in antiquity and today. Environ. Geol. 27 (2), 126–134.

Cvijic, J., 1895 [2005]. Karst, a geographic monograph [Serbian version of "Das Karstphänomen"]. Reprinted In: Stevanovic, Z., Mijatovic, B. (Eds.), Cvijic and karst/Cvijic et karst. Spec. ed. of Serb. Acad. Sci. and Arts, Belgrade, Serbia, pp. 57–146.

Denic-Jukic, V., Jukic, D., 2003. Composite transfer functions for karst aquifers. J. Hydrol. 274, 80–94.

Eftimi, R., 1971. Discharge regime of Selita Spring [in Albanian]. Permbledhje Studimesh 2, 65–76.

Eftimi, R., 1998. Some data about the hydrochemistry of karst water of Dajti Mountain [in Albanian]. Studime Gjeografike 11, 60–65.

Eftimi, R., Amataj, S., Zoto, J., 2007. Groundwater circulation in two transboundary carbonate aquifers of Albania; their vulnerability and protection. IAH Selected Papers on Hydrogeology. In: Witkowski, A., Kowalczyk, A., Vrba, J. (Eds.), Groundwater Vulnerability Assessment and Mapping, Ustron, vol. 11. Taylor & Francis Group, London, pp. 199–212.

Eftimi, R., Zoto, J., 1997. Isotope study of the connection of Ohrid and Prespa lakes. Proceedings of Intern. Symp, Towards Integrated Conservation and Sustainable Development of Transboundary Macro and Micro Prespa Lakes, Korçë, Albania, pp. 32–37.

Elektroprojekt, Zagreb, 1999. Design proposal of Ombla underground dam. Techn. Doc, Zagreb, Croatia (unpublished).

Gospodaric, R., Habic, P., 1976. Underground water tracing. Proceedings of the Third Intern. Symp. of Underground Water Tracing (3. SUWT), Ljubljana, Slovenia.

Habic, P., 1980. Karst springs of Vipava [in Slovenian]. Guide exc. Eighth Yugoslav. Symp. of hydrogeol. and eng. geol, Portoroz, pp. 9–13.

Iurkiewicz, A., Oraseanu, I., 1997. Karstic terrains and major karstic systems in Romania. In: Gunay, G., Johnson, I. (Eds.), Karst waters and environmental, impacts. Balkema, Rotterdam, The Netherlands, pp. 471–478.

Kekic, A., 1982. About groundwater in karstic terrains [in Serbian]. Proceedings of the Seventh Yugoslav. Symp. of hydrogeol. and eng. geol., vol. 1, Novi Sad, pp. 123–133.

Kolbezen, M., Pristov, J., 1998. Surface streams and water balance of Slovenia. MOP: Hidrometeorološki zavod RS. Ljubljana, Slovenia.

Kranjc, A. (Ed.), 1997. Karst hydrogeological investigations in South-Western Slovenia. Acta Carsologica, 26/1, Ljubljana, Slovenia.

Kresic, N., 1984. Hydrogeology of karst terrains of Drina catchment upstream of Bajina Basta, Serbia. MS thesis, Fac, Min. Geol. Univ. of Belgrade, Belgrade, Serbia.

Kresic, N., 1991. Quantitative Karst Hydrogeology with Elements of Groundwater Protection [in Serbian]. Naucna Knjiga, Belgrade, Serbia.

Krivic, P., Bricelj, M., Zupan, M., 1989. Podzemne vodne zveze na področju Čičarije in osrednjega dela Istre (Slovenija, Hrvatska, NW Jugoslavija). Acta Carsologica 18, Ljubljana, Slovenia, pp. 265–295.

Mijatovic, B., 1968. A method of studying the hydrodynamic regime of karst aquifers by analysis of the discharge curve and level fluctuations during recession. Bull of Inst. For Geol. and Geophys. Res. Ser., B. Belgrade, Serbia, pp. 43–81.

Mijatovic, B., 2005. The groundwater discharge in Mediterranean karst coastal zones and freshwater tapping: Set problems and adopted solutions—Case studies. In: Stevanovic, Z., Milanovic, P. (Eds.), Water Resources and Environmental Problems in Karst—CVIJIĆ 2005. Spec. ed, FMG, Belgrade, Serbia, pp. 259–266.

Milanovic, P., 1979. Hidrogeologija karsta i metode istrazivanja [Karst hydrogeology and research methods]. HE Trebisnjice, Trebinje, Yugoslavia.

Milanovic, P., 1996. Ombla spring, Croatia. Environ. Geol. 27 (2), 105–107.

Milanovic, P., 2005a. Water potential in southeastern Dinarides. In: Stevanovic, Z., Milanovic, P. (Eds.), Water Resources and Environmental Problems in Karst—CVIJIĆ 2005. Spec. ed., FMG, Belgrade, Serbia, pp. 249–257.

Milanovic, P., 2006. Karst of eastern Herzegovina and Dubrovnik littoral. ASOS, Belgrade, Serbia.

Milanovic, S., 2005b. Hydrogeological characteristics of some deep siphonal springs in Serbia and Montenegro karst. In: Stevanovic, P., Milanovic, P. (Eds.), Water Resources and Environmental Problems in Karst—CVIJIĆ 2005. Spec. ed., FMG, Belgrade, Serbia, pp. 451–458.

Milanovic, S., 2005c. Perucac underground spring and storage. In: Stevanovic, Z., Milanovic, P. (Eds.), Water Resources and Environmental Problems in Karst—CVIJIĆ 2005. Spec. ed. FMG, Belgrade, Serbia, pp. 16–20.

Monopolis, D., Mastopis, K., 1969. Hydrogeological investigation of Almiros spring. Report IGME, Athens (unpublished).

Oraseanu, I., 1993. Hydrogeological regional classification of the Romanian karst. In: Theoretical and applied karstology, vol. 6. Acad. Rom., Bucharest, Romania, pp. 175–180.

Petric, M., 2002. Characteristics of recharge-discharge relations in karst aquifers. Založba ZRC, Ljubljana, Slovenia.

Peric, J., Simic, M., Milivojevic, M., 1980. An idea to storing part of Beli Drim water within underground reservoirs for water supply and irrigation of Metohija (in Serbian). In: Transactions of FMG, vol. 22. Belgrade, Serbia, pp. 251–292.

Radulovic, M., 2000. Karst hydrogeology of Montenegro. Sep. issue of Geological Bulletin, vol. 18. Spec. ed. Geol. Survey of Montenegro, Podgorica, Montenegro.

Radulovic, M., Radulovic, V., Stevanovic, Z., Komatina, M., Dubljevic, V., 2005. General geology and hydrogeology of Dinaric karst, in Montenegro and Serbia. In: Milanovic, P., Stevanovic, Z., Radulovic, M. (Eds.), Water resources and environmental problems in karst, Excursion Guide of IAH Intern. Conf. KARST 2005. Belgrade-Kotor, Belgrade, Serbia, pp. 16–20.

Spachos, T., Voudouris, K., Drosos, D., Dimopoulos, G., Soulios, G., 2006. Groundwater levels fluctuation in aquifer systems of borehole fields of Thessaloniki waterworks. Proceedings of 10th Pan-Hellenic Sci. Cong. Xanthi, on CD session 13.4.

Stevanović, Z., 1994. Karst ground waters of Carpatho-Balkanides in Eastern Serbia. In: Stevanović, Z., Filipović, B. (Eds.), Ground waters in carbonate rocks of the Carpathian-Balkan mountain range. Spec. ed. of CBGA, Allston, Jersey, UK, pp. 203–237.

Stevanovic, Z., 1997. Resources and potential of karst ground waters in Serbia. Rec. de rapp. du Comité pour le Karst et spéléologie Acad, Serbe des Sciences et des Arts VI, vol. 72, Belgrade, Serbia, pp. 55–73.

Stevanovic, Z., Mijatovic, B., 2005. Cvijic and karst/Cvijic et karst. Spec. ed. Board on karst and spel. Serb. Acad. of Sci. and Arts, Belgrade, Serbia.

Touloumdjian, C., 2005. The springs of Montenegro and Dinaric karst. In: Stevanovic, Z., Milanovic, P. (Eds.), Water Resources and Environmental Problems in Karst—CVIJIĆ 2005. Spec. ed. FMG, Belgrade, Serbia, pp. 443–450.

CHAPTER 10.2

Case Study: Kläffer Spring—the major spring of the Vienna water supply (Austria)

Lukas Plan[1] Gerhard Kuschnig[2] and Hermann Stadler[3]

[1]Formerly University of Vienna, Department for Geodynamics and Sedimentology; now at Natural History Museum Vienna, Department for Karst and Caves, Vienna, Austria.
[2]Vienna Waterworks, Vienna, Austria.
[3]Joanneum Research, Institute of Water Resources Management, Hydrogeology and Geophysics, Graz, Austria.

10.2.1 INTRODUCTION

For the country of Austria, water from karstic catchment areas is a very important drinking water resource. About 22 percent (Kralik, 2001) of the nation's surface consists of karstic rocks (Figure 10.2–1). Most of the large cities like Vienna, Salzburg, Graz, Linz, and Innsbruck are significantly supplied by karst waters. In total, about 4 million inhabitants or about 50 percent of the population is served from areas with karstic carbonate rocks (Kralik, 2001).

We present the biggest spring of the eastern Alps, the so-called Kläffer Spring, or Kläfferquelle in German. It is also the most important spring that supplies the city of Vienna, where a water supply from developed karst springs dates back to the 19th century, when the venturesome plans of the geologist Eduard Suess to construct a 120 km long pipeline were accepted by the city council. Originally called the Kaiser Franz Josefs Water Main, it was completed in 1873 and captured major springs in the Rax and Schneeberg area south of Vienna. The growing demand for water led to the construction of the 200 km long Second Vienna Water Main, finished in 1910, which develops the Kläffer Spring and smaller springs in the Hochschwab area southwest of Vienna.

Today, Vienna is the classical example of a metropolis with 1.7 million inhabitants that is almost entirely supplied with high-quality freshwater from karstic catchment areas. In 2007, a total of 142.07 million m^3 of water was supplied to Vienna, of which 96.6 percent originated from developed karst springs (Rumpold, 2008). These springs are located at the foot of four major karst massifs in the eastern part of the northern Calcareous Alps (Figure 10.2–1). About 57 percent of the water derives from the Hochschwab massif, where the Kläffer Spring is the biggest and most important one.

Detailed research studies on the springs and their catchment areas have been carried out within the framework of the European Union projects KATER I and II, which were initiated by the Vienna Waterworks to assure quality and reliance of long-term water supply (Kuschnig, 2006).

Copyright © 2010, Elsevier Inc. All rights reserved.

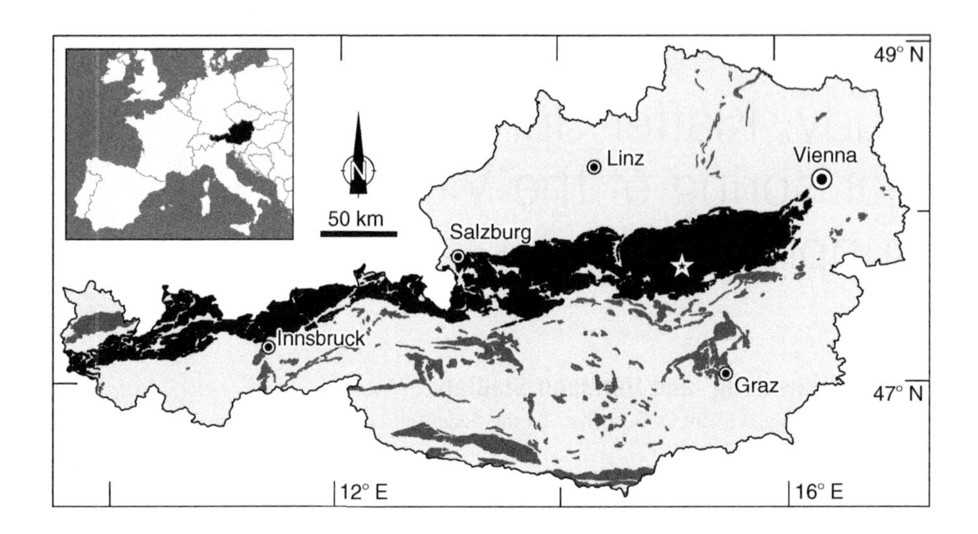

FIGURE 10.2–1 Distribution of karst rocks in Austria (after Schubert, 2003). The northern Calcareous Alps (NCA) are shown in black; other units with karstic rocks in dark gray. Locations of major cities that are significantly supplied with water from karstic catchment areas are labeled. The star indicates the position of the Kläffer Spring.

This chapter focuses on the geological and morphological settings of the catchment area and the complex arrangement of the Kläffer Spring itself, which actually consists of several outlets. Further, quantitative and qualitative hydrological parameters are given and interpreted and measurement techniques are explained. The historic development of the Viennese Water Main and actual data on the consumption and distribution are presented. The last section describes water quality management and protection strategies.

10.2.2 GEOGRAPHICAL AND GEOLOGICAL SETTING

10.2.2.1 Topography, climate, soils, and vegetation

The Hochschwab, one of the large karst plateaus of the northern Calcareous Alps (Figure 10.2–1), is located in the north of the Austrian province of Styria (Figure 10.2–2). The whole mountain range covers an area of 650 km^2 (Stummer and Plan, 2002), whereas some 85 km^2 belong to its main plateau system, which significantly drains to the Kläffer Spring. Several other, smaller plateaus are separated from it by valleys of mainly glacial origin. The plateau area ranges from about 1400 m elevation up to the Hochschwab summit at 2277 m (Figure 10.2–3). The whole massif is encircled by rather deeply incised valleys with floors at 500 to 700 m.

The alpine climate is characterized by low temperatures with high amounts of precipitation. In winter, snow piles up some meters. The mean temperature at Wildalpen (610 m; Figure 10.2–4) is 5.9°C (Steiermärkirsche Landesregierung, 2008). At the Schiestl Refuge (2153 m), close to the summit, the mean temperature is –0.2°C. In a north-south section, the precipitation ranges about 1463 mm/a (Wildalpen), 2200 mm/a (Schiestl Refuge; Wakonigg, 1980), and 1213 mm/a at St. Ilgen (880 m), which lies south of the plateau.

Up to an elevation of approximately 1600 m, vegetation is dominated by forests with spruce and beech trees in the lower parts and larches in the higher parts. Above the timberline, mugo pines (*Pinus mugo*) occur up to about 1900 m. Grassy vegetation persists up to the summits, where soil coverage has not been washed out due to karstification or erosion. The plateau is uninhabited, with some gravel roads for forestry in the

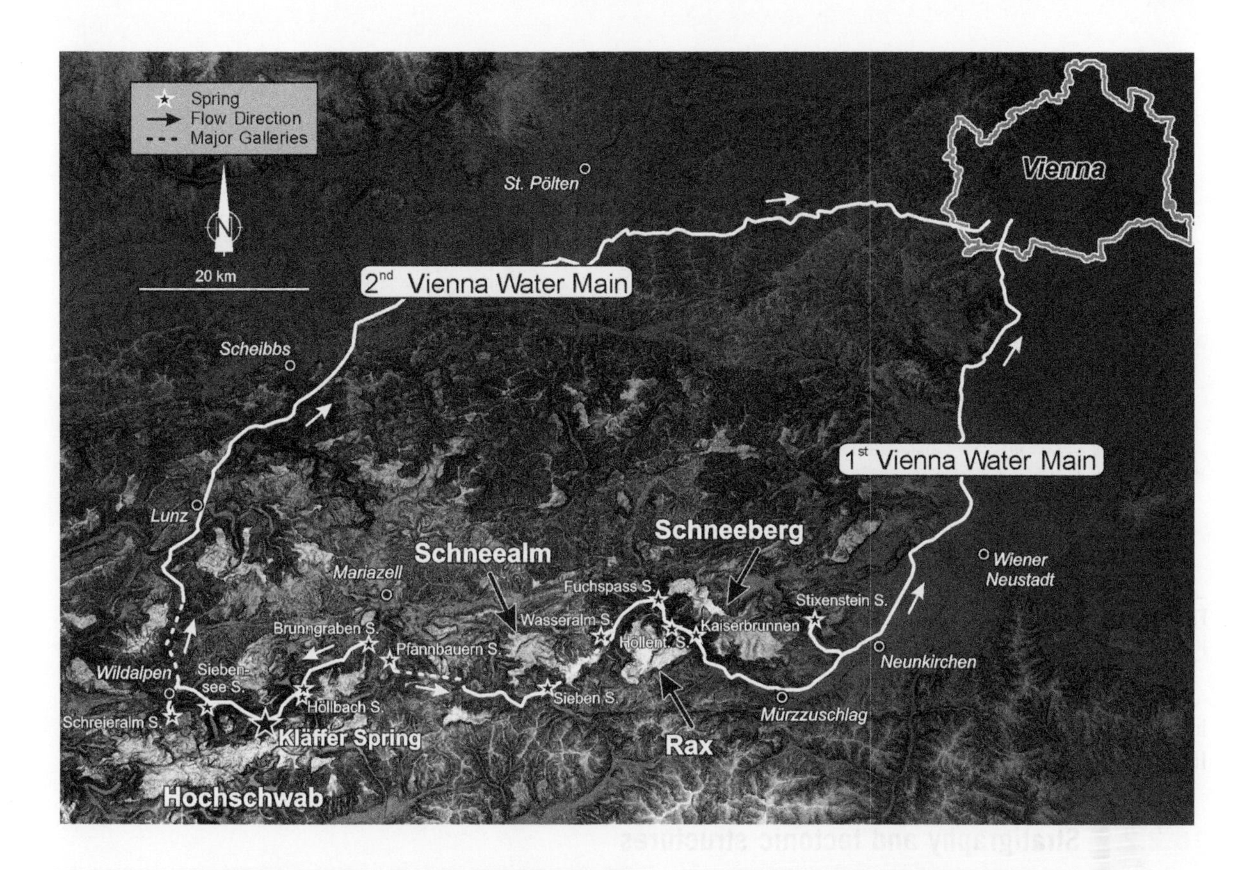

FIGURE 10.2–2 The catchment areas and springs of the First and Second Vienna Water Mains.

FIGURE 10.2–3 View from the Hochschwab summit (2277 m) over the Aflenzer Staritzen, which are part of the plateau and catchment area.

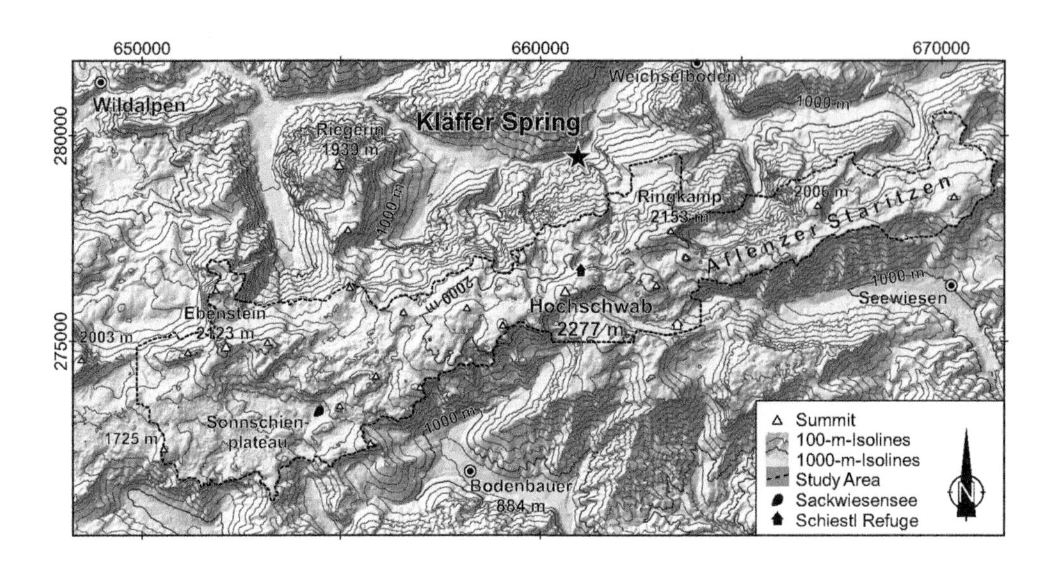

FIGURE 10.2–4 Shaded digital elevation model showing the Hochschwab karst massif and the position of the Kläffer Spring.

lower parts and some tourist facilities like hiking trails and mountain huts. During the summer, there is limited pasture (some hundred cattle) on the lower plateau areas.

10.2.2.2 Stratigraphy and tectonic structures

The Hochschwab is part of the northern Calcareous Alps (Figure 10.2–1), with the Kläffer Spring and its catchment area located in the Juvavic Mürzalpen Nappe. The stratigraphic sequence (Mandl et al., 2002) comprises mainly up to about 2000 m thick Permian to Upper Triassic sediments with Middle and Upper Triassic carbonates forming a huge karst aquifer (Figure 10.2–5).

The sandstone and shale of the Lower Triassic Werfen Formation underlying these calcareous rocks form the only significant aquiclude and nonkarstic rocks in the area. They are exposed along the southern and eastern slopes of the massif, along faults on the northern side, and in some parts of the plateau, where they were elevated along major faults.

Middle Triassic carbonates include up to 1500 m platform limestone and dolostones of the Steinalm and Wetterstein Formations in lagoonal and reef facies. The Kläffer Spring emerges from well-bedded intraplatform basin limestones, which are referred to as *Sonderentwicklung*. To the south, the platform carbonates pass into fore reef, slope, and basinal limestones (Wetterstein Formation, slope; Grafensteig Formation; Sonnschien Formation) with about 600 m total stratigraphic thickness.

The Upper Triassic is dominated by up to several hundred meters thick platform limestone (Dachstein Formation), which covers only the elevated part of the plateau southwest and southeast of the summit. Sediments of the Cretaceous to Paleocene Gosau group locally overlie the Triassic units with a major unconformity but have no significance in the catchment area of the Kläffer Spring.

In the Oligocene and Early Miocene periods, up to 1 km of crystalline sediments of the Augenstein Formation were deposited over an Eocene hilly paleo-karst landscape, which is referred to as *Dachstein Paleosurface* (Frisch et al., 2000, 2002). They were removed in the Middle Miocene period and only

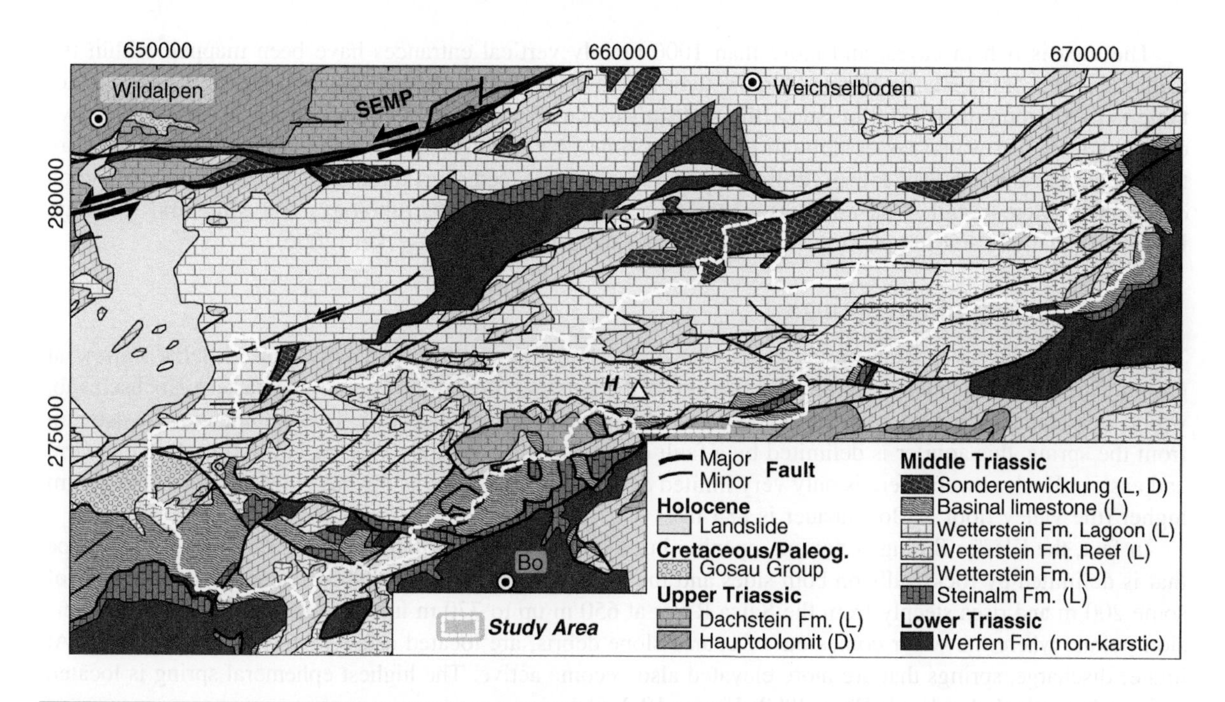

FIGURE 10.2–5 Simplified geological map of the Hochschwab. (Modified after Mandl et al., 2002.)

redeposited crystalline pebbles ("Augenstein") are found today, but weathering products remain as widespread paleo soils. Quaternary sediments have only local significance on the plateau.

The Hochschwab was created by a complex polyphase tectonic evolution that comprises permeable (leaking) faults as well as faults with low-permeable fault rocks (fault seals), such as dolomite cataclasites. With this background, tectonic analyses in the Hochschwab massif address the identification and characterization of structures relevant for hydrogeological and karstification processes. Analyses include regional fault mapping, the establishment of a deformation model, and descriptions of the kinematic groups of faults. For details, see Decker and Reiter (2001); Decker, Plan, and Reiter (2006); and Plan and Decker (2006).

10.2.2.3 Karst features

The entire plateau and most of its slopes are well karstified and show subsurface drainage. Among surface karst features, dolines are the most frequent, and karst morphological field mapping within the catchment area of the spring located some 7000 features. The distribution is inhomogeneous, mainly due to glacial erosion. Very high densities of small dolines are found in formerly glaciated areas and few, but big, features (up to 500 m diameter) occur on paleo surfaces that were not significantly reshaped by glaciers. Doline morphology is mostly funnel, bowl, or pit shaped; and most features are observed to be of solutional origin. Several polje-like features are found. On the one hand, they can be classified as structural poljes, as they developed above the aquiclude Werfen Formation, which is uplifted in relation to the surrounding Triassic carbonates at major tectonic faults. On the other hand, these features are polygenetic, as glacial erosion also played a significant role in their development. They show a surface drainage system infiltrating into ponors. Polygenetic glaciokarstic depressions are found in areas with massive glacial erosion and reach diameters of more than 500 m and depths of 60 m. All closed depressions make up more than 11 percent of the plateau catchment area.

The area is rich in caves, and more than 1000 mainly vertical entrances have been mapped within the catchment area of the Kläffer Spring. A more than 700 m deep vadose canyon shaft shows that the karst water table is partly more than 1000 m below the plateau surface (Plan, 2004). Observations in these water active caves indicate that water passes through the vadose zone very rapidly. Springs, streams, and ponors are rare throughout the karst plateau except for the poljes. Seepages that feed short minor streams are mostly found on dolostones. More detailed descriptions concerning karst morphology, lithology, and structural geology, including detailed karst morphological maps, are presented in Plan and Decker (2006).

10.2.2.4 Setting of the springs

The Kläffer Spring is situated on the north side of the Hochschwab, in the deeply incised Salza Valley, at 650 m. The position is almost at the lowest point of the karst aquifer of the central and eastern Hochschwab, which is a characteristic of a mature karst system (Ford and Williams, 2007). Two kilometers downstream from the spring, this aquifer is delimited by a fault contact to the aquiclude Werfen Formation. In the central and eastern Hochschwab, there is only very limited drainage to the south, as the base level is more than 200 m higher (the valley floor at Bodenbauer is 884 m).

In fact, the Kläffer Spring is not one opening but a couple of bigger and smaller springs on a steep slope that is delimited by rock walls on both sides and the upper end. This circle-shaped feature has a diameter of some 200 m and rises steeply from the Salza River at 650 m up to 770 m in the south. The permanent emersion points, where the water comes out of coarse slope debris, are located a few meters above the Salza. At higher discharge, springs that are more elevated also become active. The highest ephemeral spring is located 105 m above the Salza level (Plan, 2002; Figure 10.2–6).

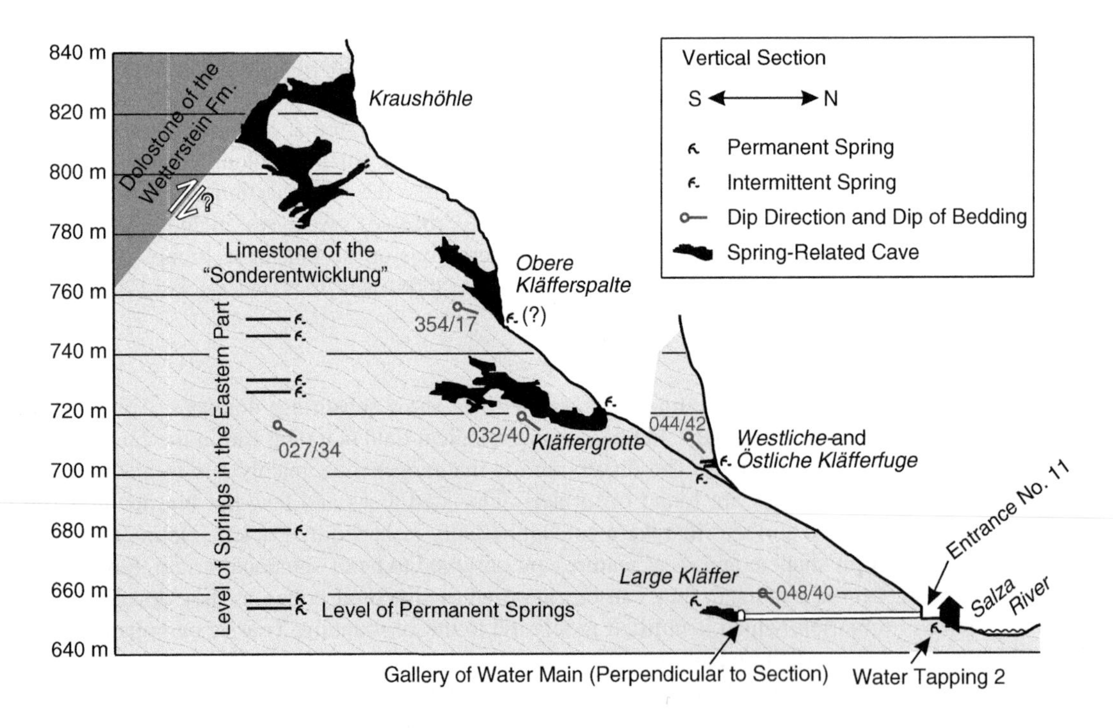

FIGURE 10.2–6 Vertical geological section of the Kläffer Springs and related caves. (Modified after Plan, 2002.)

To develop the spring, a tunnel was driven underneath the surface openings and two water-bearing cavities were intersected, which directly drain into the tunnel of the water main (Drennig, 1988). However, hydrochemical and isotopic data reveal that all these springs are fed by one karst water body.

Three of the intermittent springs are accessible caves, whereas the Kläffergrotte is the biggest one, having a length of 125 m. Higher on the slope, two additional caves exhibit former outlets of the spring system (Plan, 2002). The partly vertical Kraushöhle is the most elevated at 817 m, having a surveyed length of 196 m. Eroded speleothems and conglomerates reveal at least one period where the cave was reactivated due to Pleistocene changes within the area. All caves are terminated by massif boulder chocks, but a major cave system has to be expected behind these springs.

Detailed geological observations at the surface, in the caves, and in the artificial gallery reveal that the Kläffer Spring developed along a system of north-south trending strike slip faults that cut through older north northeast-south southwest trending ones. The lithology is the so-called Sonderentwicklung, where the slightly folded north-northeast dipping bedding also influenced conduit development. To the south, a major south-dipping reverse fault forms the contact to dolostones of the Wetterstein Formation in the hanging wall (Figure 10.2–6).

10.2.3 HYDROGEOLOGY OF THE SPRING

The Kläffer Spring is a typical limestone karst spring with a dynamic outflow. Since exact measurements started in April 1995, the highest discharge was reached on August 7, 2006, with more than 45 m^3/s. The highest daily mean discharge was over 34 m^3/s, also during a summer thunderstorm event. The lowest discharge was measured on December 20, 1999, with 389 L/s; the lowest daily mean discharge was 444 L/s. The mean discharge in this period (1995–2006) was calculated by means of 15 minute measurements with 5146 L/s. The general hydrographical situation (Figure 10.2–7) shows that the snowmelt processes drive

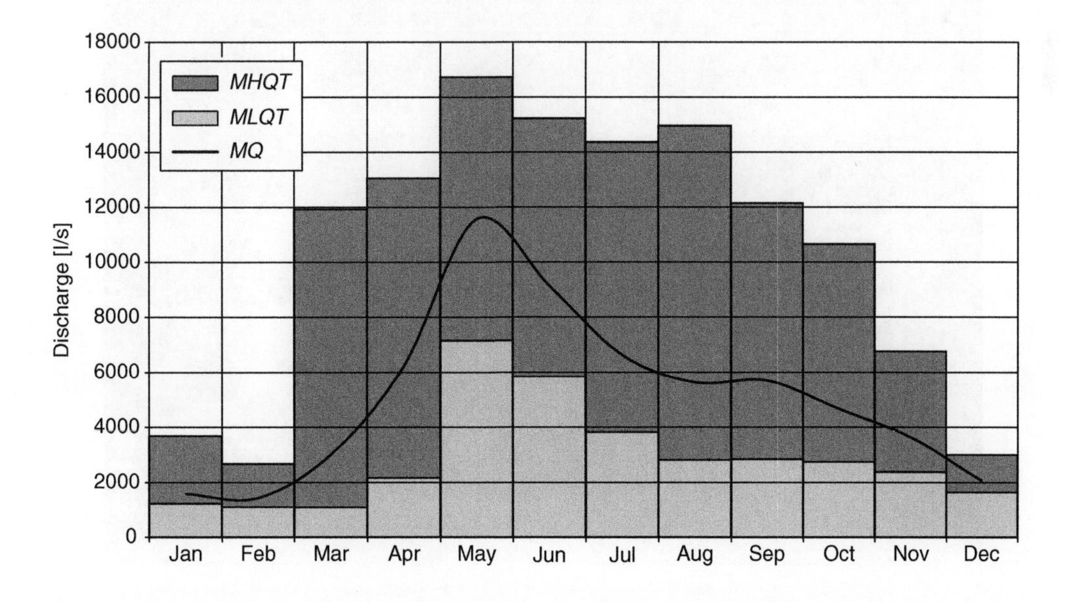

FIGURE 10.2–7 Time series of mean discharge of the Kläffer Spring from 1995 to 2006: MHQT, daily mean of highest discharge per month; MNQT, daily mean of lowest discharge per month; MQ, mean discharge.

the whole seasonal dynamic of the springs. The retardation of the decline of the discharge in August and September is caused by heavy rain events during thunderstorms.

The seasonal course of conductivity and temperature is also strongly influenced by snowmelt processes, which reach their culmination at the end of May or beginning of June. During this time, conductivity declines to 145 μS/cm (at 25°C), while the temperature reaches a minimum of 4.76°C. The highest values of conductivity are reached before the snow starts to melt, with about 215 μS/cm and for temperatures of about 5.73°C. The annual mean values are calculated with 193 μS/cm and 5.30°C.

Using discharge and electric conductivity measured every quarter hour, a mass balance was calculated for the catchment of the spring. It indicates a loss of 2.12×10^7 kg of carbonate rock per year (October 2000 to October 2001), which gives an average carbonate dissolution rate of 112 μm/a for an assumed catchment area of about 70 km^2 (for details on the method and values obtained by other methods, see Plan, 2003).

During the period of intensive investigation (1996–2003), isotopic measurements also were carried out at the Kläffer Spring. They concerned the mean residence time, the mean altitude of the catchment area, and the homogeneity of the springwater at the different outlets of the Kläffer Spring during flood events (Figure 10.2–8; Stadler and Strobl, 2006).

The mean residence time was calculated on the basis of an exponential model with tritium data giving 0.8 to 1.5 years. On the basis of attenuation of the seasonal cycle of oxygen-18, a residence time of 0.5 to 0.8 years was calculated (Stadler and Strobl, 1997).

FIGURE 10.2–8 Part of the Kläffer Spring during a flood in August 2002: (Upper right) waterfall out of the Kläffergrotte; (front) Salza River during flood conditions.

During the investigations concerning the mean altitude of the catchment by means of oxygen-18 fractionation processes, it could be shown that, for all springs that drain relevant parts of the Hochschwab plateau, the isotopic height-decreasing effect is overlain by a special fractionating process on the plateau, which is driven mainly by the buildup of the snow cover (Stadler and Strobl, 2006). With this knowledge, the mean altitude of the catchment area was calculated to be approximately 1700 m. This correlates well with morphometric analyses and geological mapping results. A determination of the catchment area of the Kläffer Spring by means of geological mapping and methods of structural geology shows a maximum size from this point of view of about 60–70 km^2. Great problems with a hydrological delineation occur because of missing precipitation stations at the Hochschwab plateau, problems of discharge measurements at the springs itself, and possible shifting of the boundaries due to different hydrological situations in this karstic catchment.

As a typical Alpine karst spring, the Kläffer Spring is also affected by different events of turbidity. They are mainly caused by heavy thunderstorms during summertime. A typical phenomenon during snowmelt can be monitored at Kläffer Spring. This is shown in Figure 10.2–9. The gauge height (shown is the gauge height of one of the outlets to the Salza River of the tapping; Stadler and Strobl, 1996) recognizes the starting of a snow melting period. The changing gauge height or discharge, respectively, is caused by the daily cycle of the air temperature in the catchment area around 0°C. It can be seen that the turbidity, especially at the spring outlet Large Kläffer, goes inverse to the turbidity, showing a cycle of about 24 hours. As the turbidity is caused mainly by increased shearing force in the karst conduits by mobilization of sediments, it comes to a dilution by clear melt waters.

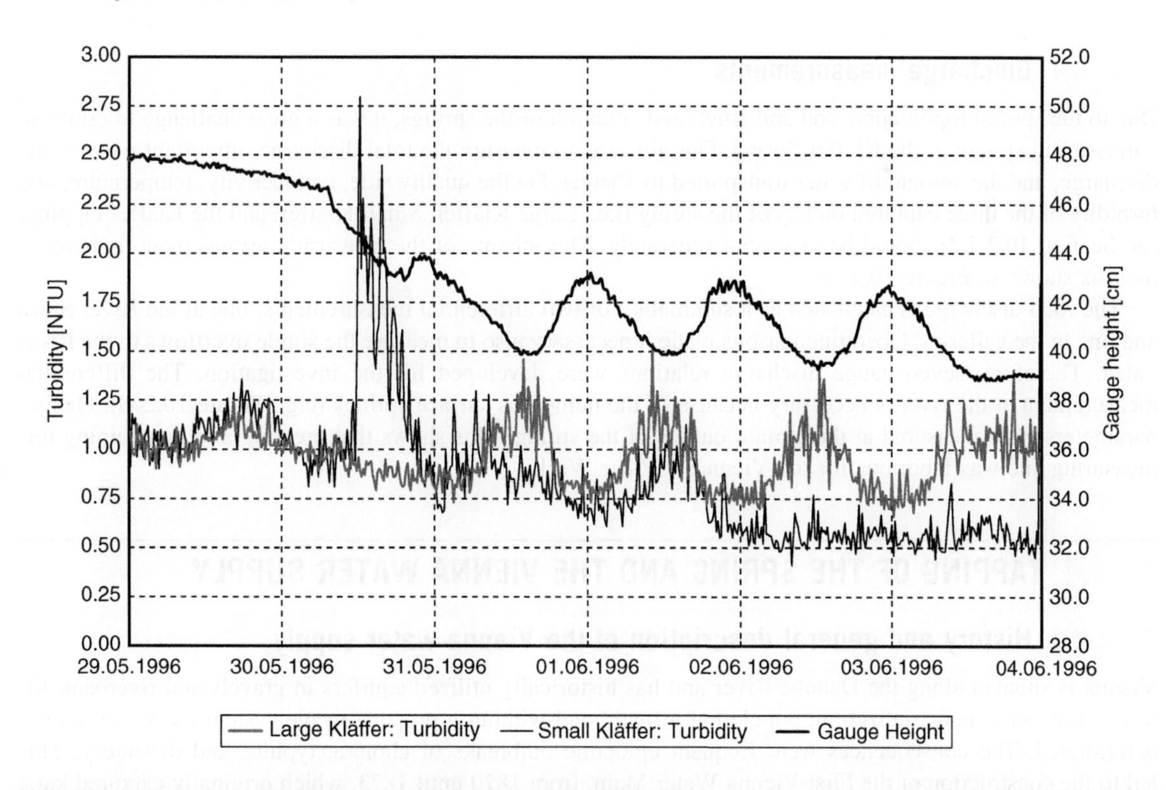

FIGURE 10.2–9 Turbidity during starting of snowmelt at Kläffer Spring.

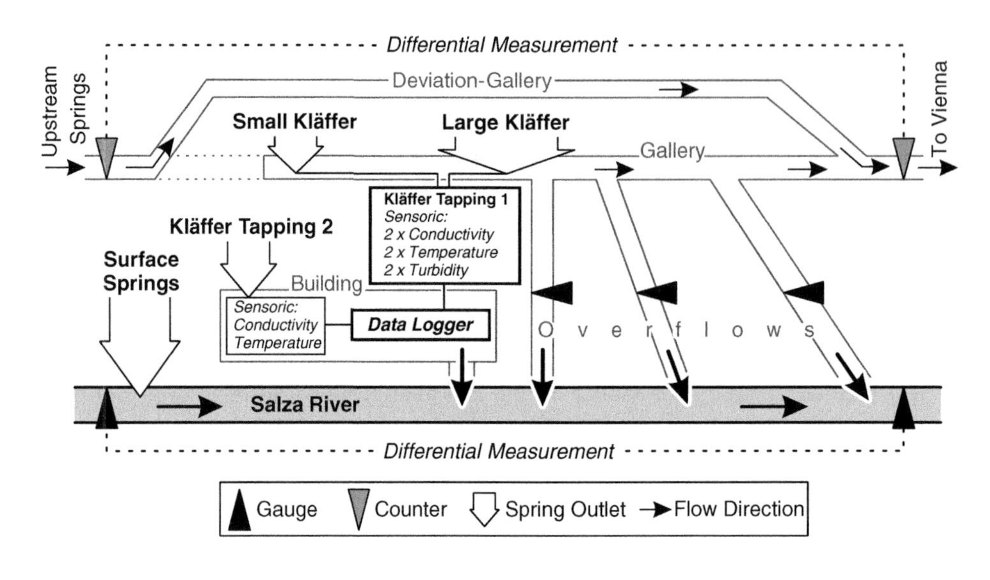

FIGURE 10.2–10 Scheme of the measuring arrangement and equipment at Kläffer Spring.

10.2.3.1 Discharge measurements

Due to the special topographic and constructional situation of the springs, it was a great challenge to establish a measuring system at the Kläffer Spring. Our aim was to measure the total discharge, important parts of the discharge, and the amount of water transported to Vienna. On the quality side, conductivity, temperature, and turbidity of the three captured outlets of the spring (i.e., Large Kläffer, Small Kläffer, and the Kläffer tapping; see Section 10.2.4.2) should be measured separately. The scheme of the measuring arrangement and equipment is shown in Figure 10.2–10.

The total discharge is calculated as a summation of two differential measurements: one at the River Salza and one in the galleries. Operating reasons made it necessary also to measure the single overflows to the River Salza. Therefore, seven gauge discharge relations were developed for the investigation. The differential measurement at the river is necessary because of the numerous surface springs (e.g., Figure 10.2–8). Quality parameters were measured at three main outlets of the spring. This shows that erecting and maintaining this measuring site was laborious for the Viennese Water Works.

10.2.4 TAPPING OF THE SPRING AND THE VIENNA WATER SUPPLY

10.2.4.1 History and general description of the Vienna water supply

Vienna is situated along the Danube River and has historically utilized aquifers in gravels and riverbank filtrates. But, with industrialization, population growth, and without a proper sewage system the water quality deteriorated. The consequences were frequent epidemic outbreaks of cholera, typhus, and dysentery. This led to the construction of the First Vienna Water Main, from 1870 until 1873, which originally captured karst waters from the Rax and Schneeberg massifs (Drennig, 1973).

Due to the continuing growth of the city and the demand for higher quantities of water, the First Water Main did not suffice. So the Vienna City Council decided on March 27, 1900, to build the Second Water Main from the Salza Valley to Vienna, with the originally called Kläfferbrünne as the main spring (Drennig, 1988).

Later on, because of the rising consumption but especially to account for the high differences between minimal and maximal discharge of the karstic springs additional springs (e.g., from the Schneealm massif, Figure 10.2–2) were developed and connected to the First and Second Water Mains and incorporated into the Viennese water supply system (Drennig, 1973, 1988).

In 2007, the First Water Main provided 61.87 million m^3 water to Vienna, which is 43.5 percent of the demand, and the Second Main provided 75.40 million m^3 or 53.1 percent. The remaining 3.4 percent were developed from groundwater wells in Vienna (Rumpold, 2008). Figure 10.2–11 shows one of the Vienna Waterworks reservoirs storing springwater in the city.

The drinking water is tapped at the springs and flows, driven by gravity, to Vienna. Along the course of the water main, especially in the catchment areas, power stations are incorporated, which produce hydropower for some 50,000 households. There are also several water gates to empty the galleries for cleaning and construction work.

10.2.4.2 Tapping of the Kläffer Spring and water distribution

As described, the spring consists of several outlets that do not allow capturing at the surface. Therefore, a gallery was built into the slope to find a water-bearing crevice, which succeeded after some 30 m (Figure 10.2–12). Two cavities (Large and Small Kläffer) were intersected that directly drain into the gallery

FIGURE 10.2–11 Reservoir at Vienna Rosenhügel with a volume of 120,500 m^3.

FIGURE 10.2–12 Large Kläffer draining into the gallery of the water main.

(Figure 10.2–13). Later on, in 1947, a constant surface outlet was also developed, referred to as Kläffer Tapping No. 2. As it lies below the other outlets, it has to be pumped into the gallery. Only 0.6 to 2 m^3/s of the discharge of these three inputs from the Kläffer Spring can be used for water supply. The rest of the water captured inside the gallery extrudes through overflows (Figure 10.2–14) into the Salza River, like the original surface springs.

The pipeline has a length of about 200 km and the water needs some 36 hours from the Kläffer Spring to the first reservoir in Vienna. The mean cross sections of the pipeline vary from 1.2 to 1.9 m in width and from 1.6 to 2.1 m in height.

In Vienna, the water pipe network has a length of approximately 3100 km. Except for about 10 percent of the households, especially in the western, more elevated part of Vienna, the water is supplied to the consumers without pumping, just driven by gravity. This is facilitated by a system of 32 reservoirs with a total volume of approximately 1.5 million m^3, which equals a demand for 3.5 days. The water is marginally chlorinated and needs no further treatment.

FIGURE 10.2–13 Large Kläffer (background) drains directly into the gallery. Photo taken during high water condition.

FIGURE 10.2–14 Partly flooded overflow gallery, the Kläffer Spring.

10.2.5 WATER QUALITY, CONTAMINATION PROBLEMS, AND PROTECTION STRATEGIES

10.2.5.1 Contamination hazards in the spring catchment and vulnerability mapping

Apart from airborne pollution, potential hazards to the karst aquifer originate mainly from pasture, tourism, forestry, and an unnaturally high game population (mainly, chamois and deer). The inputs from pasture, tourism, and the game population are mainly feces with their microbial hazards.

To get data on the morphology of the catchment area, the plateau portion of the catchment area was field mapped in detail (1:5000) by Plan and Decker (2006). Data and results are summarized in a karst morphological GIS, which covers 59 km^2. Entities are linked to an extensive database comprising information on 12,700 karst features, including karst landforms (dolines, poljes, polygenetic glaciokarstic-depressions, dry valleys, caves, and karren), hydrological items (springs, surface streams, ponors, and ponds), geologic items relevant as protective covers (Cenozoic clay covers and glacial deposits), and anthropogenic features as possible sources of pollution (mountain huts, hiking trails, gravel roads, meadows used for pasture).

Using the other data achieved within the KATER I and II project (see Section 10.2.1), Plan et al. (2008) assessed the resource vulnerability (for definitions, see Zwahlen, 2004) utilizing the COP method (Vías et al., 2006) for the plateau area of the Hochschwab. In general, carbonates are very well karstified throughout and overlying layers are mostly absent, not protective, and sometimes even enhance point recharge. Vulnerability according to the COP method (Figure 10.2–15) is high or very high in 82 percent of the area (35 percent has *very high* vulnerability, 47 percent *high* vulnerability, 17 percent *moderate* vulnerability, 1 percent low vulnerability, no area has *very low* vulnerability). The resulting map is reasonable but does not differentiate vulnerabilities to the extent that the results can be used for further protective measures in this already well-protected area.

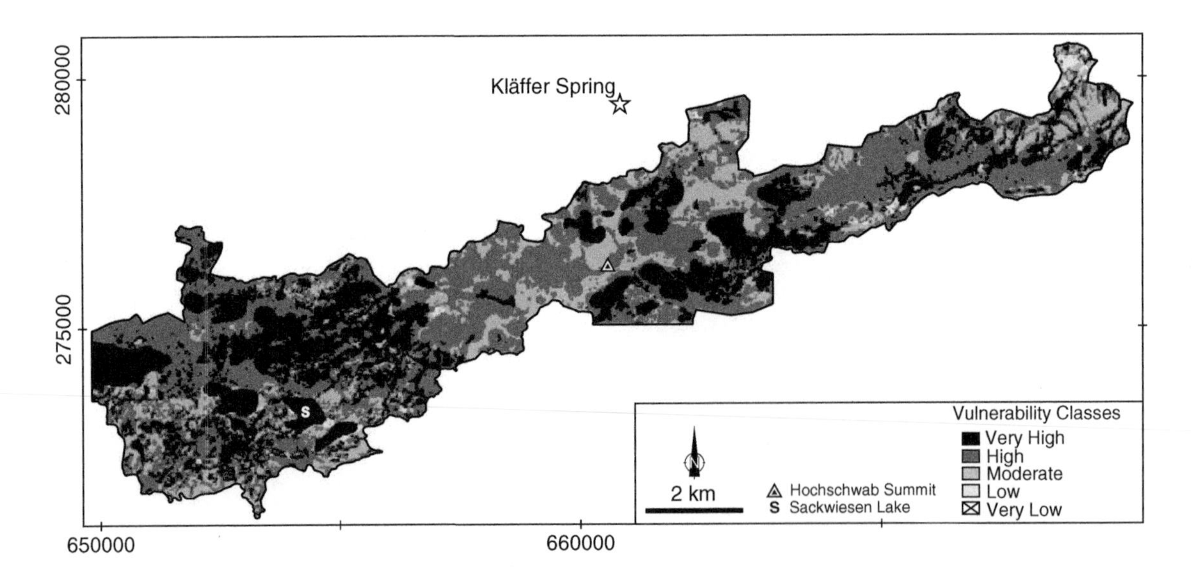

FIGURE 10.2–15 COP vulnerability map of the plateau area of the catchment of the Kläffer Spring. (Modified after Plan et al., 2008.)

Although these areas are generally uninhabited and protected by law, land-use conflicts arise from pasture, tourism, forestry, and hunting. Complete protection of all highly vulnerable areas is not possible, and only a small and manageable number of extremely vulnerable sites can be subject to rigid protective measures. Therefore, an extension for the upper end of the vulnerability scale was developed that allows identifying so-called ultravulnerable areas (Plan et al., 2008). This enhancement of the conventional approach points out that infiltration conditions are of key importance for vulnerability. The method accounts for karst genetical and hydrologic processes using qualitative and quantitative properties of karst depressions and sinking streams, including parameters calculated from digital elevation models. Using this approach 1.7 percent of the Hochschwab plateau area is delineated as *ultravulnerable*. The resulting vulnerability map highlights spots of maximum vulnerability, and in combination with a hazard assessment, some critical areas can be protected. For example, ponors of major poljes and their catchment can be fenced off to prevent fecal contamination by cattle (Plan et al., 2008). For the whole catchment area, neither the COP nor any other vulnerability assessment method has been applied as yet.

10.2.5.2 Chemical and microbial springwater quality

Due to the remote location with limited land-use activities, there is almost no hazard of chemical contamination. The highest measured nitrate concentration was 5 mg/L. For forestry, there is also no use of fertilizers and pesticides in pasture activities and no additional concentrated feed with potential contents of hormones or heavy metals. Tourism is only hiking, climbing, and backcountry skiing in wintertime. At the tourist huts, the solid waste, except for compost, is disposed of in the valley outside of the catchment areas. The fluid waste (feces and wastewater) is disposed of by sewage pipes down to the valley or at least treated with chlorinated lime where sewage pipes are not possible before being spread on meadows.

A slight temporary microbial contamination derives from the undisposed-of human feces and the feces from cattle and deer. The Kläffer Spring, which shows a very dynamic discharge, is especially prone to microbial contamination during strong precipitation events or snowmelt, whereas the base flow is almost free of microbial contamination. Event water boosts turbidity, the spectral absorption coefficient (SAC), and correlating with it, the microbial contamination. The method in use to prevent bringing microbially contaminated water to the consumers is to distract the water to the fore flood in the case of rising turbidity and SAC, which are measured continuously.

10.2.5.3 Protection zones and land-use restrictions

To minimize the input to the aquifers, protection zones have been decreed by law, where the defined areas are dedicated primarily for water supply. Several activities are not allowed or have to be approved by the government body in charge. Additionally, the city of Vienna pursued the policy of acquiring the catchment areas since 1870. But often, older rights still exist, like rights to the pasture. In case of conflicts, consensual solutions are the aim. This policy is supported by financial incentives from the city of Vienna.

To impose or even suggest land-use restrictions, the potential hazards and risks have to be known exactly. The contamination path and its likelihood have to be shown and explained. For acquiring all those data, the Vienna Waterworks started a systematic and area-covering karst research program, including two projects funded by the European Union, in 1994. The results of the KATER (karst water research program) and KATER II, like maps, data, and studies, are integrated into a geographical information system (GIS) based decision support system, which helps develop transparent and comprehensible land-use plans.

Another important aspect is the way of cultivating forests, which cover two thirds of the catchment areas of springs utilized for Vienna. Natural, diverse, and structured forests do not constitute hazards; on the contrary, they fulfill a vital and important protection function. It depends on their silvicultural management practices. Large parts of the total catchment areas (approximately 970 km^2) of the Vienna water supply are owned (325 km^2) and managed by the city of Vienna. The forest stands owned by the city of Vienna are markedly dedicated as spring protection forests. Clear cuts and the use of fertilizers as well as pesticides or any other chemicals are forbidden. Silvicultural measures that diminish the threats of wind throws and beetle diseases are applied.

REFERENCES

Decker, K., Reiter, F., 2001. Strukturgeologische Methoden zur Charakterisierung von Karstgrundwasserleitern im Hochschwabmassiv. In: Mandl, G. (Ed.), Geologische Bundesanstalt Arbeitstagung 2001. Geologische Bundesanstalt, Vienna, pp. 206–212.

Decker, K., Plan, L., Reiter, F., 2006. Tectonic Assessment of Deep Groundwater Pathways in Fractured and Karstified Aquifers, Hochschwab Massif, Austria. Proceedings of All about Karst and Water, Vienna. pp. 138–142.

Drennig, A., 1973. Die I. Wiener Hochquellenwasserleitung. Magistrat der Stadt Wien, Abteilung 31, Wasserwerke, Vienna.

Drennig, A., 1988. Die II. Wiener Hochquellenwasserleitung. Magistrat der Stadt Wien, Abteilung 31, Wasserwerke, Vienna.

Ford, D., Williams, P., 2007. Karst Hydrogeology and Geomorphology. Wiley, Chichester, UK.

Frisch, W., Székely, B., Kuhlemann, J., Dunkl, I., 2000. Geomorphological evolution of the Eastern Alps in response to Miocene tectonics. Zeitschrift für Geomorphologie 44 (1), 103–138.

Frisch, W., Kuhlemann, J., Dunkl, I., Székely, B., Vennemann, T., Rettenbacher, A., 2002. Dachstein-Altfläche, Augenstein-Formation und Höhlenentwicklung—Die Geschichte der letzten 35 Millionen Jahre in den zentralen Nördlichen Kalkalpen. Die Höhle 53 (1), 1–36.

Kralik, M., 2001. Strategie zum Schutz der Karstwassergebiete in Österreich. BE-189. Umweltbundesamt GmbH, Vienna.

Kuschnig, G., 2006. KATER II Projekt Web Portal. Available at: www.kater.at, accessed December 2008.

Mandl, G., Bryda, G., Kreuss, O., Moser, M., Pavlik, W., 2002. Erstellung moderner geologischer Karten als Grundlage für karsthydrogeologische Spezialuntersuchungen im Hochschwabgebiet. Unpublished final report to the Viennese Waterworks, Geol, Bundesanstalt, Vienna.

Plan, L., 2002. Speläologisch-tektonische Untersuchungen der Karstwasserdynamik im Einzugsgebiet der bedeutendsten Quelle der Ostalpen (Kläfferquelle, Hochschwab). Speldok 11. Verband Österreichischer Höhlenforscher, Vienna.

Plan, L., 2004. Speläologische Characterisierung und Analyse des Hochschwab-Plateaus, Steiermark. Die Höhle 55 (1–4), 19–33.

Plan, L., Decker, K., 2006. Quantitative karst morphology of the Hochschwab plateau, Eastern Alps, Austria. (Int. Karst Atlas No. 19). Zeitschrift für Geomorphologie, Supplement 147, 29–56.

Plan, L., Decker, K., Faber, R., Wagreich, M., Grasemann, B., 2008. Karst morphology and groundwater vulnerability of high alpine karst plateaus. Env. Geol. Special Issue, doi:10.1016/j.geomorph.2008.09.011.

Rumpold, A., 2008. Wiener Wasser—Statistik. Available at: www.wien.gv.at/wienwasser/statistik.html, accessed December 2008.

Schubert, G., 2003. Hydrogeological Map of Austria 1: 500.000, Geologische Bundesanstalt, Vienna.

Stadler, H., Strobl, E., 1996. Karstwasserdynamik und Karstwasserschutz Hochschwab (STA28K). Endbericht, 1. Arbeitsjahr. Unpublished report, Joanneum Research, Graz, Austria.

Stadler, H., Stobl, E., 1997. Karstwasserdynamik Zeller Staritzen. Endbericht. Unpublished report, Joanneum Research, Graz, Austria.

Stadler, H., Strobl, E., 2006. Hydrogeologie Hochschwab Zusammenfassung. Unpublished final report, Joanneum Research, Graz, Austria.

Steiermärkische Landesregierung, 2008. Aktuelle Niederschlags- und Lufttemperaturwerte. Available at: www.wasserwirtschaft.steiermark.at, accessed December 2008.

Stummer, G., Plan, L. (eds.), 2002. Speldok-Austria—Handbuch zum Österreichischen Höhlenverzeichnis. Speldok-10. Verband Österreichischer Höhlenforscher, Vienna.

Vías, J.M., Andreo, B., Perles, M.J., Carrasco, F., Vadillo, I., Jiménez, P., 2006. Proposed method for groundwater vulnerability mapping in carbonate (karstic) aquifers: The COP method. Hydrogeology Journal 14 (6), 912–925.

Wakonigg, H., 1980. Die Niederschlagsverhältnisse im südlichen Hochschwabgebiet. In: Fabiani, E., Weißensteiner, V., Wakonigg, H. (Eds.), Grund- und Karstwasseruntersuchungen im Hochschwabgebiet, Part II. Amt der Steiermärkischen Landesregierung, Graz, Austria, pp. 65–141.

Zwahlen, F. (Ed.), 2004. Vulnerability and risk mapping for the protection of carbonate (karst) aquifers, EUR 20912. Final Report COST Action 620, European Commission, Directorate-General XII Science, Research and Development, Brussels.

CHAPTER

Case Study: Characterization, exploitation, and protection of the Malenščica karst spring, Slovenia

10.3

Metka Petric
Karst Research Institute, Scientific Research Centre, Slovenian Academy of Sciences and Arts, Postojna, Slovenia

10.3.1 INTRODUCTION

The Malenščica karst spring is situated at the southern border of the Planina karst polje in southwestern Slovenia. At present, it is captured for the water supply of the municipalities of Postojna and Pivka, and it is one of the most important water resources in Slovenia. Its capacity, with a minimum discharge exceeding 1 m^3/s, is more than sufficient for the present needs of the population within these two municipalities. Since the 1950s, the Malenščica spring has been included in different plans for local and regional water supply, but various doubts regarding its suitability have always been present. Due to the low position of the spring, water has to be pumped up 200 m into the water supply system. Plans for the construction of a reservoir for a hydropower station on the Planina karst polje, which would result in the flooding of the spring, were discussed at different times. Furthermore, the water quality has been in the past and is at present increasingly endangered. Therefore, the Malenščica spring was first used by only a nearby village, but after the long droughts at the end of the 1960s and in the absence of better solutions, it was decided to capture it for the water supply of the broader Postojna region. It remains so today, when the main problem is the risk of pollution due to various human activities in its extensive recharge area. At present, the main task is implementing suitable protection measures within this area. The basis for efficient protection is a clear understanding of the characteristics of water flow and the transport of substances within the karst water system, which has been achieved through various geological, geomorphological, speleological, and hydrogeological studies applying different research techniques.

10.3.2 GEOLOGICAL SETTING

The studied area is a part of the karst basin of the Ljubljanica spring, one of the largest karst springs in the world. Its recharge area comprises a system of interconnected karst poljes, of which the Planina Polje is one of the most typical. Several karst springs are located at its southern border, the most important being the Unica and Malenščica Springs (Figure 10.3–1). The Unica River flows from Planina Cave, and the

Copyright © 2010, Elsevier Inc. All rights reserved.

FIGURE 10.3–1 Malenščica and Unica Springs at the southwestern edge of Planina Polje.

Malenščica is its right tributary. Their catchments with surface and underground water flows overlap. They can be divided into three separate but hydrologically connected parts (Figure 10.3–2). The central part is the karst massif of Javorniki and Sneznik, which borders the eastern side of the valley of the Pivka River and its tributaries. On the eastern and northern sides of Javorniki is a string of karst poljes (the biggest of these is Cerknica Polje) that are distributed gradually in a southeast-northwest direction. These three areas can be named the Javorniki, Pivka, and Cerknica parts of the catchment. In the Javorniki part, the underground flow is dominant, while in the other two parts, surface streams are also present. They are recharged mainly by karst waters, and after a certain distance of surface flow, they sink underground again.

In the regional sense, this area is a part of southwestern Slovenia, for which a nappe structure is characteristic (Placer, 1981). The Hrusica nappe (most of the Cerknica part of the catchment) is thrust over the Sneznik thrust sheet (Pivka and Javorniki parts of the catchment), which farther to the southwest is thrust over the Komen thrust sheet. These nappes are intersected with numerous faults, the most important among them being the Idrija and Predjama faults, both of which extend in the Dinaric northwest-southeast direction. The Predjama fault zone crosses the Pivka Valley and Javorniki, and the Idrija fault zone, the system of karst poljes.

The oldest rocks in the catchment are Norian-Rhetian dolomites found between Planina and Cerknica Poljes (Buser, Grad, and Plenicar, 1967). They are intersected by the Idrija fault zone, where the dolomites are crushed into millonite and act as a hydrological barrier. Along it, underground water is forced to flow to the surface through several springs, including the Malenščica and Unica Springs. Jurassic limestone and dolomite compose the northeastern border of karst poljes from Cerknica Polje to Babno Polje. The area of Javorniki and Sneznik is composed of Cretaceous carbonate rocks, mostly limestone, which forms the central karst aquifer of the area. A surface drainage network has developed on the very poorly permeable Eocene flysch in the Pivka valley. Quaternary alluvial sediments are deposited along the surface streams and on karst poljes.

10.3.3 THE MALENŠČICA SPRING

Based on known geological, geomorphological, and speleological characteristics, it can be concluded that the Pivka River sinks into the Postojna Cave and emerges again in the Planina Cave, where it flows along the so-called Pivka branch of the cave (Figure 10.3–2). In the cave, it is joined by the stream along the Rak branch, which collects groundwater mainly from the areas of Rakov Skocjan and Cerknica Polje.

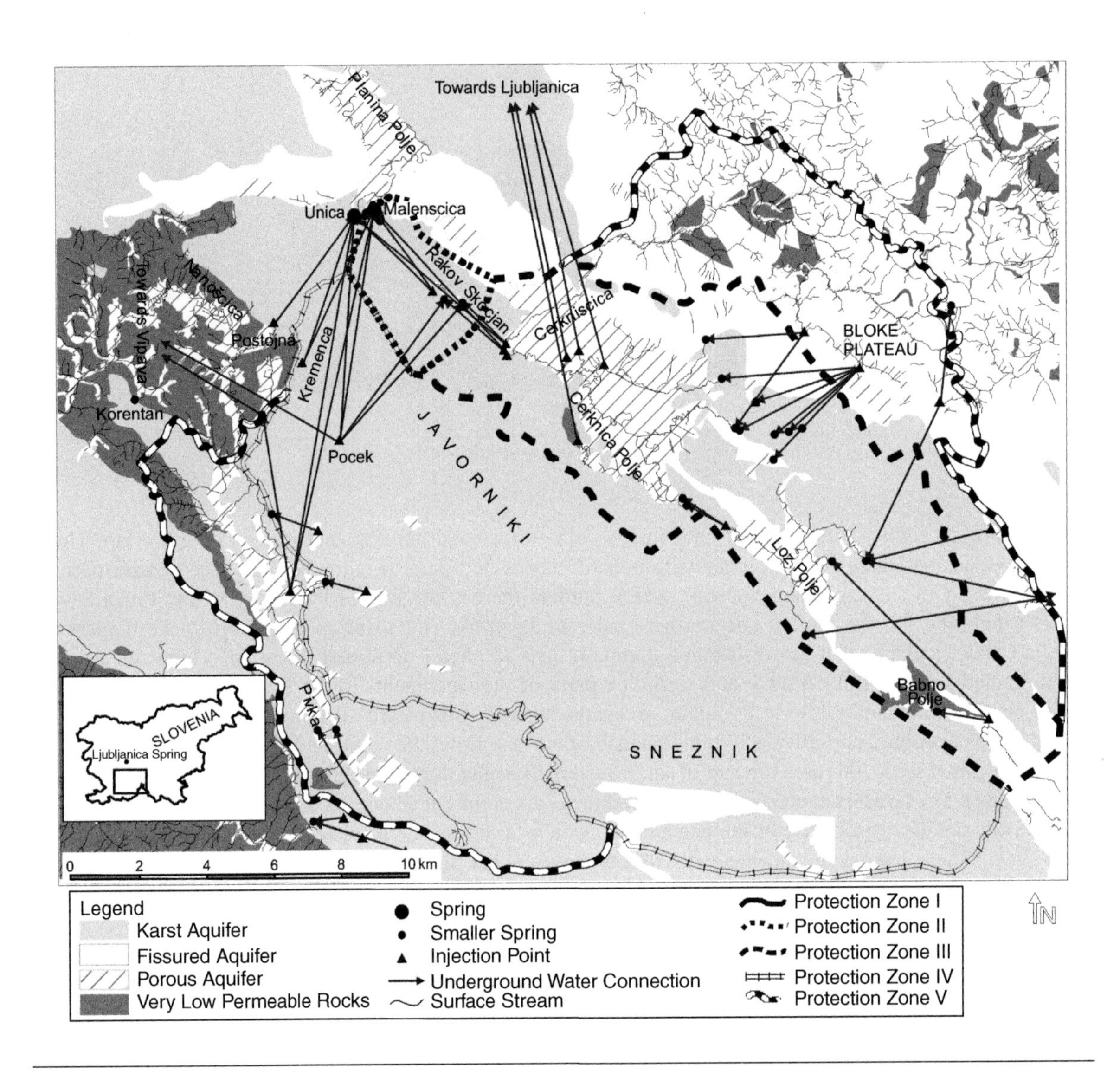

FIGURE 10.3–2 Hydrogeological map of the recharge area of springs on Planina Polje.

The common outflow from the cave is the Unica Spring (Figure 10.3–3). The length of the cave, which is especially interesting due to this underground confluence of two cave streams, is 6.6 km.

In the geological past, the main outflow from this karst system was the Malenščica Stream, which is situated in an approximately 1 km long and 200 m wide collapsed valley at the southwestern edge of Planina Polje and is a right tributary of the Unica River. Due to the collapses of karst channels over time, the main groundwater flow was redirected toward Planina Cave, and only limited amounts of water now flow in a dispersed outflow through the several springs contributing to the Malenščica stream. The largest of these springs is the lowest one, in which the water level oscillates between 448 and 449.5 m above sea level (asl) and which

FIGURE 10.3–3 The Unica River emerges from the Planina Cave.

is permanently active (Figure 10.3–4). Upward along the valley, three intermittent springs are situated at altitudes between 455 and 470 m asl. (Habic, 1987).

The common daily discharge of the permanent and intermittent springs of the Malenščica is regularly measured by the Environmental Agency of the Republic of Slovenia. According to data for the 1961–1990 period, the lowest discharge was 1.1 m^3/s, the mean discharge 6.7 m^3/s, and the maximum discharge 9.9 m^3/s (Kolbezen and Pristov, 1998). Still higher discharges were measured in 2002, with a maximum daily value of 11.2 m^3/s on October 24, 2002.

Comparison with the Unica Spring, with discharges from several hundreds of liters per second to almost 100 m^3/s, confirms that the maximum flow rates of the Malenščica Spring are limited. At low water, Malenščica is the main outflow of the karst system, but at high water, only one tenth of the common discharge on Planina Polje comes through this spring (Figure 10.3–5). Already at medium water, the discharge of Malenščica Spring exceeds 8 m^3/s, but only at very high water does it increase above 9 m^3/s.

The immediate hinterland of the Malenščica Spring is composed of Lower Cretaceous dark gray micritic limestone (Car and Gospodaric, 1984). The valley cuts into the anticline apex, which is intersected by a strong fault in the Dinaric direction (Figure 10.3–6). Parallel to it are several other fissured zones and faults on both sides of the valley, which direct the flow of groundwater. Diving in the only larger fissure (30 to 50 cm wide, 1.5 m deep) in the main Malenščica Spring showed that water flows upward through collapsed blocks and rubble along a rocky wall (Habic, 1987). The position and direction of the fissure indicate the proximity of a subvertical fault in the Dinaric direction, along which groundwater flows to the surface.

FIGURE 10.3–4 The Malenščica Spring is captured for water supply.

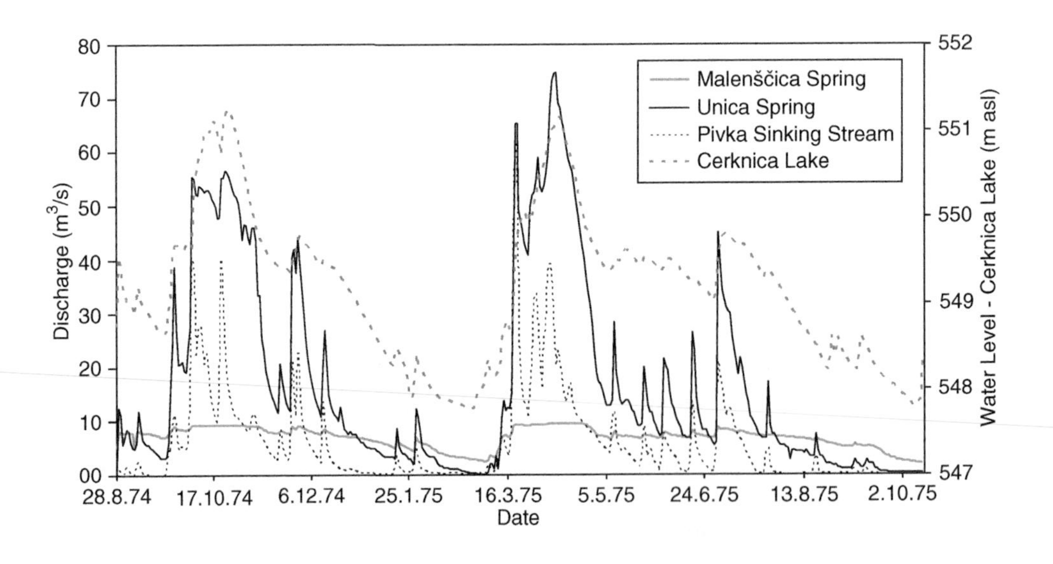

FIGURE 10.3–5 Comparison between discharges of Malenščica and Unica Springs, and water flow in their catchment for selected hydrological years.

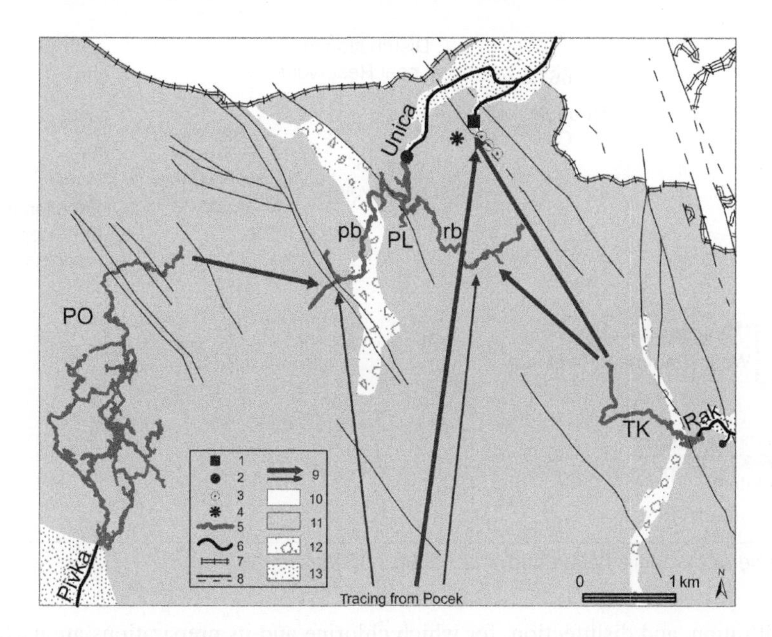

FIGURE 10.3–6 Geological structure of the immediate hinterland of the Malenščica Spring: (1) Malenščica Spring, captured; (2) permanent spring; (3) intermittent spring; (4) water treatment plant; (5) karst caves (PO, Postojna Cave; PL, Planina Cave; pb, Pivka branch; rb, Rak branch; TK, Tkalca Cave); (6) surface stream; (7) thrust line; (8) fault, visible and covered; (9) main and secondary directions of groundwater flow proven by tracer tests; (10) upper Triassic dolomite; (11) Cretaceous limestone and dolomite; (12) Cretaceous breccia; (13) Quaternary alluvial deposits. (Adapted from Car and Gospodaric, 1984.)

10.3.4 DEVELOPMENT OF THE WATER SUPPLY SYSTEM IN THE POSTOJNA AREA

In the Postojna area, smaller springs that allowed gravitational distribution were used for the water supply in the past. A pumping station was built at the Korentan Spring in 1954 (Figure 10.3–2), and even at low water, it was possible to pump 20 L/s, which was sufficient for the needs of the time. However, a long drought at the end of the 1960s demonstrated the necessity of a more productive source, and the Malenščica Spring seemed to be the best choice. Studies were carried out to test the possibility of pumping the groundwater in the Javorniki karst massif closer to Postojna, but the results of geological, geomorphological, speleological, and hydrogeological analyses, including test pumping in three exploration boreholes, were not promising enough to continue the research.

At that time, the Malenščica Spring was captured for the water supply of the nearby village of Planina. One of the lateral springs was captured, because a house blocked access to the main spring. At low water, only 6 L/s was available, and therefore geophysical research, drilling, and a tracer test were employed to find a main karst conduit but with no success. Based on additional detailed structural-lithological mapping and speleological research, including diving, it was decided to build a new water intake at the main spring. The area of the spring was deepened by 3 m to the natural dam of the Malenščica by the Quaternary sediments of the Planina Polje. There, a new pumping station was built, which is still active.

The system is managed by the Kovod Water Supply Company of Postojna. From the intake at the spring, water is first pumped 75 m to the water treatment plant where the raw water is treated through the processes

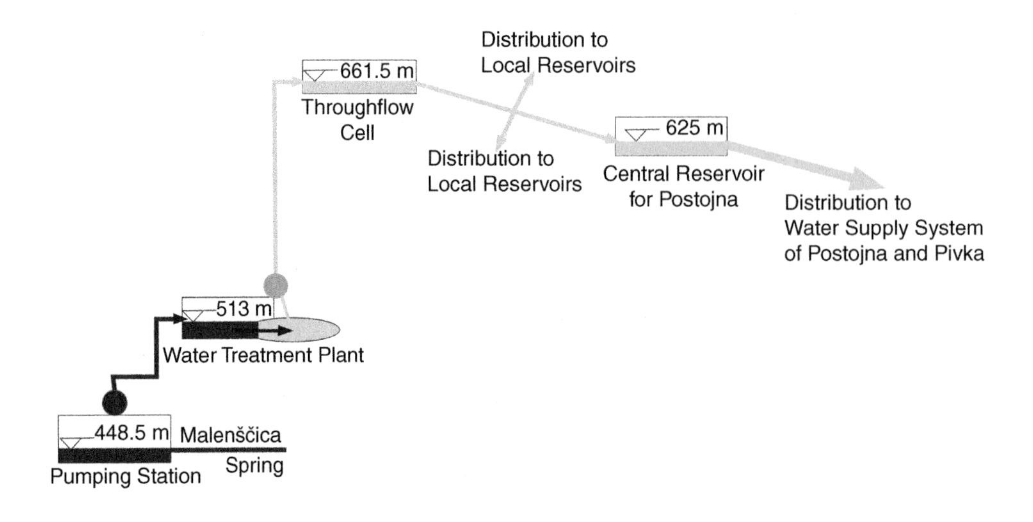

FIGURE 10.3–7 Scheme of pumping station and water supply system.

of sedimentation, filtration, and disinfection, for which chlorine and its preparations are used (Figure 10.3–7). The capacity of the plant is 250 L/s. The treated water is pumped 150 m up to the throughflow cell and, from there, distributed gravitationally toward numerous reservoirs. The water is used to supply the approximately 21,000 inhabitants of the municipalities of Postojna and Pivka. Only a low proportion (less than 10 percent) of the water of this central water supply system is contributed by smaller local springs. The yearly consumption of water is approximately 1.4 million m^3.

10.3.5 PAST AND CURRENT STUDIES OF THE SPRING AND ITS CATCHMENT

Underground water connections between the springs and surface flows within their catchment were proven by several tracer tests (Figure 10.3–2). In the Pivka part of the catchment, tracer tests were performed in 1928, 1974, and 1977. Tracers were injected in the Pivka River at the ponor into the Postojna Cave, and the underground water connection with the Pivka branch of the Planina Cave was confirmed (Habic, 1987). At various hydrological conditions, apparent flow velocities between 1.3 and 4 cm/s were determined. However, no tracer was detected in the Malenščica Spring.

In the Javorniki part of the catchment, three tracer tests were carried out. In 1955, the tracer from the stream sinking into the Brezno na Kremenci Cave and, in 1988, the tracer from the ponors in the riverbed of the upper Pivka River were detected at the Malenščica and Unica Springs (Habic, 1987). Apparent flow velocities were around 0.5 cm/s. More information was gathered by the tracing at the Pocek military training area in June 1997, when the tracer was injected into the rocky bottom of a doline (Kogovsek and Petric, 2004). The groundwater flow toward the Malenščica Spring and the Rak and Pivka branches in the Planina Cave was proven, and secondary directions toward some other springs were confirmed. The highest concentrations of tracer were detected in the Malenščica Spring, where the amount of recovered tracer was 55 percent and the apparent dominant flow velocity was 0.7 cm/s. Approximately 26 percent of the injected tracer flowed with the same velocity below the flysch of the Pivka valley toward the Vipava Spring.

In the Cerknica part of the catchment, the ponor caves at the northwestern edge of Cerknica Polje were chosen as injection points six times between 1939 and 1971 (Gams, 1970; Habic, 1987). Underground water flow toward the Rak Spring (at different hydrological conditions, apparent flow velocities between 2.2 and 18 cm/s) and other springs in Rakov Skocjan (apparent flow velocities between 1 and 7 cm/s) was proven. Further observation of these tracers and five other tests between 1928 and 1967, in which tracers were injected in the sinking Rak River at the northwestern edge of Rakov Skocjan, confirmed the existence of an underground water connection with the springs on Planina Polje. Apparent flow velocities toward the Rak branch in the Planina Cave were from 3.5 to 6.3 cm/s and toward the Malenščica Spring between 3.9 and 7.1 cm/s. An extremely low velocity of 0.2 cm/s was calculated for the flow toward Malenščica at very low water in August 1967 (Gams, 1970). During one of the speleological expeditions in the Rak branch of the Planina Cave in August 1950, a tracer was injected at low-water conditions into the rapids in front of the last siphon, approximately 2600 m inside the cave (Michler, 1955). Water flow toward the siphon, deeper into the cave, and farther toward the Malenščica Spring with the apparent flow velocity of 1.3 cm/s was proven. Unfortunately, no calculations of the proportion of recovered tracer were made in any of these tests.

Through four tests with the injection of tracer in the ground ponors on Cerknica Polje, a direct connection with the Ljubljanica Spring was confirmed (Gospodaric and Habic, 1976). Figure 10.3–2 shows the results of tracings on the Bloke plateau, the Babno and Loz Poljes, and the western border of the Javorniki massif. Based on these results, it was possible to define the borders of the recharge area of the springs on Planina Polje more accurately.

In addition to tracings with artificial tracers, natural tracers were monitored occasionally at the Malenščica Spring. A survey of the published data (Habic, 1986, 1987; Kogovsek, 2001, 2004) and the results of continuous measurements performed by the Karst Research Institute since autumn 2007 give the following values: temperature between 3°C and 19°C, pH between 7.2 and 8.3, specific electrical conductivity (SEC) between 220 and 424 μS/cm, and Ca/Mg ratio between 3.5 and 8.5. During selected water waves, various parameters (discharge, temperature, SEC, pH, turbidity) have been continuously measured in the Malenščica Spring, and chemical analyses of water samples have been performed since 1997 (Kogovsek, 2001, 2004). Based on the results obtained, it was possible to deduce the different proportions of contribution from different parts of the recharge area to the Malenščica outflow. For example, the inflow from the area of Cerknica Polje, on which an intermittent lake forms at high water, is reflected in significant oscillations of water temperature (the lake temperatures depend on air temperatures) and lower values of the Ca/Mg ratio (larger proportion of water from dolomite areas). Waters from the Cerknica part of the catchment have a relatively higher influence on the Malenščica Spring, especially when the intermittent lake is active. At low water, the influence of waters from the Javorniki part of the catchment is more important.

A deficiency of the research described was that only parameters in the Malenščica Spring were regularly monitored, and the characteristics of water flows in the recharge area were measured only occasionally. It was therefore decided, in autumn 2007, to set up a monitoring network with measurements of discharges, temperature, and SEC at 30 minute intervals at the Malenščica and Unica Springs, in the Rak branch of the Planina Cave, on the Pivka River at the ponor in Postojna Cave, on the Rak River at the ponor in Tkalca Cave, at the Kotlici and Rak Springs in Rakov Skocjan, and at one ponor cave at the southwestern edge of Cerknica Polje. Additionally, during selected water waves, samples for chemical analysis are taken at these points. As an example, selected data for a short period are presented in Figure 10.3–8. The SEC is the most stable in the Malenščica Spring. The reactions to changes in hydrological conditions are the most intensive in the Pivka River, and they result in changes of SEC in the Unica Spring with a certain time lag as well as smaller intensity due to additional inflows from other parts of the recharge area of this spring. The differences and changes of the Ca/Mg ratio also seem to be a useful indicator for the more detailed analysis of the karst system in the catchment of the Malenščica Spring, but this research project is still in its initial phase and more results will be available in the future.

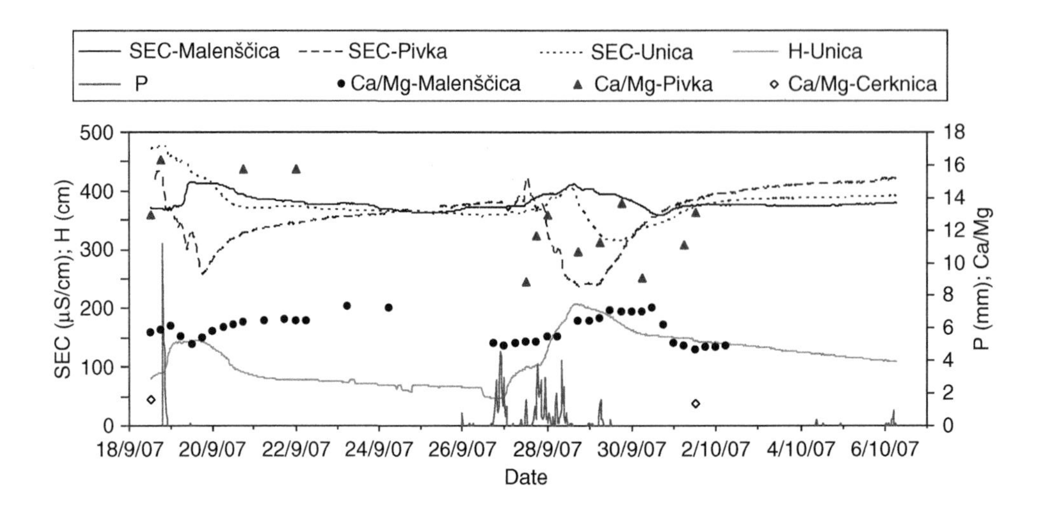

FIGURE 10.3–8 Comparison of precipitation in the catchment and water level, specific electrical conductivity, and Ca/Mg ratio in the Malenščica Spring and water flow in its recharge area.

Environmental isotopes were analyzed between 1981 and 1985 (Pezdic and Urbanc, 1987). Monthly samples were taken in water flows and precipitation in different parts of the catchment. It was demonstrated that, depending on hydrological conditions, different proportions of inflows from different parts of the catchment recharge the Malenščica Spring. In dry winter periods, recharge from the Javorniki part prevailed, but at high water, the isotopic compositions of Malenščica and Rak were practically the same, indicating an important recharge from the Cerknica part of the catchment. Significant differences in the isotopic compositions of Unica and Malenščica were observed.

Based on all the described results, we can make some inferences about the functioning of the karst system in the recharge area of the Malenščica Spring at low and high water (Figure 10.3–9). The Malenščica Spring is recharged mainly from the Cerknica and Javorniki parts of the catchment, and there is no direct connection with the ponor of the Pivka River in Postojna Cave. At high water, inflows from the Cerknica part dominate and the intermittent springs are active. The Rak branch in Planina Cave is recharged from both the Cerknica and Javorniki parts, and the Pivka branch mostly from the Pivka part.

At low water, after the emptying of the intermittent Cerknica Lake, the proportion of inflows from the Javorniki part to the Malenščica spring is more important. Interesting was the result of tracing within the Rak branch of Planina Cave, in which flow deeper into the cave and farther toward the Malenščica Spring was proven.

Interesting results were obtained by the tracer tests. From the catchment of the Malenščica Spring, groundwater flows toward the Vipava Spring (which belongs to the Adriatic basin) and directly toward the Ljubljanica Spring as well. These facts additionally complicate the delineation of the recharge area of the Malenščica Spring. In spite of a relatively extensive database, this remains a very difficult task. According to the calculation of hydrological balance, the extent of the catchment is significantly underestimated. Considering the mean discharge of 6.7 m^3/s and the annual effective precipitation of 1050 mm (Kolbezen and Pristov, 1998), the minimum extent of the recharge area is 200 km^2. However, according to the results of tracer tests, this area is significantly larger. A calculation of hydrological balance was therefore done for the common discharge of the springs on Planina Polje. The mean annual discharge is 22.6 m^3/s, which makes the minimum

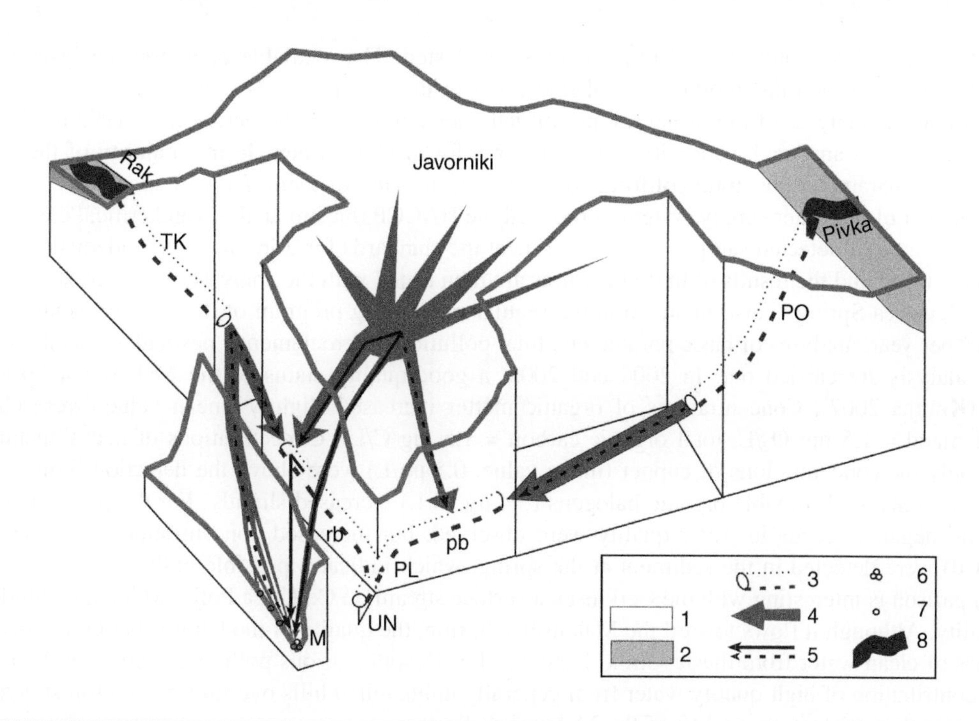

FIGURE 10.3–9 Conceptual model of the functioning of the karst system in the recharge area of the Malenščica Spring at low and high water: (1) Karst aquifer, (2) less permeable rocks, (3) karst caves with permanent water flow (PO, Postojna Cave; PL, Planina Cave: pb, Pivka branch; rb, Rak branch; UN, Unica Spring; TK, Tkalca Cave), (4) directions and shares of groundwater flow (indicated by the thickness of the arrow) from different parts of recharge areas at high water, (5) main and secondary directions of groundwater flow at low water, (6) permanent Malenščica Spring, (7) intermittent Malenščica Spring, (8) sinking stream.

extent of the recharge area 680 km^2. We should additionally consider the groundwater flow from this catchment toward the Vipava and Ljubljanica Springs. Based on all the described characteristics and known hydrogeological conditions, the full extent of the recharge area of the springs on Planina Polje was estimated to be 746 km^2. Considering the effective precipitation and mean discharge of the Malenščica Spring, we can calculate that only 27 percent of water from the catchment flows through this spring.

10.3.6 QUALITY OF THE WATER RESOURCE

According to legal provisions, a basic analysis of physical and chemical parameters was performed on the raw water of the Malenščica Spring once per month and an extended analysis once per year. In the period from 1999 to 2006, 67 of 73 analyzed samples were within the standards for drinking water. Basic analysis includes assessment of color, temperature, turbidity, pH, SEC, and concentration of ammonium. Until 2004, the consumption of added $KMnO_4$ was also regularly measured (in 27 samples, values between 0.02 and 7.9 mg $KMnO_4$/L were measured; in one sample the maximum allowed value of 10 mg $KMnO_4$/L was exceeded). In individual samples, the concentrations of nitrates (2.4–2.9 mg NO_3/L), chlorides (2–6.7 mg Cl/L), and total organic carbon (0.78–2.9 mg C/L) were measured. All values were within the standards for drinking water.

In the extended analysis, more than 100 parameters are tested. The available data from analyses in 2001, 2002, and 2005 confirmed the good quality of raw water in the Malenščica Spring.

Bacteriological analyses of raw water are performed once a month. In the period from 1992 to 2006, only 5 of the 160 samples analyzed were within the standards for drinking water. In the majority of the samples, the problematic parameter was total coliform bacteria and, in many of them, *Escherichia coli*.

The manager of the water supply system introduced the HACCP (hazard analysis and critical control point) system. In the case of detected samples that do not meet the standards for drinking water, adequate corrective measures are taken and the results of these measures are then tested with the analysis of control samples.

The Malenščica Spring is also included in the regular monitoring program of the Environmental Agency. Four times per year, analyses of basic parameters, total pollution, microelements, pesticides, metabolites, and bacterial analysis are carried out. In 2004 and 2005, a good-quality status of the Malenščica Spring was reported (Kranjc, 2007). Concentrations of organic matter increased slightly (mean values were chemical oxygen demand = 1.5 mg O_2/L, total organic carbon = 1.5 mg C/L). Concentrations of heavy metals were low, and only the concentrations of copper (mean value, 0.5 μg/L) were above the detection limit.

Concentrations of absorbable organic halogens (3.4 μg Cl/L) increased slightly. For the period from 1998 to 2005, no negative trends in water quality were observed, but increased concentrations of heavy metals (Pb, Cr, Cd) were detected in the sediment at the spring, which indicates possible pollution.

A comparison is interesting with the Cerkniscica surface stream on Cerknica Polje, which has significantly lower quality. Although it flows toward the Malenščica Spring, the quality of the latter is better, mainly due to the inflows of clean water from the Javorniki karst aquifer. Despite various pollution sources in the recharge area, the contribution of high-quality water from generally uninhabited hills overgrown with forest is decisive for preserving the satisfactory quality of the Malenščica Spring.

However, one probable reason for the constantly good results of water quality analysis is that these controls are not frequent enough. The water quality of karst springs changes very rapidly, and therefore more representative results would be obtained with more frequent analyses. Additionally, the characteristics of flow should be considered and the monitoring plan should be adapted to hydrological conditions. Some suggestions are offered in the Conclusion.

10.3.7 STATE OF PROTECTION AND SOURCES OF POLLUTION

An expert basis for the protection of the Malenščica Spring was prepared in 1987 (Habic, 1987). Based on the existing data on geological, geomorphological, speleological, and hydrogeological characteristics and on the results of tracer tests, the extent of the recharge area was defined. Within it, five protection zones, with the most rigorous regime in zone I, were proposed (Figure 10.3–2):

- Zone I is the area of the spring within a fence.
- Zone II comprises approximately 22 km^2 of karst aquifer in the immediate hinterland, including Rakov Skocjan.
- Zone III has an extent of 198 km^2 in the area of karst poljes.
- Zone IV has an extent of 313 km^2 in the area of the Javorniki and Sneznik karst massif.
- Zone V has an extent of 213 km^2 in the area of the surface basin of the Cerkniscica River, the Bloke plateau, and the upper part of the Pivka valley.
- Unfortunately, no protection decree has been legally adopted. According to the legislation at the time, the municipalities were responsible for adopting such decrees. In the case of the Malenščica Spring with its extensive catchment, this would involve several municipalities. Some of them do not use this resource for their water supply, and this is probably one of the main reasons they have not adopted a protection decree.

According to new Slovene legislation, the state is now responsible for such decrees, but the problem of protecting the Malenščica Spring has not yet been solved and remains an important task for the future. In recent years, a new methodology for defining protection zones has been developed and new data about the characteristics of the karst system in the recharge area of the Malenščica Spring have also been gathered. It would therefore make sense to revise first the existing expert basis using modern techniques, then define adequate protection measures and adopt a suitable protection decree.

The Malenščica Spring has an extensive recharge area, and the number of potential sources of pollution is correspondingly high. Only Postojna and Cerknica have sewage systems with a water treatment plant. Industry in bigger settlements additionally burdens the environment. Agriculture is more developed in the area of karst poljes, but in general, traditional extensive practices are employed. Slovenia's central military training area is situated in the area of Pocek in the Javorniki massif, and nearby is the landfill for nonhazardous waste from the Postojna and Pivka municipalities. According to the new Slovene legislation, landfills in karst areas have to be closed. Even before this, however, the monitoring of negative impacts on the environment, including measurements of the parameters of groundwater contamination by hazardous substances, must be organized (Petric and Sebela, 2005). The traffic routes (expressway, roads, railway) that cross the recharge area are an additional source of pollution. The Karst Research Institute has monitored runoff water from the expressway (Kogovsek, 1995), and various contaminants (indicated by the parameters for chemical oxygen demand, biochemical oxygen demand, Pb, Cd) have been detected in concentrations above the maximum allowed values. This is permanent pollution, and cumulative concentrations can be more harmful; therefore, runoff water from roads should be treated before it flows into the karst aquifers. Heavy metals were found in higher concentrations in the sediments at the Malenščica Spring, but due to the variety of potential sources of pollution, it is difficult to determine reliably the source responsible. All these findings confirm the importance of regularly monitoring the influences of different sources of pollution on karst waters.

10.3.8 PLANNED FURTHER RESEARCH

The Malenščica Spring remains for the foreseeable future an important water resource for the supply of the wider Postojna area, and further research is therefore aimed primarily toward its better protection. An important starting point is the study of relations between various contribution areas within the catchment, which are strongly dependent on meteorological and hydrological conditions, which change very rapidly. To better understand these relations, the meteorological, hydrological, and hydrochemical parameters at the springs and at various locations within the recharge area are being monitored. The gathered data will be compared with the results of a combined tracer test. A simultaneous injection of two artificial tracers in the ponor of the Pivka River in Postojna Cave and in one of the ponor caves at the northwestern edge of Cerknica Polje at high water is planned to assess the characteristics of inflow from the Pivka and Cerknica parts of the catchment. The results obtained about the characteristics of flow and transport of substances through the karst water system will provide valuable information for the planning of protection. Modern techniques of vulnerability and risk mapping will be applied and tested. The ultimate goal is to use these findings as an expert basis for the proper protection of the Malenščica Spring as an important source of drinking water.

10.3.9 CONCLUSION

In Slovenia, karst springs are the most important sources of water supply. In general, their capacity is sufficient, and most problems are related to their protection against various sources of pollution. The presented case study of the Malenščica Spring shows a typical example of a karst water source in Slovenia, but similar conclusions could be drawn for numerous karst springs all around the world. Three main problems discussed in the text should be emphasized:

- The specific nature of karst aquifers, especially the large extent of catchments and the unknown characteristics of groundwater flow.
- Implementation of proposed protection measures in practice.
- Inadequate approach to the monitoring of karst water quality.

On the one hand, the implementation of strict protection measures over large catchment areas ensures the preservation of a good-quality water resource, but on the other, it limits the activities of people living within these areas as well as the development of their communities. An effective solution for protection involves finding a reasonable balance between the two options. A prerequisite for this is a better understanding of the functioning of karst water systems, which can be achieved through various research methods, such as tracing with natural and artificial tracers. In this way, a more detailed delimitation of areas with different degrees of vulnerability can be prepared and adequate protection measures within them suggested.

The next step is to implement the suggested protection measures in practice. Slovene experience with the assignment of this task to local administrations has not been very good. Municipalities within the protection zones that do not use the drinking water resource in question often ignored the need to adopt a protection decree, and no protection measures have been implemented on their territory. With the new legislation adopted, the transfer of this task to the state level should bring a necessary improvement.

An important measure for water protection is the regular monitoring of water quality, which can detect possible pollution. However, the results of the presented case study indicate that the monitoring of water quality performed on karst springs is often inadequate and certain problems with water quality can be overlooked. The state of the water quality of karst springs changes very rapidly following changes in hydrological conditions, and therefore a monitoring plan should be based on a good understanding of the characteristics of water flow and the transport of substances in karst water systems. When only occasional sampling is done, it should be carried out during various hydrological conditions. When possible, a more detailed sampling during a selected water wave, after the first, more intensive precipitation following a dry period, is recommended. Such precipitation events cause a more intensive washing out of contaminants and higher possible pollution can be detected at the springs.

REFERENCES

Buser, S., Grad, K., Plenicar, M., 1967. Osnovna geoloska karta SFRJ, list Postojna 1:100000. Zvezni geoloski zavod, Belgrade, Serbia.

Car, J., Gospodaric, R., 1984. O geologiji krasa med Postojno, Planino in Cerknico. Acta carsologica 12, 91–106.

Gams, I., 1970. Maksimiranost kraskih podzemeljskih pretokov na primeru ozemlja med Cerkniskim in Planinskim poljem. Acta carsologica 5, 171–187.

Gospodaric, R., Habic, P., 1976. Underground water tracing. Investigations in Slovenia 1972–1975. Institute for Karst Research, Ljubljana, Slovenia.

Habic, P., 1986. Pomen speleohidroloskih raziskav pri zajemanju kraskih vodnih virov. Nas krs 12, 17–30.

Habic, P., 1987. Raziskave kraskih izvirov v Malnih pri Planini in zaledja vodnih virov v obcini Postojna. Report, Karst Research Institute, Postojna, Slovenia.

Kogovsek, J., 1995. Podrobno spremljanje kvalitete vode, odtekajoce z avtoceste in njen vpliv na krasko vodo. Annales 5 (7), 149–154.

Kogovsek, J., 2001. Monitoring the Malenščica water pulse by several parameters in November 1997. Acta carsologica 30 (1), 39–53.

Kogovsek, J., 2004. Physico-chemical properties of waters in the Malenščica recharge area (Slovenia). Acta carsologica 33 (1), 143–158.

Kogovsek, J., Petric, M., 2004. Advantages of longer-term tracing—Three case studies from Slovenia. Env. Geol. 47, 76–83.

Kolbezen, M., Pristov, J., 1998. Povrsinski vodotoki in vodna bilanca Slovenije. MOP-Hidrometeoroloski zavod Republike Slovenije, Ljubljana, Slovenia.

Kranjc, M., 2007. Porocilo o kakovosti podzemne vode v Sloveniji v letih 2004 in 2005. Agencija Republike Slovenije za okolje, Ljubljana, Slovenia.

Michler, I., 1955. Rakov rokav Planinske jame. Acta carsologica 1, 73–90.

Petric, M., Sebela, S., 2005. Hydrogeological research as a basis for the preparation of the plan of monitoring groundwater contamination—A case study of the Stara vas landfill near Postojna (SW Slovenia). Acta carsologica 34 (2), 489–505.

Pezdic, J., Urbanc, J., 1987. Sledenje kraskih tokov z uporabo stabilnih izotopov kisika v vodi. Nase jame 29, 5–15.

Placer, L., 1981. Geoloska zgradba jugozahodne Slovenije. Geologija 24 (1), 27–60.

CHAPTER

Case Study: Hydrogeology and exploitation of Izvarna Spring, Romania

10.4

Adrian Iurkiewicz
Environmental Geology and Geophysics Research Department, University of Bucharest, Romania

10.4.1 INTRODUCTION

The size (amount) and stability of water flowing out from Izvarna-Celei swamps bordering the calcareous slopes of the Valcan Mountains for centuries probably impressed the local inhabitants and the pilgrims passing through the northern Oltenia region. This could also have been the reason why the swamps were mentioned as "piscinae de Cheley" within the *Diploma* granted in 1247 by Bela IV, king of Hungary, to the religious order of the Malta Knights (Conea, 1937). The first information related to a centralized water supply in the area dates back to the 14th century and is related to the water supply of Tismana Monastery, which is one of the oldest, most important, and most beautiful monasteries in Romania.

Integrated within the western group of the southern Carpathians and oriented west southwest-east northeast, the Valcan Mountains extend over more than 900 km^2 from Motru valley to Jiu deep valley. The altitude ranges mainly between 400 (500) and 1800 m, with a longitudinal ridge lowering abruptly to the north northeast (Petrosani Depression) and in cascades toward the south. Important calcareous areas cover the southern slopes of these mountains, where all representative karst features, such as dolines, ponors, caves, and sinkholes, occur.

10.4.2 PREVIOUS INVESTIGATIONS

Based on investigations and regular discharge monitoring conducted over the years by different government institutions, the main group of springs of Izvarna was tapped in 1966 for the water supply of the city of Craiova. Information on main karst systems of the southern Valcan can be found in Ilie (1970, 1973); Sencu (1967); T. Constantinescu (1975); and Vintilescu, Constantinescu, and Diaconu (1970). Other data were gathered and included in unpublished reports or investigations conducted for the implementation of the Cerna-Motru-Tismana hydropower project. Key underground connections between the main surface water inflows and karst springs of the southern Valcan Mountains were described by Radulescu et al. (1987).

Overall hydrogeologic surveys of the karst systems on the southern slope of the Valcan Mountains were conducted by a Prospectiuni Company team (Iurkiewicz, Slavoaca, and Slavoaca, 1991/1992) and concluded by Iurkiewicz and Mangin (1994), Iurkiewicz et al., (1996) and eventually by Iurkiewicz (1994, 2004).

Copyright © 2010, Elsevier Inc. All rights reserved.

10.4.3 GEOLOGY

As mentioned previously, the Valcan Mountains are included in the group of the western ranges of the southern Carpathians, all of which have a similar tectonic pattern, dominated by the overthrust of the Getic Domain over the crystalline and sedimentary formations of the Danubian Domain. According to the model proposed by Pop (1973), Pop et al. (1975), Marinescu et al. (1989), and Stan et al. (1979) for the geologic map of Romania (1:50,000), sheets of Tismana, Peştişani şi Câmpu lui Neag, the overall geological structure of the Valcan Mountains (Figure 10.4–1) includes a crystalline basement, consisting of crystalline schists and granites of Precambrian age, and a sedimentary cover, starting with Lias formations (sandstone, shale, and conglomerate).

The limestone, the thickness of which may reach 1000 m, is of Dogger-Aptian age and partly layered (the lower sequence) and partly massive (the upper sequence). It is unconformably covered by a flyschoid series of

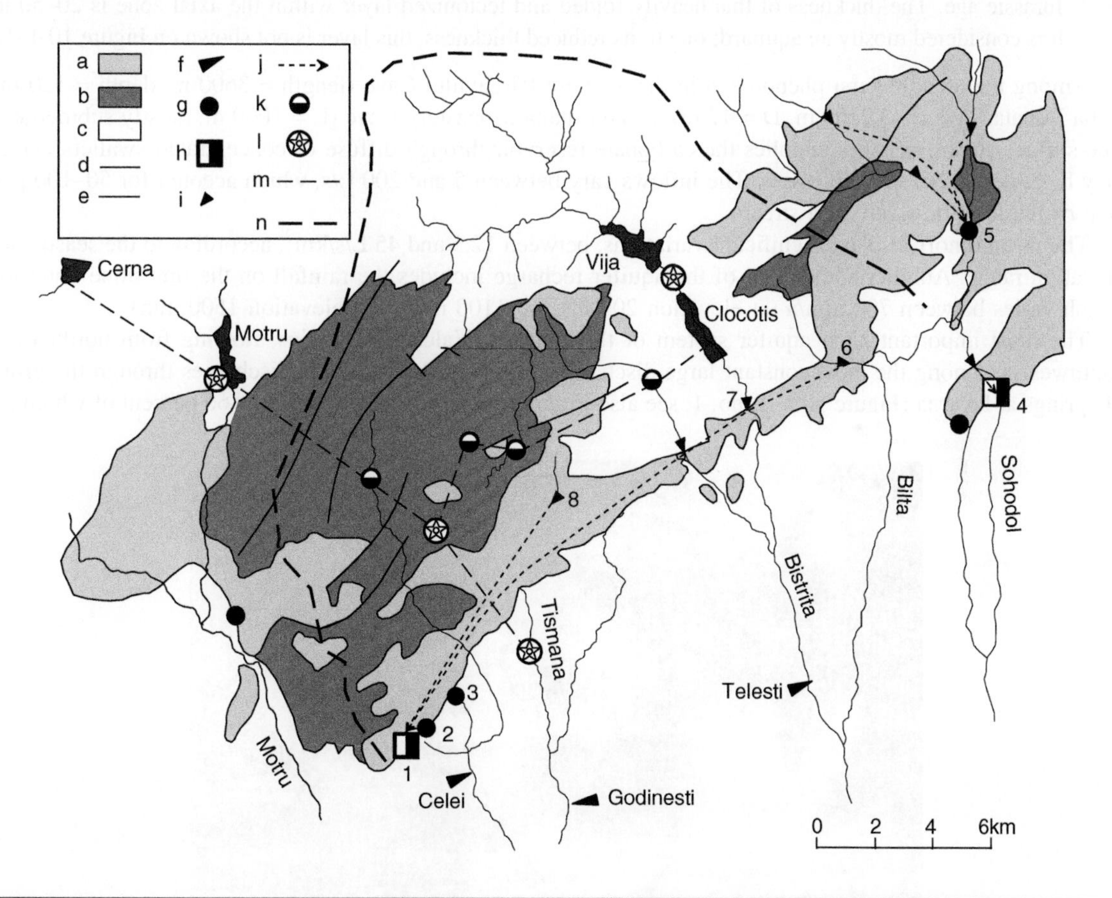

FIGURE 10.4–1 Izvarna karst system: (a) Limestone, (b) granitoids, (c) metamorphic or sedimentary rocks (mainly impermeable), (d) geologic boundary, (e) fault, (f) hydrometeorological station, (g) karst spring, (h) water intake structure, (i) swallet, (j) proved underground drainage, (k) local surface stream intake, (l) hydropower plant, (m) water transport galleries, and (n) estimated catchment of the system. Karst springs: (1) Izvarna (tapped); (2) mill; (3) Bolborosu; (4) Sohodol (Valceaua, tapped); (5) Picuiel (group). swallets (ponors); (6) Bâlta Creek; (7) Bistrita valley; and (8) Pargavu Creek.

Cenomanian-Senonian age. Part of the Jurassic-Cretaceous sediments, limestones occupy a normal position between the Liassic sandstones and the Upper Cretaceous flysch, while other parts are thrusted over the flysch and largely occur as relatively thin bodies. Towards the south southwestern part of the massif, the limestone of Urgonian facies frequently display a massive, reef character and build a low and elongated barrier, crossed by many surface streams. The existing fault systems of regional amplitude and the overall high degree of fracturing favored the organization of remarkable underground drainages. Apart from limestone, two other formations likely play an important role to the recharge of the carbonate aquifers:

- Granites and granitoids of the plutonic Tismana body to the north of the carbonate complex. These rocks largely occur from Bistricioara valley up to Motru valley, displaying good fissure permeability, which favors the storage and circulation of the groundwater (Figure 10.4–1).
- The package of sedimentary rocks that starts with arkosian (feldspathic) "Schela Formation" of Lower Jurassic age. The thickness of this heavily folded and tectonized layer within the axial zone is 20–50 m. It is considered mostly an aquitard; due to its reduced thickness, this layer is not shown on Figure 10.4–1.

Among the notable karst phenomena in the area are Pârgavului Cave (length = 3600 m, depth = 120 m), Râpa Vânata Cave (L = 2100 m, D = 47 m), and Tismana Monastery Cave (L = 1000 m, mostly submerged). The surface stream network supplies the carbonate reservoir through diffuse or concentrated swallets, which may be either partial or total losses. The inflows vary between 5 and 200 L/s, which account for 50–100 percent of the total flow rate of a stream.

The mean runoff displays significant variations, between 12.5 and 45 L/s/km^2, according to the season and elevation range. Another component of the aquifer recharge includes the rainfall on the limestone outcrops, which varies between 700 mm/a (at elevation 200 m) and 1100 mm/a (at elevation 1500 mm).

The most important karst aquifer system of the southern Valcan Mountains, flowing from northeast to southwest, is among the most constant large discharge springs in Romania and discharges through the group of springs at Izvarna (Figure 10.4–1, no. 1; see also Figures 10.4–2 and 10.4–3), some 60 percent of which are

FIGURE 10.4–2 Izvarna Spring before tapping.

FIGURE 10.4–3 Old mill at the spring in 2000. (Photo by N.Tomoniu with permission.)

tapped and piped for water supply of the city of Craiova (300,000 inhabitants). The system is developed in a low, elongated limestone body, dissected by many surface streams. There is morphologic and hydrologic evidence (Iurkiewicz and Mangin, 1994) that an upper underground karst drainage level overlies the main deep drainage. It has been also inferred that the karst system receives an additional underground supply provided by a fissured aquifer hosted by a Tismana granitoid body located to the north (Radulescu et al., 1987).

10.4.4 SYSTEM DISCHARGE AND WATER INTAKE STRUCTURES

The decision to tap the springs of Izvarna for the public water supply of the city of Craiova (115 km from the springs) was made following complex investigations carried out during 1957–1960. The amount of water for capture and the design of the intake structure were selected based on the results of geologic studies, installation of exploratory wells, geophysical investigations, and the spring flow monitoring.

The detailed geology in the spring area was clarified on the basis of several boreholes of different depths, also used for the design of the intake structure. Vertical electric soundings (VESs) also contributed to the identification of an important dip-slip fault (or system of faults) with a displacement of more than 150 m (Figure 10.4–4). To this extent, the inferred fault represents the main control for the spring occurrence. A few shallow wells drilled through the overburden layer (around 9 m thick) surprisingly proved no lateral leakage; therefore, the building (Figure 10.4–5) was founded to –2.5 m in gravel and not in limestone.

As previously mentioned, during the initial monitoring interval (1957–1960), each group of springs was measured separately. The results are included in Table 10.4–1, where the third column shows the minimum

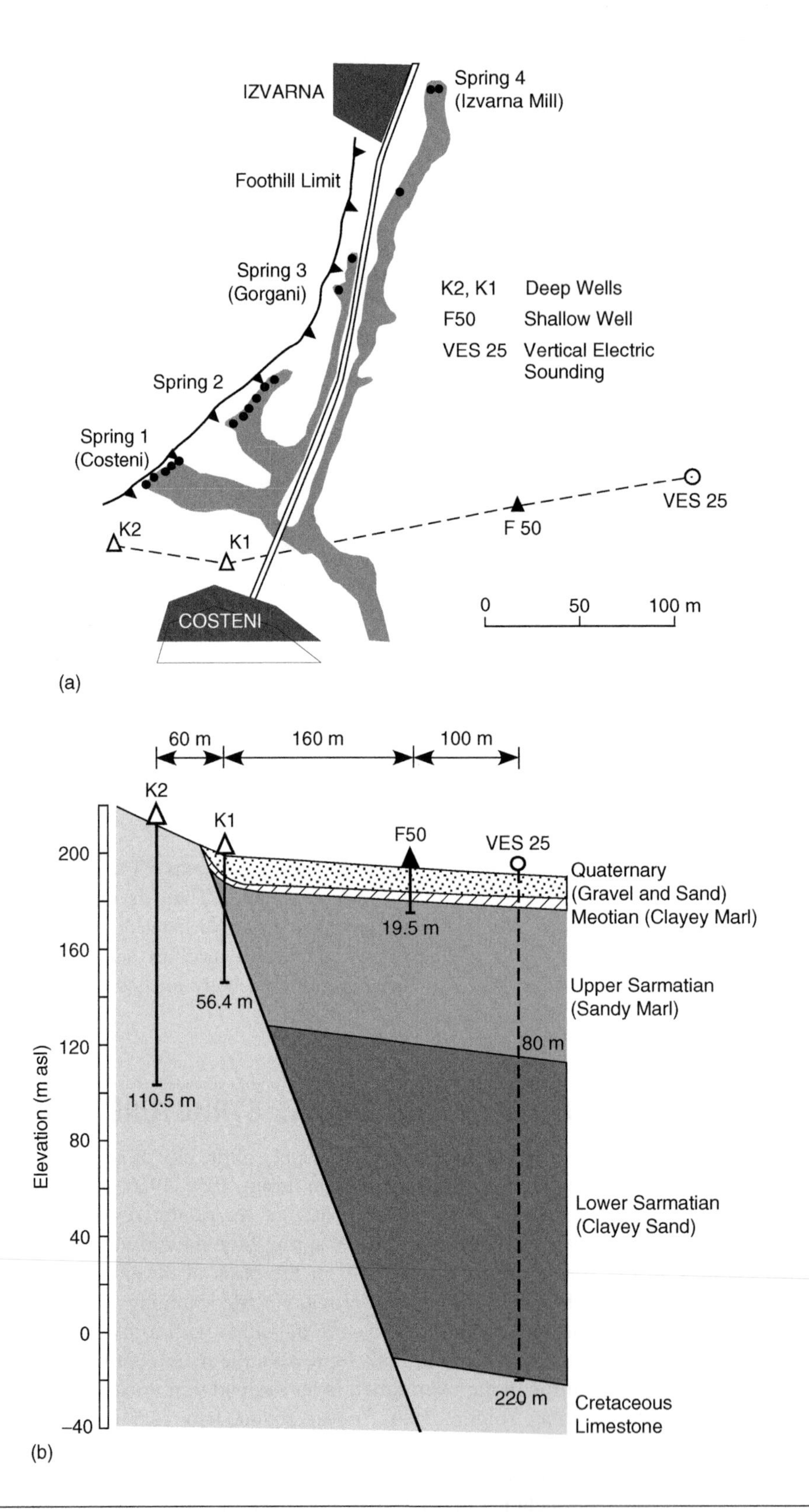

FIGURE 10.4–4 (a) Map and (b) cross section of the Izvarna Spring area. (Simplified from Constantinescu, 1980, and Teodorescu, 1965.)

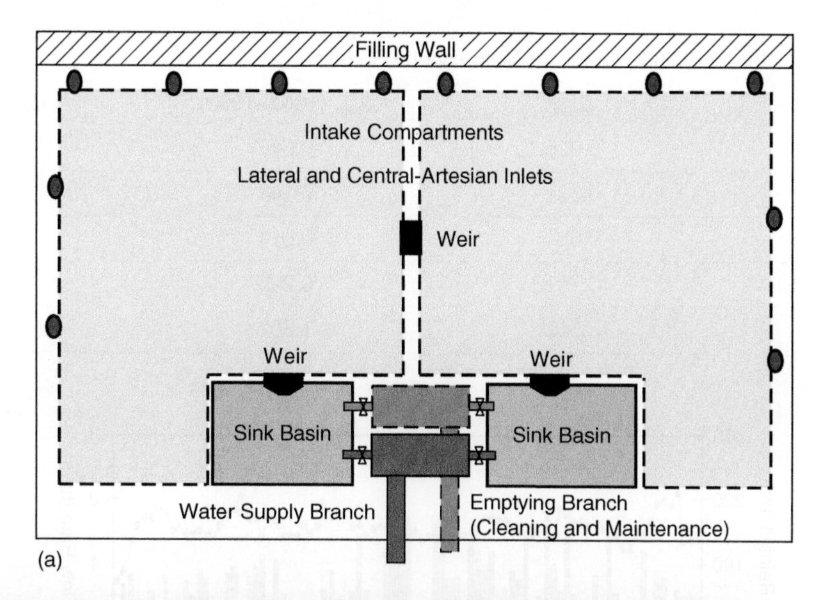

FIGURE 10.4–5 (a) Schematic of the main intake system of the Izvarna Spring (not to scale); (b) main intake basin. (Simplified from Constantinescu, 1980, and Teodorescu, 1965.)

flow rates with a probable frequency of 6.6 percent ($^1/_{15}$), based on the correlation with the time series (1921–1960) of the neighboring Tismana stream. It is worth mentioning that neither the maximum nor minimum flow rates were recorded simultaneously for all the springs. The cumulative discharge from all springs was gauged for four subsequent years (1961–1964) and the whole hydrograph (1957–1964) is presented in Figure 10.4–6.

Table 10.4–1 Characteristic Values (m^3/s) for Main Izvarna Springs

Spring Name	Q_{max}	Q_{min} (1957–1960)	Q_{min} (6.6%)
Spring 1 Costeni	0.925	0.645	0.405
Spring 2	0.231	0.148	0.095
Spring 3 Gorgani	0.654	0.514	0.315
Spring 4 Izvarna (Mill)	0.701	0.295	0.185
Total	2.511	1.602	1.000

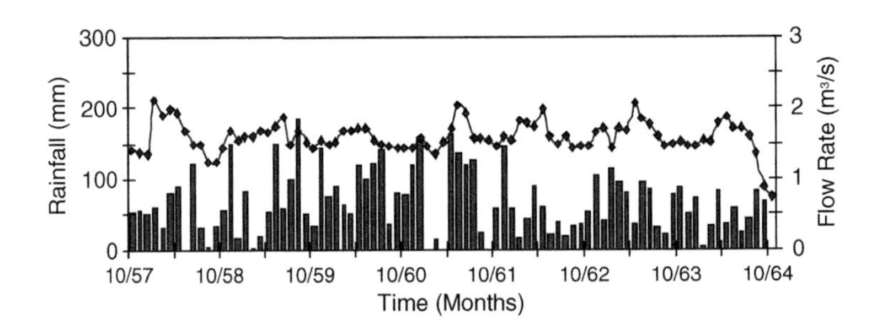

FIGURE 10.4–6 Flow rate (monthly average) versus rainfall (1957–1964) of the Izvarna Springs.

10.4.5 RECHARGE OF THE SYSTEM

The main inflows of the system include, from east to the west, Bâlta valley (10–30 L/s; no. 4 on Figure 10.4-1), Bistrita valley (200–400 L/s; no. 7 on Figure 10.4–1), and Pargavu valley (10–50 L/s, no. 8). Losses were also identified along the valleys of Bistricioara (50–70 L/s) and Pocruia (10–15 L/s), most probably part of the same system (Table 10.4–2).

Table 10.4–2 Morphohydrographic Characteristics of the Izvarna System

Spring		Main Losses			
		Valley	Bâlta	Bistrita	Pargavu
		Elevation (m)	420	325	370
		Q (L/s)	10–30	200–400	10–40
Name	Elevation (m)	Q (L/s)	Aerial Distance Ponor–Spring (km)		
Izvarna (Caught)	200	1400–2200	19.7	15.6	10
Izvarna (Mill)	202	200–400	19.4	15.3	9.7

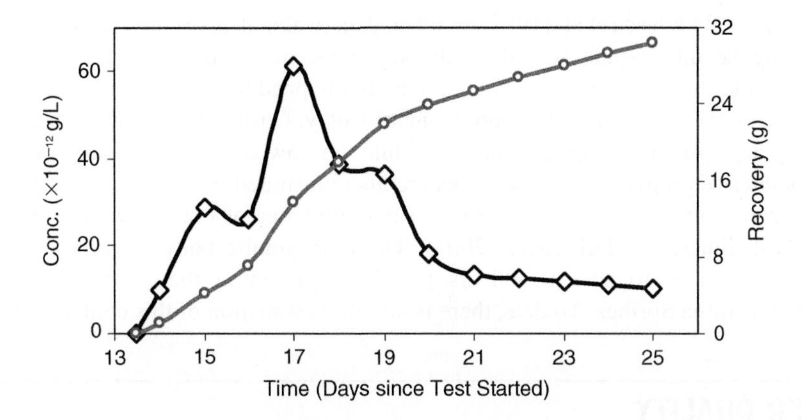

FIGURE 10.4–7 Tracer experiment of Bâlta stream losses.

Above the longitudinal drainage, which covers a distance of 20 km between the extreme points that tracing experiments proved to be connected (Figure 10.4–7), several other systems exist, apparently independent of the main one. The previously mentioned important cave systems in the area are in fact associated with these independent systems.

Clear delineation of the system catchment poses many problems. Hence, the water balance computations (Radulescu et al., 1987) performed on the 1953–1980 data showed major discharge deficits at gauging stations on the Tismana and Bistrita rivers, as well as surplus discharges measured at Orlea (including Izvarna Springs piped to Craiova, Table 10.4–3). Most likely this surplus is even higher due to usual underestimation of evapotranspiration when using the Turc formula (Kullman, 1990). On the other hand, the hydrographic basin of valleys, upstream of the limestone bar, cannot be entirely attributed to the Izvarna catchment. A more complicated situation is represented by the eastern limit of the catchment, disputed with another important aquifer structure of the southern Valcan Mountains, the so-called Jales-Valceaua karst system. It is assumed that part of the latter is already captured via a partial ponor located on the Sohodol valley. Along the northern limit of the system the inferred contribution of the granite body to the recharge of the karst system has been

Table 10.4–3 Elements of Water Budget (Simplified from Radulescu et al., 1987)

	Area	Average Elevation	Rainfall		Evapotransp. (Turc)		Runoff	Water Budget
Basin	km^2	m	mm	m^3/s	mm	m^3/s	m^3/s	m^3/s
Tismana	126	501	1040	4.16	426	1.70	1.6	(+) 0.85
Bistrita	270	721	1120	9.59	430.7	3.69	4.95	(+) 0.95
Orlea	61.8	573	1060	2.08	427.4	0.84	2.68	(–) 1.44
Total	759.8		1073.3	15.82	428	6.23		(+) 0.36

Note:
(+) Deficit: Rainfall > Evapotranspiration + Runoff (Input > Output).
(–) Surplus: Rainfall < Evapotranspiration + Runoff (Input < Output).

averaged to 8 L/s/km^2 by Radulescu et al. (1987) based on data (water affluxes through active fissures) from the galleries mined during the implementation of the hydropower system (Cerna-Motru-Tismana). A comparatively similar value of 10 L/s/km^2 resulted for a small granite body (mapped to 2.8 km^2) occurring eastward from the Izvarna karst system, on the neighboring Sohodol (Runcu) valley. During the low-water period, a sector of the valley is completely dry upstream of the granite area, while the flow at its downstream limit is on the order of 25–30 L/s. The conductive properties of these rocks are also illustrated by diffuse losses of 18–20 L/s on the granite bed (15 percent from the total river flow), in a fractured area of Pocruia valley 300–400 m upstream from the contact with limestone (Iurkiewicz, 2004). The total granite body extends to an area of around 40 km^2, that roughly leads to around 350–400 L/s, possibly representing the granite recharge component to the total discharge of Izvarna Springs. To date, there is no other estimation of this component.

10.4.6 WATER QUALITY

Complete chemical analyses performed within SC Prospectiuni laboratories indicated the presence of calcium bicarbonate and calcium-sodium water types with a very little chemical change. Sampling the karst springs for two subsequent years (low-water season) showed insignificant changes of chemical load, with a similar trend for most karst springs from the southern Valcan Mountains.

Additional data for the surface-groundwater relationship and other aspects concerning regional hydrogeology resulted from monthly sampling of isotope contents (^{3}H, ^{2}H, and ^{18}O) of the main emergences. Electric conductivity and flow rate measurements (Izvarna Mill spring) complemented the isotope monitoring. In the short term (one year), the temporal variation of the isotopic contents is less significant, while the electric conductivity and discharge exhibit broadly similar behavior. Based on stable isotopes contents, two main karst zones were identified (Tenu, Iurkiewicz, and Davidescu, 2008):

- Upper karst zone (ZCS), extending mainly to the northeastern Valcan Mountains.
- Lower karst zone (ZCI), along the southern border, starting from Runcu springs and passing through Izvarna up to Baia de Aramă.

For both zones, point distribution on graph (Figure 10.4–8) also suggests a gradual transition eastward-westward of isotope content toward more negative content. This could be indicative of a longer groundwater transit, resulting from a differentiated recharge from areas of higher altitude.

The bacteriological content of the springs draining the upper part of the Izvarna system is remarkable high. Conversely, the low bacteriological content of the main Izvarna spring (tapped springs), draining the bottom part of the system, substantiates again the distinct origin of this component of the spring's discharge.

10.4.7 SYSTEM ANALYSIS

Use of relevant system analysis methods, as described by Mangin (1975, 1982, 1984), on the series of rainfall and discharge data recorded during 1956–1964 (courtesy of INMH and ISPIF, Teodorescu, 1964) lead to the following results:

- The low-water regime commenced after 1.2–1.3 m^3/s discharge values; the upper drainage level reacted with additional contributions during high-water seasons, when system discharges were over 2.4 m^3/s.
- The two components of the recession curves obtained for the Izvarna system separately describe the functioning manner of the subsystems. Thus, the initial very steep postrainfall decrease features the response of the shallow component of the discharge, that is, the fast depletion of the highly karstified

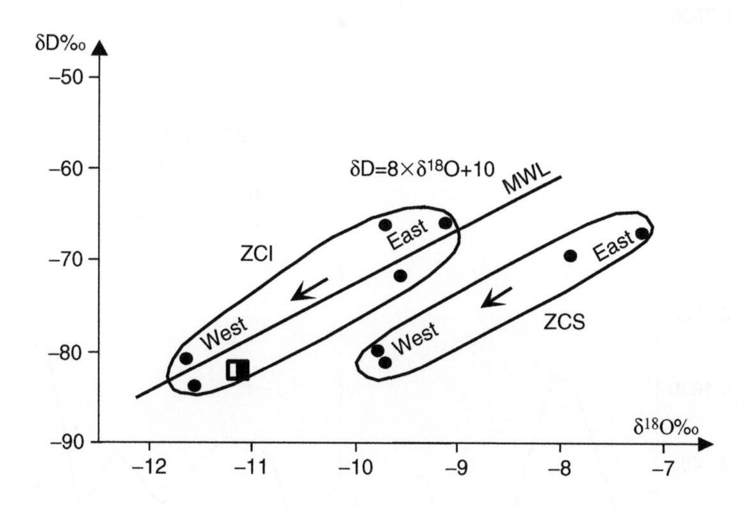

FIGURE 10.4–8 δD–δO diagram (1990) for some karst springs from southern Valcan Mountains. (Modified from Tenu et al., 2008.)

but less extended unsaturated zone; during the subsequent long depletion periods, the discharge slowly decreases, the water provided mainly by the deep component of the system. Interpretation of the α, V_{dyn}, *i*, and *k* parameters (Mangin, 1975) is awkward, since depletion occurs in a rather nonkarstic flow regime: The computed α values are abnormally low (0.0003 d^{-1}), while V_{dyn} is correspondingly large. In any case, according to the *i* and *k* (1) parameters, the system behavior better resembles that of a porous aquifer than a karstic one. At the same time, sudden oscillations, on the order of 200–500 L/s, which occur in the discharge series during the depletion period, seem to suggest that Izvarna Spring is only the overflow of a deep drainage that extends farther southwestward.

- The dynamic reserves identified through recession curve analysis were very high, between 90 and 425 million m^3.

The separate examination of each hydrologic cycle within the general frame of the correlative and spectral analyses indicates that the reserves of the system are affected by rainfall events for long periods, which may extend up to two years. In this respect, it has been observed that the recurrence of comparatively "dry" hydrologic cycles every one or two years results in drought flow rates lower than 500 L/s and, implicitly, in the occurrence of the sudden drops in the discharge series. However, this also explains the presence of high resources, even during medium-term drought seasons.

Thus it may be stated that the sudden drops of the discharges identified on the system hydrograph are connected with the occurrence of two very poor hydrologic cycles (low rainfall amount) that may succeed at maximum two year intervals. This is clearly visible on the long-term cross-correlogram (rainfall-flow rate) using filtered data (to eliminate the incidental correlation shorter than seasonal) displaying no seasonal correlation but a maximum correlation at two year intervals (750 days; Figure 10.4–9).

Based on the values of the descriptive parameters, it was concluded that the Izvarna karst aquifer system exhibits a complex organization, consisting of subsystems with various degrees of interconnectiveness. Systemic analysis identified the relationships between subsystems and provided an overall characterization of each system behavior. Thus, the Izvarna system as a whole consists of parallel subsystems that interact only within the global function of the system (Figure 10.4–10).

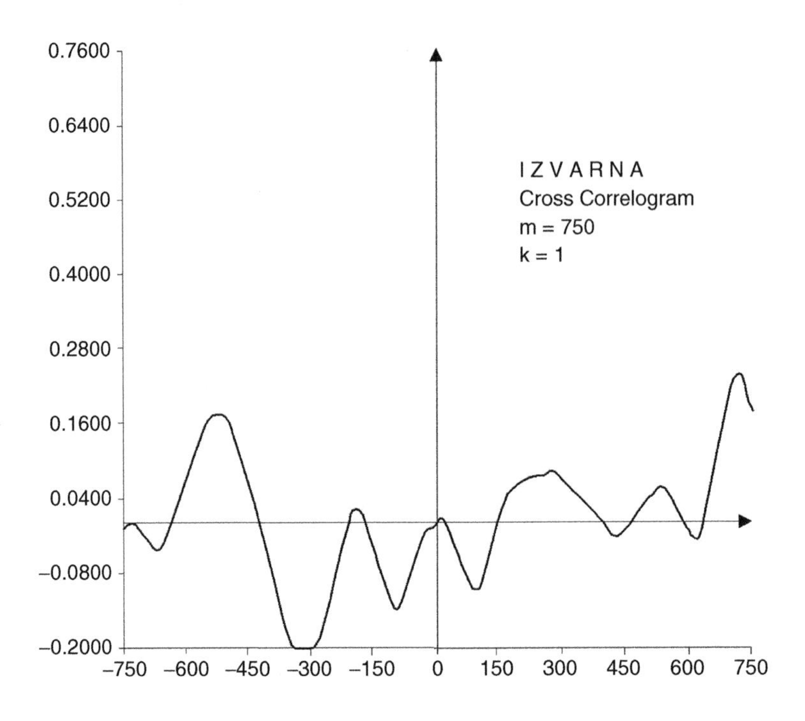

FIGURE 10.4–9 Long-term cross correlogram (rainfall-flow rate, filtered data).

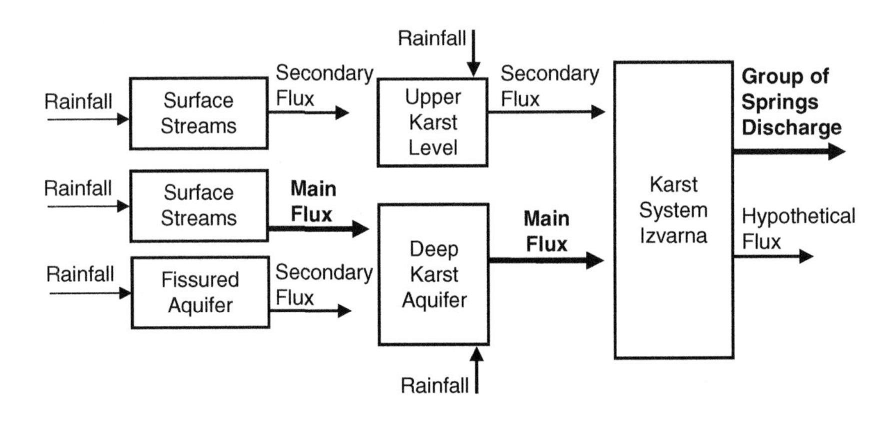

FIGURE 10.4–10 Izvarna karst system.

10.4.8 INTERFERENCE WITH HYDROPOWER SYSTEM AND TREND ANALYSIS

The hydropower system of Cerna-Motru-Tismana was designed to convey minor perennial streams through a network of secondary pipes, galleries, and intercalated small power plants that eventually emerge at a large underground power plant. The skeleton of the system consists of four hydropower plants and five main surface water bodies (lakes), out of which four are represented in Figure 10.4–1. Some five smaller intake structures built on tributaries (Tismana, Tismanita, and Bistricioara) will be added to the system. The work associated with the development of the hydropower system induced significant changes in the regime of surface streams and groundwater drainages. Large volumes of water otherwise continuously supplying the karst system via main ponors are transferred outside their catchments to supply the power system. Radulescu et al. (1987) already pointed out that these works might strongly affect the discharge of the karst springs. Consequently, the project of increasing the yield tapped by the Izvarna intake was canceled.

Trend analysis on Izvarna Springs discharges was conducted over a period of more than 15 years to predict the available discharge and possible anthropogenic influence on the tapped springs (Iurkiewicz et al., 1996). The filtered hydrograph of Orlea valley (supplied mainly by the Izvarna group of springs) was compared with the discharges recorded on Şuşiţa Verde valley (located eastward), as a nonkarstic catchment area preserving its natural flow regime (Figure 10.4–11, filter amplitude of 365 days).

The main results of the analysis are as follows:

- Incidence on both charts of a newly outlined five year periodicity, due to the long-term climate features.
- Decreasing trend of both series in the 1978–1983 range.
- Subsequent to 1983, a decreasing trend associated only to the Orlea series; during the same period, the Şuşiţa series maintained a cyclic evolution, without dropping below a virtually constant value (1.1 m^3/s); hence, 1983 can be considered as the start of the degradation of the Orlea system dynamic resources.
- A steeper decline of the Orlea series after 1992; it might reflect the Izvarna system "inertia" subsequent to the drought of that year and/or the filling of the Vija reservoir; the latter is part of the Cerna-Motru-Tismana hydropower system, to which it provides important flow rates, being situated upstream with respect to the main swallet that supplies the karst aquifer ($Q_{ins.}$ = 300–400 L/s; Figure 10.4–1).

The recession curve analysis performed, starting from 1983, indicates a dramatic reduction of the stored dynamic water volume (V_d), from 123×10^6 m^3 in 1984 to 9.16×10^6 m^3 in 1994 (Figure 10.4–12). Although during 1993–1994 the considered parameters appear to be stable and the recorded flow rate even increased at the end of the recession period (Q_r), the analysis of the cyclic (five years periodicity) behavior indicates that, at the end of the 1993–1998 cycle, the flow rates of the karst springs might drop below the yield required for the water supply of the city of Craiova (800 L/s).

The significance of all the previously discussed parameters lies in the manner in which they are used. Due to different water supply and hydropower projects, the hydrodynamic behavior of the Izvarna system has been radically changed compared with main data series (1957–1964) previously presented.

However, some general remarks extracted from the analyses of the previously mentioned data remain valid. An obvious application is to rely on the evaluation of the dynamic volume of the flooded zone in assessing the available amount of water that can be supplied to consumers. The vulnerability to pollution may be estimated as well from information on the internal structure and the hydrodynamic behavior of the system:

- The deep structure of the Izvarna system shows a strong regulation capacity, mainly due to nonkarstic hydrodynamic behavior. Accordingly, the base component of the source will be influenced less by physical pollution (immiscible pollutants) and, within the first stage (10–14 days), even by a chemical pollution (miscible pollutants). On the other hand, even in the case of a more or less accidental pollution, the component will suffer the consequences over a very long period, most likely a few years.
- The other discharge component (upper karst drainage level) is much smaller quantitatively, but its transit through the system is virtually instantaneous, which makes it highly vulnerable to pollution.

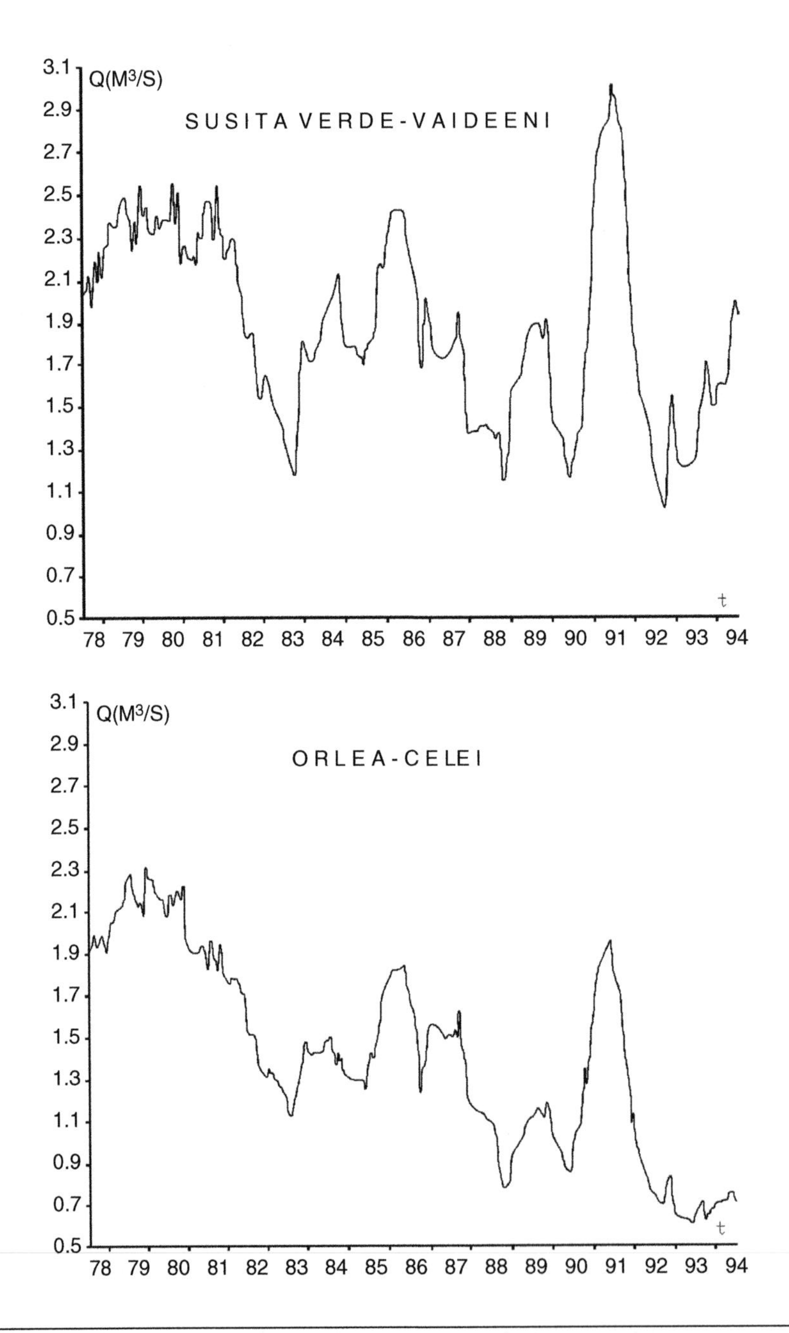

FIGURE 10.4–11 Trend analysis of the Şuşiţa Verde valley (top) and Orlea valley (bottom).

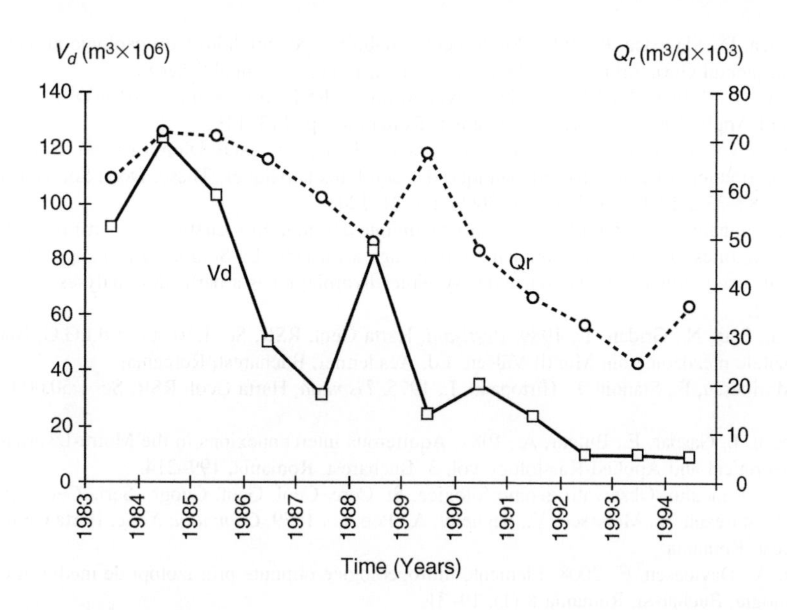

FIGURE 10.4–12 Dynamic stored water volume evolution in Orlea catchment basin: V_d, dynamic stored water volume; Q_r, end of depletion flow rate.

The data presented here only partially reflect the actual situation of the aquifer resources. There are no recent data to quantify the real impact of the interactions between stress factors, among which the most important are climate change and the full operation of the hydropower system. The highly delayed response of the aquifer to natural or anthropic stresses requires careful planning of any operation touching the karst aquifer in one way or another. Efforts have been made to collect new data and draw attention to the possible danger. Recommendations to integrated surface-groundwater management capable of satisfying the industrial and potable water demand in the area are also envisaged.

REFERENCES

Conea, I., 1937. Basinul Izvarna-Celei, cuib de veche viata romaneasca, Corectari geografice in Diploma Ioanitilor (1247). Buletinul Societatii Regale Romane de Geografie LVI, 246–270.

Constantinescu, G., 1980. Captarile de apa subterana din Romania, Ed. Tehnica. Bucharest, Romania.

Constantinescu, T., 1975. Considérations sur les grottes situées entre les rivieres Susita Verde et Sohodol (Monts Vâlcan-Carpates Meridionales). Trav. Inst. Speol. “Emil Racovita” XIV, 169–188.

Ilie, D.I., 1970. Carstul din nordul Olteniei, Teza de doctorat. Archives of the University of Bucharest, Romania.

Ilie, D.I., 1973. Resurgenţele din zona Tismana—Izvarna, (NV Olteniei). Annals of the University of Bucharest, Geografie XXII, 27–40.

Iurkiewicz, A., 1994. L’interet de l’analyse systémique dans l’evaluation de la vulnerabilité à la pollution des systémes karstiques des Monts Valcan. Proceedings of the International Symposium on the Impact of Industrial Activities on Groundwater, May 23–28, 1994, Constantza, Romania, pp. 291–299.

Iurkiewicz, A., 2004. Analiza sistemică în investigarea hidrodinamică a acviferelor carstice (exemple semnificative din România). PhD thesis, University of Bucharest, Romania.

Iurkiewicz, A., Mangin, A., 1994. Utilisation de l’analyse systémique dans l’étude des aquiferes karstiques des Monts Vâlcan. In: Theoretical and Applied Karstology, vol. 7. Ed. Academiei Române, Bucharest, Romania, pp. 9–96.

Iurkiewicz, A., Slavoaca, D., Slavoaca, R., 1991. Studii pentru stabilirea potentialului hidrogeologic al depozitelor carbonatice din muntii Vâlcan, judetul Gorj. Arhiva S.C. Prospectiuni-S.A., Bucuresti (unpublished).

Iurkiewicz, A., Voica, M., Bulgăr, A., 1996. Indirect evaluation of the Izvarna Karst System Discharge Trend (Romania). In: Theoretical and Applied Karstology, 9, Bucharest, Romania, pp. 113–119.

Kullman, E., 1990. Krasovo-puklinove vody. Karst-fissure waters. Geologicky ustav Dionyza Stura, Bratislava, Slovakia.

Mangin, A., 1975. Contribution a l'étude hydrodynamique des aquiféres karstiques. These. Doct. Sci. Nat. Dijon. In: Ann. Spéléol 29(3), pp. 283–332, 29(4), pp. 495–601, 30(1), pp. 21–124.

Mangin, A., 1982. Determination du comportement hydrodynamique des aquiferes karstiques a partir de l'étude des informations fournies par leurs exutoires. In: Colloque national en hommage a Castany. La Source, Orléans, pp. 397–403.

Mangin, A., 1984. Pour une meilleure connaissance des systémes hydrologiques a partir des analyses correlatoire et spectrale. J. Hydrol. 67, 25–43.

Marinescu, F., Pop, G., Stan, N., Gridan, T., 1989. *Peştişani*, Harta Geol. RSR, Sc. 1:50.000, Ed.I.G.G, Bucharest, Romania.

Pop, G., 1973. Depozitele mezozoice din Munţii Vâlcan. Ed. Academiei, Bucharest, Romania.

Pop, G., Berza, T., Marinescu, F., Stănoiu, I., Hîrtopanu, I., 1975. *Tismana*, Harta Geol. RSR, Sc. 1:50.000, Ed.IGG, Bucharest, Romania.

Radulescu, D., Stanescu, I., Gaspar, E., Bulgar, A., 1987. Aquiferous interconnexions in the Motru-Izvarna-Tismana-Bistrita karst area. In: Theoretical and Applied Karstology vol. 3. Bucharest, Romania, 199–214.

Sencu, V., 1967. Cheile Runcului, Observaţii geomorfologice. St. Cerc. Geol. Geof. Geogr., Seria Geografie 19 (1), 81–94.

Stan, N., Stănoiu, I., Năstăseanu, S., Moisescu, V., Seghedy, A., Pop, G., 1979. *Câmpu lui Neag*, Harta Geol. RSR, Sc. 1:50.000, Ed.I.G.G, Bucharest, Romania.

Tenu, A., Iurkiewicz, A., Davidescu, F., 2008. Elemente hidrogeologice obţinute prin izotopi de mediu in carstul munţilor Valcan. Hidrogeologia, Bucharest, Romania 8 (1), 19–31.

Teodorescu, M., 1964. Alimentarea cu apa a orasului Craiova. In: Hidrotehnica, Gospodarirea apelor, Meteorologia, vol. 9/3. Bucharest, Romania. pp. 109–117.

Vintilescu, I., Constantinescu, T., Diaconu, G., 1970. Grottes, phenomenes karstiques et situation hydrologiques dans la valée de la Şuşiţa Verde (Monts de Vâlcan). Trav. Inst. Speol. 9, 9–34.

CHAPTER

Case Study: Intake of the Bolje Sestre karst spring for the regional water supply of the Montenegro coastal area

10.5

Zoran Stevanovic
Department of Hydrogeology, School of Mining and Geology, University of Belgrade, Serbia

10.5.1 INTRODUCTION

Even though the coastal region of Montenegro is of crucial importance to the tourism industry and the main generator of its economy, its long-standing water shortage problem is still unsolved. Over 150,000 inhabitants and tourists connected to the public water supply system in the coastal area experience water shortages and regular interruptions in supply during the summer months. Yet Montenegro is one of the most water-rich countries in Europe (see Chapter 10.1). It follows that one will ask why such a situation exists and what engineers have done about it.

First, Montenegro is a country where karstic rocks are dominant. Almost 90 percent of the population consumes and depends on water solely from karst aquifers. An unstable discharge regime, fast circulation, a very deep groundwater table, and the presence of water limited to discharge zones and in restricted karstic pathways are typical characteristics of its karst aquifers. Therefore, in Montenegrin land with such a high development of karst, the problem of finding and successfully tapping precious groundwater is increasing. Climatic and hydrological conditions during Pleistocene and even during recent times were not favorable for groundwater sources. Many springs along the shorelines of the Adriatic Sea and Skadar (Scutari) Lake were submerged as a result of a general rise in the seawater level, hydrographic piracy of some rivers, and intensive karstification that deepened the water table below the regional erosion base (Mijatovic, 1997; M. Radulovic et al., 2005).

Over time, the water shortage was understood as one of the main obstacles to further development, which is why, in the last few years, several synchronized mitigation actions have been undertaken. A program for urgent rehabilitation of the distribution network (losses in some cases exceeded 70 percent) was conducted in the waterworks of coastal municipalities, many intakes were improved, and several successful projects for karstic springs regulation undertaken (Radulovic, 2000; Stevanovic and Radulovic, 1997). All this was followed by the reactivation of an old idea: to tap and direct water from the continental part and water-rich Skadar basin to the coast. Therefore, construction of some crucial elements of a new regional system, such as the main reservoir on the coast, the hydrotechnical tunnel between the continent and the coast (5 km long),

Copyright © 2010, Elsevier Inc. All rights reserved.

and the coastal parts of the pipeline was completed. The issues of the source, intake structure, water treatment, and exact position of the pipeline have been left for the later project stages. Such an approach may not seem entirely reasonable, but it is the result of an attempt to speed up the project completion. Apart from different technical opinions as to how to select the best source, the long delay in this project is caused by difficulties and complicated conditions for tapping and transferring water from the Skadar basin.

10.5.2 REGIONAL WATER SUPPLY SYSTEM—CONCEPTION AND DISPUTES

The Montenegro coastal Regional Water Supply System (RWSS) will cover the water supply of the following municipalities: Herceg Novi, Kotor, Tivat, Budva, Bar, and Ulcinj (Buric and Radulovic, 2005). The scheme is based on conceptual designs prepared in 1990, when a *Water Supply Master Plan for the Coastal Area* was prepared by the Energoprojekt Belgrade and Aqua Emit Italy.

The RWSS is planned for a maximum capacity of 1.5 m^3/s in two stages, the first of which will ensure delivery of 1.0 m^3/s. The total pipeline length is about 140 km, of which the continental part from Skadar Lake to the coast is 33 km (IK Consulting Engineers, 2006). The northern coastal line from Djurmani to Herceg Novi is 73 km long, while the southern line of 34 km will supply Bar and Ulcinj (Figure 10.5–1). The cost-benefit analysis prepared in 2005 suggested that the preferred source would be the spring Bolje Sestre (Malo Blato Gulf). The World Bank and European Bank for Reconstruction and Development approved the study and its conclusions, paving the way for the design and construction.

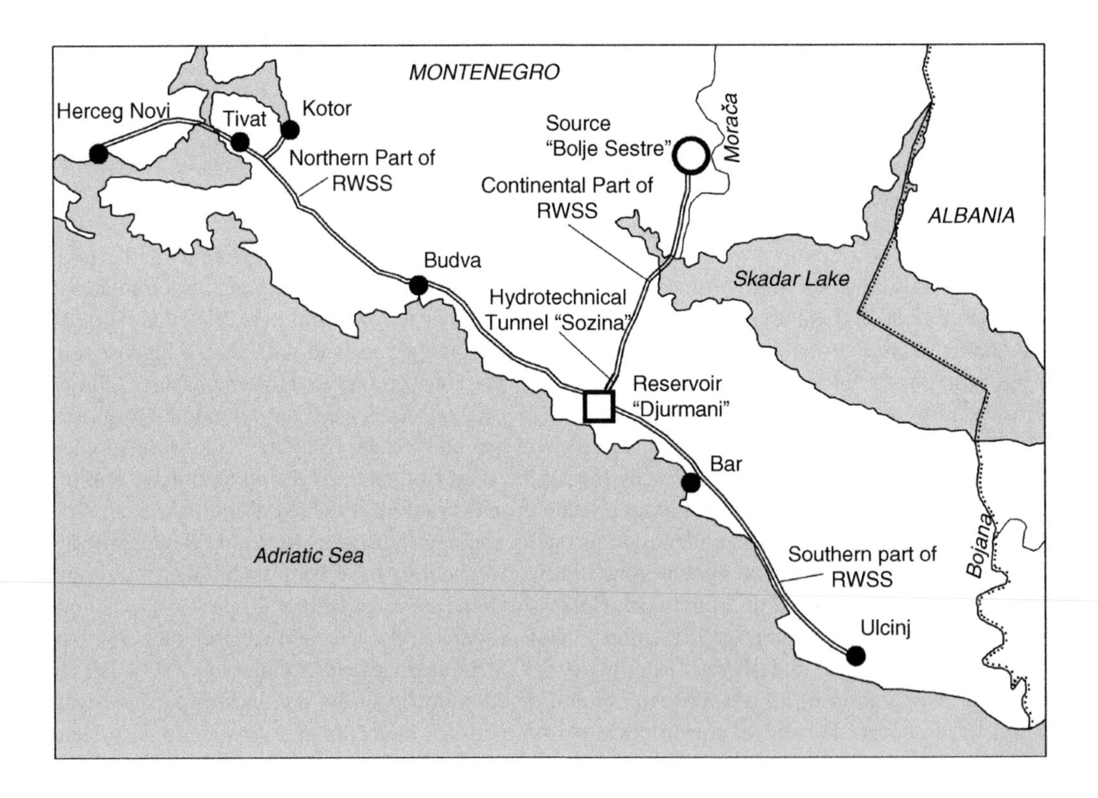

FIGURE 10.5–1 Continental and coastal water pipelines from Bolje Sestre source.

The Bolje Sestre karstic spring discharge area is located within the Skadar (Shkoder, Skutari) Lake system, the largest on the Balkan peninsula and an important biodiversity site (included in the Ramsar list). The project implementation would be coordinated with the National Parks of Montenegro by ensuring minimum environmental impact during construction and operation. The design pipelines should remain above the maximum lake water level but some parts will also be impounded. This requires particular attention to material selection and installation concerning local seismic and liquefaction conditions. An environmental impact study (ITSC Ltd. Montenegro, 2005) concluded that Bolje Sestre intake will reduce the volume of lake water by only 0.03 percent, or 2 mm of the water column.

Skadar Lake (Figures 10.5–2 and 10.5–3) is situated in the Skadar-Zeta depression, in the karstic terrain of the southeastern Dinaric Alps. The mountains Lovćen, Sutorman, Rumija, and Tarabosh lie on the southern

FIGURE 10.5–2 One of the small islands in Skadar Lake (southwestern margin).

FIGURE 10.5–3 Location of Bolje Sestre source.

side of the lake, while the Shkoder city lowlands lie on the eastern shore. Its northern coast is flat, gradually descending toward the lake, and covered with lush vegetation. The southern coast is steep and rugged. Skadar Lake is relatively shallow and the deepest part of the lakebed is below sea level, making the lake a cryptodepression. The altitude of the water table is 4.6–9 m above sea level (asl); the lake's depth near the shore is 5–9 m on average, and the maximum depth is more than 60 m. Around 60 percent of the lake is in Montenegro, while 40 percent is in Albania. The average lake surface area is 475 km^2. During the summer season, it reduces to 370 km^2, while during the winter season it increases to about 540 km^2 (Hrvacevic, 2004). Skadar Lake has a peculiar water regime, with water level fluctuations of up to 5 m. Noticeable oscillation of the water level of Skadar Lake results in long-term flooding of vast areas of the Donja Zeta. This is a major problem in littoral regions. The total lake volume is 1.9 km^3. Its catchment area occupies around 5000 km^2. Average annual precipitation reaches 2000 mm.

The Moraca River, with its two tributaries, Zeta and Cijevna/Cemi, contributes 62 percent of the lake's water. About 30 percent of it comes from many sublacustrine springs called *eyes*. The rest comes directly from the mountains or rainfall. The Bojana/Buna and Drim/Drini Rivers play an important role. The Bojana/Buna River flows from Skadar Lake (near the Albanian city of Shkoder), with an average flow rate of 320 m^3/s. Combined with the flow of the Drim/Drini River, the lake drains into the Adriatic Sea at an average rate of 682 m^3/s.

Numerous sublacustrine springs are registered along the rim of the lake. The largest are Bazagur (elevation 9.15 m), Karuc (19.45 m), and Volac (10.60 m) at the northwestern rim, and Kaludjerovo Oko (13.0 m), and Radusko Oko at the southwestern rim. The Radusko Oko is the best-known "eye," with its bottom below the middle of the lake for approximately 60 m, which means that it lies below sea level for about 53 m (V. Radulovic, 1997; V. Radulovic and M. Radulovic, 1997). The "eyes" drain a rich karstic aquifer and discharge a very large amount of water. Their average summary yield is assumed to be over 40 m^3/s, but their tapping and separation from the lake water, which is essential, is a very difficult problem to solve. In fact, many of today's sublacustrine springs had been discharged above or around the lake level until the middle of the 19th century. Then, after a major flood, significant hydrographic piracy caused the Bojana/Buna River to start to receive part of its water from the neighboring Drim/Drini River, and tail lake water generally increased by submerging most of the existing springs.

Some earlier preliminary studies identified the Karuc sablacustrine spring (Figure 10.5–4), which lies west of the Bolje Sestre in the neighboring gulf of Skadar Lake, as an optimal solution for the regional water supply (Energoprojekt, 1994; 1995). But tapping water from a depth of some 20 m without having a mixture of groundwater and lake water is very sensitive and a difficult problem to solve (Zikic, 1997). The quality of the lake water is variable, and the eutrophication process can be intensive over the summer months and in zones far from the groundwater "eyes"; nonetheless, the capital Podgorica and its industrial pollutants, led by the Alumina Plant, contribute to its further deterioration (Filipovic, 1981).

10.5.3 INVESTIGATION OF BOLJE SESTRE SOURCE AS A BASE FOR THE RWSS FINAL DESIGN

The recommendation of a cost-benefit analysis to study alternative sources, including the Bolje Sestre Spring, just opened a new chapter in this water story but, at the same time, offers a new prospective for the region (Zikic et al., 2005). Some earlier studies identified the potential of the Bolje Sestre for a regional water supply but did not provide a full argument; it is known that the spring is characterized by a significant capacity and better topographical and hydrogeological conditions for placing the intake than many other Skadar "eyes" (Radulovic, 2000). However, the identified set of major problems that require adequate responses includes the following:

FIGURE 10.5–4 Sublacustrine spring Karuc at Skadar Lake shore.

- Source minimal capacity.
- Source intake.
- Source water quality.
- Source protection against pollution.
- Water treatment.
- Water pumping and transfer to the other side of the lake.

Since 2005, the Malo Blato basin, as a northwestern part of the lake, and the Bolje Sestre Spring became targets of systematic hydrogeological research. During 2005 and 2006, detailed surveys, including hydrological measurements of the spring discharge, hydrogeological mapping (Figure 10.5–5), geophysical survey (geoelectric tomography and electromagnetic very low frequency (VLF) method; Figure 10.5–6), remote sensing analyses, drilling of investigation boreholes, tracing tests on neighboring streams, and permanent sampling and analyses of the water quality (biological, chemical, and radiological) were carried out. The obtained and evaluated results fully complied with the previous assessment that the spring Bolje Sestre could be the optimal option for regional supply.

The catchment area of the Bolje Sestre consists of highly karstified limestone of the Upper Jurassic and Cretaceous age, which are intersected by numerous faults (mostly northeast-southwest and east-west oriented). These rocks are outcropped at Kolozub and several other hills along the Malo Blato shoreline (one of the gulfs of Skadar Lake). Numerous small dolines, caverns and a few caves, and buried siphons

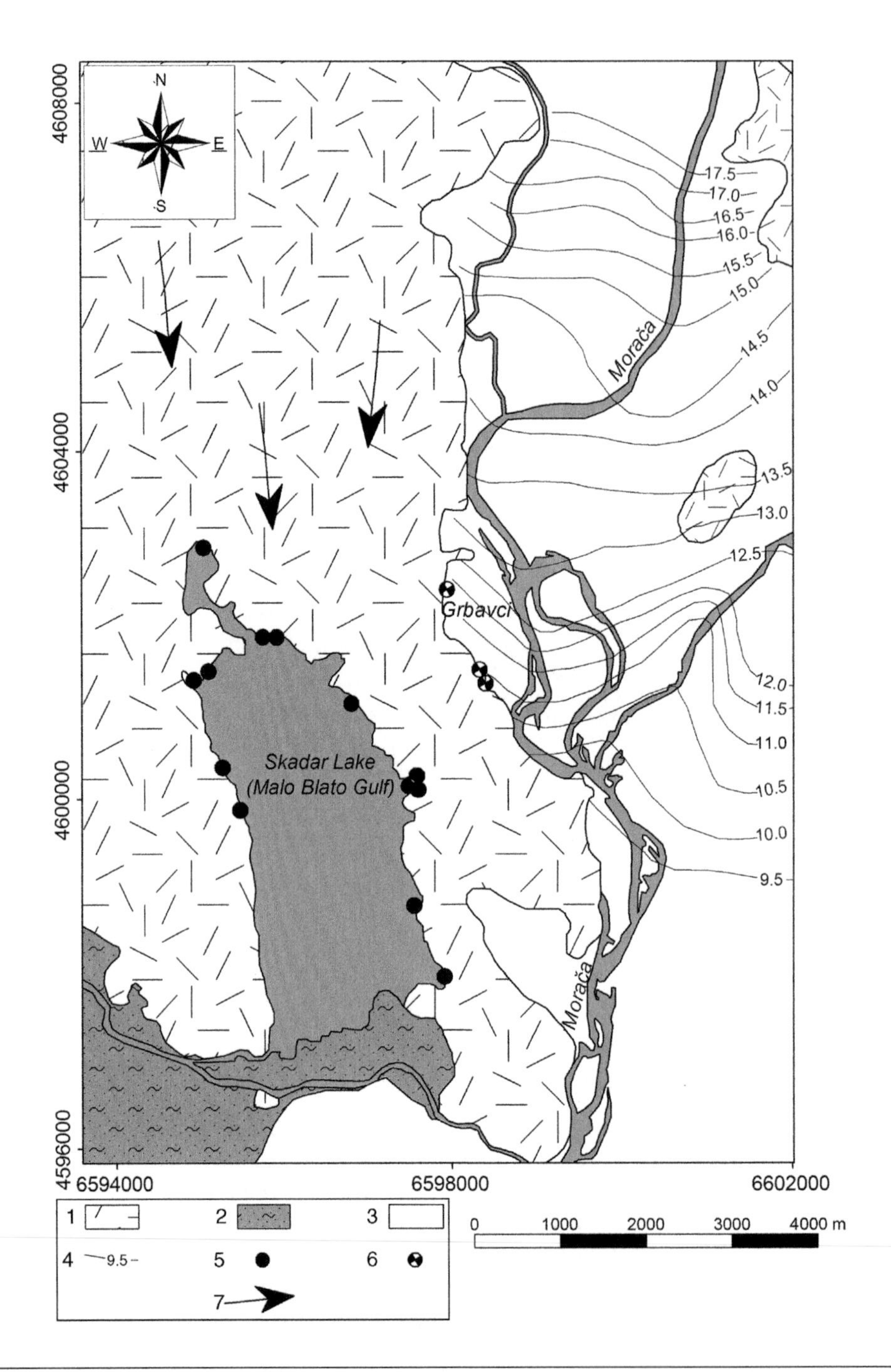

FIGURE 10.5–5 Hydrogeological map of the Bolje Sestre source: (1) Karstic aquifer, (2) low-permeable lacustrine sediments (saturated or dry depending on lake level fluctuations), (3) intergranular aquifer in fluvial and glacial sediments, (4) hydraulic head contour lines, (5) spring, (6) borehole, (7) groundwater flow direction. (From Radulovic and Radulovic, 2005.)

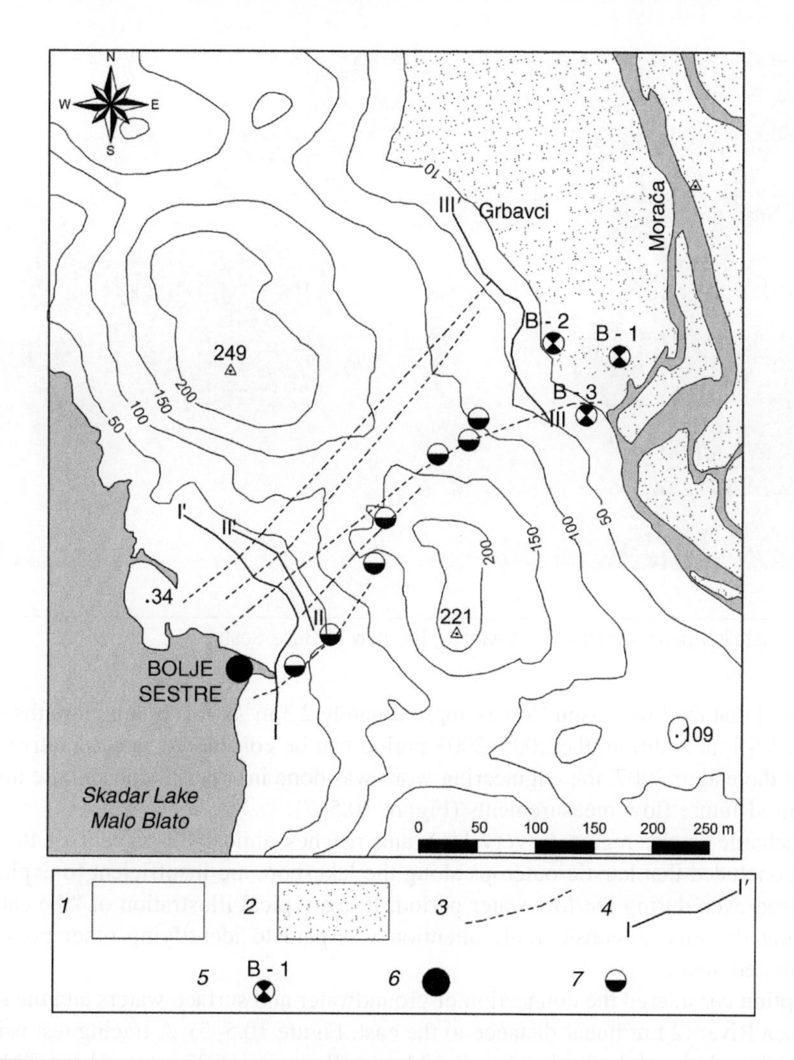

FIGURE 10.5–6 Electromagnetic VLF exploration of Bolje Sestre catchment: (1) Karstic aquifer, (2) intergranular aquifer of the Morača alluvial sediments, (3) assumed groundwater pathways, (4) VLF profiles, (5) borehole, (6) spring, (7) registered ponors (sinks, swallow holes).

are registered in this zone (Besic, 1969; V. Radulovic and M. Radulovic, 1997). Toward the east, the rich karstic aquifer descends and is overlain by recent fluvial-glacial and alluvial sediments (Figure 10.5–5).

To assess the tectonic pattern and dominated groundwater pathways within the karstic aquifer, field geophysical survey and remote sensing analysis were undertaken. They included the application of geoelectrical scanning (two-dimensional tomography) and electromagnetic survey by the VLF method. In total, the length of the geophysical profile covers some 3000 m. Remote sensing analysis of tectonic elements was conducted on Landsat satellite images.

The Bolje Sestre source arises from a typical karst siphonal discharge, which is sublacustrine, and extends across an area of approximately 10 × 5 m^2, both on the bed of the lake and along the shore edges almost at the lake level. In such a situation, the yield measurements were extremely difficult: The measurements during

FIGURE 10.5–7 The installed panel system for streaming the flow of Bolje Sestre.

2005–2007 indicated that the low season flow is approximately 2.3 m^3/s. It is much more than required for the second stage of RWSS, in addition, the 2005–2007 period can be considered representative due to very low annual rainfall. At the end of 2007, the engineering work was done in order to concentrate turbulent flow and facilitate the required future flow measurements (Figure 10.5–7).

The annual recharge in the region is very high and reaches almost 80 percent of the annual rainfall. However, it was concluded that karstic outcrops along the lakeshore are insufficient to explain the very high discharges registered even during the low-water period. It is a typical illustration of "the catchment deficit." This is why, during the survey, considerable attention was paid to identifying other possible sources and delineating contributed areas.

One studied option considered the connection of groundwater and surface waters and the seepage from the neighboring Moraca River (2 km linear distance to the east. Figure 10.5–5). A tracing test using sodium fluorescein was undertaken during July 2005, when the Moraca flow was of 32 m^3/s. About 20 kg of fluorescein was introduced into the river some 2 km upstream of village Grbavci and monitored at 12 observation points (springs around the lake, downstream sites along the riverbanks, and local water points). Although all colored water disappeared very fast after the test was started, the tracer was not registered at any of the points for the next 17 days (Hydrometeorological Survey of Montenegro, 2005). Three possible reasons were evaluated: (1) small tracer quantity, (2) slow velocity and large storage (most probable) and consequently short observation period, and (3) instruments too imprecise to detect very low concentrations.

The results of all the aforementioned activities were evaluated and taken into consideration for the next stage of the survey, which included drilling and testing of investigation boreholes. The first group of three boreholes were located in the village of Grbavci in the alluvial plain of the Moraca River (Figure 10.5–5). All were drilled to a depth of 30 m. The borehole B-1 did not reach carbonate bedrock. The hydraulic conductivity of highly permeable alluvial and underlying fluvial-glacial sediment is in the range of 2×10^{-2} to 1×10^{-4} m/s. An air lift test confirmed rich groundwater reserves. The boreholes B-2 and B-3 were drilled within the karstic aquifer at the eastern edge of the Kolozub massif, which was assumed to belong to the catchment of the Bolje Sestre Spring. Although B-2 passed through highly karstified and saturated carbonate rocks, B-3 was dry and found only compact or slightly fissured limestone. Thus, the vertical and lateral variation in permeability of this nonhomogeneous karstic aquifer was confirmed.

To assess the hydraulic connection of porous and karstic aquifers as well as confirm whether there is a "short connection" between surface and groundwater, another tracing test was undertaken at the drilled borehole B-2 (Figure 10.5–6). Five kilograms of sodium fluorescein was injected into the most karstified interval at a depth of 8 m and registered after approximately 17 hours at Ckanjak Spring (Montenegro Geological Survey, 2007), the first neighboring spring north of Bolje Sestre. The calculated tracer velocity is 0.025 m/s, which complies with the average values obtained from similar tests conducted in highly karstified Montenegrin carbonate rocks (M. Radulovic, 2000). However, the tracer was not identified at the Bolje Sestre, which confirms that, even if communication with the river water exists, it does so in a very limited manner, going through a buffer represented by very thick intergranular aquifer of alluvial and fluvial-glacial sediments and moving slowly toward the underlying karst.

This variation is further confirmed by numerous chemical analyses of surface waters (Skadar Lake, Moraca River) and groundwater sampled from the Bolje Sestre and other springs in its vicinity, as well as from investigation boreholes and shallow wells of the Moraca alluvium. The concentration of the main components and microconstituents identifies a very different origin of the waters. The same results are obtained by correlation of the biological content and the river and lake sediments. The Moraca River water, despite its different origin, is also of generally good quality with a slightly increased concentration of phenols, nitrites, and ammonia (A 2 class according to Montenegro standards).

Water samples collected from the Bolje Sestre have consistently shown the water to be of very good—and even excellent—quality with the potential for bottling. The Bolje Sestre groundwater is low mineralized (TDS ~ 200 mg/L), and HCO_3 and Ca are the dominant ions (HCO_3 is 128 mg/L, Ca is 48 mg/L, Mg is 16 mg/L, Cl is 10 mg/L), while the hardness is a little above moderate values. All analyzed microconstituents are far below maximum permitted levels or, as in the case of As and cyanides, even below detection limit of the laboratory equipment used. Microbiological analyses confirmed water A 1 class (I class Kohl; i.e., I B class of index of phosphate activity, 0.07).

The research enabled the confirmation of the suitability of the Bolje Sestre Spring and movement to the next stage of investigation and design of RWSS (Stevanovic, 2007). Of the six previously noted major problems, the two most critical and most professionally challenging are the intake and the source protection. Some other problems include

1. A more reliable flow measurement system was required to make the sublacustrine discharges from the spring area converge better. Along with fixing the panels (baffles) to the bed of the lake, the Hydrometeorological Survey of Montenegro installed the gauge, limnigraphs, and other equipment for continual measuring of the Bolje Sestre flow (Figure 10.5–7). Given that the work was just recently completed, it is still not possible to evaluate the collected data, but it is known that no values smaller than 2.3 m^3/s were registered during the last hydrologic cycle (2007–2008). The maximum flow was much more difficult to measure due to a wide dispersion of the streamlines, but it exceeds 8 m^3/s. However, a water budget and stochastic analyses of groundwater regime are planned and will be carried out before project completion (Stevanovic and Radulovic, 2007).
2. Although no serious problem with turbidity or microbiological contamination was identified in the springwater, the treatment is designed to prevent such obstructions, which are so typical where the karst is concerned. Therefore, water from the intake will be pumped out to the nearby water treatment plant, where the treatment will include two major processes: fast filtration and ultraviolet disinfection (Scott Wilson and IK Consulting Engineers, 2007).

3. Two reservoirs (Bolje Sestre 1 in the first phase for $Q = 1.1\ m^3/s$, $V = 500\ m^3$; and Bolje Sestre 2 in the second phase for $Q_{final} = 1.5\ m^3/s$, $V = 700\ m^3$) will be located on the hill over the source (at altitudes of 56–61 m asl) and collect clean water after the treatment.
4. The main gravity pipeline will follow the eastern margin of Malo Blato Gulf, cross Skadar Lake using facilities of the existing embankment and the bridge, and reach the hydrotechnical tunnel Sozina, where the continental part of RWSS ends (Figure 10.5–1). This will minimize problems of seismicity and liquefaction, which would be more problematic if the pipeline were situated on the bed of the lake.
5. After passing the Sozina tunnel, the water will be stored in the large Djurmani reservoir on the seaside, from which it will be diverted to the north and south (Figure 10.5–1).

10.5.4 HOW TO TAP AND PROTECT BOLJE SESTRE WATER

From the beginning of this project, it was clear that the main problem is how to tap the Bolje Sestre water in such a manner to ensure the required demands of $1.5\ m^3/s$ and prevent the mixing of karstic groundwater and lake water. Whereas the amount of water is not considered to be a big problem, minimizing the influence of lake water requires very careful planning of intake as well as optimizing future exploitation.

To reduce the risk of these hazards, a series of fieldwork activities was conducted in narrow discharge zones along the shore (Montenegro Geological Survey, 2007; Stevanovic and Radulovic, 2007; Figure 10.5–8).

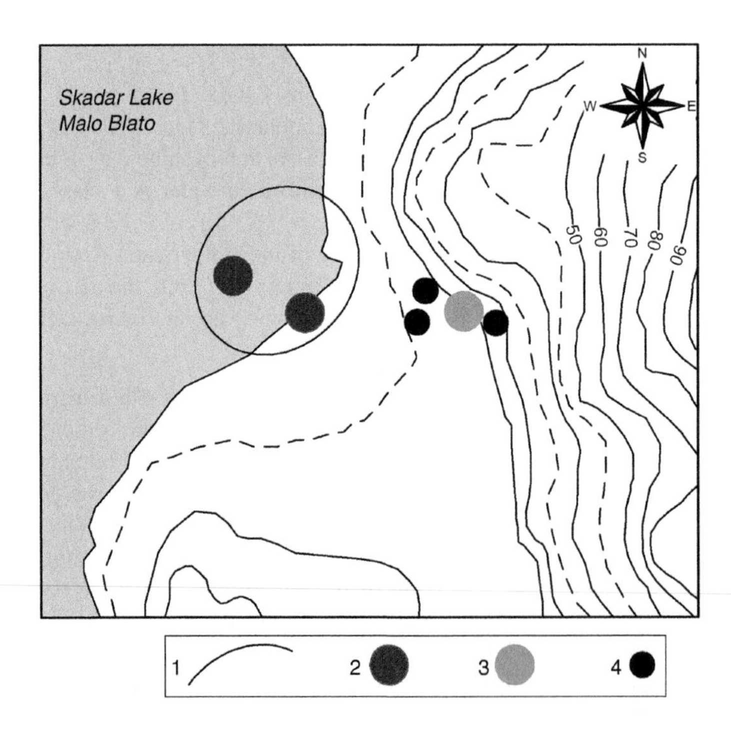

FIGURE 10.5–8 Narrow discharge zone of Bolje Sestre: (1) Elliptical shape zone where option A with coffer dam is evaluated, (2) discharge points, (3) location of borehole B-8 and potential shaft as per option C, (4) proposed observation boreholes.

Three optional designs of the source works intake were evaluated (Stevanovic and Radulovic, 2007; Scott Wilson and IK Consulting Engineers, 2007):

- **Option A**. By enclosing the area of the source discharge and using a small coffer dam, the discharging water can be captured and pumped into the scheme.
- **Option B**. Custom-built metallic pipes can be jacked into the ground of the lake and anchored by underwater cementation, creating intake shafts for pumping.
- **Option C**. A shaft or large diameter well system, located on land, approximately 5–6 m above the lake level, cased off with steel casing, enclosed in a suitable pump-control house, with appropriate electrical and mechanical controls, can be operated remotely or manually as required (Figure 10.5–9).

Option C was selected as the preferable one by the hydrogeologists and given priority when an additional detail survey started. The basic idea was to tap the water before it reaches the submerged discharge points. Simply, this should be the best way to avoid mixture with waters of the lake and the threat of groundwater quality deterioration. However, that the pumping capacity from the shaft-well must not result in significant drawdown and exaggerate lake water intrusion has been taken into consideration (Stevanovic et al., 2008).

The detail surveys in the two zones along the shore aiming to provide clarification to the preceding dilemmas included the following activities:

- Hydrologic measurements of the source discharge.
- Water quality analyses of the source, other water points, lakes, and streams.
- Photogeologic (stereoscopic) analysis of the main lineaments.
- Detailed geophysical surveying and logging in concordance with the exploratory drilling program.
- Borehole drilling and testing (exploratory drilling is carried out to depths no greater than 20 m in a configuration decided on the basis of the geophysics).
- Tracing tests at the newly drilled boreholes.
- Hydrogeologic mapping at scales 1:1000–1:2500.
- Geodetic surveying in potential intake area.

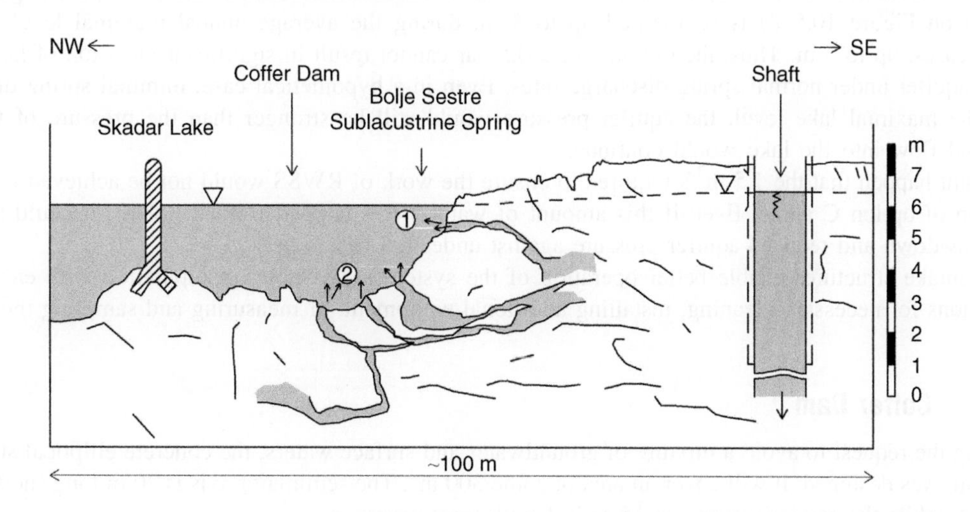

FIGURE 10.5–9 Draft intake design of Bolje Sestre (combination of dam founded on the rocks of the lake bed and the shaft on B-8 borehole location).

- Creating a three-dimensional elevation model of the terrain.
- Mapping potential pollutants and assessing risk (survey for sanitary protection zones and preventive measures).
- Creating the monitoring system for the source.

The essential results were obtained by drilling, pumping, and tracing tests. Eight boreholes were drilled, four in each of two zones. Although most of the boreholes were drilled in fissured or even compact limestone blocks, borehole no. 8, located some 45 m south of (behind) discharge zone 1 of the Bolje Sestre (Figure 10.5–8), reached very karstified rocks and cavernous intervals. The main caverns are registered at the depths of 6–6.8 m, 7.6–10.8 m, and 12.1–12.2 m. An air lift test, a pumping test, and eventually a tracing test confirmed that borehole B-8 is located directly on the main underground conduit that diverts water to the Bolje Sestre discharge points. This is an additional argument in favor of the theory of very concentrated flows within the Dinaric karst of Montenegro.

With a pumping capacity of 5 L/s, no single millimeter of drawdown is recorded. Despite the small pumping rate (the limited borehole diameter did not permit use of larger pumps), it was concluded that the main karstic "artery" has been found. Infused tracer sodium fluorescein passed the 45 m distance from B-8 to the discharge zone close to the shore in 52 minutes, at a velocity of 0.014 m/s (IK Consulting Engineers, 2007).

10.5.5 BOLJE SESTRE INTAKE

Some investigation and work continued in winter 2008–2009 (when this book was written) and some will continue along with civil work in 2009 and 2010. However, up to the present, the research results obtained enabled the formulation of a final design and solution for tapping this source. Of the three options discussed, the design considered optimal is a combination of options A and C (Scott Wilson and IK Consulting Engineers, 2007, Figure 10.5–9).

What were the basic arguments in favor of such a solution?

The lake level fluctuates throughout the year between 4.6 and 9 m asl. Bolje Sestre water (discharge zone 2, indicated on Figure 10.5–8) is submerged up to 3 m, during the average annual maximal level, and, in extreme cases, up to 5 m. Thus, the pressure of a 0.5 bar cannot result in significant intrusion of lake water into the aquifer under normal spring discharge rates. Even in a hypothetical case, minimal spring discharge during the maximal lake level, the aquifer pressure would still be stronger than the pressure of the lake water, and flow into the lake would continue.

It might happen that the 1.5 m^3/s required to ensure the work of RWSS would not be achieved by implementation of option C alone. Even if this amount of water were pumped from the shaft, it could result in larger drawdown and reduced aquifer pressure against undesired lake water.

Two intake structures enable better operation of the system as a whole: manipulation with energy and interruptions for necessary cleaning, installing additional equipment, or measuring and sampling the water.

10.5.5.1 Coffer Dam

Following the request to avoid a mixture of groundwater and surface waters, the concrete elliptical structure-coffer dam was designed. It will cover an area of some 300 m^2. The semimajor axis is 20 m long (north-south direction), while the semiminor axis is 15 m in length (east-west).

For a proper dam foundation and to assess possible leakage below the dam, two investigation boreholes were drilled 14 m from the shore. They confirmed the existence of 2.5–2.9 m thick mud on the lake floor, which must be cleaned out before the construction begins. Additional geomechanical exploration and laboratory tests confirmed the suitability of the carbonate rocks for the excavation and dam construction.

Considering the importance and influence of lake water fluctuations on the intake, a special removable spillway section (rubber gate) was designed for installation on the dam crest. The idea is to allow the uncaptured springwater to overflow easily into the lake and to manipulate the height of the water level inside the coffer dam.

The following criteria are considered for the spillway design:

1. To control the difference between the springwater and lake water levels.
2. To enable free springwater flow from the dam to the lake until the lake water altitude is lower than 8 m asl.
3. To have inundated springwater flow from the dam to the lake at lake water altitudes between 8 and 9 m asl.

Scott Wilson and IK Consulting Engineers (2007) projected the following technical elements for this engineering solution (Figure 10.5–10):

- Concrete side walls (nonflow part).
- Concrete dam with rubber gate spillway.

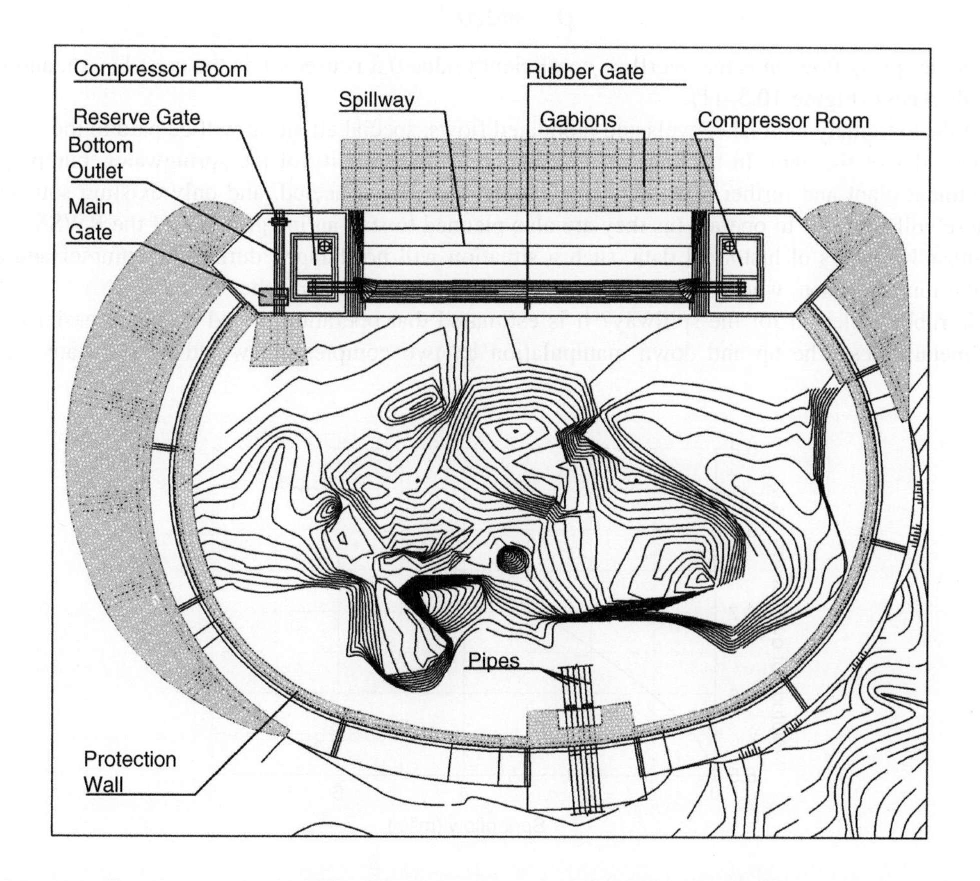

FIGURE 10.5–10 The coffer dam intake of Bolje Sestre designed by IK Consulting Engineers and Scott Wilson. (Scott Wilson and IK Consulting Engineers, 2007; courtesy of V. Letica and B. Trajkovic.)

- Fixed concrete blocks from both sides of the spillway.
- Bottom outlet.
- Orifice for the pipe.

The concrete side walls (nonflow part) are of an elliptical shape, 5.1–7.0 m high with a crest at an altitude of 9.5 m asl. The width of the wall is 0.6 m.

The concrete dam with a rubber gate spillway is 24 m long, and its task is to evacuate nontapped water. In fact, the gate has the form of a wide tube made from special resistant rubber that can be filled with compressed air. The dam foundation altitude is at 2 m asl (0.5 m below bedrock) while the crest is at 6.5 m asl. The internal part of the dam has a slope of 3:1. The rubber gate, when filled with air, is 3.0 m wide and 1.5 m high. It is fixed by being anchored to the concrete dam crest. The rubber gate is empty and inactive when the water level of the lake is below 6.5 m asl. For the altitudes of the lake's water table, 6.50 m asl $\leq Z \leq$ 8 m asl, the spillway activates at 8 m asl (the compressor fills the air into the rubber tube). For extreme water levels of Skadar Lake over 8 m asl, the rubber gate remains active, and the inundated overflow continues from the spring toward the lake.

To calculate the water thickness over the wide dam crest, the following equation is applied:

$$Q = mb2gH^{3/2}$$

where Q is the spring flow, m is the overflow coefficient (value 0.3 is used), b is the crest length, and H is the overflow thickness (Figure 10.5–11).

During the extremely high lake levels and inundated flows, special attention will be paid to the water quality on both sides of the dam. In the case of deterioration of the quality of the springwater, pumping to the water treatment plant and further distribution will be temporarily canceled, and only existing sources along the seashore will continue to operate (as they are also planned to stay an integral part of the RWSS). According to almost 100 years of historical data, such a situation will not happen during the summer and autumn months, the tourist season, when the local water consumption regularly reaches its peaks.

Why is rubber selected for the spillway? It is estimated that operation would be much easier than with classical metal gates. The up and down manipulation by two compressors would be fully automatic and

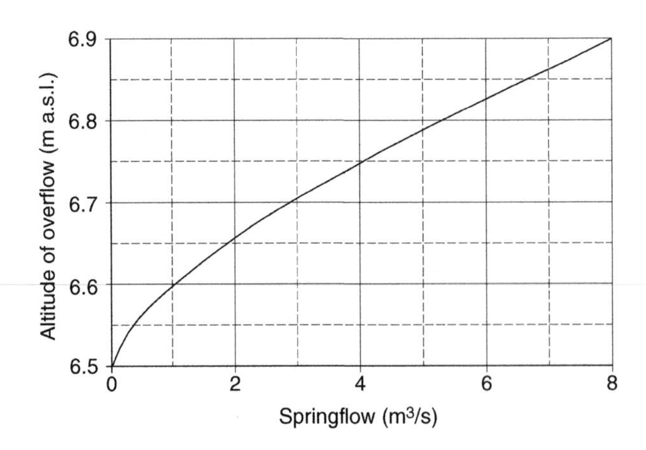

FIGURE 10.5–11 The curve shows different elevations of water level over the dam crest versus the spring flow values for an inactive rubber gate (6.5 m asl).

remotely controlled. The rubber tube is easily transported, mounted, and removed for repair, if necessary. Its maintenance should also be inexpensive. And, last but not least, there is no negative impact on the environment and water inside and outside the coffer dam.

For both concrete side walls and the dam with the rubber gate, "Larsen" sheet piles are temporarily mounted to prevent saturation by lake water during the building. Permanent pumping of springwater from the reservoir part will also be conducted for the same purpose.

The gate manipulation facilities are placed on fixed concrete blocks located on both sides of the spillway (Figure 10.5–12). The purpose of the bottom outlet is to divert water from the internal reservoir to the lake in emergency cases or enable inspection or intervention on the spillway (rubber gate). It will be controlled by two panel gates, manipulated from the dam. The outlet dimensions are 0.70 m × 0.70 m, and its length is 8.8 m. On the eastern side of the intake structure are two orifices for the main pipes connected with the pumping station.

To amortize the effects of the hydraulic jump, a gabion carpet 5 m wide will be fixed to the dam's foundation (Figure 10.5–13), to accommodate the differences when the lake's minimal water table is at 5.6 m asl and the water level in the reservoir is at a 6.9 m asl crest. The gabions, as a protective measure, also reduce the effects of raising the mud deposited on the lake bed.

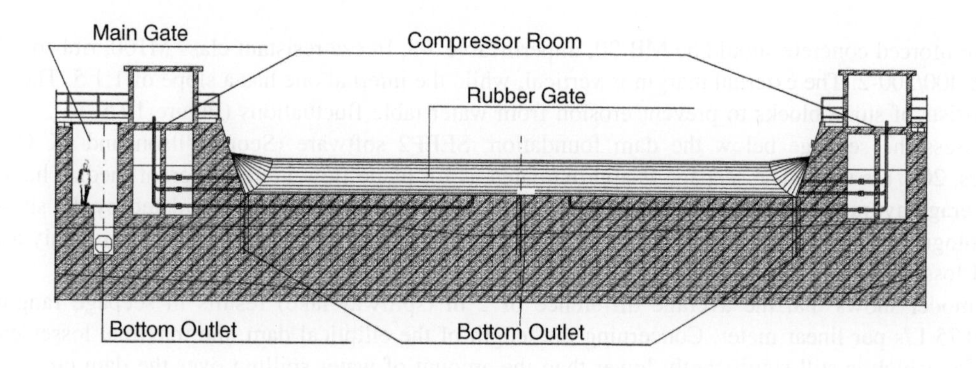

FIGURE 10.5–12 Bolje Sestre coffer dam, spillway details (From Letica and Trajkovic, in IK Consulting Engineers and Scott Wilson, 2007; courtesy of A. Tucovic and M. Zikic.)

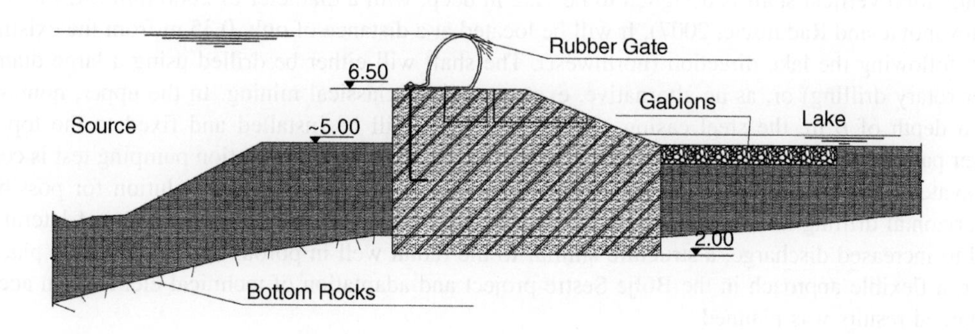

FIGURE 10.5–13 Bolje Sestre coffer dam section, rubber gate and gabions details. (From Letica and Trajkovic, in IK Consulting Engineers and Scott Wilson, 2007; courtesy of A. Tucovic and M. Zikic.)

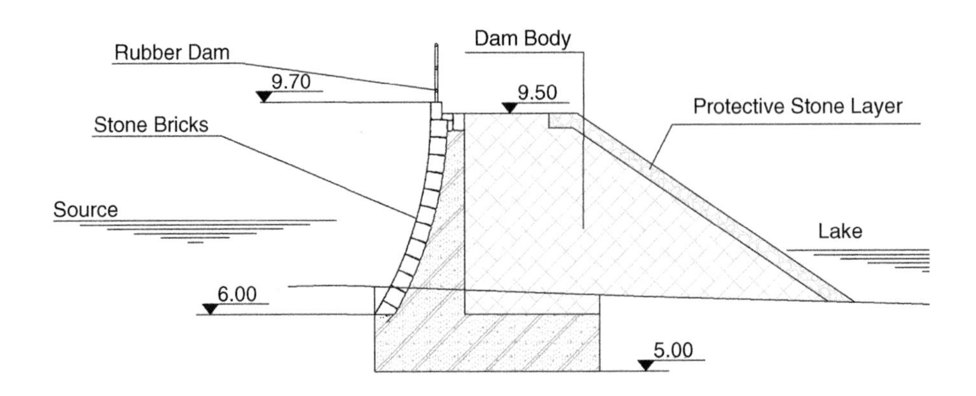

FIGURE 10.5–14 Bolje Sestre coffer dam section, protective wall details in a fixed part. (From Letica and Trajkovic, in IK Consulting Engineers and Scott Wilson, 2007; courtesy of A. Tucovic and M. Zikic.)

The reinforced concrete should be MB 30, impermeable V8, freeze resistant class M100, ribbed reinforcement RA 400/500-2. The external margin is vertical, while the internal one has a slope of 1:1.5. The embankment consists of stone blocks to prevent erosion from water table fluctuations (Figure 10.5–14).

To assess the seepage below the dam foundation, SEEP2 software (Scott Wilson and IK Consulting Engineers, 2007) is used. The seepage is approximated as for plane flow in noncoherent media characterized by an average hydraulic conductivity of 10^{-4} m/s. This method cannot be fully validated by the site-specific hydrogeological conditions, where intensively karstified rocks are present, but it is used to roughly assess the potential losses.

The model shows that the average difference of 2 m (spillway-lake) results in seepage ranging from 0.07 to 0.75 L/s per linear meter. Concerning the length of the elliptical dam, the maximal losses can reach 0.075 m^3/s, which is still significantly lower than the amount of water spilling over the dam crest.

10.5.5.2 Shaft

The exploitation vertical shaft is designed to be 12.2 m deep, with a diameter of 2000 mm (Stevanovic et al., 2007; Stevanovic and Radulovic, 2007). It will be located at a distance of only 0.35 m from the existing borehole B-8 following the lake direction (northwest). The shaft will either be drilled using a large diameter bit (hammer-rotary drilling) or, as an alternative, excavated as in classical mining. In the upper, nonproductive part, to a depth of 6 m, the steel casing pipe of 1500 mm will be installed and fixed to the top surface. The lower part, 6–12.2 m, is an open hole (Figure 10.5–15). Once the long duration pumping test is completed and excavated conduits recorded by geophysical logging and video camera, the solution for possible additional horizontal drilling lateral from the main cavities will be reevaluated. The expansion of lateral cavities can lead to increased discharge, a structure similar to the radial well in porous aquifer could be placed here. However, a flexible approach in the Bolje Sestre project and adaptation of technical elements in accordance with obtained results was planned.

The groundwater extraction rates, pumping capacity, and working regime will be determined after the construction of the intake structures and their individual and simultaneous testing. The work will be closely monitored, however, and tailored to ensure the best possible water quality.

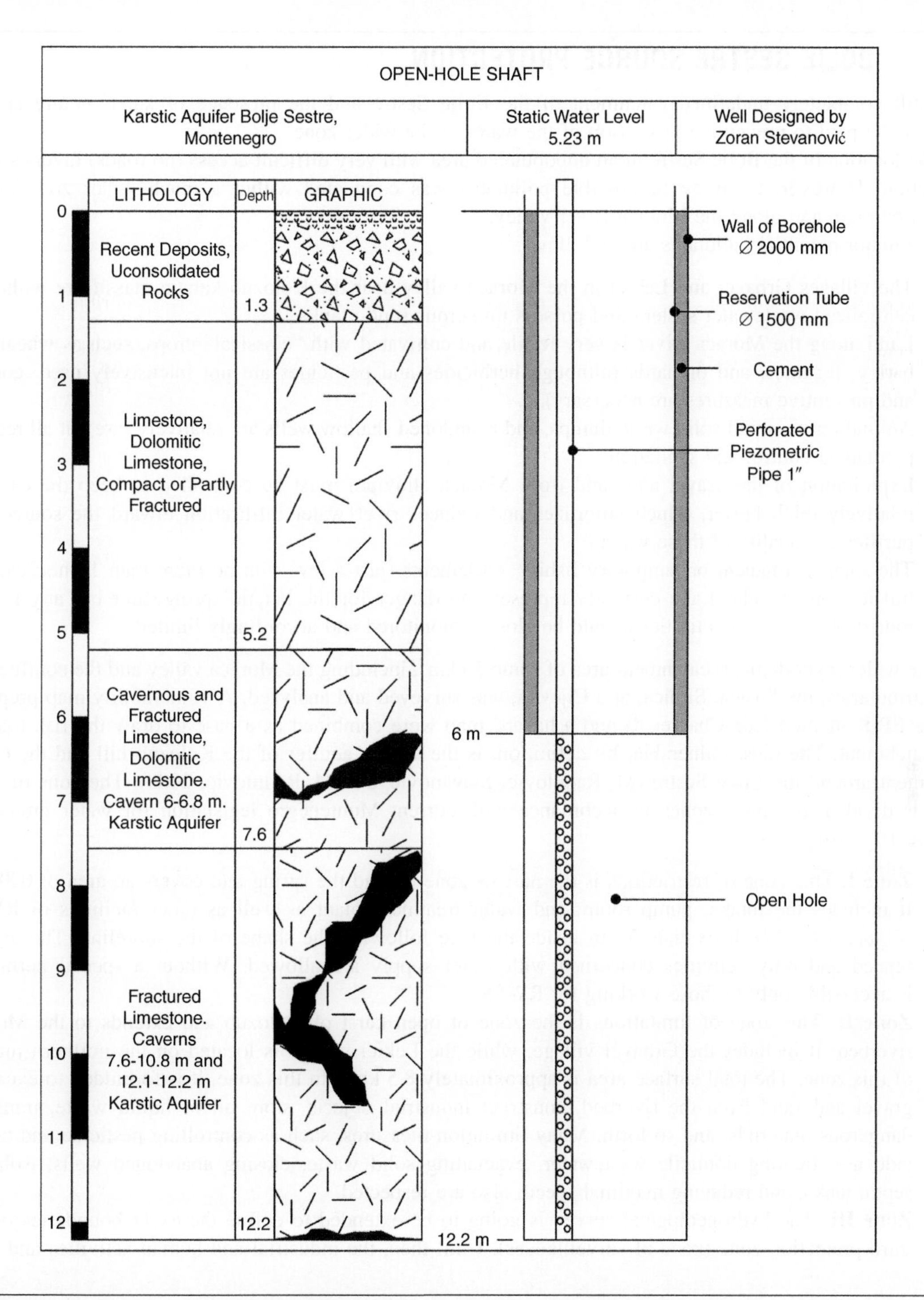

FIGURE 10.5–15 Design of exploitation shaft at B-8 location. (From Stevanovic et al., 2007.)

10.5.6 BOLJE SESTRE SOURCE PROTECTION

The still completely undefined catchment of the Bolje Sestre and the presence of karst require special attention be paid to preventive protection of the water in the wider zone.

The location of the Bolje Sestre in an unpopulated area with very difficult access (no roads) favors source protection. However, a survey of possible pollutants was conducted within a much wider zone in the Podgorica plain and along the Moraca River valley.

The major potential pollutants are as follows:

1. The villages Grbavci and Lekici in the Moraca valley near the Kolozub karstic massif are without a centralized wastewater system and possess numerous septic tanks.
2. Land along the Moraca River is very fertile and cultivated with "classical" crops, such as wheat and barley, legumes, and orchards (although herbicides and pesticides are not intensively used, control and preventive measures are necessary).
3. Animal farms, small solid waste dumps, and abandoned shallow wells are relatively rare but all require permanent control and sanitation.
4. Exploitation of the gravel and sand from Moraca alluvium must be controlled to keep the existing relatively thick buffer, which amortizes and reduces river water infiltration toward the source and purifies the quality of these waters.
5. The small permanent or temporary fishing settlements (just a few with no more than 10 houses) and traffic along Skadar Lake currently represent no danger for the karstic springwater but any further tourist or recreation activities should be closely monitored and accordingly limited.

The wider (hypothetical) catchment area of some 50 km^2, including the Moraca valley and the confluences of its tributaries, the Susica, Sitnica, and Cijevna, was surveyed and analyzed. A vulnerability map prepared by the EPIK method (see Chapter 8) and a hazard map were combined as a base to draw the risk map of this catchment. The most vulnerable, by definition, is the karstic aquifer of the Kolozub hill and the entire catchment around the Bolje Sestre (M. Radulovic, Stevanovic, and M. Radulovic, 2008). The zone of open karst is divided into three zones in accordance with current Montenegro legislation for water protection (Figure 10.5–16):

1. **Zone I**. This zone of restrictions is the narrow zone around the spring and covers an area of 0.38 ha. It includes the intake, pump room, and water treatment plant as well as other facilities of RWSS (Figure 10.5–17). It extends 50 m inside the lake following the shape of the shoreline. The area is fenced and only activities concerned with water supply are allowed. Without a special permit, it is accessible only to those working for RWSS.
2. **Zone II**. This zone of limitations is the zone of open karst of Kolozub and extends to the Moraca riverbed. It includes the Grbavci village, while the Lekici village is located on the northern margin of this zone. The total surface area is approximately 8.5 km^2. In this zone, it is forbidden to excavate gravel and sand from the riverbed, construct industrial objects, store oil or liquid waste, transport dangerous materials, and so forth. Many limitation measures, such as controlling pesticide and herbicide use, treating domicile wastewater, evacuating solid waste, closing abandoned wells, isolating septic tanks, and reducing maximal speed, also are respected.
3. **Zone III**. The hydrogeological survey is going to be extended to define the exact boundaries of the third protection zone (zone of surveillances). Until then, the industrial and human activities and their

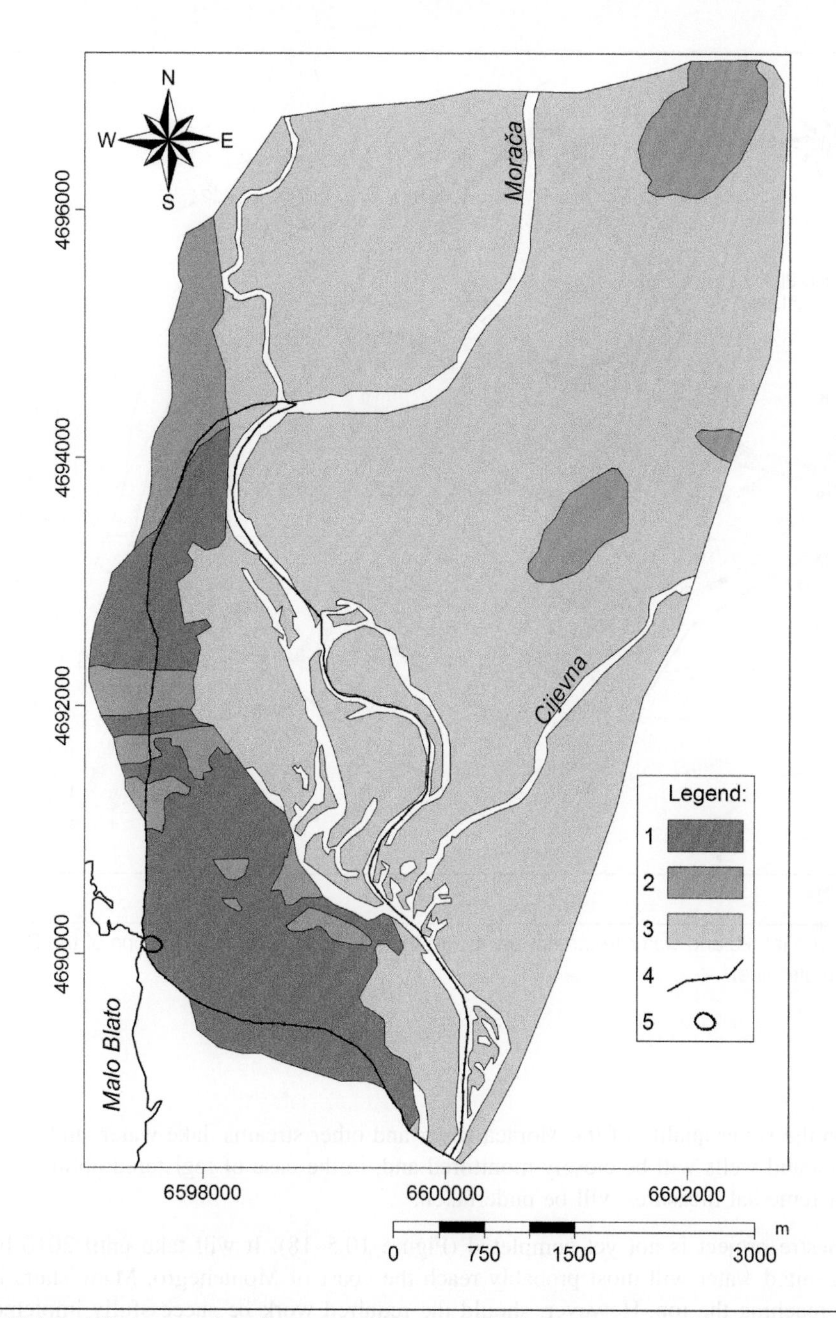

FIGURE 10.5–16 Vulnerability map and sanitary protection zone of Bolje Sestre source: (1) Very high risk from pollution, highly karstified rocks; (2) high risk from pollution, karstified and fissured rocks; (3) moderate risk from pollution, alluvial and fluvioglacial deposits; (4) zone II of sanitary protection of the source, zone of limited activities; (5) narrow protection zone I around source, restrictive access. (From Radulovic et al., 2008.)

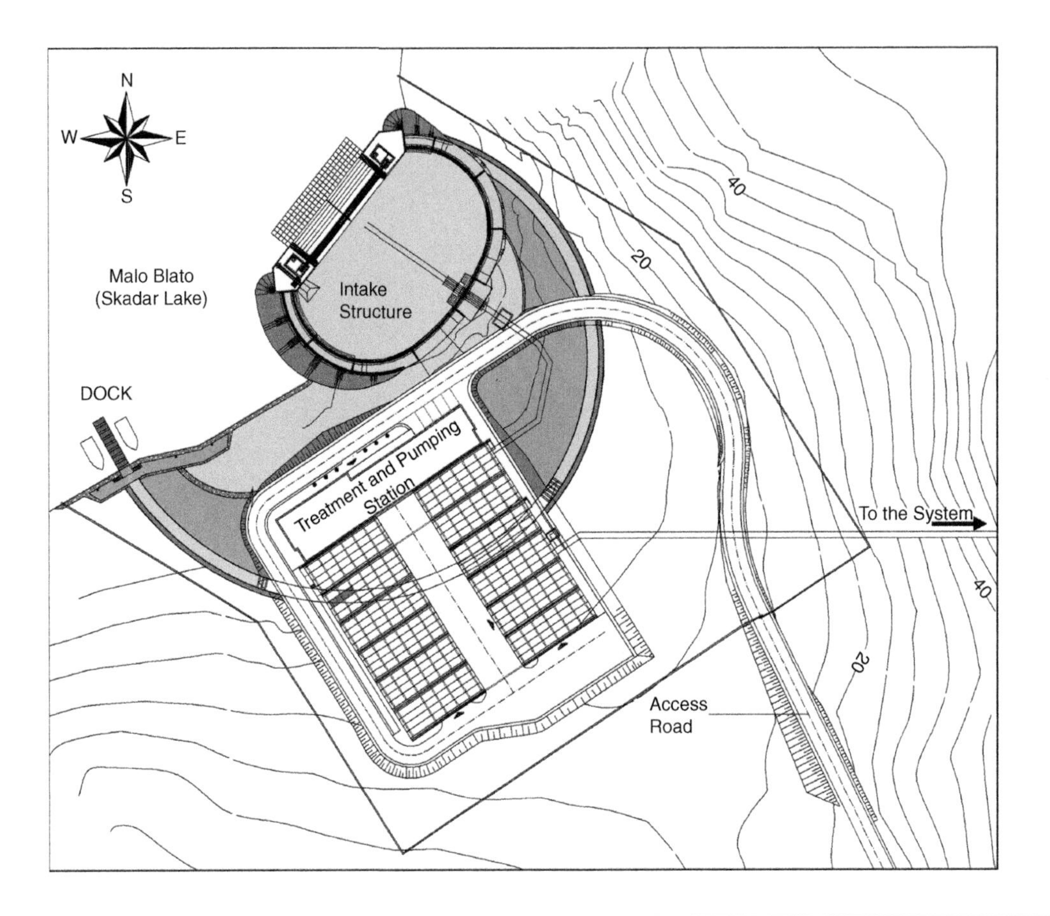

FIGURE 10.5–17 The intake and water treatment plant with the first sanitary protection zone of the Bolje Sestre Spring (bold perimeter line).

impact on the water quality of the Moraca River and other streams, lake water, and groundwater from the springs and wells will be closely monitored and, in the case of registered pollution, adequate protective or remedial measures will be undertaken.

The Bolje Sestre project is not yet completed (Figure 10.5–18). It will take until 2010 before the first drops of long-awaited water will most probably reach the coast of Montenegro. Many stairs have yet to be climbed before reaching the top. However, should the required work be successfully implemented, there is a high possibility that the Bolje Sestre source can provide good-quality water and a reliable and sustainable yield for the next 30 years, without interruptions and failure.

One final fact is worthy of special attention: The current cost of the survey and design of the two intakes and their future construction is calculated to be in the range of 1–2 percent of the total investment for the implementation of this project.

FIGURE 10.5–18 Bolje Sestre tapping structure under construction in December 2008. The curved line approximates the position of the future coffer dam.

ACKNOWLEDGMENTS

I gratefully acknowledge the data obtained and friendship of the colleagues with whom I worked closely on this project for more than two years. Special thanks go to Andra Tucovic, Miomir Zikic, Vicko Letica, Biljana Trajkovic, Aleksandar Balatov (IK Consulting Engineers), Micko Radulovic (Faculty of Civil Engineering, Podgorica), Vladan Dubljevic and his staff from the Geological Survey of Montenegro, Tomasz Krawczyk, Shammy Puri (Scott & Wilson), Milan Radulovic ("Geoprojekt," Podgorica), Darko Novakovic (Hydrometeorological Survey of Montenegro). And my warmest thanks for their full support and understanding go to the officials and colleagues from the regional waterworks Crnogorsko Primorje, Budva, headed by M. Pejovic and P. Pavicevic.

REFERENCES

Besic, Z., 1969. Geology of Montenegro (Geologija Crne Gore). In: Vol. 2. Karst of Montenegro (Karst Crne Gore) Spec. Ed. Geol. Soc. Montenegro, Podgorica, Montenegro.

Buric, M., Radulovic, M., 2005. A projection of long-term water supply of the settlements along the Montenegrin coast with a regional water system. In: Milanovic, P., Stevanovic, Z., Radulovic, M. (Eds.), Water resources and environmental problems in karst, Exc. Guide of IAH Intern. Conf. KARST 2005. Spec. ed. FMG, Belgrade-Kotor, Serbia and Montenegro, pp. 48–50.

Energoprojekt, Belgrade, 1994. Technology Process Project for WTP Karuc, Study. Regional waterworks Crnogorsko Primorje, Budva, Montenegro (unpublished).
Energoprojekt, Belgrade, 1995. Report on Karuc water source. Third phase of water quality investigations, regional waterworks Crnogorsko Primorje, Budva, Montenegro (unpublished).
Filipovic, S., 1981. Effects of pollution on Lake Skadar and its most important tributaries. The biota and limnology of Lake Skadar. Univ. of Michigan, Ann Arbor/Biological institute, Podgorica, Montenegro.
Hrvacevic, S., 2004. Resources of surface waters of Montenegro. Spec. ed. Electro system of Montenegro, Podgorica, Montenegro.
Hydrometeorological Survey of Montenegro HMSM, 2005. Tracing test of Lower Moraca in Malo Blato area. [Report in Serbian]. Doc. Regional waterworks Crnogorsko Primorje, Budva, Montenegro (unpublished).
IK Consulting Engineers, 2006. Conception and design of source Bolje Sestre for regional water supply of Montenegro coast, Vols. 1 and 2. Regional waterworks Crnogorsko Primorje, Budva, Montenegro (unpublished).
IK Consulting Engineers, 2007. Source Bolje Sestre—Report of undertaken studies. Regional waterworks Crnogorsko Primorje, Budva, Montenegro (unpublished).
ITSC Ltd. Montenegro, 2006. Environmental impact assessment study of Bolje Sestre source, regional waterworks Crnogorsko Primorje. Budva, Montenegro (unpublished).
Mijatovic, B., 1997. Etude de certains caracteres et de couplage des phenomenes karstiques dans l'ensemble Dinarides—Hellenides. In: Rec. de rapp. du Com. pour le karst et spéléologie Acad, Serbe des Sciences et des Arts, VI, vol. 72. Belgrade, Serbia, pp. 39–53.
Montenegro Geological Survey, 2007. Report on drilling and testing of investigation boreholes (Bolje Sestre, Malo Blato, Skadar Lake). Geol. Surv. Monte, Podgorica, Montenegro (unpublished).
Radulovic, M., 2000. In: Karst hydrogeology of Montenegro. Sep. issue of Geological Bulletin, vol. XVIII. Spec. Ed. Geol. Survey of Montenegro, Podgorica, Montenegro.
Radulovic, M., Radulovic, Mi., 2005. Hydrogeological survey of terrain between Grbavci and Bolje Sestre. [Report in Serbian]. Geoprojekt, Podgorica, Montenegro (unpublished).
Radulovic, M., Radulovic, V., Stevanovic, Z., Komatina, M., Dubljevic, V., 2005. General geology and hydrogeology of Dinaric karst, in Montenegro and Serbia. In: Milanovic, P., Stevanovic, Z., Radulovic, M. (Eds.), Water resources and environmental problems in karst, Exc. Guide of IAH Intern. Conf. KARST 2005. Spec. ed. FMG, Belgrade-Kotor, Serbia and Montenegro, pp. 16–20.
Radulovic, M., Stevanovic, Z., Radulovic, Mi., 2008. Methodology of determination of the zone of sanitary protection of source Bolje Sestre planned for regional water supply of Montenegrin coast. In: Proceedings of Intern. Conf. on Civil Engineering—science and practice, Zabljak, Montenegro, pp. 1081–1086.
Radulovic, V., 1997. The Skadar Lake water and of brim springs as source for water supply. In: CANU (Montenegro Acad. of Sci.), 44: Natural values and protection of Lake Skadar. Podgorica, Montenegro.
Radulovic, V., Radulovic, M., 1997. Karst of Montenegro. In: Stevanovic, Z. (Ed.), 100 years of Hydrogeology in Yugoslavia. Spec. ed. Fac. Min. Geol. (FMG), Belgrade, Serbia, pp. 147–185.
Scott Wilson and IK Consulting Engineers, 2007. Main Project of intake, water treatment plant and pumping stations on the source of regional water supply of coastal area of Republic of Montenegro. Vol. 2, Intake Bolje Sestre, books 1 and 2, hydroengineering and construction parts. Regional waterworks Crnogorsko Primorje, Budva, Montenegro (unpublished).
Stevanovic, Z., 2007. Project of hydrogeological exploration for the main design of Bolje Sestre source for regional water supply of Montenegrin coastal area (Malo Blato, Skadar Lake). Fac. Min. Geol. (FMG). University of Belgrade, Belgrade, Serbia (unpublished).
Stevanovic, Z., Radulovic, M., 1997. Regulation of karst aquifer as a base for optimal exploitation of water potential—Case studies from Serbia and Montenegro [in Serbian]. In: Proceedings of Conf. on Groundwater Sources. Society for Water Technology and Sanitary Engineering, Belgrade, Serbia, pp. 445–465.
Stevanovic, Z., Radulovic, M., 2007. Report on hydrogeological survey of Bolje Sestre source, period May–August 2007. Regional waterworks Crnogorsko Primorje, Budva, Montenegro (unpublished).
Stevanovic, Z., Radulovic, M., Puri, S., Radulovic, Mi., 2008. Karstic source Bolje Sestre—Optimal solution for regional water supply of Montenegro coastal area. In: Rec. de rapp. du Com. pour le karst et spéléologie Acad, vol. IX. Serbe des Sciences et des Arts, Belgrade, Serbia, pp. 33–64.
Zikic, M., 1997. Final design of water intake Karuc for regional water supply of Montenegrin coast. Energoprojekt, Belgrade, Serbia (unpublished).
Zikic, M., Buric, M., Radulovic, M., 2005. Cost-benefit analyses of potential water sources for long term solution of the Coastal Region of Republic of Montenegro. Regional waterworks Crnogorsko Primorje, Budva, Montenegro (unpublished).

CHAPTER

Case Study: Geological and hydrogeological properties of Turkish karst and major karstic springs

10.6

Gültekin Günay
Retired from Ankara Hacettepe University, Turkey

10.6.1 INTRODUCTION

Turkey is situated entirely in the Mediterranean sector of the Alpine orogenic belt. This belt lies between the Russian belt in the north and the African and Arabic belts in the south. Most of the rocks and sediments of Turkey represent geosynclines and orogenic facies and differ from craton rocks and the shelf zones. Alpine orogeny and younger epiorogenic movements had an effect throughout the country. Tectonic structures are clearly recognized in the morphology. If the tectonic map of Turkey and the map that shows the distribution of the caves in Turkey with respect to their occurrences are studied together, the following conclusion can be drawn: The major tectonic structures of Turkey are closely related to the karst features of the country.

In general, the rocks that show karstification belong to the geosynclines and epeiric regions of the continent. In other words, they correspond to pelagic and benthonic facies. In Turkey, with the strong folding and uplifting of the Mediterranean-Alpine geosyncline and underthrusting of the geosyncline sediments by basement blocks to the north and south, two major Alpine structures have been exposed.

10.6.2 KARST WATER RESOURCES OF TURKEY AND RELATED PROBLEMS

The water resources of the karst terrains of Turkey are relatively rich and as such are very important for the economic development of the country. Some important and beneficial characteristics of these water resources are as follows (Eroskay and Günay, 1980):

1. Large average annual precipitation over most of the high mountains produces a large runoff per unit area.
2. Large differences in level toward the sea and relatively short distances give to these water resources an especially high hydroelectrical potential.
3. The melting of accumulated snow at high elevations sustains the low flows well into the first half of the dry summer season in many areas of karst terrains.

Copyright © 2010, Elsevier Inc. All rights reserved.

4. High mountains have little water evaporation and relatively short periods of water available for evaporation, so that runoff most often represents a large portion of the total precipitation.

However, the karst terrains of Turkey are also responsible for some detrimental characteristics of these water resources in several regions.

5. The surface concentration of these waters into streams is of a much lower density than is the case with the noncarbonate terrains, so that some developments and uses of water resources become less convenient and often marginal.
6. Water from rain and snowmelt infiltrates relatively fast at high elevations and appears at very low levels, thus making difficult both the access to it and the utilization of its hydroelectrical power potential, except in a limited number of cases of high-level lakes and rivers.
7. The water resources often appear as large springs at low levels, close to the sea, where the limited areas for irrigation and the relatively small hydroelectrical power heads of the potential power plants preclude the use of their full potential.

10.6.3 BASIC FACTORS RESPONSIBLE FOR LARGE KARSTIFICATION IN TURKEY

The following are the basic reasons for the large karstification of most carbonate rocks of Turkey:

1. The strong orogenic movements involving the carbonate rocks lifted them much above the sea level and so created significant differences in level and powerful energy gradients for surface and underground water circulation.
2. The intense orogeny of folded, faulted, upthrusted, overthrusted, and highly fractured rocks provided both openings for the initial water circulation and opportunities for subsequent large rock solution and the creation of secondary porosity.
3. Orographic processes and the resulting high mountain ranges represented barriers for the movement of air masses, forcing them to rise significantly and precipitate rain and snow, and so provide large amounts of water for the rapid infiltration, circulation, and solution of carbonates.

Many external and internal factors are responsible for the type and the degree of karstification of an area of carbonate rocks. The basic fact is, however, that the geological structure, the orogeny, and the connected tectonics provide the basic framework that permits, enhances, or impedes the processes of karstification. The Alpine orogeny and the following epiorogenic movements in Turkey have been important factors in karstification (Figure 10.6–1). This karstification of carbonate rocks is spread almost everywhere in Turkey. It is found particularly in the regions of the Taurus Mountain range, in western and southwestern Anatolia, in the Konya closed basin, in eastern Anatolia (Keban Reservoir region), and in southern Anatolia (a large Ceylanpinar plain).

The karst features of western Turkey, bordering the Aegean and Mediterranean Seas, dramatically demonstrate the tectonic, lithologic, and climatic controls over the occurrence, movement, and chemical character of groundwater. The Taurus Mountains were formed by folding and overthrust faulting during the Alpine orogeny (Back and Gunay, 1992). This orogenic system also contains the karstified mountains of Greece and Yugoslavia; consequently, the western extension of the Taurus Mountains is submerged beneath the Mediterranean. At different times in geological history, the Aegean Islands have been connected among themselves as well as to the lands of Turkey and Greece. Since the late Tertiary period, western Turkey has experienced a downward movement that has been largely controlled by faults. The highly crenulated and dissected Aegean coastline of Turkey is due to this tectonic subsidence.

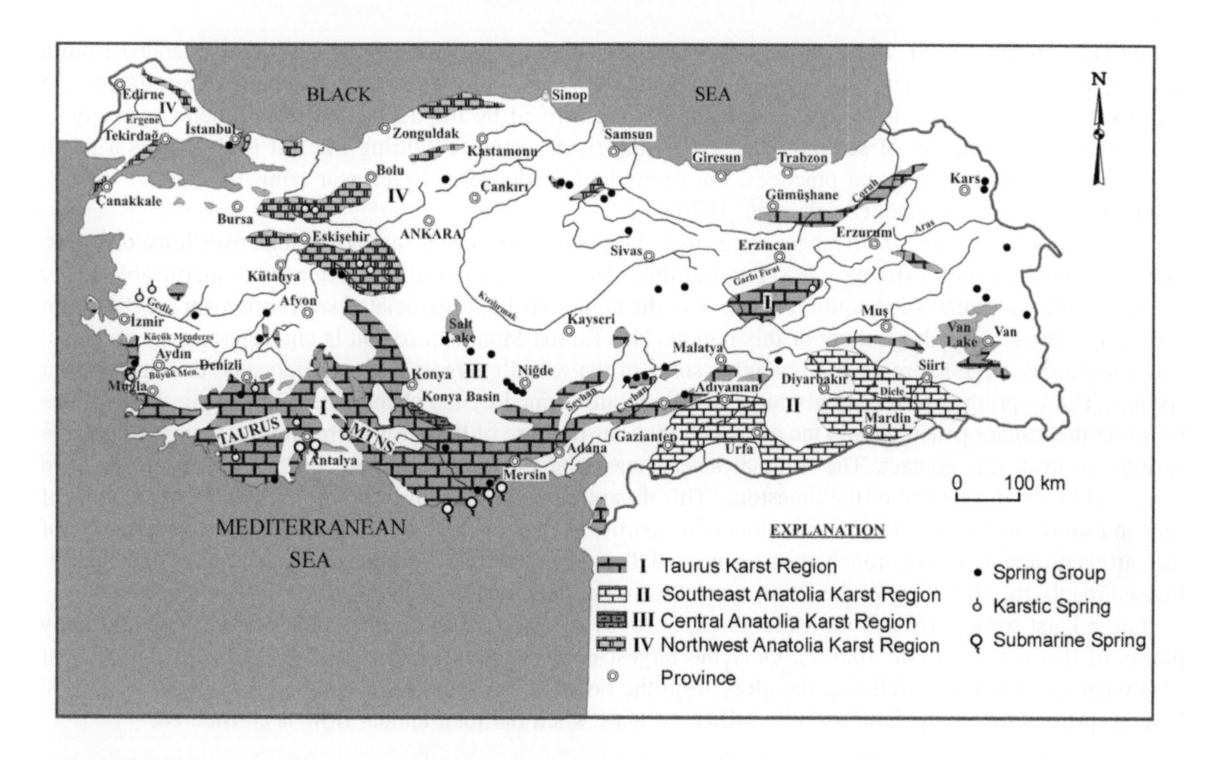

FIGURE 10.6–1 Major tectonic units of Turkish karst. (After Eroskay and Günay, 1980.)

The epiorogenic coastal movements continued from late Tertiary into the Quaternary period. The Quaternary tectonism is characterized by normal faults and, in this western part, block faults. Many of the coastal bays are in structural grabens that extend inland as major valleys. At approximately the middle of the early Pleistocene epoch, structural activity changed the river systems. In the late Tertiary period, the river drainage from the highlands was directed to the north and the south. The early Pleistocene faulting diverted these rivers to a westerly direction. With uplift, karstification increased, and the surface drainage of many lakes and rivers was transformed into subsurface drainage.

During the Ice Age, the climate changed drastically several times. Although the Taurus Mountains are now free of ice, they were covered with glaciers during that period. Throughout Turkey, the snowline during the last glacial epoch was 1000 to 1200 m lower than at present. Hence, the temperature in July would have been 6 to 8°C lower than at present. Those regions that were below the snowline were subjected to the effects of the periglacial climate during the Ice Ages. The climatic fluctuations also had effects at the lower elevations. During the glacial epochs, the lakes expanded due to reduced evaporation. For example, the level of Lake Burdur (near Antalya) rose by 90 m, Lake Tuz (near Ankara) by 110 m. East of Konya, a lake that was 125 × 25 km disappeared entirely. These climatic changes were controls on the extensive dissolution of limestone and precipitation of the travertine terraces of Antalya.

At the end of the last Ice Age, approximately 10,000 years ago, the temperature rose rapidly to its present value. As evaporation increased, groundwater levels declined and many springs ceased to flow. Freshwater lakes either turned into saline lakes or dried up entirely. The climate did not change much during the Holocene epoch, with the exception of the period of 5500–2500 BC, when it was somewhat warmer and more humid (Brinkmann, 1976). At the beginning of the Holocene period, the people of Turkey changed from a hunting and gathering society to a planned agricultural society with farming and ranching. Even though the tectonic subsidence continues today, new land is being formed by the deposition of sediments in many of the estuaries. The amount of sediment carried by the rivers increased during the last millennia as a result of deforestation and agricultural practices. For example, the Menderes River near İzmir has had an average delta advance of about 6 m/a (Brinkmann, 1976).

Much of the limestone, which ranges in age from Mesozoic to Holocene, has been either overthrust or deposited on formations with extremely low permeability. The combination of this extensive impermeable lower boundary of the regional karst aquifers along with the numerous faults associated with tectonism are the major controls on the karst hydrogeology of this region. The Taurus Mountain region is characterized by abundant water resources, large hydroelectric potential, some of the world's largest karst aquifers, and the largest karst springs. These springs of exceedingly large discharge are formed by the channelized flow along the fractures by water that cannot penetrate into the deeper formations because of their low permeability; many of the large springs occur at this contact. The original heterogeneity of the aquifer formed by the tectonism has been increased by the dissolution of the limestone. This dissolution results from infiltration of the great amounts of rain and snow that occur in the Taurus Mountains, particularly at the higher elevations. If the transmissivity of the carbonate aquifers were more homogeneous and the aquifers were hydrologically connected to the underlying sediments, the storativity of the entire system would greatly increase and diffuse flow would occur.

Large karst springs issue from Mesozoic limestones and Tertiary limestone and conglomerates in many places of this region (Table 10.6–1). Only the largest ones are mentioned here. The Dumanli Spring near Manavgat (Antalya) may well take the glory from the Fountain of Vaucluse in southern France as the largest known spring discharging from a single orifice, at least when the total annual flow is considered.

10.6.4 DUMANLI SPRING (MANAVGAT RIVER, ANTALYA)

By 1982, the enormous Dumanli Spring in the Mediterranean region of Turkey was already submerged by about 120 m of head produced by the Oymapinar reservoir. The spring contributes one third to the annual discharge of the Manavgat River, which will be dammed at Oymapinar. The mean discharge of the spring is estimated at about 50 m^3/s; therefore, its total annual outflow is about 1.6×10^9 m^3. In October 1978, the spring discharge was measured by the dye-dilution technique (Karanjac and Gunay, 1980). The flow rate of about 35.6 m^3/s at the very end of the dry period (at the end of the spring's discharge recession) motivated the authors to declare the Dumanli the largest karstic spring in the world issuing from a single orifice (Figure 10.6–2).

The results of the dye-dilution technique applied in this experiment were compared with staff-gauge readings upstream and downstream of the Dumanli Spring. To the best of the authors' knowledge, this is the first reported dye-dilution experiment that accurately measured the discharge of a river with an annual flow of about 50 m^3/s.

The spring discharges from a cave in a narrow gorge. Its outlet point is at an elevation of about 62 m asl (above sea level), less than 5 m higher than the river surface and no more than 10 m from the riverbank. Indirectly, by subtracting the flow upstream of the Dumanli from the flow recorded at a limnigraph station downstream of the Dumanli and with appropriate corrections made for intervening inflows, it was speculated that the minimum flow of the Dumanli could be about 25 m^3/s in a very dry year, its normal minimum (in a normal hydrologic year) about 30–35 m^3/s, and its mean annual flow about 50 m^3/s. Thus, its contribution to the Manavgat River flow (Figure 10.6–3), which has an order of magnitude of about 1.6×10^9 m^3, accounts for one third of the total annual river discharge.

Table 10.6–1 Distribution of Large Karst Springs in the Taurus Karst Region, and Their Average Flow Rate: Location of Karst Springs from West to East

Province	Spring	Q_{av} (m^3/s)
İzmir	Halkapınar Spring	1.2
	Bakırçay plain springs	1.2
Manisa	Göksu, Göldeğirmeni, Palamut, Sarıkız, Gölmarmara Spring	5.0
Aydin	K. Menderes plain springs	2.0
	B. Menderes plain springs	1.0
Muğla	Marmaris, Gökova Spring	
	Fethiye Karaçay, Akçay, Yaka, Kargıçayı, Yuvarlakçay, and Eşençayı Springs	35.0
Antalya	Finike-Tekke and Salur Springs	3.0
	Elmalı-Akçay-Demre plains springs	7.0
	Boğaçay plain springs	2.5
	Kırkgöz Springs	20.0
	Düdenbaşı Spring (underground river)	10.0
	Köprüçayı River valley, Olukköprü Springs	25.0
	Manavgat River valley springs (Dumanlı and others)	60.0
Isparta	Hoyran, Gelendost-Yalvaç plains springs	1.0
Afyon	Akarçay basin springs	1.5
Içel	Gilindire-Soğuksu spring and Gözce plain springs	2.0
	Silifke and Erdemli Springs	5.0
Maraş	Maraş plains springs	8.0
	Göksun plain springs	8.0
Hatay	Asi basin springs	3.0
Muş	Muş plain springs	0.8

To permit a better appreciation of the size of the aquifer drained by the Dumanli, a recession hydrograph of differences in the flow downstream and upstream of the Dumanli is presented. This hydrograph is a cumulative one of several springs between two gauging stations, but the Dumanli itself contributes at least 90 percent.

The recession hydrograph shown in Figure 10.6–4 is a cumulative one, containing the discharges of the Dumanli Spring and of Sevinç Moizi, and Yarpuzlu Springs. The latter two may have a minimum discharge of several cubic meters per second. The recession of the discharges of these springs is exceptionally slow.

FIGURE 10.6–2 Dumanli Spring before impounding.

When an equation of the Maillet (1905) type, namely, $Q_t = Q_0^{-\alpha t}$, is fitted to the regression line, the following hydrodynamic characteristics of the springs are obtained:

$$\alpha = 0.0026 \text{ day}^{-1}$$

$$Q_0 = 60.0 \text{ m}^3/\text{s on April 19, 1977}$$

$$Q_t = 40.9 \text{ m}^3/\text{s on September 13, 1977}$$

$$V_0 = 2 \times 10^9 \text{ m}^3 \text{ on April 19, 1977}$$

$$V_{dr} = 0.63 \times 10^9 \text{ m}^3$$

$$V_{rem} = 1.37 \times 10^9 \text{ m}^3$$

where α is the recession coefficient; Q_0 and Q_t are the flow rates of the springs at the beginning and end of recession, respectively; V_0 is the volume of water stored in the aquifer above the outlet elevations of the

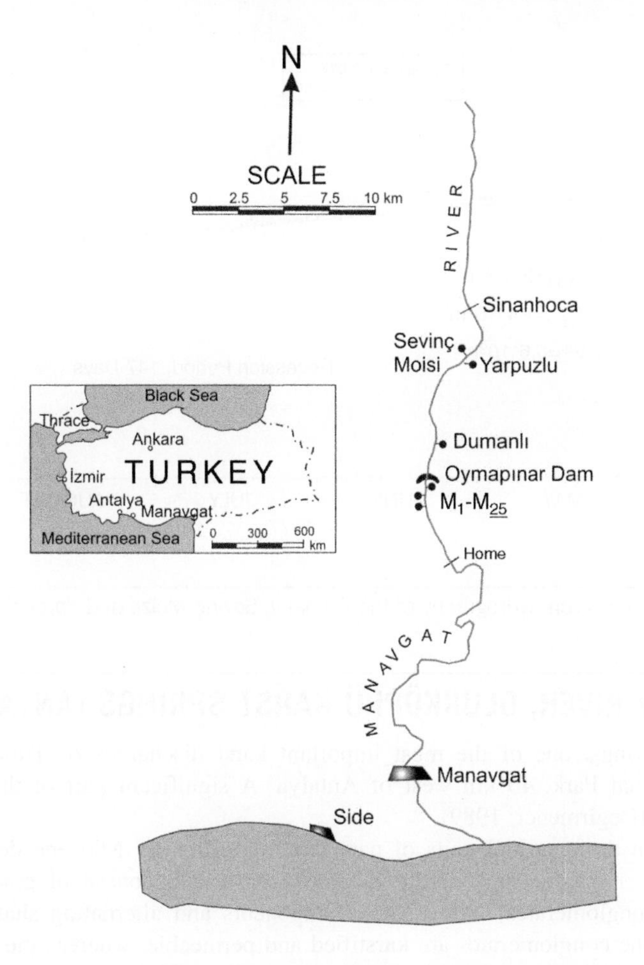

FIGURE 10.6–3 Location map of Manavgat River, Oymapinar reservoir, and Dumanli Spring.

springs at the beginning of recession; and V_{dr} and V_{rem} are the volume drained throughout the 147 day recession period and the volume still in storage at the end of recession, respectively.

The coefficient of variation of the Dumanli Spring, defined as the ratio of its maximum to minimum flow rate, is probably between 2 and 2.5. Since most of the area that might eventually contribute to the Dumanli aquifer has an annual precipitation on the order of 1500–2000 mm, the catchment area of Dumanli Spring is estimated to be between 1200 and 1500 km^2.

Based on this analysis, the following can be concluded: (1) The Dumanli plus several smaller karstic springs discharge, in the five dry months of the year, about 630×10^6 m^3; (2) near the end of the recession period, about 1.370×10^6 m^3 still remain in storage; (3) the coefficient of recession points to a very large drainage system, high storage volume, and slow drainage; and (4) the drainage area contributing to the Dumanli appears to be large, certainly more than 1000 km^2.

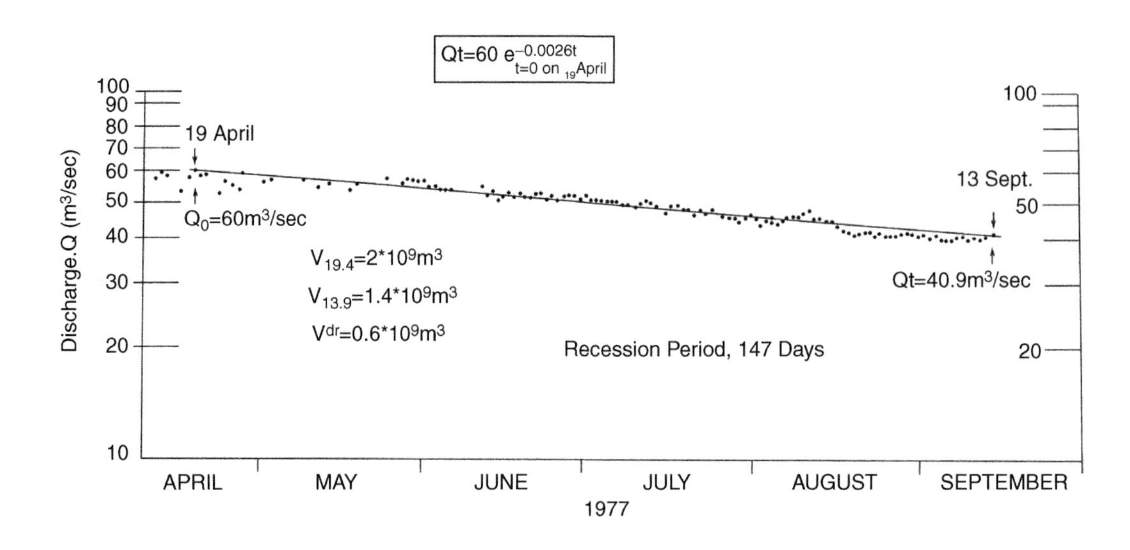

FIGURE 10.6–4 Cumulative recession hydrography of the Dumanlı, Sevinç Moizi, and Yarpuzlu Springs in 1977.

10.6.5 KÖPRÜÇAY RIVER, OLUKKÖPRÜ KARST SPRINGS (ANTALYA)

The Olukköprü karst springs, one of the most important karst discharges of Turkey, are located in the Köprüçay Canyon National Park, 40 km west of Antalya. A significant part of the recharge takes place through adjacent basins (Degirmenci, 1989).

All the older rocks in the area and units of nappes are overlain by Miocene deposits, which are over 1000 m thick. They cover a large area of the Köprüçay basin and consist of gradually alternating conglomerates (Köprüçay conglomerates) with coarse components and alternating shale-sandstone sequences (Beşkonak formation). The conglomerates are karstified and permeable, whereas the Beşkonak formation is impermeable.

Due to its dominant calcareous components and carbonate cement, the Köprüçay conglomerate is karstified along fractures, joints, and fault zones and is thus permeable. Numerous dolines, ponors, and caves have been found in the basin. The Kuruköprü cave with a total length of 530 m in the Beşkonak reservoir area is a typical example of the karstic features in the Köprüçay conglomerates.

The Olukköprü Springs are named after the Roman stone bridge constructed in the narrow canyon through which the Köprüçay River flows. Along the right bank of the river, numerous springs occur slightly above the river level. The occurrence of paleo-outlets at the higher levels clearly indicates the evolution of the karst erosion base. The springs discharge through bedding planes dipping 5–10 degrees toward the valley and through joints and fractures.

The Olukköprü Springs, together with the Böğrümköprü, Alabalik Pool, and Oğlanuctuğu Springs in the south, are discharges from the same aquifer. This conglomerate aquifer is surrounded by the impermeable Beşkonak formation from the east and south. Thus, all the groundwater discharges in the area take place between the Kikgeçit Creek in the north and the Oğlanuçtuğu fault in the south (Figure 10.6–5).

Based on drilling data obtained from one of the boreholes south of the Oğlanuçtuğu fault, the thickness of the Beşkonak formation (sandstone-shale), which acts as an impermeable barrier in the region, is over 240 m.

FIGURE 10.6–5 Beşkonak-Olukköprü karst springs.

Due to the location of the Olukköprü Springs between the Bulasan and Beşkonak flow gauging stations and to the negligible contribution from the tributaries especially in the dry season, a relatively correct analysis of the springs' discharge rates is possible based on comparison of Bulasan and Beşkonak flow records.

In the year of minimum flow (1964), a significant flow rate of 30 m^3/s was observed at the Beşkonak flow gauging station, even though the river was completely dry at the Bulasan flow gauging station. This flow can be regarded as the minimum contribution of the Olukköprü, Böğrümköprü, Alabalik Pool, and Oğlanuçtuğu Springs. The contribution of these springs to the Köprüçay River was evaluated from the monthly flow records between June and September 1982. Accordingly, the average flow rates for these springs are

- Olukköprü Springs: 31 m^3/s
- Böğrümköprü Springs: 2.5 m^3/s
- Alabalik Pool + Oğlanuçtuğu Springs: 6 m^3/s

Based on the same flow gauging records, it can be concluded that, in dry periods, about four fifths of the total flow (48 m^3/s) recorded at the Beşkonak flow gauging station are supplied by the Olukköprü and other springs in the region, whereas one fifth of the flow comes from the catchment area to the north of the Bulasan flow gauging station (Degirmenci and Gunay, 1993).

The long-term (1941–1987) annual average flow of the Köprüçay River is 86.4 m^3/s. About 68 percent of this amount is met by base flow (groundwater). For a year of minimum flow, this ratio can increase to as much as 75 percent. The abundance of the base flow in comparison with the surface flow can be explained by the discharge of the Olukköprü karst springs. This also gives rise to the existence of a regular flow regime, which is an important factor where the construction of dams, especially river-type power plants, is concerned.

To determine the recharge areas of the springs, water budget calculations based on the rainfall-evaporation-surface runoff have been made for the subbasins of Köprüçay and adjacent basins, where flow data have been recorded. The evaluations gave a groundwater contribution of 20 m^3/s from the adjacent basins to the Köprüçay basin. Based on the budget calculations and hydrogeological evaluations, the following conclusions were drawn:

- About 10 m^3/s of this groundwater comes from the Büyük Çandir subbasin of the Aksu basin.
- The remaining 10 m^3/s is the groundwater inflow to the Köprüçay basin that may be supplied by the karst terrains located between the Dumanli Mountain and Kartoz.

The Olukköprü Springs discharge water in the deep Köprüçay River valley from these gently dipping conglomerates along the bedding planes, fractures, and joints. Most springs are at altitudes of 160–170 m asl. Several individual springs yield more than 4 m^3/s. The old spring outlets are visible at higher elevations on the valley slopes. As the Köprüçay River lowered its channel, the spring levels also fell. Waters of these springs on the two banks may be connected by siphonlike channels. Conglomerates are limited by the interfingering shales to the east in the vicinity of these springs. Gently dipping shales appear under the conglomerates at the right bank 1.5 km downstream of the springs. Therefore, these springs are forced to discharge water at the area where the conglomerates encounter the impervious formation both in the east and at their bottom.

The total minimum flow of all the Olukköprü Springs north of Beşkonak in a dry year was found to be 30 m^3/s. This is based on the analysis of the discharge of the Köprüçay River. The maximum flow of all the springs may be twice or three times the minimum flow. If one takes the position that these springs emerge as a spring zone from a unique aquifer, an attractive hypothesis that is supported by various facts, then the Olukköprü Springs by their minimum total flow of about 30 m^3/s may exceed the minimum flow of the Dumanli Spring (estimated to have a minimum discharge of 20–30 m^3/s). Therefore, all factors indicate that the Olukköprü Springs may be among the largest karst springs in the world.

The mean discharge of the Köprüçay River, measured at a gauging station 20 km from the Mediterranean Sea, is 117 m^3/s. This corresponds to a runoff of about 1560 mm precipitation for the surface drainage area of the Köprüçay River basin. Because the mean annual precipitation of the surface basin is only about 1070 mm, it is obvious that water comes to the springs from outside the surface watershed as well. The largest contribution to the Köprüçay River flow is from the Olukköprü Springs. Their contribution was 37 m^3/s in 1970. The lack of conformity between the surface drainage area and the underground drainage area must be accepted. The surplus water may be coming from the area of the Eğridir and Kovada Lakes but more likely from the Beyşehir Lake through the Kirkkavak (Kepez) fault. An additional drainage area may be in the high mountains between these lakes. It is likely that two thirds of the springs' water come from the surface drainage area, while a third comes from the adjacent areas through karst subaquifers and from the lake percolations.

Discharge measurements at the Bulaşan and Beşkonak stream gauging stations show that the river flow is at its minimum in September and October, at the beginning of the rainy season. The mean annual flow at Beşkonak was 2.9 km^3 for the period of 1963–1971. The mean base flow is 60 percent of the total flow, which is approximately 1.7 km^3. About 70 percent of the Köprüçay River flow comes through the Olukköprü Springs. The recession coefficient α is very small, even smaller than expected for a mature karst aquifer. It varies between 0.003 and 0.0082 for the period of analysis.

If the water of the Beyşehir Lake and some water of the Eğridir and Kovada Lakes watershed areas feed the aquifer of the Olukköprü Springs, then they have a long route of about 50 km or more. Since the recession curve has a low value of α, say, on average, $\alpha = 0.005$, then at the minimum flow of $Q = 30\ m^3/s$, the "dynamic" storage (water volume above the level of the Olukköprü Springs) may be about 6 km^3. Since the limestone conglomerates are very deep (possibly 1000 m) and the comprehensive limestone that feeds water into the conglomerates are also deep, the "static" storage volume must also be large for these springs, as has already been inferred for the Dumanli Springs. This conclusion coincides with the results of studies of the natural isotopes of these springs.

10.6.6 BEYAZSU AND KARASU KARST SPRINGS (MARDIN-NUSAYBIN AREA, SOUTHEAST TURKEY)

The mean annual precipitation in southeastern Anatolia ranges between 550 and 850 mm. Farther to the north, the annual precipitation exceeds 1200 mm at the Gercus and Haberli meteorological stations. Kiziltepe Plain, with an area of about 600 km^2 at the Turkish-Syrian border in southern Turkey, is of great importance from the standpoint of the socioeconomic development of the region. Although the topography and soil structure favor agriculture, shortage of water is the major factor hindering the development of the plain.

The multiaquifer system in the plain consists of three individual aquifers whose water might be available for irrigation and domestic use. The upper aquifer is made of unconsolidated clastics and basal conglomerate. The static water level at this aquifer ranges between 5 and 30 m from the surface, and groundwater is available through shallow dug wells. This aquifer is separated from the underlying limestone aquifer by an impermeable unit made up of marl and clay. The lower aquifer, which is located at greater depth, is made of fractured and karstified limestone and dolomite of Upper Cretaceous age (Eroskay and Günay, 1980). The hydrogeological properties of deep carbonate aquifers differ significantly, since they are highly karstified. The data of this deep paleokarstic aquifer are obtained mainly from oil drilling.

The most important water resources are in Eocene Midyat limestone, which is highly karstified (Altinli, 1966). The limestone outcrops at very limited areas (80 km^2) in the west, whereas it has a much larger areal extent in the east. Karstic springs discharging from the Midyat formation are located mainly in the eastern part, where this formation outcrops over large areas. The infiltration rate of the Midyat limestone is calculated as 80 percent, which provides an important groundwater reservoir. The discharge rates of the Beyazsu and Karasu karst springs were measured in February 1985 as 4.25 m^3/s and 4 m^3/s, respectively (DSI, 1987; Figure 10.6–6). The existence of these springs at high altitudes (600–650 m) may reflect that the karstification base does not extend to great depths. The discharging water is of the typical calcium-bicarbonate class. The water of the Beyazsu Springs is tapped and distributed to Mardin city.

10.6.7 MUĞLA-GÖKOVA KARST SPRINGS (AKYAKA-AKBÜK AREA)

The springs of Gökova consist of the springs that are discharged from the coastline between Akyaka-Akbük, situated in the north of Gökova Bay. The springs are related to the Gökova fault system. In the 60 km long fault system, which is a graben still in the process of opening, the southern blocks have descended while the northern blocks have risen. In relation to the faults of Quaternary or younger age, the limestone at Haticeana and Babadağ and the underground system of Köprüçay conglomerates that are commonly related to the limestone have descended (Kurttaş, Günay, and Gamalmaz, 2000).

Under such conditions, the karst system as well as overlain alluvial sediments have been affected by saltwater. The springs that are recharged from the north and northeast rise to the surface, passing through

FIGURE 10.6–6 Beyazsu Spring.

the saltwater zone as they flow along the fault lines. The spring that surfaces from underneath the coastal alluvium in the north of Akbük Bay is the discharge from Haticeana limestone situated there, again in relation to the faults. It is presumed that, in the west of Akbük Bay, the Yatağan formation, by being situated on the block that descended due to fault formation, constitutes a barrier before the karstic limestone and is effective in the formation of the springs (Eroskay et al., 1992).

The Azmak spring group is composed of discharges from the slope of the fault line, located along the road between Akyaka and Çaydere, to the east of Akyaka village. The springs surfaced at elevations near sea level, and the channel into which they are discharged forms a group of springs (Azmak). In the measurements taken by the State Hydraulic Works (DSI) in August 1990, the total flow rate of the spring group was calculated as 11.18 m^3/s. The results of the salinity measurements applied to the spring group were observed to vary, depending on place and time (Eroskay et al., 1992).

The Akbük Bay Spring is discharged from alluvium at an elevation close to sea level, situated in the northwest of Akbük Bay. It formed an Azmak of a length of 300 m. The spring is controlled by a fault, behind which is situated the Haticeana formation. Most probably, the spring is recharged by the Yilanli and Haticeana limestone. In July 1991, 20°C of temperature, 0.007 of salinity, and 11,000 μmhos/cm of electrical conductivity (EC) were measured (Eroskay et al., 1992).

10.6.8 NIĞDE-POZANTI ŞEKERPINARI SPRINGS (SOUTH TURKEY)

The Şekerpinari Springs are in the central Taurus Mountain region located near Alpu town in Pozanti county (near the Ankara-Adana highway and railway). The geology of the Pozanti and surrounding area is decribed by Demirtaşli (1981). The Bolkar Mountains form the backbone of the eastern part of the Inner Taurus belt, which is composed mainly of slightly metamorphic limestone and slate of the Bolkar group. The age of the Bolkar group rocks ranges from Permian to Late Cretaceous, and they were thrusted over formations of Late Cretaceous-Paleocene age of the Ereğli-Ulukişla basin.

Due to the high elevation (1000–3500 m asl), the climate of the Bolkar Mountains in the central Taurides region shows more continental characteristics than that of the Mediterranean climate type. The summers are hot and dry. The winters are cold and wet. Most of the precipitation falls in the form of snow, and at high elevations, the snow cover stays even until early summer. On the northern slopes of the Bolkar Mountains, the snow does not melt even in summer. The average precipitation for the period 1980–2000 was 678 mm/a.

The Şekerpinari Spring discharges are between 6 and 12 m^3/s (Figure 10.6–7). The spring's regression curve analysis resulted in a coefficient of $\alpha = 0.0178$ day^{-1}. Storage capacity of the spring is $V_s = 38{,}831 \times 10^6$ m^3.

The recorded values of calcium in the Şekerpinari water are between 24.0 and 43.0 mg/L. The values of magnesium are 4.7–10.9 mg/L, sodium 1.8 mg/L, potassium 0.3 mg/L, bicarbonate 109.8 mg/L, sulfates 0.8 mg/L, chloride 1.4 mg/L, pH 7.9, TDS 130 mg/L, EC 200 μS/cm, temperature 9.0°C.

FIGURE 10.6–7 Pozanti-Şekerpinari Spring.

10.6.9 SAKARYABAŞI KARSTIC SPRINGS (ÇIFTELER)

Upper Sakarya basin drains mainly through the Sakaryabaşi karstic springs, from which the Sakarya River emerges in the Eskişehir-Çifteler area. The main aquifer is formed within the Gökçeyayla formation, which is composed primarily of shelf-type carbonates of Triassic-Upper Cretaceous ages. Dolomitic limestone is dominant in the lower section of the unit, while the upper section is mostly cherty limestone. The uppermost section of the unit is composed of shale and the radiolarite, which indicate a deepening transition to the Çöğürler complex. The Çöğürler complex has been defined as an ophiolitic olistostrome (Figure 10.6–8).

Most of the spring discharge occurs from the Gökçeyayla aquifer (Figure 10.6–9). To estimate the hydrological characteristics of the Sakaryabaşi Springs, a hydrological drainage area of ~4300 km^2 was examined. The results of the hydrological, isotope hydrological, and karst hydrogeological studies conducted to determine the recharge area of the Sakaryabaşi Springs are briefly discussed next.

For the basin in general, the mean precipitation is 376.8 mm/a (period 1967–1995) and the mean temperature value is 10.5°C. The annual evaporation in the basin is 334.5 mm/a. According to the calculations, 89 percent of the precipitation that occurs in the basin evaporates. Using the dye dilution method, in December 1995, the total drainage of the Sakaryabaşi Springs was calculated as 5.6 m^3/s. In the study area, water budget calculations have been done for the basin in general. The basin is recharged from the surface

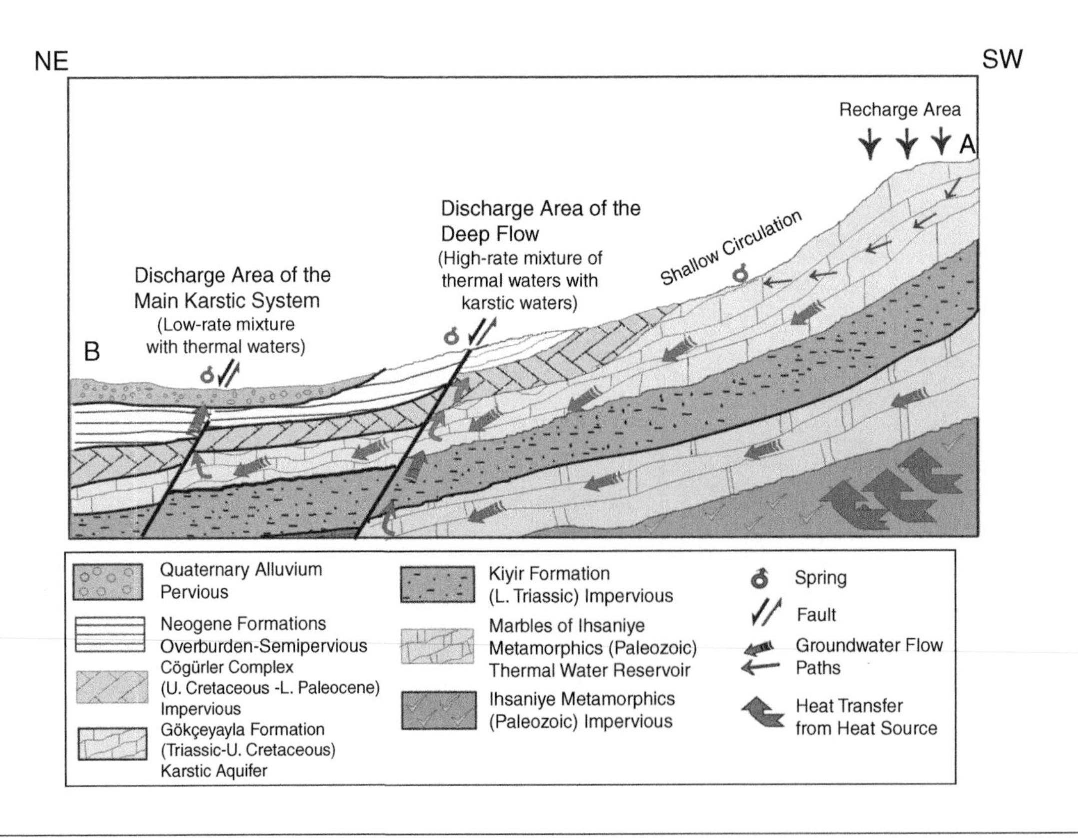

FIGURE 10.6–8 Cross section of Eskişehir-Çifteler area.

FIGURE 10.6–9 Sakaryabaşi discharge zone.

drainage area. All water samples studied are situated on and between the $\delta D = 8\ \delta^{18}O + 14.5$ Ankara Meteoric Line and the $\delta D = 8\ \delta^{18}O + 10$ Global Meteoric Line. The elevation-dependent change in the oxygen-18 content of precipitation is found to be –0.33 for each 100 m increase in elevation (–0.33/100 $\delta^{18}O$). In the isotopic composition of the springwater, such deviations as evaporation and hydrothermal interaction have not been detected (Guven, 1996; Gunay, 2006).

The Sakaryabaşi Springs, which have low tritium and high EC values, are deep circulating karstic geothermal springs. The possible recharge area of the Sakaryabaşi Springs has been defined as the Gökçeyayla formation of Triassic dolomite limestone located in the southwest of the study area. The additional area required for balancing the recharge and discharge in the basin corresponds to an area in the south where limestone is densely distributed. According to Ergeneli (1994), with the exception of Ilicabaşi Spring, the Sakaryabaşi, Başkurt, and Sadiroğlu Springs are on the tectonic unit that recharges the Sakarya River.

In this study, it was found that Ilicabaşi Spring recharges the Sakarya River through an output of 0.45 m^3/s. Since schists hydrogeologically classified as Paleozoic are in the vicinity of Afyon, the Sakaryabaşi Springs are not recharged by this area. It seems also unlikely that the Sakaryabaşi Springs are recharged by the Akşehir and Eber Lakes, approximately 100 km away and at an elevation of 950 m. Another reason for this assertion is that in the oxygen-18/deuterium graph, the Sakaryabaşi Springs and other springs are situated on the Global Meteoric Line, and an evaporation effect on the springs has not been detected. If there were a recharge from Akşehir and Eber Lakes into the Sakaryabaşi Springs, the springs would be situated on the evaporation line in the oxygen-18/deuterium graph. When the chemical analysis results of the springs and the lake waters are compared, there are no relationships between the two waters. The water of Akşehir and Eber Lakes contain $NaCO_3 \cdot HCO_3$, whereas the Sakaryabaşi springs contain $CaCO_3 \cdot HCO_3$ (Table 10.6–2).

Table 10.6–2 Hydrochemical Results of Field Measurements

Sample Name	Temperature (°C)	pH	EC (μs/cm)	TDS (mg/L)	Hardness (°Fr)
Sakaryabaşi	18.7–23	7.28	750–820	460–480	41.45–47.9
Havuzbaşi	19–20	7.05–7.77	833–900	430–550	42.6
Kirkgiz Lake	18.7–23	7.31–7.74	780–1000	420–560	38.1
Hamampinari	23.8–2	6.35–6.9	820–900	460–630	38.4–51.8
Karaburgu	20.5–22	7.17–7.4	841–890	450–460	43.05

10.6.10 KARST HYDROGEOLOGY OF PAMUKKALE KARST SPRINGS (DENIZLI)

The main aquifer supplying hot water to the Pamukkale thermal springs is the highly fractured and karstic marble of Paleozoic age. The marble is confined by the thick, impervious-semipervious units of Pliocene age from the top. These units form the overburden layer needed for the occurrence of the thermal springs.

The catchment area extending along the Yenice Horst is covered mainly by limestone of Pliocene age. Recharge of the aquifer is provided by infiltration of precipitation through the outcrops of marble and limestone, which have a high secondary porosity and permeability.

The annual mean areal precipitation was found by the isohyetal method to be 563 mm. Four major springs exist at the edge of the travertine area (Figure 10.6–10). All these outlets are located along the major fault line extending northwest-southeast. The total discharge of the springs is about 385 L/s on average. This figure does not change significantly during the year. Recession curve analysis revealed that the annual active reservoir capacity of the aquifer is about 16×10^6 m^3. This suggests that about 30 percent more can be abstracted safely from the aquifer to be utilized in travertine formation or for other purposes (UKAM, 1994 and 1995).

FIGURE 10.6–10 Thick travertine deposits at Pamukkale Springs, world-famous geoheritage site.

Chemical and environmental isotopic evaluations revealed that hot water is mainly of meteoric origin with a long (about 20–30 years) turnover time. The chemical composition of the waters is almost constant. They are of calcium bicarbonate character representing karstic aquifer. Figure 10.6–11 depicts the ^{18}O-*D* relation for the hot springs of Pamukkale and other hot springs existing in the adjacent hydrological sites. A geothermal exchange effect is not observed in the hot waters. They are affected, to a great extent, by meteoric waters.

To explain the recharge and flow regime of the hot springs, regular and systematic observation, measurement, and sampling programs were performed on a weekly basis. This knowledge assisted in developing a conceptual model for the regional hydrodynamic structure. However, precipitation kinetics of the travertine were also studied in this framework. In this context, the required physical, chemical, and hydraulic conditions for the maximum deposition of travertine were investigated for the sake of conservation of the white travertine area. Precipitation kinetics and mass transfer calculations revealed that the hot water starts travertine deposition when the saturation with respect to calcite attains a value about four times its initial value at the outlet. The maximum rate of deposition is attained when this figure is about sixfold the initial value

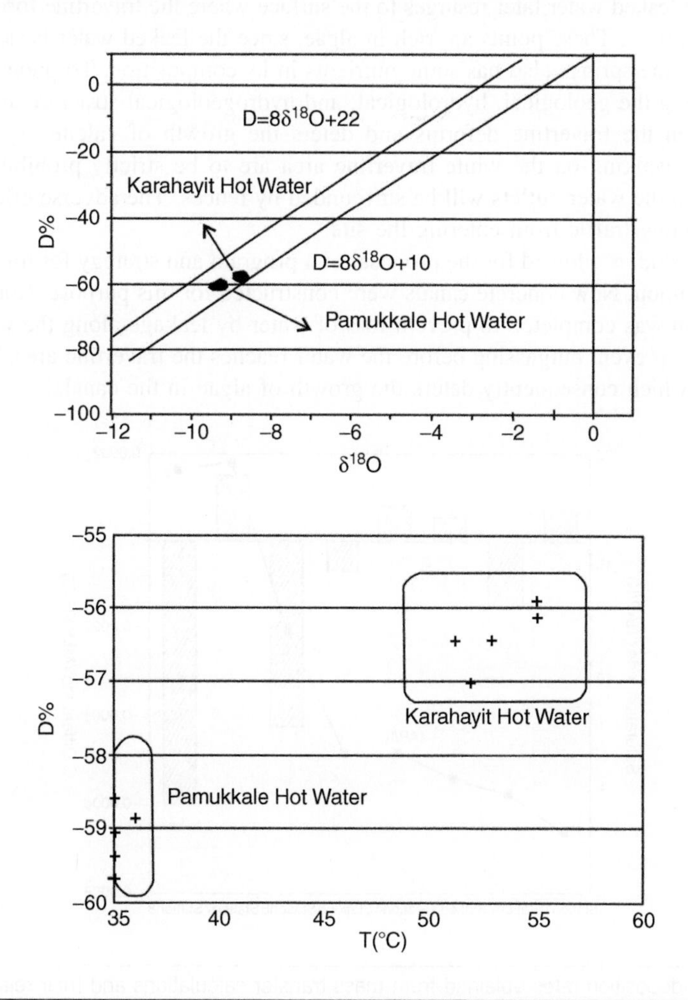

FIGURE. 10.6-11 Oxygen-18/deuterium relation in thermal springs of Pamukkale.

(Figure 10.6–12). Precipitation is found to be enhanced by the hydraulic agitation. Travertine deposition is maximum at the rim of terraces, where outgassing of CO_2 is enhanced by a turbulent flow. However, deposition is also significantly high when the water flows at high velocity and the flow depth is low. These hydraulic controls were investigated using the relation between the pH value and saturation index.

The white travertine has been becoming darker, yellowish, and brownish since the establishment of tourist sites and hotels on the travertine area. The hotels take the hot water from the spring outlets directly to their swimming pools before they release it onto the travertine. The pools have had more than one adverse effect on the properties of the water. First, because of the high rate of outgassing of CO_2, the capacity of the water in the pools to deposit travertine has decreased. Second, the people swimming in the pool leave some organic relics, which cause a rapid growth of algae. The algae in turn cause a change in color of the travertine from white to greenish, yellowish, and brownish.

The lack of a sewage system and presence of septic tanks for every hotel are the other major sources of pollution. The septic tanks are dug in the travertine, and though they are lined with cement, they leak wastewater into the travertine. The leaked water later resurges to the surface where the travertine forms a steep topography at the bottom of the terraces. These points are rich in algae, since the leaked water is rich in nutrients.

The hot water from the springs also has some nutrients in its composition. To prevent this pollution, protection areas considering the geological, hydrological, and hydrogeological structure are delineated.

Because walking on the travertine deforms and deters the growth of calcite crystals, entrance to the travertine terraces and walking on the white travertine area are to be strictly prohibited. For this purpose, the white travertine and the water outlets will be surrounded by fences. The adverse effect of traffic will also be removed by prohibiting traffic from entering the site.

Hydrogeochemical studies allowed for the proposal of a program and strategy for travertine deposition and protection against pollution. New concrete canals were constructed for this purpose. Construction of concrete canals of about 2400 m was completed to prevent loss of water by leakage along the water path. The canals are covered with lids to prevent outgassing before the water reaches the travertine area. The lids serve also as obstacles to sunlight, which consequently deters the growth of algae in the canals.

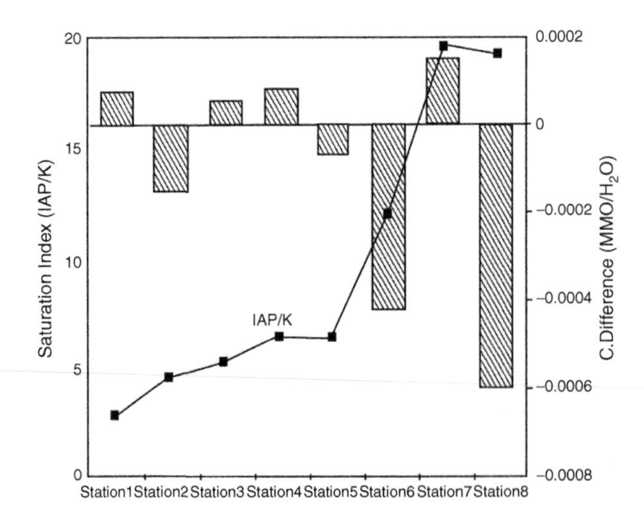

FIGURE 10.6–12 Calcite deposition rates obtained from mass transfer calculations and their relation to the saturation index with respect to calcite.

The old asphalt road crossing the travertine is abandoned and new terraces imitating the natural morphology have been constructed. Tourist activities (hotels, shopping centers, etc.) will be reduced or prohibited in some fragile places. Walking on the travertine and traffic on the site will be strictly prohibited. Septic tanks will be removed; mobile toilets will be provided instead. A special security team will be composed for the area, and the water distribution system will be operated by authorized technical persons like hydrogeologists.

Special regulations and laws should be enacted to conserve the travertine area based on the scientific findings and evaluations. The delineation of protection zones against pollution should be enforced and the requirements for implementation should be developed immediately (Gunay and Simsek, 1997).

REFERENCES

Altinli, I.E., 1966. Geology of Eastern and Southeastern Anatolia. Bulletin No. 66, MTA, Ankara, Turkey.

Back, W., Günay, G., 1992. Tectonic influences on groundwater flow systems in karst of the southwest Taurus Mountains, Turkey. In: International Contributions to Hydrogeology Vol. 13. Verlag Heinz Heise, Hannover, Germany.

Brinkmann, R., 1976. Geology of Turkey. Ferdinand Enke Verlag, Stuttgart, Germany.

Demirtaşli, E., 1981. Summary of the Paleozoic stratigraphy and Variscan Events in the Taurous Belt. Newsletter, IGCP Project No. 5, Correlation of Variscan and Prevariscan Events in the Alpine Mediterranean Belt 3 (1), 44–57.

Değirmenci, M., 1989. Köprüçay ve dolayinin (Antalya) karst hidrojeoloji incelemesi [Karst hydrogeological investigation of the Köprüçay basin and its vicinity, Antalya]. PhD thesis, Hacettepe University, Ankara, Turkey (unpublished).

Değirmenci, M., Günay, G., 1993. Origin and catchment area of the Olukköprü karst springs. Hydrogeological Process in Karst Terranes. Proceedings of the Antalya Symposium and Field Seminar, October 1990. IAHS Publication, no. 207.

DSI, 1987. Beyazsu and Karasu karst springs measurements. Report No. 87. Ankara, Turkey.

Ergeneli, N., 1994. Sakaryabaşi kaynaklari ve dolayinin karst hidrojeolojisi incelemesi [Karst hydrogeology of the Sakaryabaşi springs area]. MS thesis, HÜ Fen Bilimleri Enstitüsü, Ankara, Turkey.

Eroskay, S.O., Günay, G., 1980. Tecto-genetic classification and hydrogeological properties of the karst regions in Turkey. In: Günay, G. (Ed.), Karst Hydrogeology. Oymapinar-Antalya, Turkey UNDP Project TUR/77/015; reprinted in: Günay, G. (Ed.), Proc., Intern. Symp. and Field Seminar on Karst Hydrogeology. Oymapinar, Hacettepe University, Ankara, Turkey, pp. 1–41 (1979).

Eroskay, S.O., Gözübol, A.M., Gürpinar, O., Şenyuva, T., 1992. Muğla Gökova ile Milas-Savran ve Ekinambari Karst Kaynaklarinin Jeolojik ve Hidrojeolojik İncelemesi, Sonuç Raporu. DSİ Genel Müdürlüğü, Ankara, Turkey.

Günay, G., 2006. Hydrology and hydrogeology of Sakaryabaşi karstic springs, Çifteler, Turkey. Environ. Geol. 51, 229–240.

Günay, G., Şimşek, Ş., 1997. Karst hydrogeology and environmental impacts of Denizli-Pamukkale springs. In: Günay, Johnson, (Eds.), Karst Waters and Environmental Impacts. Balkema, Rotterdam, the Netherlands.

Güven, F., 1996. Sakaryabaşi kaynaklarinin çevresel izotop hidrojeolojisi incelemesi [Environmental Isotope Hydrology of the Sakaryabaşi Springs]. MS thesis. Hacettepe University, Ankara, Turkey.

Karanjac, J., Günay, G., 1980. Dumanli Spring, Turkey—The Largest Karstic Spring in the World? J. Hydrol. 45, 219–231.

Kurttaş, T., Günay, G., Gemalmaz, A., 2000. Karst Hydrogeology of the Gökova, Guide Book. In: Karst 2000 symposium. September 17–26, 2000, Marmaris, Turkey.

Maillet, E., 1905. Essais d'hydraulique souterraine et fluviale. (Hermann, A. Ed.), Libraire Scientifique, Paris.

UKAM, 1994 and 1995. Projects Phase-I and Phase-II: Report on conservation and development of Pamukkale travertines. Hacettepe University, Int. Res. Centre for Karst Water Res., Ministry of Culture, Turkey [in Turkish, unpublished].

CHAPTER

Case Study: Springs of the Zagros mountain range (Iran and Iraq)

10.7

Ezzat Raeisi[1] **and Zoran Stevanovic**[2]
[1]Department of Earth Science, Shiraz University, Iran;
[2]Department of Hydrogeology, Faculty of Mining and Geology, University of Belgrade, Serbia

10.7.1 INTRODUCTION

The Zagros Mountain range represents the southern, Asian branch of the Alpine geosynclines. Most of the Zagros Mountains (also known as the Iranian Mountains) are in the western and southern part of the Iranian territory. The geological boundary with the Taurides Mountains on the east is the largest regional fault along the Great Zab valley in Iraq. Towards the southwest, the Zagros Mountains foothills extends to the Upper Mesopotamia plain, Khuzestan plain, and the Shaat El Arab marshland. In the south, the elevation of the Zagros Mountains gradually declines toward the Persian Gulf. The highest peaks in Iran are over 4000 meters above sea level (m asl); that is, Zard Kuh is 4548 m asl and Mount Dena is 4359 m asl. A considerable amount of snow falls during the winter and rainfall is often well over 500 mm per year. The precipitation represents the main recharge of the aquifers. The Zagros Mountains are composed mainly of karstic carbonate rocks. A high percentage of precipitation recharges the karstic aquifers. Water of the karst aquifer emerges mainly through the springs along the contact with nonkarstic formations or alluvium.

10.7.2 HYDROGEOLOGY AND KARSTIC AQUIFERS OF THE ZAGROS RANGE

Systematic geological investigations in the Middle East started in the early 1920s and were aimed at the assessment of oil resources. Although the regional geology is well explored (Dubertret, 1959; James and Wynd, 1965; Stocklin, 1968; Stocklin and Setudehnia, 1971; Falcon, 1974; Buday, 1980; Darvishzadeh, 1991; Alavi, 2004), many hydrogeological investigations were undertaken for practical purposes: to tap groundwater for drinking or for irrigation (Noble, 1926) or to assess the conditions for building dams, which has a long tradition in the region (Mohammadi, Raeisi, and Bakalowicz, 2007a, 2007b).

The Zagros Mountains make up Iran and Iraq's largest mountain range. They have a total length of 1600 km from northeast Iraq to the Straits of Hormuz in the Persian Gulf (Figure 10.7–1). The Zagros Mountains were formed by a collision of two tectonic plates, the Eurasian and Arabian. The Zagros orogenic system is divided longitudinally into the three major tectonic zones: (1) thrust zone (in the north), (2) simply folded zone (or high folded zone) in the center, and (3) plain (or low folded zone) in the south (Stocklin, 1968). The thrust zone is a narrow zone of thrusting bound on the north by the main Zagros thrust fault. This zone contains the oldest

Copyright © 2010, Elsevier Inc. All rights reserved.

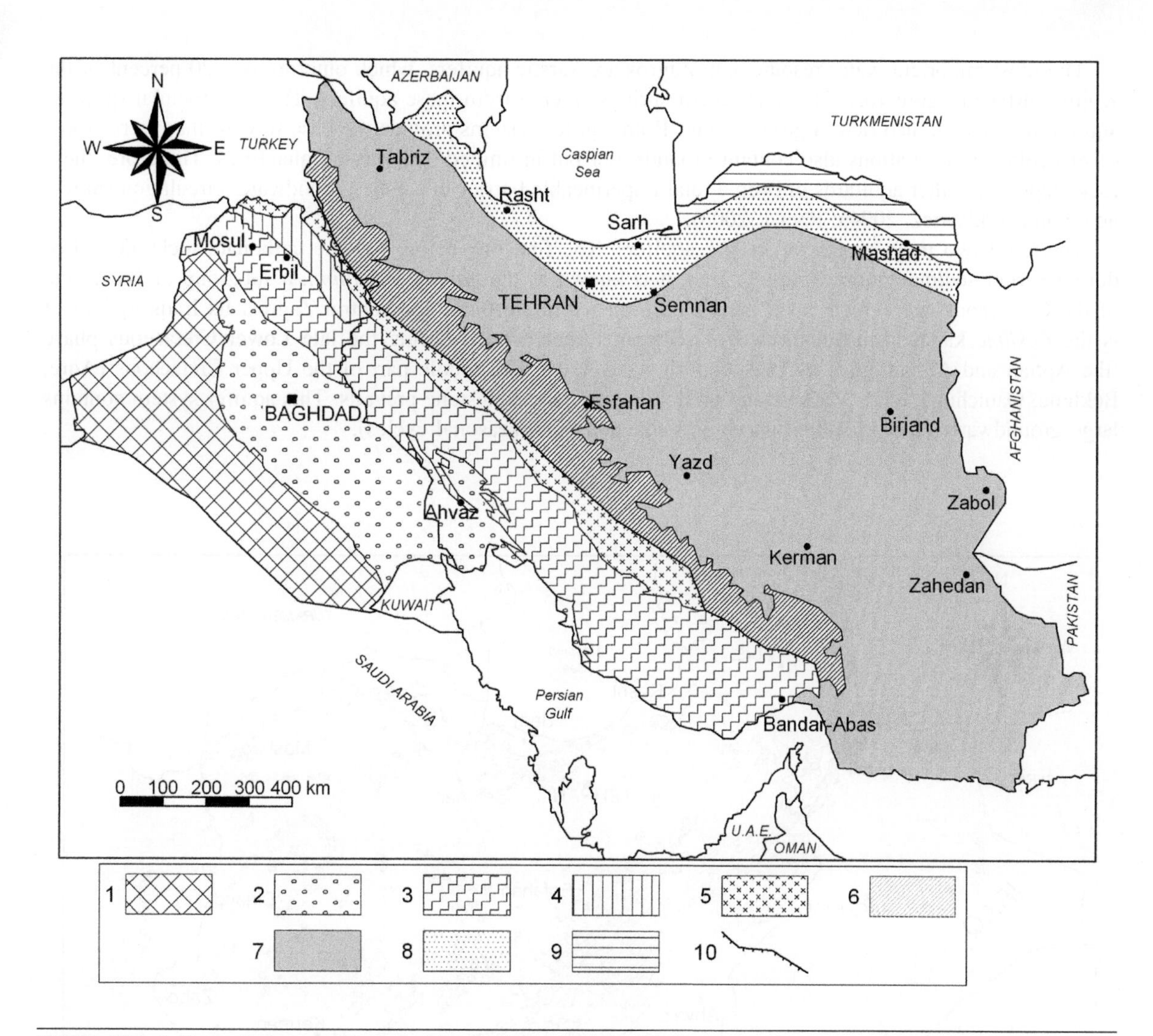

FIGURE 10.7–1 Geological map of Iran and Iraq: (1) Stable shelf, (2) Mesopotamian zone and Khuzestan plain, (3) low folded zone, (4) high folded zone, (5) thrust zone, (6) Sanandaj-Sirjan zone, (7) central Iran tectonic units, (8) Khazar Talesh Ziveh and Makran zone, (9) Kopet Dagh zone, (10) Zagros thrust fault.

geological formations, of the pre-Triassic age, and formations of the late Tertiary period. The rocks of this zone are crushed and intensively faulted. In the simply folded zone, a sequence of Precambrian to Pliocene shelf sediments, about 8–10 km thick, has undergone folding from the Miocene period to recent time. Intensive uplifting and folding of the sedimentary complex sequence are the result of Alpine orogenic polyphase deformation (Stocklin, 1968; Buday and Jassim, 1987; Stevanovic and Markovic, 2004). The simply folded zone consists of long, linear, asymmetrical folds. Anticlines are well exposed and separated by broad valleys (Miliaresis, 2001). The numerous anticlines topographically represent mountain ridges, while between them wide synclines were formed. On the plain (or low folded zone), the alluvial and terrace sediments cover the older formations. Salt domes and salt glaciers are a common feature of the Zagros Mountains, especially in southern Iran.

The most important water resources in Zagros are karstic aquifers, which outcrop over 20 percent of the Zagros surface (Figure 10.7–2). Hundreds of springs emerge from these aquifers. The intergranular (porous) aquifer systems are also rich in groundwater. Both aquifer systems provide the base flow of the rivers. Some other geological formations also contain groundwater but in limited and varying quantities. Therefore, these rocks represent either aquitards or even a total impermeable barrier to karstic groundwater circulation (Stevanovic and Iurkiewicz, 2004a; Raeisi, 2008).

There are two main groups of karstic aquifers. The first one belongs to the carbonate rocks deposited during earlier sedimentation cycles. In Iran, it is known as the *Sarvak formation*, a subgroup of Bangestan of the Cretaceous age, which has a great extension within northern and central Zagros. Its Iraqi equivalent is the *Bekhme* karstic aquifer, formed of sediments accumulated during the late Lower Cretaceous phase (the Aptian and Albian ages) as well as during the lower and middle part of the Upper Cretaceous (Aqre, Bekhme, Qamchuga, etc.). Massive and predominantly pure limestone prevails. This aquifer system contains large groundwater reserves, albeit varying in size and time (seasonal cycles).

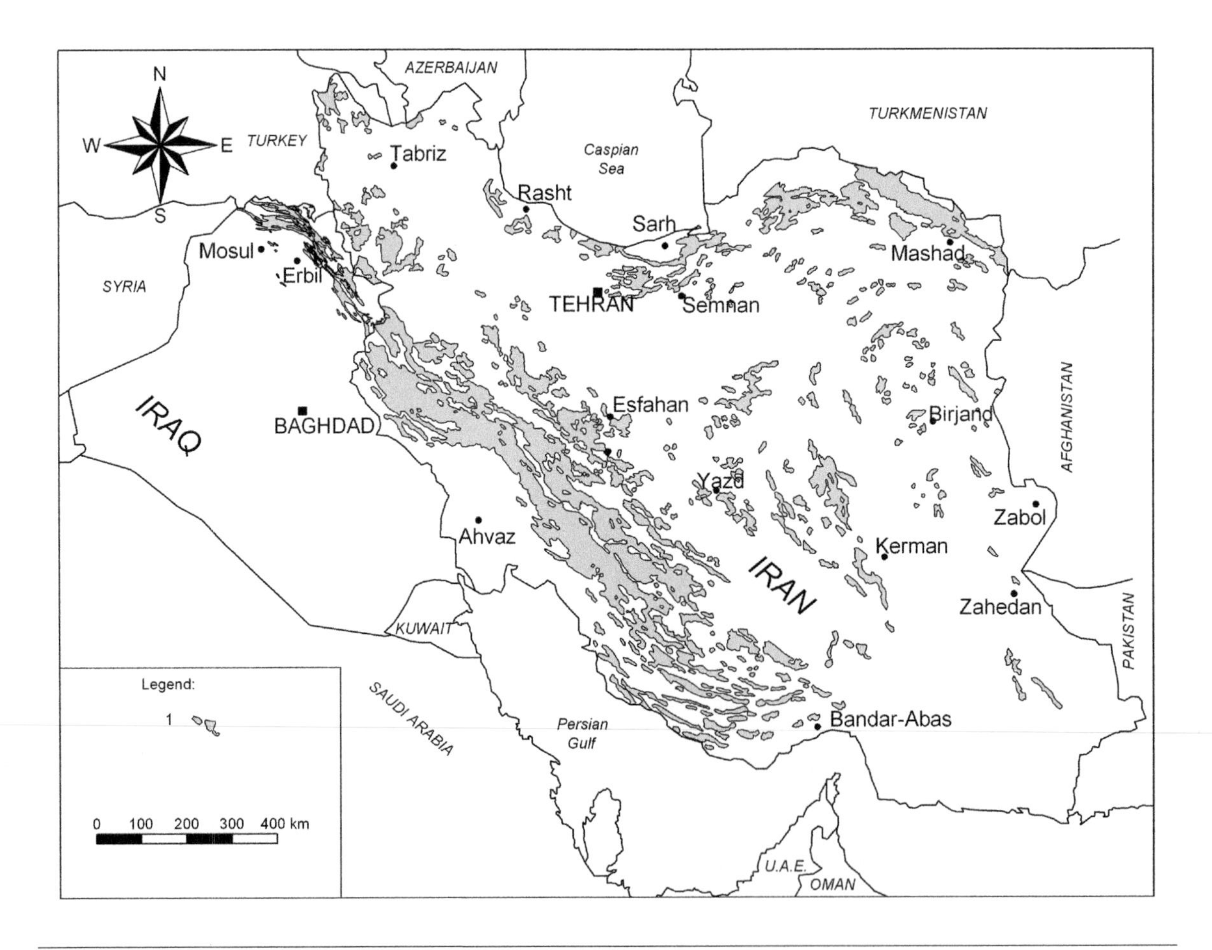

FIGURE 10.7–2 Distribution of karstic aquifers in Iran and Iraq: (1) Karstified rocks.

The second, younger karstic aquifer system is of the Oligo-Miocene age. In Iran this is Asmari limestone or undifferentiated Asmari-Jahrum, while its Iraqi stratigraphic equivalent is known as *Pila Spi*. It is present predominantly in the southern parts of the Zagros Mountains. This aquifer also contains large reserves; in some cases, the well capacity might achieve 60 L/s with high artesian pressure (Stevanovic and Iurkiewicz, 2004b; Mohammadi et al., 2007). The other karstic aquifers are Cretaceous Darian-Gadvan-Fahliyan, Cretaceous Ilam, Eocene Tarbur, and Miocene Guri, which are locally the main water resources in some parts of Iranian Zagros.

Ashjari and Raeisi (2006a) and Raeisi (2008) showed that most of the karstic aquifers in the Zagros folded zone form a broad highland aquifer with the following characteristics:

1. They are sandwiched between two thick, impermeable formations, where the hydrogeological connections between karstic formations do not exist except on rare occasions when circulation is possible along the major faults.
2. The outcropped karstic formations build up the high mountains. The impermeable overlying formations are eroded from the high elevation parts of the anticlines, and they are mostly exposed at the foot of the anticlines or buried under a thin alluvium. The impermeable formations act as a local base of erosion, and karst water discharges through the springs at the contact zone.
3. Flow from one karstic anticline to the parallel, adjacent karstic anticlines is unlikely because the karstic formations in synclines are normally buried under very thick overburdens. These characteristics imply that karstic formations in the Zagros anticlines are mostly independent aquifers. The aquifer boundary is limited to the overlying and underlying impermeable formations or adjacent alluvium.

10.7.3 FACTORS CONTROLLING SPRING DISCHARGE AND CHEMISTRY

Ashjari and Raeisi (2006a) presented a conceptual model for flow direction in Zagros based on the 72 anticlines. The anticlines were divided into two main groups based on the presence or absence of the hydraulic connectivity between the limbs. The geological and tectonic settings are the main controlling factors within these two groups. Each group was further classified into four subgroups based on the location of the discharging zones, namely, one or both down-plunge noses, limbs, traversing river and combination of down-plunges, limbs, and river. The discharging zones may be located in adjacent or successive anticlines. The discharging zones are controlled mainly by the local base level. In most cases where there is no hydraulic connection between the limbs, the direction of flow is initially along the bedding plane dip and finally parallel to the strike at the foot of the anticline. In most cases having connections between two limbs, the regional direction of flow, in the connection part, is in the opposite direction to the bedding plane dip and eventually parallel to the strike. The results show that the principal controlling factor of predominate regional flow is the anticlinial structure of aquifers and geometry of bedrock.

Ashjari and Raeisi (2006b) studied the lithological control of water chemistry based on hydrochemical data from 195 karst springs in south-central Iran. The springs emerge from the Fahliyan-Gadvan-Dariyan, Sarvak, Tarbur, Asmari-Jahrum, and Guri karstic formations. The springs were classified into four groups based on the specific conductance (SC) and water type (Table 10.7–1). The type of water was bicarbonate, sulfate, and chloride, and the SC ranged from 364 to 13,500 μs/cm. Of 195 springs, 155 were allocated to Group B, the SC less than 500 μS/cm and the type of water bicarbonate. The quality of karst springs is controlled mainly by the lithology of the water route inside the karst aquifer. As the marl or marly limestone on the karst formation increases, the water type changes from bicarbonate (Group B1) to bicarbonate sulfate or bicarbonate chloride (Group B2). If gypsum and anhydrate are the contact lithology of the neighboring formations, the water type is mainly sulfate (Group S). Intrusion of saline water from the neighboring lakes

Table 10.7–1 Classification of Karst Springs Based on Electroconductivity and Water Type (Ashjari and Raeisi, 2006b)

Group	SC (μScm^{-1})	Water Type	Number of Springs	Q (L/s)
B1	<500	HCO_3	112	5–3112
B2	500–1160	HCO_3	43	5–2855
		$HCO_3{\cdot}SO_4$		
		$HCO_3{\cdot}Cl$		
S	1160–1850	SO_4 $SO_4{\cdot}Cl$	26	5–163
C	2300–13,500	Cl	14	5–98

or alluvium aquifers into karst aquifers and the mixing of saline water from salt domes with the karst water are the main sources of chloride-type waters (Group C).

Similar to the affirmation made earlier by Stevanovic and Iurkiewicz (2008) that groundwater quality is simply the result of the geology and hydrology of the area, source rock composition, weathering in the source area, and final mineral composition of the sediments are the main factors controlling the chemical composition of the water. When pure carbonate rocks prevail in the mountains, water pH values range from 6.5 to 8.0, and water has a low mineral content, generally less than 500 ppm.

The factor that directly reflects the regime of karstic aquifers in the area is related to the high variability in terms of time, area, and quantitative distribution of recharge. The rainy season in the lower Zagros usually ends in April and no rainfall event occurs until late September. Therefore, during a recession period, the spring hydrographs display typical monotone depletion of the accumulated resources. The underground drainage development depends on many factors but mostly on effective porosity, the base of karstification, and accumulated nonrenewable (static) groundwater reserves. Therefore, while many springs drastically reduce their yield, others slowly discharge by slightly reducing their spring flow during recession periods.

10.7.4 MAJOR SPRINGS AND THEIR CHARACTERISTICS

10.7.4.1 Iranian springs

Milanovic and Agili (1993) studied seven karstic aquifers in the Kazerun region in southern Iran. The anticlines are mountain ridges and the synclines are valleys and plains. The karstic aquifers are composed mainly of the Asmari formation and a limited outcrop of the Sarvak formation. The main resurgence of the karst aquifers is several springs in contact with the impermeable formation or alluvium. The total annual discharge of this study area is 21 m^3/s. The main springs are Dadin (4.4 m^3/s), Renjan (0.8 m^3/s), and Sarab Doukhtaran (0.3 m^3/s).

Sasan Spring is the biggest karstic spring of this region. It emerges from the Dashtak aquifer at an elevation of 814 m asl (Figure 10.7–3). The average annual discharge is 6.3 m^3/s. The SC of this spring ranges from 450 to 900 μS/cm. Sasan Spring is the main water supply of the city of Busher, located on the coast of the Persian Gulf. It is a 60 m walk from the famous Sassanid rock relief and less than 500 m from the city of Bishapur and the Anahita Temple. The famous and touristy Shapour cave is located in the catchment area of this spring.

FIGURE 10.7–3 Sasan Spring.

The Maharlu basin, with an area of 4029 km^2, is located in southern Iran. It is a closed basin with several perennial, ephemeral, and intermittent streams draining into Maharlu Lake. The karst aquifer of the Maharlu basin in southern Iran was studied in detail by the Karst Research Centre, the Ministry of Energy of Iran (1993), with the cooperation of the Strojexport Company, of the former Czechoslovakia. The karst outcrop was divided into eight anticline aquifers, consisting of Sarvak, Tarbur, and mainly Asmari-Jahrum formations. The annual average precipitation of Maharlu basin is 435 mm. The karst water emerges as karst springs at the foot of anticlines, flowing into adjacent alluviums and karst qanats (gently sloping underground galleries first developed in ancient Persia). The average discharge of the springs, qanats, and pumping wells is 2.4, 2.6, and 2.2 m^3/s, respectively. Two salt plugs, which intrude in the eastern part of Maharlu basin, lessen the quality of the adjacent karst aquifers.

The Alvand River basin, with an area of 2700 km^2, is situated in northwestern Iran. The average annual rainfall is 530 mm. Approximately 36 percent of the Alvand River basin is composed of carbonate aquifers, mainly Asmari limestone and the local Ilam limestone. The hydrogeology of this region was studied by Karimi et al. (Karimi, Raeisi, and Zare, 2003; Karimi, Raeisi, and Bakalowicz, 2005b). The Alvand basin consists of seven anticlines, separated by synclines. Most of the aquifers are characterized by autogenic recharge. The total groundwater of the Alvand basin discharges at 12 main and 23 minor karst springs. The annual discharge of the main springs varies between 140 and 3300 L/s. Two groups of springwater were identified: the first with a low ion concentration, especially sulfate, low temperature, light isotope concentration, and high elevation of the recharged area, and the second with moderate to high mineralization, especially sulfate, a higher temperature, heavy isotope composition, and a low-altitude recharge area. The main factors controlling the groundwater composition and its seasonal variations are the leaching of evaporate formations.

Gilan Spring is the biggest spring in the Alvand basin. The maximum, minimum, and mean annual discharges of Gilan Spring were 1120, 730, and 940 L/s, respectively. The catchment area of this spring is Tertiary karstic Asmari formation in the core of the Saravan anticline. The electrical conductivity varied from

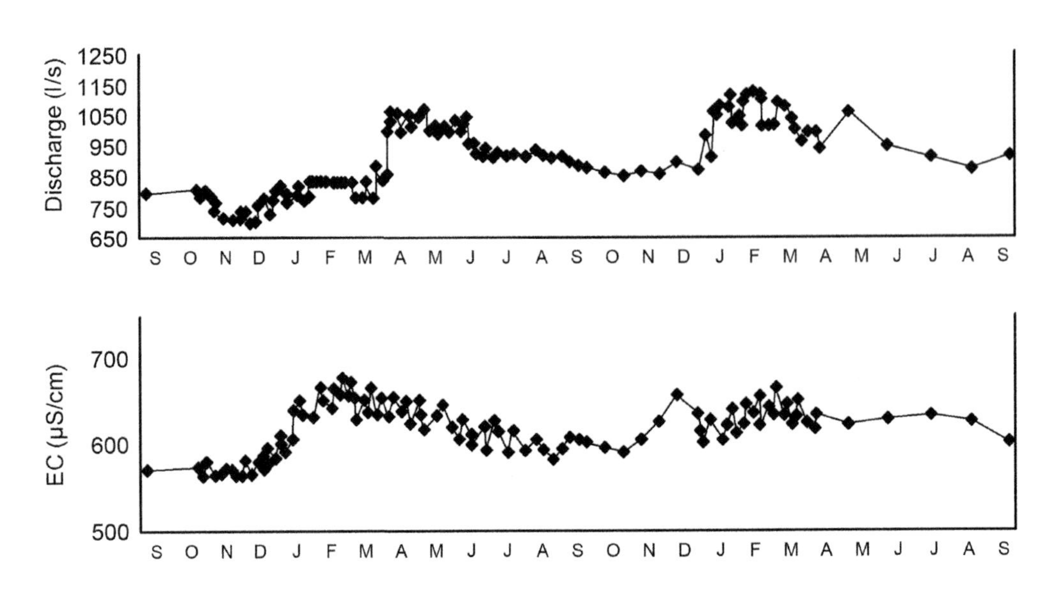

FIGURE 10.7–4 Gilan Spring hydrograph and electroconductivity (EC) values versus time.

583 to 677 µS/cm. The recession coefficient, percentage of base and quick flow by local minimum method, variation coefficient of the SC and discharge, and the standard deviation of temperature indicate that the flow system in the Gilan aquifer is characterized by a dominant diffuse flow. The SC decreases gradually during the dry period and the minimum value occurs seven months after the peak discharge (Figure 10.7–4). The flow velocity based on the lag time between peak discharge and minimum SC indicates that the conduit flow contributes to the spring discharge during the base flow. The catchment area of Gilan Spring is 37 km long and 5 km wide; therefore, it takes a long time for a more distant conduit flow to reach the spring. Peak discharge is provided mainly by the conduit flow within parts of the aquifer in the vicinity of the Gilan Spring.

To determine the source of the high-pressure artesian karstic aquifer beneath the Khersan III dam site (Mohammadi et al., 2007a, 2007b), Keshavarz (2003) studied the hydrogeology of three anticlines in Iran: Rig, Laki, and Shorum in south Zagros, in ChahrMahah-Bakhtiari province. These anticlines are composed mainly of Asmari formation. The number of springs emerging from the Rig, Laki, and Shorum aquifers is 12, 14, and 2 with an annual discharge of 3030, 930, and 42 L/s, respectively. Most of these springs emerge not at the local base level (Khersan River) but at higher elevations in the middle of the Asmari formation. The drainage pattern is predisposed by red marly layers outcropped in the upper part of the Asmari formation. The two biggest springs of the Rig aquifer are the Parvaz and Atashgah Springs, with an average annual discharge of 975 and 963 L/s, respectively. The karst water of the Laki anticline discharges mainly into the adjacent Khersan River and two small springs. The biggest spring of the Shorum aquifer is Gakadeh, with an average annual discharge of 600 L/s. The electrical conductivity of most of the springs is less than 400 µS/cm and the type of water is bicarbonate.

The Khersan III dam, with a height of 148 m, is under construction in a narrow valley, where the Khersan River cuts the plunge of the Laki anticline. The reservoir is located at the foot of the highly elevated Rig and Shorum anticlines. The length of reservoir will be about 32 km with a width of less than 2 km. The normal water level of reservoir will be 1415 m asl. All the karst springs of the southern flank of Rig anticline and the northern flank of the Shorum anticline are located beside the reservoir above the normal water level. The Khersan River flow rate, at the dam site, ranges from 20 to more than 80 m^3/s with an average of about 40 m^3/s.

Atashgah Spring is a tourist attraction in Iran (Figure 10.7–5). This spring emerges from the contact of the middle Asmari formation with the marly interlayers of the upper Asmari at an elevation of 1710 m asl (Keshavarz, 2003). This spring flows toward the Khersan River at an elevation of 1280 m asl, creating several waterfalls over a short distance. There are no sinkholes or caves in the catchment area of the spring. Melting snow is one of the main recharges. The average SC is 283 μS/cm and the water temperature 12°C. The time series of discharge and SC are presented in Figure 10.7–6. The flow regime is mainly diffuse. This spring will be accessible only by boat after Khersan dam construction.

FIGURE 10.7–5 Atashgah Spring.

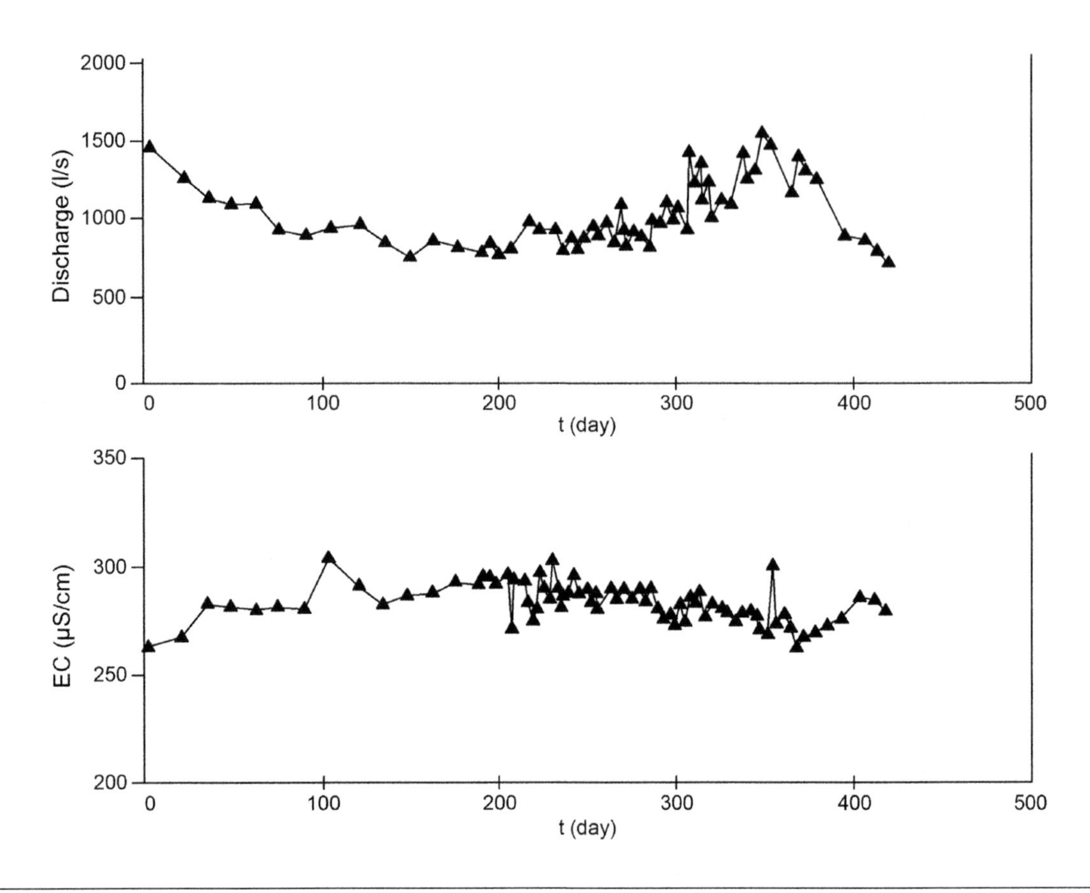

FIGURE 10.7–6 Atashgah Spring electroconductivity (EC) and discharge values versus time.

Fourteen springs, collectively called the *Atashkadeh Springs*, emerge from the southern flank of the Podenow anticline in Fars province, southern Iran. This anticline, 120 km long and 7.5 km wide, is located in the Zagros simply folded zone. It is composed mainly of Tertiary karstic limestone-dolomite Asmari formation. The Atashkadeh Springs are concentrated in an area less than 0.5 km^2. The total average discharge of these springs is 1700 L/s. The Ghomp, the largest of the Atashkadeh Springs, is an ascending picturesque spring with a diameter of 30 m (Figure 10.7–7) and a mean annual discharge of 1400 L/s. The Castle of Ardeshir e Babakan (in Persian known also as the Atashkadeh), built in 224 AD by Ardashir I of the Sassanian Empire, is located beside the Ghomp Springs (Figure 10.7–7). The catchment area of the Atashkadeh Springs is the part of the southern and northern flank of the Podenow anticline based on the geological settings, water balance, and dye tracing (Karimi et al., 2005a; Asadi, 1998). The Firozabad River flows through the U-shaped Tangab valley in the catchment area of the Atashkadeh Springs. The Tangab dam, under construction, is located at the beginning of the Tangab valley in the northern flank of the Podenow anticline.

These springs have a common main conduit and the same catchment area because (1) the hydrographs of the 14 Atashkadeh Springs have similar trends, (2) the SC time series of the springs overlap in most parts of the curve, and (3) the uranine dye tracer injected in the right abutment of the Tangab dam on the northern flank of the Podenow anticline was detected in all the Atashkadeh Springs (Asadi, 1998). The dye concentration curves of all these springs overlap, resulting in only one peak. This implies that only one major

FIGURE 10.7–7 Atashkadeh Spring.

conduit system transfers water to these 14 springs. The main conduit branches into 14 springs only near the Atashkadeh area.

Konarsiah Spring emerges from the Sarvak formation in the core of the Aghar anticline, south Zagros (Figure 10.7–8). The Sarvak formation is extensively fractured. The water balance and geological setting indicate that the Sarvak formation is the source of Konarsiah Spring. Part of the Sarvak formation is in direct contact with the Konarsiah salt dome, which with an area of several square kilometers and 2500 sinkholes, is the main source of deterioration of karst water in the study area. The minimum and maximum discharges are 140 and 267 L/s, respectively. The average temperature is 24.4°C (Sharafi, Raeisi, and Farhoodi, 1996). The average SC is 5500–7000 μS/cm, and the type of water is chloride. The percentage of chloride and sodium ions is 88 percent.

Margoon waterfall spring is one of the most beautiful tourist attractions in Fars province, Iran (Figure 10.7–9). It emerges from a bedding plane with a length of 60 m at an elevation of 2130 m asl The height of the waterfall is 58 m. The catchment area of the spring is the Jahrum formation. The SC is 309 μS/cm, and the type of water is bicarbonate. No systematic measurement has been done on the Margoon Spring, but the average discharge is assessed to be around 500 L/s.

Gasre Gomshe karst qanat spring has two galleries connecting to the foot of a small Gasre Gomshe anticline (Figure 10.7–10). This anticline is connected to the Puk anticline. The catchment area has a length of 48 km, a maximum width of 2.5 km, and an area of 61 km^2. The minimum, maximum, and average discharges of this qanat are 447, 843, and 607 L/s respectively, with an average SC of 500 μS/cm. The type of water is calcium magnesium bicarbonate. The water is used mainly for irrigation of the famous Gasrodasht gardens (northwestern of Shiraz). The water is distributed among tens of farms and gardens under a traditional gravity system. Part of the catchment area of Gasre Gomshe qanat spring is in direct contact with the alluvium, and it is crossed by the main road connecting Shiraz to Sepidan (Karst Research Centre of Iran, 1993). A protection program is recommended for pollution control.

FIGURE 10.7–8 Sampling of Konarsiah's water.

FIGURE 10.7–9 Margoon Spring.

FIGURE 10.7–10 Gasre Gomshe qanat spring.

10.7.4.2 Iraqi springs

During 2001 (considered one of the driest years for decades), about 100 monitored springs in Iraq that drain both karstic aquifers, Bekhme and Pila Spi, discharged a combined minimum of about 14 m^3/s (Table 10.7–2). Three major Zagros springs, the Bekhal (Figure 10.7–11), the Saruchawa, and the Sarchinar (Figure 10.7–12), had a minimal discharge of over 0.5 m^3/s. The summary maximal flow of all springs during the winter of 2002 reached 130 m^3/s. The average Q_{min}:Q_{max} for all monitored springs was about 1:8 (Stevanovic and Iurkiewicz, 2004a).

The famous and impressive *Bekhal Spring*, in the easternmost part of Zagros in northern Iraq, is also a tourist attraction. The spring is located at the lithostratigraphical contact between the Qamchuga massive limestone (Lower Cretaceous) and the overlaid Bekhme thick-bedded limestone (Upper Cretaceous). The limestone layers were highly deformed and intersected by two faults, which directly predisposed the discharge point. From the Bekhal, there is a very short and steep valley up to the confluence with the Rawanduz River.

Bekhal Spring was included in the list of potential UNDP projects for small hydropower plants (Figure 10.7–11). The measurements in 2000–2002 indicated a minimum of 0.75 m^3/s, while the maximum was over 17 m^3/s (March 2002). The spring drains a very large catchment, assumed to be at least 65 km^2 of the northern Korak Mountain and surroundings.

Table 10.7–2 Discharge of Iraqi Karstic Springs in 2001 (Stevanovic and Iurkiewicz, 2004b)

Aquifer	Number of Springs	Total Q_{max} (L/s)	Total Q_{min} (L/s)	Total Q_{max}:Q_{min}
Karst aquifer Bekhme	67	108,600	10,600	1:10
Karst fissured aquifer Pila Spi	31	22,500	3150	1:7

FIGURE 10.7–11 Bekhal Spring in Rawanduz area.

FIGURE 10.7–12 Typical intake for ascending springs in Iraq.

This spring plays an important role in future water use and management, for both the central and northern part of the Erbil governorate of Iraq. Even at its base flow in the low-water season, Bekhal Spring discharges more water than the Rawanduz stream flow, with its much larger catchment area (including the Iranian part). Currently Bekhal's water is partly tapped (at an average amount of 0.3 m^3/s) and distributed through a gravity pipeline to the city of Dyana (Soran), where it is used exclusively for human consumption. The very steep water slope of this gravity spring enables it to introduce a very simple tapping structure, which can often be seen in similar topographic conditions. Water from a discharge point flows directly into several large-diameter pipes and is further diverted to the city (no chlorination is applied at the site).

The *Saruchawa Springs* near Dokan Lake in Iraq is related to the boundary between Kometan (Bekhme aquifer) and the impervious Shiranish formation. The plunging of the limb of the anticlinal structure of Mackok Mountain to the plain dictates the location, a classical situation for the category of the largest springs in northern Iraq. The discharge zone includes several strong ascedending springs emerging in a diffuse manner and several other springs discharging with a concentrated flow. The upper group of springs is tapped into a concrete structure, and three pumps supply the village of Saruchawa with water. The minimal recorded discharge in 2001 was 1.75 m^3/s, while in the rainier year of 2002, the minimum discharge was three times greater (5.50 m^3/s).

Sarchinar Spring is well known in the region for its very large discharge and the great resort area formed around the spring. It is considered one of the local natural beauties, but at the same time, it is the main source of drinking water for the entire city of Sulaimani (center of one Iraqi governorate). The authors of the Parsons Company study were the first to describe Sarchinar Spring. It was mentioned as "one of the larger springs, [which] supplies municipal water for the 43,000 inhabitants of Sulaimani. In June 1957, the yield of the spring was calculated to be 56,000 US gallons per minute. Water is also diverted from this spring and used extensively for irrigation. . . ." (Anonymous, 1957). Half a century later, the city is 10 times bigger and the springwater is insufficient to meet all demands. Therefore, during the summer months, the surface waters are pumped out from the Dokan reservoir to the town, a distance of some 50 km. Like many ascending springs in the area, this spring is tapped by a protective wall, which surrounds a diffuse discharge zone (Figure 10.7–13).

In the case of the Sarchinar, 57 centrifugal pumps are installed to pump the water from the constructed pool ("lake") to the chlorination unit and farther to the pipelines (diameter from 200 to 900 mm) which divert the water to the different city zones (Figure 10.7–13). The discharge of this important spring is in the range of 0.60 to 7.5 m^3/s.

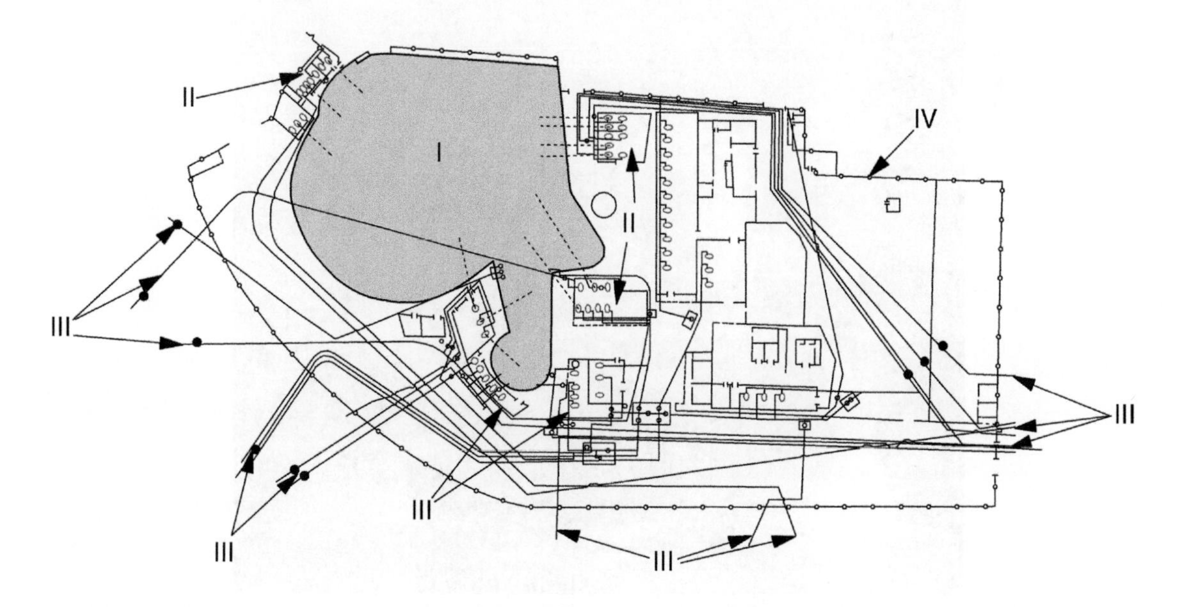

FIGURE 10.7–13 Sarchinar Spring "lake" and pumping stations: (I) pool with ascending springwater, (II) pump houses, (III) pipelines diverting water toward the different consumers, (IV) protective fence.

Zulum, the most impressive spring in the Sulaimani governorate of Iraq, draining the Bekhme aquifer, is located at the head of the Ahmad Awa valley, northeastern of the Khurmal thrust zone, at an elevation of around 900 m asl. It is locally also called *Zalum* or *Zalm*. It appears very close to the Iranian border, and a significant part of its catchment is in Iran. The spring emerges from a large cave directly to the right of a steep slope and cascades to the bottom of the valley some 30 m below, making a very inspiring waterfall. The majority of the catchment area consists of Triassic massive or well-bedded limestone (Avroman formation).

At the foothills, a small hydropower plant (with its penstock parallel to the cascade) has been built, using part of the flow and 24 m head (Figure 10.7–14). During the drought of 2000–2001, according to UNDP data, the spring discharges varied from 0.35 m^3/s to 5.98 m^3/s. The waters of Zulum are diverted into two irrigation channels and into pipelines installed for the gravity water supply of the towns of Halabja (0.4 m^3/s) and Khurmal (0.2 m^3/s), and the village of Ahmed Awa. Ali and Ameen (2005) found the water quality so good that it could be used as bottled drinking water.

The flow mechanism of this spring was analyzed through a time series analysis of daily spring discharge and daily rainfall data for the period from October 2004–July 2006 (Figure 10.7–15). The behavior of the spring hydrograph shows that this spring reacts strongly to the first rainfall events within the period continuing from the first steps through 15 days.

The autocorrelogram signifies a large storage with a high correlation coefficient and statistical significance of 73 days. Annual cyclicity of the system recharging-discharging is verified by the spectral density analysis. The dynamic reserves of the spring calculated on the Maillet recession curve are 62.7×10^6 m^3 per 277 days. This is equivalent to 2.62 m^3/s. The aquifer requires about five years to be exhausted if no replenishment occurs (Ali, 2008).

FIGURE 10.7–14 Zulum Spring (Khurmal area, Sulaimani governorate).

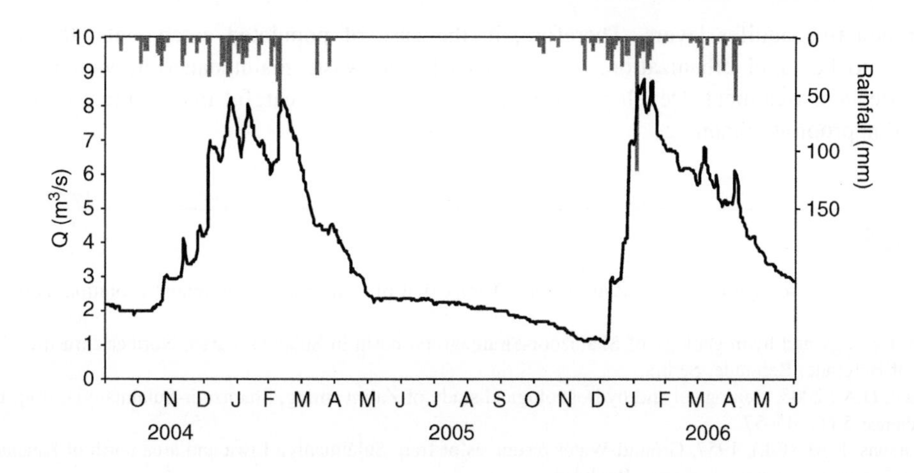

FIGURE 10.7–15 Zulum Spring hydrograph for the period October 2004–June 2006.

Some other springs are important simply because of their mythical, historical, or tourism significance. One such case is the Jundian spring (near the Rezan River and the ancient city of Rawanduz), which is drying out and is characterized by intermittent fluctuations throughout the year. The local Kurds believe that Jundian has existed "since the beginning of time."

Also famous for their "strange" characteristics are the Awa Spi ("White Water") Spring and Qaytool and Mamlaha, all located in the Low folded zone (Sangaw area, Sulaimani governorate, Iraq). The presence of karstified carbonate layers of Pila Spi and tectonized contact with evaporitic Fars formations enable the mixture of fresh and acidic water components. Thus, the Qaytool water can be so highly saline that deposited salt is consumed by the local villagers. This water is not even used as livestock supply.

10.7.4.3 Utilization and protection of the springwater

The water of most karst springs is used directly for the water supply of nearby cities (the first priority), irrigation purposes, and recently, as bottled water and for fish production. It is also the main source of the river base flow, providing agricultural water in the far downstream regions. Some karst springs provide the river's base flow, which is stored extensively in the dam reservoir during the cold season. In Iran, the catchment area of most karst springs is the unpopulated high mountains; therefore, it is naturally protected, except in a few cases around the big cities, where houses were built on the catchment area of the springs. Fortunately, a sewage network is under construction in these cities. The natural beauty of some of the tourist springs is affected by the housing and water supply pipes. In Iran, the water of a few karst springs has deteriorated due to the mixing with saline water of salt domes and intrusion of saline lake and evaporate formations. Several research projects to prevent the effect of natural hazards on the high quality of water from karstic springs are under consideration.

Current knowledge indicates that, in northern Iraq, the effects of pollution are much less than in industrialized countries. However, some of the springs are located inside rural and even urbanized zones. Another major problem for most of the cities is canalization: There are often more septic tanks than centralized systems with wastewater treatment facilities, and the constructed septic tanks are sometimes just over the underlying productive aquifer layers. Therefore, in the case of populated catchments, the most serious problems seem to be rapid urbanization, waste disposal, and waste maintenance as well as industrial or municipal wastewater treatment. Development of the region requires careful management of water and the ecosystem and appropriate planning.

REFERENCES

Alavi, M., 2004. Regional stratigraphy of the Zagros Folds-Thrust Belt of Iran and its proforeland evolution. Am. J. Sci. 304, 1–20.

Ali, S.S., 2008. Geology and hydrogeology of Sharazoor-Piramagroon basin in Sulaimani area, Northeastern Iraq. PhD thesis, University of Belgrade, Belgrade, Serbia.

Ali, S.S., Ameen, D.A., 2005. Geological and hydrogeological study of Zalim spring, Sharazoor-Sulaimanyia, Iraq. Iraqi Journal of Earth Science 5 (1), 45–57.

Anonymous Parsons, R.M. (Ed.), 1957. Ground-Water resources of Iraq. Sulaimaniya Liwa and area north of Khanaqin, Vol. 12 Development board, Government of Iraq, Baghdad.

Asadi, N., 1998. The study on watertightness problem in the Tangab dam using uranine dye tracer. M. Sc. thesis, Shiraz University, Shiraz, Iran.

Ashjari, J., Raeisi, E., 2006a. Influences of anticlinal structure on regional flow, Zagros, Iran. Journal of Cave and Karst Studies 68 (3), 118–129.

Ashjari, J., Raeisi, E., 2006b. Lithological control on water chemistry in karst aquifers of the Zagros Range, Iran. Cave and Karst Science 33, 111–118.

Buday, T., 1980. The regional geology of Iraq. Strartigraphy and paleogeography. 1, SOM, Baghdad, Iraq.

Buday, T., Jassim, S.Z., 1987. The regional geology of Iraq. Vol. 2, Tectonism, magmatism and metamorphism. Print. Dep., S.E. Geol. Surv. and Mineral Invest, Baghdad, Iraq.

Darvishzadeh, A., 1991. Geology of Iran. Amirkabir, Tehran, Iran.

Dubertret, L., 1959. Lexique stratigraphique international. Asie, Fascicule 10a, Iraq, (International Stratigraphic Glossary, Asia, Vol. 10a, Iraq), Centre National de la Recherche Scientifique, Paris.

Falcon, N.L., 1974. Southern Iran: Zagros Mountains in Mesozoic-Cenozoic orogenic belts. Geologica: Society of London, Spec. Publ. 4, 199–211.

James, G.A., Wynd, J.G., 1965. Stratigraphic nomenclature of Iranian Oil Consortium Agreement area. Bulletin of American Association of Petroleum Geologists 49 (12), 2182–2245.

Karimi, H., Raeisi, E., Zare, M., 2003. Hydrodynamic Behavior of the Gilan Karst Spring, West of the Zagros, Iran. Journal of Cave and Karst Science 30 (1), 15–22.

Karimi, H., Raeisi, E., Zare, M., 2005a. Physicochemical time series of karst spring as a tool to differentiate the source of spring water. Carbonates Evaporites 20 (2), 138–147.

Karimi, H., Raeisi, E., Bakalowicz, M., 2005b. Characteristics the main karst aquifers of the Alvand Basin, Northwest of Zagros, Iran, by a hydrogeochemical approach. Journal of Hydrogeology 13, 787–799.

Karst Research Centre of Iran, 1993. Comprehensive study and research in water resource of the Maharlu Basin (Fars). Vols. 1–4 (unpublished).

Keshavarz, T., 2003. Study on the karstic artesian aquifers in Khersan 3 dam site. M. Sc. thesis, Department of Geosciences, Shiraz University, Shiraz, Iran.

Milanovic, P., Agili, B., 1993. Hydrogeological characteristics and groundwater mismanagement of Kazerun karst aquifer, Zagros, Iran. In: Proceedings of the Symp. and Field Seminar: Hydrogeological Process in Karst Terrains. IAHS publ., International Association of Hydrological Sciences, Wallingford, Oxfordshire, UK, pp. 163–171.

Miliaresis, G.C., 2001. Geomorphometric mapping of Zagros Ranges at regional scale. Comput. Geosci. 27, 775–786.

Mohammadi, Z., Raeisi, E., Bakalowicz, M., 2007a. Evidence of karst from behavior of the Asmari limestone aquifer at the Khersan 3 dam site, southern Iran. Journal of Hydrological Science 52, 206–220.

Mohammadi, Z., Raeisi, E., Bakalowicz, M., 2007b. Method of leakage study at the karst dam site, a case study: Khersan 3 dam, Iran. Env. Geol. 52, 1053–1065.

Noble, A.H., 1926. Subsurface water resources of Iraq. The Government Press, Baghdad.

Raeisi, E., 2008. Groundwater storage calculation in karst aquifers with alluvium or no-flow boundaries. Journal of Cave and Karst Studies 63, 62–70.

Sharafi, A., Raeisi, E., Farhoodi, G., 1996. Contamination of Konarsiah karst spring by saltdome. International Journal of Engineering 9.

Stevanovic, Z., Markovic, M., 2004. Hydrogeology of northern Iraq, Vol. 1, Climate, hydrology, geomorphology and geology. Spec. Ed. FAO (Spec. Emerg. Prog. Serv.), Rome.

Stevanovic, Z., Iurkiewicz, A., 2004a. Hydrogeology of northern Iraq, Vol. 2, Regional hydrogeology and aquifer systems. Spec. Ed. FAO (Spec. Emerg. Prog. Serv.), Rome.

Stevanovic, Z., Iurkiewicz, A., 2004b. Karst of Iraqi Kurdistan—Distribution, development and aquifers. Rec. trav. VIII, Spec. ed. Board on Karst and Speleol. Serb Acad. Sci. and Arts, DCLVI, Belgrade, Serbia, pp. 31–53.

Stevanovic, Z., Iurkiewicz, A., 2008. Groundwater management in northern Iraq. Hydrogeology Journal 17, 367–378.

Stocklin, J., 1968. Structural history and tectonics of Iran: a review. AAPG Bull. 52 (7), 1229–1258.

Stocklin, J., Setudehnia, A., 1971. Stratigraphic Lexicon of Iran. Geological Survey of Iran: Report 18-1971, Tehran, Iran.

CHAPTER

Case Study: Sheshpeer Spring, Iran

10.8

Ezzat Raeisi
Department of Earth Science, Shiraz University, Shiraz, Iran

10.8.1 INTRODUCTION

The Sheshpeer Spring is located in the Zagros region, 80 km northwest of Shiraz (south central Iran). It emerges from the karstic Sarvak formation. Extensive research, including geology, geomorphology, hydrogeology, dye tracing, hydrochemistry, and hydrology was done in this region by experts and students of Shiraz University (Marandi, 1990; Kasaeyan, 1990; Pezeshkpoor, 1991; Azizi, 1992; Porhemat, 1993; Karami, 1993; Eftekhari, 1994). The Sheshpeer Spring is a tourist attraction, well known for fresh, cold water (Figure 10.8–1). The biggest ski area in southern Iran is located on the catchment area of this spring.

FIGURE 10.8–1 Sheshpeer Spring.

Copyright © 2010, Elsevier Inc. All rights reserved.

10.8.2 GEOLOGICAL SETTING OF THE CATCHMENT AREA

The catchment of Sheshpeer Spring is the northern and part of southern flanks of the Barm-Firooz anticline and northern flank of the Gar anticline, and it is located at the border of the simple folded zone and thrust zone (Figure 10.8–2) between their elevations of 3714 m asl (above sea level) and 2330 m asl, respectively. Geological formations are the Hormuz (Paleozoic), Khami group (lower Jurassic-Albian), Kazhdomi (Albian-Cenomanian), Sarvak formation (Albian-Turonian), Pabdeh-Gurpi (Santonian-Oligocene),

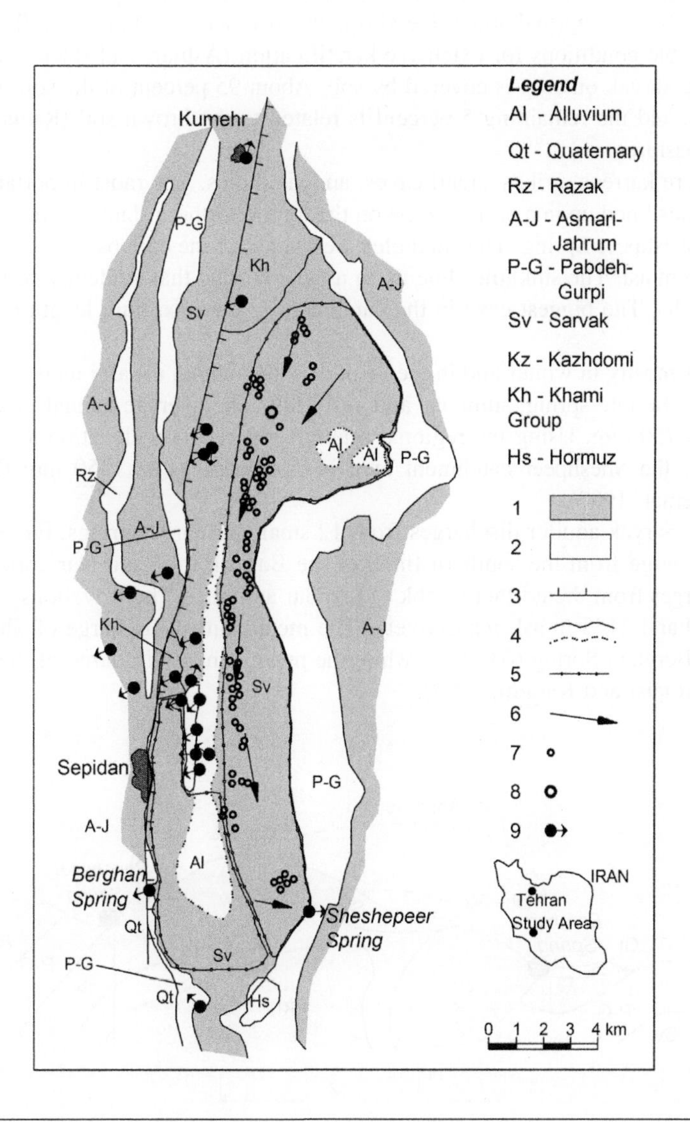

FIGURE 10.8–2 Geological map of Sheshpeer Spring catchment area: (1) Karst, (2) nonkarstic formations, (3) thrust, (4) geologic boundary, (5) catchment divide, (6) groundwater flow direction, (7) dye injection point, (8) spring. (After Raeisi, Pezeshkpoor, and More, 1993.)

Asmari-Jahrum (Paleocene), Razak (Early Miocene), and Bakhtiari (late Pliocene-Pleistocene) (Figure 10.8–2). The detailed lithology of these formations is described by James and Wynd (1965), Falcon (1974), and Alavi (2004).

The Barm-Firooz and Gar anticlines are connected by a saddle-shaped plunge. The exposed cores of these anticlines are made predominantly of calcareous Sarvak formation, which is both underlain and overlain by impermeable shale of Kazhdomi and Pabdeh-Gurpi formations. The main tectonic feature is the presence of a major thrust fault (Figure 10.8–2). The northern flank of the anticlines has been brought up by the tectonic forces, and the southern flank is so shattered that it either is completely removed or seen as large slide blocks (Figure 10.8–3). Several normal and strike-slip faults also occur and the overall tectonic setting of the area has produced suitable conditions for extensive karstification (Ashjari and Raeisi, 2006).

Forty percent of the Sarvak outcrop is covered by soil. About 95 percent of the soil belongs to the regosol and lithosol categories, and the remaining 5 percent is related to the brown soil (Raeisi and Karami, 1996). The area is a natural pasture.

The karst features are karrens, grikes, small caves, and sinkholes. The most important karst feature is the presence of 250 sinkholes lined up in a narrow zone on the top of northern flanks from the beginning of catchment area to near the Sheshpeer Spring. The sinkholes are mainly of the collapse type. The highest sinkhole is at an elevation of 3245 m asl. The sinkholes line up in a narrow zone that evidently coincides with the direction of longitudinal faults. The biggest cave in the catchment has a maximum length of 20 m and is located along a fracture.

Precipitation occurs mostly in winter and in the form of snow, which usually melts in early April. There is no precipitation during the late spring, summer, and early fall. The average annual precipitation at Berghan station (2110 m asl) is 750 mm. Using the regional relationship between elevation and rainfall, the average annual precipitation of the Sheshpeer catchment area is calculated to be 1350 mm (Raeisi, Pezeshkpoor, and More, 1993; Porhemat, 1993).

Groundwater of the Sarvak aquifer discharges from 12 small and large springs. Eleven of them, including the Berghan Spring, emerge from the southern flank of the Barm-Firooz and Gar anticlines, while only the Sheshpeer Spring emerges from the northern flank of the Gar anticline. The elevations of Sheshpeer and Berghan Springs are 2330 and 2145 m asl, respectively. The mean annual discharge of Sheshpeer is five times (3247 L/s) that of the Berghan Spring (632 L/s), while the mean annual discharge of the other springs ranges from 1.4 to 68.3 L/s (Raeisi and Karami, 1997).

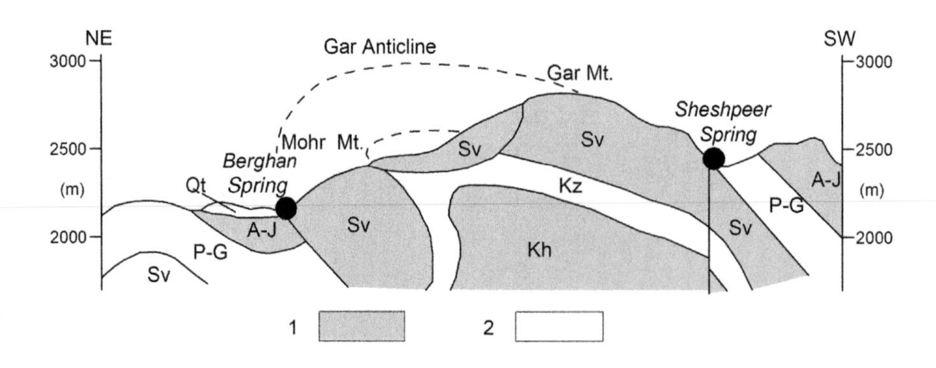

FIGURE 10.8–3 Cross section of Berghan-Sheshpeer Spring: (1) Karst, (2) nonkarstic formations, other indexes of geological units are the same as in Figure 10.8–2. (From Raeisi, Pezeshkpoor, and More, 1993.)

10.8.3 DYE TRACING

Thirty kilograms of sodium fluorescein was injected into the sinking stream of a sinkhole in the northern flank of the Barm-Firooz anticline, 18 km from the Sheshpeer Spring (Figure 10.8–2). The sampling sites were 12 springs of the Sarvak formation and 21 more springs from the neighboring Asmari-Jahrum formations. The tracer was detected only in the Sheshpeer Spring. The tracer concentration curve and hydrograph of Sheshpeer Spring are shown in Figure 10.8–4. The Sheshpeer Spring is most probably the only resurgence point, in spite of only 33 percent dye recovery. The low recovery is due to tracer absorption by the rock masses, deposit of sediment in the conduits, dye dilution by huge aquifer dynamic volume, and long dye residence time. The results of three experiments conducted at the Shiraz University laboratory reveal that 18 percent of sodium fluorescein was lost after 200 hours, and after 24 hours, 8, 11, and 25 percent of sodium fluorescein was adsorbed by limestone, marl, and clay, respectively (Raeisi, Zare, and Eftekhari, 1999).

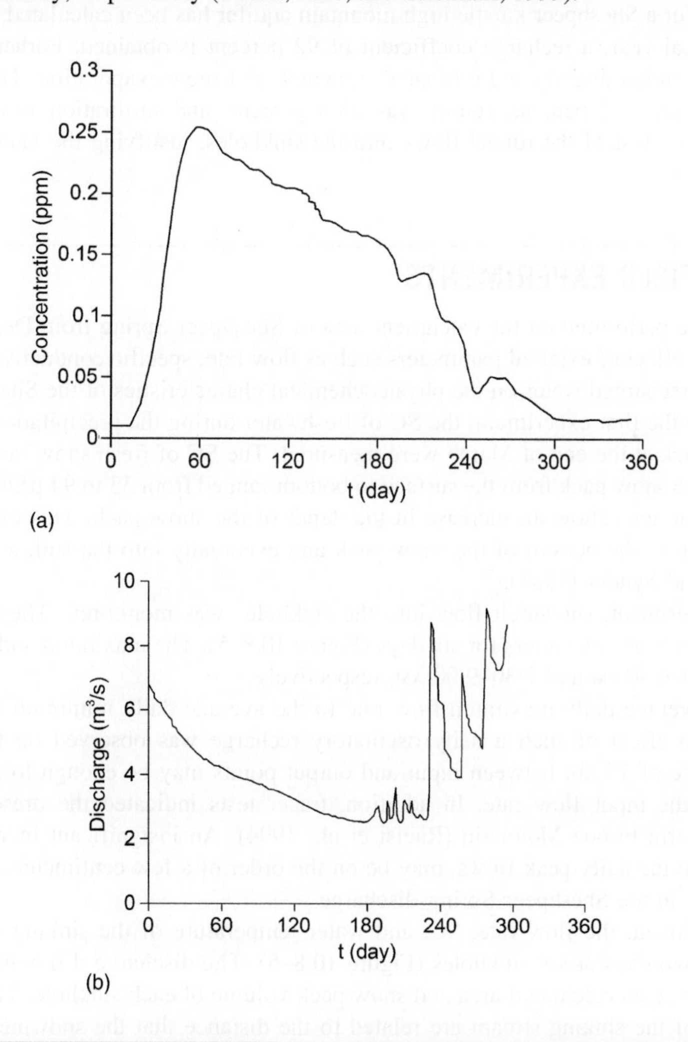

FIGURE 10.8–4 (a) Sodium fluorescein concentration curve and (b) hydrograph of Sheshpeer Spring.

The sinking stream is muddy during the intense snow melting period. The turbidity of the Sheshpeer springwater was occasionally measured and can reach 33 NTU after a heavy rain. Therefore, the deposition of sediment inside the conduit of the Sheshpeer system enhances the dye adsorption. The dynamic volume of stored water in the Sheshpeer aquifer is about 200×10^8 m^3 (Raeisi et al., 1993). Therefore, the extensive contact surface of water with bedrock adsorbs large amounts of tracer. The times to the leading edge, peak, and trailing edge of the dye concentration curve were 19, 64, and 287 days, respectively. The direct distance between the injecting sinkhole and Sheshpeer Spring is 18 km. The flow velocity in the karst system is 39 m/h, 12 m/h, and 6 m/h based on the time to the leading edge (19 days), peak (64 days), and centroid (134 days) of the concentration curve, respectively.

Based on the results of this test, geological setting, sinkhole trend, and hydrochemistry of Sheshpeer Spring and other springs including Berghan Spring (Raeisi and Karami, 1997), the catchment area of the Sheshpeer Spring is assessed to be 81 km^2 (Pezeshkpoor, 1991).

The water budget for a Sheshpeer karstic high mountain aquifer has been calculated by Raeisi (2008). For an average hydrological year, a recharge coefficient of 92 percent is obtained. Porhemat (1993) measured runoff and snow evaporation directly and estimated snowmelt and snow evaporation. The average percentage of snow evaporation was 5.2 percent, runoff was 25.3 percent, and infiltration from melting snow was 69.5 percent. However, most of the runoff flows into the sinkholes, justifying the value estimated from the water budget.

10.8.4 IN-THE-FIELD EXPERIMENTS

Four experiments were performed on the catchment area of Sheshpeer Spring from December 1991 to April 1992 to determine the effect of external parameters such as flow rate, specific conductivity (SC), temperature, and dissolved ions of recharged water on the physicochemical characteristics of the Sheshpeer Spring (Raeisi and Karami, 1996). In the first experiment, the SC of freshwater during the precipitation period and the various depths of snow pack at the end of March were measured. The SC of fresh snow ranged from 7 to 20 μS/cm, while the SC of the snow pack from the surface to bottom ranged from 33 to 94 μS/cm. All the major ions and SC except chloride ions show an increase in the depth of the snow pack. This effect is due to the ion migration from the top to the bottom of the snow pack and eventually into the soil, a process that was also reported by Jeffries and Synder (1981).

In the second experiment, snowmelt flow into the sinkholes was monitored. The flow rates of sinking streams were measured every two hours for six days (Figure 10.8–5). The maximum and minimum discharges were observed at 4:30–6:30 PM and 7:30–9:00 AM, respectively.

The ratio of the average daily maximum flow rate to the average daily minimum flow rate varied from 5 to 10. However, no effect of such a daily oscillatory recharge was observed on the Sheshpeer Spring hydrograph. A distance of 15 km between input and output points may be enough to suppress the effect of daily oscillations of the input flow rate. In addition, tracer tests indicated the presence of an extensive reservoir under the Barm-Firooz Mountain (Raeisi et al., 1994). An insignificant increase in the hydraulic head, corresponding to the daily peak flows, may be on the order of a few centimeters, incapable of causing a measurable increase in the Sheshpeer Spring discharge.

In the third experiment, the flow rate, SC, and water temperature of the sinking water were measured every two hours for two days at six sinkholes (Figure 10.8–5). The discharge differences between the sinkholes depend on the unequal catchment area and snow pack volume of each sinkhole. The diversities in water temperature and SC of the sinking stream are related to the distance that the snowmelt water flows on the ground surface to reach the sinkhole.

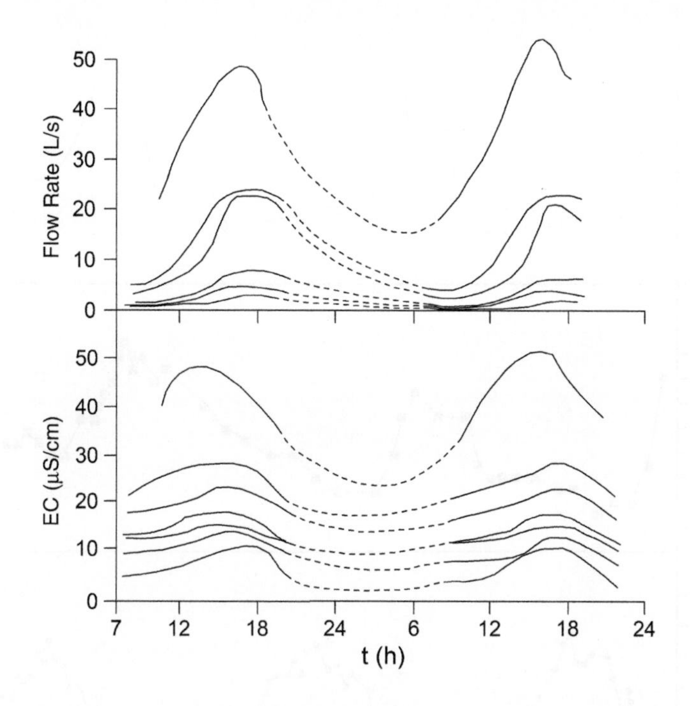

FIGURE 10.8–5 Hourly values of flow and electroconductivity of sinking water in experimental sinkholes.

In the fourth experiment, the temperature and SC of snowmelt water at the dripping point on the surface and at 40 cm below the ground surface were measured at five sites. The SC of water at 40 cm below the surface was 2.5 times higher than the SC of the water on the surface. It was concluded that, in the presence of thick soil cover, water may be close to saturation, reducing the role of the aquifer on the spring chemograph. The results of these experiments indicate that, if the physicochemical characteristics of a karst spring are going to be used to determine the characteristics of a corresponding aquifer, the effect of external factors on the outflow should be counted first, then the characteristics of the karst aquifer should be determined.

10.8.5 SHESHPEER DISCHARGE AND CHEMISTRY

The Sheshpeer Spring discharge regime was observed daily during the wet season and once every three weeks during the rest of the study period from March 1990 to November 1992 (Figure 10.8–6). The recession coefficients (α) were evaluated using the Maillet equation (1905). The first and second recession periods resulted with α equal to 0.0082–0.015 and 0.0028–0.0038.

The hydrograph may be divided into the first recession, second recession, and the precipitation time periods (Figure 10.8–6). The first recession (α_1) starts with a decrease in the discharge of the spring and coincides with the period after snowmelt infiltrates the aquifer. During this period, the groundwater level is relatively high, and due to the steep hydraulic gradient, water still discharges at a higher rate. The second recession period (α_2) coincides with the dry season, when no recharge from rain occurs. The precipitation period (w) starts with the beginning of rainfall and an increase in discharge. The base flow and quick flow of both regimes were determined from the hydrograph using the conventional separation method. Almost 40 percent

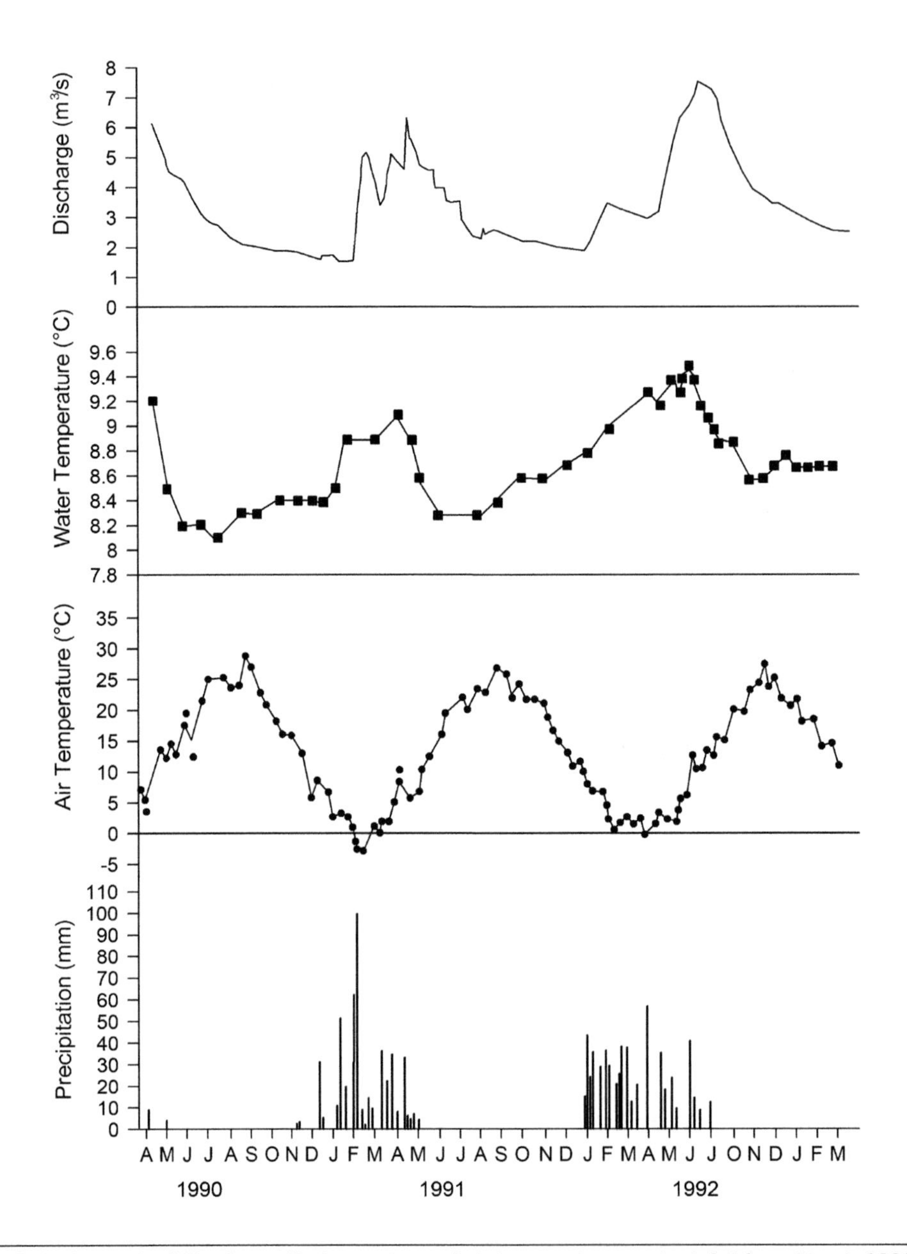

FIGURE 10.8–6 Hydrograph of Sheshpeer Spring, water and air temperature, and rainfall from March 1990 to November 1992.

of the α_1 regime is produced by the quick flow through sinkholes and large fissures and the rest comes from the stored water in the pore space (base flow). In the α_2 regime, which coincides with the dry season, when no recharge from snowmelt occurs, the stored water in the pore space and small joints gradually discharges into large conduits to build the base flow. The large conduit in the second regime acts as a reservoir for the stored water and merely provides a water transportation medium.

The Sheshpeer water is the typical calcium bicarbonate type. The measured SC ranges from 243 to 295 μS/cm. The SC is relatively low in the first recession period. This is due to the melting snow, which quickly reaches the spring and has little time for dissolution. In addition, Raeisi and Karami (1996) show that the snowmelt in the first recession period contains insignificant amounts of dissolved ions. During the second recession period, the SC increases as there is no dilution by recharge water, and flow from small pores with longer residence time contributes to spring discharge. In the precipitation period, however, the SC is relatively high and contradicts the expected higher flow rates. The increase in the SC is due to the flushing of water with a long residence time in the deeper phreatic zone and the downward ion migration, which causes the initial winter snowmelt to have higher amounts of dissolved ions. The calcite saturation index (SI_c) shows that the springwater is almost supersaturated during the three periods, which suggests a diffuse flow system. The dolomite saturation index indicates undersaturation, as would be expected, because the aquifer is dominantly calcareous. The minimum, maximum, and annual average temperatures are 8.4°C, 9.5°C, and 9°C, respectively. Therefore, it can be concluded that the springwater discharges from depths that are not affected by external temperature.

The following facts support a diffuse flow regime in the Sheshpeer aquifer:

1. Low variation was found in temperature, total hardness, SC, and dissolved ions.
2. The dye recovery curve exhibits an obtuse peak and a long tail. It implies a unique conduit with a large underground "lake."
3. The average discharge during the dye elapsed time (287 days) was 4.27 m^3/s. The dilution of dye by at least 106×10^6 m^3 of karst water implies an extensive underground reservoir.
4. Velocities less than 18 m/h involve long underground retentions (Ford and Williams, 2007). The velocity based on the dye tracing is 12 m/h, also confirming a large underground reservoir.
5. The conventional separation method indicates that 60 percent of the flow is contributed to by the base flow (diffuse).
6. If a velocity of more than 3.6 m/h is accepted as the conduit flow, it takes four months for the conduit water to flow out of the Sheshpeer aquifer and the remaining stored water in the aquifer is diffuse flow.

A conceptual model is presented to justify the share of both conduit and diffuse flow in the Sheshpeer aquifer. The sinkholes were developed at the higher base level. They are relict and, at present, are located in the vadose zone. Part of the recharge water flows into the sinkholes and reaches the water table in a short time. The other part of the water percolates directly into the wide openings of the grikes, being stored in the small pores of the epikarst and eventually reaching the water table. The water table is located 300 m below the ground surface in most parts of the aquifer; therefore, the recharged water mainly stored in small openings raises the water table and gradually enters the conduit system and provides the share of diffuse flow in Sheshpeer discharge.

10.8.6 WATER USE

The Sheshpeer Spring is an attractive tourist site; throughout the year, people picnic on both sides of the Sheshpeer River, enjoying the fresh, cold water. During the dry season, the Sheshpeer water is used mainly for irrigation of the adjacent plain. The water eventually joins the Persian Gulf during the wet season. A mineral water factory and fish farm were recently constructed a few hundred meters downstream of the spring. In 1803, an attempt was made to transfer the water of Sheshpeer Spring to Shiraz via an open canal, but only its first part, which is 26 km long, is used today to irrigate the adjacent Homayjan plain. The canal irrigated hundreds of small traditional farms using a very precise water division system with no sophisticated instruments. The Sheshpeer springwater was occasionally muddy after a heavy rain, due to 250 sinkholes in the catchment area; therefore, it is recommended to protect the sinkhole area. A dam is under study on the Sheshpeer River, 8 km downstream of the Sheshpeer Spring, located on the plunge of the Gar anticline at the beginning of a karstic valley. The main water resource of the dam would be the Sheshpeer Spring during the winter, late fall, and early spring, in addition to the runoff from the 63 km^2 catchment area of the dam site. The annual river flow on the dam site, based on 41 years of data, ranges from 37 to 132 $\times$ 10^6 m^3, respectively. The dam should be 66 m high; the elevation of a normal water level should reach 2302 m asl, while the reservoir volume would be 50 million m^3. The dam would provide 38.5 $\times$ 10^6 m^3 of drinking water for the city of Shiraz and 20.3 $\times$ 10^6 m^3 of irrigation water for the adjacent plain. The water would be transferred to Shiraz by 65 km of pipe under pressure. A detailed karst hydrogeology study including two dye tracers has recently been started to optimize the grout curtain.

ACKNOWLEDGMENTS

I thank the Research Council of Shiraz University for its financial support. I express my sincere gratitude to Dr. Zoran Stevanovic for his assistance in the revision of this manuscript.

REFERENCES

Alavi, M., 2004. Regional stratigraphy of the Zagros Folds-Thrust Belt of Iran and its proforeland evolution. Am. J. Sci. 304, 1–20.

Ashjari, J., Raeisi, E., 2006. Anticlinal structure influences on regional flow, Zagros, Iran. Journal of Cave and Karst Studies 68 (3), 119–127.

Azizi, M., 1992. Snow hydrology in a small catchment area, Barm-Firooz mountain. Unpublished MSc thesis, Shiraz University, Shiraz, Iran.

Eftekhari, A., 1994. Hydrogeological characteristic of Sepidan karstic region using dye tracing. MSc thesis, Shiraz University, Shiraz, Iran.

Falcon, N.L., 1974. Southern Iran: Zagros Mountains in Mesozoic-Cenozoic orogenic belts. Geologica: Society of London, Special Publ. 4, 199–211.

Ford, D.C., Williams, P.W., 2007. Karst geomorphology and hydrology. Unwin Hyman, Winchester, MA.

James, G.A., Wynd, J.G., 1965. Stratigraphic nomenclature of Iranian Oil Consortium Agreement area. Bulletin of American Association of Petroleum Geologists 49 (12), 2182–2245.

Jeffries, D.S., Synder, W.R., 1981. Variations in the chemical composition of the snowpack and associated waters in central Ontario. Dwin Quebec City, Quebec, Canada.

Karami, G., 1993. Relationship of physiochemical characteristics of recharged water with the karstic springs in Gar and Barm-Firooz anticline. MSc thesis, Shiraz University, Shiraz, Iran.

Kasaeyan, A., 1990. Morphology of Barm-Firooz sinkholes and its effect on the general flow direction. Unpublished MSc thesis, Shiraz University, Shiraz, Iran.

Maillet, E., 1905. Essais d´hydraulique souterraine et fluviale. Hermann, Paris.

Marandi, K., 1990. Morphology of Gar mountain sinkholes and its effect on the general flow direction. Unpublished thesis, Shiraz University, Iran.

Pezeshkpoor, P., 1991. Hydrogeological and hydrochemical evaluation of Kuh-e Gar and Barm-Firooz springs. MSc thesis, Shiraz University, Shiraz, Iran.

Porhemat, J., 1993. Evaluation of hydrological balance parameters in karstic highland catchment area. MSc thesis, Shiraz University, Shiraz, Iran.

Raeisi, E., 2008. Groundwater storage calculation in karst aquifers with alluvium or no-flow boundaries. Journal of Cave and Karst Studies 63 (3).

Raeisi, E., Karami, G., 1996. The governing factors of the physical and hydrochemical characteristics of karst springs. Carbonate and Evaporate 11 (2), 162–168.

Raeisi, E., Karami, G., 1997. Hydrodynamic of Berghan karst spring as indicators of aquifer characteristics. Journal of Cave and Karst Studies 59 (3), 112–118.

Raeisi, E., Pezeshkpoor, P., More, F., 1993. Characteristics of karst aquifer as indicated by temporal changes of the springs physico-chemical parameters: Iranian Journal of Science and Technology 17, 17–28.

Raeisi, E., Zare, M., Eftekhari, P., 1999. Application of dye tracing for determining characteristics of Sheshpeer karst spring, Iran. Theoretical and Applied Karstology, Bucharest 11–12, 109–118.

CHAPTER

Case Study: Protection of Edwards Aquifer Springs, the United States

10.9

Gregg Eckhardt
www.edwardsaquifer.net, Texas

10.9.1 INTRODUCTION

In south central Texas, the Edwards Aquifer gives rise to some of the largest and most important springs in the United States. Bursting forth as fountains from Cretaceous limestone, they sustained native populations for over 11,000 years. In the 1600s, Spanish colonial outposts on the New World frontier were established where ample spring flows created lush oases on the edge of the vast Chihuahuan desert. Major American cities, like Austin and San Antonio, grew up along the Texas "spring line," where Indian and buffalo trails connecting natural artesian discharges became today's major highways.

As the decades passed, thousands of Edwards Aquifer wells (Figure 10.9–1) and a lack of legal pumping restrictions caused steep declines in spring flow rates, and some began to flow only intermittently. By the 1980s, a number of highly adapted endangered species had been identified that rely on natural artesian spring flows and aquatic habitats just downstream. Even as water resources became increasingly strained by the needs of over 2 million people, there was a growing awareness of the importance of spring flows for both environmental and economic concerns. In addition to endangered species, important regional economies also depend on spring flows, and large commercial fisheries in the coastal bays and estuaries depend on freshwater inflows.

In the last decade, several engineering solutions have been applied to the problem of spring flow maintenance. At San Antonio's Aquifer Storage and Recovery facility, retrieval of Edwards Aquifer water placed in underground sands allows deferral of pumping during critical summer months, when spring flows are most at risk. Another approach to deferral of pumping during critical times was the construction of America's largest recycled water distribution system, capable of replacing 20 percent of San Antonio's aquifer pumpage with tertiary treated wastewater effluent. Looking forward, an innovative recharge and recirculation strategy envisions construction of structures to hold enhanced recharge in storage until needed for protection of spring flows and downstream flows.

10.9.2 HYDROGEOLOGIC SETTING

The Edwards Aquifer is one of the largest and most prolific in the world, stretching for 160 miles in an arch shaped curve through central Texas (Figure 10.9–2). Its formation involved the alternating transgression and regression of ancient seas. Limestone laid down in thick beds was extensively eroded when exposed during

Copyright © 2010 by Gregg Eckhardt.

FIGURE 10.9–1 The first large Edwards Aquifer wells, drilled in San Antonio. (From Hill and Vaughan, 1896.)

shoreline regressions, becoming highly porous and permeable due to karstification. This honeycombed limestone, capable of storing and transmitting large amounts of water, was covered over again with clays and relatively impermeable layers that formed a confining unit.

Subsequently, the mountain-building episode that rejuvenated the present-day Rocky Mountains also resulted in the deposition of thick sediments across Texas by wind and water. As mountains rose to the northwest, sediments were deposited toward the southeast, with their thickness increasing toward the present-day coast. The tremendous weight of these sediments caused a series of parallel faults (Figure 10.9–3) to form and created the Balcones escarpment, which separates the uplifted and rocky Texas hill country from the deep sediments of the low-lying coastal plain. Vastly different land uses on either side of the escarpment are evident in satellite imagery. Below the escarpment is the blackland prairie of the Gulf Coast, where agriculture and urban uses are

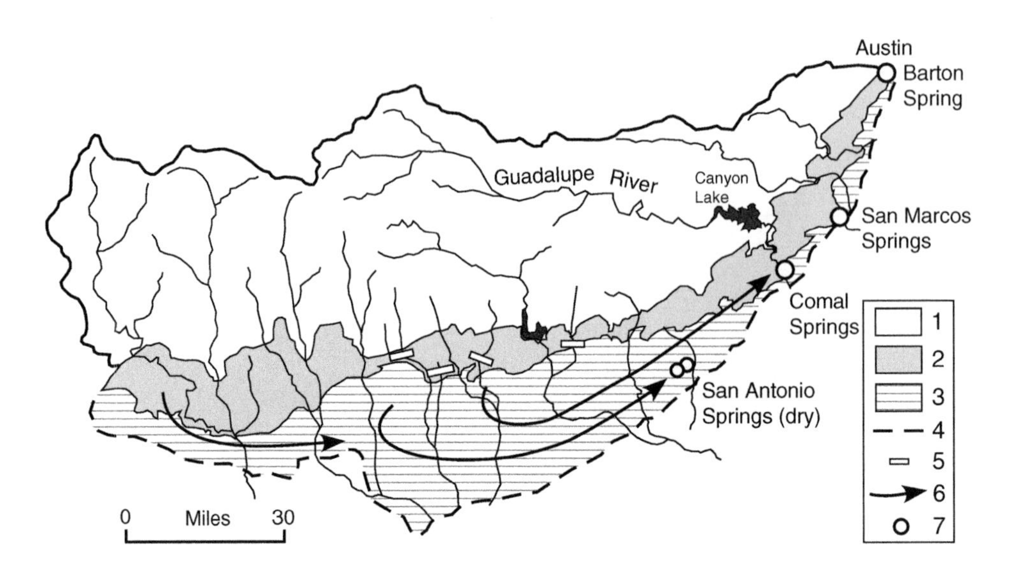

FIGURE 10.9–2 Edwards Aquifer in Texas with main flow paths: (1) Drainage area; (2) recharge area (carbonate rock outcrops); (3) confined zone of the aquifer; (4) "bad water" line, east and south of this line groundwater has high salinity and is not potable; (5) recharge dam; (6) general direction of potable groundwater flow in the confined zone of the aquifer; and (7) large karst spring. (Modified from Maclay and Small, 1986.)

dominant, while above the escarpment lies the Texas hill country, where ranching and grazing are common. Property on the Balcones escarpment, especially where it runs through northern San Antonio and western Austin, is highly valued for its rugged hill country beauty and lovely views of the city lights.

During formation of the Balcones escarpment, the limestone layers that had been laid down flat became tilted, forming narrow bands at the surface. This is the present-day geology. The Edwards limestone is between 300 and 700 ft thick, outcrops at the surface in a narrow band below the escarpment, is tilted downward toward the south and east, and is overlain by younger limestone layers and thousands of feet of sediments.

Where the Edwards limestone is exposed, over an area of approximately 1250 square miles, the highly faulted and fractured outcrop can accept large quantities of recharge. Water drains toward the recharge zone from the contributing zone, an area of about 5400 square miles that is known both as the Edwards Plateau and the Texas hill country. On the plateau, elevations range between 1000 and 2300 ft above sea level, and the rugged, rolling topography is covered with thick woodlands of oak and cedar. Runoff from rainfall collects in streams or infiltrates the shallow water table aquifers of the plateau. Water table springs discharge at the base of the Balcones escarpment and combine with surface runoff to feed the streams and rivers that cross the Edwards recharge zone. About 75–80 percent of total recharge occurs when these waterways cross the permeable formation and go underground. A small percentage occurs when precipitation falls directly on the outcrop. A surface water reservoir built partly on the recharge zone, Medina Lake, contributes large amounts of water to the aquifer. Most of the annual average recharge of about 711,600 acre-feet (for the period 1934–2006) occurs in the western counties of Medina and Uvalde, where the Edwards outcrop is wider at the surface. Rainfall in the region averages about 30 in. per year, but it is highly variable, so annual recharge amounts also vary widely. The variable nature of rainfall has large impacts for water availability in any given year.

In the recharge zone, there are no other rock formations overlying the Edwards: It is exposed at the surface. So the aquifer here is unconfined and has a water table that rises and falls in response to rainfall.

FIGURE 10.9–3 This limestone fault can be seen from U.S. Highway 281 about 20 miles north of San Antonio. The Balcones fault zone contains a complicated series of such faults and fractures. Slickensides can be observed here, the smooth striated surface produced on rock by movement along a fault. (Copyright Gregg Eckhardt, printed with permission.)

However, the major portion of the Edwards, the artesian zone, is confined between the Glen Rose limestone below and the Del Rio clay on top, and it has no water table.

The sheer weight of new water entering the aquifer in the recharge zone puts tremendous pressure on water that is already deeper down in the formation. Flowing artesian wells and springs exist where hydraulic pressure is sufficient to force water up through wells and faults to the surface. Major natural discharge occurs at San Marcos Springs and Comal Springs in the northeast (Figures 10.9–4 and 10.9–5, respectively), where a total of nine threatened or endangered species depend on habitats created by spring flows. In San Antonio, two large clusters of springs formed the headwaters of the San Antonio River. However, as the pumping from the Edwards increased, the springs dried up. For almost 100 years, the base flow of the San Antonio River was provided by potable Edwards wells drilled specifically for flow augmentation (Figure 10.9–6).

FIGURES 10.9–4 San Marcos Springs. In San Marcos, Texas, more than 200 springs issue from three large fissures and many smaller openings. In 1849, the springs were inundated when General Edward Burleson, who served as vice-president of the Texas Republic, built a dam just downstream of the springs to operate a gristmill. About half the combined discharge of the springs are seen here, leaving Spring Lake over a spillway. (Copyright Gregg Eckhardt, printed with permission.)

FIGURES 10.9–5 Comal Springs. (Copyright Gregg Eckhardt, printed with permission.)

FIGURE 10.9–6 This Edwards well is in San Antonio's Brackenridge Park, just below San Antonio Springs, and was used for almost 100 years to augment the flow of the city's namesake river. It was shut off in 2000, when San Antonio's completion of the nation's largest recycled water distribution system allowed the flow to be replaced by discharge of water from the city's treatment plants. (Copyright Gregg Eckhardt, printed with permission.)

Water in the Edwards moves generally from southwest to northeast, and a number of barrier faults make it difficult for waters in various limestone units to mix together. These faults, along with the varying porosities and permeabilities of the limestone, control the movement of water in the aquifer. Several index wells are used to monitor the amount of pressure that water in the artesian zone is under. Changing pressure is reflected in rising or falling well levels. These well levels are also used to trigger drought restrictions that require users to reduce pumping during critical periods.

In the San Antonio segment of the Edwards Aquifer, water quality is exceptionally good. The waters are alkaline, with hardness ranging from 250 to 300 mg/L. Low concentrations of regulated metals occur well below drinking water standards. Municipal water purveyors provide chlorine disinfection prior to distribution; otherwise, no treatment is required. Interestingly, most of the major spring complexes produce water from several underground flow paths, and water from spring outlets only a few feet apart can be quite distinct in color and flavor. Most have an emerald green hue, and some are aquamarine or blue. Figure 10.9–7 shows two polls containing the four largest of the San Pedro Springs. San Antonio was founded on the nearby banks in 1718, and the site was the social and recreational center of the young frontier town for many decades. In 1729, the lands around the springs were declared to be a public place by King Philip V of Spain, making it one of the oldest public parks in the United States. Today, due to heavy pumping demands on the Edwards Aquifer, the San Pedro Springs flow only during times of heavy rainfall.

10.9.3 LEGAL FRAMEWORK

Until the 1990s, withdrawal of waters from the Edwards Aquifer was unregulated. A fundamental flaw in Texas water law meant that for most of Texas's history, groundwater was treated as if it were completely separate and different from surface water. In the 1800s, the movement of groundwater was deemed by courts to

FIGURE 10.9–7 San Pedro Springs near San Antonio, Texas. (Copyright Gregg Eckhardt, printed with permission.)

be unknowable and in the realm of the "occult," so legislators refused to make any laws regarding its use. Today we know that surface and groundwater are interconnected and inseparable, but for over a century the "separation myth" was a major hurdle in the development of an integrated and conjunctive body of water law in Texas.

The differences in how surface water and groundwater were regulated were profound. Surface water is the property of the state and its use is highly regulated, while Edwards groundwater was deemed to be the property of whomever's land it was under. To use surface water, a user must apply for a state permit, then available water rights are assigned using a "first in time, first in right" method. With groundwater, however, the "right of capture" prevailed for many decades. This was also called the *law of the biggest pump*, because all could pump as much water from under their land as they wanted, as long as they put it to a beneficial use. So, just one person could legally use *all* the aquifer water if he or she could pump it out and put it to use. A person could do this without regard for the impact on anybody else, and the idea of unlimited rights to groundwater became deeply ingrained in Texas culture, which strongly leans toward private property rights over state controls.

The fact that the law did not recognize any connection between surface water and groundwater only complicated matters. In the 1900s, as use of groundwater from the Edwards became greater over time, water that used to come out of the ground at springs was now being pumped from wells and eventually placed back in the rivers by wastewater treatment plants, mainly those owned and operated by the San Antonio Water System. Under Texas law, privately developed groundwater is owned by the developer, and there is no obligation to discharge or release any water downstream after use. For decades, the state had been assigning surface water rights to people whose water had long ago originated as spring flow but now originated as wastewater treatment plant discharges. Water rights holders might not have access to a single drop of water if the city of San Antonio decided to reuse its wastewater instead of discharging it downstream, and this was later an important consideration in the development of reuse plans.

In the 1950s, the region experienced the worst drought in recorded history. Afterward, numerous water planning studies were undertaken but little progress was made toward management of the Edwards Aquifer or toward development of other water resources. As a response to the 1950s drought, the Edwards Underground Water District was created in 1959 and it was charged with conserving and protecting water in the aquifer. However, it had no authority to restrict groundwater pumping, and for over 40 years, it was mainly a data collection agency. In 1961, the state released the Texas Water Plan, which discouraged overreliance on the Edwards and recommended several new reservoirs. An update to the 1961 plan was produced in 1966, with a final version in 1968; and it outlined an ambitious program of statewide reservoir construction with those around San Antonio being in Phase I. Reservoirs to meet San Antonio's needs were the Cuero I, Cuero II, Goliad, Cibolo, and Cloptin Crossing. None of these has been built, and none is currently contemplated. The most significant contribution of the 1968 Texas Water Plan was the determination that, based on historical rates of recharge and discharge, withdrawals from the Edwards should not exceed 400,000 acre-feet per year. As we shall see, this number stood the test of time, at least until recently.

San Antonio responded to the 1950s drought with innovation and foresight. The city pioneered the large-scale use of treatment plant effluent for cooling electrical generating plants, constructing two large lakes for that purpose. San Antonio also brought court action seeking to become a regional partner in the development of Canyon Lake reservoir, 30 miles north of town. The courts ruled, however, that San Antonio did not need water from elsewhere because it had not fully exploited the resources in its own basin, including the Edwards Aquifer. For the next three decades, water resource development languished, and San Antonio got a reputation as a do-nothing city.

In 1991, a startup catfish farm near San Antonio made all of south Texas painfully aware the right of capture was no longer a workable approach to regional water supply. Ronnie Pucek drilled the world's largest water well (Figure 10.9–8) and began using more than 50 million gallons per day, as much as one fourth of the entire San Antonio metropolitan area. Moreover, his well was located in a spot that intercepted flows before they reached San Antonio. Farther up the underground flow path, spring flows at Comal and San Marcos dropped like a rock. The city realized that, unless the law changed, it was possible that San Antonio could be left without water, and environmentalists realized that spring flows supporting endangered species were in imminent danger (Figure 10.9–9).

Environmentalists brought court action that forever altered the legal landscape of Edwards withdrawals. The Lone Star Chapter of the Sierra Club filed a lawsuit against the U.S. Fish and Wildlife Service, claiming the service was not adequately protecting endangered species that depend on the aquifer. The Sierra Club argued that Comal and San Marcos Springs could dry up if unrestricted pumping continued, and that would constitute a "taking" as defined by the Endangered Species Act. The Sierra Club asked that the service be required to ensure minimum spring flows to protect the endangered species. After a two year trial, in January 1993, Federal Judge Lucius Bunton of the U.S. District Court in Midland ruled in favor of the Sierra Club and others who had joined the suit along with the club. The court found that, if unrestricted withdrawals continued, endangered and threatened species would be "taken" as defined by the act. The court also found that the Fish and Wildlife Service had failed to implement a recovery plan for San Marcos and Comal Springs and had caused risk or jeopardy to the endangered species. Judge Bunton ordered that spring flow must be maintained, even during the most severe drought, such as in the 1950s. He directed the Texas Water Commission to prepare and submit a plan to ensure spring flows, and he directed the service to determine spring flow levels that would result in "take" or "jeopardy" of the species. The service subsequently determined that level was 150 ft^3/s at Comal Springs.

FIGURE 10.9–8 The world's largest water well drilled by Ronnie Pucek. The well, capable of producing more than 50 million gallons per day, was used for a catfish farm near San Antonio. (Copyright Gregg Eckhardt, printed with permission.)

FIGURE 10.9–9 The Texas blind salamander is a sightless, cave-dwelling creature that reaches a mature length of about 5 in. It spends its life in complete darkness and is very sensitive to changes in water quality. Biologists know of only one population, living in Edwards Aquifer caves around San Marcos. It is one of seven federally endangered species that depend on the Edwards Aquifer. (Copyright Gregg Eckhardt, printed with permission.)

In the 1993 ruling, Judge Bunton also announced the Texas Legislature had to enact a regulatory plan to limit withdrawals from the aquifer, or he would implement his own plan. In May 1993, the Texas Legislature passed Senate Bill 1477, which abolished the Edwards Underground Water District, created the Edwards Aquifer Authority, and authorized the new agency to issue permits and regulate groundwater withdrawals from the Edwards. This essentially ended the right of free capture in the Edwards region and it laid out the legal framework for assigning ownership of water to people who had been using it for many years. The bill also created means to market groundwater rights by making permits transferable (with some restrictions), and it set a cap on permits at 450,000 acre-feet annually, to be reduced to 400,000 acre-feet in 2008, the number identified in the 1968 Texas Water Plan. It also provided for short-term permits for additional use when rainfall and recharge are high, required the authority to adopt a Critical Period Management Plan to reduce pumping during droughts, and addressed the question of preserving endangered species habitats by requiring the authority to provide continuous minimum spring flows.

While the new legislation established a pumping cap, it also directed the new authority to issue minimum pumping rights to persons who could prove their use during the prior 21 years. Those rights turned out to be far in excess of the legislative pumping cap, which created a difficult conundrum for the agency. After several attempts to solve the issue by rule making, it was finally addressed by the Texas legislature in 2007. The pumping cap had been scheduled to be reduced to 400,000 acre-feet in 2008; instead, it was raised to 572,000 acre-feet. In many years, withdrawal of the entire permitted volume would not affect spring flows to the extent that endangered species are harmed. To ensure adequate spring flows during dry years, new critical period management rules were also adopted, which set more stringent trigger levels at the several monitoring wells used to index aquifer pressure.

With Edwards supplies now limited by law and a new legal directive in place to protect endangered species habitats by ensuring spring flows, the region had to quickly get into the business of water supply management and development. In 1998, San Antonio adopted a 50-Year Water Supply Plan, which addressed the city's critical need to develop and manage water resources for a growing population while protecting regional environments. During the plan development, there was a strong consensus that, before San Antonio could seek new water sources from outside its own basin, it ought to demonstrate to its neighbors that it was making the best use possible of its own resources. Long criticized as a do-nothing city, several key components of the new plan established San Antonio as a leader in developing water management strategies that protect the water environment of Texas. In the last decade, two major projects have been brought on line, as described next.

10.9.4 TWIN OAKS AQUIFER STORAGE AND RECOVERY PROJECT

Unlike the Edwards Aquifer, with its high transmission rates, water in a sand aquifer tends to stay in place or move very slowly. Water injected into unconfined sand forms a stationary dome, and if there are confining layers, then water spreads out horizontally. Either way, it is possible to store water in sand and come back years later and extract that same water. This is the concept behind aquifer storage and recovery (ASR). Because rainfall in south Texas and Edwards Aquifer recharge are both highly variable, the region swings from being water rich to water poor, sometimes within the same year. So ASR technology was viewed as a way to address one of the region's biggest problems: There are hardly any storage locations where water can be put in times of plenty for later use. With an ASR facility available, Edwards water could be injected into locally occurring sand aquifers and extracted during times of shortage. Water from other sources could also be transported and stored in the sand aquifer for later use. Some advantages of storing water in sand aquifers instead of a reservoir are that no water evaporates and there is far less potential for contamination.

Another major advantage is that land is not lost. The many environmental impacts of reservoirs are avoided, and land can remain undisturbed and productive. Also, it is possible to ensure the people who pay for the project are the ones who benefit.

In September 1996, the San Antonio Water System and Bexar Metropolitan Water District were awarded a $200,000 state grant to study the possibility of storing water in the Carrizo-Wilcox and Glen Rose Aquifers, the saline zone of the Edwards Aquifer, and the Austin Chalk and Anacacho limestone formations. The study included looking at availability of water for storage, whether the source waters were compatible with water in the destination aquifers, and the quality and movement of water in each aquifer. The Carrizo-Wilcox Aquifer was identified as having the characteristics necessary for aquifer storage and recovery. The Carrizo-Wilcox is composed mainly of sand interbedded with gravel, silt, clay, and lignite. It extends from the Rio Grande in south Texas northeastward into Arkansas and Louisiana, passing through southern Bexar, Wilson, and Atascosa counties. In some places, the water has a high iron content, and hydrogen sulfide and methane also occur. Carrizo water is easily treated by conventional methods, and lots of people currently use it without treatment.

In September 1999, the San Antonio Water System (SAWS) purchased a 261 acre farm over the Carrizo-Wilcox, near the Atascosa-Bexar county line, and added two more large adjacent tracts in February 2000 for a total of over 3200 acres. SAWS developed an aggressive project timetable to have the facility on line and producing water quickly, and plans included leaving the farms in agricultural production.

In February 2001, the SAWS's board approved $7.53 million in engineering design contracts to move the project forward. Engineers began preparing plans for a facility capable of storing about 30,000 acre-feet of water per year, enough for about 60,000 families.

On July 16, 2002, the SAWS Board of Trustees approved $110 million in construction and engineering contracts to build a 30 mgd treatment facility, 17 ASR wells, and 29 miles of pipelines and pumps to deliver and integrate the water into the city's distribution system. A groundbreaking ceremony was held on July 31, construction began on August 1, and the facility opened in June 2004. A second phase added an additional 17 wells. Overall, the project cost about $215 million.

During wet years, the plant operates in recharge mode, storing excess available Edwards waters. State law requires that, before water can be injected, it has to meet drinking water quality standards, so there is no chance that water already in the ground could be contaminated. Withdrawal of these stored waters during dry periods reduces Edwards Aquifer pumpage and helps maintain the natural spring flows that provide critical habitat for endangered species.

By the end of 2006, over 20,000 acre-feet had been stored for later use. This water came in very handy during a drought later that year, when more than 6400 acre-feet of stored water was produced, deferring Edwards pumping and protecting spring flows (Figure 10.9–10). When the rains returned, the facility went back into recharge mode and began storing excess Edwards waters throughout the very rainy year of 2007.

In the summer of 2008, the facility was called back into action after an extremely dry 10 month period resulted in declining index well levels and spring flows. Drought restrictions are triggered when the level of the J-17 index well remains below 660 ft for 10 consecutive days, and restrictions stay in place for 30 days after the well rises back above the trigger level. When the trigger was reached in June, Twin Oaks switched to recovery mode and began producing about 15 cfs of stored Edwards water, about one third of its pumping capacity. As seen in Figure 10.9–11, the J-17 well level rose in response to reduced pumping and calls for conservation. Extremely heavy demands later in the month caused the well level to begin declining again, and facility operators contemplated increasing the pumping rate; however, Hurricane Dolly soaked the region, drought restrictions were canceled, and the Twin Oaks plant went on standby, having successfully maintained well levels and spring flows during a critical time.

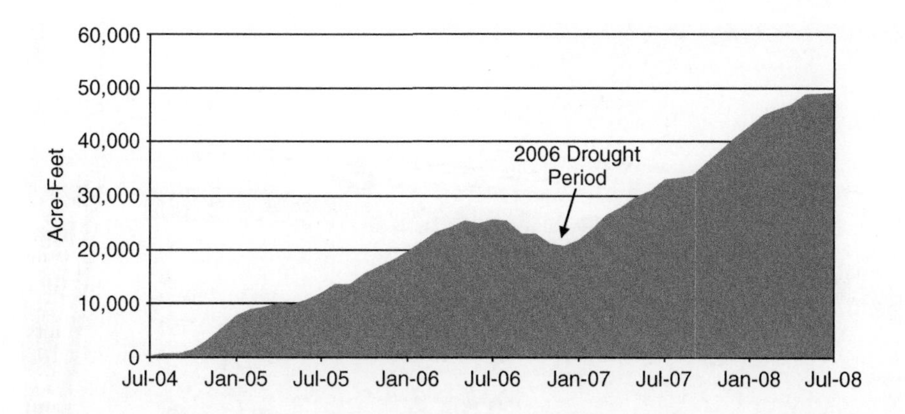

FIGURE 10.9–10 Twin Oaks ASR volume in storage (acre-feet) from July 2004 to June 2008. (Data from San Antonio Water System.)

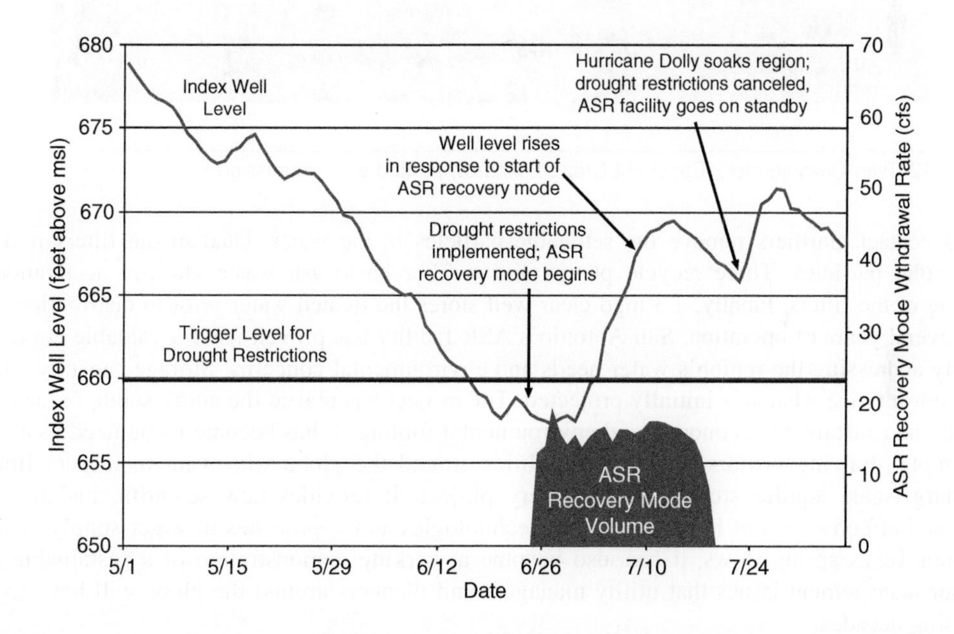

FIGURE 10.9–11 Index well J-17 and Twin Oaks recovery mode from May to July 2008. (Copyright Gregg Eckhardt, printed with permission.)

The Twin Oaks facility also has several wells designated for recovery of native Carrizo Aquifer water and a treatment plant designed to make Carrizo water compatible with Edwards Aquifer supplies. Carbon dioxide and lime are added to raw Carrizo water to increase the pH, hardness, and carbonate alkalinity. A step-feed aeration process removes any remaining carbon dioxide, provides oxidation of iron and hydrogen sulfide, and increases the dissolved oxygen concentration (Figure 10.9–12). Polymer is then added to the aerated water to assist coagulation of suspended solids into large, settleable particles. Potassium permanganate is also added to oxidize manganese into an insoluble form that can be removed by sedimentation and filtration.

FIGURE 10.9–12 Twin Oaks aerator. (Copyright Gregg Eckhardt, printed with permission.)

Next, solid contact clarifiers remove the settleable particles in the water. Dual media filters remove any remaining solid particles. Three recycle pumps and a filter backwash waste clarifier accommodate the backwashing of the filters. Finally, a 3 mgd clear well stores the treated water prior to distribution.

After several years of operation, San Antonio's ASR facility has proven itself a valuable component of successfully addressing the region's water needs and environmental concerns. Storage capacity has turned out to be almost twice what was initially projected. The project has placed the entire south Texas region on a more solid and sustainable economic and environmental footing. It has become recognized as an international example, drawing visitors from water utilities around the globe who want to inspect firsthand a working, large-scale aquifer storage and recovery project. It provides new scientific insights into the feasibility and effectiveness of imaginative new technologies and approaches to water supply. While protecting south Texas spring flows, it has also become a working demonstration of a sustainable solution to the water management issues that utility managers and planners around the globe will have to address in the coming decades.

10.9.5 RECYCLED WATER DISTRIBUTION

A second major project aimed at managing south Texas water resources while protecting spring flows was construction of America's largest recycled water distribution system (Figure 10.9–13). A primary goal was to reduce San Antonio's pumpage from the Edwards Aquifer, thereby protecting endangered species habitats and critical ecosystems in Comal and San Marcos Springs. A second primary goal was to accomplish local aquatic ecosystem enhancement and restoration by providing stream flow augmentation at four newly established discharge locations on the San Antonio River and Salado Creek. Both streams were originally fed by natural spring flows and, as springs dried up, potable flows from artesian wells. In the case of Salado Creek,

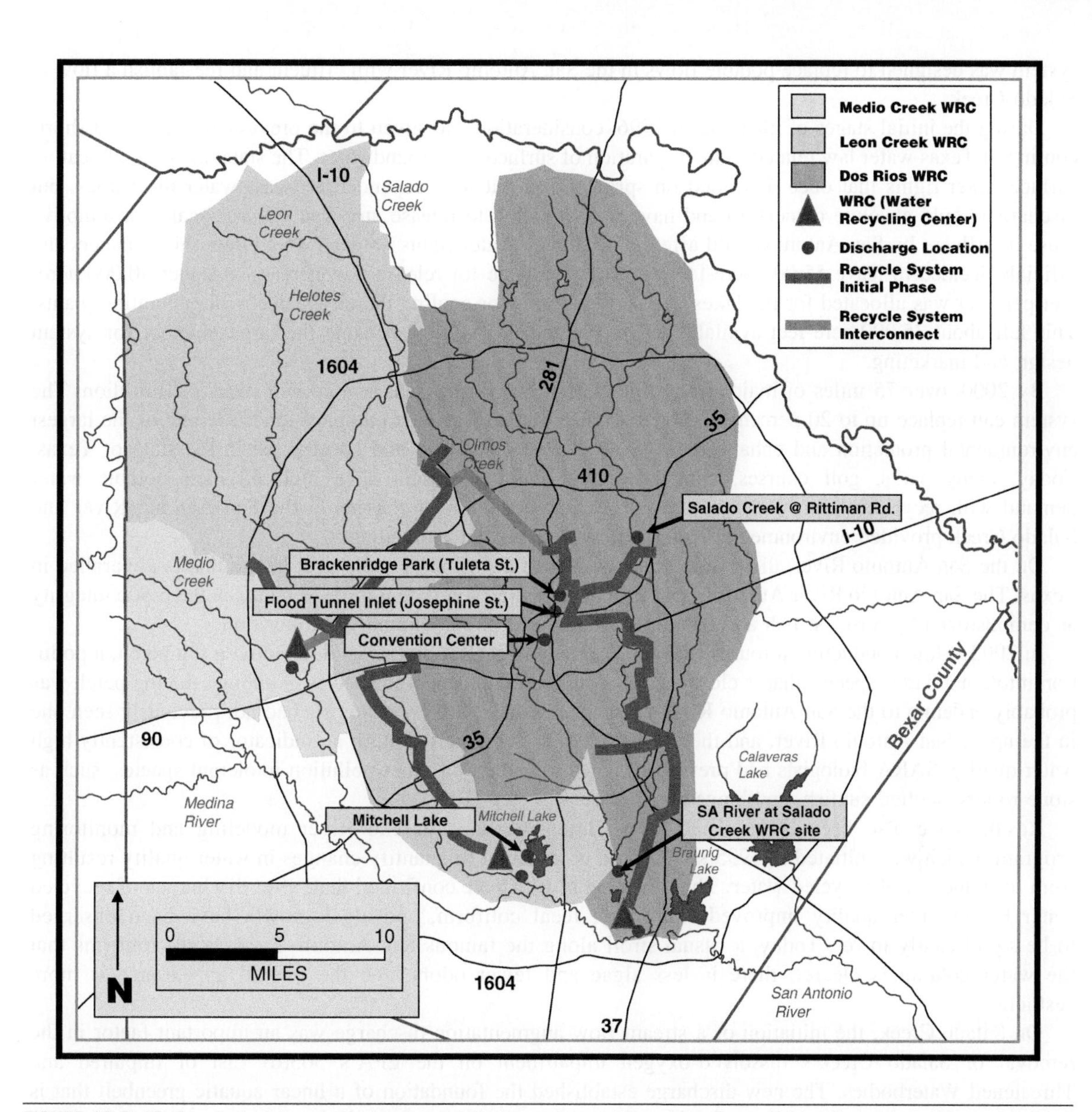

FIGURE 10.9–13 San Antonio's recycled water distribution system. (Copyright Gregg Eckhardt, printed with permission.)

a large well that had supplied base flow for almost a century was deemed by authorities to be a "waste," and it was ordered plugged in the early 1990s. Subsequently, the stream was listed by the U.S. Environmental Protection Agency on its 303(d) List of Impaired and Threatened Waterbodies because of low dissolved oxygen levels. In downtown San Antonio, potable Edwards wells continued to provide base flow in the famous River Walk, one of the state's largest economic generators and tourist destinations. In addition to delivering water to customers for nonpotable industrial uses and applications such as landscape irrigation, the recycled water system was designed to replace potable flows in the San Antonio River with effluent and reestablish a flow in Salado Creek.

During the initial stages of planning in 1996, consideration was given to the previously mentioned shortcoming of Texas water law related to the separation of surface and groundwater. The state had been allocating surface water rights that once depended on spring flows but now depended on wastewater discharges, but dischargers hold absolute ownership and have no obligation to release any water. Large-scale consumptive reuse of effluent by San Antonio could affect downstream water rights holders. To address this, planners and officials decided to leave 55,000 acre-feet per year available for release downstream. Another 40,000 acre-feet per year was allocated for the lakes built in the 1960s for cooling the city's electrical generation plants. This left about 35,000 acre-feet available for reuse, and this volume became the target number for system design and marketing.

By 2000, over 75 miles of major trunk lines had been constructed, at a cost of over $120 million. The system can replace up to 20 percent of San Antonio's annual Edwards pumpage, and it is one of the largest environmental protection and enhancement projects ever conceived and constructed in the state of Texas. Today, many parks, golf courses, cemeteries, and industrial users have replaced their potable water demand with recycled water, and the stream flow augmentation discharges to the San Antonio River and Salado Creek provided environmental benefits that have proven astounding.

On the San Antonio River, the program has witnessed a revival of ecological health that is unparalled in Texas. The San Antonio River Authority (SARA) documented significant improvements in the biotic integrity as demonstrated by a robust fish and benthic macroinvertebrate community composition.

In 2002, while conducting a routine habitat assessment, SARA biologists discovered a log perch, a pollution intolerant darter species that is closely related to several endangered species. Although the log perch was probably endemic to the San Antonio River in its natural state, SARA biologists had not previously seen one in the upper San Antonio River, and they considered the log perch's return an indicator of consistently high water quality. SARA biologists had previously noted the return of other pollution intolerant species, such as stone rollers, spotted sunfish, and longear sunfish.

Results have also been documented by an intensive San Antonio River modeling and monitoring program, which was initiated in 1995 so it would be possible to quantify changes in water quality resulting from introduction of recycled water. The sampling results have confirmed that, after discharges of recycled water began, river quality improved. Turbidity, fecal coliform, and algal growth have been observed to be significantly lower. Today, a casual stroll along the famous San Antonio River Walk confirms that the water column is clearer, there is less algae and fewer odors, and the overall appearance is more aesthetic.

On Salado Creek, the initiation of a stream flow augmentation discharge was an important factor in the removal of Salado Creek's dissolved-oxygen impairment on the EPA's 303(d) List of Impaired and Threatened Waterbodies. The new discharge established the foundation of a linear aquatic greenbelt that is quickly becoming a cherished natural refuge for the San Antonio community.

The construction and operation of such a large-scale recycled water distribution system has been perceived as demonstrating a commitment to leadership in water resource development, environmental improvement, aquatic resource protection, and stewardship. The project has also served as a model for others; San Antonio is now recognized internationally as a leader in conservation and aquatic ecosystem protection. Many water utilities have expressed a high level of interest in the lessons learned by the San Antonio Water System while planning, developing, and operating the recycled water distribution system. The utility is actively sharing its experiences and all it has learned at professional conferences and through community outreach presentations.

10.9.6 RECHARGE AND RECIRCULATION

A third approach to manage and ensure spring flows has been dubbed *recharge and recirculation*. It is a collection of water management strategies that would add enhanced recharge to the Edwards Aquifer and hold that enhanced recharge in aquifer storage until needed for water supply and protection of spring flows and downstream flows in critical times.

Two types of projects are under consideration for providing enhanced recharge to the Edwards (Figure 10.9–14). Type I projects involve an engineered structure placed on the contributing zone that would capture flows and hold them for release to the recharge zone. Type II projects include an engineered structure placed directly on the recharge zone, and water would be held for direct infiltration. Potential sources of water are storm water flows, unused Edwards permits that might be purchased or leased, and diversions from other regional rivers and streams. New firm yield water supplies would come from issuance of recharge recovery rights and diminished critical period pumping reductions. Some of the additional water recharged would be recirculated by removing it through wells down-gradient from where it recharged and pumping it back to recharge areas.

The first phase of the project involved using a model of the Edwards Aquifer prepared by the U.S. Geological Survey to evaluate feasibility. Two recharge scenarios were modeled, and both scenarios predicted long-term storage benefits of up to several years when 25,000 acre-feet were placed in the aquifer.

A second phase involved using the same model to simulate aquifer responses to recharge at eight locations. Numerous combinations of recharge timing, volume, and location were simulated. The simulations predicted that, by introducing approximately 149,000 acre-feet of enhanced recharge and applying critical period management rules, Comal Springs could be kept from going dry during a repeat of the 1950s drought of record.

The third phase of the project, currently under way, is evaluating potential operational parameters, water sources, and costs for the various scenarios. The optimum placement of recirculation wells is being determined based on factors like impact to spring flows and length of recirculation pipelines. Potential water sources are also being evaluated, such as diversions from the Guadalupe River and several area reservoirs.

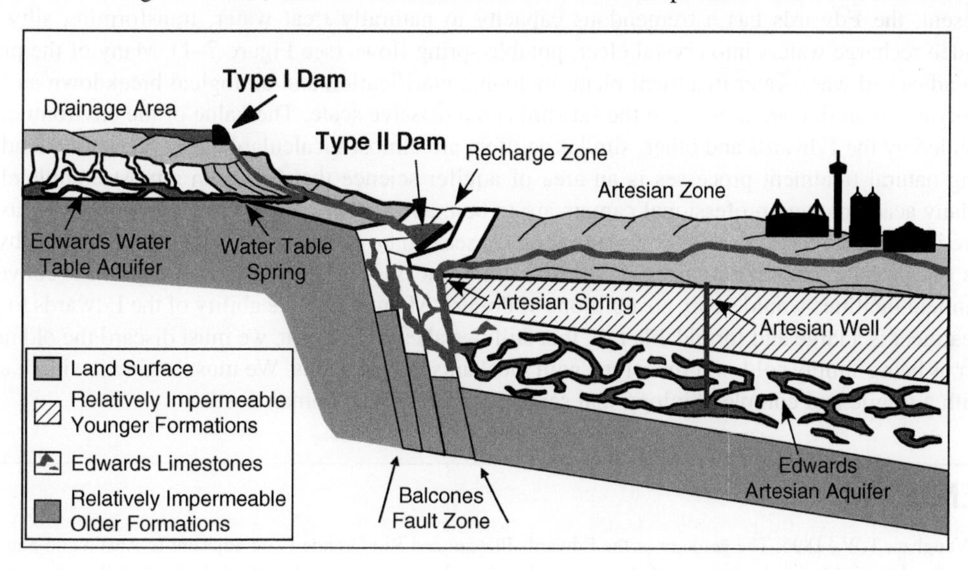

FIGURE 10.9–14 Projects for providing enhanced recharge to the Edwards Aquifer. (Copyright Gregg Eckhardt, printed with permission.)

10.9.7 SUMMARY AND OTHER CONSIDERATIONS

Tremendous progress has been made toward ensuring the long-term survival of central Texas springs. The first step was revamping of unworkable laws that gave all users infinite access to a resource that is, in fact, finite. A new agency was established to allocate Edwards Aquifer pumping rights based on historical use. In the last decade, engineering solutions have delivered innovative strategies for storage, reuse, and conservation, and ongoing efforts may provide additional recharge and recirculation management tools that will help ensure spring flows. Today, the volume issue is largely settled, and it appears likely that many future generations will benefit from the technological and engineering solutions that have been applied.

Managing recharge water quality in the Edwards mainly involves controlling land use, but Texas is a state where politically powerful landowners and developers view environmental controls and restrictions as a seizure of private property. As growth exploded over the recharge and contributing zones around San Antonio, effective rules aimed at maintaining recharge water quality have remained completely inadequate. On the contributing zone, land use will ultimately determine runoff and recharge water quality, but ranchers and residents of the hill country are not Edwards Aquifer users, and they are generally not very enthused about restricting their own development to protect Edwards water supplies. There is also disagreement about whether the Edwards Aquifer Authority or the state's primary environmental agency, the Texas Commission on Environmental Quality, should make and enforce water quality rules. Many political subdivisions are involved and dozens of groundwater conservation districts, all of which have different views and goals.

Meanwhile, elected officials that dare to attempt managing development for protection of water quality run the risk of being drummed out of office by developer-financed candidates who are willing to sacrifice common resources for short-term profits. In Helotes, for example, development around an important recharge stream rallied concerned residents to elect conservation-minded representatives, but they were subsequently removed when powerful development interests rallied a private property rights backlash. With shortsighted politicians guiding the development around critical recharge locations, the long-term outlook for Edwards water quality is not good.

At present, the Edwards has a tremendous capacity to naturally treat water, transforming silty brown, organic laden recharge waters into crystal clear, potable spring flows (see Figure 7–1). Many of the processes used in an advanced wastewater treatment plant, including clarification and biological breakdown and stabilization of organic material, are at work in the Edwards on a massive scale. The value of the environmental services provided by the Edwards and other, similar aquifers are almost incalculable, and yet, understanding and quantifying natural treatment processes is an area of aquifer science that has been almost completely overlooked. Many academic and professional careers are to be made in this area. As in a conventional wastewater plant, there is a delicate balance, and these natural processes can be disrupted in many ways, such as by excessive loads of inorganic or organic material and toxic substances. Replacing these services with conventional water treatment facilities would easily cost billions of dollars. Maintaining the ability of the Edwards to provide natural treatment is a large and formidable task that still lies ahead. To begin, we must discard the old notion of the aquifer as being simply cold, wet limestone with mechanistic flow paths. We must begin viewing it as living system with a fragile and complex biology that serves our most basic human needs.

REFERENCES

Hill, R.T., Vaughan, T.W., 1896. The geology of the Edwards Plateau and Rio Grande Plain adjacent to Austin and San Antonio, Texas, with references to the occurrence of underground waters. U.S. Geological Survey 18th Annual Report, part 2-B, Washington, DC, pp. 103–321.

Maclay, R.W., Small, T.A., 1986. Carbonate hydrology and hydrology of the Edwards aquifer in the San Antonio area, Texas. Texas Water Development Board Report 296, Austin, Texas.

Case Study: Utilization and protection of large karst springs in China 10.10

Qiang Wu[1], **Liting Xing**[1], **and Wanfang Zhou**[2]
[1]Institute of Water Hazard Prevention and Water Resource, China University of Mining and Technology, Beijing, China;
[2]Earth Resource Technology, Huntsville, Alabama

10.10.1 INTRODUCTION

Karst regions occupy approximately 25 percent of the land surface of the Earth. Karst aquifers are often major sources of water supply in these areas. However, karst aquifers are generally considered to be particularly vulnerable to pollution and anthropogenic impacts. Large withdrawals of water in karst areas for municipal, agricultural, and industrial use may competitively affect the water supply in surrounding areas; deterioration of surface water and groundwater quality from agricultural, industrial, or private development may occur; and improper injections of waste into a karst system may contaminate the water supply. To utilize karst groundwater sustainably, measures for elimination of pollution sources and protection of groundwater are necessary. Typical preventative strategies include land-use control and establishment of groundwater protection zones.

Springs are characteristic of most karst systems. There is no exception in the widely distributed karst areas of China, where the soluble rocks encompass approximately 3.44×10^6 km^2, of which 9.1×10^5 km^2 is composed of bare carbonate rocks (Yuan, 1998). Figure 10.10–1 shows the different types of springs that often occur in China. Because of the differences in climate, topography, lithology, and geologic structure, karst terrain is generally concentrated in two regions of China. One of the regions encompasses the Shanxi Plateau and neighboring provinces in north China, an area of about 470,000 km^2 in a semiarid climate zone. The other karst region is located in the southwest region of China, with an area of approximately 500,000 km^2 in a humid climate zone (Yuan, 1994).

Karstification is highly influenced by precipitation and terrain, which can cause large regional differences in karst spring flow. The karst systems in humid southwest China are characterized by well-developed caves and highly connected underground flow channels (Yuan, 1994). The ratios of maximum to minimum spring flow are generally 10:1000, which is indicative of a rapid response to precipitation in the humid regions. In this region, the total thickness of carbonate rock ranges from 3,000 to 10,000 m. Most karst development is in the Devonian, Permian, and Middle and Low Triassic systems. The abundant rainfall, orogenic tectonic activities, folding and faulting, and dynamic circulation of water dramatically facilitated karstification in this area. Many world-known tropical karst landscapes are located in this region. The water resource is well utilized through drilling supply wells into the underground rivers, augmenting springs, and constructing

Copyright © 2010, Elsevier Inc. All rights reserved.

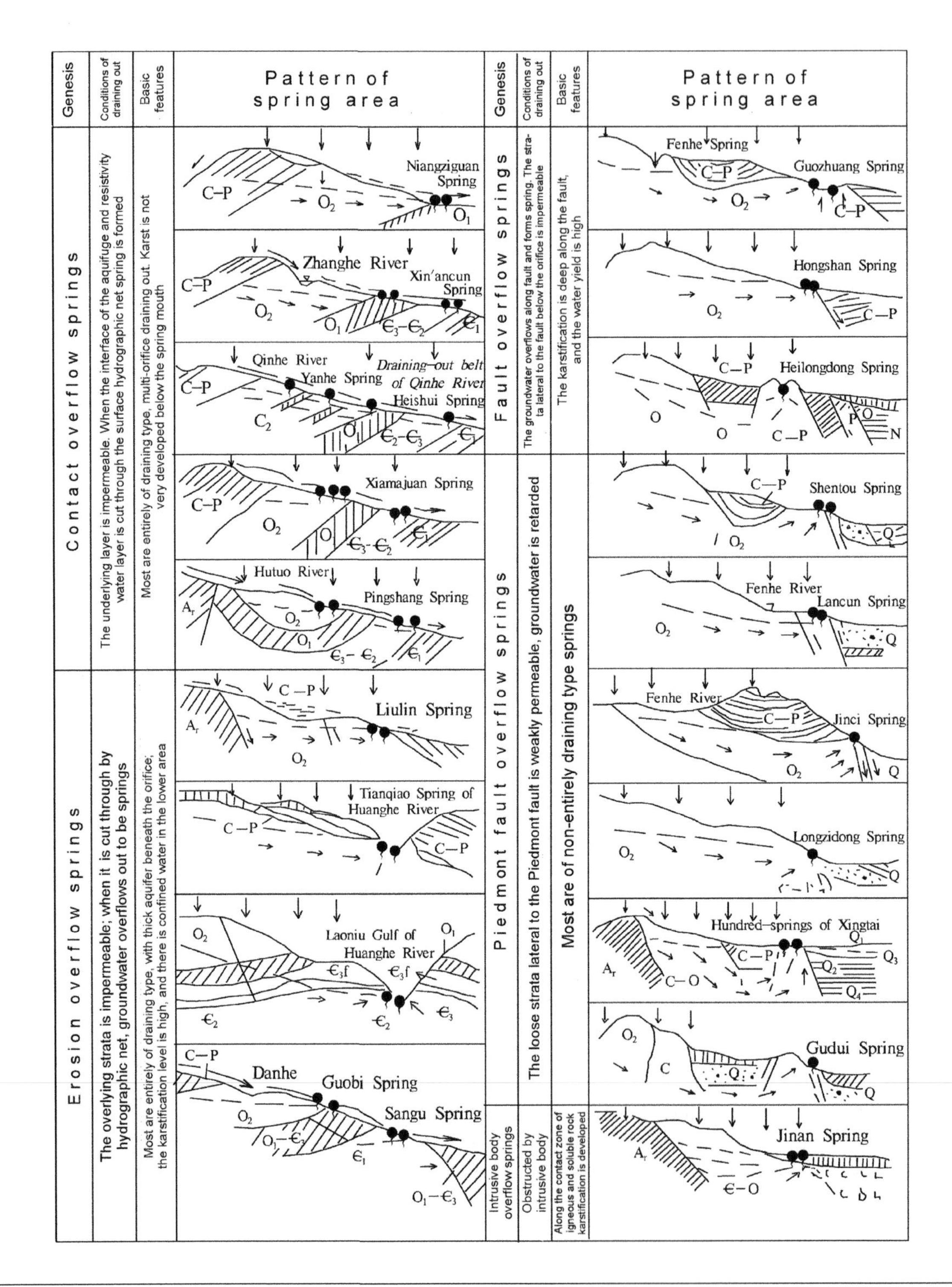

FIGURE 10.10–1 Classification of karst springs in China. (From Yuan, 1994.)

underground water dams and reservoirs. For example, in Guizhou province, approximately 2680 municipalities use springs as their potable water supply sources. Many hydropower stations, such as those along Maotiao River, were constructed in underground rivers or at the orifice of large springs for electric generation. Hot and mineral springs are used for irrigation, sprout cultivation, bathing, and medical purposes. Karst landscapes, coupled with the rich water resources, have become tourist attractions, such as Li River of Guangxi, Huangguoshu Waterfalls of Guizhou, and Jiuzhaigou of Huanglong, Sichuan.

In contrast, in semiarid areas of China, most carbonate aquifers are overlain by thick Permian carboniferous-sandstone and shale, and Quaternary sediments. Karstification in these areas is generally not so well developed on a regional scale because of low precipitation and thick overburden. Karst features are characterized by widespread, dissolution-enlarged fractures or conduits rather than subsurface rivers. The geological structures are composed primarily of large-scale gentle folds and large fault blocks, which result in large karst basins. There is usually a unified flow field in each karst system. The presence of large springs is one of the most important hydrogeolologic characteristics of these karst basins. At least 60 springs have discharges of more than 1 m^3/s (Yuan, 1994). As shown in Figure 10.10–2, distribution of the large springs is clustered. Table 10.10–1 lists 18 large springs in Shanxi province. The areal extent of a karst groundwater system is usually up to several thousands of square kilometers. The multiple types of porosity in the carbonate rocks provide extensive storage for precipitation. Groundwater flows preferentially through discrete larger conduits while the fracture-pore system stores the majority of water. Many large springs drain the entire basin

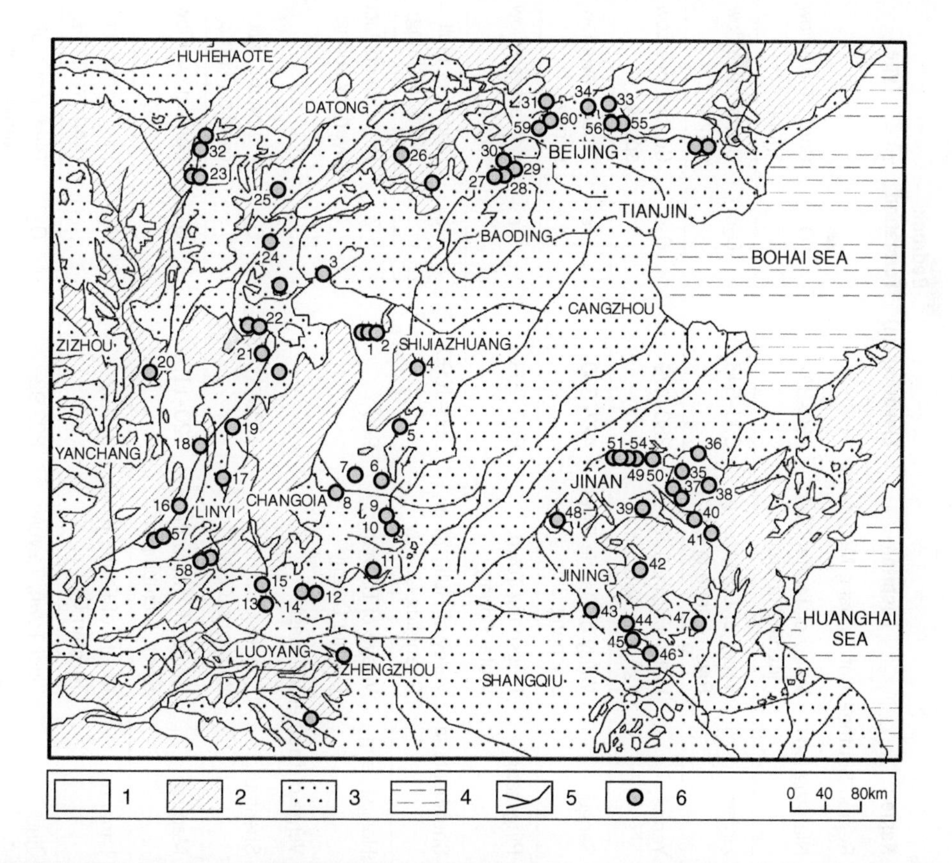

FIGURE 10.10–2 Large springs in north China. (From Yuan, 1994.)

Table 10.10–1 Characteristics of 18 Large Springs in Shanxi, China (From Guo, Zhang, and Yu, 2004)

Name	Location	Elevation (m)	Bedrock Formation	Type	Catchment Area (km^2)	Limestone Area (km^2)	Average Discharge (m^3/s)	Utilization Rate (%)
Region Southeast of Taihang Mountains								
Niangziguan	Valley of Taohe River, Pingding County	360–392	O_1	Contact overflow spring	4467	2218	12.1	44.9
Xin'an	Valley of Dongzhang River, Lucheng City	600–643	$O_2\varepsilon$	Contact overflow spring	13,000	2600	10.1	26.3
Yanhe	Valley of Qinhe River, Yangcheng County	380–479	$O_2\varepsilon$	Contact overflow spring	2575	1375	11.9	9.6
Sangu	Valley of Danhe River, Jincheng City	427–560	ε	Erosion overflow spring	2813	1008	7.2	30.9
Pingshang	Pingshang Town, Wutai County	640–703	ε	Contact overflow spring	2817	750	4.5	2.8
Shuishentang	Valley of Huliu River, Guangling County	960–1090	ε	Piedmont fault overflow spring		153	0.85	11.5
Hongshileng	Village of Dashahu, Lingqiu County	940	Z	Intrusive contact overflow spring	74	58	0.2	1.4
Verge of the Central Basin								
Shentou	Shentou Town, Wangzhou City	1059–1063	O_2	Piedmont fault overflow spring	4500	2410	7.17	92.0
Maquan	Gorge of Yangwuh River, Yuanping City	120–1150	O_2	Contact overflow spring	754	329	1.23	92.4
Lancun	Mouth of Fenhe River, Taiyuan City	810.92	O_2	Piedmont fault overflow spring	2344	1440	6.32	111.9
Jinci	Jinci Town, Taiyuan City	802.92	O_2	Piedmont fault overflow spring	1528	391	2.18	186.6
Hongshan	Hongshan Town, Jiexiu City	916	O_2	Fault overflow spring	650	260	1.23	63.7

Guozhuang	Valley of Fenhe River, Huozhou City	516	O_2	Fault overflow spring	5000	1400	7.59	66.3
Guangshengci	Foot of Guangshengsi Mountain, Hongtong County	581.5	O_2	Fault overflow spring		948	3.89	99.3
Longzici	Front of mountain, southwest of Linfen City	465–479	O_2	Piedmont fault overflow spring	3250	1340	5.48	21.7
Gudui	Sanquan village, Xinjiang County	450	ε	Piedmont fault overflow spring			1.19	61.6
Region West of Luliang Mountains								
Liulin	Valley of Sanchuan River, Liulin County	790–801	O_2	Erosion overflow spring	5100	830	3.19	11.2
Tianqiao	Gorge of Huanghe River, Baode County	816–830	O_2	Erosion overflow spring	13,974		8.50	10.0

with only 5–10 percent underflow. Discharge at the majority of springs is relatively steady. The ratio between the maximum and minimum discharges within a hydrological year is no more than 5 for most springs, with the common range from 2 to 3. Spring discharges in these semiarid regions generally lag behind precipitation by 2–10 years. The springs in semiarid areas can be classified as slow-response springs.

In Shanxi province, the eastern side of Taihang Mountains, and the area of Jinan-Xuzhou-Huaiyin where large springs exist, the karst water has been fully utilized because of the excellent water quality and accessibility. Karst groundwater is the primary water supply in Taiyuan, Siquan, Changzhi, Xingtai, Handan, Anyang, Jiaozuo, Jinan, Zibo, Xuzhou, Laiwu, Zaozhuang, and other cities. The largest springs have become the most important water supply for the adjacent cities and energy bases. In Shanxi province, for example, many springs, such as Shentou Spring, Liuquan Spring, Guozhuang Spring, Niangziguan Spring, and Xinan Spring, have been augmented to provide water supply.

Table 10.10–2 lists 8 large karst spring systems in Hebei province. Their total drainage area is approximately 1.9×10^4 km^2, and the exploitable water resource is estimated to be approximately 16×10^8 m^3/a. The current utilization rate is more than 80 percent. Overexploitation has been reported in several spring watersheds. In Jici Spring and Lancun Spring of Shanxi province, Shigu Spring, Xingtai Spring, Dongfenghu Spring, Helongdong Spring of Hebei province, Yuquanshan Spring of Beijing, Huixian Hundred-Springs, Jiulishan Spring of Jiaozuo, Henan province, and Jinan Springs of Shandong province, the flow rates decreased because of inappropriate exploitations, dewatering in mines, and other factors. A recent survey of 29 springs of Shandong province indicates that the discharges in 20 of them decreased significantly. Figures 10.10–3 and 10.10–4 show two of the Jinan Springs, illustrating their great cultural and historic importance to generations of Chinese.

Table 10.10–2 Utilization of 10 Large Karst Spring Systems in Hebei Province (8 of 10 shown here) (From Chen and Ma, 2002)

System Name	Main Water Sources	Exploitable Resource (10^4 m^3/a)	Remaining Resource (10^4 m^3/a)	Utilization Rate (%)
Hot Spring-Nanshan Spring of Yuxian	Outside of Yuxian county town	7100	5588	21
Laiyuan Spring		13,875	6197	55
Shuimocao Spring of Quyang		14,569	10,323	29
Weizhou Spring	Kuangshi town, Kengkou power plant, Shangan power plant, coal mine	34,633	6687	81
Shigu Spring		2745	0	100
Hundred-Springs of Xingtai	Xishimen, Yushikuangshan village, Zhongguanxiyao, Xingtai city, coal mine	21,814	57	100
Dongfenghu Spring	105 power plant, She county	6559	0	100
Heilonghe Spring	Yangjiaopu, Feng-Feng, Niuji, Handan power plant, drainage of coal mine	37,141	0	100

FIGURE 10.10–3 Baotu Spring in Jinan.

FIGURE 10.10–4 Black Tiger Spring in Jinan.

Discussed in the following sections are two large springs in north China: the Jinan Springs and Niangziguan Spring. Jinan Spring, in Shandong province, consists of springs 51 through 54 in Figure 10.10–2. Niangziguan Spring, in Shanxi province, is listed as spring 1 in Figure 10.10–2.

10.10.2 THE JINAN SPRINGS

Jinan, the capital city of Shandong province on China's east coast, is a regional, political, economic, cultural, scientific, and educational center. Its location is shown in Figure 10.10–2. Jinan is near the Tai Mountains to the south and neighbors the Yellow River on the north. The terrain of the territory of Jinan slopes down from the south to the north and its landforms vary from hills and inclined plains in front of the mountains to the alluvial plains of the Yellow River. Jinan is known for its beautiful springs and it is often referred to as the *spring city*. Generally, the Jinan Springs are composed of 10 spring groups: Baotuquan spring group, Zhenzhuquan spring group, Heihuquan spring group, Wulongtan spring group, Baiquan spring group, Yongquan spring group, Yuhequan spring group, Baimaiquan spring group, Jiashaquan spring group, and Hongfanchi spring group. Table 10.10–3 lists the individual spring orifices in each spring group. There are 733 natural fountains in the area, among which 219 are in an urban district, 217 in Licheng district, 156 in Zhangqiu city, 104 in Changqing district, and 137 in Pingyin county. The most famous four spring groups are Baotu (Baotuquan), Black Tiger (Heihuquan), Pearl (Zhenzhuquan), and Five Dragon (Wulongquan) (see Figures 10.10–3 and 10.10–4).

Since the 1980s, the groundwater levels in the aquifer that feeds the springs have declined significantly because of overirrigation and overexploitation. Many of the springs have become intermittent, and some ceased to flow. Many studies have been performed to protect the springs. Successive large-scale link experiments on the spring sources were made to determine the characteristics of the groundwater flow in the karst area (Zhang, Wang, and Zhai, 2004). The study on controlling parameters of the Jinan springs indicated the maximum available amount of groundwater (Li, Hu, and You, 2004). The development of geographical information system (GIS)-based models showed that the landscape spatial pattern had changed since 1995. Several studies indicated that the integrated water resource analysis and management in a given jurisdiction were capable of identifying the feasible water allocations (Wu and Xu, 2005). The support of the province government and local regulatory agencies led to a series of measures, including diversion of water from the Yellow River to Jinan, artificial recharge of groundwater, rational exploitation of groundwater, and better utilization of surface water.

10.10.2.1 Hydrogeological settings

The Jinan Spring area covers 1486 km^2 and constitutes a karst water system. The stratigraphic sequence consists of powdery clay, conglomerate and clay of the Quaternary age, grit of the Permian age, diorite of the Carboniferous age, limestone of the Ordovician age, as well as the Fengshan, Gushan, and Zhangxia formations of the Cambrian age. The zone forms a monocline structure dipping slightly to the north. The major stratum exposed in the study area is the Ordovician limestone. The Ordovician limestone is the most important aquifer, which is under the entire city, with groundwater levels 10–100 m below the surface.

The average annual precipitation in the spring area is 647 mm, most of which occurs between June and September. The annual maximum precipitation was 1194.5 mm in 1962, and the minimum precipitation was 340 mm in 1989. In the last 20 years, the regional precipitation has shown a general trend of decrease. Several surface streams flow through the spring basin, including the Yellow River, Beisha River, and Xiaoqing River. The Yellow River is the main inflow water source of Jinan city. There is no evidence that the river has a hydraulic connection with the underlying karst groundwater. However, the Yufu River and Beisha

Table 10.10–3 General Descriptions of 10 Spring Groups in Jinan City

Spring Group	Main Springs
Baotuquan	Lies in the southwest of the old city. This group has 28 spring orifices, among which 27 are in Baotuquan Park: Baotu Spring, Jinxian Spring, Huanghua Spring, Liuxu Spring, Woniu Spring, Suyu Spring, Mapao Spring, Wuyou Spring, Shiwan Spring, Zhanlu Spring, Manjing Spring, Dengzhou Spring, Dukang Spring, Wangshui Spring, Laojinxian Spring, Qianjing Spring, Xibo Spring, Hunsha Spring, Jiu Spring, Donggao Spring, Luosi Spring, Shangzhi Spring, Cang Spring, Huaqiangzi Spring, Baiyun Spring, Quantingchi Spring, Bailongwan Spring, and Yinhuchi Spring
Wulongtan	The area is 6 hectares and consists of 28 spring orifices: Wulongtan Spring, Guwen Spring, Xianqing Spring, Tianjing Spring, Yueya Spring, Ximizhi Spring, Guanjiachi Spring, Huima Spring, Qiuxi Spring, Yu Spring, Lian Spring, Dongmizhi Spring, Xixin Spring, Jingshui Spring, Jingchi Spring, Dongliu Spring, Beixibo Spring, Lexi Spring, Tanxi Spring, Qishisan Spring, Qing Spring, Jing Spring, Jingming Spring, Xianmingchi Spring, Yuhong Spring, Cong'er Spring, Chi Spring, and Li Spring
Heihuquan	The area is 1.5 hectares and consists of 16 spring orifices: Hehu Spring, Pipa Spring, Manao Spring, Baishi Spring, Jiunv Spring, Jinhu Spring, Nanzhenzhu Spring, Douya Spring, Wulian Spring, Ren Spring, Yinsi Spring, Huibo Spring, Duibo Spring, Yihu Spring, Gujian Spring, and Shoukang Spring
Zhenzhuquan	This group lies in adjacent parks and consists of 21 spring orifices: Zhenzhu Spring, Sanshui Spring, Xiting Spring, Chu Spring, Zhuoying Spring, Yuhuan Spring, Furong Spring, Shun Spring, Tengjiao Spring, Shuangzhong Spring, Ganyingjing Spring, Hui Spring, Zhiyu Spring, Yunlou Spring, Liushi Spring, Zhusha Spring, Bukui Spring, Guangfu Spring, Shanmian Spring, Xiaogan Spring, and Taiji Spring
Baiquan	This group consists of Bai Spring, Hua Spring, Yinma Spring, Hua Spring, Hui Spring, Yayahulu Spring, Cao Spring, Leng Spring, Tuan Spring, and Ma Spring
Yongquan	This group lies in Liubu, Jinxiuchuan, and Xiying: Yong Spring, Kuju Spring, Bishu Spring, Tu Spring, Niyu Spring, Da Spring, Shengshui Spring, Duanhua Spring, Li Spring, Shengchi Spring, Nan Spring, Boluo Spring, Mujia Spring, Xilao Spring, Xuan Spring, Nanganlu Spring, Pipa Spring, Liu Spring, Che Spring, Yinyang Spring, Liangshui Spring, Siqing Spring, Baihua Spring, Basu Spring, Zhigong Spring, Zaolin Spring, Sheng Spring, Huanglu Spring, Hudong Spring, Xuehua Spring, Ouchi Spring, Xizhang Spring, Shuiliandong Spring, Shenyi Spring, Dishui Spring, Fengle Spring, Qianggan Spring, Zahu Spring, Dahua Spring, Shicha Spring, Wolong Spring, Bingbing Spring, Shui Spring, Liangwan Spring, Lupao Spring, Kuli Spring, Sanlongtan, and Yundou Spring
Yuhequan	This group consists of Yuhe Spring, Tangdou Spring, Yulou Spring, Dongliu Spring, Laoyuhe Spring, Xianghulu Spring, Dong Spring, Huanglu Spring, Zhugong Spring, Humen Spring, Zhong Spring, Huangxie Spring, Lujing Spring, Yihe Spring, and Heihu Spring
Baimaiquan	This group lies in Zhangqiu city and is the biggest spring group in the east of Jinan district: Baimai Spring, Dongma Spring, Mo Spring, Meihua Spring, Ximawan, Jingming Spring, Suyu Spring, Longwan Spring, Jinjing Spring, Lingxiu Spring, Hehua Spring, Yanming Spring, Dalongyan Spring, Xiaolongyan Spring, Fantang Spring, Shaizidi Spring, Yule Spring, Xiejia Spring, Pan Spring, and Bai Spring
Jiashaquan	Jiasha Spring, Zhuoxi Spring, Qingleng Spring, Tanbao Spring, Xiaolu Spring, Dishui Spring, Ganlu Spring, Shuanghe Spring, Baihe Spring, Shangfang Spring, Langgong Spring, Niubi Spring, Longju Spring, Shuang Spring, Wangjia Spring, Changshou Spring, Wolong Spring, Duanjia Spring, Baihu Spring, Runyu Spring, Kanggou Spring, Hui Spring, Yuzhu Spring, Qinglong Spring, Shengtian Spring, and Mashan Spring
Hongfanchi	This group lies in Pingyin County: Hongfan Spring, Shuyuan Spring, Hu Spring, Riyue Spring, Jiangnv Spring, Tianchi Spring, Mochi Spring, Tianru Spring, Baiyan Spring, Bajian Spring, Lianhua Spring, Ding Spring, Lang Spring, Changgou Spring, and Baisha Spring

River, originating from north of the Taishan Mountains, have strong hydraulic connections with the karst groundwater. Construction of reservoirs, including Wohushan, Jinxiuchuan, and Yuqingwu reservoirs, upstream has reduced the recharge to the karst groundwater from both rivers. The Xiaoqing River, originating from the Muli village, a west suburb of Jinan, has been severely polluted by various activities in the city since the 1960s, when the river was clean.

Generally, concealed faults are the primary geologic structure, such as the north northwest Qianfushan fault, Mashan fault, Dongwu fault, and Wenhuaqiao fault; north northeast Ganggou fault; and north-south Chaomidian fault (Figure 10.10–5). Lithology, faults, and topography control the boundaries of the spring drainage basin. The southern boundary of the drainage basin consists of lithology and a surface watershed divide, which includes Gangxinzhuang-Taohuayu-Momoding in the west; Huangshanding, Xianghuoluzi Mountain to Changchengling in the southwest; Dagaojianshan Mountain in the northeast; and Wenfengshan Mountain, Paomaling Peak, and Dongwu fault in the southeast. The northern boundary is defined by the

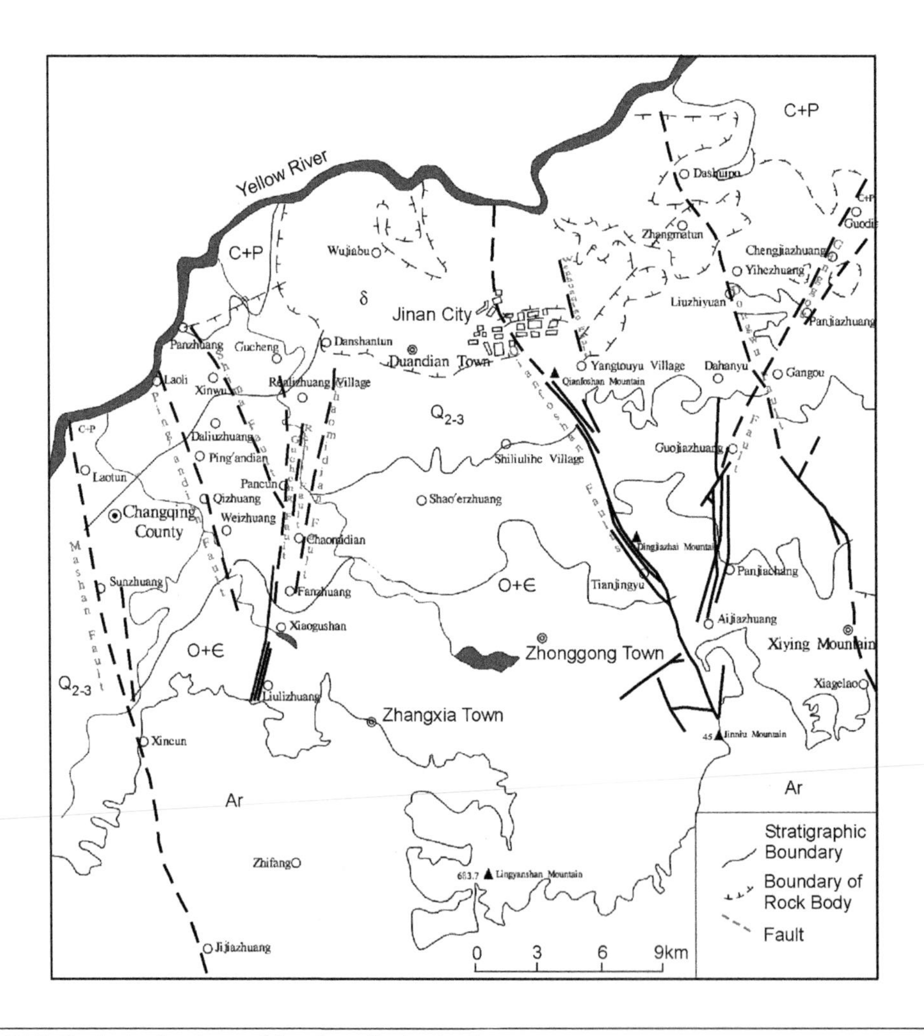

FIGURE 10.10–5 Geologic structures of the Jinan Spring basin.

intrusive rock mass. The eastern boundary is bounded by Dongwu fault. The western boundary is bounded by Mashan fault. The total drainage area is 1486 km^2. The most significant hydrogeological units are Cambrian and Ordovician carbonate rocks characterized by karst features at different scales.

Precipitation is the primary water source to the spring system. It recharges the subsurface through four modes:

- **Direct infiltration on bedrock outcrops.** This is probably the most important mode of recharge. Perennial dynamic observational data show that the change of water level and artesian flow is correlated closely with atmospheric precipitation. In the rainy season, the karstic water level rises universally, but in dry months, such as April, May, and June, the water level is much lower. Isotope analysis of groundwater and surface water samples shows that the karstic water of the spring area comes from atmospheric precipitation.
- **Riverbed leakage.** South of the spring area, the karstic water is recharged by the river leakage. The Wohushan reservoir is also a recharge source.
- **Indirect recharge from Quaternary sediments.** Along the middle and upper reaches of the Yufu River and Beisha River, coarse sand with pebbles is present and overlies the limestone directly. Precipitation recharges the porous aquifer then the karstic water.
- **External recharge.** The eastern and western boundaries connect the aquifers beyond the spring basin.

10.10.2.2 Hydrodynamics of the Jinan Springs

The Jinan Springs result from unique topographic, geologic, and hydrogeological conditions. The Ordovician limestone is the main source of water and outcrops in the south. The topography in Jinan is higher in the south than in the north. Springwater comes from the mountainous area of the south. The more than 260 m elevation difference between south and north makes the surface water and groundwater flow northward along multiple porosities in the formation, which then is blocked by impermeable diorite. Consequently, the springwater discharging from the crevices among the rocks helps maintain a certain groundwater level in the aquifer. In other words, there could be enough water in the spring area to keep a certain groundwater level. According to testing data from the Baotu Spring (Meng, 2003), which is known not only as the largest one in Jinan but also as one of the most famous springs in China, the threshold groundwater level to maintain the spring flow is 27 m above mean sea level. Baotu Spring flows only when the groundwater levels are above this value.

The close relationship between the spring discharge and the groundwater level in the aquifer is supported by historical data at the springs (Figure 10.10–6). From 1959 to 1967, the annual average groundwater level in the urban area ranged from 28.75 to 32.85 m, the corresponding spring discharges were between 3.5 and 5.8 m^3/s. The average amount of water utilized was only 1 m^3/s. When the utilization increased from 1.7 to 3.2 m^3/s, from 1968 to 1975, the spring discharges varied from 1.6 to 1.8 m^3/s, while the groundwater level changed from 28.06 to 28.75 m. During the period 1975–1981, when the groundwater utilization reached the level of 3.6 m^3/s, the groundwater level decreased from 28.16 m to 26.78 m. Between 1982 and 2002, the utilization reached a peak at 6.4 m^3/s. As a result of overexploitation, the spring flow ceased discharging several times. Starting in 2002, efforts were made to protect the spring. By 2008, the spring discharge remained at a reasonably high level of 2.1 m^3/s.

Sustainable discharge at the spring is not the only concern. Increased land development and activities in the spring basin have led to groundwater contamination. As shown in Figure 10.10–7, the water quality at Baotu Spring has been deteriorating since 1958. In 1958, the SO_4^{2-} and Cl^- concentrations were 8.47 mg/L and 9.06 mg/L, respectively. In 2007, their concentrations increased to 60.18 mg/L and 41.55 mg/L, respectively.

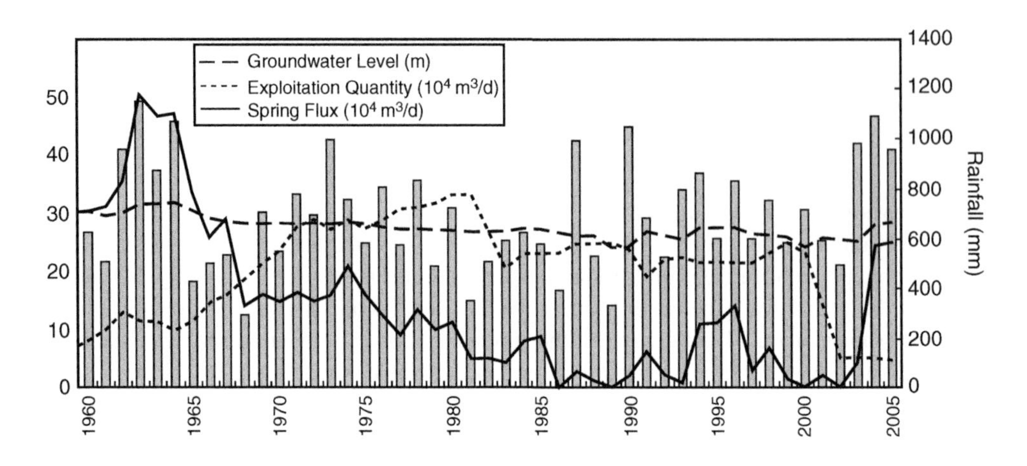

FIGURE 10.10–6 Spring discharge, precipitation, and groundwater level in the Jinan Spring basin.

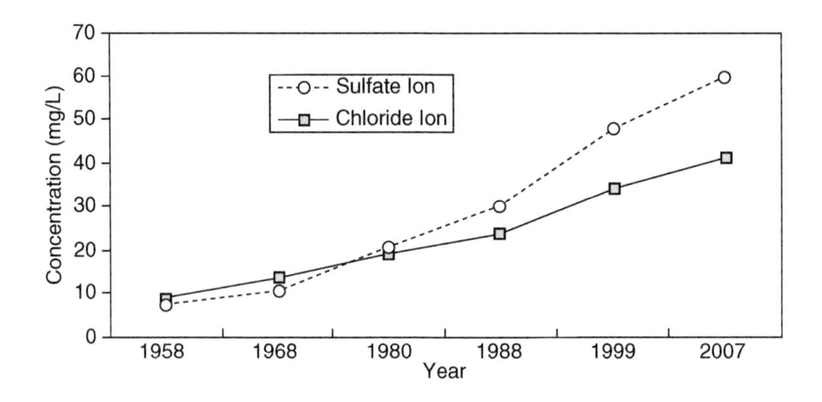

FIGURE 10.10–7 Chemographs of sulfate and chloride at Baotu Spring.

10.10.2.3 Application of a three-dimensional hydrogeological model to the protection of the Jinan Springs

Through the support of province government and local regulatory agencies, a series of measures, including diversion of water from the Yellow River to Jinan, artificial recharge of groundwater, rational exploitation of groundwater, and better utilization of surface water have been proposed to protect the Jinan Springs. To evaluate the effectiveness of the measures for spring protection, a three-dimensional model that integrates the data from various sources and represents the current understanding of the geologic and hydrogeological conditions of the spring area was constructed for the spring watershed. The three-dimensional model combines environmental geography, GIS, groundwater modeling, remote sensing, computational geometry, data mining, visualization, and virtual reality simulation. The simulation provides the necessary visual accuracy to make assessments quickly with minimum financial outlay. It is a valuable tool for policy makers to develop effective and practical strategies for spring protection.

Data on the regional hydrogeology, engineering geology, and environmental geology were accumulated through various studies over the years. To utilize these available data for three-dimensional simulation, source-oriented integration is used. The source-oriented integration offers a visual, physical, and uniform channel in three-dimensional space. It aims to support the interoperability of various data sources and allow any possible form of geological data to be stored, processed, displayed, and manipulated in the same coordination system.

The geological data are classified into three types: direct data, indirect data, and assistant data. Direct data, such as borehole data, property data, and spring measurements, are original sampling data obtained by direct observations and survey and are highly accurate. They are utilized directly in the three-dimensional modeling system and managed and stored with databases in MS Access, Oracle, and other formats. Indirect data are also original but have different precisions with different resolutions of graphs, such as boundaries, faults, geological maps, topographic and structural geology maps, two-dimensional and three-dimensional seismic reflection data, exploring data, as well as contours of the water table, gullies, lakes, and so on. This type of data should be stored as files after being digitized. However, this kind of data cannot be used as direct inputs for the three-dimensional modeling system. To integrate the indirect data, they need to be digitized using the existing software systems, like AutoCAD, ArcView GIS, or other interpretative software; and these digitized data need to be converted into the three-dimensional modeling system by employing the CAD/GIS Data Conversion Interface (DCI) as well as DEM DCI. Assistant data are used in the process of three-dimensional modeling as icons like two-dimensional and three-dimensional primitives, and texture maps including remote sensing images, and scanned maps. Once the diverse kinds of geological data are integrated into the three-dimensional modeling system, geometrical objects abstracted from these data are analyzed and mapped into different abstract levels of space. Point objects are at the bottom level then line and polygon objects.

Construction of the correct three-dimensional fault geometric models is one of the crucial problems in modeling and visualization of the geological processes. Fault data, both original and those interpreted from boreholes and cross sections, are scarce in the Jinan Spring area. It is not often feasible that the complicated geological structure in the area can be completely approximated using only these data and commonly used methods. According to the characteristics of the faults in the Jinan Spring area, a modeling framework is presented that aims to establish an efficient fault model. The process mainly involves four steps: (1) determining the property parameters of faults, such as dip, direction, and displacement, and building a database of them; (2) digitizing the fault lines extracted from the structural geology maps and converting them to the three-dimensional modeling system through data transformation DCI; (3) importing the scatter fault points acquired from boreholes and cross sections, which is used as the base data for faults simulation; and (4) deducing the geometry shape of the faults by applying the methodologies for the mathematical description and computer simulation of faults (Wu and Xu, 2005).

The three-dimensional modeling project significantly improved the understanding of the spring basin, including the causes of springs drying up, soil erosion, and groundwater overexploitation. The water and soil erosion in the upper reaches of the spring area is especially serious. Many factors contribute to water and soil erosion, including the unscientific cultivation and deforestation. The mountainous area in the south is the main source for the Jinan springwater. The recharge area of the springs covers 1500 km^2, which comprises 550 km^2 of direct recharge area and 950 km^2 of indirect recharge. Both are in the mountainous area in the south. The remote sensing image analysis indicates that the forest coverage rate is lower, and the ecological situation tends to get worse in the mountainous area in the south. The increase of the impermeable area and decrease of the cultivated land and vegetation cover resulted in a significant change in the surface conditions in recent years. This change influences the rate and volume of the rainfall runoff and leads to a reduction in the amount of groundwater available in the spring area. The result of the simulation shows the severe groundwater overexploitation in the lower reaches of the spring area, which poses a major threat to the water ecosystems. The expansion of Jinan city, coupled with the rapid population growth and the even faster growing consumption of water, means an increased demand for water (Xing, 2006).

10.10.3 NIANGZIGUAN SPRING

Niangziguan Spring (E115°, N37° 50′), the largest karst spring in north China, is located in the Mianhe River valley, Taihang Mountains of Shanxi province (Figure 10.10–8). The average discharge over the period 1959–2006 was 10 m^3/s. The maximum annual recorded spring flow was 15.75 m^3/s in 1964 and the minimum was 5.73 m^3/s in 1995. Niangziguan Spring does not represent just one spring orifice, rather it consists of

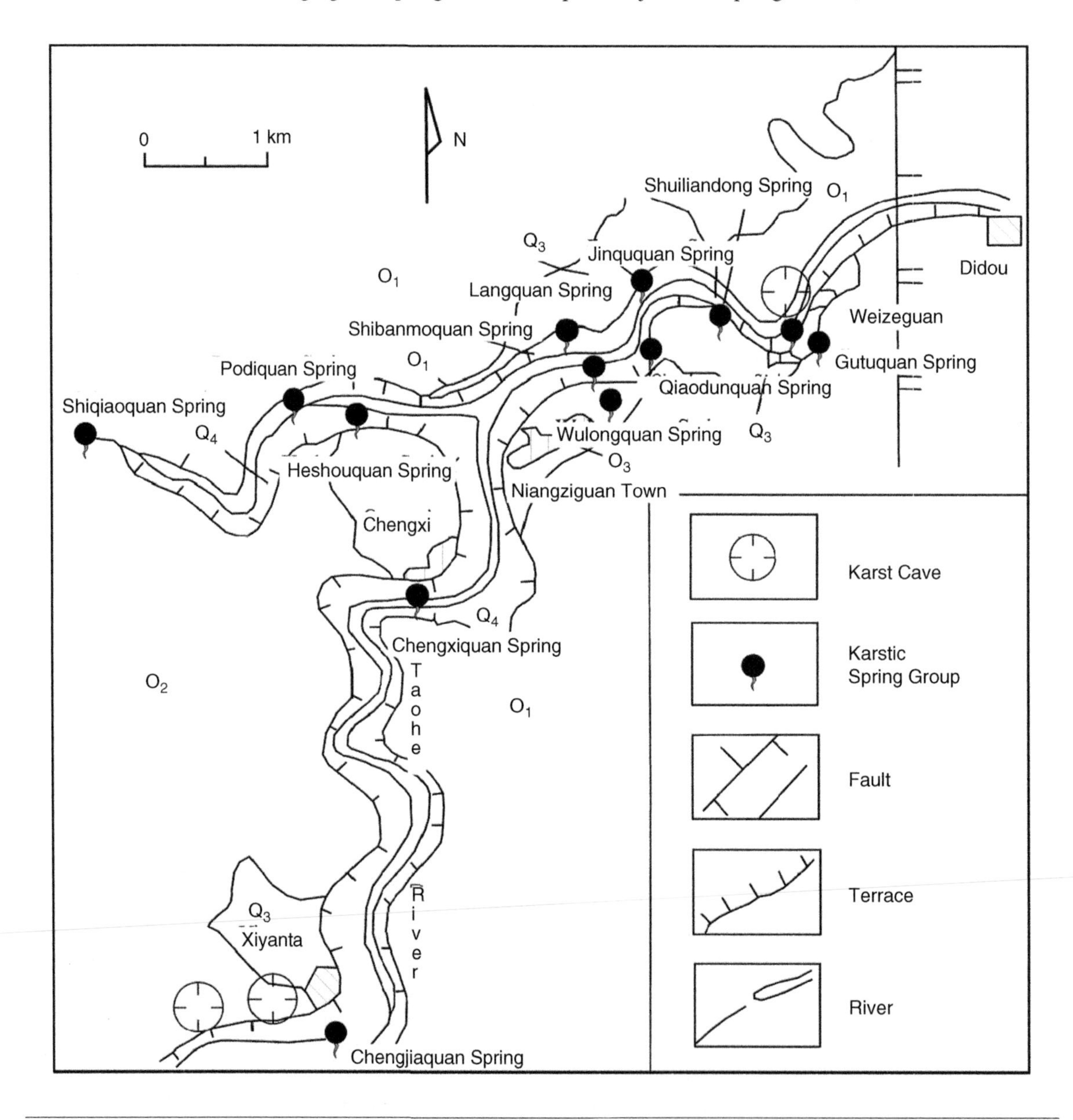

FIGURE 10.10–8 Location of Niangziguan Spring. (From Han, Lu, and Li, 1993.)

Table 10.10–4 Springs in the Niangziguan Spring Group (Han, Lu, and Li 1993)

Spring	Maximum Flow (m^3/s)	Elevation (m)	Physiographical Character of Discharge Point	Remarks
Weizeguan Spring	1.08	371	Second terrace	
Chengxi Spring	1.57	379	River floodplain	
Podi Spring	0.953	377	Lower part of second terrace	
Wulong Spring	2.1	369	Second bottom	
Shibanmo Spring	0.178	364.9	Upper river floodplain	
Gun Spring	0.335	386	Under part of second bottom	
Jinqu Spring	0.383	360	Upper river floodplain	
Qiaodun Spring	0.575	364	River floodplain	
Hebeicun Spring	0.067	381.5	Second terrace	Flow ceased
Shuiliandong Spring	3.15	386	Second terrace	Flow ceased
Chengjia Spring	1.5	392	River floodplain	Flow ceased

11 individual springs: Podi Spring, Chengjia Spring, Chengxi Spring, Wulong Spring, Shibanmo Spring, Gun Spring, Hebeicun Spring, Qiaodun Spring, Jinqu Spring, Shuiliandong Spring, and Weizeguan Spring. The springs are distributed along the Mianhe River and the zone of discharge stretches approximately 7 km. Table 10.10–4 lists the elevations of their discharge points.

In the Niangziguan Spring basin, the annual mean air temperature is 10.9°C. The highest recorded temperature is 40.2°C and the lowest is −28°C. The annual mean potential water surface evaporation is 1202 mm. Surface water in the study area include Wenhe River, Taohe River, Mianhe River, Songxihe River, Qingzhangdongyuan River, and Qingzhangxiyuan River. The Wenhe River and Taohe River converge to form the Mianhe River at Niangziguan town.

From 1956 to 2006, the annual average precipitation was 528 mm. The largest recorded annual precipitation was 847 mm in 1963 and the smallest 288 mm in 1972. Approximately 60–70 percent of the precipitation occurs in July, August, and September. As shown in Figure 10.10–9, the average rainfall in the Niangzigaun Spring area has had a general decreasing trend since the 1960s. The average annual rainfall was 595.0 mm from 1954 to 1959, 607.9 mm from 1960 to 1969, 495.2 mm from 1970 to 1979, 501.9 mm from 1980 to 1989, 485.7 mm from 1990 to 1999, and 499.4 mm from 2000 to 2006.

Because precipitation is the most important recharge source to Niangziguan Spring, the decrease in precipitation has contributed, among other factors, as discussed in this chapter, to the decrease in the spring discharge. As shown in Figure 10.10–9, the average annual flow in the 1960s was 13.7 m^3/s, 11.4 m^3/s from 1971 to 1980, 8.76 m^3/s from 1981 to 1989, and 7.3 m^3/s from 1991 to 2004. The calculated attenuation rate of the annual flow was 0.16 m^3/s. The Shuiliandong Spring, Chengjia Spring, and Hebeicun Spring dried up. The striking degradation of the karst water system poses a significant threat to the sustainable economic development of this region.

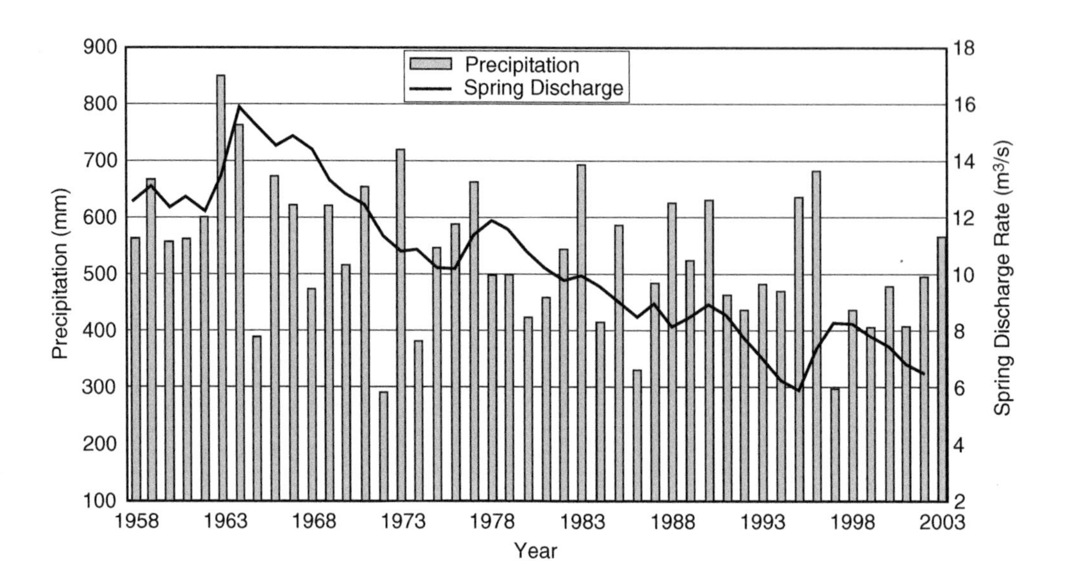

FIGURE 10.10–9 Recorded precipitation and discharge hydrograph of Niangziguan Spring.

10.10.3.1 Geological settings

The lithology in the Niangziguan Spring area consists of Archaeozoic antiquity metamorphic rocks, Middle Proterozoic quartz sandstone, Paleozoic Cambrian-Ordovician carbonate rocks, Carboniferous and Permian coal-bearing carbonate and clastic rocks, Mesozoic Triassic clastic rocks, and Cenozoic Tertiary and Quaternary unconsolidated layer and basalt (Figures 10.10–10 through 10.10–12):

- **Cambrian (ε).** The lower part is aubergine shale and mudstone intercalated with 120 m dolomite. This is the aquifuge of the regional flow system. The middle upper layer is thick oolitic limestone, dolomitic limestone, coarse crystalline dolomite, and edgewise limestone. The karst crannies developed into an aquifer, which outcrops at the east and north fringe out of the system.
- **Lower Ordovician series (O_1).** It is dolomite with flint nodule and argillaceous dolomite, most of which is buried under the middle Ordovician with the developing of karst crannies and the runoff is slow. The karst developed well near the discharge area and the fracture zone on the fault structure. Most of the Niangziguan Spring group outcrop among the lower Ordovician dolomite.
- **Middle Ordovician series (O_2).** It is a typical mixture creation of sulfate rocks and carbonate rocks, and the karstification has stratification. The lithology characteristics of the Fengfeng group, upper Majiagou group, and lower Majiagou group from upper to lower are limestone-angle conglomeratic marlite or marlite, and there is gesso marlite or angle conglomeratic marlite in the lower part of each group. These stratums distribute around the whole spring area and the development of surface joints and fissures in them provides good channels for the infiltration of rainfall and river water. In the deep ground, prime abundance corrosion fissures and honeycomb structure corrosion pores with a few karst caves provide a huge water storage space for composing the aquifer in the system. This stratum is the main aquifer for the Niangziguan Spring area.

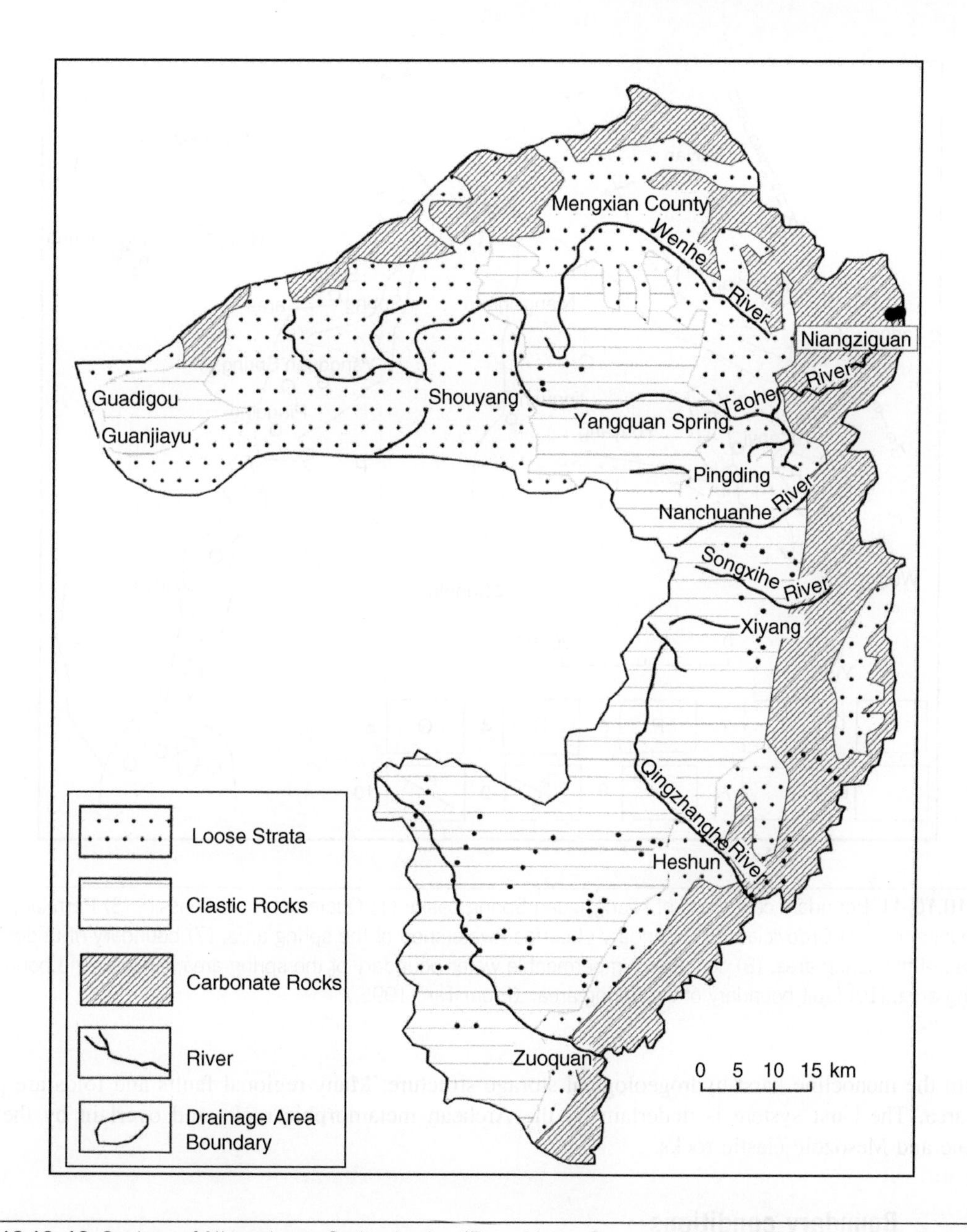

FIGURE 10.10–10 Geology of Niangziguan Spring basin. (From Liang, Gao, and Zhang, 2005.)

- **Carboniferous-Permian (C-P)**. This stratum is a coal series stratum composed of shale, sandstone, a coal bed, and limestone.
- **Cenozoic**. This stratum includes the Tertiary Pliocene (N_2) and Quaternary (Q), which is mostly composed of clay, alluvial-diluvial sand, sandy gravel, and sandy loam.

Structurally, the Niangziguan Spring area is on the northeast limb of the Xinshui syncline and is a big syncline tilting northeastward. The lower Paleozoic carbonate rocks tilt toward the sag center from north to east

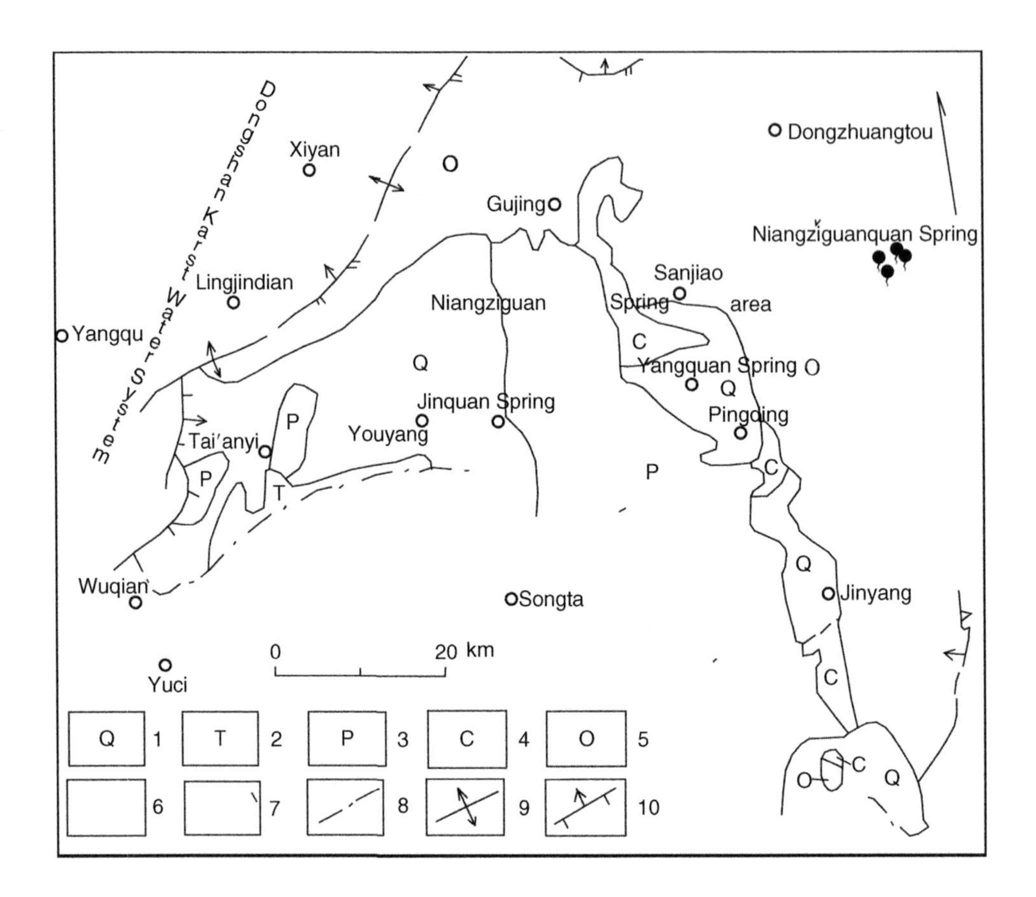

FIGURE 10.10–11 Boundary conditions of Niangziguan Spring basin: (1) Quaternary, (2) Triassic, (3) Permian, (4) Carboniferous, (5) Ordovician, (6) boundary of surface watershed of the spring area, (7) boundary of underground watershed of the spring area, (8) concealed impermeable water boundary of the spring area, (9) anticline boundary of the spring area, (10) fault boundary of the spring area. (From Tan, 1995.)

and form the monocline karst hydrogeological storage structure. Many regional faults and folds are present in the area. The karst system is underlain by the Archean metamorphic rocks and overlain by the upper Paleozoic and Mesozoic clastic rocks.

10.10.3.2 Boundary conditions

The Niangziguan Spring basin includes Pingding, Mengxian, Yangquan outskirts, and Xiyang areas and part of Heshun and Shouyang. Figure 10.10–11 shows the general boundary conditions of the spring basin. The eastern and northern boundaries are composed of lower Ordovician dolomite. The southern boundary is the gentle karst subsurface watershed on Heshun Poli and the Xin'an Spring area. South of the western boundary is the topographic high formed by the Triassic and Permian clastics, and north is the East Hill anticline. North of the western boundary is a 20 km gentle undulant fault named the Niutounao-Xitougou-Yangzhuang fault. The fault is a normal fault, along which the middle Ordovician limestone is

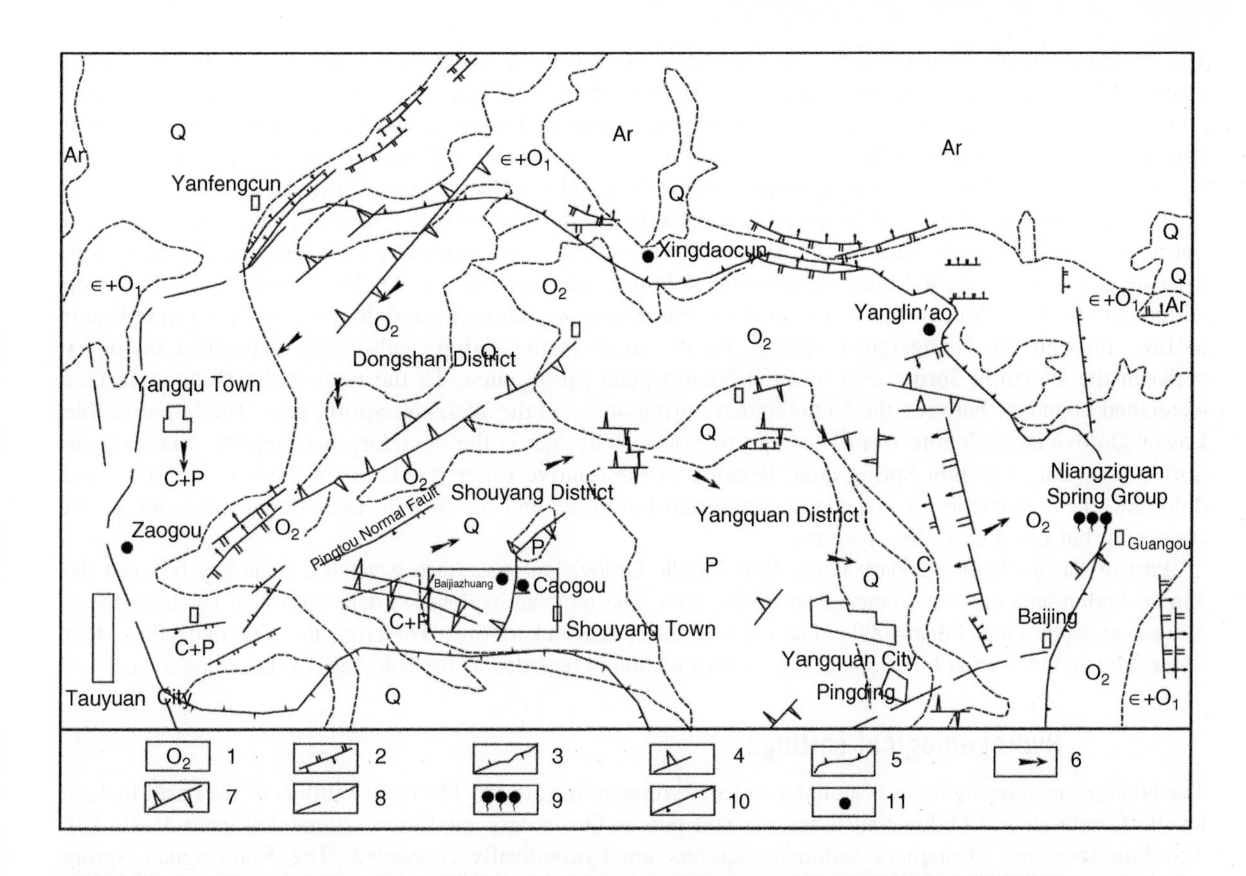

FIGURE 10.10–12 Hydrogeology of Niangziguan Spring basin: (1) Strata, (2) reverse faults, (3) retention boundary, (4) syncline, (5) underground watershed, (6) groundwater flow direction, (7) anticline, (8) impermeable boundary, (9) spring group, (10) normal faults, (11) sinkhole ("collapse column"). (From Jiang and Ji, 2007.)

in direct contact with the Carboniferous-Permian sandstone. This prevents the eastward flow of karst groundwater of the Dongshan system and forms the southwestward runoff. The fault throw at the south part of the fault varies gradually, but the bilateral middle Ordovician limestone is still hydraulically connected. South of the western boundary is the Dayukou fault and the inferred Xushagou-Niedian fault. The Dayukou fault is oriented in the northeast direction, and the Xushagou-Niedian fault is oriented in the northwest direction.

The northwest boundary is the Taiyuan Dongshan anticline, Dafangshan-Wenjiashan fault, Yangpingwang anticline, and Heishiyao-Xingdao reverse fault. It consists of a series of mountain peaks and ridges on topography and physiognomy. The anticline axis coincides with the surface watershed. The core is the Ordovician and Cambrian formations. The main bilateral aquifers of karst water contact the lower Ordovician dolomite and the thick Cambrian shale layer. At the south Dafangshan-Wenjiashan fault, the lower Ordovician, Cambrian, and Archaeozoic metamorphic rocks were uplifted and have direct contact with the middle Ordovician aquifer.

The northern boundary consists of the Shuilingdi-Shenquan Cambrian subsurface watershed and the Waicun-Pifunao-Zihejian surface watershed from west to east. The Shuilingdi-Shenquan Cambrian subsurface watershed is the boundary line between the Niangziguan Spring area and the Xingdao Spring area.

The Waicun-Pifunao-Zihejian surface watershed is the boundary line between the Yuxian basin and the Huduohe drainage area, which is composed of Middle Ordovician limestone with a slope angle larger than 30 degrees. The impermeable layer of the Zhuangzhi section in Lower Ordovician dolomite forms the local boundary of the Niangziguan Spring area. The Xijiazhuang section, a Cambrian subsurface watershed, is the boundary line between the Lianggouqiao Spring area and the Niangziguan Spring area.

The eastern boundary is the Weizeguan reverse fault, an impermeable boundary. This fault is along a north-south direction, crosscutting the Mianhe River valley, which broadens after crossing the fault from west to east. The western Lower Ordovician dolomite was uplifted by the Weizeguan fault. The deep entrenchment of the Mianhe River and the impermeable Lower Ordovician dolomite force the groundwater to flow through the Niangziguan Spring. To the south is the Baijing subsurface watershed boundary between the Weizhou Spring area and the Niangziguan Spring area. To the west is the Hulutao surface watershed boundary between the Niangziguan Spring area and the Weizhou Spring area. The impermeable Lower Ordovician dolomite from Dawaqiu to South Hengshan is the boundary between the Niangziguan Spring area and east Gubi Spring area. Because of the relative water resistance of the Lower Ordovician dolomite, the water level of the middle Cambrian limestone and the Middle Ordovician limestone is not connected but has a cascading pattern.

East of the southern boundary is the Boli Middle Ordovician subsurface watershed boundary between the Xin'an Spring area and the Niangziguan Spring area. The regional exploration data show that karstification is so weak at depths greater than 1000 m that it can be taken as a nonkarst area. Therefore, the 1000 m depth contour of the Middle Ordovician limestone at the western section is regarded as the boundary of the spring area.

10.10.3.3 Hydrogeological settings

The Niangziguan Spring basin is an independent hydrogeological unit. The main aquifers of the basin include karstic Cambrian and Ordovician limestone and porous Quaternary sandstone sediment (Figure 10.10–12). The limestone and Quaternary sediment aquifers are hydraulically connected. The Niangziguan Spring receives water from a 7394 km^2 basin that includes the city of Yangquan and the counties of Pingding, Heshun, Zuoquan, Xiyang, Yuxian, and Shouyang. Small basins and gentle sloping river valleys are the primary physiographic features, but extensive areas of the Niangziguan Spring basin consist of rough hilly terrain, where the altitude ranges from 1200 to 1600 m. The western part of the basin is higher than the eastern part, with the general topography of the basin inclining to the east. The Mianhe Valley, where the Niangziguan Spring discharges, has the lowest altitude in the Niangziguan Spring basin, ranging from 360 to 392 m.

In general, the farther to the west, the thicker is the overlying strata. In the westernmost area of the Niangziguan Spring basin, where the overlying Permian and Triassic sandstone and shale is several hundreds of meters thick, karst is the least developed. There is basically no feed flow from surface water to the karst groundwater in this area. The surface water generally flows to the east, and much of it recharges to groundwater. In this area, karst groundwater storage is small, and groundwater flow is very slow. This anthropogenic area includes Shouyang county, the southwest of Yangquan city, the west of Pingding county, and parts of Xiyang county, Heshun county, and Zuoquan county.

The recharge area can be divided into northern and southern parts, bounded by the Taohe River and Wenhe River. This area includes the south of Yuxian and Pingding counties, the north of Shouyang county, together with the counties of Xiyang, Heshun, and Zuoquan, and occupies an area of over 3000 km. Groundwater recharge occurs mainly from infiltration of precipitation into outcropping limestone areas, riverbeds, sinkholes, and seepage from overlying strata. The aquifers in the recharge area have a varying thickness from 10 m to several decameters. The groundwater level is 100–500 m below the surface. The groundwater hydraulic gradient ranges from 0.76 to 0.9 percent. The velocity of groundwater flow is still relatively slow.

The triangular area between the Taohe River and the Wenhe River is the confluence area of the basin. In the confluence area, karstification has reached a mature stage, with extensive dissolution of the limestone and full development of an arterial underground drainage system with flows in zones of high permeability. The groundwater hydraulic slope is generally less than 0.3 percent. Aquifers in this zone have a thickness of 200–400 m. This area is very vulnerable to pollution.

The discharge area is located at Niangziguan town, in the northeast of the spring basin, where the karst groundwater flow is blocked and discharges to the surface as a result of the incision of the Mianhe River and the elevation of a low permeability bedrock of clayey dolomite and covering an area of 50 km^2. Karstic conduits are well developed and are the major paths of groundwater flow to springs. The karstic conduits have an average diameter of 0.02 m, with 4.57 m being the largest known diameter. Cobbles have been reported in some of the conduits. The conduits are distributed mainly from near the surface to 40 m deep. Regional groundwater is confined. The hydraulic gradient is about 0.35 percent. Groundwater flow in some parts of the area is turbulent. Because water moves from the recharge area to the discharge area through long distances over a period of years, the spring discharge has been closely related to precipitation over the past seven years. Therefore, once the karst groundwater is polluted, contaminants linger for a long time and diffuse over a large area.

10.10.3.4 Utilization of the karst groundwater in the Niangziguan Springs

Since the early 1970s, karst groundwater in the Niangziguan basin has been exploited for irrigation, municipal use, and industrial water supply. Usage has accelerated since the 1980s. Nowadays, the karst groundwater is a major water source for the Niangziguan Springs basin, one of the heavy industrial regions in China, for coal mining, power generation, chemical engineering, and metallurgy. Karst groundwater in the Niangziguan Springs basin is utilized from spring discharge and wells. In 1998, the total volume of karst groundwater use was 149 million m^3, 82 percent of which came from spring outlets and 18 percent from wells. From 1980 to 2001, the annual volume of abstracted water from wells increased from 6.43 million to 30.8 million m^3. This enormous exploitation of groundwater reaped large socioeconomic benefits for the area but caused water quality and quantity problems. Overexploitation, uncontrolled urban and industrial discharges, and agricultural intensification are causing increasingly widespread degradation of the aquifers and reducing the discharge of springs. The discharge of Niangziguan Springs has steadily decreased during the period from 1956 to 2003. Attenuation of spring discharge has become a major concern for local economic development. The spring system is further deteriorated by such factors as dewatering in coal mines and lack of land-use control. The coalfields in the Niangziguan Spring basin extract the Carboniferous-Permian coal seam, which is underlain by the carbonate strata. With the exploitation of the coal, a great quantity of water was pumped out of the mines, which destroyed that natural balance between recharge and discharge. As a result of uncontrolled land use, the forest coverage in the spring area has decreased by 4 percent, which has changed the local runoff conditions. More surface runoff tends to flow out of this region, thus the infiltration rate into the subsurface decreases (He and Wang, 2001a; 2001b).

In the Niangziguan Springs basin, concentrated infiltration points, such as faults, fractures, fissures, and conduits, are connected to the karst network. Hazardous substances from industries and disposal of waste products are potential threats to the karst aquifer. Currently, the detected rate of ammonia nitrogen in the karst water is 97 percent, with 20 percent exceeding the drinking water standard. The hydroxybenzene detected rate is 37 percent. The situation at the springs is worse. Monitoring of five springs in the Niangziguan group indicates that the detected ammonia nitrogen rate is 100 percent and the hydroxybenzene rate is 80 percent. In addition, other chemical constituents are showing upward trends. Figure 10.10–13 shows the chemographs of sulfate concentrations at two springs.

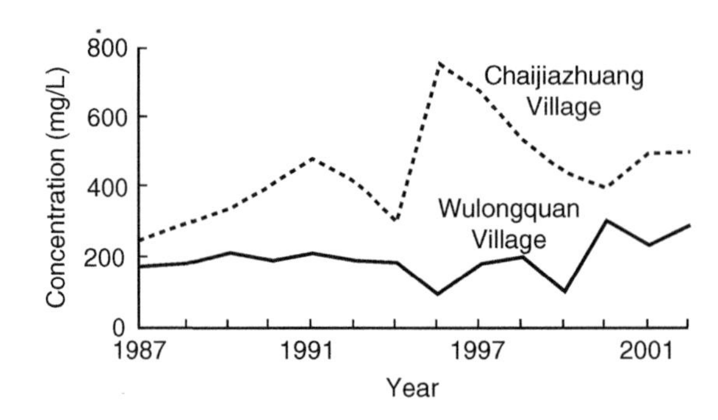

FIGURE 10.10–13 Sulfate chemographs at two springs. (From Zhang, 2007.)

10.10.3.5 Management strategies

Improved management of karst groundwater resources in the Niangziguan Spring basin is urgently needed to mitigate actual and prevent potential degradation caused by excessive exploitation and inadequate pollution control. Unless the groundwater is protected, scarcity of water and escalating water supply costs will be accompanied by potentially negative impacts on human health. Different management measures should be taken in accordance with the hydrodynamic properties of the karst system (Yan, 2005).

The discharge and confluence areas are susceptible to a greater range of environmental impacts, due to their hydraulic and hydrogeological characteristics. Some pollution threats are the direct leaching of sewage flow into the groundwater. Stricter water environmental quality standards should be established for this area. Protecting groundwater resources against pollution demands (1) regulations for the planning, location, and construction of settlements, houses, buildings, roads, and industrial plants; (2) prohibition of the direct or indirect disposal of untreated liquid wastes and wastewater into the ground; (3) prohibition of solid waste storage in this area; (4) prohibition of the construction of sewage and wastewater collection systems or pollutant treatment plants within this area unless high standards are used; (5) prohibition or restriction of the use and transportation of hazardous and toxic materials; and (6) prohibition of the overexploitation of the karst groundwater without a license for water withdrawal.

In the recharge area, accelerated development of the coal industry has brought about great social and economic benefits but also many environmental problems. Dewatering for coal mining has created problems of groundwater subsidence that affected industrial, municipal, and agricultural areas. When the underground coal mining compartments were dewatered, nearby wells went dry and a large number of sinkholes formed. Polluted waters drained from coal mining areas may cause groundwater pollution. Direct discharge of mining wastewater should be forbidden. Stacking of gangues or mine waste materials should be forbidden. Appropriate land development policies should be put in place. Construction projects that produce pollutants should not be permitted without safeguards. Discharges originating from industrial factories should not be permitted unless the discharge water quality is in compliance with national environmental standards. Proper location, construction, and operation of fertilizer and pesticide storage facilities; correct application of fertilizers and pesticides; and proper collection storage and treatment methods for solid wastes are also required.

The westernmost area, which appears to have greater relative immunity to pollution, is the more suitable zone for urbanization and development. Construction of projects should be allowed, as well as the enlargement of urban areas; but even in this zone, discharge drainage must be restricted. Chemical industrial plants and their warehouses, wastewater leakage into the ground through open seepage pits, and injection of untreated industrial wastewater into dry or decommissioned wells should not be permitted to avoid further environmental degradation.

ACKNOWLEDGMENTS

This research was supported by the China National Natural Science Foundation (grants 40572149 and 40772162), the National Key Project of Scientific and Technical Supporting Programs (2007BAK24B01 and 2006BAB16B04), and the "973" Project (grant 2006CB202205).

REFERENCES

Chen, M., Ma, F., 2002. Groundwater resources and environment in China. Earthquake Publishing House, Beijing, China.

Guo, Z., Zhang, H., Yu, K., 2004. Multiple causes of attenuation of karst spring discharge in Shanxi Province. Journal of Geotechnical Investigation 2, 22–24.

Han, X., Lu, R., Li, Q., 1993. Karst water system—investigation of large karst springs in Shanxi. Geological Publishing House, Beijing, China.

He, Y., Wang, X., 2001a. The research of karst water resources conservation in Niangziguan Spring basin. Systems Engineering—Theory and Practice (4), 137–138.

He, Y., Wang, X., 2001b. The study of grey system models of Niangziguan Spring. Journal of System Engineering 6 (1), 40–43.

Jiang, X., Ji, S., 2007. Application of karstic water hydrochemical characteristics in hydrogeological condition analysis—A case study of Niangziguan Springs in Shouyang area. Coal Geology of China 19 (4), 45–47.

Li, C., 2005. Dynamic character of Niangziguan karst spring. Shanxi Hydrotechnics 1, 44–45.

Li, C., Hu, A., You, Q., 2004. The research into the evolvement trend of karst water quality for fountain areas in Jinan. Land and Resources in Shandong Province 20 (1), 35–38.

Liang, Y., Gao, H., Zhang, J., 2005. Preliminary quantitative analysis on the causes of discharge attenuation in Niangziguan Spring. Casologica Sinica 24 (3), 228–229.

Tan, K., 1995. Application of satellite images to analysis of structural control of karst groundwater in Niangziguan Springs. North China Geology and Mineral Resource 10 (4), 612–615.

Wu, Q., Xu, H., 2005. A three-dimensional model and its potential application to spring protection. Env. Geol. 48 (4–5), 551–558.

Xing, L., 2006. Present situation and protection strategies for environmental problems of karst groundwater in Jinan Spring region. Journal of University of Jinan (Science and Technology) 20 (4), 345–349.

Yan, M., 2005. Discuss of the utilization and protection countermeasure of Niangziguan Spring. Shanxi Water Resource 2, 41–42.

Yuan, D., 1994. China karstology. Geological Publishing House, Beijing, China.

Yuan, D., Cai, G., 1998. Karst environmentology. Chongqing Publishing House, Chongqing, China.

Zhang, J., 2007. The recognition of karst water system evolution in Niangziguan Spring. Journal of Taiyuan University of Technology 38 (5), 421–423.

Zhang, Y., Miu, Z., Mao, J., 1985. Applied karstology and speleology. Gui Zhou People's Publishing House, Guiyang, China.

Zhang, D., Wang, T., Zhai, H., 2004. Carbonate water contamination and genetic analysis of Niangziguan spring area. Safety and Environmental Engineering 11 (1), 5–8.

Index

Note: Page numbers followed *f* indicates figures and *t* indicates tables.

I

K

L

M

O

P

T

U

W

Printed and bound by CPI Group (UK) Ltd, Croydon, CR0 4YY
08/05/2025
01864850-0005